Pharmacokinetic/Pharmacodynamic Data Analysis:

Concepts and Applications

3rd edition

Revised and expanded

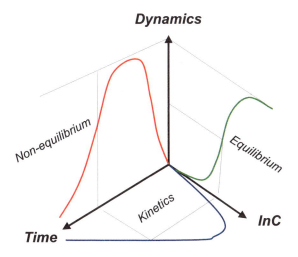

APOTEKARSOCIETETEN

SWEDISH PHARMACEUTICAL SOCIETY

This book is dedicated to Barbro Gabrielsson for boundless patience and understanding during preparation of this book.

They were "so intent on making everything numerical" that they frequently missed seeing what was there to be seen.
Barbara McClintock
Nobel Prize Laureate

Pharmacokinetic and Pharmacodynamic Data Analysis:
Concepts and Applicaiontions, 3rd edition by J Gabrielsson and D Weiner

© 2000 J Gabrielsson, D Weiner and Swedish Pharmaceutical Society,
Swedish Pharmaceutical Press, P.O. Box 1136, S-111 81 Stockholm, Sweden

ISBN 91 8627 492 9

Digital original förkopiering ABA Kopiering AB

Kristianstads Boktryckeri AB, Sweden 2000

Foreword

Pharmacokinetics is that branch of science, which deals with the time course of drug in the body. Specifically it is the study of drug absorption, distribution and elimination. The companion subject of pharmacodynamics deals with the time course of drug action and is intimately linked to pharmacokinetics. As Lord Kelvin said:

"When you can measure what you are speaking about, and express it in numbers, you know something about it: but when you cannot measure it, when you cannot express it in numbers, your knowledge is of a meager and unsatisfactory kind: it may be the beginning of knowledge, but you have scarcely, in your thoughts, advanced to the stage of science."

This is especially true of pharmacokinetics/pharmacodynamics (*PK/PD*) and modern *PK/PD* has developed into a relatively sophisticated mathematical discipline.

The importance of *PK/PD* in drug development is becoming increasingly recognized and now permeates the program from preclinical development through to Phase IV clinical trials. Specifically in preclinical studies, *PK/PD* is used to support drug discovery, interpret toxicokinetic experiments and via physiological modeling to extrapolate from animal to man. In the clinical program *PK/PD* is used to support dose-finding and dose-escalation studies and there is at least one instance where it has been used to recommend a dose which was not originally studied in the efficacy-safety studies. More recent applications include the concentration-controlled clinical trial and population *PK/PD*.

However like all biological experiments, *PK/PD* data is noisy and one has to use sophisticated data analysis techniques to estimate parameters of interest. Therefore a scientist working with *PK/PD* data has frequently to fit a model to experimental data. One has to be careful about the term model as applied in *PK/PD*. There are a number of so-called model independent methods that have become popular recently. However in a *PK/PD* context model independent implies fewer assumptions about the structural *PK/PD* model. Even with these methods one is often faced with fitting some empirical expression to the data, for example a sum of exponentials, and when I talk about fitting models to data, it is meant in this sense.

Most models used in *PK/PD* are nonlinear functions of the parameters of interest and consequently severe data analysis problems arise. Therefore nonlinear regression and maximum likelihood techniques have to be used which are much more computer intensive than their linear counterparts. In addition there are several theoretical problems specific to nonlinear models, such as nonuniqueness of the solution and the estimation of confidence intervals. Weighting schemes are a particular problem for *PK/PD* data as it is generally impossible to get replicate measurements.

Professional scientists but amateur statisticians often carry out *PK/PD* data analysis. Consequently the scientist working in the *PK/PD* area is dependent on the availability of good software packages. Despite the lack of a complete understanding of the methodology the user of such packages needs to be convinced of their reliability and accuracy. Although it is extremely difficult to completely *validate* a nonlinear regression program some, perhaps even extensive, testing is required. As a final caveat I would note that it is incumbent on the *PK/PD* scientist to obtain a reasonable understanding of the methodology behind a software package if he or she is going to be able to correctly interpret the output and diagnostics. It is also desirable that the software producer should provide adequate user support.

Therefore it is timely that the current book has appeared. Although it is not intended to be a validation exercise it is reassuring that the package tested behaved well over a wide range of problems. It will provide a valuable introductory text to new researchers and be a useful reference for established scientists.

Leon Aarons
Manchester, 2000

Preface to the third edition

It has now been three years since the second edition of *Pharmacokinetic and Pharmacodynamic Data Analysis: Concepts and Applications* was published. During that time we have developed and run a number of both basic/intermediate and advanced courses in pharmacokinetic and pharmacodynamic data analysis at Madingley Hall, Cambridge. In parallel with that project we have also rewritten the older text, updated equations with a professional formula editor, and have re-drawn all figures with a splash of colour added to some of them. A substantial effort has been invested into the new pharmacodynamics sections including new exercises.

The book has two functions. Its primary role is to serve as a repository of kinetic and dynamic datasets and models, and in addition to serve as a text in biological data analysis. As a service to our readers we have expanded the previous sections on pharmacokinetic and pharmacodynamic theory. The length of such a section is still a matter of balance since we do not aim to write a comprehensive book on only *PK/PD* concepts. Wherever possible, examples from our own data bank will be used throughout the text to illustrate concepts.

This edition is a major revision of the second. We have tried to improve the graphical presentation of the material, although it is far from perfect. We have searched for and tried to stress the *trail from data to insight*. The third edition centres on how to use the power of graphics to visualise, manipulate, summarise and extract information from regression data. Several sections, such as derivation of parameter estimates are now discussed more thoroughly in the applications. Therefore, some of the previous datasets have been deleted and some new ones have been added, primary in the area of pharmacodynamic models. Experimental design issues have been gathered under a separate section to include discussions on *variance inflation factors*, *partial derivatives* with respect to the parameters and the *delta function*. Our aim is both to cover these areas from a practical point of view, and to give references to more theory.

The sections on turnover concepts and inter-species scaling are expanded and integrated. The sections on *in vitro/in vivo* and *in vivo/in vivo* extrapolation are illustrated by means of new examples. Turnover concepts are integrated with the first-order one-compartment constant rate input model and the indirect response model. The pharmacodynamics section now also includes kinetics of drug action, scaling of pharmacodynamic parameters, synergy via turnover or hyperbolic models, and a set of asymmetric binding models, the operational model, as well as multiple dose dynamics. A great deal of work has been invested in updating the theory of dynamic turnover (indirect response) models and tolerance models.

The book still supports the use of *WinNonlin*, although users of other programs may benefit from the provided data sets and theory. A diskette with all command files is enclosed with the book.

We have had a great opportunity to study the seminal work of Professor William Jusko and his colleagues. Their work have heavily influenced our own interest to write the third edition of this book.

We would like to personally acknowledge a number of individuals who made this book possible. We have had the opportunity to collaborate with Jenny Watson, Dunedin, New Zealand, during the technical, pedagogic and linguistic revision of the manuscript. Her tremendous insight and never-ending enthusiasm has been a wonderful source of inspiration. Our sincere thanks goes to Professor Leon Aarons, University of Manchester and Dr. Jan Lundström, AstraZeneca R&D Södertälje, who volunteered to review draft copies of this book. However, if there are still any inconsistencies, typos or other flaws, they remain the sole responsibility of the authors. Dr. Piet-Hein van der Graaf, Pfizer, has been a first-class teacher and instrumental for introducing us to receptor binding theory and operational models. Professor John Urquhart, Palo Alto, continuous to be a great source of inspiration, and has led JG to invest more time on studying *"God's drugs"*. Dr. Åke Norberg, Huddinge Hospital, Stockholm, generously shared his rich data bank of ethanol kinetics. Professor Lennart Hellspong, Södertörn University College, kindly reviewed some chapters with respect to their (lack of) rhetoric content. Colleagues and management at Preclinical Development, AstraZeneca R&D Södertälje, have been tremendously supportive and are greatly acknowledged for making a stimulating scientific environment. Appreciation goes to Nils Granelli who did the cover illustration, Anders Forsman, ABA, and to Maud Sundén and the Swedish Pharmaceutical Press for agreeing to publish the third edition of this book.

JG wants to forward his sincere gratitude to Professors Douwe Breimer and Meindert Danhof for giving him the opportunity to annually visit and lecture at the Leiden/Amsterdam Center for Drug Research (L.A.C.D.R.).

Johan Gabrielsson and Daniel Weiner
Stockholm and Cary, 2000

Table of Contents

Pharmacodynamic models

CHAPTER 1 – The meaning of modeling

Objectives

◆ To introduce the reader to the book

◆ To explain why we model biological systems

◆ To discuss constants, parameters and variables

1.1   Where Do You Want to Begin?

This is a revised and very expanded version of the previous edition of the book. For readers who wish a more in-depth discussion of underlying *PK/PD* (pharmacokinetic-pharmacodynamic) theory, receptor pharmacology or general regression analysis, we recommend text books such as *Applied Therapeutics 3rd ed.* edited by Evans, Schentag and Jusko [1992], *Clinical Pharmacokinetics: Concepts and Applications 3rd ed.* by Rowland and Tozer [1995], *Pharmacological Analysis of Drug-Receptor Interaction, 2nd ed.* by Kenakin [1993] and *Applied Regression Analysis 3rd ed.* by Draper and Smith [1998]. In addition, an extensive bibliography is included at the end of the book.

This book offers the means to perform a complete, critical and hopefully a more accurate analysis of your kinetic and dynamic data. It aims both to give you some theoretical background in an easy-to-understand format, and more importantly to expose you to a large selection of different modeling situations that will give you the practical experience of how to perform a successful analysis of kinetic and dynamic data.

The easiest way of using this book is to first read this chapter. It will help to open paths through the material. Throughout this book we emphasize how to critically evaluate the output from fitting a model to data and to apply statistical techniques for model selection and discussion. We will address questions such as

◆ What needs to be done and how can this be accomplished with available tools such as pencil, paper and software?

◆ How are initial estimates obtained graphically?

◆ What model should be proposed?

◆ How can the adequacy of the fit be assessed?

◆ What data are needed to distinguish between e.g., the one and two-compartment models or the clearance and volume parameters?

◆ How are the parameters in a model interpreted?

Chapters 1 to 5 make up the concept section of this book, and chapters PK1 to PK45 and PD1 to PD27 are the applications sections. Chapter 2 gives an overview of regression theory including objective functions, search algorithms and weighting. Chapter 3 is an introductory text to pharmacokinetics (one and multi-compartment systems, absorption and disposition kinetics, clearance concepts, turnover, nonlinearities, non-compartmental analysis and inter-species scaling including tables of physiological variables (e.g., organ weights and blood flows) of laboratory animals). Chapter 4 introduces pharmacodynamics (law of mass-action, receptor binding models, pharmacodynamic models, kinetics of drug action, time-delays, turnover, dose-response-time data analysis, tolerance, baseline models, logistic models and modeling synergistic responses). Chapter 5 is intentionally placed after the kinetics and dynamics chapters, since it covers general modeling strategies including experimental design. This chapter starts with a section on *Exploratory Data Analysis* (*EDA*) and how to understand your experimental data better. It discusses how complicated model you can get out of your data and how to obtain initial estimates by means of graphical methods and convolution and deconvolution techniques. The latter techniques are covered from a pragmatic point of view. We teach these techniques by means of pictures, since we believe that this may help the novice to get started within a complex area. We believe that these techniques are of substantial value when studying absorption processes independently from regression modeling. Then we cover minimization algorithms, iterations and assessing goodness-of-fit. Discrimination between rival models is presented by means of F-test and Akaike and Schwarz criteria. Experimental design issues as the delta function, the variance inflation factor and partial derivatives are also dealt with. Chapter 5 also contains a brief discussion of outliers and finishes with a checklist for assessing goodness-of-fit.

The applications section is aimed to make you familiar with the *art of modeling* by means of real modeling problems. Start with something simple such as an intravenous bolus data set of a compound that obeys first-order one-compartment plasma kinetics. We recommend that beginners make themselves familiar with the first exercise (PK1). This provides an introduction to kinetic data analysis and contains different representations (models) of the same mono-exponential decline curve. Extravascular dosing is then explored (PK2 and PK3). We continue with an exercise covering four different first-order kinetic models with different parameterizations that are fit to multiple dose data (PK4). If you want to increase the complexity to include multi-compartment kinetics, read the section (e.g., PK7) on fitting mono, two and three-exponential models to data.

A number of examples cover various routes of administration such as intravenous bolus (PK1, PK7), intravenous infusion (PK13), oral (PK11, PK14), transdermal (PK33) and inhalation (PK25). We will make you familiar with how to write your own routine for repeated dosing (e.g., PK4). This can be most useful when you have specified your model in terms of differential equations. Differential equations are also covered in-depth with respect to different parameterizations of linear (PK8) and nonlinear (PK17 to PK24) models. Toxicokinetics is introduced in several exercises (PK15, PK26, PK27). Allometric models are discussed for one (PK28) and multi-compartment (PK29) models. Simultaneous use of several sources of data is stressed throughout this book. Besides numerous practical examples, including plasma and urine

(PK5, PK6 and PK16), several doses (PK17, PK19), subjects (PK20), species (PK28, PK29), *in vitro* systems (PK38, PK44), parent compound and metabolite (PK19, PK45), routes (PK10, PK12), and multiple source dynamic data (e.g., PD2, PD6, PD7, PD11, PD18) are included. Toxicokinetic data has been expanded with a 12-month dog study (PK27) that demonstrates the power *of EDA*. Experimental design issues are elaborated even more now (PK39, PK44, PK45, PD2, PD3, PD6, PD25 and PD26). Reversible metabolism (PK34), Bayesian forecasting (PK35), drug delivery (PK36), *in vitro/in vivo* extrapolation (PK37, PK38) are some commonly requested exercises. We have also included enterohepatic recirculation (PK40), *NCA versus* regression analysis (PK41), and saturable (PK42) and multiple routes of absorption (PK43).

Receptor binding models (PD1) are expanded to also include a discussion on operational models (PD2). Response measurement after repeated dosing (PD6) is now included, since it has been requested by numerous students and readers of previous editions of our book. Dose-response-time data analysis (PD22 to PD25) is expanded to also include body temperature data (PD25) and data exhibiting tolerance (PD11). The derivation of initial parameter estimates has been expanded throughout all problem-solving sections.

You will also get exposed to different parameterizations such as using micro- or macro-constants. Some readers are more used to inter-compartmental micro-constants such as k_{12} and k_{21}, while others prefer inter-compartmental distribution terms Cl_d (also called distribution clearance). Both parameterizations are covered (PK8).

The section on nonlinear pharmacokinetics includes capacity (PK17 to PK20), flow (PK23, PK24), and time (PK21, PK22) dependent kinetics such as saturable elimination, changes in blood flow and enzymatic induction processes, respectively. Nonlinear (PK19) drug-metabolite models are analyzed for multi-compartment systems.

Turnover concepts (PK30 to PK32) of endogenous compounds (*IgG, estradiol* and *hyaluronan,* respectively) are covered separately with three examples. How to gain information about production (synthesis or secretion) and loss (catabolism or excretion) of the endogenous compound is illustrated by introducing an external dose.

In Chapter 4 a more in-depth discussion of pharmacodynamic models is given, covering the law of mass-action, binding and saturation models (PD1, PD2, PD8, PD14, PD15), indirect response (turnover) models (PD4 to PD7) of different origin, tolerance/rebound (PD9 to PD12) models, and dose-response-time data analysis (PD22 to PD25). Synergy is discussed using turnover and hyperbolic (PD26) functions. We have also added an incomplete concentration-effect dataset (PD27). At the end of the pharmacodynamics section, we have added a discussion on pulsative (oscillating) functions, which relates to both dynamics (PD12) and kinetics (PK35).

1.2   Why Model the Data?

There is a growing recognition of the importance of pharmacokinetics (*PK*) and pharmacodynamics (*PD*) in all phases of drug development (Breimer and Danhof, [1997]). The primary objective is to characterize drug specific factors (e.g., EC_{50}, E_{max}) and system specific factors (e.g., k_{in}, k_{out}) that govern the 'on/off' response-time courses. The effects of a drug depends usually on the drug-receptor interaction (primary effect) and on the responses of the physiological system and its control mechanisms to the disturbance caused by the primary effect (Struyker-Boudier *et al* [1990]).

Drug actions trigger the onset of various kinds of physiological responses, which have their own time-scales of operation (e.g., baroreceptor activation occurs in seconds whereas baroreceptor adaptation occurs in hours or days; vascular remodeling may take weeks or months). In ignoring 'off' responses, which are often markedly asymmetric to 'on' responses, one miss a powerful constraint on *PD* modeling, which is that the model has to be able to simulate both the 'on' and the 'off responses (personal communication with Dr. John Urquhart [2000]).

In preclinical studies, *PK/PD* is used to interpret toxicokinetic data, and via physiological modeling and allometric scaling, to extrapolate results from animal to man. During early clinical testing, *PK/PD* is used to aid in the interpretation of dose-response and dose escalation studies. There are now several instances where *PK/PD* modeling has been used by regulatory agencies to recommend a dose and/or regimen that was not originally studied as part of the clinical program.

The analysis of *PK/PD* data can be complex and time consuming, partly because modeling is as much an art as it is a science (Shahin *et al*, [1984]). As in all biological experiments, biological, assay and other noise contaminate *PK/PD* data and one often has to use sophisticated data analysis and modeling techniques to estimate parameters of interest and to separate the relative sources of variation.

Historically, many scientists were hampered from performing *PK/PD* modeling because of a lack of training and a lack of ease in using *PK/PD* modeling software. Fortunately, in recent years several *PK/PD* modeling programs (e.g., *Kinetica*, *WinNonlin*) have been released that employ good statistical algorithms and are relatively easy to use. (Gabrielsson and Weiner [1999]).

The emphasis throughout this book is on how to apply biological reasoning and statistical techniques to model selection and discussion, and how to critically evaluate the output from fitting a model to data. In particular, we focus on the analysis of *rich* data sets, rather than on modeling problems associated with *sparse* data. However, we sometimes combine several sources of *sparse* data and fit them simultaneously.

Modern software enables scientists and non-professional modelers/statisticians to analyze their own data. Good analysts let the underlying biology and the data that were collected drive the analysis. It is important to take time to understand the data and the results of the modeling properly. As Barbara McClintock has stated (Keller [1983]) they were "so intent on making everything numerical" that they frequently missed seeing what was there to be seen. We will illustrate these concepts via the modeling of numerous datasets, several containing concentration-time data, concentration-response data and the others response-time data.

1.3   The Road to Successful Modeling

In order to be successful in modeling, we recommend that you take a look at the modeling carousel depicted in Figure 1.1. One usually has an idea about the kinetics and dynamics of a drug prior to designing the study (Step 1). This notion about how the system will behave includes hypotheses to be tested (e.g., drug A has an effect on blood pressure which is optimal at dose B), ideas about the drug's behavior (by increasing the dose, the response to the drug will increase), and a tentative model of the system (the response *versus* concentration relationship can be modeled by an E_{max} model).

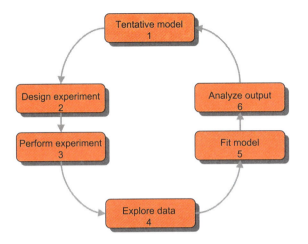

Figure 1.1 The modeling carousel including tentative model (Step 1), experimental design (Step 2), experiment (Step 3), exploratory data analysis (Step 4), model fitting (Step 5) and analysis and model discrimination (Step 6).

A successful modeling project is dependent upon conducting a well-designed study. The selection of doses to be administered, and the times that plasma concentrations and responses should be measured, are an intrinsic part of the design step (Step 2). This step includes both the actual design of the trial and simulations to understand the range of results that might be seen. *CATD* is a powerful tool for designing clinical trials and, in our opinion, will become increasingly an integral aspect of the drug development process.

In Step 3, the actual experiment is performed to collect the data. After data are gathered, we recommend that an exploratory (graphical) data analysis (*EDA*) be performed to confirm or suggest modifications to the tentative model(s). This is Step 4. Unfortunately, this important exercise of graphical exploratory analysis is often overlooked. The data, when collected from a well-designed trial, will lead one to better understand the behavior of the system/model.

A powerful way of evaluating possible nonlinearities in the data involves dose normalizing the plasma concentration curves following several dose levels, and then plotting the dose-normalized concentration profiles against time. Alternatively, derived variables such as Area Under the plasma concentration-time Curve *AUC* may be plotted against dose. This type of design is frequently used in preclinical studies. Toxicokinetic analyses, for example, effectively utilise *EDA* methods by plotting plasma concentrations and *AUCs* (exposure) against dose, gender or time to ascertain if the kinetics of a drug is nonlinear over the range studied.

In *EDA,* one commonly starts with plots of concentration *versus* time, response *versus* time and response *versus* concentration. These three relationships / plots will reveal attributes of the data that require special care during modeling. These include time delays between concentration and response, the relationship between concentration and response (saturation, hysteresis, and adaptation), and the need for further data to fully characterize the pharmacodynamic behavior of the drug.

The next step is the selection of a model and the fitting of that model to the experimental data (Step 5) using regression (usually non-linear) analysis. This is done to estimate the model parameters and the precision of the parameter estimates. The selection of the model is usually suggested from prior studies, and from using the graphical displays discussed above.

Before fitting a model to data, the *EDA* should have suggested not only information about the structure of the model but also the initial estimates of the model parameters. Note that algorithms for fitting non-linear models generally require initial estimates of the parameters. Most good computer programs, will derive the initial parameter estimates for you. However, we recommend you carefully assess the appropriateness of the estimates. There are numerous examples where computer programs converge to a local minimum (Figure 1.2).

WRSS

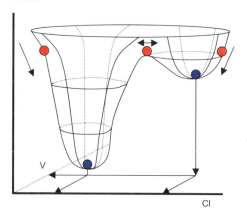

Figure 1.2 *Schematic diagram of the parameter space for a 1-compartment intravenous bolus model containing clearance (Cl) and volume of distribution (V) as model parameters. WRSS denotes the weighted residual sum of squares.*

Errors such as this can often be avoided by having used good initial estimates. To ensure that a program has not converged to a local minimum, we suggest rerunning the program with different initial estimates, and possibly using algorithms that employ different approaches to model fitting. Examples are the simplex and Levenberg-Marquardt algorithms.

Step 6 involves the evaluation of the program output and diagnostics, such as goodness of fit, correlation between parameters, residual analysis, and parameter accuracy and precision. This evaluation forms the basis of an assessment as to how well the model explains the data, and is used for comparisons of competing models. We generally look at plots of observed and predicted data superimposed on a linear and semi-logarithmic scale, or at residual plots such as absolute and weighted residuals vs. the independent variable (e.g., time) or the predicted variable (e.g., concentration). Do the plots demonstrate any trends? Residual plots are probably the most efficient means to search for both the best fitting structural model (e.g., mono- or bi-exponential decay) or variance model (e.g., constant absolute error or constant relative/proportional error).

Step 6 gives the final answers to assess how appropriate the model(s) is (are). However, you still need to challenge your model(s). A good way to challenge a model is to test the model on a new set of experimental data. This process is often called the model validation step. The model can either be used for design purposes (simulation of new studies) of a new experiment (Step 2), or to fit a new set of data (Step 5). In our experience, the experimenter seldom views model building as an iterative process, which it should be.

Multiple dose levels, repeated dosing, and multiple input rates and routes of administration are often needed to answer questions such as "Does the response reach a maximum or threshold?", "Are there active metabolites?" and "Is there evidence of development of tolerance or sensitisation to the drug?". Thus, in order to extract maximum information from a study, it is critical that the study be properly designed.

The goals of model building are to reduce data (estimate the model parameters and their precision) and to test hypotheses (about these parameters) and compare competing models. However, posing sharp questions about the model (challenging the model) is an important aspect of model building that is often overlooked.

As stated above, model building and fitting can be a complex and lengthy process. Thus, we strongly encourage scientists only to model when they need to. For example, in situations where all that is needed is a measure of drug exposure (including bioavailability), or perhaps total clearance, *NCA* will usually suffice. While one might argue that even *NCA* is a form of modeling, the assumptions involved are less restrictive than fitting compartmental models, and the analyses can be completed very quickly. Conversely, one may be required to fit a model if one is using single dose data to predict multiple dose results, or if the kinetics are known to be nonlinear in the range being studied, or if one needs to quantify the relationship between *PK* and *PD*. This would usually be a compartmental model or a physiologically based pharmacokinetic (*PBPK*) model.

1.4   Constants, Parameters and Variables

We have already introduced the expressions *constant, parameter* and *variable*. They are all part of the modeling jargon. We will explain what they mean and how we treat them in this book.

Assume that you have done an experiment and want to characterize the kinetics of the drug with a simple model. The model may be written as

$$\hat{C} = \frac{D}{V} e^{-K \cdot t} \qquad\qquad (1:1)$$

The dose D is a true *constant*, which has a known value and which stays unchanged. The volume of distribution V and the elimination rate constant K are *parameters* to be estimated by fitting the model to the data. Note that in this model D, V and K are *constant* values since they do not vary with time, concentration or dose, whereas $\hat{C}$ and t are *variables*. Time t is the *independent variable* and $\hat{C}$ is the predicted value of the *dependent variable* based on the model. The latter depends on D, V, K and t. $\hat{C}$ is also called the *model predicted* concentration in comparison to the *observed* concentration, denoted as C. The relationship between the observed and predicted dependent variables and the measurement *error (ε)* is given below

$$C = \hat{C} + \varepsilon \qquad\qquad (1:2)$$

$\hat{C}$ is assumed to be the true or mean concentration value, whereas C is the observed concentration value.

For the purpose of our discussion, we denote the parameters used in the model specification, such as the volume and the rate-constant as being *primary* model parameters. Often though, of equal or perhaps more importance, are parameters which are themselves functions of the (primary) model parameters. These functions of the model parameters will be denoted as *secondary* parameters, and include (for this model) clearance ($Cl = K \cdot V$), area under the curve ($AUC = D/Cl$) and half-life ($t_{1/2} = ln\ (2)/K$). Thus, the words *primary* and *secondary* relate to the model parameterization and not to the relative importance of the parameters.

A *data descriptor,* such as $\hat{C}$, is the concentration or response predicted from the estimated parameters. Since data descriptors are calculated from estimated parameters, they also have an accuracy and precision. The partial cancellation of parameter covariances can cause data descriptors to be estimated with higher precision and accuracy than the sometimes poorly estimated, highly correlated parameters from which they are derived (Ebling [1995]). That is, concentrations predicted by a particular model may have greater precision than the estimated model parameters.

The primary goal in regression is to understand how a *response variable* (e.g., concentration or effect) depends on one or more *predictors* (e.g., dose, clearance, time, age, weight, and even concentration). The response variable is often called the *dependent* variable, and the predictor variable may be called the *independent,* *explanatory* variable or *covariate.*

CHAPTER 2 - Parameter estimation

Objectives

♦ **To introduce linear and nonlinear regression models**

♦ **To demonstrate criteria for best fit and minimization methods**

♦ **To discuss the choice of weights**

♦ **To review methods for obtaining least squares estimates of nonlinear models**

♦ **To discuss constraints on the parameter space**

♦ **To discuss estimation of functions of parameters**

♦ **To discuss software validation**

2.1   Background

A number of algorithms have been proposed for estimating the parameters in a general nonlinear model. None of the algorithms is universally best, as they can all fail in certain situations. Unfortunately, it is not always apparent when an algorithm has failed to find the *best* solution.

One measure of a good computer program is whether or not the program provides information that will enable the user to determine if it converges to the correct solution. Computer programs that provide parameter estimates but do not provide any diagnostics should be avoided.

In this chapter, several of the most commonly employed estimation algorithms are discussed along with their relative strengths and weaknesses. Criteria for *best* fit are presented, along with diagnostic statistics. This chapter is not intended to provide the user with an in-depth discussion of the mathematics of nonlinear estimation. However, references are provided which cover this topic in more depth. Rather, the intent is to provide the user with enough information to be able to intelligently assess how well the model fits the data, and what to do when something is wrong.

To facilitate the discussion of parameter estimation, we first focus on the estimation of parameters for linear models. These methods are then extended to the estimation of parameters for nonlinear models.

2.2   Linear and Nonlinear Models

Let us assume that we have measured a series of concentrations, often denoted by Y_i, where i denotes the ith measurement. For notational convenience we will omit the '*i*' in situations where its use is obvious. Assume that N concentrations were obtained; that is, *i* takes on values from *1* to *N*. Actually, Y need not denote only concentrations, it could represent any measured quantity such as a change in blood pressure, urinary excretion rate, etc. Let us further assume that the concentration data can be modeled by some set of independent (controlled) variables such as *time, pH, temperature*, etc. We denote these independent variables as $X_1, X_2, ...$, where the subscript denotes the values of each independent variable corresponding to the ith observation. We further assume that

$$Y = f(\underline{X}, \underline{\beta}) + \varepsilon \qquad (2{:}1)$$

where

$$\underline{X} = (X_1, X_2, \dots X_n) \qquad (2{:}2)$$

and

$$\underline{\beta} = (\beta_1, \beta_2, \dots \beta_n) \qquad (2{:}3)$$

That is, $\underline{X}$ denotes the vector of independent variables corresponding to the observation and $\underline{\beta}$ denotes the vector of model parameters, which must be estimated. The function $f(\underline{X},\underline{\beta})$ represents the true deterministic concentration corresponding to $\underline{X}$ and $\underline{\beta}$, and ε denotes a random unobservable error. We assume that the distribution of the errors (ε) is normal, and has a mean of zero and a variance of σ^2.

$$\varepsilon = N(0, \sigma^2) \qquad (2{:}4)$$

Note that we are assuming that, given values of $\underline{X}$ and $\underline{\beta}$, we can compute the *exact* value of Y, except for the magnitude of the random error ε. Models of this type are said to be deterministic. A model is said to be linear if

$$f(X, \beta) = X\beta \qquad (2{:}5)$$

that is, if f can be written as the product of a matrix $\underline{X}$ and a vector $\underline{\beta}$. Note that *all* linear models can be expressed by Equation 2:5. The following are examples of linear models

$$Y = \beta_0 + \beta_1 X \qquad (2{:}6a)$$

$$Y = \beta_0 + \beta_1 X + \beta_2 X^2 \qquad (2{:}6b)$$

The first model is a simple linear regression model and, if one were to draw a graph of this model, it would appear as a straight line (with intercept β_0 and slope β_1). The second model is a quadratic regression model and, if one were to draw a graph of this model, it might not appear as a straight line. However, this second model is still linear in the sense that it is a linear function of the parameters β_1 and β_2. One way to determine if a model is linear in the model parameters is to compute partial derivatives with respect to each of the parameters in the model. If none of the partial derivatives involve model parameters, then the model is said to be linear.

$$\frac{dY}{d\beta_1} = X \qquad (2{:}7)$$

$$\frac{dY}{d\beta_2} = X^2 \qquad (2{:}8)$$

For the second model above (Equation 2:6b), the partial derivative with respect to β_0 is 1, the partial derivative with respect to β_1 is X (Equation 2:7) and the partial derivative with respect to β_2 is X^2 (Equation 2:8). Since none of these derivatives involve β_0, β_1 or β_2, the model is said to be linear.

The following model (Equation 2:9) is nonlinear, however, as the partial derivative with respect to β_1 clearly depends on β_2 (Equation 2:10).

$$Y = \beta_0 + \beta_1 e^{-\beta_2 \cdot X} \tag{2:9}$$

$$\frac{dY}{d\beta_1} = e^{-\beta_2 \cdot X} \tag{2:10}$$

Note that the above model, being nonlinear, cannot be expressed as a product of two matrices as in Equation 2:5. Examples of linear and nonlinear regression models are shown in Table 2.1.

Table 2.1   Comparative examples of linear and nonlinear regression models

	Linear models		Nonlinear models	
Function	$Y = K \cdot X + L$	$Y = aX + bX^2$	$Y = Ae^{-\alpha \cdot t} + Be^{-\beta \cdot t}$	$E = \dfrac{E_{max} C}{EC_{50} + C}$
Derivative	$\dfrac{dY}{dK} = X$	$\dfrac{dY}{db} = X^2$	$\dfrac{dY}{d\alpha} = -At \cdot e^{-\alpha \cdot t}$	$\dfrac{dE}{dEC_{50}} = -\dfrac{E_{max} \cdot C}{\left(EC_{50} + C\right)^2}$

NOTE: Linear pharmacokinetic models, such as first order kinetic models, are generally nonlinear regression models with respect to K and V vis-à-vis the dependent variable C.

$$C = \frac{D}{V} e^{-K \cdot t}$$

C and K are not directly proportional to each other. This also holds true for nonlinear kinetic and dynamic models. Although this model could be made linear by taking the logarithm of both sides of the equation, as written it is nonlinear.

2.3   Criteria for Best Fit - Minimization Methods

2.3.1   Least squares methods - *OLS, WLS* and *ELS*

We mentioned earlier that data are needed to estimate the model parameters. Our objective is to determine an estimate of β, such that the differences between the observed and predicted concentrations are in some sense *small*. Three of the most commonly employed criteria are ordinary least squares (*OLS*), weighted least squares (*WLS*), and extended least squares (*ELS* is a maximum likelihood procedure). These criteria are achieved by minimizing the following quantities (often called objective functions, e.g., O_{OLS}, O_{WLS} or O_{ELS})

$$O_{OLS} = \sum_{i=1}^{n} (C_i - \hat{C}_i)^2 \tag{2:11a}$$

$$O_{WLS} = \sum_{i=1}^{n} W_i (C_i - \hat{C}_i)^2 \tag{2:11b}$$

$$O_{ELS} = \sum_{i=1}^{n} \left[W_i (C_i - \hat{C}_i)^2 + \ln(\text{var}(\hat{C}_i)) \right] \tag{2:11c}$$

$\hat{C}_i$ denotes the predicted value of C_i based on the model and W_i denotes the weight.

The correct criteria for best fit depends upon the assumption underlying the functional form of the variances of the dependent variable C. The most commonly employed variance model is

$$\text{var}(C) = \sigma^2 f(X, \beta)^\lambda \tag{2:12}$$

The theoretically correct weight to be assigned to the observation, based on maximum likelihood considerations, is the reciprocal of the variance of the observation (Draper and Smith [1998]). Thus, observations associated with *large* variances get *less* weight when computing the above objective functions.

For purposes of this discussion, λ is assumed to be known. Note that if $\lambda = 0$, the variances of the C_i are identical and constant. In this case, the *OLS, WLS* and *ELS* criteria are equivalent. In addition, if the weights are constant valued (that is, do not involve any parameters that must be estimated), then estimation of parameters using *WLS* and *ELS* is equivalent.

If the variance model is some function of the independent variables $\underline{X}$, but does not depend on any parameters which must be estimated, then *WLS* is the appropriate criteria to use. If the variance model also depends on one or more of the model parameters (β), then a variation of *WLS* is appropriate. In this instance, the procedure is to first assume some value for the weights (W_i) and perform *WLS*. The resultant estimated model parameters are then used to obtain an updated estimate of W_i, and *WLS* is again performed using these updated values. This process is continued until the process converges. This modification of *WLS* is denoted as iterative reweighted least squares (*IRLS*).

It has been proposed that rather than arbitrarily assign a value of *1* or *2* to λ, λ should be estimated as a parameter using *ELS* (Sheiner and Beal [1985]). *ELS*, like *IRLS*, is an iterative procedure. However, in the authors' experience, the resultant plots of observed versus predicted values obtained by *ELS* and *IRLS* are generally indistinguishable. For the types of data and models usually encountered in pharmacokinetics, simulation studies have shown that *ELS* offers no practical advantage over *IRLS*, and can often produce parameters which are more biased or have greater variability than those produced by *IRLS* (Metzler [1987], van Houweligen [1988]).

The theoretically correct weight to assign to an observation when fitting a model to data is the reciprocal of the variance of the observation. Usually the *true* variance of an observation is not known and must be assumed to take on some functional form. For example, many researchers often assume that the variance of a measurement is proportional to the observed value of the concentration or to the square of the observed value of the observation. For this latter case, Equation 2:12 is assumed to take on the following form

$$var(C) \propto C^2 \qquad (2:13)$$

Note that for this example the variances depend only on the observed data and do not depend on any parameters which must be estimated; thus *WLS* would be the appropriate criteria to use to estimate the model parameters. However, assuming the model is correct, the predicted concentration is a better estimate of the true concentration than is the observed concentration. Thus the equation would take on the following form

$$var(C) \propto \hat{C}^2 \qquad (2:14)$$

where the *hat* denotes a predicted value. Note that predicted concentrations are based on the model, and as such are functions of the estimated parameters in the model. Thus *IRLS* would be the appropriate criteria to estimate the model parameters for this variance model.

2.3.2   Alternative methods - Generalized Least Squares

If the error variance is not normally distributed and/or the variance applied is not correct, then the *Maximum Likelihood (ML)* and *IRLS* methods may provide biased or poor estimates. To overcome the limitations of the latter methods the following simple approach has been proposed (Seber and Wild [1989]).

---

1.   **Use *OLS* to get an estimate of the parameters $\hat{O}_{OLS}$**

2.   **Set weight $w_i = var\left(\hat{O}_{OLS}^{-1}\right)$ or $w_i = var\left(\hat{O}_{OLS}^{-2}\right)$**

3.   **Re-fit model using *WLS* with weights obtained from step 2**

The following example is intended to illustrate the *Generalized Least Squares (GLS)* fit of a 2-compartment first-order input model with a constant CV error model. In step 1, we start by fitting the models (in this case a two-compartment model with first order input) to data by means of *OLS*, *i.e.*, weights set to 1 or $W_i = 1$. Table 2.2 contains information on time, observed concentration, predicted concentration, derived weights, and the predicted concentration by means of *GLS*.

Table 2.2   Data obtained from a *GLS* example

Time	C_{obs}	$\hat{C}_{OLS}$	$w_i = \hat{C}_{OLS}^{-2}$	$\hat{C}_{GLS}$
0.0833	13.9	12.07	0.006864	13.89
0.167	152	161.7	0.0000382	157.6
0.25	226	213	0.0000220	209.5
0.5	204	211.8	0.0000240	212.1
1	149	145.7	0.0000471	146.8
1.5	100	99.35	0.0001013	100.1
2	66	68.37	0.0002137	68.85
3	36	33.77	0.0008769	33.71
4	17.7	18.13	0.003042	17.76
6	6.9	7.405	0.018237	6.908
8	3.96	4.573	0.047819	4.21
12	2.89	2.716	0.13556	2.655
24	0.9	0.757	1.74460	0.9456
25	0.9	0.681	2.1563	0.8682

From the first fit we then use the calculated concentrations to predict the individual weights for step 2. This is done by taking the squared reciprocal concentration (assuming a constant CV error model). That weight is then used in the *WLS* fit in step 3. The calculated concentrations are then *GLS*. The resulting *OLS* and *GLS* fits are shown in Figure 2.1 below.

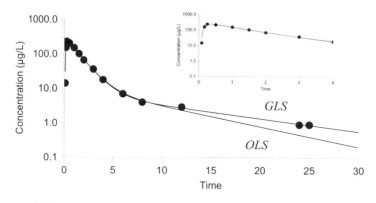

Figure 2.1 *Concentration-time plot of observed (●) and predicted (solid lines) by means of OLS and GLS. The inset shows the observed and predicted (OLS and GLS superimposed) concentration-time course within the first four hours.*

As you can see in this particular example, the difference between the *OLS* and *GLS* fit is marginal. As a general rule, we would recommend the use of *IRLS*. In our experience it is a very robust method.

2.4 Considerations in the Choice of Weights

2.4.1 Why weight?

Weighting schemes exist to account for heterogeneity in the variance of the data. From principles of maximum likelihood it is generally proposed that the weight W_i, should be inversely proportional to the variance σ_i^2 of the observed value Y_i.

$$W_i = \frac{1}{\sigma_i^2} \tag{2:15}$$

We want observations that contain little (or less) error to influence the parameter estimates and their precision more than observations prone to larger error. This is schematically depicted in Figure 2.2.

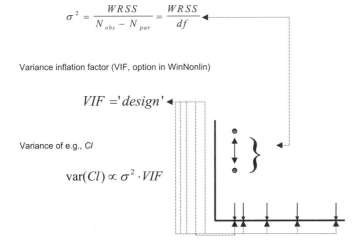

Preditced mean variance (σ^2)

$$\sigma^2 = \frac{WRSS}{N_{obs} - N_{par}} = \frac{WRSS}{df}$$

Variance inflation factor (VIF, option in WinNonlin)

$$VIF = 'design'$$

Variance of e.g., *Cl*

$$var(Cl) \propto \sigma^2 \cdot VIF$$

Figure 2.2 Schematic diagram of the determinants of the precision (variance σ_i^2) of a model parameter. The variance is determined by experimental error and the design point, i.e. at what time points the concentrations are measured. The Y axis is concentration and the X axis time.

If correct weighting is applied, the weighted residuals will be comparable for all data included in the analysis. The experimental error *(ε)* is generally assumed to be normally distributed with mean zero and a variance of σ_i^2, and is mathematically expressed as $\varepsilon = N(0, \sigma^2)$.

2.4.2 Constant absolute error

If variance is (assumed to be) constant for all observations, then the weights are typically set to a constant value, usually 1.

$$W_i = \frac{1}{constant} \tag{2:16}$$

Figure 2.3 displays the behavior of a constant absolute variance (error).

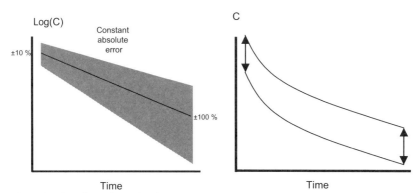

Figure 2.3 *Semi-logarithmic plot of concentration versus time including the variability in data (shown as shaded area) when we have a constant absolute error (left). Concentration versus time with the same error distribution on a linear scale (right).*

A semi-logarithmic plot of pooled data gives a cone-formed shape as shown in Figure 2.3 (left). On a linear scale the error behaves as shown in Figure 2.3 (right). A value of *100 ± 10 (mean ± SD)* at high plasma concentrations corresponds to a *10%* coefficient of variation *CV (100·SD/mean)* and *10 ± 10* at low concentrations to a *100% CV*, respectively. We have an absolute error of *10* in both instances although the relative error, as measured by the *CV*, is ten times higher in the latter case.

2.4.3   Poisson error

When the error variance is proportional to the mean of the predicted value, e.g., the concentration,

$$\sigma_i^2 \propto \hat{C}_i \tag{2:17}$$

then the weights are set to

$$W_i = \frac{1}{\hat{C}_i} \tag{2:18}$$

This weighting scheme assigns relatively larger weights to lower concentrations. Since the weights involve the model parameters (i.e., $\hat{C}_i$ is a function of the model parameters), one would generally use *IRLS* for estimation purposes.

2.4.4   Constant relative error - Proportional error

The error is said to be proportional when the coefficient of variation is constant over the observed concentration range. With many experimental protocols, you expect the experimental scatter to be a constant percentage of the *C* value. In such a case, observations with high *C* values will have more scatter than observations with low *C* values.

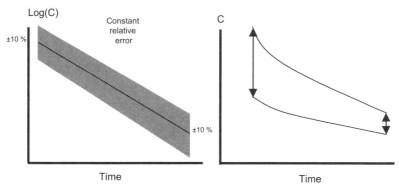

Figure 2.4 *Semi-logarithmic plot of concentration versus time (left) including the variability in data (shown as shaded area) when we have a constant relative error. Concentration versus time (right) with the same error distribution on a linear scale.*

When the unweighted residual sum of squares (*RSS*) is minimized, observations with high *Y* values will have a larger influence, while those with smaller *Y* values will be relatively ignored. However, this problem can be avoided by minimizing the sum of the relative distances. In this case, weight by the reciprocal of the concentration squared. Figure 2.4 depicts the situation where there is a constant relative error. A semi-logarithmic plot of pooled data exhibits a cone-shaped form. Values of *100 ± 10 (mean ± SD)* in the high concentration range and *10 ± 1* in the lower concentration range correspond to a coefficient of variation of *10%*. There is an absolute error of *10* and *1*, respectively, while the relative error is constant.

We recommend that one also plot data from all individuals in the same plot to observe what type of error structure there is in the study population. This gives an idea about the sum of intra- and inter-individual variability. Figure 2.5 shows data in a semi-logarithmic scale that displays constant absolute (*homoscedastic*) error and constant relative (*proportional*) error. We include a more thorough discussion of error distribution in the section on residual analysis (see 5.7.1, 5.7.2).

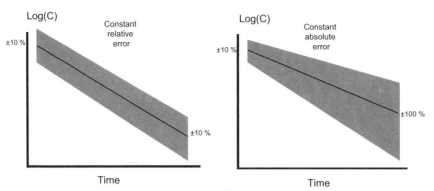

Figure 2.5 *Semi-logarithmic plots illustrating the difference between constant relative (proportional) error (left) and constant absolute error (right).*

Although it is not often discussed, *OLS* is inherently biased towards fitting *larger* values relatively more closely than *smaller* values. This is a consequence of the nonlinearity of the data and is also accentuated by a large range in the data. To see this, consider the following hypothetical example. Suppose we are trying to fit a model to a dataset, which contains the observed values of 10 and 10,000. Recall that *OLS* obtains parameter estimates, which minimize the sum of squared residuals. Thus it is the residual value associated with an observation, and not the observation itself, which has an impact on *OLS*. For our hypothetical example, the best *OLS* fit might be obtained when the predicted values are 5 and 9995, so that the corresponding residuals are both 5. Note that although both data points are associated with the same magnitude of residual, the larger data value is fit much more closely in terms of percentage deviation; the corresponding values are –50% and –0.05%. When dealing with assay data, it is not uncommon for the data to span a few or several orders of magnitude. Thus, in these instances, it is often necessary to assign weights to the observations, which result in parameter estimates that enable the model to fit *low* values as well as *high* values. If a model systematically overestimates *large* concentrations and systematically under-estimates *small* concentrations, then weighting by the reciprocal of the concentration will generally improve the fit.

Traditionally, many scientists have chosen to weight each observation by the reciprocal of the observation itself, or by the square of the observation, in order to achieve a more consistent fit to low and high values. While this idea is not without merit, a better procedure is to weight by the reciprocal of the predicted value. The reasons for this are two-fold. First, if the true variance of a value is proportional to the value itself, then the theoretically correct weight is the reciprocal of the predicted value, as was discussed in the previous section. Second, *OLS* is known to be very sensitive to outliers. That is, if one or more of the data values are incorrect or contaminated in some way, *OLS* may produce parameter estimates that are highly biased. *WLS*, with weights defined as the reciprocal of the observed values, will be of little or no value in reducing the bias as one or more of the weights are also based on outlier values. *IRLS*, on the other hand, is more robust against outliers, because the weights are the reciprocals of the predicted values from the model and not of the outlier values themselves. Thus, the recommendation is that the predicted value be used in place of the observed value when it is desirable to weight by some function of the *true* response.

2.4.5 Graphical estimation of weights

In addition to the iterative methods that are employed to assess the type and magnitude of weights, there are more robust approaches. We will demonstrate a graphical method (Cook and Weisberg [1994]). If the residual is positive it means that the observation is above the predicted curve. If the residual is negative, the observation is below the predicted curve. We assume that $var(C_i)$ is proportional to the predicted concentration raised to a factor λ. Then

$$var(C) \propto var(residual) \propto \hat{C}_i^{\lambda} \qquad (2:19)$$

We recommend that you start your estimation procedure by means of an *OLS* fit assuming a constant absolute error (constant absolute variance). From that fit, a scatter plot of the natural logarithm of the squared residuals, $ln(residual^2)$, can be plotted against the natural logarithm of the predicted value, $ln(\hat{C}_i)$. From that plot, one can obtain the slope of the regressed line. The slope is proportional to λ. Data in Table 2.3 have been used to exemplify a graphical method for derivation of the weighting exponent, λ. Some of the small residuals were deleted (sic!) since these will produce large negative $ln(residual^2)$ values and therefore influence the linear regression a disproportionate amount.

Table 2.3 Data used to exemplify graphical estimation of weights.

Time	C_{obs}	$\hat{C}_{OLS}$	$\ln(resid^2)$	$\ln(\hat{C}_{OLS})$
0.0833	13.9	12.07	1.21	2.49
0.167	152	161.7	4.54	5.09
0.25	226	213	5.13	5.36
0.5	204	211.8	4.11	5.36
1	149	145.7	2.39	4.98
1.5	100	99.35	–	4.60
2	66	68.37	1.73	4.22
3	36	33.77	1.60	3.52
4	17.7	18.13	–1.69	2.90
6	6.9	7.405	–1.37	2.00
8	3.96	4.573	–0.98	1.52
12	2.89	2.716	–3.50	1.00
24	0.9	0.7571	–	–
25	0.9	0.681	–	–

In Figure 2.6 below, the slope of the fitted line is 1.6, which we would use as the exponent in the weighting function λ. In a later section we will use this data set for *GLS* estimation and then apply an exponent λ equivalent to 2.

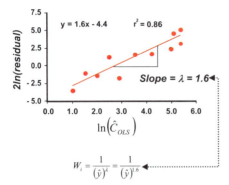

Figure 2.6 Natural logarithm of the squared residual versus the natural logarithm of the predicted value using OLS. The slope (1.6) of the regressed line corresponds to the weighting exponent.

This slope λ will be equivalent to the exponent of the weighting function below. Figure 2.7 illustrates schematically the relationship between log-residual as a function of log-predicted concentration.

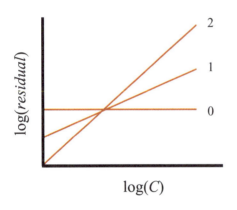

Figure 2.7 Schematic representation of Log-residual versus log-predicted C. The number in the margin corresponds to the weighting exponent λ.

$$W_i \propto \frac{1}{var(\hat{C})} \propto \frac{1}{\hat{C}^\lambda} \qquad (2:20)$$

For a constant CV model (proportional error or constant relative variance), λ is equivalent to 2 and for a constant absolute error model, λ is equivalent to 0, i.e., error is independent of the magnitude of C. For a Poisson error distribution, λ is 1.

2.5   Application of Least Squares to Linear Models

Recall that the general linear model can be written as

$$Y = f(\underline{X},\underline{\beta}) + \varepsilon \qquad (2:21)$$

where

$$f(X,\beta) = X\beta \qquad (2:22)$$

$\underline{X}$ denotes the design matrix, and $\underline{\beta}$ denotes a vector of model parameters. It can be shown (Draper and Smith [1998]) that the *OLS* estimate of $\underline{\beta}$ is

$$\hat{\beta} = (X'X)^{-1} X'Y \qquad (2:23)$$

The variance of $\hat{\beta}$ is given by

$$\text{var}(\hat{\beta}) = \hat{\sigma}^2 (X'X)^{-1} \qquad (2:24)$$

where $\hat{\sigma}^2$ is the predicted mean variance

$$\hat{\sigma}^2 = \sum_{i=1}^{n} \frac{(Y_i - \hat{Y}_i)^2}{N_{obs} - N_{par}} \qquad (2:25)$$

N_{obs} is the number of data observations and N_{par} is the number of parameters to be estimated. $(X'X)^{-1}$ is also called the matrix of variance inflation factors *VIF*, which will be discussed more in depth in Section 5.9.3 (Pharsight [1999]).

2.6 Application of Least Squares to Nonlinear Models

2.6.1 Background

As we saw in Section 2.2, $f(\underline{X},\underline{\beta})$ cannot be written as $\underline{X}\underline{\beta}$ for nonlinear models. Thus the technique that was used to obtain the least squares parameter estimates for linear models is not directly applicable to nonlinear models. A variation of the linear model method can be used, however, as can a number of other methods.

To better understand the difficulty in fitting nonlinear models as compared to linear models consider the following example. Figure 2.8 (A) displays the weighted residual sum of squares *WRSS* surface associated with fitting a linear model like $Y = \alpha - \beta x$ to a particular data set, while Figure 2.8 (B) displays the corresponding contour plot.

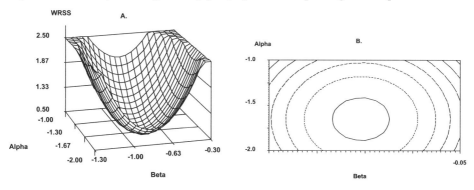

Figure 2.8 *Weighted residual sum of squares WRSS surface for various values of alpha and beta (A) and contour plot of the weighted residual sum of squares surface (B).*

The contours show values of α and β that generate the same *WRSS*. Note that the *best* solution is unique.

Next, consider the following nonlinear model (example adapted from Seber and Wild [1989])

$$Y = \alpha \cdot e^{-\beta \cdot x} \qquad\qquad (2{:}26)$$

The *WRSS* obtained by fitting this model to a particular data set is displayed in Figure 2.9 (A), and the corresponding contour plot in Figure 2.9 (B).

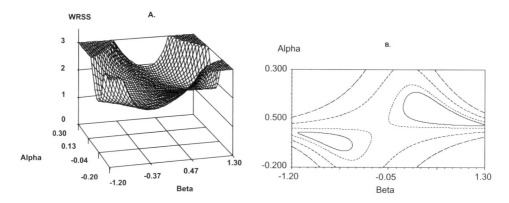

Figure 2.9 *Weighted residual sum of squares WRSS surface for various values of alpha and beta (A) and contour plot of the weighted residual sum of squares surface (B).*

Note that for this example there are two relative minima. Thus, different software packages or algorithms may well provide differing estimates of the *best* solution. The actual shape of the *WRSS* surface and resulting contour plot will vary depending on the model fitted to the data. However, these examples serve to illustrate the inherent difficulty in fitting nonlinear models. Below the following methods will be discussed for obtaining least squares estimates for nonlinear models:

◆ **Random search methods**
◆ **Stripping or peeling methods**
◆ **Linearization methods**
◆ **Simplex methods**

2.6.2   Random search methods

Random search methods require knowledge of the range of plausible values that parameters can take on. That is, we assume that for each of the i parameters

$$L_i \le \beta_i \le U_i$$

For some increment h_i, compute the following values

$$L_i, L_i+h_i, L_i+2h_i, \dots, U_i$$

Next, compute the weighted residual sum of squares for all possible combinations of values $\beta_1, \beta_2, \dots \beta_n$. Note that the number of possible combinations to try is the product of $h_1, h_2, \dots, h_p$. The *best* estimate of β is taken to be the values of the parameters that correspond to the smallest value of *WRSS*.

The use of search methods should be limited to obtaining *initial estimates* for the more complex estimation procedures.

Comments on random search

◆ **The random search procedure can be very slow**

◆ **The pattern of search is inefficient and independent of previously computed *WRSS***

◆ **Knowledge is required of L_i and U_i for each of the parameters**

◆ **The accuracy of the final estimates depends on the magnitude of h_i**

◆ **No measure of precision of the parameter estimates is readily available**

2.6.3   Stripping or peeling methods

If the kinetics of the system is assumed to be first order, then the model expressing concentrations as a function of time can be written as a sum of exponentials as follows

$$C = \sum_{i=1}^{n} A_i \cdot e^{-\lambda_i \cdot t} \tag{2:27}$$

An example of Equation 2:27 is the one-compartment open model. This model can be written as

$$C = \frac{K_a F D_{po}}{V(K_a - K)} \left[e^{-K \cdot t} - e^{-K_a \cdot t} \right] \tag{2:28}$$

If one assumes that the absorption rate is greater than the elimination rate, $K_a \gg K$, then at some point t_i in time, Equation 2:28 reduces to

$$C = \frac{K_a F D_{po}}{V(K_a - K)} e^{-K \cdot t} \tag{2:29}$$

Taking the natural logarithm (*ln[]*) of both sides, one obtains

$$\ln[C] = \ln\left[\frac{K_a F D_{po}}{V(K_a - K)} \right] - K \cdot t \tag{2:30}$$

This corresponds to the terminal slope in Figure 2.10.

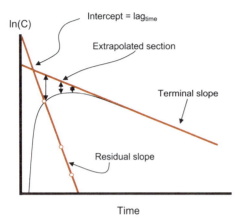

Figure 2.10 *Schematic representation of the terminal and residual slopes obtained from a first-order absorption elimination curve.*

Thus for large enough t, an estimate of K can be obtained from the slope of the graph of Equation 2:30.

Now that we have obtained an estimate of K, it remains to obtain estimates for K_a and V. To do this, we must first subtract the concentrations obtained using Equation 2:29 from those obtained from Equation 2:28 for those points corresponding to times prior to the terminal phase. The equation for these residuals is given as

$$C_{residual} = \frac{K_a F D_{po}}{V(K_a - K)} e^{-K_a \cdot t} \qquad (2:31)$$

Again, taking the natural logarithms of both sides of Equation 2:31, one obtains

$$\ln\left[C_{residual}\right] = \ln\left[\frac{K_a F D_{po}}{V(K_a - K)}\right] - K_a \cdot t \qquad (2:32)$$

which corresponds to the residual slope in Figure 2.10. Hence K_a can be obtained from a graph of Equation 2:32. Then since we have estimates of K_a and K, and since we know the oral dose D_{po}, we can estimate the quantity V/F from the intercept of Equation 2:31 or 2:32. Note that estimation of V, rather than V/F, requires intravenous data in order to estimate the fraction of the dose that is absorbed. Note that for drugs whose absorption is slower than elimination, the terminal slope provides an estimate of K_a and not K. This is known as the flip-flop effect.

Curve stripping is used by some programs to derive initial estimates for models defined as sums of exponentials.

Comments on stripping methods

◆ **Only applicable for models which can be expressed as a sum of exponentials**

◆ **Method fails when rate constants are approximately equal**

◆ **Inefficient since only a portion of the data is used to estimate each parameter**

2.6.4   Linearization methods

Most of the nonlinear regression routines implemented into commercially available software packages employ linearization techniques. Recall that in Section 2.5 we noted that the parameters and their variances for a general linear model could be estimated as in Equation 2:21 and 2:22. In order to apply these results to nonlinear models, we must *linearize* the nonlinear model by taking a series expansion of the nonlinear model (Bates and Watts [1988]). This enables us to express a general nonlinear model in the following linearized form

$$f(\underline{X},\underline{\beta}) = f(\underline{X},\underline{\beta}^0) + D(\underline{\beta} - \underline{\beta}^0)'$$ (2:33)

where

$$D = \left[\frac{df(X,\beta)}{d\beta_i} \right]_{\beta_i = \beta_i^0}$$ (2:34)

and β^0 are initial parameter estimates obtained from curve stripping, prior data or some other means. The matrix D is the matrix of partial derivatives of the model parameters evaluated at some initial estimated values of the parameters. A superscript 0 denotes these initial values. Further, since

$$C = f(\underline{X},\beta) + \varepsilon$$ (2:35)

we can say that

$$R = D\Delta\underline{\beta}' + \varepsilon$$ (2:36)

where R is the matrix of residuals

$$R = (C - f(X,\beta^0))$$ (2:37)

and

$$\Delta\underline{\beta} = \beta - \beta^0$$ (2:38)

Note that Equation 2:36 is in the form of the general linear model (Equation 2:21) where D plays the role of $\underline{X}$ and $\Delta\beta$ plays the role of β. Thus, $\Delta\beta$ can be estimated as in Equation 2:36 and as denoted below

$$\Delta\underline{\beta} = (D'D)^{-1}D'R$$ (2:39)

We can then set

$$\underline{\beta}^1 = \underline{\beta}^0 + \Delta\underline{\beta} \tag{2:40}$$

and repeat the process. This process is repeated until the estimated parameters are unchanged or until the weighted residual sum of squares *WRSS* remains unchanged, and is known as the Gauss–Newton method. One nice feature of this method is that the variances of the estimates can be readily computed as in Equation 2:24 with the matrix D playing the role of X.

Hartley [1961] suggested a modification of this method, which can greatly speed convergence. This modification can be briefly described as follows. Compute the *WRSS* at $\beta^0 + \upsilon \Delta\underline{\beta}$ for $\upsilon=0,\ 0.5,\ 1$, and denote the values as *WRSS(0)*, *WRSS(0.5)* and *WRSS(1)*. Then

$$\underline{\beta}^1 = \underline{\beta}^0 + \upsilon_{min}\Delta\underline{\beta} \tag{2:41}$$

where

$$\upsilon_{min} = 0.5 + 0.25WRSS(0) - \frac{WRSS(1)}{WRSS(1) - 2WRSS(0.5) + WRSS(0)} \tag{2:42}$$

Other linearization methods, such as the method of steepest descent, are variations of the above methodology. How well the nonlinear model can be approximated by a linear model is dependent on several things, including the *intrinsic curvature* of the linearized model and how close the initial parameter estimates are to the true values. While linearization methods usually perform well, they do sometimes fail to converge or converge very slowly. Often this is due to difficulty in computing $(D'D)^{-1}$. In these instances, algorithms proposed by Marquardt or Levenberg which do not involve the direct computation of $(D'D)^{-1}$ are often employed to estimate the parameters (Fletcher [1987]).

Comments on linearization methods

◆ **$D'D$ and hence $(D'D)^{-1}$ are often ill-conditioned**
◆ **The iterative procedure can sometimes converge very slowly or not at all**
◆ **Results are asymptotic; that is, as $N \to \infty$ in such a way that a finer and finer mesh is taken on the time domain, the estimated parameters will converge to the *true* values of the parameters and the estimate of sigma squared will converge to the true value of sigma squared**
◆ **Many computer programs estimate the derivative matrix D using difference equations**

2.6.5   Simplex methods

The Simplex method has been around for many years and is a very powerful optimization tool (Fletcher [1987]). The methodology has not been widely used, as it is computationally intense. However, the advent of faster and cheaper computers makes this methodology more attractive. One of the most widely used simplex methods is the Nelder-Mead simplex. The simplex algorithm is difficult to visualize when estimating several parameters, but can be readily described and displayed graphically for the two-parameter case.

The first step is to compute the *WRSS* at three equidistant points spanned by the two parameters. This equilateral triangle is referred to as a simplex. Note that for three parameters, the simplex would take the shape of a regular tetrahedron. The next step is to determine the vertex of the triangle at which the *WRSS* is a maximum. Then reflect the simplex through the centroid of (the line connecting) the remaining two vertices of the triangle. The next step is to compute the *WRSS* associated with the new vertex. We have now defined a new triangle, and the above process is repeated.

If the *WRSS* associated with a newly defined vertex is greater than the previous *WRSS*, then the simplex is reflected about the centroid associated with the next largest value of the *WRSS*. If one of the vertices remains unchanged for some specified number of iterations, then the size of the triangle is reduced and the search procedure restarted. Convergence is assumed when the size of the simplex has been reduced by a specified factor.

Nelder and Mead [1965] proposed an improvement on the basic simplex algorithm. This allows for expansion and contraction of the triangle based on the relative values of the *WRSS* at the different vertices. A hypothetical illustration of the simplex algorithm is displayed in Figure 2.11 for a model requiring estimation of two parameters, which are denoted as *Cl* and *V*. This method, like the linearization methods described above, requires initial parameter estimates. These estimates are used in the construction of the initial equilateral triangle denoted by the vertices 1-3.

Cl

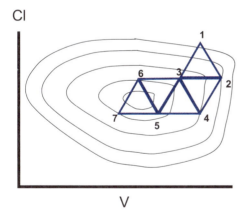

Figure 2.11 Contour plot of the weighted residual sum of squares surface versus clearance (Cl) and volume of distribution (V).

V

The contours on the plot identify values of the parameters, which are associated with similar values of *WRSS*. Since the *WRSS* associated with vertex 2 is the largest, the simplex is reflected about the line connecting the vertices 1 and 3, which yields a new vertex, 4. Our *working* triangle is now defined by vertices 2, 3 and 4. Since vertex 2 is now associated with the largest value of the *WRSS*, the simplex is now reflected about the line connecting vertices 3 and 4. The process is continued as described above.

Generally, the simplex algorithm is less sensitive to a poor choice of starting values than the linearization methods. One drawback of the method is that it doesn't directly yield any estimates of variability of the estimated parameters. Computer programs can circumvent this problem by using the simplex algorithm to obtain parameter estimates and then estimate the variances of the parameters using linearization methods.

Comments on the simplex methods
◆ **One of the most robust minimization algorithms**
◆ **Generally slower to converge than other algorithms but faster computers are making it more attractive**
◆ **Generally less sensitive to poor initial estimates since it is less likely to take a *huge* step**
◆ **The method does not directly yield estimates of variability**

2.7   Constraints on the Parameter Space

Often an estimation algorithm may compute the values of the parameters that give the best fit of the model to the data, even though one or more of the parameter values may not be biologically meaningful. For example, a transfer rate between two compartments in a compartmental model may be estimated to be negative. Thus, we may wish to obtain the parameter estimates which produce the best fit, but within some bounded parameter space. Many methods have been proposed for handling such constraints, and two of these methods are discussed below.

It is worth noting that constraints can easily be incorporated into some algorithms, such as simplex methods, by simply setting the weighted residual sum of squares to infinity if a constraint is violated. However, this method generally will not work with linearization methods. A change of variables can sometimes change a constrained problem into an unconstrained problem. For example, estimating k_{10} with the constraint $k_{10} > 0$ is equivalent to estimating $K^* = (k_{10})^2$ with K^* free to assume any value.

Some nonlinear regression programs can handle constraints, which are specified by defining lower and upper bounds on each of the parameters as follows

$$L_i \leq \beta_i \leq U_i \tag{2:43}$$

L and U are user-specified constants (Pharsight [1999]). The procedure is to first define a new parameter, which is defined on $[0,1]$ as in Equation 2:44.

$$\beta_i^* = \frac{\beta_i - L_i}{U_i - L_i} \qquad 0 < \beta_i^* < 1 \tag{2:44}$$

The next step is to use a normit transform to transform the parameter to a new parameter Z_i, which is unconstrained.

$$\beta_i^* = \int_{-\infty}^{Z_i} \frac{1}{\sqrt{2\Pi}} e^{-y^2/2} dy \qquad -\infty < Z_i < \infty \tag{2:45}$$

Transformation of parameters is often useful in situations other than putting bounds on the parameter space. For example, in situations where the magnitudes of the parameters differ substantially, the estimation algorithm may fail to converge. The reason is that a small change in a parameter with a relatively large magnitude may have minimal impact on the *WRSS*, while a small change in a parameter with a relatively small magnitude may have a substantial impact on the *WRSS*. For this reason, we recommend that transformations of the parameters in a model be employed when the parameters vary greatly in magnitude. Again, this is easily accomplished in *WinNonlin* by specifying upper and lower boundaries on the parameters.

2.8   Estimating Functions of Parameters

In some instances, our primary interest is not in the actual model parameters per se, but in some function of the model parameters. For example, let us assume that we have a drug whose kinetics are first order, so that plasma concentrations at any point in time can be expressed as a sum of three exponentials as described by Equation 2:27. Our primary interest may not be in the A_i and λ_i, but in the area under the concentration curve (*AUC*), which can be computed as follows

$$AUC = \sum_{i=1}^{n} \frac{A_i}{\lambda_i}$$
(2:46)

Clearly, an estimate of *AUC* is easily obtained by substituting the estimated values of A and λ into Equation 2:46. But how can we get an estimated standard error of the *AUC*? Let $g(\beta)$ denote the function of parameters we wish to estimate. Next compute the matrix of partial derivatives of g with respect to each of the β_i and denote it as G.

$$G = \left[\frac{dg(\beta)}{d\beta_i} \right]_{\beta_i = \beta_i^0}$$
(2:47)

Using the linearization technique that was discussed in Section 2.6, the variance of the estimated value of the function can be computed as

$$Var[g(\beta)] = G'(D'D)^{-1}G\sigma^2$$
(2:48)

The standard error is then the square root of Equation 2:48. Note that one must use caution when interpreting these standard errors. To facilitate their computation several levels of approximations were used. Firstly, a linear model approximated the nonlinear model, and the variances of the model parameters were approximated as well. Next, the function g was approximated by linearization and the corresponding variances approximated as in 2:48. How accurately these standard errors estimate the *true* standard errors of the model parameters and functions of the model parameters can only be evaluated through Monte-Carlo methods. In the authors' experience, the standard errors computed in the fashion described above generally provide reliable estimates. *WinNonlin* is capable of estimating functions of model parameters and their standard errors. These are denoted as *secondary* parameters in the program.

2.9   Validation of Software

2.9.1   What do we mean by software validation?

We consider software validation to comprise three areas:

◆ Systems development - usually employs a system development life cycle (*SDLC*) methodology. This is a process which includes (but is not limited to) development of user requirements, functional specifications and user *acceptance testing* (Chamberlain [1994], Double and McKendry [1994], Stokes *et al* [1994]). *Acceptance testing* is a procedure in which the same dataset run with the same software package on the same computer at two or more occasions give *correct* and *consistent* results. This can be accomplished by comparing the results from manually calculated datasets with that from the actual software package. Such a procedure checks that the results are *correct*. To check the *consistency* between two or more calculations, the same dataset can be run over and over again. *A priori* expectations and deviations should be documented. E.g., a larger difference (between manually obtained and program calculated results) than say 5% will generally not be accepted. No difference (identical number with respect to all output) should be accepted between two runs using the same dataset, the same program (version), and the same computer.

◆ System installation - this part of the validation process, usually performed by an end user, involves running test data sets through the systems. The output is then compared to known results. The test runs may come from the literature, or may be included as part of the documentation provided as part of the system.

◆ Testing of user applications - many systems, such as *Excel*™ or *WinNonlin*, allow the user to write their own programs or models within the confines of the system.

2.9.2   Computer Systems Validation (*CSV*)

WinNonlin was validated using a *SDLC* methodology. The methodology utilises Standard Operating Procedures (*SOPs*), which were reviewed and approved by an independent *CSV* consultant. In addition, the final *CSV* package, which includes all components of the *SDLC* methodology, was also sent to an outside *CSV* consultant for review, and the consultant provided the software vendor with a certificate stating that the program was developed in conformance with company *SOPs*.

The testing of the modeling component involved testing of the compiled libraries, *ASCII* libraries and the modeling language (which allows users to write their own models). Data sets were employed for single and multiple doses, and with and without missing values.

Algorithms were evaluated for robustness by fitting a wide variety of data sets (multiple peaks, low information during the absorption phase, etc.), and using varying values for the initial estimates and parameter boundaries.

Testing of the *NCA* routines was more straightforward, as the algorithms are relatively simple and the results can be compared to hand compilations. However, the testing included the handling of *tough* data sets, including relatively flat profiles, profiles where λ_z cannot be reliably estimated, and profiles containing one or more zero concentrations. Test data should also include two or more concentration peaks or two or more consecutive equal concentrations. Extrapolation of the concentration-time course (and the area) within and outside a dosing interval at steady state is also recommended.

2.9.3   Testing of user models

We recommend that users routinely include as part of the model statements, comments that describe the following:

◆ A description of the model and when it is appropriate for it to be used
◆ The name of the developer
◆ The creation date
◆ The modification date(s) and changes which were made
◆ A description of model parameters (and order)
◆ A description of secondary parameters
◆ A description of any constants which are required
◆ If applicable, a reference to a published description/validation of the model

The user should test the model using examples from the literature or other sources, where the solution is known. Test data sets should include one or more zero responses (if plausible), as these values often cause problems with modeling code. For example, both

$$W_i = 1 / Y$$
$$W_i = Y ** -1$$

will fail if Y takes on the value zero.

If appropriate (as is the case for all linear compartment models) the model output (when written in integrated form) should be compared to output from writing the model as a system of differential equations. Although the output will not match exactly, the differences should be small enough to be attributed to computational or rounding errors.

All of the test runs should be kept on file, particularly if the model is being used in conjunction with a regulatory submission. A test run with the vendor supplied data sets should always be done when the program is installed. This checks that the program is correctly installed. The kineticist is then highly recommended to perform his own *acceptance test* for some of his own data sets prior to e.g., the *NCA*, depending on the nature of the data and the study. This should be done mainly due to two reasons; the kineticist is always responsible for the analysis; the kineticist is less biased prior to the program run than after. There are no right or wrong answer to what method should be used for the acceptance test.

CHAPTER 3 - Pharmacokinetic concepts

Objectives

◆ **To review pharmacokinetic concepts**

◆ **To add insight into experimental design for the estimation of parameters**

◆ **To provide a wide range of pharmacokinetic systems**

3.1   Background

The transfer of a drug from its absorption site to the blood and the various steps involved in the distribution of the drug in the body are shown in schematic form in Figure 3.1. The boxes represent a compartmentalisation of the body and may include one, two, or multi-compartment systems.

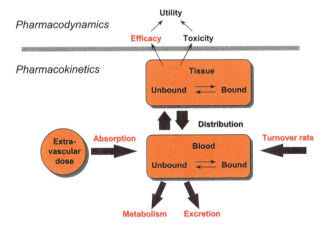

Figure 3.1 *Diagram showing the interrelationship between absorption, distribution, binding and elimination kinetics as well as dynamics of a drug. The turnover occurs throughout the body.*

A basic understanding of drug absorption, distribution and elimination, and the relationship between kinetics and dynamics, as well as the underlying mathematics, is a fundamental aspect of pharmacokinetic/pharmacodynamic (*PK/PD*) modeling. In the following, we give a background to some of the kinetic and dynamic models that are discussed in later chapters. However, it is not our aim to cover the theory of pharmacokinetics in too much detail; rather, we provide an introduction to applications. We start with traditional compartmental analysis applied to various inputs (Sections 3.2 and 3.3) and then gradually move on to clearance models (Section 3.4), turnover concepts (Section 3.5), and nonlinear models (Section 3.6) including such topics as capacity, flow, time, and binding dependent models. Thereafter we elaborate on non-compartmental analysis (Section 3.7 - computational methods, strategies for λ_z, pertinent pharmacokinetic estimates, issues related to steady state) and then inter-species scaling (Section 3.8 - exposure, allometry, time scales, parameters, Dedrick plots).

For further reading on basic pharmacokinetic principles, we refer the reader to Benet [1972], Benet and Galeazzi [1979], Gibaldi and Perrier [1982], Nakashima and Benet [1988, 1989], Jusko [1992], and Rowland and Tozer [1995]. Houston [1994] and Pang [1985, 1998] provide excellent texts on metabolite kinetics. Van Rossum and Lingen [1983], Labrecque and Belanger [1985], and Levy [1982] have reviewed phenomena such as time and dose dependencies. Concepts and theory on inter-species scaling are ellegantly reviewed by Boxenbaum [1992] and Schmidt-Nielsen [1996].

3.2   One-compartment Models

3.2.1   Intravenous bolus administration

In general, the transfer rate of a drug from one compartment to another is governed by first-order kinetics. This means that the rate of change of drug from a specific compartment to another is proportional to its concentration within the source compartment. Figure 3.2 shows a schematic presentation of a one-compartment model. C_p and V are the concentration in plasma and the volume of distribution, respectively.

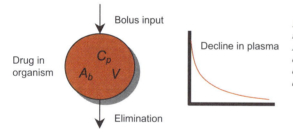

Figure 3.2 The one-compartment model with first-order elimination. Amount of drug in the body, concentration in plasma and volume of distribution are denoted A_b, C_p and V, respectively.

Volume of distribution (V) is the apparent space (volume) which a drug distributes into. V may be viewed as a proportionality factor between the total amount of drug in the body A_b and the concentration of drug in plasma C_p, which is the reference space.

Figure 3.3 shows schematically how one could interpret the volume of distribution. Two physically equal buckets of 10 L are filled with water. Some active charcoal is dispersed in one of the buckets. A 100 unit dose of a drug is put into each bucket. When the drug has dissolved in the water the concentration of drug in each bucket is measured. The concentration in the left hand bucket is 10 units/L and in the right hand bucket 1 unit/L. By dividing the dose of 100 units with each of the measured concentrations one would obtain apparent volumes of 10 and 100 L, respectively. The reason why the concentration of drug in the right hand bucket is so low is because the drug not only dissolves into the water but also binds to the charcoal, which makes it inaccessible to the site of measurement (water). However, we still have 100 units of drug in each bucket, because no drug has been eliminated. The apparent volumes estimated above relate the total amount of drug in the system (bucket or body) to the site of measurement (water or plasma).

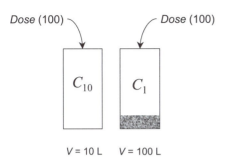

Figure 3.3 Schematic illustration of the volume of distribution, V. A dose of 100 units is dissolved in each of the buckets, resulting in a concentration C of 10 and 1 units/L since the apparent bucket volumes are 10 and 100 L, respectively. The greyish area at the bottom of the right hand bucket represents active charcoal.

If one administers a bolus dose of a drug that behaves like the one-compartment system presented schematically in Figure 3.2, and measures the decline in a reference volume, such as plasma, one would obtain a plasma concentration-time course as depicted in Figure 3.4. Note the different characteristics of the two plots.

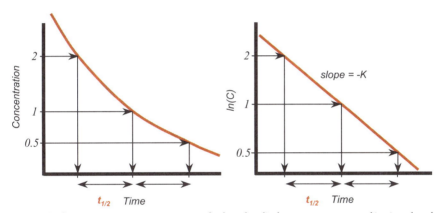

Figure 3.4 Plasma concentration-time courses of a drug that displays one-compartment kinetics, plotted linearly (left) and semi-logarithmically (right). $t_{1/2}$ denotes the half-life.

The relationship between the concentration C and the rate of decline dC/dt in concentration may be expressed mathematically for a one-compartment model with first-order kinetics when drug is administered as a bolus dose as follows

$$\frac{dC}{dt} = -K \cdot C = -\frac{Cl}{V} \cdot C \tag{3:1}$$

dC/dt is the rate of change of the plasma concentration, C is the plasma concentration and K is the first-order rate constant associated with the elimination process. Clearance (Cl) relates the rate of elimination (dX/dt) to C as follows

$$V \cdot \frac{dC}{dt} = \frac{dX}{dt} = -Cl \cdot C \qquad\qquad (3:2)$$

This can be represented graphically as a proportionality (slope) factor as in Figure 3.5.

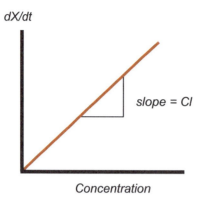

dX/dt

slope = Cl

Concentration

Figure 3.5 *Elimination rate as a function of plasma concentration. The slope corresponds to clearance, Cl. Note that the slope has a positive sign which contrasts the minus sign in Equation 3:2. The latter indicates a decline process.*

Clearance (Cl, e.g., in Equation 3:2 and Figure 3.5*)* is the volume of blood or plasma that is totally cleared of its content of drug per unit time (e.g., mL/min or L/min).

Elimination rate constant (K), also called the fractional rate constant, relates to the amount of drug in the body which is eliminated per unit time (time^{-1}). *-K* is equivalent to the slope of the line resulting from a plot of *ln(C) versus* time (see Figure 3.4). The biological half-life of the drug $(t_{1/2})$ is the time required for reducing by half the amount of drug remaining in the body. *K* may also be directly estimated from $K = ln(2)/t_{1/2}$. As can be seen from Equation 3:1, *K* is equal to clearance *(Cl)* divided by volume *(V)*. *Cl, K* and $t_{1/2}$ relate to first-order processes.

Since *Cl* and *V* are the primary pharmacokinetic parameters of interest, we will often use them instead of *K*. Equation 3:1 can be expressed in explicit (integrated) form as

$$C = C^0 \cdot e^{-K \cdot t} = C^0 \cdot e^{-\frac{Cl}{V} \cdot t} \qquad\qquad (3:3)$$

C is the dependent variable (plasma concentration) and C^0 is the concentration at time zero (Figure 3.6). Figures 3.6 and 3.7 show the typical one-compartmental behavior of a drug following a 10 mg bolus dose with a C^0 of 1000 µg/L and a half-life of about 34 minutes. The semi-logarithmic scale (relative scale) reveals more clearly the number of declining phases of the concentration-time course, and a linear scale (absolute scale) demonstrates the time it takes to reduce the concentration by say 90%.

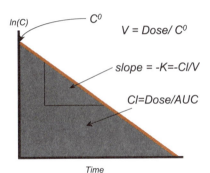

$V = Dose/ C^0$

slope = -K=-Cl/V

Cl=Dose/AUC

Figure 3.6 *Semi-logarithmic plot of plasma concentration-time data. The data show mono-exponential decline described by Equation 3:2. The intercept of the concentration axis is C^0, slope is the elimination rate constant K, volume V is the ratio of Dose-to-C^0, and clearance Cl is obtained from the ratio of Dose-to-area under the curve AUC.*

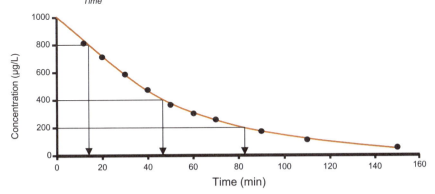

Figure 3.7 *Cartesian plot of plasma concentration-time data following an intravenous bolus dose equivalent to 10 mg. The data show mono-exponential decline described by Equation 3:2.*

Most computer programs will present data in log-linear plots as in Figure 3.6. To estimate K, one would perform a regression analysis on the *ln(concentration)* versus time; the resultant slope is $-K$ and intercept is $ln(C^0)$, which are the model parameters in this case. Initial estimates of K and $t_{1/2}$ can be derived from the slope. Using the data in Figure 3.7

$$slope = \frac{y_2 - y_1}{x_2 - x_1} = -K = \frac{\ln(800) - \ln(400)}{13 - 47} = -0.02\,\text{min}^{-1} \qquad (3:4)$$

The half-life becomes

$$t_{1/2} = \frac{\ln(2)}{K} = \frac{0.693}{0.02} \approx 34\,\text{min} \qquad (3:5)$$

The intercept of the concentration axis is 1000 µg/L. According to Figure 3.6, the volume of distribution V can be calculated as

$$V = \frac{Dose}{C^0} = \frac{10000}{1000} = 10L \qquad (3:6)$$

The total area under the curve, $AUC_{0-\infty}$ is either obtained by direct integration of Equation 3:3 from zero to infinity

$$AUC_0^\infty = \int_0^\infty C^0 e^{-K \cdot t} dt = C^0 \left[\frac{e^{-K \cdot \infty}}{-K} - \frac{e^{-K \cdot 0}}{-K} \right] = C^0 \left[\frac{0}{-K} - \frac{1}{-K} \right] = \frac{C^0}{K} = \frac{D_{iv}}{VK} \qquad (3:7)$$

or by means of the trapezoidal rule by summing the incremental areas from t equal to 0 to t equal to 150 and then adding the extrapolated area from $t = 150$ to $t = \infty$.

$$AUC_0^\infty = \sum_{i=1}^n AUC_i + \frac{C_{150}}{K} \qquad (3:8)$$

Equation 3:3 may be written to include the dose D_{iv}, plasma clearance Cl and volume of distribution V as follows

$$C = C^0 \cdot e^{-K \cdot t} = \frac{D_{iv}}{V} \cdot e^{-\frac{Cl}{V} \cdot t} \qquad (3:9)$$

For a compound that exhibits one-compartment linear kinetics and is given as a bolus dose, the area under the concentration curve (AUC also called zero moment) and first moment ($AUMC$) curve, clearance Cl, volume of distribution V, and mean residence time MRT are calculated as follows

$$AUC_0^\infty = \int_0^\infty C dt = \int_0^\infty C^0 \cdot e^{-K \cdot t} dt = \frac{C^0}{K} \qquad (3:10)$$

The area under the first moment curve from time zero to infinity, $AUMC_{0-\infty}$

$$AUMC_0^\infty = \int_0^\infty t \cdot C dt = \int_0^\infty t \cdot C^0 e^{-K \cdot t} dt = \frac{C^0}{K^2} \qquad (3:11)$$

Clearance is the dose divided by the total area under the curve

$$Cl = \frac{D_{iv}}{AUC_0^\infty} = \frac{D_{iv}}{\left[\dfrac{C^0}{K} \right]} \qquad (3:12)$$

The volume of distribution V, is calculated as D_{iv} divided by the intercept at time zero

$$V = \frac{D_{iv}}{C^0} \qquad (3:13)$$

Otherwise volume V can be expressed as the total amount of drug in the body A_b at any time divided by the corresponding total plasma concentration, or by means of free or unbound concentrations

$$A_b = V \cdot C = V_u \cdot C_u \qquad (3:14)$$

Therefore V can be derived from unbound volume of distribution V_u and the free fraction f_u, as

$$V = V_u \frac{C_u}{C} = V_u f_u \qquad (3:15)$$

Note that f_u and V_u are the independent variables and V is a function of the two. The number of binding sites and the affinity between drug and the binding plasma protein determines f_u. V_u is independent of plasma binding but is dependent on tissue binding.

Mean residence time MRT is the average time a molecule stays in the body (unit time). *MRT* is commonly expressed as the ratio of *AUMC*-to-*AUC* following an intravenous bolus only

$$MRT = \frac{AUMC}{AUC} \qquad (3:16)$$

3.2.2   Constant rate infusion

If we infuse drug at a constant rate over a period of time, the plasma concentration will rise as shown in Figure 3.8. Fifty percent of steady state is reached after one half-life (30 min), 75% after 2 half-lives, and 87.5% after 3 half-lives etc. From a practical point of view, we say that 90% of steady state is reached after 3 - 4 half-lives.

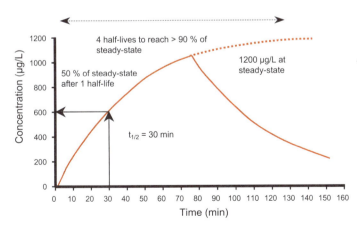

Figure 3.8 The plasma concentration-time course following a constant rate infusion. Infusion during one half-life establishes 50% of steady state. 3-4 half-lives are needed to reach about 90% of the steady state concentration.

When the infusion is stopped, the plasma concentration will decrease (exponentially). If drug is given as a constant rate infusion R_{in}, the rate is equal to the dose divided by the length of infusion T_{inf}, as long as t is less than or equal to T_{inf}. During infusion, the differential equation for the one-compartment drug would be

$$\frac{dC}{dt} = \frac{R_{in}}{V} - \frac{Cl}{V} \cdot C \qquad (3{:}17)$$

The analytical solution of the differential equation during the constant rate infusion is

$$C = \frac{R_{in}}{Cl} \cdot \left[1 - e^{-\frac{Cl}{V} \cdot t} \right] \qquad (3{:}18)$$

The concentration at steady state C_{ss}, may also be calculated as R_{in} divided by Cl, hence

$$C_{ss} = \frac{R_{in}}{Cl} \qquad (3{:}19)$$

Cl and V are model parameters to be estimated by fitting this model to the observed concentration-time data. R_{in} is generally a known constant. This is the simplest example of deconvolution. The method of deconvolution is discussed in section 5.4.2. By fitting a one-compartment disposition model (called the *weighting function*) to the observed plasma data following a constant rate of infusion (called the *response function*), the disposition function (Cl, V) can be estimated since the *input function* is known. The input function is the infusion regimen. If the infusion is stopped at time T_{inf}, the concentration time-course post-infusion will have the following form

$$C = \frac{R_{in}}{Cl} \cdot \left[1 - e^{-\frac{Cl}{V} \cdot T_{inf}} \right] e^{-\frac{Cl}{V} \cdot t'} \qquad (3{:}20)$$

where $t' = t - T_{inf}$. Equations 3:21 to 3:23 and Figure 3.9 show how data obtained from the infusion phase can be used to estimate K.

$$C = C_{ss} \cdot \left[1 - e^{-\frac{Cl}{V} \cdot t} \right] \qquad (3{:}21)$$

$$C_{ss} - C = C_{ss} \cdot e^{-\frac{Cl}{V} \cdot t} \qquad (3{:}22)$$

$$\ln\left[\frac{C_{ss} - C}{C_{ss}} \right] = \ln(R) = -\frac{Cl}{V} \cdot t = -K \cdot t \qquad (3{:}23)$$

R denotes $(C_{ss}-C)/C_{ss}$. $ln(R)$ plotted *versus* t gives a slope equal to $-K$ (Figure 3.9).

In(R)

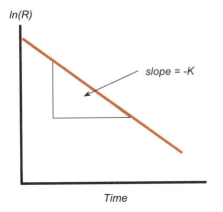

slope = -K

Time

Figure 3.9 Relationship between ln(R) and time. The slope of that line gives the elimination rate constant K.

Under the assumption that equal doses are given, we can derive a relationship for converting C^I to C^0 (for explanations of these parameters see Figure 3.10).

$$C^I = \frac{R_{in}}{Cl}\left[e^{\frac{Cl}{V}\cdot T_{inf}} -1 \right]$$
(3:24)

$Dose = R_{in}\cdot T_{inf}$ which rearranged gives $R_{in} = Dose/\ T_{inf}$. Hence

$$C^I = \frac{Dose}{T_{inf}}\cdot\frac{1}{Cl}\left[e^{\frac{Cl}{V}\cdot T_{inf}} -1 \right]$$
(3:25)

$Dose = C^0\cdot V$ after a bolus dose. Thus

$$C^I = \frac{C^0 V}{T_{inf}}\cdot\frac{1}{Cl}\left[e^{\frac{Cl}{V}\cdot T_{inf}} -1 \right]$$
(3:26)

and rearrangement gives

$$C^0 = \frac{C^I T_{inf}\frac{Cl}{V}}{e^{\frac{Cl}{V}\cdot T_{inf}} -1}$$
(3:27)

This relationship applies to all mammillary compartment models (Gibaldi and Perrier [1982]) as shown in Equation 3:28

$$C^0 = \frac{C^I T_{inf}\lambda_z}{e^{\lambda_z\cdot T_{inf}} -1}$$
(3:28)

where λ_z is the terminal slope. Figure 3.10 shows schematically the relationship between the length of infusion and the back-extrapolated intercept C^I. By increasing the length of infusion from *a* to *b*, one will increase the back-extrapolated intercept according to the dashed lines.

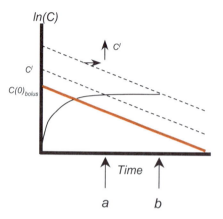

Figure 3.10 *Relationship between C^I and C^0 after a bolus dose (red line) and a constant rate infusion (black line) of different lengths (a or b). The longer the infusion proceeds (b compared to a), the higher is the back-extrapolated intercept (C^I). This is evident from the dashed extrapolations for a and b.*

Steady state is that state where successive concentrations are the same over time because the rate into the system equals the rate out. For drugs administered at constant rate such as an infusion, this means that concentrations are no longer changing with time. For other dosing regimens, the concept is somewhat weaker, meaning that equivalent time-concentration profiles occur after successive administration of the same dose and frequency. The concept can also be applied to extravascular administration. Again, the only requirement is that dose size and dose frequency are identical when comparing two consecutive intervals.

It is possible for a system to reach a *steady state* while not being genuinely at *equilibrium*. *Equilibrium* is thermodynamically defined as a closed system at zero entropy. However, we use *steady state* and *equilibrium* interchangeably in this book.

3.2.3   Integration of disposition concepts

Figure 3.11 illustrates schematically the kinetics of six different systems. It compares how *a)* equal volumes but different clearances, *b)* equal clearances and different volumes, and *c)* different clearances and different volumes, affect the time course of drug in plasma after bolus dosing and constant rate intravenous infusion. Note that only clearance affects the steady state concentration, but both clearance and volume determines the time to steady state (via the half-life). The volume determines the maximum and minimum plasma concentration at steady state when drug is given intermittently.

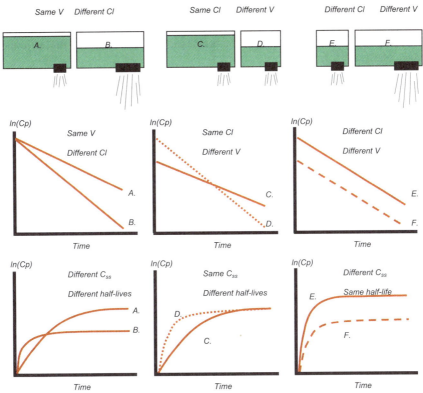

Figure 3.11 *Diagram showing how equal volumes and different clearances (A, B), equal clearances and different volumes (C, D), different clearances and different volumes (E, F) affect the time course of drug in plasma after intravenous bolus dosing (equal doses, upper row of graphs) and constant rate intravenous infusion (equal dosing rates, lower row of graphs).*

3.2.4   Constant rate infusion – computer code

If you want to implement an infusion model with an input that starts at time t_1 and lasts to t_2 we propose the following program code. Figure 3.12 displays the behavior of the infusion rate or input function *Input*, between time t_1 and time t_2 and the resulting plasma concentration time course. The total infusion dose is denoted *Dose*. Before t_1, *Input* is equal to zero. *Input* is equal to $Dose/(t_2-t_1)$ between t_1 and t_2, and then after t_2 *Input* is again zero. This can be expressed as an *IF . . . THEN . . . . ELSE . . . ENDIF* statement as follows. If *time* is less than or equal to t_1 or *time* is greater than or equal to t_2 then *Input* is equal to zero. Else, *Input* is equal to $Dose/(t_2 - t_1)$.

```
IF T LE T1 OR T GE T2 THEN
        INPUT = 0.
ELSE
        INPUT = DOSE/(T2 – T1)
ENDIF
```

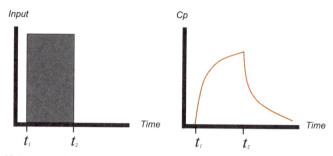

Figure 3.12 *Diagram illustrating the plasma concentration (Cp) time course for a constant rate intravenous infusion (Input) of duration (t_2-t_1). The total infused dose is the integral (shaded area) of the input function versus time course.*

3.2.5　　　Extravascular administration

The administration of a drug by all routes (e.g., oral) other than the systemic (e.g., intravenous) route will introduce an absorption step. The change in body drug content or plasma drug concentration will now be more complex, since the rate of change of drug will be a function of both the absorption rate and the elimination rate (Figure 3.13).

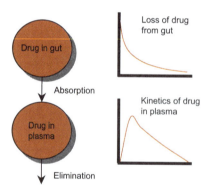

Figure 3.13 *The one-compartment model with absorption from the gut compartment.*

A typical plasma concentration-time plot after extravascular administration is shown in Figure 3.14. Data are obtained after administration of a controlled-release (*CR*) dosage form. Note how late the peak concentration appers. The terminal half-life (approximately 8h, blue line) is due to absorption-rate limited elimination. The true half-life of this compound is represented by the red line (4h).

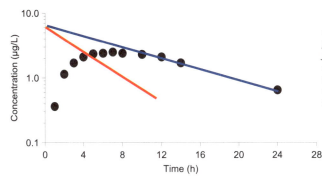

Figure 3.14 *Semi-logarithmic plot of plasma concentration-time data following an extravascular dose equivalent to 100 μg. The data show mono-exponential decline. An explanation of how the slopes were derived can be found in Figure 3.16.*

Typically, one would estimate the elimination rate constant from the terminal slope of the extravascular concentration-time profile (i.e., after an instant release tablet). In some cases, however, the absorption process will be the rate-limiting step (e.g., controlled release tablet). This is known as the *flip-flop* situation, where the ascending limb of the concentration-time curve will contain information about the elimination rate constant and the descending portion will contain information about the absorption rate constant.

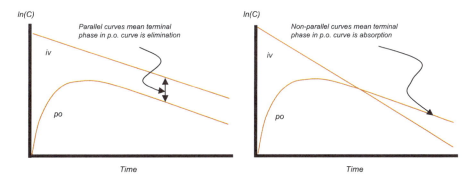

Figure 3.15 *Plasma concentration-time data following an extravascular dose. The left hand figure demonstrates a system where the elimination rate constant is obtained from the terminal portion of the oral curve since it declines in parallel with the intravenous curve. The intravenous curve represents only elimination. The right hand figure demonstrates the flip-flop situation for a system which displays absorption rate limited elimination. Therefore, the terminal portion of the oral curve will not run in parallel with the intravenous curve. The latter reflects the true elimination of drug.*

In order to discriminate between the two phases, the true elimination phase of the drug after an intravenous dose needs to be calculated. That is shown for the *flip-flop* situation in Figure 3.15. Note that the slope of the terminal portion of the extravascular curve is not parallel to the slope of the intravenous curve. The latter slope gives the true elimination rate constant and the slope of the former gives the absorption rate constant.

The differential equation for a drug that obeys a one-compartment model, given orally, will take the following form

$$\frac{dC}{dt} = \frac{K_a FD_{po} e^{-K_a \cdot t}}{V} - K \cdot C \qquad (3:29)$$

where K_a denotes the apparent first-order absorption rate constant and F denotes the bioavailability. The rate of loss from the gut compartment after an oral dose becomes

$$Rate\ of\ loss = K_a D_{po} e^{-K_a \cdot t} \qquad (3:30)$$

and rate of input into plasma becomes

$$Rate\ of\ input = K_a FD_{po} e^{-K_a \cdot t} \qquad (3:31)$$

The integrated form of Equation 3:29 is

$$C = \frac{K_a FD_{po}}{V(K_a - K)} \left[e^{-K \cdot t} - e^{-K_a \cdot t} \right] \qquad (3:32)$$

The slope of the linear portion of the terminal phase of the curve when plotted on a semi-loge scale is equal to $-K$, assuming that $K_a \gg K$. The slope of the linear residual portion of the initial phase of the curve is equal to $-K_a$. The rate constants K_a and K may be determined by the curve stripping method, also called the method of residuals (see also section 2.6.3 and Gibaldi and Perrier [1982]). However, unlike the situation found after intravenous administration, C^0 does not correspond to the term (D/V). Rather, C^0, extrapolated from the terminal portion of the curve, is a complex function of the amount of drug eliminated during the absorption phase, the time required for absorption, the dose, and the volume of distribution.

The area under the plasma concentration-time curve can be obtained by integrating Equation 3:32

$$AUC_0^\infty = \int_0^\infty \frac{K_a FD_{po}}{V(K_a - K)} \left[e^{-K \cdot t} - e^{-K_a \cdot t} \right] dt \qquad (3:33)$$

which gives

$$AUC_0^\infty = \frac{K_a FD_{po}}{V(K_a - K)} \left[\frac{1}{K} - \frac{1}{K_a} \right] \qquad (3:34)$$

and after rearrangement

$$AUC_0^\infty = \frac{FD_{po}}{VK} = \frac{FD_{po}}{Cl} \qquad (3:35)$$

The time course of concentration in plasma of certain compounds suggests a time-delay between oral administration and the apparent onset of absorption (Figure 3.16). The dissolution of the drug particles and the absorption of the dissolved drug into the systemic circulation take a certain amount of time until the concentration of drug is measurable in plasma (Equation 3:36).

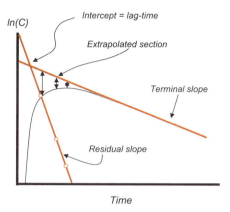

Figure 3.16 Observed and model predicted plasma data following an oral dose of a drug displaying some lag-time. See also Figures 3.14 and 3.20. The intercept between the extrapolated terminal slope and the residual slope is called the lag-time.

If we deconvolute extravascular data with intravenous data from the same subject we will be able to observe a time delay (often 10-20 min) of the flux of drug into the body. This time delay is often called the *lag-time*.

Figure 3.17 shows a series of typical input rates (top) and their corresponding plasma concentrations (bottom). Bolus (A) and zero-order (B) inputs are commonly applied regimens for systemic dosing. First-order input (C) is usually assumed to occur after extravascular dosing (such as with intra-duodenal, intra-peritoneal, oral, and subcutaneous dosing). A multiple zero-order input (D) system is encountered with the transdermal patch or the intravaginal ring. Multiple first-order input (E) systems are usually seen when the drug is given by the sublingual route, with some of the drug being absorbed locally from the buccal cavity and some being swallowed and absorbed via the gastrointestinal tract. Combinations of logarithmic and zero-order input (F) are required when one wants to establish a steady state rapidly.

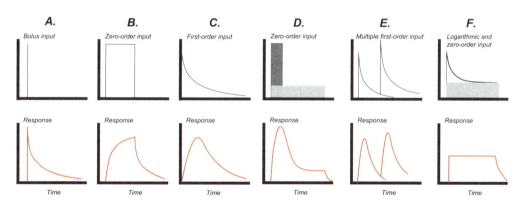

Figure 3.17 Input functions versus time (upper figures) and their corresponding response (plasma concentrations) versus time curves (bottom figures), assuming the weighting function to be a unit disposition function. The scale is arithmetic.

The resulting first-order input function (curve C in Figure 3.17) declines linearly on a semi-log$_e$ scale as shown in Figure 3.18.

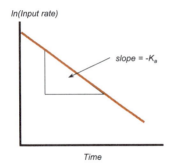

In(Input rate)

slope = -K_a

Time

Figure 3.18 *Semi-logarithmic plot of input function for a pure first-order input versus time.*

The fraction of the cumulative input is also called bioavailability *F*. This is simply the integral of the fractional input rate and is unitless. After having given an oral tablet, the tablet must disintegrate and dissolve. Then the drug must cross the gut wall before the drug appears at the sampling site. Clearly, this takes some time. This is often referred to as the *lag-time*. A lag-time may easily be fit into the model and subsequently estimated as an additional parameter as follows

$$C = \frac{K_a FD_{po}}{V(K_a - K)} \cdot e^{-K \cdot (t - tlag)} - \frac{K_a FD_{po}}{V(K_a - K)} \cdot e^{-K_a \cdot (t - tlag)} \tag{3:36}$$

when $t \geq t_{lag}$, and $C = 0$ when $t < t_{lag}$. The model parameters will then be K_a, K, V/F and t_{lag}. The reason for the ratio V/F is simply due to the inability to determine F and V separately. This is an inherent limitation of the model since neither F nor V appears elsewhere in the model, and without information following an intravenous dose, unique values for F and V cannot be determined.

3.2.6   Multiple dosing

We have discussed single dose administrations in the previous sections. Now, it is time to expand that discussion to also include kinetics after multiple dosing, since that is often the case in drug therapy.

When a multiple extravascular dosing regimen is initiated, plasma concentrations will increase, reach a maximum, and then decline. Generally, a second dose is administered before the first dose is completely eliminated (Figure 3.19). Therefore, plasma concentrations resulting from the second dose will be higher than those from the first dose. This accumulation will continue to occur until steady state is reached.

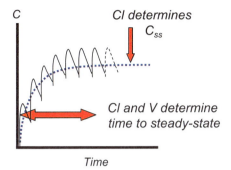

C

Cl determines

Css

*Cl and V determine
time to steady-state*

Time

Figure 3.19 *Schematic diagram showing
difference between constant rate input (heavy
dotted curve) and intermittent extravascular
dosing (solid curve). If the dosing rate is the
same per unit time for the two regimens, then
the steady state concentrations will be equal. Cl
governs the steady state concentration (together
with the dosing rate) and both Cl and V govern
the time to steady state via the half-life.*

The extent to which a compound will accumulate relative to the first dose can be quantified by an accumulation factor R which is dependent on the dosing interval τ and the half-life $t_{1/2}$ of the drug (Gibaldi and Perrier [1982]). Thus

$$R = \frac{1}{1 - e^{-\varepsilon \cdot \tau}} \qquad (3\text{:}37)$$

where ε is $ln2/t_{1/2}$. The smaller the ratio $\varepsilon/t_{1/2}$ the greater will be the extent of accumulation. When $\varepsilon = t_{1/2}$ the average concentration at steady state will be about twice the average concentration after the first dose (Gibaldi and Perrier [1982]).

The time to 90% of steady state is only determined by the half-life, whereas the average steady state concentration is determined by the dosing rate and clearance

$$C_{ss} = C_{average} = \frac{FD_{po}}{\tau} \cdot \frac{1}{Cl} = \frac{AUC}{\tau} \qquad (3\text{:}38)$$

AUC is either $AUC_{0\text{-}\infty}$ after a single dose or $AUC_{0\text{-}\tau}$ (area within the $0\text{-}\tau$ dosing interval) at steady state, assuming linear kinetics.

3.2.7   Estimation of pertinent parameters

We start the analysis by estimating the rate constants of the absorption and elimination phase. The rate constants are obtained from the slopes

$$K = -\frac{\ln(C_1) - \ln(C_2)}{t_1 - t_2} \qquad (3\text{:}39)$$

The terminal slope is then back-extrapolated to the concentration axis (abscissa). By means of the method of residuals (see section 2.6.3 and Gibaldi and Perrier [1982]) we subtract the observed concentrations on the upswing of the curve (usually called absorption phase) from the extrapolated line (see Figure 3.20).

In(C)

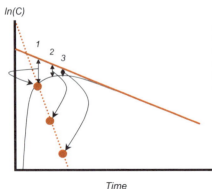

Time

Figure 3.20 *Plot illustrating method of residuals for estimation of the absorption rate constant. Filled circles represent the difference between the back-extrapolated curve and the observed values on the up-swing of the concentration curve. See also Figure 3.16.*

This difference ($C_{residual} = C_{extrapolated} - C_{observed}$) is then plotted in the same figure and a straight line is drawn through the data. The slope of the new line gives the absorption rate constant, K_a[1]. You can also observe that the back-extrapolated terminal slope and the absorption slope intersect at about 5 minutes (Figure 3.21). This suggests that there is a lag-time involved before you start to observe drug at the site of sampling.

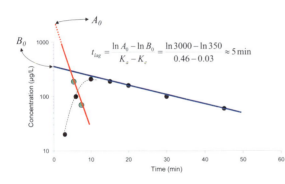

Figure 3.21 *Method of estimating lag-time t_{lag} graphically and by means of Equation 3:41. A_0 and B_0 are the back-extrapolated intercepts with the Y axis. K_e denotes K.*

The residual and back-extrapolated lines intersect when t is equal to t_{lag}, which gives

$$A_0 \cdot e^{-K_a \cdot t_{lag}} = B_0 \cdot e^{-K \cdot t_{lag}}$$

(3:40)

Rearranging this, shows that t_{lag} can be expressed as a function of K_a, K and the two intercepts A_0 and B_0.

$$t_{lag} = \frac{\ln A_0 - \ln B_0}{K_a - K}$$

(3:41)

What does a K of 0.1 or 0.99 h^{-1} mean? Since K is also called the fractional rate

---

[1]Intravenous data are required to determine if the terminal portion of the curve corresponds to the absorption or elimination phase (see Figure 3.15). The slope of the intravenous curve will parallel the true elimination phase after the oral dose.

constant, it means in this case that approximately 10% and 63%, respectively, is eliminated per hour. For example, the fractions remaining when K is 0.1 and 0.99, respectively, are

$$e^{-0.1} = 0.9048 \tag{3:42}$$

$$e^{-0.99} = 0.3716 \tag{3:43}$$

and the fractions lost for K equal to 0.1 and 0.99, respectively, are

$$1 - e^{-0.1} = 0.09516 \approx 0.10 \tag{3:44}$$

$$1 - e^{-0.99} = 0.6284 \approx 0.63 \tag{3:45}$$

Provided steady state is established, a value of K of 0.1 and 0.99 h^{-1}, means that 10 and 99%, respectively, of the amount at steady state is being removed per hour. At steady state, this equals the amount being replenished per hour, i.e., the input rate. The initial estimates of the model parameters can be derived according to Table 3.1.

Table 3.1   Pertinent kinetic estimates after extravascular dosing (see Table 3.2)

Parameter	Calculation	Description
Elimination rate constant (t^{-1})	$slope = -K = \dfrac{\ln C_1 - \ln C_2}{t_1 - t_2}$	(C_1, t_1) and (C_2, t_2) are obtained from the terminal portion of the curve
Absorption rate constant (t^{-1})	$residual\ slope = -K_a = \dfrac{\ln R_1 - \ln R_2}{t_1 - t_2}$	(R_1, t_1) and (R_2, t_2) are obtained from the residual curve
Volume/F (volume)	$\dfrac{V}{F} \approx \dfrac{Dose}{C_{max}}$	This gives an initial approximation of V/F
t_{lag} (t)	$t_{lag} = \dfrac{\ln A_0 - \ln B_0}{K_a - K}$	The time where the extrapolated terminal phase intersects with the residual curve. A_0 is the residual intercept with the y axis and B_0 is the back-extrapolated intercept of the terminal slope.

The peak plasma concentration C_{max}, and the time to reach the peak t_{max}, can be calculated as follows according to Gibaldi and Perrier [1982]. The first-order input/output model is written as

$$C = \frac{K_a F D_{po}}{V(K_a - K)} \left[e^{-K \cdot t} - e^{-K_a \cdot t} \right] \tag{3:46}$$

Multiply each exponential term by means of the constant to obtain

$$C = \frac{K_a FD_{po}}{V(K_a - K)} \cdot e^{-K \cdot t} - \frac{K_a FD_{po}}{V(K_a - K)} \cdot e^{-K_a \cdot t} \tag{3:47}$$

Then take the derivative of that expression

$$\frac{dC}{dt} = \frac{K_a^2 FD_{po}}{V(K_a - K)} \cdot e^{-K_a \cdot t} - \frac{K_a K FD_{po}}{V(K_a - K)} \cdot e^{-K \cdot t} \tag{3:48}$$

Set the derivative to zero

$$\frac{dC}{dt} = 0 \tag{3:49}$$

and rearrange

$$\frac{K_a^2 FD_{po}}{V(K_a - K)} \cdot e^{-K_a \cdot t_{max}} = \frac{K_a K FD_{po}}{V(K_a - K)} \cdot e^{-K \cdot t_{max}} \tag{3:50}$$

$K_a \cdot FD_{po}/V \cdot (K_a-K)$ cancels out and one obtains

$$\frac{K_a}{K} = \frac{e^{-K \cdot t_{max}}}{e^{-K_a \cdot t_{max}}} \tag{3:51}$$

Take the natural logarithm of both sides and solve for t_{max}

$$t_{max} = \frac{1}{K_a - K} \ln\left[\frac{K_a}{K}\right] \tag{3:52}$$

The resulting plasma concentration C_{max} at t_{max} then becomes

$$C_{max} = \frac{K_a FD_{po}}{V(K_a - K)}\left[e^{-K \cdot t_{max}} - e^{-K_a \cdot t_{max}}\right] = \frac{FD_{po}}{V} \cdot e^{-K \cdot t_{max}} \tag{3:53}$$

If drug absorption is delayed by means of a lag-time, then t_{max} becomes

$$t_{max} = \frac{1}{K_a - K} \ln\left[\frac{K_a}{K}\right] + t_{lag} \tag{3:54}$$

and C_{max}

$$C_{max} = \frac{FD_{po}}{V} \cdot e^{-K \cdot (t_{max} - t_{lag})} \tag{3:55}$$

If t_{max} and K are known, K_a can be estimated from the expression above. Other approaches to derive initial estimates of K_a are by means of curve stripping, the residual method shown above and the methods by Pidgeon and Pitlick [1980] and Vaughan

[1976]. Pidgeon and Pitlick proposed Equation 3:56 which is depicted in Figure 3.22

$$\frac{K_a}{F} = \frac{C_{max}}{\int\limits_{t_{max}}^{\infty} Cdt - \frac{C_{max}}{K}}$$

(3:56)

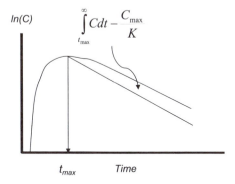

Figure 3.22 *Diagram illustrating the method of Pidgeon and Pitlick [1980]. C_{max} divided by the area between the down-swing of the concentration-time curve and the line from C_{max} parallel to the terminal slope, gives K_a/F.*

Vaughan derived Equation 3:57 for K_a

$$K_a = \frac{K}{1 - \frac{I_{iv}}{I_{po}} \cdot \frac{FD_{po}}{D_{iv}}}$$

(3:57)

I_{iv} and I_{po} are the back-extrapolated intercepts with the concentration axis of the terminal intravenous and oral phases. These are shown in Figure 3.23. λ_z is the terminal slope.

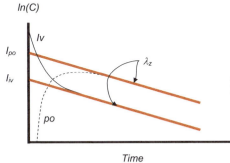

Figure 3.23 *Diagram illustrating the method of Vaughan [1976].*

A difficulty sometimes encountered in least-squares fitting of a one-compartment model with first-order absorption is that estimated values of K_a and K are almost identical. This anomaly is explained by the existence of a class of data sets for which least-squares estimates of the rate constants are complex quantities. Such data sets may arise either from an unfortunate combination of random (e.g., assay) errors in the concentration values if K_a and K are sufficiently similar in magnitude, or from delayed absorption. The usual recommendation in such cases, given without discussion of the cause, has been to use the special equation

$$C = \frac{K'FD_{po}}{V} \cdot t \cdot e^{-K' \cdot t} \tag{3:58}$$

which is easily derived from Equation 3:29 as the limiting form with $K_a = K = K'$. We will demonstrate the applicability of this model (Equation 3:58) in a later section (PK4). Note that a semi-logarithmic plot of concentration *versus* time yields a nonlinear relationship for this model. There will be no apparently linear segments (phases) since one adds the logarithm of time *(ln t)* to the overall function.

$$\ln(C) = \ln\left[\frac{K'FD_{po}}{V}\right] + \ln[t] - K' \cdot t \tag{3:59}$$

One starts by differentiating Equation 3:58 in order to obtain t_{max} and C_{max}, which gives

$$\frac{dC}{dt} = \frac{K'FD_{po}}{V} \cdot e^{-K' \cdot t} - \frac{K'^2 FD_{po}}{V} \cdot t \cdot e^{-K' \cdot t} \tag{3:60}$$

At t_{max}, $C = C_{max}$ and

$$\frac{dC}{dt} = 0 \tag{3:61}$$

Rearranging Equation 3:60 gives

$$\frac{K'FD_{po}}{V} \cdot e^{-K' \cdot t_{max}} = \frac{K'^2 FD_{po}}{V} \cdot t \cdot e^{-K' \cdot t_{max}} \tag{3:62}$$

Some terms cancel out and one obtains

$$t_{max} = \frac{1}{K'} \tag{3:63}$$

To get C_{max} at t equals t_{max}, t in Equation 3:62 is replaced by $1/K'$ which gives

$$C_{max} = \frac{K'FD_{po}}{V} \cdot \frac{1}{K'} \cdot e^{-K' \cdot \frac{1}{K'}} \tag{3:64}$$

This expression of C_{max} can be simplified to

$$C_{max} = \frac{FD_{po}}{V} \cdot e^{-1} = \frac{0.37 \cdot FD_{po}}{V} \qquad (3:65)$$

The absorption of chemicals is complex and involves several processes as mentioned above. In general, we assume that absorption follows a first-order process, but there are exceptions such as zero-order absorption. The equation that describes drug concentration in plasma following zero-order input is given by

$$C = \frac{R_{in}}{V \cdot K} \left[1 - e^{-K \cdot T_{abs}} \right] \cdot e^{-K \cdot t'} \qquad (3:66)$$

K may be expressed as Cl/V. Equation 3:66 then becomes

$$C = \frac{R_{in}}{Cl} \left[1 - e^{-\frac{Cl}{V} \cdot T_{abs}} \right] \cdot e^{-\frac{Cl}{V} \cdot t'} \qquad (3:67)$$

R_{in} is the apparent zero-order absorption rate constant and t' the time after drug absorption, which means actual time minus the absorption time, T_{abs}. During absorption T_{abs} is equal to the actual time t, and after cessation of absorption T is a constant. T_{abs} is a parameter to be estimated just like K. Note that this model is identical to the constant infusion model, except that we estimate the duration of input as a parameter. However, we want to make it clear that there are many exceptions, where the absorption process cannot be approximated by a first or zero-order process, but rather a more complex mixture of the two. In some cases drugs are also exsorbed from the blood across the gastrointestinal membranes by passive or active processes. The extent of this process will be determined by blood protein binding, ionization, lipophilicity, cacophilicity[2] (extent of sorption into membrane cells), molecular size, transporters (e.g., P-glycoprotein) and perfusion rates. On the other hand, the transdermal therapeutic system, and the estradiol vaginal ring are typical examples of controlled zero-order input. We will demonstrate examples of these two systems in the applications section of the book (PK3 and PK4). For further reading on modes, rates and routes of absorption, see Creasy and Jaffe [1991], Robinson [1991], Andersson *et al* [1993], Yoshida *et al* [1993], de Boer [1994], and Arimori and Nakano [1998].

3.2.8   Parameter ambiguities

Ideally, one should obtain an independent estimate of K to discriminate between the rate constants and to avoid ambiguity in interpreting the results from the curve fitting procedure. The absorption rate constant K_a, obtained from fitting Equation 3:36 to data is a best estimate of the first-order loss of drug from the gastrointestinal tract, not of the first-order appearance of drug in plasma (Figure 3.24).

---

[2] Cacophilicity denotes uptake and storage of chemicals into Caco-2 cells and was originally termed by D. McCarthy, AstraZeneca R&D Wilmington, with coworkers at Johns Hopkins, Baltimore.

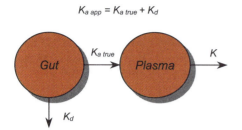

$$K_{a\ app} = K_{a\ true} + K_d$$

Figure 3.24 *Rate constants for the one-compartment model with first-order input from the gut compartment. Relationship between the apparent absorption rate constant, $K_{a\ app}$, and the true absorption rate constant, $K_{a\ true}$. K_d is the rate constant for loss of drug via the feces or chemical degradation within the gut region.*

In the following, we will assume that K_a obtained from regression is the true absorption rate constant. Remember that the rate of loss of drug from the gut compartment can be written as

$$Rate\ of\ loss = K_a D_{po} \cdot e^{-K_a \cdot t} \tag{3:68}$$

and the rate of input in plasma as

$$Rate\ of\ input = K_{a\ true} F D_{po} \cdot e^{-K_a \cdot t} \tag{3:69}$$

These are illustrated in Figure 3.25.

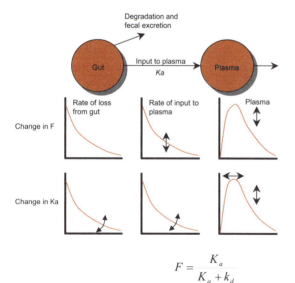

Figure 3.25 *Illustration of the first–order loss from the gut and first-order input into plasma. Bioavailability sets the rate into plasma in relation to rate of loss from gut. Rate of input (input) convoluted with the unit disposition function (not shown) gives the response function. The Y axes are either amount or concentration and X axes time. The arrows denote the changes of the curves when either F or K_a is changed.*

$$F = \frac{K_a}{K_a + k_d} \tag{3:70}$$

The availability needs to be measured parallel to apparent $K_{a\ app}$ in order to obtain a correct estimate of the true absorption rate constant. *AUC* depends on the bioavailability *F*, which is determined by absorption across the gastrointestinal membranes and hepatic extraction. Degradation of drug in gut and fecal excretion also affects *F*.

3.2.9   Estimation of bioavailability

The conventional way of estimating the bioavailability F is by sequential administration of the systemic (intravenous) and extravascular (e.g., oral) doses with an interval of say one day or a week between administrations. First the intravenous dose is given and plasma samples are collected and then the oral dose is given some time later and plasma samples are collected (The order of the intravenous and oral doses may be randomized). For estimation of bioavailability, the underlying assumption is that clearance is constant between the two occasions. However, one can never be sure of that. Some compounds induce or inhibit e.g., the liver enzymes, and that will affect the time course of the second dose.

   If one administers the two different regimens by means of a much shorter dosing interval (e.g., a few hours), nonstationarity in clearance is largely avoided. This can be seen in Figure 3.26 (bottom).

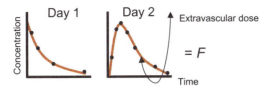

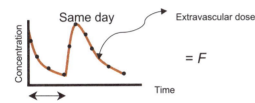

Figure 3.26 Effect on plasma concentration-time curve of varying the interval between successive intravenous (iv) and extravascular doses. Sequential administration of reference dose (iv) and test dose (e.g., po) where the doses are separated by e.g., a day (top). The bottom part of the figure shows the reference (iv) and test (po) doses being given the same day but separated by a short interval indicated by the horizontal arrow. '=F' denotes how one obtains the bioavailability.

   The next step would be to fit a model (Equation 3.71) which is the sum of the two doses to the experimental data.

$$C_{iv} = C^0 \cdot e^{-K \cdot t} = \frac{D}{V} \cdot e^{-\frac{Cl}{V} \cdot t} \tag{3:71}$$

C_{po} includes a term t_{dose}, which is the time for the oral dose.

$$C_{po} = \frac{K_a F D_{po}}{V(K_a - K)} \cdot \left[e^{-K \cdot (t - t_{dose})} - e^{-K_a \cdot (t - t_{dose})} \right] \tag{3:72}$$

   The function for the sum of the two doses is then fitted to the total plasma concentration-time curve

$$C_{total} = C_{iv} + C_{po} \tag{3:73}$$

3.2.10   First-order input – implementation of computer code

If you want to implement a first-order input rate model that starts at time t_1 we propose the following program code. Figure 3.27 displays the behavior of the input rate or input function, *Input*, starting at time t_1 and the resulting plasma concentration-time curve.

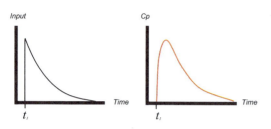

Figure 3.27 *The first-order input function (Input) following an extravascular dose and the resulting plasma concentration-time curve (C_p). The total administered dose is the integral (area under the input function) of the input function versus time. t_1 is either a lag-time or the time at which the extravascular dose is given.*

The total dose is denoted *Dose*. *F* is the bioavailability. *Input* is equal to zero before t_1. Otherwise, *Input* equals $K_a \cdot FD_{po} \, e^{-Ka \cdot (t-t1)}$ starting at t_1. The process is first-order with a rate constant equal to K_a. A plot of the *ln(Input rate)* gives a straight line with a slope of K_a. This can be expressed as an IF . . . THEN . . . ELSE . . . ENDIF statement as follows:

If time is greater than or equal to t_1 then *Input* is equal to $F \cdot Dose \cdot Ka \cdot e^{-Ka(t-t1)}$. Else, *Input* is equal to zero.

```
IF T GE T1 THEN
       INPUT = F*DOSE*KA*EXP(-KA*(T-T1))
ELSE
       INPUT = 0.
ENDIF
```

3.2.11   Absorption from multiples sites

It is not too uncommon to observe multiple peaks in the plasma concentration after extravascular dosing of a drug. Multiple peaks may not necessarily mean enterohepatic recirculation. It can be as simple as absorption of the drug from multiple sites. Figure 3.28 shows a schematic model of a drug that is placed sublingually. Rapid absorption occurs from the buccal cavity of a fraction (*frct*) of the dose. This corresponds to the first peak in the concentration-time curve. Some of the drug (1-*frct*) is swallowed together with the saliva and is then taken up by the gastrointestinal tract. This corresponds to the second peak.

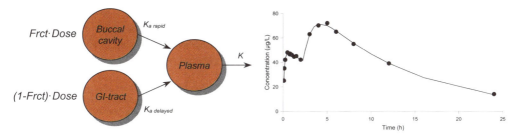

Figure 3.28 *A two-site absorption model (left), with a typical plasma concentration-time curve shown (right). Note the late second peak appearance at about 5 hours.*

The input rate from the buccal cavity is given by means of Equation 3:74.

$$Rapid\ input = frct \cdot Dose \cdot K_{a\,rapid} \cdot e^{-K_{a\,rapid} \cdot t} \tag{3:74}$$

The delayed absorption of the fraction (1-*frct*) of drug that escapes buccal absorption but is taken up by the gastrointestinal tract is written as

$$Delayed\ input = (1 - frct) \cdot Dose \cdot K_{a\,delayed} \cdot e^{-K_{a\,delayed} \cdot (t - t_{lag})} \tag{3:75}$$

The total input rate in Equation 3:76 is then the sum of the rapid (buccal) and delayed (gastrointestinal) rates. Note that the rapid $K_{a\,rapid}$ and delayed $K_{a\,delayed}$ uptake have different absorption rate constants. The differential equation model of the plasma kinetics is written as total input rate minus rate of elimination, which becomes

$$V \cdot \frac{dC}{dt} = Total\ input\ rate - Cl \cdot C \tag{3:76}$$

This generic model is very useful for other first-order input sites and can be extended to also include multiple zero-order input such as burst and maintainance infusion from a transdermal device such as the patch.

It seems that multiple peaks after oral administration of a drug could have several causes, such as reduced gastric motility (Wang *et al* [1999]), absorption differences caused by pH dependent permeability, transporter densities, solubility differences due to pH, efflux mechanisms, enterohepatic recycling and the presence of absorption windows along the gastrointestinal tract etc. Example PK43 demonstrates partial buccal absorption and partial absorption via the gastrointestinal tract (due to swallowing) of the drug substance in a sublingual dosage form.

3.2.12   Plasma and urine data

One may also want to discuss clearance as the proportionality factor between rate of elimination (*amount/time*) and plasma concentration (*amount/volume*).

$$V \frac{dC}{dt} = \frac{dX}{dt} = -Cl \cdot C \qquad\qquad (3{:}77)$$

where X denotes the amount of drug in plasma at time t. If one then integrates this equation from zero to infinity then one obtains

$$\int_0^\infty \frac{dX(t)}{dt} = \int_0^\infty I(t)dt - Cl \int_0^\infty C(t)dt \qquad\qquad (3{:}78)$$

which integrated becomes

$$X(t)\Big|_0^\infty = dose - Cl \cdot AUC \qquad\qquad (3{:}79)$$

The total amount eliminated is equal to the administered dose and the integral of the plasma concentration-time curve is simply the AUC. As t goes to infinity, Equation 3:79 becomes

$$Cl = \frac{Dose}{AUC} \qquad\qquad (3{:}80)$$

The elimination (excretion) rate of drug into the urine (dX_u/dt) is

$$\frac{dX_u}{dt} = Cl_R \cdot C \qquad\qquad (3{:}81)$$

where Cl_R is the renal clearance (see Figure 3.29). In practice, renal clearance is estimated by dividing the average urinary excretion rate (dX_u/dt) by the drug concentration in plasma at the time corresponding to the midpoint of the urine collection period (Gibaldi and Perrier [1982]).

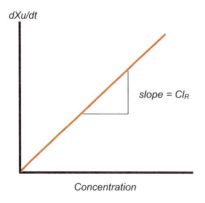

dXu/dt

slope = Cl$_R$

Concentration

Figure 3.29 *Excretion rate as a function of plasma concentration. The slope corresponds to renal clearance, Cl_R.*

The total amount of drug eliminated into urine X_{u0} is then equal to the product of

Cl_R and AUC according to

$$X_{u\,0-\infty} = Cl_R \cdot AUC \tag{3:82}$$

From this relationship it is obvious that one can calculate renal clearance from either urine or plasma data. Information about the administered dose is not necessary. One may also choose to collect urine over a certain time interval, e.g., t_1 to t_2. If one then relates the amount excreted into urine to the corresponding area under the plasma concentration-time curve within that time interval, one may then estimate renal clearance from the following expression

$$X_u(t)\Big|_{t_1}^{t_2} = Cl_R \cdot \int_{t_1}^{t_2} Cdt = Cl_R \cdot AUC_{t_1}^{t_2} \tag{3:83}$$

In some cases it is possible to determine the elimination kinetics of a drug from urinary excretion data. The model in Figure 3.30 mimics the disposition of a one-compartment drug after an intravenous bolus dose and parallel urinary and metabolic elimination.

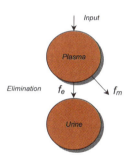

Figure 3.30 Schematic illustration of the one-compartment model with partial urinary excretion. The parameters f_e and f_m stand for fraction of input dose excreted into urine and being metabolized, respectively. The urinary excretion rate constant k_u is obtained from the product of f_e and K.

3.2.13 Analysis of urinary excretion

Several methods have been proposed for analysis of urinary data. The two most common methods are the *cumulative excretion* and *urinary excretion rate* (volume of urine collected within a time interval times the concentration of drug in the urine) plotted *versus* the midpoint time of the collection interval.

The *excretion rate* plot is of limited value when the drug half-life is short relative to the urine collection interval, and when incomplete emptying of the bladder occurs. The cumulative excretion is used to produce the *Amount Remaining to be Excreted (ARE* plot). The *ARE* plot uses actual time and not the midpoint of the collection interval. The use of the midpoint is necessary since the excretion-rate is averaged over the collection interval. The *ARE* plot required a correct estimate of the total amount (X_{u0-}) excreted via the urine. These values are obtained by summing up the amount for each collection interval. The smoothed *ARE* plot is therefore limited since absorption problems, cumulative assay errors and incomplete urine collection will produce biased results.

Changes in renal clearance due to pH and urine flow will readily be seen in the rate plot, whereas clearance is smoothed out in the *ARE* plot (Rowland and Tozer [1993]).

Nonlinear regression of the cumulative excretion of drug in urine plotted *versus* the actual time is, in our experience, a robust method provided data are available over at least one half-life and it is superior over the *ARE* plot in that the regression uses all (untransformed) data simultaneously. It is not necessary to obtain the total amount excreted via the urine, nor is it necessary to have complete bladder emptying within all sampling intervals. There is not a need to collect the urine over short intervals relative to the half-life of the drug. Equation 3:84 gives the expression for use in nonlinear regression of cumulative amount of drug in urine.

$$\frac{dX_u}{dt} = Cl_R \cdot C = f_e \cdot Cl \cdot C \qquad (3:84)$$

Renal clearance Cl_R and fraction of dose excreted via urine f_e are the parameters to be estimated from the regression. Equation 3:84 is fit to urinary data simultaneously with regression of e.g., Equation 3:1 to plasma data. Recall Equation 3:1, which assumes that plasma concentrations display mono-exponential decay.

$$\frac{dC}{dt} = -K \cdot C = -\frac{Cl}{V} \cdot C \qquad (3:1)$$

Total plasma clearance Cl and volume of distribution V are the regression parameters of the plasma model. Simultaneous fitting of several sources (urine and plasma) strengthens the results in terms of parameter accuracy and precision. We refer the reader to e.g., sections PK5, PK6 and PK15 for a discussion on simultaneous fitting plasma and urine data.

Figure 3.31 illustrates observed plasma and urine data.

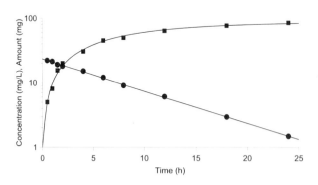

Figure 3.31 Observed plasma concentration (closed circles) and cumulative amount in urine (closed squares) following an intravenous bolus dose. Solid lines are predicted curves.

Cl_R is estimated from the product of f_e and plasma clearance Cl. The fraction of dose excreted via the urine may be viewed as the ratio of k_u (fractional rate constant for excretion into urine) over K according to Equation 3:85

$$f_e = \frac{k_u}{k_u + k_m} = \frac{k_u}{K} \tag{3:85}$$

The k_m parameter is the rate constant for the metabolic route of elimination. To solve this problem with short collection intervals, plasma and urine data can be modeled simultaneously for estimation of parameters they have in common, such as Cl and V as seen below. Then f_e has to represent substantial excretion of unchanged drug, i.e., this method often fails when $f_e \leq 0.1$. The analytical solution of Equation 3:1 is as follows

$$C = \frac{D_{iv}}{V} \cdot e^{-\frac{Cl}{V} \cdot t} \tag{3:86}$$

The corresponding function (analytical solution of Equation 3:84) for accumulated amount in urine is

$$X_u = f_e \cdot D_{iv} \left[1 - e^{-\frac{Cl}{V} \cdot t} \right] \tag{3:87}$$

The plasma equation following extravascular administration for a one-compartment model is

$$C = \frac{K_a FD_{po}}{V \cdot \left(K_a - \frac{Cl}{V} \right)} \cdot \left[e^{-\frac{Cl}{V} \cdot (t - t_{lag})} - e^{-K_a \cdot (t - t_{lag})} \right] \tag{3:88}$$

The accumulated amount in urine X_u is

$$X_u = f_e \cdot K_a \cdot FD_{po} \left[\frac{1}{K_a} + \frac{e^{-\frac{Cl}{V} \cdot (t - t_{lag})}}{\frac{Cl}{V} - K_a} - \frac{\frac{Cl}{V} \cdot e^{-K_a \cdot (t - t_{lag})}}{K_a \cdot \left(\frac{Cl}{V} - K_a \right)} \right] \tag{3:89}$$

Assuming a one-compartment system, the differential equation for plasma following administration of an intravenous bolus dose is

$$V \frac{dC}{dt} = -Cl \cdot C \tag{3:1}$$

Recall that the elimination (excretion) rate of drug into the urine is equal to

$$\frac{dX_u}{dt} = Cl_R \cdot C \tag{3:84}$$

The analog of Equation 3:1 corresponding to an oral dose for plasma is

$$V \frac{dC}{dt} = In - Cl \cdot C \tag{3:90}$$

In represents the first-order input function. The excretion rate into urine is still modeled according to Equation 3:84. The beauty of specifying the model as a system of differential equations lies in the simplicity of the equations. Thus we only need to specify *In*. For a bolus dose, *In* is a *spike* at time zero, i.e., $C^0 = Dose/V$. For a first-order input, *In* becomes

$$In = K_a \cdot FD_{po} \cdot e^{-K_a \cdot t} \tag{3:91}$$

For a zero-order input when $t < T_{inf}$, *In* becomes

$$In = \frac{D_{inf}}{T_{inf}} \tag{3:92}$$

When $t > T_{inf}$, $In = 0$. The general expression of excretion rate of drug into urine is

$$\frac{dX_u}{dt} = Cl_R \cdot C \tag{3:93}$$

After a bolus dose, excretion rate becomes

$$\frac{dX_u}{dt} = Cl_R \cdot C^0 \cdot e^{-\frac{CL}{V} \cdot t} \tag{3:94}$$

If C^0 is replaced by D_{iv}/V

$$\frac{dX_u}{dt} = Cl_R \cdot \frac{D_{iv}}{V} \cdot e^{-K \cdot t} \tag{3:95}$$

Taking the natural logarithm of both sides gives

$$\ln\left[\frac{dX_u}{dt}\right] = \ln\left[Cl_R \cdot \frac{D_{iv}}{V}\right] - K \cdot t \tag{3:96}$$

Plotting *ln(excretion rate) versus* time gives a negative slope equal to the negative of the *elimination rate* constant K and an intercept of the Y axis equal to the total excreted amount of drug into the urine. Integrating Equation 3.95 gives the accumulated amount of drug excreted into the urine

$$X_u = f_e \cdot D_{iv}\left[1 - e^{-\frac{Cl}{V} \cdot t}\right] = f_e \cdot D_{iv}\left[1 - e^{-K \cdot t}\right] \tag{3:97}$$

Expanding the right side by $f_e \cdot D_{iv}$ gives

$$X_u = f_e \cdot D_{iv} - f_e \cdot D_{iv} \cdot e^{-K \cdot t} \qquad (3{:}98)$$

Replacing $f_e \cdot D_{iv}$ by

$$X_u^\infty = f_e \cdot D_{iv} \qquad (3{:}99)$$

and inserting Equation 3:101 into Equation 3:99 gives

$$X_u = X_u^\infty - X_u^\infty \cdot e^{-K \cdot t} \qquad (3{:}100)$$

Rearranging

$$X_u^\infty - X_u = X_u^\infty \cdot e^{-K \cdot t} \qquad (3{:}101)$$

and taking the natural logarithm of both sides gives

$$\ln\!\left(X_{u0-\infty} - X_u\right) = \ln\!\left(X_{u0-\infty}\right) - K \cdot t \qquad (3{:}102)$$

A plot of $ln(X_{u0-\infty}\text{-}X_u)$ versus t gives a line with a slope of $-K$ and an intercept of the Y axis of $ln(X_{u0-\infty})$. Rearrangement of Equation 3.84 gives

$$Cl_R = \frac{\dfrac{dX_u}{dt}}{C} \qquad (3{:}103)$$

Taking the integral of the rate of excretion (dX_u/dt) and the concentration (C) between t_1 and t_2

$$Cl_R = \frac{\left(X_u\right)_{t_1}^{t_2}}{AUC_{t_1}^{t_2}} \qquad (3{:}104)$$

which gives the amount of drug excreted into urine divided by the area under the plasma concentration curve for the same time interval. Note that renal clearance can be obtained from selected time intervals. This is a means of determining whether renal clearance is time independent or not. If one integrates Equation 3:105 from 0 to infinity one gets

$$Cl_R = \frac{X_u^\infty}{AUC_0^\infty} \qquad (3{:}105)$$

which is equivalent to

$$Cl_R = \frac{f_e \cdot K \cdot Dose}{C} \qquad (3{:}106)$$

3.3   Multi-Compartment Models

3.3.1   Catenerary and mammillary models

When drugs are administered rapidly and concentrations measured frequently, the mono-exponential (one-compartment) model is not very suitable for predicting the time course of drug concentrations. This is due to the fact that it takes time for the drug to distribute into the tissues and reach equilibrium. This section demonstrates how to handle situations where we do not have an instantaneous equilibrium between plasma and tissues.

Multi-compartment models are more complex than the one-compartment model, introducing additional compartments for distribution and re-distribution of the drug between a central compartment (blood and rapidly equilibrated organs) and one or several peripheral compartments (more slowly equilibrating tissues, which may either be poorly perfused or surrounded by protective membranes). In contrast to the one-compartment model, a longer time is usually required in order to achieve total distribution equilibrium between blood and tissues.

The multi-compartment catenery system is characterized by means of a number of compartments which are in a series or a chain (Figure 3.32). While flux between adjacent compartments may be either uni or bi-directional, the overall flux of substance is sequential, via successive compartments in the series.

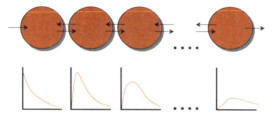

Figure 3.32 Schematic illustration of the multi-compartment catenerary system. AUC is the same for all compartments in series

The multi-compartment mammillary system is characterized by means of a number of peripheral compartments, which are connected to a single central compartment (Godfrey [1983]). The expression originates from the mammillary glands, which are organized in such a fashion (Figure 3.33). Input and irreversible loss of drug is generally assumed to occur only from the central compartment.

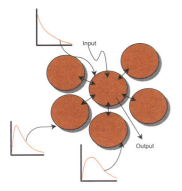

Figure 3.33 Schematic illustration of the multi-compartment mammillary system. The peripheral compartments are attached to a single central compartment.

The effect of adding a second or peripheral compartment (Figure 3.34) to the one-compartment model is to introduce a second exponential term into the predicted time course of the plasma concentration, so that it comprises a rapid and a slow phase. This pattern is often found experimentally, and is most clearly revealed when the concentration data are plotted semi-logarithmically. Figure 3.34 demonstrates schematically the time course of drug in the central and peripheral compartments after a bolus dose. The concentration declines continuously in the central compartment. This process combines irreversible elimination and distribution, which is called disposition. The concentration is zero at the start in the peripheral compartment. It then increases reaches a peak where influx is equal to efflux, and finally declines again. The increase and decline in the peripheral compartment are due to distributional processes such as perfusion and diffusion into and out of that compartment.

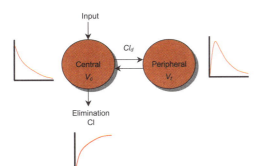

Figure 3.34 The two-compartment mammillary system with a bolus input of drug. Note how the concentration-time course of drug will vary between the central and peripheral compartments, and the accumulated urinary amount.

If we assume that drug is excreted into the urine, the time course of the accumulated amount of drug excreted is illustrated at the bottom of Figure 3.34. Hence, depending on the fluid or tissue of the body where one is measuring drug, one will observe very different concentration-time curves. The drug will display rapid kinetics in highly vascularized tissues and organs, whereas in poorly perfused tissue such as adipose tissue, the drug will display a slow turnover.

3.3.2   Intravenous bolus administration

A bi-exponential decline is associated with a two-compartment model

$$C = A \cdot e^{-\alpha \cdot t} + B \cdot e^{-\beta \cdot t} \qquad (3:108)$$

The primary model parameters in Equation 3:108 are A, α, B and β, where $-\alpha$ and $-\beta$ correspond to the initial and terminal slopes, respectively. Their corresponding half-lives are estimated as

$$t_{1/2\alpha} = \frac{\ln(2)}{\alpha} \qquad (3:109a)$$

$$t_{1/2\beta} = \frac{\ln(2)}{\beta} \qquad (3:109b)$$

Figure 3.35 shows the plasma concentration-time data of a compound that displays multi (two)-compartment characteristics following an intravenous bolus dose. The same figure demonstrates how the bi-exponential decline is dissected into its principal parts. α is obtained by means of the method of residuals, which means that the back-extrapolated β-slope is subtracted from the experimental data during the initial phase. The difference corresponds to the α-phase and the slope is $-\alpha$. The intercept of the back-extrapolated concentration-time curve on the concentration axis C^0 is equal to $A + B$.

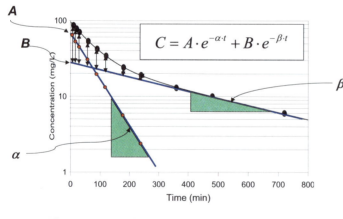

Figure 3.35 Semi-logarithmic plot of the bi-exponential decay in plasma of a drug administered as an intravenous bolus dose. A ≈ 70 and B ≈ 28 mg/L. α ≈ ln(70/10)/130 and β ≈ ln(28/10)/450.

The parameters describing the bi-exponential decline in Equation 3:108 are also called the *macro-constants*, as compared to k_{12}, k_{21}, k_{10}, and V_c, which are called *micro-constants*. We may parameterize the model in terms of clearance Cl, inter-compartmental distribution Cl_d and distribution volumes for the respective compartment V_i (Figure 3.36). For a two-compartment model, V_c basically represents plasma or blood and vascular bed, while V_t represents tissues (e.g., muscle and adipose tissues). Cl_d corresponds to the inter-compartmental distribution between plasma and tissue, which embodies perfusion, diffusion and active transport processes as well as partitioning (see section 3.3.3). Cl_d has the same units as blood flow (volume/time). The Cl_d parameter can be converted into the micro-constants.

$$Cl_d = k_{12} \cdot V_c = k_{21} \cdot V_t \qquad (3{:}110)$$

where k_{12} and k_{21} represent the inter-compartmental rate constants.

Cl is the irreversible loss of drug from the central compartment. A fuller discussion of these parameters and of how to obtain their initial estimates is given in the section on *Reparameterizations of the two-compartment model* (Section 3.3.3).

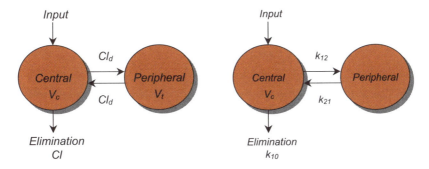

Figure 3.36 *A two-compartment model parameterized with physiological parameters (Cl, V_c, Cl_d and V_t) and micro-constants (k_{10}, k_{12}, k_{21} and V_c).*

The *macro-constants* can be estimated from the *micro-constants* and vice-versa. *A* and *B* in the *macro-constant* model are conglomerates of both the dose (D_{iv}), the macro-constant slopes (-α, -β) and a *micro-constant* (k_{21}).

$$A = \frac{D_{iv}}{V_c} \cdot \frac{\alpha - k_{21}}{\alpha - \beta} \qquad (3{:}111)$$

$$B = \frac{D_{iv}}{V_c} \cdot \frac{\beta - k_{21}}{\beta - \alpha} \qquad (3{:}112)$$

Incorporation of these in the bi-exponential model gives

$$C = \frac{D_{iv}}{V_c} \cdot \left\{ \frac{k_{21} - \alpha}{\beta - \alpha} \cdot e^{-\alpha \cdot t} + \frac{k_{21} - \beta}{\alpha - \beta} \cdot e^{-\beta \cdot t} \right\} \qquad (3{:}113)$$

α is obtained from

$$\alpha = \frac{k_{21} \cdot k_{10}}{\beta} \qquad (3{:}114)$$

and β is expressed as

$$\beta = \frac{1}{2}\left[k_{12} + k_{21} + k_{10} - \sqrt{\left(k_{12} + k_{21} + k_{10}\right)^2 - 4k_{21} \cdot k_{10}}\right] \tag{3:115}$$

Rearranging Equation 3.114 gives k_{10}

$$k_{10} = \frac{\alpha \cdot \beta}{k_{21}} \tag{3:116}$$

k_{12} is obtained from

$$k_{12} = \alpha + \beta - k_{21} - k_{10} \tag{3:117}$$

k_{21} can be estimated directly from the macro-constants

$$k_{21} = \frac{A \cdot \beta + B \cdot \alpha}{A + B} \tag{3:118}$$

V_c is estimated as

$$V_c = \frac{D_{iv}}{A + B} \tag{3:119}$$

The total area under the plasma concentration-time curve (or zero moment curve), AUC, is derived from the macro-constants as

$$AUC = \frac{A}{\alpha} + \frac{B}{\beta} \tag{3:120}$$

The respective areas under the α and β-phases are estimated as

$$AUC_\alpha = \frac{A}{\alpha} \tag{3:121}$$

$$AUC_\beta = \frac{B}{\beta} \tag{3:122}$$

The fractional areas under the α and β-phases are estimated as

$$\%AUC_\alpha = \frac{\dfrac{A}{\alpha}}{\dfrac{A}{\alpha} + \dfrac{B}{\beta}} \cdot 100 \tag{3:123}$$

$$\%AUC_\beta = \frac{\dfrac{B}{\beta}}{\dfrac{A}{\alpha} + \dfrac{B}{\beta}} \cdot 100 \tag{3:124}$$

The phase that generally contributes most to the total area is called the elimination phase. Note that the fractional area corresponds to the fraction of the dose that is eliminated during that phase. In other words, the α-phase can also be the elimination phase for some compounds (e.g., aminoglycosides). We will use the initial and terminal phase for the respective portion of the concentration-time curve in a semi-logarithmic diagram. Unfortunately, the terminal phase is often denoted the elimination phase, which may not always be true.

The total area under the first moment curve *AUMC* is derived from the *macro-constants* as

$$AUMC = \frac{A}{\alpha^2} + \frac{B}{\beta^2} \qquad (3:125)$$

Clearance is the dose divided by the total area under the curve

$$Cl = \frac{D_{iv}}{AUC} = \frac{D_{iv}}{\dfrac{A}{\alpha} + \dfrac{B}{\beta}} \qquad (3:126)$$

The volume of the central compartment V_c, is expressed as D_{iv} divided by the intercepts

$$V_c = \frac{D_{iv}}{A + B} \qquad (3:127)$$

The mean residence time *MRT* is the average time a molecule stays in the system. It is expressed as the ratio of *AUMC*-to-*AUC* (following intravenous bolus dosing only)

$$MRT = \frac{AUMC}{AUC} \qquad (3:128)$$

The volume of distribution at *steady state* V_{ss} is

$$V_{ss} = MRT \cdot Cl = \frac{AUMC}{AUC} \cdot \frac{D_{iv}}{AUC} \qquad (3:129)$$

This is the sum of the central and peripheral volume terms

$$V_{ss} = V_c + V_t \qquad (3:130)$$

For a two-compartment model V_{ss} becomes

$$V_{ss} = V_c \left[1 + \frac{k_{12}}{k_{21}} \right] \qquad (3:131)$$

and, including *Cl*, α and β, becomes

$$V_{ss} = Cl \cdot \left[\frac{1}{\alpha} + \frac{1}{\beta} - \frac{1}{k_{21}} \right]$$ (3:132)

For a three-compartment model, V_{ss} becomes

$$V_{ss} = V_c \left[1 + \frac{k_{12}}{k_{21}} + \frac{k_{13}}{k_{31}} \right]$$ (3:133)

When elimination occurs from the central compartment, a generic expression of V_{ss} containing the *micro-constants* is

$$V_{ss} = V_c \left[1 + \sum \frac{k_{1i}}{E_i} \right] = V_c \left[1 + \sum \frac{k_{1i}}{k_{i1}} \right]$$ (3:134)

Note that E_i is the sum of the exit-rate constants from the ith compartment (Benet [1972]). There are three variations of the two-compartment mammillary model, which are shown in Figure 3.37. Two of them are four-parameter models (V_c and three rate constants) and one is a five-parameter model (V_c and four rate constants).

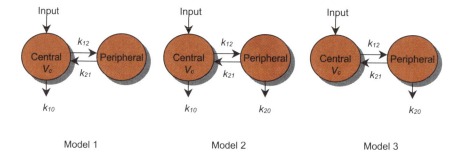

Model 1   Model 2   Model 3

Figure 3.37 *The three variations of the two-compartment model.*

V_{ss} for model 2 in Figure 3.37 is written according to Equation 3:135,

$$V_{ss} = V_c \cdot \left[1 + \frac{k_{12}}{k_{20} + k_{21}} \right]$$ (3:135)

The three models are not uniquely identifiable from a bi-exponential plasma concentration curve. One would need data from an additional source (such as the peripheral compartment).

Table 3.2 contains a summary of the *SHAM-* (*Slope, Height, Area, and Moment*) analysis of an intravenous bolus concentration curve for a drug displaying two-compartment kinetics (Lassen and Perl [1979]).

Table 3.2   Summary table of non-compartmental parameters and variables

Parameters	Equation	Dimension	Comment
Plasma curve	$A\,e^{-\alpha t} + B\,e^{-\beta t}$	amount/volume	iv bolus
Slope(S)	$A\alpha + B\beta$	amount/volume/t	S
Height (H)	$A + B$	amount/volume	H
Area (AUC)	$A/\alpha + B/\beta$	(amount/volume)t	A
Moment (AUMC)	$A/\alpha^2 + B/\beta^2$	(amount/volume)t^2	M
Centroid (MRT)	$AUMC/AUC$	t	mean residence time
Centroid^{-1}	$AUC/AUMC$	t^{-1}	fractional catabolic rate
Height/Slope	$(A + B)/(A\alpha + B\beta)$	t	mean transit time in plasma
Slope/Height	$(A\alpha + B\beta)/(A + B)$	t^{-1}	fractional transcapillary escape
Area/Height	$AUC/(A + B)$	t	mean sojourn time in plasma
Dose/Area (*Cl*)	D/AUC	volume/t	clearance
Dose/Height (V_c)	$D/(A + B)$	volume	central volume

The table is adapted from Lassen and Perl [1979].

3.3.3   Reparameterizations of the two-compartment model

It is possible to reparameterize the bi-exponential model (Equation 3:136) and eliminate *A* as a parameter by using D_{iv} and *Cl*, and to fit this new model to the data. The model derives from Equations 3:136 and 3:137

$$C = A \cdot e^{-\alpha \cdot t} + B \cdot e^{-\beta \cdot t} \qquad (3{:}136)$$

Cl is estimated as dose divided by area

$$Cl = \frac{D_{iv}}{\dfrac{A}{\alpha} + \dfrac{B}{\beta}} \qquad (3{:}137)$$

which rearranged gives

$$A = \alpha \cdot \left[\frac{D_{iv}}{Cl} - \frac{B}{\beta} \right] \qquad (3{:}138)$$

Equation 3:138 substituted into Equation 3:136 yields

$$C = \alpha \cdot \left[\frac{D_{iv}}{Cl} - \frac{B}{\beta} \right] \cdot e^{-\alpha \cdot t} + B \cdot e^{-\beta \cdot t} \qquad (3{:}139)$$

Cl, α, β and *B* are now the model parameters as compared to *A, α, B* and *β* in the original bi-exponential model.

One could either specify the model as an ordinary bi-exponential decay model as in Equation 3:136 or alternatively one could give the volume of distribution as a function of time as seen in Figure 3.38 and in Equation 3:140.

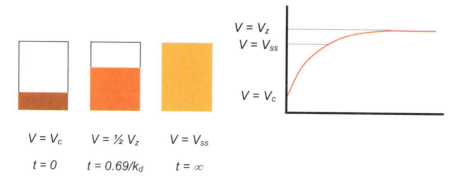

$V = V_c$ $V = \frac{1}{2} V_z$ $V = V_{ss}$

$t = 0$ $t = 0.69/k_d$ $t = \infty$

Figure 3.38 *Schematic representation of volume of distribution as a function of time. Assuming that the drug is administered as an intravenous bolus, it distributes initially in a volume corresponding to V_c and then gradually reaches pseudo-equilibrium at V_z. V_z corresponds to V_β in a two-compartment model and K_d is the first-order rate constant in Equation 3:143.*

The Takada [1981] distribution model implements V as a function of time as follows

$$C = \frac{D_{iv}}{V_c + V_t} \cdot e^{-\beta \cdot t}$$
(3:140)

where V_t is defined as

$$V_t = \frac{V_{max} \cdot t}{K_d + t}$$
(3:141)

V_{max} and K_d correspond to the maximum tissue volume and time to establish half-maximal distribution equilibrium, respectively. A time-dependent volume approach is to utilize the Colburn distribution model (Colburn [1983]). Here, the plasma concentration is defined as

$$C = \frac{D_{iv}}{V_c + V_t} \cdot e^{-\beta \cdot t}$$
(3:142)

where V_t is

$$V_t = V_{max} \cdot \left[1 - e^{-K_d \cdot t}\right]$$
(3:143)

V_{max} and K_d correspond to the maximum tissue volume and the first-order equilibrium rate constant, respectively.

Pharmacokinetics has moved away from parameterizing elimination in terms of *micro constants*. Rather, the more physiologically relevant use of clearance is now widely accepted. To do so, the model is specified in terms of Cl, V_c, Cl_d, and V_t (Benet [1972], Jusko [1980]) (Figure 3.39).

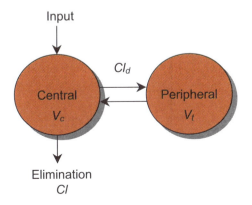

Input

Figure 3.39 A two-compartment model parameterized with physiological parameters such as Cl, V_c, Cl_d, and V_t.

These parameters correspond to plasma clearance Cl, the volume of the central compartment V_c, inter-compartmental distribution Cl_d, and the volume of the peripheral compartment V_t, respectively.

$$V_c = \frac{D_{iv}}{A + B} \tag{3:144}$$

$$Cl = \frac{D_{iv}}{\dfrac{A}{\alpha} + \dfrac{B}{\beta}} = \frac{D_{iv}}{AUC} \tag{3:145}$$

$$V_{ss} = MRT \cdot Cl = \frac{AUMC}{AUC} \cdot \frac{D_{iv}}{AUC} \tag{3:146}$$

$$V_{ss} = V_c + V_t \tag{3:147}$$

or

$$V_t = V_{ss} - V_c \tag{3:148}$$

Figure 3.40 depicts the early time course of a drug that obeys bi-exponential decline.

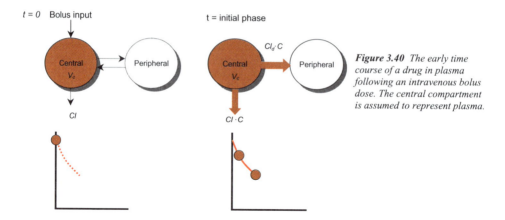

Figure 3.40 *The early time course of a drug in plasma following an intravenous bolus dose. The central compartment is assumed to represent plasma.*

A good initial estimate of the inter-compartmental distribution parameter, which can be compared to blood flow, is

$$Cl_d = \lambda_0 \cdot V_c - Cl \qquad (3{:}149)$$

where λ_0 is approximately Q/V_{blood} (Q is blood flow to the tissues which we approximate with Cl_d) since the concentration in the central compartment can, at early time points after a bolus dose, be approximated with

$$V_c \frac{dC}{dt} = -Cl \cdot C - Cl_d \cdot C \qquad (3{:}150)$$

and

$$\lambda_0 = \frac{Cl_d - Cl}{V_c} \qquad (3{:}151)$$

λ_0 is also the maximum value of α if you consider the initial loss from the central compartment to be approximately

$$C_{early} \approx \frac{D_{iv}}{V_c} \cdot e^{-\frac{Cl_d + Cl}{V_c} \cdot t} \qquad (3{:}152)$$

The inter-compartmental distribution parameter, can also be derived from the micro-constants

$$k_{12} = \frac{Cl_d}{V_c} \qquad (3{:}153)$$

$$k_{21} = \frac{Cl_d}{V_t} \qquad (3{:}154)$$

The k_{12} and k_{21} parameters are the inter-compartmental rate constants (fractional) which determine the rate of drug movement from the central to the peripheral compartment and back, respectively. These equations simply allow Cl_d to be calculated from the *micro-constants*, but should not be viewed as indicative of the determinants of Cl_d. The equation for the central or plasma compartment is

$$V_c \frac{dC}{dt} = In - Cl \cdot C - Cl_d \cdot C + Cl_d \cdot C_t \qquad (3:155)$$

In is the input function, which in this case is bolus input. The equation for the peripheral or tissue compartment is

$$V_t \frac{dC_t}{dt} = Cl_d \cdot C - Cl_d \cdot C_t \qquad (3:156)$$

At steady state, $dC/dt = dC_t/dt = 0$, and rate of input *In,* equals rate of elimination ($Cl \cdot C$). That is,

$$In = Cl \cdot C \qquad (3:157)$$

For the plasma compartment, the rate of input into the extravascular space equals the rate of removal according to

$$Cl_d \cdot C = Cl_d \cdot C_t \qquad (3:158)$$

Cl_d cancels out on both sides, which at *steady state* results in

$$C = C_t \qquad (3:159)$$

This may, of course, not always be so. We might have accumulation of drug into various tissues rendering higher tissue concentrations than those actually observed in plasma. However, using the parameterization in Equations 3:155 and 3:156, without partitioning into the extravascular compartment, the central and peripheral concentration will be equal at steady state. If they are not equal at steady state, this indicates pehaps partitioning. In such a case one can model that by two different Cl_d terms. The ratio of the two gives the partition coefficient at steady state. Figure 3.41 depicts the concentration-time course of drug in central and peripheral compartments and how they approach each other when there is no elimination from the central compartment and when the partition coefficient between the tissue and plasma compartment is equal to one.

C

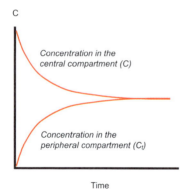

Concentration in the
central compartment (C)

Concentration in the
peripheral compartment (Ct)

Time

Figure 3.41. *The concentration-time course of drug in the central and peripheral compartments of the physiological compartment model in Figure 3.39.*

Many low-clearance drugs have a tendency to exhibit mono-exponential decline in plasma whereas high clearance drugs mostly show a multi-exponential decline. *MRT* then becomes

$$MRT = \frac{V_{ss}}{Cl} \tag{3:160}$$

The *mean transit time* through a compartment, is the average time spent by a molecule from its entry into the compartment to its exit. Mean transit time *MTT*, for the central compartment is then

$$MTT = \bar{t} = \frac{V_c}{Cl + Cl_d} \tag{3:161}$$

Sojourn time is the time required on average for a molecule to make a single pass through a tissue or an organ. Like transit and residence times, it is the quotient of essential volume and flow parameters. Table 3.3 compares transit, sojourn and residence times.

Table 3.3 Transit, sojourn and residence times

At compartment level	At tissue level	At body level
Transit time	–	–
Sojourn time	Sojourn time	–
Residence time	Residence time	Residence time

Table 3.4 contains a summary of five different parameterizations of the two-compartment model.

Table 3.4   Comparisons of model parameters

Model	Parameters	Relationship
Macro-constants	A, α, B, β	See Table 3.2
Micro-constants	V_c, k_{10}, k_{12}, k_{21}	$k_{12} = \dfrac{Cl_d}{V_c}$ $k_{21} = \dfrac{Cl_d}{V_t}$
Physiological constants	Cl, V_c, Cl_d, V_t	$Cl = k_{10} \cdot V_c$ $Cl_d = k_{12} \cdot V_c$ $Cl_d = k_{21} \cdot V_t$
Takada [1981]	β, V_c, K_d, V_{max}	$V_t = \dfrac{V_{max} \cdot t}{K_d + t}$
Colburn [1983]	β, V_c, K_d, V_{max}	$V_t = V_{max} \cdot \left[1 - e^{-K_d \cdot t}\right]$

3.3.4   Constant rate infusion

A *constant rate* infusion is preferred when a drug has a narrow therapeutic index and/or when one wants to observe and evaluate a pharmacodynamic effect. Both the bolus and extravascular input regimens are cumbersome since one does not have very much control over the input of drug when drug is administered. The longer the duration of an infusion, the smaller the amount of initial phase observed. However, if data are collected frequently enough during the early time points, one will be able to observe a certain dip corresponding to the initial loss from the central compartment (Figure 3.42).

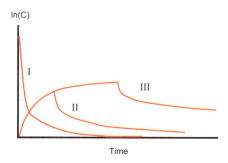

Figure 3.42 The concentration-time course of a multi-compartment drug after a bolus (I), short duration infusion (II), and wash-out after steady state has been established (III). Note the reduction of the initial phase.

This relationship applies to all mammillary compartment models. The intercept on the concentration axis of the back-extrapolated terminal phase is

$$C^0 = C(\lambda_z) = \frac{C^I T_{inf} \lambda_z}{e^{\lambda_z \cdot T_{inf}} - 1} \tag{3:162}$$

C^I, T_{inf} and λ_z are the back-extrapolated terminal phase after cessation of an infusion, the infusion time and the terminal slope, respectively. For a two-compartment system the intercept on the concentration axis of the back extrapolated β–phase becomes

$$C(\beta) = B = \frac{C^I T_{inf} \beta}{e^{\beta \cdot T_{inf}} - 1} \tag{3:163}$$

λ_z is the terminal slope and $C(\beta)$ corresponds to B for a two-compartment model (Gibaldi and Perrier [1982]). Similarly, for a two-compartment model, $C(\alpha)$ or A can be calculated if C^I of the initial phase after stop of infusion is estimated. How does one derive initial parameter estimates for this system, and can it be done by a graphical method? The derivations are given in Equations 3:164 to 3:174, and Figure 3.43.

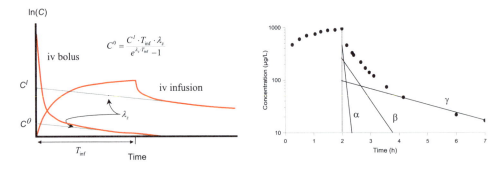

Figure 3.43 *The left hand figure displays schematically the kinetics of a multi-compartment drug after a bolus dose and constant rate infusion. This figure also relates the intercept of the back-extrapolated terminal slope (C^I) after an infusion to the intercept of the same compound after a bolus dose ($C^{\lambda z}$). In the case of the alpha-, beta- and gamma-phase, $C^{\lambda z}$ corresponds to A, B and C, respectively. The right hand figure displays observed concentration-time data on a log-linear scale and includes the alpha-, beta- and gamma-slopes.*

In the example shown in Figure 3.43, the slope of the terminal phase is

$$slope = \frac{y_2 - y_1}{x_2 - x_1} = -\lambda_z = \frac{\ln(180) - \ln(100)}{0 - 2} = -0.29 \, h^{-1} \tag{3:164}$$

The slope of the intermediate phase then becomes

$$-\beta = \frac{\ln(250) - \ln(10)}{0 - 1.75} = -1.83 \, h^{-1} \tag{3:165}$$

The slope of the initial phase is approximately

$$-\alpha = \frac{\ln(750) - \ln(10)}{0 - 0.25} \approx -13 \, h^{-1} \tag{3:166}$$

α is a value that is usually associated with a high degree of uncertainty due to the small number of observations usually collected during that phase. Consequently, the value $C(\alpha)$ or A will also be poorly estimated.

$$C^0 = \frac{C^I T_{inf} \lambda_z}{e^{\lambda_z \cdot T_{inf}} - 1} \tag{3:167}$$

C^0, C^I, T_{inf} and λ_z are the estimated bolus intercept, the back-extrapolated intercept at time zero (0) after the *constant rate* infusion, length of the infusion and the slope of the actual phase (e.g., α, β or γ), respectively. $C(\alpha)$, $C(\beta)$ and $C(\gamma)$ correspond respectively to A, B and C coefficients in a tri-exponential model.

$$C(\gamma) = C = \frac{180 \cdot 2 \cdot 0.29}{e^{0.29 \cdot 2} - 1} \approx 130 \, \mu g / L \tag{3:168}$$

$$C(\beta) = B = \frac{6500 \cdot 2 \cdot 1.83}{e^{1.83 \cdot 2} - 1} \approx 630 \, \mu g / L \tag{3:169}$$

$$C(\alpha) = A = \frac{\infty}{e^{13 \cdot 2} - 1} \approx 20000 \, \mu g / L \tag{3:170}$$

As previously stated, α is usually connected with a high degree of uncertainty and therefore the value $C(\alpha)$ will also be poorly estimated. We had to assume a reasonable value of 20000 µg/L. Total clearance is estimated from

$$Cl = \frac{D_{iv}}{AUC} = \frac{D_{iv}}{\dfrac{A}{\alpha} + \dfrac{B}{\beta} + \dfrac{C}{\gamma}} = \frac{10000}{\dfrac{20000}{13} + \dfrac{630}{1.83} + \dfrac{130}{0.29}} = 4.3 \, L/h/kg \tag{3:171}$$

The volume of the central compartment can then be estimated by means of

$$V_c = \frac{D_{iv}}{A + B + C} = \frac{10000}{20000 + 630 + 130} \approx 0.5 \, L/kg \tag{3:172}$$

and Cl_d is approximately

$$Cl_d = \alpha \cdot V_c - Cl = 13 \cdot 0.5 - 4.3 = 2.2 \, L/kg \tag{3:173}$$

However, we have two Cl_d parameters (since we are assuming a three-compartment model shown in Figure 3.44) and one of them is probably greater than plasma clearance

(4.3 L/kg). Therefore we set one of them to 2.2 L/kg (slow) and the other to 2.2 + 4.3 L/kg (rapid) which is 6.5 L/kg.

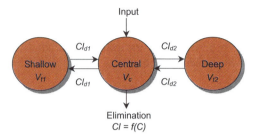

Figure 3.44. The three-compartment model with six parameters V_c, V_{t1}, V_{t2}, Cl_{d1}, Cl_{d2} and Cl. Cl is assumed to be a constant term and independent of C. The shallow compartment represents drug in rapidly equilibrated tissues (e.g., renal and hepatic tissues), and the deep compartment drug in slowly equilibrated tissues (e.g., adipose tissue or bone marrow).

The volume of distribution at *steady state* V_{ss} is predicted by means of the mean residence time *MRT* and *Cl* according to

$$V_{ss} = MRT \cdot Cl = \frac{AUMC}{AUC} \cdot \frac{D_{iv}}{AUC} \qquad (3:174)$$

$$V_{ss} = \frac{\dfrac{A}{\alpha^2} + \dfrac{B}{\beta^2} + \dfrac{C}{\gamma^2}}{\dfrac{A}{\alpha} + \dfrac{B}{\beta} + \dfrac{C}{\gamma}} \cdot \frac{D_{iv}}{\dfrac{A}{\alpha} + \dfrac{B}{\beta} + \dfrac{C}{\gamma}} = 0.79 \cdot 4.3 = 3.4 \ L/kg \qquad (3:175)$$

Figure 3.45 shows experimental data of a two-compartment model after an intravenous bolus dose and after a 6 hour *constant rate* intravenous infusion. Note how little of the initial phase is observed for the *constant rate* infusion regimen. *MRT* is 16.7 h, which also means that about 64% of drug (64% of area) is eliminated within one *MRT*. To estimate *MRT* after a *constant rate* infusion, one also has to subtract $T_{inf}/2$ from the observed value of *MRT* (i.e., *AUMC/AUC*) in order to account for the average input time, which is half of the length of the infusion.

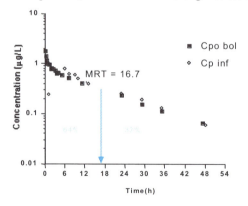

Figure 3.45 Experimental concentration-time data after a bolus (filled squares) and six-hour constant rate infusion (open diamonds). MRT refers to the intravenous bolus curve, where 64% of the dose is eliminated within the first MRT.

The concentration-time course to establish steady state for a drug with bi-exponential behavior in plasma is given by

$$C = C_{ss} \cdot \left\{ f_1 \cdot (1 - e^{-\alpha \cdot t}) + f_2 \cdot (1 - e^{-\beta \cdot t}) \right\}$$ (3:176)

where f_1 and f_2 are the fractional areas or the initial (α) and terminal (β) phases, respectively. From this relationship, one appreciates that the two phases contribute differently to the time to *steady state* (Rowland and Tozer [1995]).

3.3.5 Extravascular administration

If a drug is administered extravascularly (e.g., orally) with an absorption process that is faster than its distribution and elimination processes, one sees a concentration-time course in plasma similar to that depicted in Figure 3.46. Drug concentration in plasma is zero at the time of administration before the drug enters the circulation. The concentration then increases rapidly, reaches a maximum where the rate into the plasma compartment is equal to the rate of loss from the plasma compartment, and then decays in a multi-exponential fashion.

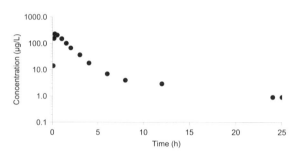

Figure 3.46 Concentration-time course of a two-compartment drug following first-order input.

The differential equation for a two-compartment model becomes

$$V_c \frac{dC}{dt} = In - Cl \cdot C - Cl_d \cdot C + Cl_d \cdot C_t$$ (3:177)

$$V_t \frac{dC_t}{dt} = Cl_d \cdot C - Cl_d \cdot C_t$$ (3:178)

The model is specified in terms of Cl, V_c, Cl_d and V_t. The input function In is

$$\text{Rate of input (no lag-time)} = In = K_a \cdot FD_{po} \cdot e^{-K_a \cdot t}$$ (3:179)

$$\text{Rate of input (with lag-time)} = In = K_a \cdot FD_{po} \cdot e^{-K_a \cdot (t - t_{lag})}$$ (3:180)

Rate of input includes two new parameters such as K_a and bioavailability F. The integrated solution for a two-compartment model with first-order input (K_a) becomes

$$C = \frac{K_a FD_{po}}{V_c} \left\{ \frac{(k_{21}-\alpha)\cdot e^{-\alpha\cdot t}}{(K_a-\alpha)(\beta-\alpha)} + \frac{(k_{21}-\beta)\cdot e^{-\beta\cdot t}}{(K_a-\beta)(\alpha-\beta)} + \frac{(k_{21}-K_a)\cdot e^{-K_a\cdot t}}{(\alpha-K_a)(\beta-K_a)} \right\} \qquad (3{:}181)$$

The k_{21}, α and β parameters are defined as in section 3.3.2. An alternative to model 3:181, which includes a lag-time, takes the form below.

$$C = \frac{K_a FD_{po}}{V_c} \left\{ \frac{(k_{21}-\alpha)\cdot e^{-\alpha\cdot(t-t_{lag})}}{(K_a-\alpha)(\beta-\alpha)} + \frac{(k_{21}-\beta)\cdot e^{-\beta\cdot(t-t_{lag})}}{(K_a-\beta)(\alpha-\beta)} + \frac{(k_{21}-K_a)\cdot e^{-K_a\cdot(t-t_{lag})}}{(\alpha-K_a)(\beta-K_a)} \right\} \quad (3{:}182)$$

when t greater than t_{lag}, and $C = 0$ when t less than or equal to t_{lag}.

3.3.6   Plasma and urine data

The beauty of specifying the model as a system of differential equations lies in the simplicity of the equations. This becomes very clear for a multi-compartment system and particularly if the input is more complex than a bolus. There is no need to complicate the equation to include the urine, since this is already taken into account in the plasma equation. The differential equations for the plasma, peripheral compartment and urine of a two-compartment system become

plasma

$$V_c \frac{dC}{dt} = In - Cl\cdot C - Cl_d \cdot C + Cl_d \cdot C_t \qquad (3{:}183)$$

peripheral compartment

$$V_t \frac{dC_t}{dt} = Cl_d \cdot C - Cl_d \cdot C_t \qquad (3{:}184)$$

urine

$$\frac{dX_u}{dt} = Cl_R \cdot C \qquad (3{:}185)$$

Cl corresponds to the total plasma clearance including Cl_R, which is the renal clearance. We only need to specify the input function *(In)* and initial conditions. For a bolus dose, *In* is a *spike* at time zero and $C^0 = Dose/V$. For a first-order input, *In* becomes

$$In = K_a \cdot FD_{po} \cdot e^{-K_a\cdot t} \qquad (3{:}186)$$

For a zero-order input when $t \le T_{inf}$, *In* becomes

$$In = \frac{D_{inf}}{T_{inf}} \qquad (3{:}187)$$

When t is greater than T_{inf}, $In = 0$.

3.4   Clearance Concepts

3.4.1   Derivation of clearance

Since clearance is such an important parameter in many kinetic models it is worthwhile knowing how to derive initial estimates of this parameter in different situations, such as those covered in the applications section. First, however, we would like to make the reader aware of fundamental texts on the clearance concept such as Riggs [1961], Wilkinson and Shand [1975], Lassen and Perl [1979], and Rowland and Tozer [1995].

Clearance may be viewed as a proportionality factor between rate of elimination (dX/dt), and plasma concentration (C)

$$\frac{dX}{dt} = Cl \cdot C \tag{3:188}$$

In a plot of elimination rate versus concentration, Cl is the slope (Figure 3.47).

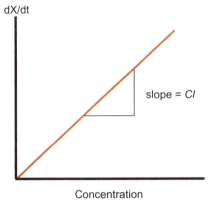

Figure 3.47 *Elimination rate as a function of plasma concentration. The slope corresponds to clearance (Cl).*

By integrating the rate-equation one obtains the relationship between dose and area under the curve. Another approach that one can use for estimation of clearance is to use dose and area ($AUC_{0-\infty}$)

$$Cl = \frac{D_{iv}}{AUC} \tag{3:189}$$

and, for oral clearance (also denoted Cl/F)

$$Cl_o = Cl/F = \frac{D_{po}}{AUC} \tag{3:190}$$

The relationship between oral clearance and systemic clearance is

$$Cl = F \cdot Cl_o \tag{3:191}$$

The *AUC* approach can be used for many types of situations, such as enterohepatic recirculation, irreversible metabolism, multiple absorption sites and peaks, etc. Clearance may also be viewed as a proportionality constant between dose and area, although clearance and dose are the primary parameters and area is a derived or secondary parameter. If the area from zero to infinity after a single dose, or the area within a dosing interval after multiple dosing, can be correctly established together with the dose, then clearance can always be estimated. The rate of elimination is the same, regardless of whether it is based on blood or plasma or unbound concentration. Thus,

$$\frac{dX}{dt} = Cl \cdot C = Cl_u \cdot C_u = Cl_{blood} \cdot C_{blood} = Cl_{ub} \cdot C_{ub} \tag{3:192}$$

Cl_{ub} and C_{ub} are the unbound clearance and unbound concentration, respectively, in blood. Rearranging Equation 3:192 gives

$$Cl = \frac{C_u}{C} \cdot Cl_u = f_u \cdot Cl_u \tag{3:193}$$

This relationship gives total clearance as a function of the free fraction f_u and unbound clearance Cl_u.

3.4.2   Extraction

The underlying assumptions in this section are that complete absorption takes place and that hepatic clearance Cl_H is the major route of elimination. Intrinsic clearance Cl_{int} is defined as the capacity of the liver to clear the blood from drug in the abscence from blood flow limitations. Hepatic organ clearance may then be derived from perfusion rate Q_H, and extraction ratio E_H, where the latter is estimated by means of the incoming C_{in} and outgoing C_{out} concentration to and from a clearing organ.

$$E_H = \frac{C_{in} - C_{out}}{C_{in}} \tag{3:194}$$

Clearance then becomes

$$Cl_H = Q_H \cdot E_H = Q_H \cdot \frac{C_{in} - C_{out}}{C_{in}} \tag{3:195}$$

Let us compare the extraction process of a compound across an organ with an assembly line (Figure 3.48). Assume that subject *A* in Figure 3.48 has the capacity of filling (cf. Q_H) 100 bottles per unit time and subject *B* of corking (cf. Cl_{int}) 1000 bottles per unit time. The result will only be 100 filled and corked bottles per unit time. If subject *A* increases or decreases production of filled bottles, the end-product will increase or decrease accordingly, since subject *B* has an over-capacity. This is flow-dependent clearance.

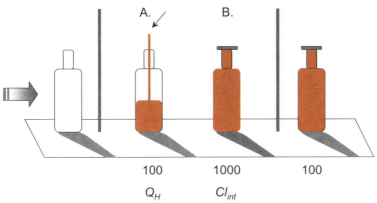

Figure 3.48 *Flow dependent kinetics symbolized by means of an assembly line. Subject A is the rate-limiting step, and will determine the number of end-products per unit time. If A increases or decreases filling capacity, the number of end-products will change accordingly.*

Now assume that subject A in Figure 3.49 has the capacity of filling (cf. Q_H) 1000 bottles per unit time and subject B of corking (cf. Cl_{int}) 100 bottles per unit time. The result will still only be 100 filled and corked bottles per unit time. If subject A increases or decreases his production of filled bottles, the end-product will not change, as long as subject B is the rate-limiting step. This situation can be compared to a flow-independent clearance.

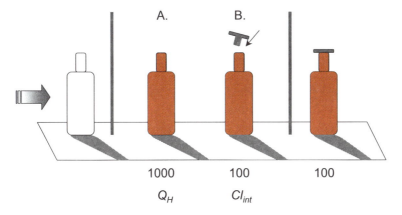

Figure 3.49 *Kinetics dependent on the metabolic capacity symbolized by means of an assembly line. Subject B is the rate-limiting step, and will determine the number of end-products per unit time. If B increases or decreases corking rate, the number of end-products will change accordingly.*

Clearance may also be derived from perfusion rate Q_H, and intrinsic clearance Cl_{int}, where the latter parameter is the maximum organ (e.g., hepatic) capacity to clear the drug with no blood flow limitations present. Consider the following expression for the liver (we assume no partitioning into the tissue)

$$V_H \frac{dC_H}{dt} = Q_H \cdot C_{in} - Q_H \cdot C_{out} - f_u \cdot Cl_{int} \cdot C_{out} \qquad (3:196)$$

$Q_H \cdot C_{in}$ is the rate of entry, $Q_H \cdot C_{out}$ is the exit rate, and $f_u \cdot Cl_{int} \cdot C_{out}$ is the elimination rate. At *steady state*

$$V_H \frac{dC_H}{dt} = 0 \qquad (3:197)$$

$$Q_H \cdot C_{in} - Q_H \cdot C_{out} - f_u \cdot Cl_{int} \cdot C_{out} = 0 \qquad (3:198)$$

$$C_{out} = \frac{Q_H}{Q_H + f_u \cdot Cl_{int}} \cdot C_{in} \qquad (3:199)$$

The extraction ratio becomes

$$E_H = \frac{C_{in} - C_{out}}{C_{in}} \qquad (3:200)$$

Equation 3:199 inserted into Equation 3:200 gives

$$E_H = \frac{C_{in} - \dfrac{Q_H}{Q_H + f_u \cdot Cl_{int}} \cdot C_{in}}{C_{in}} \qquad (3:201)$$

Rearranged

$$E_H = 1 - \frac{Q_H}{Q_H + f_u \cdot Cl_{int}} = \frac{f_u \cdot Cl_{int}}{Q_H + f_u \cdot Cl_{int}} \qquad (3:202)$$

Cl_H is blood flow Q_H times extraction E_H

$$Cl_H = Q_H \cdot E_H = \frac{Q_H \cdot f_u \cdot Cl_{int}}{Q_H + f_u \cdot Cl_{int}} \qquad (3:203)$$

We can estimate the availability F_H across the eliminating organ, provided that hepatic extraction is the only loss of availability

$$F_H = 1 - E_H = 1 - \frac{Cl_o}{Q_H + Cl_o} \qquad (3:204)$$

The relationship between oral clearance Cl_o (oral clearance also denoted Cl/F is estimated from the oral D_{po} divided by the AUC_{po}) and $f_u \cdot Cl_{int}$ might be derived from

$$Cl = F_H \cdot Cl_o \tag{3:205}$$

Substituting F with $Q_H/(Q_H + f_u \cdot Cl_{int})$ gives

$$F_H = \frac{Q_H}{Q_H + f_u \cdot Cl_{int}} \tag{3:206}$$

$$Cl = \frac{Q_H}{Q_H + f_u \cdot Cl_{int}} \cdot Cl_o \tag{3:207}$$

Then, comparing with the well-stirred model (see Section 3.4.5 for a discussion of this model),

$$Cl = \frac{Q_H}{Q_H + f_u \cdot Cl_{int}} \cdot f_u \cdot Cl_{int} \tag{3:208}$$

Thus

$$Cl_o \approx f_u \cdot Cl_{int} \tag{3:209}$$

Availability across the eliminating organ can also be estimated using dose and area

$$F_H = 1 - E_H = 1 - \frac{Cl_o}{Q_H + Cl_o} = 1 - \frac{\dfrac{D_{po}}{AUC}}{Q_H + \dfrac{D_{po}}{AUC}} \tag{3:210}$$

or

$$F_H = \frac{Q_H}{Q_H + \dfrac{D_{po}}{AUC}} \tag{3:211}$$

Under the assumption of a constant blood flow and complete absorption, this expression can be used to obtain a first approximation of extraction and bioavailability from orally administered drugs.

For highly cleared drugs, also called high extraction compounds ($E_H \geq 70\%$), $f_u \cdot Cl_{int}$ is much greater than Q_H, and $f_u \cdot Cl_{int}$ cancels out, which means that Equation 3:203 can be simplified to give Equation 3:212

$$Cl_{H(highly\ cleared)} = Q_H \cdot E_H = \frac{Q_H \cdot f_u \cdot Cl_{int}}{Q_H \ll f_u \cdot Cl_{int}} \approx Q_H \tag{3:212}$$

For lowly cleared drugs, also called low extraction compounds ($E_H \leq 70\%$), $f_u \cdot Cl_{int}$ is much less than Q_H, and then Q_H cancels out, which means that Equation 3:203 can be simplified accordingly to give Equation 3:213

$$Cl_{H(lowly\ cleared)} = Q_H \cdot E_H = \frac{Q_H \cdot f_u \cdot Cl_{int}}{Q_H \gg f_u \cdot Cl_{int}} \approx f_u \cdot Cl_{int} \tag{3:213}$$

> ☐ **Clearance of high extraction drugs is sensitive to changes in blood flow**
> ☐ **Clearance of low extraction drugs is sensitive to changes in plasma protein binding and intrinsic clearance**

For further reading on the relationship between blood flow, intrinsic clearance, and extraction ratio, see Wilkinson and Shand [1975] and Rowland and Tozer [1995].

3.4.3   Impact of route of administration

Blood flow, plasma protein binding and intrinsic clearance will influence the resulting *steady state* plasma concentration to varying degrees depending on the route of administration, e.g., intravenous or oral (Wilkinson and Shand [1975]). Let us assume that the average total drug concentration at steady state C_{ss} (also called the steady state concentration) is determined by rate of administration R_0 (also denoted R_{in} in some instances) and systemic clearance Cl according to

$$C_{ss} = \frac{R_0}{Cl} \qquad (3:214)$$

For a drug that is given by the intravenous route, the total concentration will be dependent on whether it is a highly or lowly cleared drug. A highly cleared drug has an extraction ratio larger than 0.7. A lowly cleared drug has an extraction ratio of less than 0.3. For highly cleared drugs the total concentration becomes

$$C_{ss} = \frac{R_0}{Cl} \approx \frac{R_0}{Q_H} \qquad (3:215)$$

since Cl is determined by the blood flow, and is independent of free fraction when given intravenously. The unbound concentration at steady state C_{uss} of a highly cleared drug also depends upon the free fraction (f_u) as follows

$$C_{uss} = f_u \cdot C_{ss} = \frac{f_u \cdot R_0}{Cl} \approx \frac{f_u \cdot R_0}{Q_H} \qquad (3:216)$$

For a lowly cleared drug, Cl is proportional to the intrinsic clearance times the free fraction, and total average drug concentration at steady state C_{ss} becomes

$$C_{ss} = \frac{R_0}{Cl} \approx \frac{R_0}{f_u \cdot Cl_{int}} \qquad (3:217)$$

When given by the intravenous route, the unbound concentration of a lowly cleared drug will, however, be independent of the free fraction since

$$C_{uss} = f_u \cdot C_{ss} = \frac{f_u \cdot R_0}{Cl} \approx \frac{f_u \cdot R_0}{f_u \cdot Cl_{int}} = \frac{R_0}{Cl_{int}} \tag{3:218}$$

Giving a drug by the oral route makes the situation with respect to total and free concentration of highly and lowly cleared drugs simpler. The average total drug concentration is still the rate of administration divided by clearance, but it is now multiplied by the bioavailability F.

$$C_{ss} = F_H \cdot \frac{R_0}{Cl} = \frac{Q_H}{Q_H + f_u \cdot Cl_{int}} \cdot \frac{R_0}{\dfrac{Q_H}{Q_H + f_u \cdot Cl_{int}} \cdot f_u \cdot Cl_{int}} \tag{3:219}$$

The expression $Q_H/(Q_H + f_u \cdot Cl_{int})$ cancels out, resulting in a total average concentration of

$$C_{ss} = \frac{R_0}{f_u \cdot Cl_{int}} \tag{3:220}$$

For the unbound concentration

$$C_{uss} = \frac{R_0}{Cl_{int}} \tag{3:221}$$

3.4.4   *In vitro/in vivo* comparisons of clearance

Predicting clearance from *in vitro* data can be done from simple enzyme kinetic data obtained using human tissue. The enzyme kinetics is characterized by *the maximum metabolic rate(s)* and the *enzyme-drug affinity constant(s)* (also called the *Michaelis-Menten constant*) denoted V_{max} and K_m, respectively. The approach is strengthened by comparison of preclinical *in vivo* data with predictions from animal *in vitro* data. It requires, however, assumptions concerning the main clearance mechanism, organ and enzymes. *In vitro* binding data may be used for determining free K_m. This can then be used to quantitatively assess the potential for saturation kinetics (for review see Park *et al* [1996]).

Figure 3.50A shows *in vitro* metabolic rate plotted versus substrate concentration. Figure 3.50B shows the same data converted into intrinsic Cl_{int} and organ Cl_H clearances plotted *versus* concentration for a compound that demonstrates perfusion limited extraction.

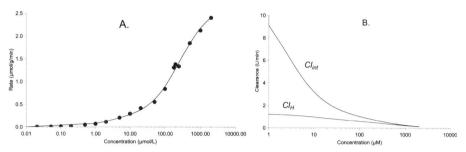

Figure 3.50 *In vitro metabolic rate versus substrate concentration (A) used for in vitro/in vivo extrapolation and prediction of intrinsic clearance and organ clearance. Figure B demonstrates how organ clearance Cl_H and intrinsic clearance Cl_{int} decreases with increasing plasma concentration of a hypothetical compound. The compound exhibits high clearance characteristics at low concentrations and then gradually changes its hepatic intrinsic clearance when the plasma concentration increases becoming a low extraction drug.*

Intrinsic clearance can also be estimated from enzyme activity by summing all parallel metabolic pathways as

$$Cl_{int} = \sum \frac{V_{max,i}}{K_{m,i} + C} \tag{3:222}$$

or

$$Cl_{int} = \frac{V_{maxA}}{K_{mA} + C} + \frac{V_{maxB}}{K_{mB} + C} + \ldots\ldots + \frac{V_{maxi}}{K_{mi} + C} \tag{3:223}$$

A, B, and *i* represent individual enzymes involved in the overall clearance of the drug (Roberts and Rowland [1985, 1988], Park *et al* [1996]). Under linear conditions, this expression can be approximated to

$$Cl_{int} = \sum \frac{V_{max,i}}{K_{m,i}} \tag{3:224}$$

However, *in vitro/in vivo* extrapolation fails under certain circumstances. One such occasion is when the drug clearance is sensitive to differences in protein binding in plasma and the incubation medium (Bäärnhielm *et al* [1986]). In such a case,

$$K_{m(free)} = f_{u(inc)} \cdot K_m \tag{3:225}$$

$$Cl_{int} \neq \frac{V_{max}}{K_m} \tag{3:226}$$

$$Cl_{int} = \frac{V_{max}}{K_{m(free)}} = \frac{V_{max}}{f_{u(inc)} \cdot K_m} \tag{3:227}$$

If clearance is approximated as follows

$$Cl \approx \frac{Q_H \cdot f_u \cdot Cl_{int}}{Q_H + f_u \cdot Cl_{int}} = \frac{Q_H \cdot f_u \cdot \dfrac{V_{max}}{f_{u(inc)} \cdot K_m}}{Q_H + f_u \cdot \dfrac{V_{max}}{f_{u(inc)} \cdot K_m}} \tag{3:228}$$

and if $f_u \approx f_{u(inc)}$, Equation 3:228 can be simplified to

$$Cl \approx \frac{Q_H \cdot \dfrac{V_{max}}{K_m}}{Q_H + \dfrac{V_{max}}{K_m}} \tag{3:229}$$

Ideally, differences in plasma protein binding should be incorporated into the extrapolation process for compounds that exhibit substantial inter-species differences in binding. This is particularly important for highly bound compounds. It is not only species differences in K_m that affect the scaling process, but also the dose, the extent of absorption (F_a), and the rate of absorption (K_a). The choice of an appropriate *in vitro* system, the contribution of non-metabolic clearance, the contribution of extra-hepatic metabolism, and the assumption that *in vitro* metabolic rates accurately reflect *in vivo* metabolic rates will govern how successful the *in vitro* and *in vivo* extrapolation and the inter-species comparisons are (Obach [1996]).

Similar to pharmacokinetic extrapolations, Levy [1993] demonstrated how one could obtain *in vitro* dynamic measures that can be used to characterize *steep* and *shallow* warfarin responders before initiation of therapy. Exercise PK38 elaborates the *pros* and *cons* of *in vitro/in vivo* extrapolations using the *single-point* method.

3.4.5 Hepatic clearance models

The *well-stirred* and *parallel-tube* models (shown schematically in Figure 3.51) have been used to describe hepatic elimination of drugs. There are also alternative but more complex models called the *dispersion* (Roberts and Rowland [1986, 1988]) and *distributed* (Bass and Keiding [1988]) models. For a review of all these models see Morgan and Smallwood [1990] and Iwatsubo *et al* [1996].

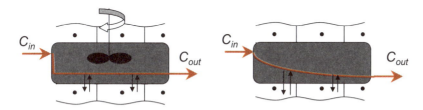

Figure 3.51 *The well-stirred (left) and parallel-tube (right) models.*

In the *well-stirred* model (also called venous equilibrium model) drug concentration is assumed to be constant throughout the hepatic compartment and equal to the outflow concentration. In the *parallel-tube* model (also called undistributed model) the uptake of drug into the hepatocytes is greates at the portal venous end of the tubes. Concentration of drug then declines exponentially along the tubes. The average concentration within the organ is the logarithmic average of C_{in} and C_{out}.

Applying the theory for the *well-stirred* model, the relationship between the hepatic venous concentration C_{out}, the incoming mixed arterial and venous splanchnic blood C_{in}, the hepatic blood flow Q_H, and intrinsic clearance Cl_{int}, is expressed as follows

$$C_{out} = C_{in} \cdot (1 - E_H) = C_{in} \cdot F_H = C_{in} \cdot \frac{Q_H}{Q_H + f_u \cdot Cl_{int}} \tag{3:230}$$

$$Cl_H = \frac{Q_H \cdot f_u \cdot Cl_{int}}{Q_H + f_u \cdot Cl_{int}} \tag{3:231}$$

E_H and F_H are the hepatic extraction ratio and availability, respectively. For simplicity we can denote $f_u \cdot Cl_{int}$ by Cl_{int}. Figure 3.52 shows the trajectory of clearance (Equation 3:231) with increasing blood flow through the clearing organ (the liver in this case).

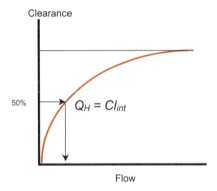

Figure 3.52 *Clearance as a function of blood flow.*

The extraction ratio is 50% when the perfusion rate Q_H is equal to intrinsic clearance Cl_{int}. The extraction ratio is greater than or equal to zero, and less than or equal to 1

$$0 \le \frac{f_u \cdot Cl_{int}}{Q_H + f_u \cdot Cl_{int}} \le 1 \tag{3:232}$$

Assuming that the *parallel-tube model* is valid for prediction of hepatic extraction we derive the following relationship between the dependent variable C_{out}, the two independent variables C_{in} and Q_H, and the intrinsic clearance Cl_{int}.

$$C_{out} = C_{in} \cdot (1 - E_H) = C_{in} \cdot F_H = C_{in} \cdot e^{-\frac{f_u \cdot Cl_{int}}{Q_H}} \tag{3:233}$$

$$Cl_H = Q_H \cdot \left[1 - e^{-\frac{f_u \cdot Cl_{int}}{Q_H}} \right] \tag{3:234}$$

Figure 3.53 illustrates the *distributed* and *dispersion* models schematically.

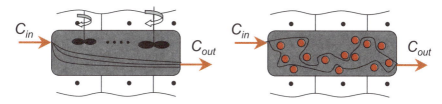

Figure 3.53 *The distributed (left) and dispersion (right) models.*

The *distributed* model can be viewed as a number of *parallel-tube* (sinusoidal) models with different geometrical characteristics with respect to their individual blood flows q_i and intrinsic clearances $Cl_{int,i}$. The higher heterogeneity of q_i and $Cl_{int,i}$ among the tubes, the closer is the resemblance of the *well-stirred* model. The lower heterogeneity of q_i and $Cl_{int,i}$ among the tubes, the more it resembles the *parallel-tube* model.

In the *distributed* model clearance Cl_H is expressed according to the following formula

$$C_{out} = C_{in} \cdot (1 - E_H) = C_{in} \cdot F_H = C_{in} \cdot e^{-\left\{ \frac{f_u \cdot Cl_{int}}{Q_H} + \frac{1}{2} \varepsilon^2 \left[\frac{f_u \cdot Cl_{int}}{Q_H} \right]^2 \right\}} \tag{3:235}$$

$$Cl_H = Q_H \cdot \left[1 - e^{-\left\{ \frac{f_u \cdot Cl_{int}}{Q_H} + \frac{1}{2} \varepsilon^2 \left(\frac{f_u \cdot Cl_{int}}{Q_H} \right)^2 \right\}} \right] \tag{3:236}$$

where ε^2 is the variance for each sinusoid in the whole liver. ε is a parameter like Cl_{int}, and can be estimated during the fitting procedure. The *distributed model* reduces to the *parallel-tube* model when either $Q_H >> Cl_{int}$ or ε is very small.

The *dispersion* model takes into account the mixing of blood within the sinusoids with respect to flow rates and path lengths. The degree of mixing is given by the dispersion number D_N. When D_N approaches infinity or zero, the *dispersion* model collapses into the *well-stirred* and *parallel-tube* models, respectively. In the *dispersion* model clearance Cl_H is expressed according to the following formula

$$C_{out} = C_{in} \cdot (1 - E_H) = C_{in} \cdot \frac{4a}{(1+a)^2 e^{\left[\frac{a-1}{2D_N} \right]} - (1-a)^2 e^{\left[\frac{a+1}{2D_N} \right]}} \tag{3:237}$$

$$Cl_H = Q_H \cdot \left[1 - \frac{4a}{(1+a)^2 e^{\left[\frac{a-1}{2D_N} \right]} - (1-a)^2 e^{-\left[\frac{a+1}{2D_N} \right]}} \right] \tag{3:238}$$

Variable a is equal to $(1 + 4R_ND_N)^{1/2}$. Cl_{int} and D_N are parameters to be estimated during the fitting procedure. The efficiency number R_N, is equal to $f_u \cdot Cl_{int}/Q$. For further discussion and analysis of the *well-stirred, parallel-tube, distributed* and *dispersion* models see PK23.

3.5   Turnover Models

3.5.1   Background

The homeostasis of the majority of endogenous compounds such as water, metals, carbohydrates, hormones, amino acids, proteins and so on, is maintained by an equilibrium between their production (synthesis, secretion) and loss (catabolism, filtration). Enzymes, like most other endogenous compounds, are synthesized by a zero-order process and removed by a first-order process, with their characteristic biological half-lives depending on the particular enzyme. An increased amount or concentration of the compound above normal values may result from increased production, diminished loss, or both. This is the area where the concept of *turnover* is applicable.

Two subjects may have identical concentrations of a compound in spite of the fact that one individual synthesizes and catabolizes the compound at twice the rate of the other. In another situation, the concentration of a protein could be 100 in one individual and 50 in the other. This difference may be due to different turnover (synthesis) rates, different elimination rates (fractional turnover rates), or both.

The turnover rate (secretion, synthesis, exogenous input) or input, the volume of the organism, and the fractional turnover rate (catabolism, filtration, excretion) or loss (Figure 3.54) govern the amount or concentration of an endogenous compound.

Figure 3.54 *This figure shows conceptually the basics of a turnover model which includes turnover rate R_{in}, fractional turnover rate k_{out} and amount of substance at steady state A_{ss}. The red arrow indicates how the input rate could be manipulated by means of exogenous input (such as intravenous infusion, or subcutaneous or per oral administration) in parallel with the endogenous turnover rate.*

The term *fractional turnover rate* means that a constant fraction is eliminated per unit time, rather than a constant amount. The larger the amount or the higher the concentration, the greater is the rate of elimination. The fractional turnover rate may not be independent of the concentration of the compound. Some metabolic processes may be saturable at higher concentrations and may be better described by for example, a Michaelis-Menten type of system. We have exemplified this by means of a practical exercise on hyaluronan turnover in PK32. See also Lebel [1988] for review.

The aims of this section are to give an overview of the theory of turnover models and then to describe different types of models such as those for hormones (estradiol E_2), enzymes (induction of CYP450), proteins (hyaluronan), and body temperature (both circadian and feedback regulated mechanisms).

3.5.2   Turnover concepts

Turnover implies steady state. To fully characterize the turnover of a compound, one needs information about the turnover rate R_{in}, the fractional turnover rate k_{out} and the amount of the compound A in the organism. The turnover rate is the amount of the compound that is secreted or synthesized per unit time. The fractional turnover rate is the fraction of the amount in the organism at steady state being turned over per unit time.

The rate of change of the compound per unit time is given by the zero-order turnover rate R_{in} (also denoted k_{in}) and the first-order loss $-k_{out} \cdot A$

$$\frac{dA}{dt} = R_{in} - k_{out} \cdot A \qquad (3:239)$$

At steady state, the rate of change is equal to zero, and this relationship can be written as

$$\frac{dA}{dt} = R_{in} - k_{out} \cdot A_0 = 0 \qquad (3:240)$$

which gives

$$R_{in} - k_{out} \cdot A_0 = 0 \qquad (3:241)$$

Rearranged, this gives an expression for the turnover rate

$$R_{in} = k_{out} \cdot A_0 \qquad (3:242)$$

The fractional turnover rate k_{out} (unit *time*$^{-1}$) is then given by the ratio of R_{in} and A_0

$$k_{out} = \frac{R_{in}}{A_0} \qquad (3:243)$$

The ratio of R_{in} and k_{out} gives the amount of the compound in the organism at steady state A_0.

$$A_0 = \frac{R_{in}}{k_{out}} \qquad (3:244)$$

A fractional turnover rate of e.g., 0.1 h^{-1}, means that 10% of the amount in the organism at steady state is turned over per hour (is produced and lost). The turnover time t_t is derived from the amount in the organism divided by the turnover rate. In other words, the turnover time is the time it takes to secrete or synthesize the same amount as there is in the system at steady state, provided there is no loss.

$$t_t = \frac{A_0}{R_{in}} = \frac{1}{k_{out}} \qquad (3:245)$$

For example, if the volume of a bucket is 10 L (A_0 then corresponds to 10 liters of fluid) and the flow of water from a tap is 1 L/min (R_{in}), then the turnover time t_t is 10 minutes. The human body contains about 42 L of water. The intake or turnover rate of water per day is about 2.5 L. The turnover time of water is then approximately 17 days (42/2.5 days). The turnover time may also be obtained from the pharmacokinetic parameters Cl and V_{ss} according to the following expression

$$t_t = \frac{A_0}{R_{in}} = \frac{C_0 \cdot V_{ss}}{C_0 \cdot Cl} = \frac{V_{ss}}{Cl} = MRT \tag{3:246}$$

Hence, turnover time may be considered as the mean residence time MRT.

$$t_t = \frac{A_0}{R_{in}} = \frac{C_0 \cdot V_{ss}}{V_{ss} \cdot \dfrac{dC}{dt}} = \frac{C_0}{\dfrac{dC}{dt}} \tag{3:247}$$

The relationship between the half-life and the turnover time is given by

$$t_{1/2} = \frac{\ln(2)}{k_{out}} = \ln(2) \cdot t_t \tag{3:248}$$

As the mean residence time is the inverse of the elimination rate constant in a one-compartment system, the turnover time is the inverse of the fractional turnover rate constant.

$$Baseline = A_0 = \frac{R_{in}}{k_{out}} \tag{3:249}$$

Figure 3.55 shows the concentration-time course of an endogenous substance when either the turnover rate or the fractional turnover rate is manipulated.

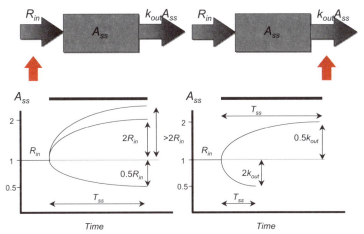

Figure 3.55 *This shows conceptually the impact of a change (red arrow) in either the turnover rate, R_{in} (left hand figures) or fractional turnover rate k_{out} (right hand figures). A change in the turnover rate (stimulus/inhibition) will only affect the steady state level (amount, concentration or response measure) but not the time to steady state T_{ss} as indicated. However, if the fractional turnover rate is affected (stimulus/inhibition), half-life also changes, with time to steady state T_{ss} increasing as half-life increases and vice versa. In addition, the actual steady state level will increase or decrease.*

Changing the turnover rate is analogous to changing the infusion rate of a drug. The time to steady state is unchanged. Changing the fractional turnover rate corresponds to changing clearance (and indirectly $t_{1/2}$) of a drug in that both the concentration at steady state changes as well as the time to the new steady state.

3.5.3   Allometric scaling of turnover parameters

If the half-life of response ($ln(2)/k_{out}$) is proportional to the turnover time of substances causing the effect, then the principles of allometric scaling are applicable to turnover parameters or pharmacodynamic response. All rate-processes like flows, clearances, secretion-rates and synthesis rates require energy, and generally scale proportionally to body weight raised to the power b of 0.6-0.8. The a parameter in Equation 3:250 is a drug-dependent proportionality constant. The turnover rate R_{in}, becomes

$$R_{in} = a \cdot BW^b \qquad\qquad (3{:}250)$$

The amount of a substance in the organism at steady state A_{ss} scales proportionally with body weight according to

$$A_{ss} = c \cdot BW^{\sim 1} \qquad\qquad (3{:}251)$$

The c parameter in Equation 3:251 is a drug-dependent proportionality constant. The fractional turnover rate k_{out} scales, like any first-order rate constant or frequency proportionally to body weight raised to the power of ~ -0.25

$$k_{out} = d \cdot BW^{-0.25} \qquad\qquad (3{:}252)$$

The d parameter in Equation 3:252 is a drug-dependent proportionality constant, and e is the ratio of a and c. The turnover time is derived as the ratio of A_{ss} to R_{in} and scales proportionally with body weight according to

$$Turnover\ time = \frac{A_{ss}}{R_{in}} = \frac{c \cdot BW^1}{a \cdot BW^{0.75}} = e \cdot BW^{0.25} \qquad\qquad (3{:}253)$$

$$t_{1/2kout} = \frac{\ln(2)}{k_{out}} = \ln(2) \cdot e \cdot BW^{0.25} \qquad\qquad (3{:}254)$$

Extended to pharmacological responses

$$t_{1/2response} = \text{constant} \cdot BW^{0.25} \qquad\qquad (3{:}255)$$

3.5.4   Feedback

We will briefly discuss three different types of feedback, namely *inherent, positive* and *negative feedback*. Feedback is also covered in section 4.9 but then specifically oriented towards responses. In the following section, feedback is discussed very broadly.

The simplest case of feedback is called *inherent* feedback. This can be seen in the following relationship

$$\frac{dC}{dt} = -K \cdot C \qquad (3{:}256)$$

which shows that the lower the concentration, the slower is the rate of decline.

Bacterial growth is an example of *positive* feedback. The more bacteria *A*, the more rapid is the rate of growth of the population dA/dt, as demonstrated in the following equation.

$$\frac{dA}{dt} = k_{growth} \cdot A - k_{kill} \cdot C(t) \cdot A \qquad (3{:}257)$$

where *C(t)*, concentration at time *t*, is a function of drug exposure.

We have adapted the hormone-glucose model, originally developed by Ackerman *et al* [1964], for our discussion of systems of *negative* feedback. Our adaptation of the model is shown in Figure 3.56. *H(t)* is the drug function stimulating or inhibiting the turnover rate of *R* (*R* could be any kind of biomarker). The fractional turnover rate k_{out} and potency of the moderator M_{50} also determine the characteristics of the model.

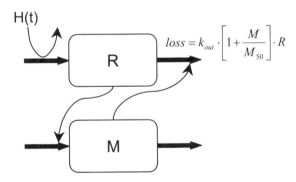

$$loss = k_{out} \cdot \left[1 + \frac{M}{M_{50}} \right] \cdot R$$

Figure 3.56 *Adaptation of the Ackerman et al's [1964] hormone-glucose model. R is the observed response compartment and M the moderator. H(t) is the drug function stimulating or inhibiting the turnover rate of R. The fractional turnover rate k_{out} and M_{50} also determine the characteristics of the model. For further discussions on this class of models see Chapter 4.*

The turnover of response *R*, is defined as

$$\frac{dR}{dt} = k_{in} \cdot H(t) - k_{out} \cdot M \qquad (3{:}258)$$

and the turnover of the moderator (*M*) of *R* is given by

$$\frac{dM}{dt} = k_{tol} \cdot R - k_{tol} \cdot M \tag{3:259}$$

The k_{tol} parameter is a first-order rate-constant that determines the production and loss of the moderator. The loss of R is a nonlinear process that also depends on the moderator M. The loss can be written as

$$loss = k_{out} \cdot M \tag{3:260}$$

This relationship shows that the higher the level of the moderator, the greater will be the rate of loss of R. In other words, in a negative feedback system the half-life of response will decrease with increasing moderator or with increasing response level. This may also explain why the time to the maximum response appears shorter if negative feedback (tolerance) is present, as a dynamic steady state is reached more rapidly. Figure 3.57 demonstrates this phenomenon.

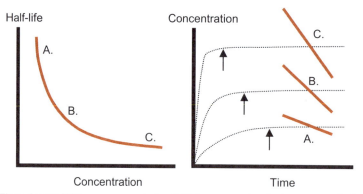

Figure 3.57 *Half-life versus concentration (left) for a system that displays negative feedback-governed elimination and simulated concentration versus time plot for three infusion rates (right) for this system. The higher the infusion rate, the shorter is the time to steady state. The half-life is determined as slopes A, B and C, that increase with an increase in concentration.*

If one kept, for example, the response or endogenous system (e.g., hyaluronan or creatinine) at a steady state level (e.g., *A, B* or *C* in Figure 3.57, left), and then administered a tracer dose as an intravenous bolus, the tracer would decay according to *A, B* or *C* in Figure 3.57 (right). The three slopes are functions of k_{out} and M, since R is equal to M at dynamic steady state in the simplest form of model 3:259 above.

$$slope_{A,B\,or\,C} = k_{out} \cdot \left[1 + \frac{R_{A,B\,or\,C}}{M_{50}}\right] \tag{3:261}$$

The half-life of for example, *IgG* shortens progressively until a limit of about 11 days is reached at a serum level of 30 mg/mL (Waldman *et al* [1970]). Further increases in the serum *IgG* concentration do not result in further shortening of the *IgG* half-life. The postulated mechanism is a saturable protection system specific for *IgG*.

3.5.5 Body temperature

Body temperature is a useful example to demonstrate turnover concepts and negative feedback. In the following section, we take a look at how body temperature can be *model*ed by means of a simple turnover model. Body temperature is characterized by means of the set point. The set point of the body temperature is the body's internal standard around which the temperature can change. In man, the set point is about 37 °C. Salicylates reduce hyperthermic states but do not reduce the set point. Acetaminophen can lower the body temperature beyond the set point (Guyton [1976]).

In an experiment, five groups of rats received a dose of 0, 30, 125, 500 or 2000 µg of *8-OH-DPAT* subcutaneously (Deveney *et al* [1999]). *8-OH-DPAT* is an agonist that binds upon the *5-HT$_{1A}$* receptor and indirectly acts as a temperature-lowering compound. The red arrow in Figure 3.54 indicates the site of inhibitory action of *8-OH-DPAT*. The decrease in body temperature from the baseline curve was obtained and is plotted in Figure 3.58 as mean values.

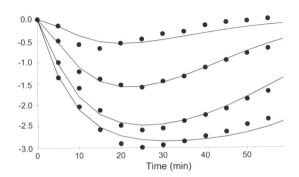

Figure 3.58 Observed and predicted response-time data after four sc doses of 8-OH-DPAT. The response (mean decrease in body temperature) of each group is expressed as change from baseline group (0 µg) (Deveney et al [1999]). See Gabrielsson et al [2000] for a more indepth review.

The inhibitory function can be expressed as

$$H(t) = 1 - \frac{I_{max} A^n}{ID_{50}^n + A^n} \qquad (3:262)$$

where I_{max}, ID_{50} and n are the maximum drug induced effect of the $H(t)$ function, the potency and the sigmoidicity factor, respectively. The turnover function thus becomes

$$\frac{dR}{dt} = k_{in} \cdot H(t) - k_{out} \cdot R \qquad (3:263)$$

The antagonistic effect of *8-OH-DPAT* on body temperature is assumed to be due to an inhibitory action on the turnover rate k_{in} rather than a stimulatory action on the loss of response. The relationship of the combined response (Equation 3:263) and inhibitory (Equation 3:264) functions at dynamic steady state therefore becomes

$$R_{ss} = R_0 \cdot H(t) = \frac{k_{in}}{k_{out}} \cdot \left[1 - \frac{I_{max} A^n}{ID_{50}^n + A^n} \right] \tag{3:264}$$

This relationship is the dose-response function for *8-OH-DPAT* with respect to body temperature in the rat. The function starts at a value equal to k_{in}/k_{out} and then decreases with increasing values of A (or *Dose*). The maximum difference from the baseline temperature (Δ_{max}) is expressed as

$$\Delta_{max} = R_0 - R_0 \cdot H(t) = \frac{k_{in}}{k_{out}} - \frac{k_{in}}{k_{out}} \cdot \left[1 - \frac{I_{max} A_\infty^n}{ID_{50}^n + A_\infty^n} \right] = \frac{k_{in}}{k_{out}} \cdot I_{max} \tag{3:265}$$

Figure 3.59 shows the dose-response curve for the experiment discussed above. From this curve one can easily obtain the ID_{50} value or the Δ-value. The latter is the maximum drug induced change from baseline, which is 3 °C (I_{max}/k_{out}).

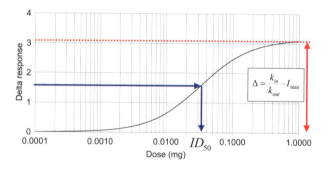

Figure 3.59 *Simulated delta response versus dose profile. The delta response corresponds to change from the baseline value. Delta (Δ) is equal to 0.41/0.13 since k_{in} is set to 1 (Gabrielsson [1996]).*

It is important to keep in mind that body temperature is a negative feedback regulated system. The analysis presented above is limited to a 60 min time interval and therefore we can approximate the turnover characteristics to a simple model without feedback regulation.

3.5.6   Immunoglobulins

A growth hormone, denoted *IgX*, was given as a subcutaneous dose of 40 µg/kg to a healthy human volunteer in order to elucidate the fractional turnover rate and turnover time. The plasma concentrations of *IgX* were measured during the experiment and are given in Figure 3.60. The mean pre-dose concentration was 32 µg/L.

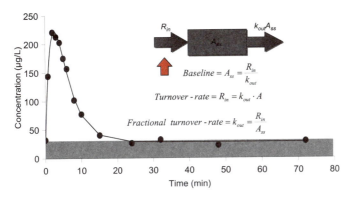

Figure 3.601 *Observed (•) and predicted (line) concentration-time data of IgX, a growth hormone, after a 40 µg/kg subcutaneous dose. R_{in} is the turnover rate of IgX and the subcutaneous dose is added on top of that input. The stippled area shows the mean pre-dose concentration.*

The basic kinetic parameters such as clearance Cl/F, volume of distribution V/F, and half-life were estimated. In addition, mean residence time MRT, which corresponds to turnover time, mean input time MIT from the subcutaneous depot and the absorption rate constant K_a were derived. The underlying differential equation that takes endogenous turnover rate R_{in} and Cl/F (denoted Cl in Equation 3:266), together with the subcutaneously administered dose, into account can be written as

$$V \cdot \frac{dC_{IgX}}{dt} = R_{in} + In_{sc} - Cl \cdot C_{IgX} \tag{3:266}$$

In_{sc} is the input function for subcutaneous administration and is defined as

$$In_{sc} = K_a \cdot FD_{sc} \cdot e^{-K_a \cdot t} \tag{3:267}$$

We assume that absorption is 100% in that bioavailability is set to 1, and no substantial feedback control occurs with this dose, regimen and time frame. Baseline kinetics are unaltered (parameter stationarity) and there is linear kinetics within the studied concentration and time range. The absorption rate of IgX from the subcutaneous site is first-order

Without the additional information obtained from the subcutaneous dose, we would not have been able to discriminate between clearance and rate of synthesis, since they form a ratio where C_{IgX} is equal to R_0/Cl. C_{IgX} is the concentration in plasma.

Cl and V were 0.03 L/h/kg and 0.10 L/kg body weight, respectively. Rate of synthesis or turnover rate was 0.78 µg/h/kg body weight. Turnover time and mean input time MIT were estimated to 3.7 h and 1.8 h, respectively. The terminal half-life was 2.5 h with a corresponding fractional turnover rate of 0.27 h^{-1}, which means that about 27% of the body load was eliminated per hour. The pool size of IgX at steady state was 3.3 µg/kg, and was calculated as the quotient between turnover rate and the fractional turnover rate. Table 3.5 contains turnover data for five immunoglobulins.

Table 3.5   Comparison of turnover characteristics of some immunoglobulins (Waldmann *et al* [1970])

Characteristics	IgG	IgA	IgM	IgD	IgE
Serum level (mg/mL)	12.1	2.5	0.93	0.023	0.0005
% of total pool in plasma	45	42	76	75	51
Total circulating pool	494	95	37	1.1	0.019
Fractional catabolic rate (% of plasma pool/day)	6.7	25	18	37	89
Half-life (days)	23	5.8	5.1	2.8	2.5
Turnover rate (mg/kg/day)	33	24	6.7	0.4	0.016

Waldmann *et al* [1970] concluded that there is no truly normal serum immunoglobulin level, because the turnover rates are not controlled by their serum concentrations or pool sizes, but are determined by the bacterial environment and the impact of antigenic exposure. In addition, age, level of antibody, genetic background and hormonal factors of the organism also function as covariates. Catabolism is governed partly by factors that affect all serum proteins (metabolic rate) and also factors that are specific for the immunoglobulin classes (Waldmann *et al* [1970]).

3.5.7   Hormones - Estradiol

The following example covers the estimation of turnover rate, fractional turnover rate, turnover time, clearance, and volume of distribution of estradiol. In postmenopausal women the endogenous level of estradiol is highly suppressed. The question is, therefore, is it due to a reduced turnover rate or an increased clearance of estradiol? The biological rhythm of estradiol in plasma is also negligible. The assessment of turnover rate is successful when basal turnover rate is unaffected by the additional input of estradiol. However, sometimes this is not so because feedback control adjusts the endogenous turnover rate to maintain equilibrium. In order to estimate the turnover and disposition parameters of estradiol, a two-compartment model was regressed using data following baseline measurements and administration of exogenous estradiol (Gabrielsson *et al* [1995]).

Estimates of clearance Cl, volume of the central compartment V_c, volume of the peripheral compartment V_t, inter-compartmental distribution Cl_d and endogenous turnover rate of estradiol R_{in}, were determined using data obtained from baseline observations and measurements after a rapid intravenous infusion iv_{inf} (Figure 3.61).

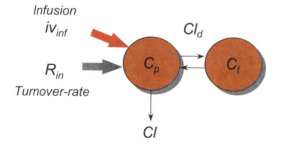

Infusion
iv_{inf}
Cl_d
R_{in}
C_p
C_t
Turnover-rate
Cl

Figure 3.61 *Schematic illustration of the endogenous and exogenous input and turnover model of estradiol.* C_p *and* C_t *are the central and peripheral estradiol concentrations.*

The predicted half-life $t_{1/2}$, turnover time and fractional turnover rate were then derived from Cl and V_{ss}. The equations for serum and extravascular space were respectively,

$$V \cdot \frac{dC}{dt} = R_{in} + In_{inf} - Cl \cdot C - Cl_d \cdot C + Cl_d \cdot C_t \qquad (3{:}268)$$

$$V \cdot \frac{dC_t}{dt} = Cl_d \cdot C - Cl_d \cdot C_t \qquad (3{:}269)$$

In_{inf} represents the constant intravenous infusion, and R_{in} the endogenous production or turnover rate of estradiol. Figure 3.62 shows the compartmental model and experimental data.

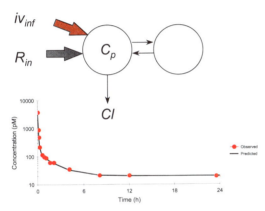

Figure 3.62 *Schematic illustration of the kinetic model taking endogenous and exogenous input into account. The red filled symbols are experimental observations and the black solid line is the model predicted concentration-time course of estradiol.*

The time course of estradiol in the subjects displayed a more rapid wash-out than expected from compiled literature data (for review, see Gabrielsson *et al* [1995]), although the predicted turnover rate of estradiol was consistent with previously reported data.

The conclusions from analyzing these data were that the turnover rate of estradiol was 20 µg/24 h and the fractional turnover rate was 0.03 min^{-1} (3%). Consequently, turnover time was 30 min and the pool size, or body load, was 0.4 µg.

In the studied post-menopausal population, the turnover rate varied between 1 and 44 µg/24 h with a mean of 20 µg per 24 h. The fractional turnover rate was on average 3% per minute, ranging between 1 to 15% per minute. This resulted in a pool size of about 0.4 µg estradiol/70 kg body weight (Lassen and Perl [1979], Gabrielsson *et al* [1995], Rowland and Tozer [1995]). Clearance was found to be 1.5-2 L/min in this population, which is similar to that in women of childbearing potential. The lowered levels of estradiol in postmenopausal women are therefore due to a reduced turnover rate rather than an increased clearance.

3.5.8   Comparison of models

Often, several parallel saturable processes are involved simultaneously in order to maintain e.g., the homeostasis. This may include saturable uptake, saturable renal reabsorption and feedback. A combination of these factors makes analysis and interpretation of turnover data (e.g., bioavailability) of endogenous compounds (e.g., drugs) more complex. Sometimes a pre-dose concentration calls for erroneous baseline subtraction, in spite of the fact that the baseline may not be constant. A constant baseline is rather the exception that proves the rule (Figure 3.63).

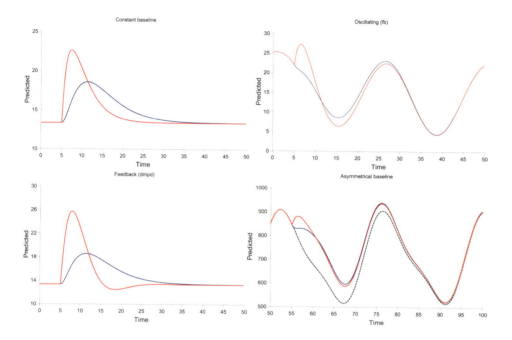

Figure 3.63 *Schematic illustration of three scenarios where we have a constant baseline (upper left), oscillating baseline (upper right), baseline level governed by feedback (bottom left) and an oscillating asymmetric baseline (bottom right). The blue line is the concentration-time relationship of compound X administered with a slow input function and the red line represents the case of rapid input into the system. The doses are the same for the rapid and slow input systems which may be appreciated from the equal AUCs in the upper left figure. Note that subtraction of a pre-dose baseline value may lead to erroneous results in the other two cases (diurnal variation and feedback). The bottom left figure representing feedback regulation of the endogenous level demonstrates also a slight rebound (dip below the expected baseline value) for the rapid dosage form.*

The turnover rates and degradation of enzymes provide important quantitative data concerning the dynamics of cellular processes (Waldmann *et al* [1970], Wenthold *et al* [1974]). Rather extensive information of this kind is already available for hepatic enzymes, although turnover rates of neuronal enzymes are somewhat lagging behind. The turnover process may be rapid in one region of the body but slow in another. A good example of this is the turnover rate of acetylcholinesterase *AChE*, which in mature rat brain has been found to be about 2-3 days (Wenthold *et al* [1974]). In comparison, the turnover rate of *AChE* in blood is probably more related to the life span of the red blood cell (60 days). Turnover may also be saturable such as for hyaluronan (see Lebel [1989] for a comprehensive review, and PK32 for a detailed analysis of experimental data on hyaluronan).

In order to demonstrate the applicability of the simple infusion (clearance model) and turnover models, we have compared them with the indirect pharmacodynamic response model. The results can be seen in Table 3.6. The models are compared with respect to parameterization, equations, determinants of steady state conditions, half-life, and baseline values. A more thorough description of the pharmacodynamic (turnover) response model is included in section 4.6.

Table 3.6   Comparisons of clearance, turnover, and response models

	Clearance model	Turnover model	Response model
Model			
Primary variable	C (amount/volume)	A (amount)	R (arbitrary units)
Parameters	V, Cl	R_{in}, k_{out}	k_{in}, k_{out}
Equation	$V \cdot \dfrac{dC}{dt} = R_0 - Cl \cdot C$	$\dfrac{dA}{dt} = R_{in} - k_{out} \cdot A$	$\dfrac{dR}{dt} = k_{in} - k_{out} \cdot R$
Determination of steady state	$C_{ss} = \dfrac{R_0}{Cl}$	$A_0 = \dfrac{R_{in}}{k_{out}}$	$R_0 = \dfrac{k_{in}}{k_{out}}$
$t_{1/2}$	$\dfrac{\ln(2) \cdot V}{Cl}$	$\dfrac{\ln(2)}{k_{out}}$	$\dfrac{\ln(2)}{k_{out}}$
Turnover time or MRT	$MRT = \dfrac{V}{Cl} = \dfrac{1}{k}$	$t = \dfrac{1}{k_{out}}$	$t = \dfrac{1}{k_{out}}$
Scale factor	$Cl \propto BW^b$	$R_{in} \propto BW^b$	$k_{in} = a \cdot BW^b$
Scale factor	$k \propto BW^{-0.25}$	$k_{out} \propto BW^{-0.25}$	$k_{out} \propto BW^{-0.25}$
Scale factor	$MRT \propto BW^{0.25}$	$t \propto BW^{0.25}$	$t \propto BW^{0.25}$
Baseline	0	$A_0 = \dfrac{R_{in}}{k_{out}}$	$R_0 = \dfrac{k_{in}}{k_{out}}$

3.6   Nonlinear Models

3.6.1   Background

This section illustrates the characteristics of nonlinear kinetics with respect to *capacity, time, flow and binding*, and how these variables may have an impact on a parameter such as clearance (Pitlick and Levy [1977], van Rossum and van Lingen [1983], Rowland and Tozer [1995]).

As an introduction to nonlinear systems, we take to liberty to cite a classical passage of Levy ([1983]). Levy wrote that *"the major distinguishing features between capacity (dose) and time dependency, is that the latter involves an actual physiological or biochemical change in the organ(s) of the body associated with the drug disposition parameters in question. For example, in time dependence of the auto- or heteroinduction type, the increase in drug intrinsic clearance results from an increase in amount of enzyme (e.g., in protein synthesis). However, in a typical Michaelis-Menten capacity (dose) dependency, drug clearance changes with concentration, and such a system should not be considered time-dependent simply because the values of pharmacokinetic parameters change with time. If that was a true time-dependent system, drug clearance should change with time while drug concentration is time invariant. It is still possible that capacity and time dependency exist simultaneously"* (Levy [1983]).

We have come to the conclusion that, from a practical point of view, one of the best ways to assess whether we have nonlinear kinetics, is by means of graphical methods. If clearance decreases or bioavailability increases, a relationship is obtained that is similar to situation A or C in Figure 3.64. If clearance increases or bioavailability decreases, situation B or D occurs.

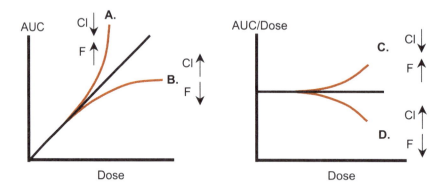

Figure 3.64 *Schematic illustration of linear compared with nonlinear absorption and disposition. The black lines are for linear kinetics, the red curves for nonlinear. Curve A displays either a decrease in clearance and/or an increase in bioavailability. Curve B displays the effect of an increased clearance and/or decreased bioavailability. Curves C and D represent the same situation for dose-normalized areas.*

If nonlinearities are observed in the half-life after intravenous administration, this is caused by changes in the disposition of drug (Cl, V_c, Cl_d and V_t). If AUC is changed, this may be due to either changes in F or Cl. If the principle of superposition is violated, we have either a change in Cl, F, or the distribution (V_c, V_t or Cl_d).

Figure 3.65 captures the relationship between dose-normalized areas *versus* dose of four systems. Example 1 is a truly linear system. Example 2 is a system that displays feedback (see also section 3.5.4 on feedback models). The higher the concentration in the central compartment, the larger is the impact of the feedback. Therefore, the dose-normalized area for some compounds decreases with an increase in dose. Example 3 displays Michaelis-Menten (*MM*) kinetics, which means saturable kinetics. The dose-normalized area increases with dose. Finally in Example 4 we have a parallel *MM* and linear elimination. At low doses, the dose-normalized area increases depending on the relative contribution of *MM* and the linear process. At very high doses the contribution from the *MM* route of elimination is marginal, with the system behaving as an apparent linear system.

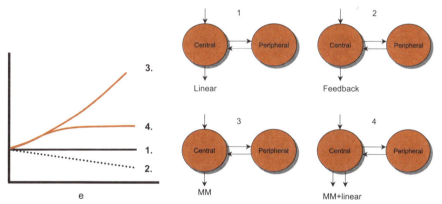

Figure 3.65 *Differences between linear (1), feedback governed (2), saturable (Michaelis-Menten) (3) and parallel linear plus saturable (4) elimination. MM denotes Michaelis-Menten kinetics.*

When the areas and concentrations cover a large enough range, the possibility of clearly comparing high and low concentrations becomes less and less practical. To circumvent this problem, the relative contribution of each concentration or area measurement is obtained by normalizing the area or concentration value by its corresponding dose. This is a commonly used approach, which we find very useful if plotted versus dose. We recommend this type of exploratory data analysis prior to any model derivation. There are situations, such as nonlinear tissue binding, *"where the volume of distribution and the apparent clearance is higher for the first dose as compared to subsequent doses, which could have implications for bioavailability and/or bioequivalence studies; where the apparent volume and clearance decrease with increasing doses; steady state infusion concentrations are reached more rapidly with increasing infusion rate; the zero time plasma concentration intercept of the true terminal phase is not dose proportional and is almost independent of dose"* (Cheung and Levy [1989]).

3.6.2   Nonlinear kinetics - Capacity

Linear kinetic models permit the application of the superposition principle, in that doubling the dose will consequently lead to a doubling of the concentrations as seen in Equation 3:270

$$C = C^0 \cdot e^{-K \cdot t} = \frac{D_{iv}}{V} \cdot e^{-\frac{Cl}{V} \cdot t} \qquad (3:270)$$

This model can be expressed in differential equation terms as a direct proportionality between the rate of change (dC/dt) and the plasma concentration

$$\frac{dC}{dt} = -K \cdot C = -\frac{Cl}{V} \cdot C \qquad (3:271)$$

At high concentrations, metabolism and/or renal secretion usually approach the capacity of the elimination system and the elimination rate is no longer strictly a constant such as K. For practical reasons, we will estimate parameters for Equation 3:271 in terms of clearance and volume of distribution. Equation 3.271 can also represent the model for Michaelis-Menten elimination if we allow Cl to take the form

$$Cl = \frac{V_{max}}{K_m + C} \qquad (3:272)$$

V_{max} and K_m are the maximum metabolic rate and the Michaelis-Menten constant, respectively. There are two limiting cases for Cl in this model. Clearance is equal to V_{max}/K_m when the concentration of drug is zero, and clearance approaches zero when the concentration goes to infinity, as shown in Figure 3.66.

Since clearance depends on plasma concentration, it is of limited utility (Tozer and Winter [1992]) and should be used carefully. Measuring the bioavailability using the AUC is also not appropriate, because clearance is not constant.

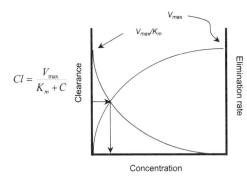

Figure 3.66 *Schematic relationship between clearance (left Y axis) and concentration and elimination rate (right Y axis) and concentration. Arrows indicate the location of K_m on the concentration axis at 50% of the maximum rate.*

Figure 3.67 displays Michaelis-Menten behavior in two subjects who received a different intravenous dose of a compound that exhibits capacity limited elimination.

Observe the nonlinear behavior at higher concentrations (e.g., a four-fold increase in dose results in more than a four-fold increase in concentration, except for $t = 0$ where the concentrations are dose proportional). Note also how the plasma half-life changes constantly during the nonlinear portion of the curve.

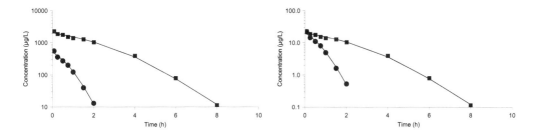

Figure 3.67 *Semi-logarithmic plots of the plasma concentration-time profiles of a compound following 25 (●) and 100 (■) unit intravenous doses to two subjects. The left hand figure shows actual data of both subjects and the right hand figure shows the dose-normalized data. Note the differences between the terminal slopes, which may be due to different K_m values in the two subjects.*

For a typical Michaelis-Menten system, the elimination rate (*Rate*) as a function of plasma concentration takes the following form

$$Rate = \frac{V_{max} \cdot C}{K_m + C} \qquad (3:273)$$

Cl will then be proportional to the ratio V_{max} over $K_m + C$ (Equation 3:272). When the concentration is much greater than K_m, *Cl* approaches zero and the elimination rate reaches a maximum value of V_{max}. There are two limiting cases of Michaelis-Menten kinetics. The first is at high plasma concentrations $C \gg K_m$ where the elimination rate approaches its limit

$$Rate \approx V_{max} \qquad (3:274)$$

Under these conditions, the elimination rate is independent of the plasma concentration and proceeds according to zero-order kinetics at a constant rate. The second is where we have low plasma concentrations ($C \ll K_m$). In this instance, the original Michaelis-Menten equation reduces to

$$Rate \approx \frac{V_{max} \cdot C}{K_m} = K' \cdot C \qquad (3:275)$$

V_{max} cannot be separated from K_m. This may also be appreciated from Figure 3.66. V_{max}/K_m is the intercept of the *Cl* axis at a concentration equal to 0. In order to estimate V_{max} and K_m accurately we need to measure plasma concentrations that cover the majority of the curvature. To estimate K_m accurately and separated from the V_{max} estimation, one needs concentration information from K_m and below. Fitting the kinetic model simultaneously to both high-dose and low-dose plasma concentration-time data is therefore recommended. V_{max} and K_m may then be estimated separately with a sufficient amount of accuracy and precision and relatively low correlation.

For a one-compartment system following intravenous dosing with Michaelis-Menten elimination, the area under the curve *AUC* can then be estimated as

$$AUC = \frac{V_{max} \cdot C^0}{2}\left[K_m + \frac{C^0}{2}\right] \qquad (3:276)$$

C^0 is *Dose/Volume* at $t = 0$. The half-life is a function of plasma concentration for the nonlinear system. It is not the time required to eliminate one-half of the drug in the body, but is rather an instantaneous value of the time required to eliminate half of the drug present, if the fractional rate of elimination were to continue at the same value (Tozer and Winter [1992]).

$$t_{1/2} = \ln(2) \cdot \frac{V_{ss}}{V_{max}}\left[K_m + C\right] \qquad (3:277)$$

Since clearance and half-life are first-order pharmacokinetic parameters, the most useful parameters for a nonlinear system are V_{max}, K_m and V_{ss}. In the case of phenytoin, Tozer and Winter pointed out the importance of further research of the time dependency, stability of V_{max}, K_m and V_{ss}. One would expect enzyme induction to increase V_{max} but to have no effect on K_m. One would expect K_m, based on the unbound concentration, to be relatively constant, because it depends on the property of the enzyme. For a comprehensive review of nonlinear (phenytoin) pharmacokinetics see Tozer and Winter [1992]. More information about how to derive initial parameter estimates of K_m and V_{max} for the one- and two-compartment models are presented in the applications section on nonlinear (capacity-limited) kinetics (exercises PK17, PK18, PK19, PK20).

3.6.3   Nonlinear kinetics - Time

3.6.3.1   Background

We will review briefly the phenomena called *induction/inhibition* and how these processes affect the time course of drug. Induction may be an effect of either increased rate of synthesis (increased turnover rate due to increased expression or synthesis) of the responsible enzyme (*[E]*) or decreased loss of enzyme (decreased fractional turnover) (Whitlock and Denison [1995], Park *et al* [1996]). Both processes increase the amount of enzyme, also called the enzyme pool. Since no information is generally available about gene transcription, translation or the fractional turnover rate of [E], k_{out} is a measure of the rate-limiting step (Figures 3.68 and 3.69). Induction may affect both the amount and/or the activity of the enzyme. The latter includes a more rapid turnover of drug molecules per unit time.

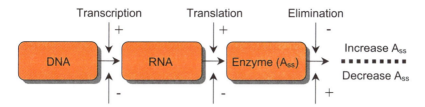

Figure 3.68 _Schematic diagram of possible causes of changes in the enzyme level A_{ss}. A positive sign indicates increased synthesis (production of enzyme) or an increased loss of enzyme and a negative sign indicates the opposite. Enhanced input or reduced loss leads to build-up of the enzyme level. Reduced input (e.g., inhibition) or enhanced loss reduces the enzyme level._

3.6.3.2   Pentobarbital induction

Pentobarbital is an inducer that affects the pool size $[E]$ of the responsible enzyme(s) for nortriptyline metabolism (Figure 3.69).

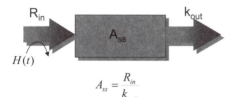

$$A_{ss} = \frac{R_{in}}{k_{out}}$$

Figure 3.69 _The turnover model. Inducer (pentobarbital) stimulates the production of enzyme. The half-life of A is not affected. We assume that an increase in R_{in}, A_{ss} and V_{max} increase oral clearance proportionately._

By increasing the turnover rate, the level of enzyme is changed (the pool size) but not the time it takes to achieve the new steady state with respect to the new pool size, since k_{out} is unaffected. The time to steady state will only be affected if k_{out} is changed. A decrease (or increase) in k_{out} will not only affect the time to steady state but also the actual steady state level. Although enzyme induction is perhaps among the most common causes of time-dependent kinetics, there may also be other reasons for time dependency, such as diurnal variations in renal function, urine pH, plasma protein concentration and hormone levels to mention just a few (Rowland and Tozer [1995]).

A study was conducted to see if the drug metabolizing enzymes of nortriptyline (_NT_) are inducible by pentobarbital (_PB_) (von Bahr _et al_ [1998]). _NT_ was administered orally as 10 mg every 8 h for a total period of 29 days (696 h). After 9 days (216 h) treatment with _PB_ (inducer) was initiated, continuing until 21.5 days (516 h) from the start of _NT_ administration. The concentration time course of _NT_ before, during and after treatment with inducer is depicted in Figure 3.70, including the predicted time course.

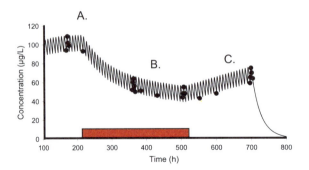

Figure 3.70 Observed (closed circles) and predicted (solid line) plasma concentrations of drug before (A), during (B), and after (C) induction. The horizontal red bar corresponds to the pentobarbital treatment period.

The model does not explicitly include the kinetics of the inducing substance *PB*. We assume that k_{out}[1] of the induced enzyme has a longer half-life than the drug. In addition, we assume that the time course from one level of enzyme activity to another will be governed by the fractional turnover rate of the enzyme. *PB* is assumed not to change the fractional turnover rate or the rate of loss of enzyme. No lag-period is included between initiation of *PB* and the appearance of altered plasma concentration of *NT*.

We will approximate the behavior of *NT* by a one-compartment model. We also assume that *PB* affects the turnover rate R_{in} of the responsible enzyme, and that the clearance of *NT* is directly proportional to the enzyme pool.

Absorption rate and lag-time of drug absorption will also be estimated. Oral clearance Cl_o for *NT* pre- and post-induction will be estimated simultaneously with mean residence time *MRT* pre- and post-induction.

Pre-induction, Cl_o is estimated as $Dose/AUC_o$, while after the start of induction (i.e., peri-induction), Cl_o is calculated as $Dose/AUC_{ss}$. The k_{out} parameter is derived from Figure 3.70 since it takes approximately 150 h to lower the drug level 50%. Dividing $ln(2)$ with 150 gives a rough estimate of k_{out}. V/F is approximated from $Dose/C_{max}$ during the pre-induction phase. We assume that an increase in R_{in}, $[E]$ and V_{max} leads to a proportionate increase of Cl_o. For a complete derivation of the relationship between the elimination rate constant of drug and time, see Gibaldi and Perrier [1982].

The pre-induction oral clearance was estimated to be 46 L/h and the induced oral clearance was about 118 L/h. This is equivalent to a corresponding increase in the enzyme level, which will have consequences for bioavailability and hepatic extraction. The volume of distribution (V/F) is 1680 L. The average plasma concentration is halved within this time frame, which is consistent with a 100% increase in the induced enzyme (Figure 3.70). The half-life of k_{out} is about 158 h, which is what one would expect from the plot of the data (Figure 3.71). The concentration-time profiles within a dosing interval pre-induction, peri-induction, and post-induction are shown in Figure 3.71.

---

[1]Note that the estimate of k_{out} is the apparent change in enzyme content, including transcription, translation and loss of enzyme.

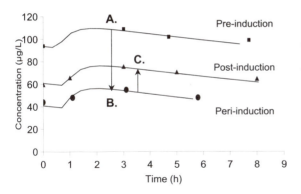

Figure 3.71 *Observed and predicted plasma concentrations of nortriptyline during a dosing interval pre-, peri- and post-induction after oral dosing. The absorption process is the rate-limiting step (flip-flop). The arrows indicate the time order of the curves.*

The half-life $t_{1/2}$, which governs the time to steady state, is determined by volume of distribution and clearance (Equation 3:278). Since the half-life is shortened during induction (*Cl* increases), and the time to establish steady state is shortened.

$$t_{1/2} = \ln(2) \cdot \frac{V}{Cl} \tag{3:278}$$

When the inducer is removed (*Cl* decreases), the half-life returns to its pre-induction value which is longer compared to the half-life during the induction phase. Therefore, the return of the induced state towards normal levels is significantly prolonged. Figure 3.72 shows the semi-logarithmic decline (left) of plasma concentration (slopes) before (A), during (B) and after (C) induction, at three different sections of the concentration-time course (right) following a constant rate intravenous infusion.

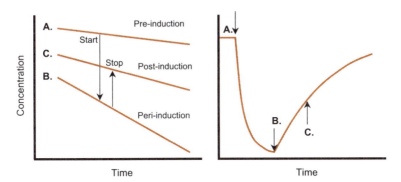

Figure 3.72 *Schematic diagram of the log-concentration time course of pre- (A), peri- (B), and post-induction (C) values after intravenous dosing (left) and the simulated concentration-time course following a constant rate input (right). The time to steady state is shortened during induction (half-life is shortened) compared to post-induction (half-life increases and returns to its pre-induction value). Note that the time scales in the left and right figures are not the same.*

Mean residence time pre- and post-induction were 36 h and 14 h, respectively. The absorption was manifested in a 50 min lag-time prior to the onset of absorption, with an absorption rate of 2.1 L/h.

However, one important limitation of this model is that neither the kinetics of the inducer nor any potential intermediate compound were taken into account. These may in part be rate-limiting. This difficulty can be overcome by measuring the kinetics of the inducing agent itself (Abramson [1986]). We can probably rule out that the inducing agent in this example, pentobarbital, with a half-life of 15-48 h was responsible for the rate-limiting step. This experiment does not predict the maximum induction possible, since only one dose level of inducer was given. Ideally, two or more dosing rates would have been chosen.

For further reading on time-dependency in kinetic variables and turnover concepts, see excellent texts such as Lassen and Perl [1979], Levy and Dumain [1979], Gibaldi and Perrier [1982] and Abramson [1986]. Abramson [1986] has demonstrated how to appropriately design and interpret results from induction studies.

3.6.3.3   Turnover of induction

Gibaldi and Perrier [1982] derived useful expressions for a time-dependent change in the enzyme level. Let us therefore assume that we can describe the turnover of the enzyme by means of Equation 3:279.

$$\frac{dE}{dt} = R_{in} - k_{out} \cdot E \tag{3:279}$$

The baseline condition then becomes

$$E_0 = \frac{R_{in}}{k_{out}} \tag{3:280}$$

For the induced enzyme we have a new steady state level E'

$$E' = \frac{R'_{in}}{k_{out}} \tag{3:281}$$

According to Gibaldi and Perrier [1982], one could derive the following expression for the time course of increase in the enzyme level E_t.

$$E_t = \frac{R'_{in}}{k_{out}} - \left(\frac{R'_{in}}{k_{out}} - \frac{R_{in}}{k_{out}} \right) \cdot e^{-k_{out} \cdot t} \tag{3:282}$$

If we assume that V_{max} is directly proportional to the enzyme level, then

$$E_0 = \frac{R_{in}}{k_{out}} = \frac{V_{max}}{k_{out}} \tag{3:283}$$

If the same holds true for the induced state, then

$$E' = \frac{R'_{in}}{k_{out}} = \frac{V'_{max}}{k_{out}} \tag{3:284}$$

Substituting all R_{in} -terms in Equation 3:282 by means of Equations 3:283 and 3:284 gives

$$V_{max_t} = V'_{max} - \left(V'_{max} - V_{max}\right) \cdot e^{-k_{out} \cdot t} \tag{3:285}$$

Replacing the V_{max}-terms by means of Cl gives

$$Cl_t = Cl' - \left(Cl' - Cl\right) \cdot e^{-k_{out} \cdot t} \tag{3:286}$$

where Cl_t is the variable for the time-dependent change in clearance. Insert Cl_t into the one-compartment constant intravenous infusion relationship

$$C_t = \frac{K^0}{Cl_t} \cdot \left[1 - e^{-\frac{Cl}{V} \cdot t}\right] \tag{3:287}$$

A parameter for the lag-time between dosing and start of induction is then included to give

$$C_t = \frac{K^0}{Cl_t} \cdot \left[1 - e^{-\frac{Cl}{V}(t-t_{lag})}\right] \tag{3:288}$$

The enzyme level during induction is dependent on the pre-induction and induced steady state enzyme pools, $[R_{in}]_{pre}/k_{out}$ and $[R_{in}]_{ss}/k_{out}$, respectively, and the first-order rate constant for enzyme degradation k_{out}. Of interest is the plasma concentration-time course of a drug that is subject to self-induction or as we have shown, a system with *drug + inducer*. The concentration-time course of drug at a constant rate of input takes the following form

$$C = \frac{R_0}{Cl(t)} \left[1 - e^{-\frac{Cl(t)}{V} \cdot (t-t_{lag})}\right] \tag{3:289}$$

when t greater t_{lag} and where $Cl(t)$ is defined as

$$Cl(t_{peri\ induction}) = Cl_{ss} - \left[Cl_{ss} - Cl_{pre}\right] \cdot e^{-k_{out} \cdot (t-t_{lag})} \tag{3:290}$$

For the post-induction state

$$Cl(t_{post\ induction}) = Cl_{pre} - \left[Cl_{pre} - Cl_{ss}\right] \cdot e^{-k_{out} \cdot (t-t_{lag})} \tag{3:291}$$

where $e^{-k_{out} \cdot (t-t_{lag})}$ is set equal to 1 if t less than or equal to t_{lag}.

The change in both steady state drug concentration, $R_0/Cl(t)$, and time to steady state, 3-4 times $\ln(2) \cdot V/Cl(t)$, becomes obvious when inspecting Equations 3:289 and 3:290. If we utilize these as the simplest form of kinetics of the induced compound to simulate the concentration-time course, we obtain the situation presented in Figure 3.73.

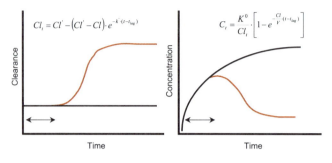

Figure 3.73 Time course of non-induced and induced clearance (left) versus time. The resulting concentration-time profiles of plasma (right) are shown for the induced and non-induced state.

3.6.3.4   Autoinduction

An alternative to the previous modeling approach is the feedback model. Let us elaborate a little on the model shown in Figure 3.74. It contains the drug model (right) and the clearance or enzyme model (left). By means of the plasma concentrations we will model this process and obtain an overall measure of the induction process. It can be seen that plasma concentrations of drug *drive* the enzyme level, which then affects the clearance of drug. This is called negative feedback, since they counterbalance each other. However, the most appropriate way to elucidate the causes of induction would have been to measure the individual processes shown in Figures 3.68 and 3.74. Simultaneous measurement of the metabolite(s) and responsible enzyme might have further strengthened our ability to precisely estimate the model parameters.

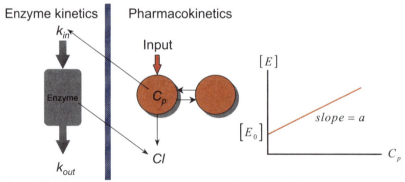

Figure 3.74 Relationship between the two-compartment drug model and the one-compartment enzyme model. Drug concentration (C) and the clearance term (Cl) link enzyme kinetics and pharmacokinetics. The right hand diagram shows the assumed linear relationship at steady state between enzyme level and plasma concentration in this example.

The differential equation for a two-compartment model specified in terms of the distribution parameter and volume of the extravascular space (Figure 3.75) then becomes

$$V_c \cdot \frac{dC}{dt} = In - Cl_E \cdot C - Cl_d \cdot C + Cl_d \cdot C_t \tag{3:292}$$

Cl_E is the inducible plasma clearance of drug. *In* is the input function, e.g., a bolus input. The equation for the peripheral compartment is

$$V_t \cdot \frac{dC_t}{dt} = Cl_d \cdot C - Cl_d \cdot C_t \qquad (3:293)$$

Autoinduction may be due to an increased transcription, translation or decreased loss of enzyme. In this example, we assume that a drug acts on the production of enzyme.

$$\frac{dE}{dt} = k_{out} \cdot [C + E_0] - k_{out} \cdot E \qquad (3:294)$$

The k_{in} parameter is substituted by $k_{out} \cdot E_0$. At baseline,

$$\frac{dE}{dt} = 0 = k_{out} \cdot [0 + E_0] - k_{out} \cdot E \qquad (3:295)$$

which implies

$$E = E_0 \qquad (3:296)$$

$$Cl_E = a \cdot [0 + E_0] \qquad (3:297)$$

The *a* parameter is a proportionality factor for clearance. In order to be able to design an appropriate dosage regimen, we also need to estimate the induced clearance and the time to induction equilibrium. Unfortunately, we only have one dose level, which may not suffice for a complete assessment of the induction phenomenon.

This exercise has shown that the feedback model (Ackerman *et al* [1964], Rescigno and Segre [1966]) also suits itself for mimicking autoinduction data. The nice feature of this model is that concentration *C* is counterbalanced by the enzymatic activity *E*. This enables us to use the model after multiple dose regimens, which is a prerequisite for mechanistically oriented models. In addition, two or more optimally spaced dose levels will probably help us to define the appropriate behavior of the induction process.

Figure 3.75 shows the fitted concentration-time profile of the aforementioned induction model together with observed data.

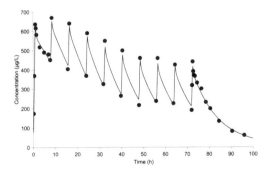

Figure 3.75 *Observed (circles) and predicted (lines) plasma concentrations versus time using the feedback model. The horizontal blue bar corresponds to the length of infusion.*

Since no information is available about gene transcription, translation or the fractional turnover rate of [E], k_{out} is a measure of the rate-limiting step. Table 3.7 contains information about the final parameter estimates of V_c, V_t, a, k_{12}, k_{21}, E_0, Cl_d and k_{out} of the two fitted models. *Model I* in Table 3.7 (Figure 3.36, left) is based on volumes and the distribution parameter and *Model II* (Figure 3.36, right) on micro-constants. The proposed model is suitable for multiple dosing and different modes of administration. The enzyme level adjusts itself with respect to drug concentration.

Table 3.7 Comparison of models

Parameter	Model I	Model II
	mean ± CV%	mean ± CV%
V_c (L)	145 ± 5	144 ± 5
a (L/h)	0.041 ± 6	0.041 ± 5
Cl_d (L/h)	–	123 ± 30
V_t (L)	–	61.7 ± 12
k_{12} (h^{-1})	0.82 ± 33	–
k_{21} (h^{-1})	1.94 ± 25	–
k_{out} (h^{-1})	0.024 ± 15	0.023 ± 12
E_0	135 ± 21	135 ± 16

The induction half-life was estimated to be 30 h; enzyme baseline 135 units. Volume of distribution is 144 L for the central compartment and volume of distribution at steady state is 206 L. Uninduced clearance is 5.5 L/h (0.041·(1+135)). Ideally, different modes of administration and two or more dose levels, as well as metabolite data, are needed to fully describe the underlying time-dependent process.

In conclusion, the amount and/or the activity (substrate turnover) of the enzyme changes during induction. Furthermore, the baseline value of the enzyme is highly variable between individuals. Therefore, the extent of induction should be compared to its pre-induction value on an individual basis. The maximum level of induction is still poorly understood. See also sections PK21 and PK22 on enzyme induction.

3.6.4   Nonlinear kinetics - Flow
A third explanation for nonlinear kinetics may be changes in the perfusion of eliminating organs such as the liver and the kidneys. We refer the reader to section 3.4 on clearance models, for a discussion of the impact of flow, and sections PK23 and PK24.

3.6.5   Nonlinear kinetics - Binding
The free fraction f_u of a drug that binds to a single site (or protein) is defined as

$$f_u(\%) = \frac{100}{1 + K_1 \cdot C_{protein}} \qquad (3:298)$$

We assume that binding occurs to only one site. C_{prot} is the plasma protein concentration and the affinity constant between drug and binding protein is denoted as K_1, the model parameter to be estimated. Figure 3.76 contains observed and model predicted binding data for a hypothetical drug.

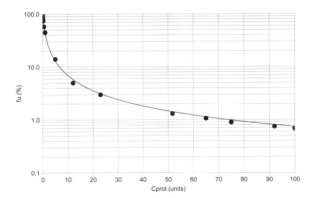

Figure 3.76 *Semi-logarithmic plot of observed (filled circles) and model predicted (solid line) free fraction (%) versus plasma protein concentration.*

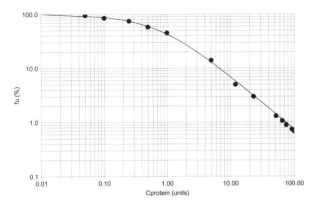

Figure 3.77 *Log-log plot of bserved (filled circles) and model predicted (solid line) free fraction (%) versus plasma protein concentration.*

Large changes in kinetic parameters which are based on total concentration and are associated with changes in plasma protein binding are frequently cited, as if they predicted some important alteration in clinical effect. When changes in binding are associated with clinical effects, it has almost always been found that this is the result of a change in unbound drug clearance caused by a mechanism quite independent of plasma protein binding (Holford [1995]). Winter [1996] concluded that there is little evidence demonstrating that monitoring unbound drug levels improves the correlation between the plasma concentration and the pharmacologic effect or therapeutic outcome. See also Levy [1986].

3.6.6   Nonlinear drug metabolite models

The plasma concentrations of drug A and metabolite B could be measured in plasma following three rapid intravenous doses of 10, 50, and 300 μmol/kg. We assume all elimination occurs from the measured plasma compartment by a Michaelis-Menten metabolic pathway. The formation of B is capacity limited, and can be approximated by the Michaelis-Menten equation.

$$V_M \frac{dC_M}{dt} = \frac{V_{max} \cdot C}{K_m + C} \tag{3:299}$$

where V_M and C_M are the volume and concentration of the metabolite, respectively. Equation 3:303 gives the complete turnover of metabolite.

Figure 3.78 depicts the experimental data, with drug concentrations in the left hand panel and metabolite concentrations in the right hand panel. Notice that the peak concentration of drug increases approximately linearly with dose, whereas the metabolite peak concentration increases less than proportionately. The slopes of the 30-180 min portion of the drug plasma concentration-time profiles decrease with dose. The 'terminal' phase of the metabolite plasma concentrations seems to run in parallel for the two highest doses.

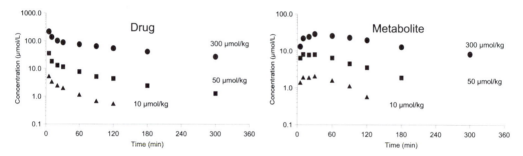

Figure 3.78 *Observed drug (left) and metabolite (right) concentration-time data. Note the change in half-life with increasing concentrations.*

The model includes a two-compartment drug model with Michaelis-Menten elimination, of which the latter is formation of metabolite, and a one-compartment metabolite model, respectively (Figure 3.79). Although this model may be rather simplistic, it may be easily expanded to encompass more complex situations encountered in metabolite kinetics.

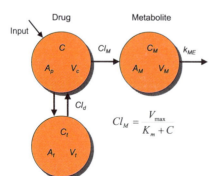

Figure 3.79 *Schematic model of drug and metabolite disposition.*

The equation for the central or plasma compartment is

$$V_c \frac{dC}{dt} = In - Cl_M \cdot C - Cl_d \cdot C + Cl_d \cdot C_t \tag{3:300}$$

where *In* is the intravenous input function, and Cl_M is equal to

$$Cl_M = \frac{V_{max}}{K_m + C} \tag{3:301}$$

The equation for the peripheral drug compartment becomes

$$V_t \frac{dC_t}{dt} = Cl_d \cdot C - Cl_d \cdot C_t \tag{3:302}$$

Metabolite concentration in plasma is

$$V_M \frac{dC_M}{dt} = Cl_M \cdot C - Cl_{ME} \cdot C_M \tag{3:303}$$

Cl_{ME} $(k_{ME} \cdot V_M)$ and V_M are the clearance and volume of distribution for the metabolite, respectively. Note that Cl_M is the metabolic clearance of drug and *formation* clearance of the metabolite. In this model the only route of elimination is the formation of the single metabolite. In Figure 3.80, the predicted data for drug and metabolite are superimposed on the observed concentration-time data that were shown in Figure 3.78.

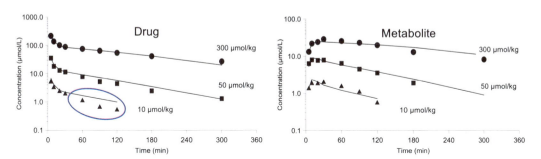

Figure 3.80 *Mean observed (filled symbols) and predicted plasma concentrations (lines) of drug and metabolite. The doses represent the amount of drug. The ellipsoid demonstrates systematic deviation between observed and predicted data.*

Although the overall fit was good, a systematic deviation was observed for observed and calculated terminal concentrations for the parent drug following the 10 μmol/kg dose (Figure 3.80, Table 3.8).

Table 3.8 Parameter estimates and their precision using both
              parent drug and metabolite data in the model

Parameter	Mean	SD	CV%
V_c (L/kg)	1.07	0.062	5.7
V_t (L/kg)	2.0	0.098	5.0
Cl_d (L/min/kg)	0.13	0.014	10
V_{max} (µmol/min/kg)	1.69	0.16	9.3
K_m (µmol/L)	57.0	8.88	16
k_{ME} (min^{-1})	0.14	0.02	14
V_{ME} (L/kg)	0.29	0.039	13

Correlation between V_{max} and K_m is 0.92.

K_m for the Michaelis-Menten elimination of parent drug was estimated to be 57 µmol/L and V_{max} 1.7 µmol/min/kg. The total volume of distribution at steady state was 3 L/kg, which is consistent with previous studies. Using data on V_{max}, K_m, and V_{ss} the half-life could be estimated for different concentrations. By means of the equation

$$t_{1/2} = \ln(2) \cdot \frac{V_{ss}}{V_{max}} \left[K_m + C \right]$$

(3:304)

the theoretically shortest half-life, when plasma concentration is zero, was calculated to be 69 min.

When the concentration was increased to 100 and 200 µmol/L, the half-lives became 198 and 326 min, respectively (Table 3.9). A flip-flop situation is seen in the metabolite data at all doses since the elimination rate constant is about 0.04 min^{-1} although the formation rate constants are 0.008, 0.003, and 0.002 min^{-1} at 10, 100, and 200 µmol/L, respectively. The formation of metabolite is the rate-limiting step, agreeing with the terminal plasma profiles for the metabolite.

Table 3.9 Model predicted *Cl*, $t_{1/2}$ and *MRT* using
              both parent drug and metabolite data in
              the model

C (µmol/L)	Cl (L/min/kg)	$t_{1/2}$ (min)	MRT (min)
1	0.029	73	106
10	0.025	84	122
50	0.016	135	195
100	0.011	198	286
200	0.007	326	468
500	0.003	707	1015

Clearance was inversely proportional to the plasma concentration of drug. The predicted half-life for metabolite was less than 10 min. This implies that the terminal half-life observed in plasma for metabolite after administration of drug was the formation half-life. No further extrapolations of the metabolite kinetics should be made based on the present results. Ideally, the metabolite should be given separately to obtain its own disposition kinetics. Figure 3.81 displays the relationship between clearance and concentration, and between half-life and concentration, respectively, for the parent drug.

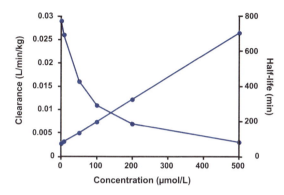

Figure 3.81 Predicted plasma clearance (left axis and curvilinear line) and half-life (right axis and straight line) of drug versus plasma concentration

One reason why the Michaelis-Menten parameters were so well defined with a high level of precision is that we had data on elimination of drug, and build-up and elimination of metabolite. It is the ideal situation to have data *on both sides* of the modeled route. The predictive strength of this model is also superior to linear elimination models. Not only does this model do a better job in interpolation of data, but it is also better for extrapolations outside the present dosing range.

A reduced model taking only the drug concentration-time profiles into account was then fit to data (Figure 3.82). This was done in order to evaluate the drug kinetics without use of the metabolite data. The results can be applied in the situation where the major concerns are an adequate systemic exposure of parent compound.

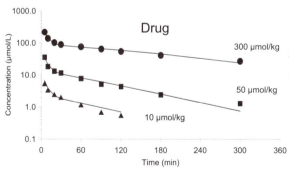

Figure 3.82 Observed and predicted plasma concentrations of drug following three doses as indicated). Only drug concentrations were modeled.

The values obtained for all parameters were consistent with results from the full model except for V_{max} and K_m, which were halved (compare Tables 3.9 and 3.10). The 40-50% reduction in V_{max} and K_m values may affect predictions at concentrations that are less than K_m, but have marginal effects at plasma concentrations of 100 μM and higher.

Table 3.10   Parameter estimates and their precision using only
                 parent drug data in the model

Parameter	Mean	SD	CV%
V_c (L/kg)	1.10	0.055	2.3
V_t (L/kg)	2.15	0.107	5.0
Cl_d (L/min/kg)	0.11	0.011	10
V_{max} (µmol/min/kg)	1.11	0.11	10
K_m (µmol/L)	27.5	5.30	19

Correlation between V_{max} and K_m is 0.91.

The model predicts a clearance of 40, 14, and 5 mL/min at drug plasma concentrations of 0, 50, and 200 µM, respectively, which is reasonably close to the full model (30, 16 and 7 mL/min, respectively).

3.7   Non-Compartmental Analysis

3.7.1   Background

Most current approaches to characterize a drug's kinetics involve non-compartmental analysis, denoted *NCA*, and nonlinear regression analysis. The advantages of the regression analysis approach are the disadvantages of the non-compartmental approach and *vice versa*. *NCA* does not require the assumption of a specific compartmental model for either drug or metabolite. The method used involves application of the trapezoidal rule for measurements of the area under a plasma concentration-time curve. This method, which applies to first-order (linear) models, is rather assumption free and can readily be automated. Benet and Galeazzi [1979], Watari and Benet [1989], and Nakashima and Benet [1989] have elaborated on the theory of *NCA,* while Gillespie [1991] discussed the pros and cons of *NCA* versus compartmental models. Figure 3.83 gives a schematic picture of the *NCA* and nonlinear regression approaches. As can be seen, *NCA* deals with sums of areas whereas regression modeling uses a function with regression parameters.

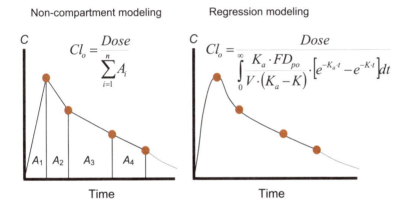

Figure 3.83 *Comparison of NCA (left) and nonlinear regression modeling (right). K_a, K and V in the right hand panel indicate the model parameters to be estimated by regressing the model to data.*

The time course of drug concentration in plasma can usually be regarded as a statistical distribution curve. The area under a plot of the plasma concentration versus time curve is referred to as the area under the zero moment curve *AUC*, and the area under the product of the concentration and time versus time curve is then called the area under the first moment curve *AUMC*. Only the areas of the zero and first moments are generally used in pharmacokinetic analysis, because the higher moments are prone to an unacceptable level of computational error.

3.7.2   Computational methods

3.7.2.1   Linear trapezoidal rule

The areas can either be calculated by means of the *linear trapezoidal rule* or by the *log-linear trapezoidal rule*. The total area is then measured by summing the incremental area of each trapezoid (Figure 3.84).

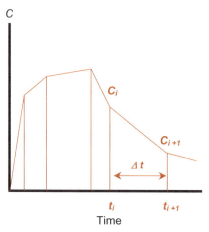

Figure 3.84 Graphical presentation of the linear trapezoidal rule. $AUC_{ti-t(i+1)}$ is the area between t_i and t_{i+1}. C_i and C_{i+1} are the corresponding plasma concentrations, and Δt is the time interval. Note that Δt may differ for different trapeziums.

The magnitude of the error associated with the estimated area depends on the width of the trapezoid and the curvature of the "true" profile. This is due to the fact that the linear trapezoidal rule overestimates the area during the descending phase (assuming elimination is *first-order*), and underestimates the area during the ascending part of the curve (Figure 3.85).

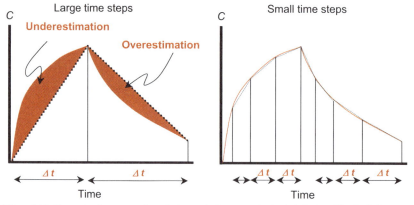

Figure 3.85 Concentration versus time during and after a constant rate infusion. The shaded area represents underestimation of the area during ascending concentrations and overestimation of the area during descending concentrations. By decreasing the time step (Δt) between observations, this under- or overestimation of the area is minimized.

Using the linear trapezoidal method for calculation of the area under the zero moment curve (*AUC*) from *0* to time t_n, we have

$$AUC_0^{t_{last}} = \sum_{i=1}^{n} \frac{C_i + C_{i+1}}{2} \cdot \Delta t \qquad (3:305)$$

where $\Delta t = t_{i+1} - t_i$ and t_{last} denotes the time of the last measurable concentration. Unless one has sampled long enough in time so that concentrations are negligible, the *AUC* as defined above will underestimate the *true AUC*. Therefore it may be necessary to extrapolate the curve out to $t = \infty$. The extrapolated area under the zero moment curve from the last sampling time to infinity (AUC_{extr}) is calculated as

$$AUC^{\infty}_{t_{last}} = \int_{t_{last}}^{\infty} C_{last} \cdot e^{-\lambda_z (t - t_{last})} dt = C_{last} \left[\frac{e^{-\lambda_z (t - t_{last})}}{-\lambda_z} \right]_{t_{last}}^{\infty} = C_{last} \left[0 - \frac{1}{-\lambda_z} \right] = \frac{C_{last}}{\lambda_z} \qquad (3{:}306)$$

where C_{last} and λ_z are the last measurable (non-zero) plasma concentration and the terminal slope on a log_e scale, respectively. One may also use the predicted concentration at t_{last} if the observed concentration for some reason deviates from the terminal regression line.

The λ_z parameter is graphically obtained from the terminal slope of the semi-logarithmic concentration-time curve as shown in Figure 3.86, with a minimum of 3-4 observations being required for accurate estimation (See section 3.7.3). The Y axis $ln(C)$ denotes the natural logarithm (log_e) of the plasma concentration C.

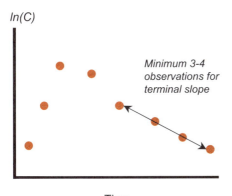

Minimum 3-4 observations for terminal slope

Figure 3.86 *Semi-log plot demonstrating the estimation of λ_z. The terminal data points are fit by log-linear regression to estimate the slope.*

The linear trapezoidal method for calculation of the area under the first moment curve (*AUMC*) from 0 to time t_{last}, is obtained from

$$AUMC^{t_{last}}_{0} = \sum_{i=1}^{n} \frac{t_i \cdot C_i + t_{i+1} \cdot C_{i+1}}{2} \cdot \Delta t \qquad (3{:}307)$$

Remembering that $\int x \cdot e^{-a \cdot x} dx = -\frac{x \cdot e^{-a \cdot x}}{a} - \frac{e^{-a \cdot x}}{a^2}$

the corresponding area under the first moment curve from time t_{last} to infinity ($AUMC_{extr}$) is computed as

$$AUMC_{t_{last}}^{\infty} = \int_{t_{last}}^{\infty} t \cdot C dt = \int_{t_{last}}^{\infty} t \cdot C_{last} e^{-\lambda_z (t - t_{last})} dt$$

$$= C_{last} \cdot e^{\lambda_z t_{last}} \left[\frac{t \cdot e^{-\lambda_z t_{last}}}{-\lambda_z} + \frac{e^{-\lambda_z t_{last}}}{-\lambda_z^2} \right]_{t_{last}}^{\infty} = \frac{C_{last} \cdot t_{last}}{\lambda_z} + \frac{C_{last}}{\lambda_z^2} \qquad (3:308)$$

3.7.2.2   Log-linear trapezoidal rule

An alternative procedure that has been proposed is the log-linear trapezoidal rule. The underlying assumption is that the plasma concentrations decline mono-exponentially between two measured concentrations. However, this method applies only for descending data and fails when $C_i = 0$ or $C_{i+1} = C_i$ (in these instances one would revert to the linear trapezoidal rule). The principal difference between the linear and the log-linear trapezoidal method is demonstrated in Figure 3.87.

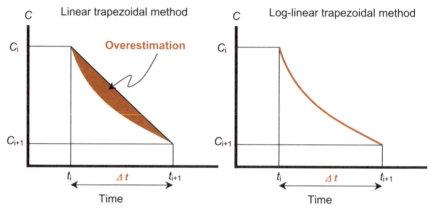

Figure 3.87 *The principal difference between the linear (left) and the log-linear (right) trapezoidal methods. The shaded region represents the over-predicted area with the linear trapezoidal rule. Note that the log-linear approximation is only true if the decay is truly mono-exponential between t_i and t_{i+1}.*

Remember that when the concentrations decline exponentially

$$C_{i+1} = C_i \cdot e^{-K(t_{i+1} - t_i)} = C_i \cdot e^{-K\Delta t} \qquad (3:309)$$

where $t_{i+1} - t_i$ is the time step (Δt) between two observations and K is the elimination rate constant for a one-compartment system. Otherwise, λ_z should be used as the slope. The above expression when rearranged gives the elimination rate constant K

$$K = \frac{\ln(C_i / C_{i+1})}{\Delta t} \qquad (3:310)$$

The *AUC* within the time interval Δt is the difference between the concentrations divided by the slope K

$$AUC_i^{i+1} = \frac{C_i - C_{i+1}}{K} = \frac{C_i - C_{i+1}}{\ln(C_i / C_{i+1})} \cdot \Delta t \qquad (3:311)$$

Using the log-linear trapezoidal method from time zero to t_n

$$AUC_0^{t_n} = \sum_{i=1}^{n} \frac{C_i - C_{i+1}}{\ln(C_i / C_{i+1})} \cdot \Delta t \qquad (3:312)$$

while the corresponding equation for *AUMC* from time zero to t_n with this method yields

$$AUMC_0^{t_n} = \sum_{i=1}^{n} \frac{t_i \cdot C_i - t_{i+1} \cdot C_{i+1}}{\ln(C_i / C_{i+1})} \cdot \Delta t - \frac{C_{i+1} - C_i}{[\ln(C_i / C_{i+1})]^2} \cdot \Delta t^2 \qquad (3:313)$$

The extrapolated area under the zero moment curve from the last sampling time to infinity (AUC_{extr}) is calculated as

$$AUC_{t_{last}}^{\infty} = \frac{C_{last}}{\lambda_z} \qquad (3:314)$$

where C_{last} and λ_z are as defined earlier. The corresponding area under the first moment curve from time zero to infinity ($AUMC_{extr}$) is

$$AUMC_{t_{last}}^{\infty} = \frac{C_{last} \cdot t_{last}}{\lambda_z} + \frac{C_{last}}{\lambda_z^2} \qquad (3:315)$$

As previously pointed out, the linear trapezoidal method gives approximate estimates of *AUC* during both the ascending and descending parts of the concentration-time curve, although the bias is usually negligible for the up-swing. The log-linear trapezoidal method may also give somewhat biased results, though to a lesser extent. Some people argue that the log-linear trapezoidal method may therefore be preferable for drugs with long half-lives. From a practical point of view this still needs to be proven. However, our own experience is that the difference between the two methods is negligible as long as a reasonable sampling design has been used. We generally use a mixture of the two methods, which means that the linear trapezoidal method is applied for increasing or equal concentrations (e.g., at the peak or a plateau) and the log-linear trapezoidal method for decreasing concentrations. This is demonstrated in Figure 3.88.

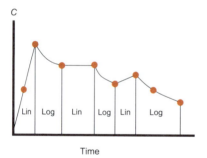

Figure 3.88 *NCA using a combination of the linear and log-linear trapezoidal methods.*

Note that *NCA* is often used in crossover studies comparing two formulations and 12 to 36 subjects. Thus, since the error associated with an individual patient's *AUC* is generally small, the (average) error associated with the average *AUC* for a formulation will generally be negligible, regardless of the method used. The choice of method is thus up to the discretion of the modeler, as long as one can explain why a particular method provides a more accurate estimate of *AUC*.

Direct integration of the function for the drug's kinetics in plasma is discussed under the introductory section on mono- and multi-exponential models and will therefore not be further elaborated here.

3.7.3   Strategies for estimation of λ_z

When estimating λ_z, we recommend that data from each individual are first plotted in a semi-log diagram. Ideally, to obtain a reliable estimate of the (terminal) slope, 3-4 half-lives would needed to have elapsed. However, sometimes this is not possible. A minimum requirement is then to have 3-4 observations for the terminal slope (Figure 3.89). By means of log-linear regression of those observations, the estimate of λ_z is obtained. This is then used for calculation of the extrapolated area as shown below

$$AUC_{t_{last}}^{\infty}\,(observed) = \frac{C_{last}}{\lambda_z}$$ (3:316)

or

$$AUC_{t_{last}}^{\infty}\,(predicted) = \frac{\hat{C}_{last}}{\lambda_z}$$ (3:317)

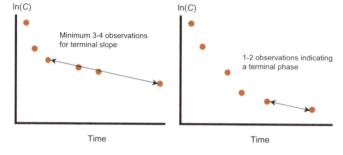

Figure 3.89 *The ideal situation (left) for estimation of the terminal slope λ_z. Another and perhaps more commonly encountered situation (right) is where one only has an indication of an additional slope.*

In Figure 3.90 the last observed concentration deviates somewhat from the regression line. The extrapolated area, if based on this concentration, would be disproportionately large as compared to the area based on the predicted concentration.

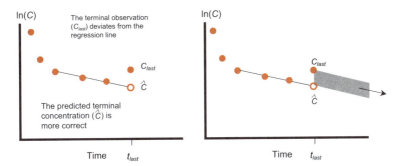

Figure 3.90 *Impact on the extrapolated area of using observed terminal concentration versus predicted concentration. The shaded area from t_{last} to infinity symbolizes the overestimation that would result. Note that if the observed terminal concentration lies below the predicted terminal concentration, then the extrapolated area would be underestimated. The open circle is the predicted concentration at t_{last}. The last observation is not included in the regression.*

The total area is obtained by summing the individual areas obtained by means of the trapezoidal rule to the last time (t_{last}), and adding the extrapolated area according to

$$AUC_{total} = AUC_0^\infty = AUC_0^{t_{last}} + AUC_{t_{last}}^\infty \qquad (3:318)$$

The fraction of AUC_{extr} to $AUC_{0-\infty}$ is calculated as

$$\% \; extrapolated \; area = \frac{AUC_{t_{last}}^\infty}{AUC_0^\infty} \cdot 100 \qquad (3:319)$$

The extrapolated area should ideally be as small as possible in comparison to the total area. We believe that AUC_{extr} should not exceed 20-25% of AUC_{total}, unless it is only used as a preliminary estimate for further study refinement.

3.7.4   Pertinent pharmacokinetic estimates

Moment analysis has become widely used in recent years as a non-compartmental approach to the estimation of clearance Cl, mean residence time MRT, steady state volume of distribution V_{ss}, and volume of distribution during the terminal phase V_z (also called $V_{d\beta}$ for a bi-exponential system). Nakashima and Benet [1988, 1989] present a general treatment for the aforementioned parameters, which includes the possibility of input/exit from any compartment in a mammillary model. They also define *exit site dependent* and *exit site independent* parameters. We will, however, assume in the following examples that input/output occurs to the central compartment.

Assuming a simple case with a one-compartment bolus system, the shape of the concentration-time profile will take the form depicted in Figure 3.91. The extrapolated area from the last sample at t_{last} to infinity is in this case small. However, the corresponding area under the first moment curve has an altogether different shape. Clearly, the extrapolated area from last sampling point to infinity will generally contribute to a much larger extent under the first moment curve as compared to the area under the zero moment curve.

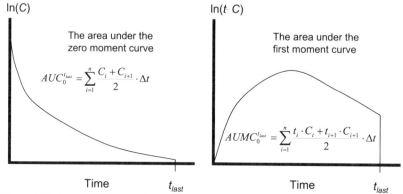

ln(C)   ln(t C)

The area under the   The area under the
zero moment curve   first moment curve

$$AUC_0^{t_{last}} = \sum_{i=1}^{n} \frac{C_i + C_{i+1}}{2} \cdot \Delta t$$

$$AUMC_0^{t_{last}} = \sum_{i=1}^{n} \frac{t_i \cdot C_i + t_{i+1} \cdot C_{i+1}}{2} \cdot \Delta t$$

Time t_{last} Time t_{last}

Figure 3.91 *Areas under the zero (left) and first (right) moment curves. The extrapolated areas beyond t_{last} will differ dramatically for the two curves.*

Before we continue with how to use different area measures, we would like to bring up some thoughts that were presented by Holford [1995]. Pharmacokinetics has moved almost completely from parameterizing elimination in terms of rate constants, with the more physiologically relevant use of clearance now being widely recognized. To put even more focus on clearance, Holford suggested that *AUC* no longer be used as a pharmacokinetic parameter. Clearance (*Cl*) or clearance over bioavailability (*Cl/F* also denoted Cl_o) is easily computed from *AUC* and dose, and *Cl* and *CL/F* can immediately be interpreted in a physiological context. On the other hand, *AUC* can be viewed as a parameter that confounds clearance and dose, and that has no intrinsic merit. While we agree with those ideas, *AUC* is still useful as a measure of exposure in toxicological studies and when dose is unknown.

Clearance is calculated from the dose and the area under the zero moment curve

$$Cl = \frac{D_{iv}}{AUC_0^{\infty}} \quad\quad\quad\quad (3:320)$$

Oral clearance (Cl_o or *Cl/F*) is calculated from the oral dose and the area under the zero moment curve

$$Cl_0 = \frac{Cl}{F} = \frac{D_{po}}{AUC_0^{\infty}} \quad\quad\quad\quad (3:321)$$

Using the areas obtained from systemic (e.g., intravenous) and extravascular (e.g., oral) dosing, the bioavailability *F* is calculated, after dose-normalization, according to

$$F = \frac{AUC_{ev}}{AUC_{iv}} \cdot \frac{D_{iv}}{D_{ev}} \qquad (3:322)$$

If the drug is given at a constant rate over a period of T_{inf}, then one also needs to adjust mean residence time MRT for the infusion time by means of subtracting $T_{inf}/2$ (infusion time/2) as follows

$$MRT = \frac{AUMC_0^\infty}{AUC_0^\infty} - \frac{T_{inf}}{2} \qquad (3:323)$$

$T_{inf}/2$ originates from the average time a molecule stays in the infusion set (e.g., syringe, catheter, line). Half of the dose is infused when the piston has travelled half of the intended distance. $T_{inf}/2$ is the mean input time MIT. Similarly for *first-order* input

$$MRT = \frac{AUMC_0^\infty}{AUC_0^\infty} - \frac{1}{K_a} \qquad (3:324)$$

Remember that K_a is the apparent *first-order* absorption rate constant derived from plasma data. This parameter may also contain processes parallel to the true absorption step of drug in the gastrointestinal tract, e.g., chemical degradation (k_d). Consequently, the mean absorption time MAT is the sum of several processes including absorption and chemical degradation

$$MAT = \frac{1}{K_{a(apparent)}} = \frac{1}{K_{a(true)} + k_d} \qquad (3:325)$$

The mean residence time of the central compartment is the sum of the inverse of the initial (α) and terminal (β) slopes corrected for the inverse of the sum of the exit rate constants from the peripheral compartment

$$MRT_{iv}(1) = \frac{1}{\alpha} + \frac{1}{\beta} - \frac{1}{E_2} \qquad (3:326)$$

Assuming that there is only one exit rate constant from the peripheral compartment, which then is k_{21}, the MRT_{iv} is

$$MRT_{iv} = \frac{1}{\alpha} + \frac{1}{\beta} - \frac{1}{k_{21}} \qquad (3:327)$$

The observed mean residence time after extravascular dosing becomes

$$\frac{AUMC_{0measured}^\infty}{AUC_{0measured}^\infty} = MRT + MIT \qquad (3:328)$$

which is the sum of the true mean residence time and mean input time. Mean input time can also be obtained from the input function according to Equation 3:329 below.

$$MIT = \frac{\int_0^\infty input\ function \cdot t\ dt}{\int_0^\infty input\ function\ dt} = \frac{\int_0^\infty input\ function \cdot t\ dt}{F \cdot Dose} \tag{3: 329}$$

Provided the input function is known

$$A_{gut} = F \cdot D_{po} \cdot e^{-K_a \cdot t} \tag{3:330}$$

and mean input time can be derived

$$MIT = \frac{\int_0^\infty F \cdot D_{po} \cdot e^{-K_a \cdot t} \cdot t\ dt}{\int_0^\infty F \cdot D_{po} \cdot e^{-K_a \cdot t}\ dt} = \frac{F \cdot D_{po} / K_a^2}{F \cdot D_{po} / K_a} = \frac{1}{K_a} \tag{3:331}$$

The volume of distribution at steady state (V_{ss}) is computed as

$$V_{ss} = MRT \cdot Cl = \frac{AUMC_0^\infty}{AUC_0^\infty} \cdot \frac{D_{iv}}{AUC_0^\infty} = \frac{D_{iv} \cdot AUMC_0^\infty}{\left[AUC_0^\infty\right]^2} \tag{3: 332}$$

The volume of distribution during the terminal phase (V_z) is computed as

$$V_z = \frac{Cl}{\lambda_z} = \frac{D_{iv}}{AUC_0^\infty} \cdot \frac{1}{\lambda_z} \tag{3: 333}$$

The same volume for a bi-exponential system (function) is computed as

$$V_{d\beta} = \frac{Cl}{\beta} = \frac{D_{iv}}{AUC_0^\infty} \cdot \frac{1}{\beta} \tag{3:334}$$

The terminal half-life $t_{1/2z}$ is readily estimated from the slope λ_z as

$$t_{1/2z} = \frac{\ln(2)}{\lambda_z} \tag{3:335}$$

The half-life of the initial α-phase is

$$t_{1/2\alpha} = \frac{\ln(2)}{\alpha} \tag{3:336}$$

The half-life of the terminal β-phase of a bi-exponential function is

$$t_{1/2\beta} = \frac{\ln(2)}{\beta} \tag{3:337}$$

Note that the $t_{1/2z}$ parameter is referred to as $t_{1/2\beta}$ in a bi-exponential system and simply $t_{1/2}$ in a mono-exponential system.

3.7.5   Issues related to steady state

Analysis can also be extended to calculation of *Cl*, *MRT*, and V_{ss} during multiple dosing (dosing interval τ). Clearance is calculated from the dose and the area under the zero moment curve within a dosing interval

$$Cl = \frac{D_{iv}}{AUC_0^{\tau}}$$
(3:338)

which is based on the fact that $AUC_{0-\infty}$ after a single dose is equivalent to $AUC_{0-\tau}$ at steady state. At steady state, $AUC_{0-\tau}$ represents the dose administered within that dosing interval since there are equal residual areas of drug from the previous and present doses as shown below (Figure 3.92).

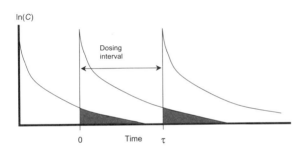

Figure 3.92 Area under the curve at steady state. The contribution of the residual area (shaded area left) from the previous dose is equal to the truncated area (also shaded, right) after τ. The total area under the concentration-time curve within the 0-τ dosing interval is thus equal to $AUC_{0-\infty}$ after a single dose.

Oral clearance (*Cl$_o$* or *Cl/F*) is calculated from the dose and the area under the zero moment curve

$$Cl_0 = \frac{Cl}{F} = \frac{D_{po}}{AUC_0^{\tau}}$$
(3:339)

In the case of incomplete sampling within a dosing interval and if one wants to extrapolate to τ or infinity, we propose the following technique

$$C_{24} = C_{last} \cdot e^{-\lambda_z \cdot (24 - t_{last})}$$
(3:340)

which is the predicted concentration at 24 h and t_{last}. The area from 0 to t_{last} is estimated by means of the linear trapezoidal method.

$$AUC_0^{t_{last}} = \sum_{i=1}^{n} \frac{C_i + C_{i+1}}{2} \cdot \Delta t$$
(3:341)

The area from t_{last} to 24 h is the difference between the area from t_{last} to infinity and the area from 24 h to infinity according to the following expression

$$AUC_{t_{last}}^{24} = AUC_{t_{last}}^{\infty} - AUC_{24}^{\infty}$$ (3:342)

$$AUC_{t_{last}}^{24} = \frac{C_{last}}{\lambda_z} - \frac{C_{last}}{\lambda_z} \cdot e^{-\lambda_z \cdot (24 - t_{last})} = \frac{C_{last}}{\lambda_z} \left[1 - e^{-\lambda_z \cdot (24 - t_{last})} \right]$$ (3:343)

which is the shaded area in Figure 3.93.

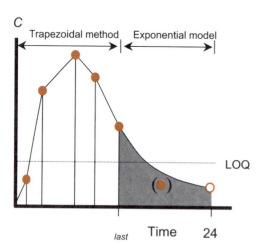

Figure 3.93 Schematic illustration of the predicted area between t_{last} and τ. In this case τ is 24 h. LOQ represents the limit of quantification. The observation within parentheses (●) is below LOQ and is therefore not used in the calculations. We generally treat values below LOQ as missing values. However, approaches that take values below LOQ into account have been proposed. The open circle (○) at 24 h denotes the predicted plasma concentration.

If dosing occurs with unequal dosing intervals at steady state (Figure 3.94), clearance can still be calculated even if plasma concentrations are only obtained during the first of the two (unequal) dosing intervals at steady state.

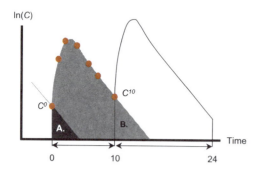

Figure 3.94 A dosing schedule with repeated but unequal dosing intervals of 10 and 14 hours, respectively. C^0 corresponds to the trough value prior to the morning dose and C^{10} is the trough value prior to the night dose. A corresponds to C^0/K and B to C^{10}/K. Ln(C) denotes the natural logarithm (log_e) of the plasma concentration.

Assume that the oral morning dose is given at 9:00 am (time 0) and the evening dose at 7:00 pm (time 10 h). The dosing intervals are then 10 and 14 h. A number of observations are obtained during the 0-10 h interval. In order to calculate oral clearance (Cl_o), the following relationship might be useful.

$$AUC_0^\infty = AUC_0^{10} \text{ (trapezoidal method)} - \frac{C^0}{K} + \frac{C^{10}}{K} \qquad (3:344)$$

and oral clearance

$$Cl_0 = \frac{Cl}{F} = \frac{D_{po}}{AUC_0^\tau} \qquad (3:345)$$

where $AUC_{0-\infty}$, AUC_{0-10} (trapezoidal method), C^0, and C^{10} are the area from zero to infinity related to the 9:00 am dose, the area between 0-10 h calculated by means of the trapezoidal method, the concentration at 9:00 am, and the concentration at 7:00 pm, respectively. If K (or λ_z) is not well-established, regression modeling might be necessary in order to obtain an accurate estimate of Cl_0.

To estimate *MRT* from steady state data requires, however, that information about the terminal slope is known since the method involves an estimate of the extrapolated area

$$MRT = \frac{AUMC_0^\tau + \tau \cdot AUC_{ssN\tau}^\infty}{AUC_0^\tau} \qquad (3:346)$$

where $\tau AUC_{N\tau-\infty}$ is derived from $\tau [AUC_{0-\infty} - AUC_{0-\tau}]$. The volume of distribution at steady state is calculated as

$$V_{ss} = Dose \cdot \frac{AUMC_0^\tau + \tau \cdot AUC_{ssN\tau}^\infty}{\left[AUC_0^\tau\right]^2} \qquad (3:347)$$

where τ, $AUC_{0-\infty}$, $AUC_{\tau-\infty}$, $AUMC_{0-\tau}$ are the dosing interval, the (zero moment) area within a dosing interval at steady state, the (zero moment) area from the last dose to infinity, and the (first moment) area within a dosing interval, respectively (Cheng and Jusko [1991]).

3.8   Interspecies Scaling

3.8.1   What is exposure?

Section 3.8 covers how to estimate, assess and extrapolate *exposure* within and between different species. Our definition of *exposure* to a drug is not primarily the dose administered, but rather the systemic plasma concentration and in some special cases tissue concentrations. Examples of systemic exposure is the peak plasma concentration (C_{max}), the integral of plasma concentration over time (AUC) or even the duration of a plasma concentration above a defined concentration level (e.g., plasma concentrations greater than e.g., *10* μmol/L over a period of 12 h). Beside the use of exposure measures such as AUC and C_{max}, we advocate the application of pharmacokinetic *parameters* such as unbound clearance Cl_u, unbound concentration C_u and free fraction f_u. Other derived parameters such as bioavailability F and half-life $t_{1/2}$ may be of value for interpretation. There are also situations with low systemic (plasma) concentrations of parent compound but high levels of parent compound or metabolites in specific organs and tissues (e.g., brain or fetus). Therefore, the tissue-to-blood partition coefficient at steady state (K_{pi}) may even be considered. Figure 3.95 illustrates different *measures* of *exposure* that have been used recently.

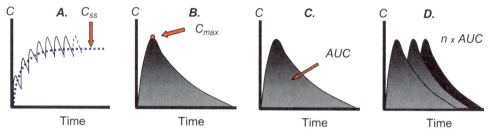

Figure 3.95 *Schematic illustration of different exposure measures. Plot A illustrates the steady state or average concentration, plot B the maximum plasma concentration (C_{max}), plot C the area under the curve (AUC) and plot D a multiple of AUCs, which have been used as a replacement of the dose.*

Small animals (e.g., rodents) generally metabolize and excrete chemicals more rapidly than humans. Such differences between man and small animals appear to be the rule rather than the exception. An important consequence of this (e.g., short half-life) is that a regimen of daily single administration of drugs produces potentially high peaks followed by a rapid decline to low or undetectable concentrations. One way to avoid this problem is to administer (input) the drug in a controlled manner (e.g., by infusion), another is to reduce the dose and dosing interval, keeping the daily dose the same. For drugs with short half-lives, the latter method would result in impractical and stressful handling of the animals. The *Alzet®* osmotic minipump is a device that will enable a constant more optimal exposure to the drug, avoiding drug holidays and missing sensitive time windows (Figure 3.96).

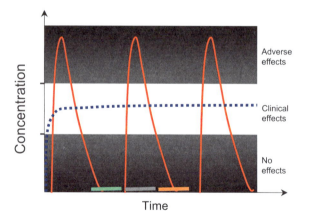

Figure 3.96 *Schematic comparison of traditional once-a-day (intermittent, red line) dosing and controlled input (blue dashed line) of a compound. The intermittent dosing will result in high peaks and low undetectable troughs with drug holidays. The green, grey and orange bars at the bottom of the figure illustrate sensitive time windows during which exposure has to be established. The intermittent dosing misses the green and orange time windows. The exposure during the grey time window is highly variable.*

Methadone, salicylic acid and valproic acid have been used as model substances to investigate whether embryonic effects on rat fetuses vary between controlled input and traditional once daily input. A marked increase in fetal adverse effects was observed at similar daily doses with a once-a-day regimen compared to the controlled input (Nau *et al* [1981], Gabrielsson *et al* [1985]). Figure 3.97 illustrates the shift in the adverse effect *versus* exposure (e.g., plasma concentration) relationship when changing the mode (intermittent *versus* controlled input) of administration.

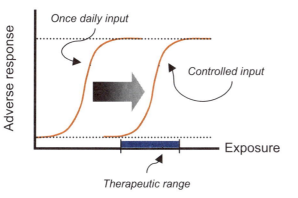

Figure 3.97 *Schematic illustration of the impact of mode of administration (e.g., controlled versus once-a-day regimens) on the adverse response outcome (e.g., fetal weight). Data adapted from Gabrielsson and Larsson [1985].*

The results from these and several other studies advocate the use of systemic exposure (controlled plasma concentrations) in combination with controlled input, when designing and assessing toxicological studies (see also *ICH Guidelines, ICH S3A*). Exposure to a chemical is generally related to *dose, gender, duration* and *time*. Figure 3.98 illustrates a commonly encountered situation where one has to tackle both capacity (dose)-limited elimination and time (induction)-dependent clearance.

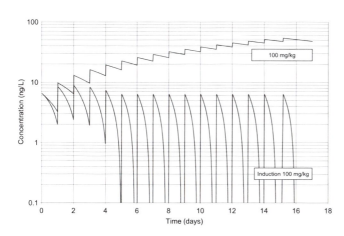

Figure 3.98 *Schematic illustration of the impact of simultaneous dose- (upper curve) and time- (induction, lower curve) dependent kinetics Accumulation is seen in the upper curve although simultaneous overlayed induction reduces the exposure substantially (lower curve). Data adapted from Gabrielsson and Larsson [1985].*

We propose that one discriminates between *extent* of exposure (C_{max} and AUC) and *duration* of exposure (treatment period, or plasma concentration-time course above a predefined concentration during Y h). The latter embodies physiological time that differs, of course, between different species. By so doing, one might be able to predict the duration of exposure in e.g., man, based on data obtained in laboratory animals.

A common misconception is that *toxicokinetics* is a special well-defined scientific area such as pharmacokinetics. *Pharmacon* comes from Greek for a drug or a toxin. Unfortunately, some people believe that *toxicokinetics* is the substitute for pharmacokinetics when we deal with high dose toxicity data. The primary objective of *toxicokinetics* is to describe the systemic exposure achieved in animals during a toxicity study, and the relationship of exposure to the *dose* level and *time* course of the study, and in some cases also *gender*. In other words, exposure could be used synonymously with *toxicokinetics*. Secondary objectives of *toxicokinetics* are (1) to relate the exposure achieved in toxicity studies to toxicological findings and contribute to the assessment of the relevance of these findings to clinical safety, (2) to support the choice of species and treatment regimen in non-clinical toxicity studies, and (3) to provide information which, in conjunction with the toxicity findings, contributes to the design of subsequent non-clinical toxicity studies.

3.8.2   What is allometry?

In the previous section, we discussed assessment of exposure with respect to *AUC* and C_{max}. These measures are obtained in the actual species of interest. In this section, we evaluate how to predict the exposure or pharmacokinetics of a drug in humans, from data obtained in another species e.g., the rat or monkey. Remember, that many drugs are well predicted in man from animal data, and that some are poorly predicted. However, what is interesting to know is why some compounds fail in the scaling exercise. What makes them different and how can we improve the predictability?

Numerous physical and physiological parameters (Y) vary according to some mathematical function of body weight (BW). For a given parameter, this function appears to be reasonably uniform across a broad range of animal species (for review see Boxenbaum [1982], McMahon and Bonner [1983], Peters [1983], Schmidt-Nielsen [1984, 1996]). This regularity results in the ability to predict with a high degree of accuracy the parameters in a given species, including humans, based on the values found for other mammalian species. The present evidence suggests that this approach is particularly useful for substances that are eliminated primarily by physical transport processes, e.g., biliary, renal, or pulmonary excretion, or when the disposition of the substance is largely independent of plasma protein binding. For qualitative differences between different species in metabolism and/or excretion of chemicals (e.g., different metabolic routes, heterogeneity in phase I and phase II drug metabolism reactions, differences in CYP450 isozyme makeup), allometric scaling may fail.

Energy consumption is of extreme importance in understanding interspecies variation in pharmacokinetics, because it influences in one or another way the amount of energy to be directed towards xenobiotic and drug elimination (Boxenbaum [1982], Boxenbaum and D'Souza [1990]). An increase in body mass is generally associated with a progressive reduction of the metabolic rate per unit mass, which then influences the turnover rate at the cellular level (Günther [1973]). In this regard, the way in which oxygen consumption varies among species depending on size is illustrated in Figure 3.99.

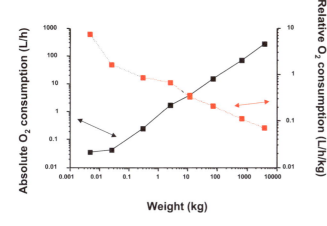

Weight (kg)

Figure 3.99 Observed rates of specific oxygen consumption of various mammals on the basis of body mass. The oxygen consumption per unit body mass (red symbols) increases rapidly with decreasing body size (Adapted from Schmidt-Nielsen [1984]). The absolute oxygen consumption is shown with black symbols). A log-log scale is used to increase readability of the large range of body weights and oxygen consumption rates.

The tremendous increase in relative oxygen consumption of the small animal necessitates that the oxygen supply, and hence blood flow, to 1 g tissue be about 100 times greater in the shrew, e.g., than in the elephant. Other physiological variables such as heart rate, respiration, food intake, turnover of drug etc., are similarly affected. Kleiber [1975] has shown that the metabolic rate of an organism is proportional to the 0.75 power of body weight

$$Metabolic\ rate = a \cdot BW^{0.75} \tag{3:348}$$

where a is a constant of appropriate units. The turnover time is proportional to $BW^{0.25}$ according to

$$Turnover\text{-}time = \frac{Energy\ content\ of\ organism}{Metabolic\ rate} = \frac{a_1 \cdot BW^{1.0}}{a_2 \cdot BW^{0.75}} \tag{3:349}$$

and hence

$$Turnover\text{-}time \propto a_3 \cdot BW^{0.25} \tag{3:350}$$

How can we then relate a discussion on pharmacokinetic parameters to size? Take a look at the following three relationships

$$surface \propto length^2 \qquad S \propto L^2$$

$$volume \propto length^3 \qquad V \propto L^3$$

$$surface \propto volume^{2/3} \qquad S \propto V^{2/3}$$

The last relationship states that as the volume of a body is increased, its surface area does not increase in the same proportion, but only in the proportion of the two-thirds power of the volume (Schmidt-Nielsen [1984]). We will now take a similar approach with regression models to relate kinetic parameters to a more easily measured variable, namely body weight.

When extrapolating within and across species, the use of regression analyses across body weight has been recommended as the most accurate methodology. This is because of difficulties inherent in measuring body surface parameters (as opposed to body weight) at the individual level. As a result, body surface parameters usually require the use of a reference criterion. Brody [1945] and Calabrese [1991] suggested several theoretical possibilities as to why surface area may have a poorer predictability across species than body weight.

The surface area of living organisms is difficult to determine and verify. For example, various species increase their coat, subcutaneous fat etc., at different times of year and in response to environmental factors, thereby changing their effective surface area to affect heat regulation. While surface area changes with the 2/3 power of body weight, this is only true for geometrically similar bodies of constant specific gravity. In fact, it is a gross oversimplification to accept that surface area determines the metabolic rate. Brody suggested that it is more rational to use BW^b instead of $BW^{2/3}$. The exponent b has revealed considerable variation in different species (0.5 - 0.8). The calculation of body weight is direct, simple and often performed with data from individual animals other than the standard reference mouse, rat, dog, monkey, and man (see Tables 3.14 and 3.15).

The body weight rule offers the investigator a very powerful tool with which to make reasonable interspecies predictions. There are compounds for which this general rule fails. In the real world, it is highly recommended that data be available from four species under similar, standardized conditions when performing regression analysis (Boxenbaum [1982], Schmidt-Nielsen [1984]).

Numerous biological vaiables have been found to be mathematical functions of body weight. If these parameters are unknown in one species, they can often be reasonably accurately predicted from known values in other species. This phenomenon is considered a *general biological regularity* rather than a *natural law* since there are some exceptions to the rule such as human lifespan, the relative weight of the brain and the amount of oxygen consumed by the brain (Boxenbaum [1982]).

Examples of the allometric relationship between four physiological variables and body weight are shown in Figure 3.98 and Equations 3:351 and 3:352 (Schmidt-Nielsen [1984]).

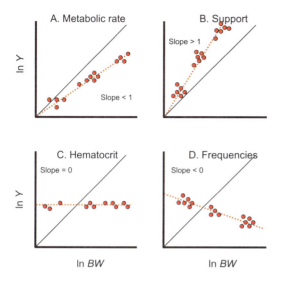

Figure 3.98 *Allometric plots with different exponents. Solid black lines illustrate theoretical relationships with an exponent of unity (adapted from Schmidt-Nielsen [1984]). Figure (A) shows how the metabolic rate increases with body weight ($b \approx 0.75$). Figure (B) illustrates how support (e.g., cross-sectional area of skeletal bones) of mammals increases out of proportion to body weight ($b \approx 1.08$). Figure (C) shows how some variables, like hematocrit, is independent of body weight ($b = 0$). Figure (D) shows how e.g., heart rate decreases with increase in body weight ($b \approx -0.25$). ln Y and ln BW denote the natural logarithm ($\log_e$) of the respective variable.*

$$Y = a \cdot BW^b \tag{3:351}$$

$$\ln(Y) = \ln(a) + b \cdot \ln(BW) \tag{3:352}$$

Body weight is denoted *BW* and the physiological or kinetic variable of interest e.g., blood flow, is denoted *Y*. The constants *a* and *b* give the intercept at a body weight of unity (1) and the slope in a log-log plot, respectively (e.g., see Figure 3.99).

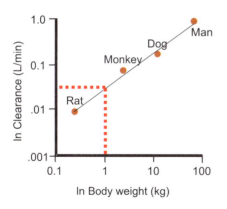

Figure 3.99 *A log-log plot of clearance versus body weight for rat, monkey, dog and man. The dotted line that originates at a body weight of one kg gives the intercept on the clearance axis that corresponds to the allometric constant 'a' in Equation 3:352, while the slope gives the constant 'b'. ln denotes the natural logarithm ($\log_e$).*

In Figure 3.99, allometric data of plasma clearance for drug *X* are plotted for rat, monkey, dog and man. It is apparent how well *Cl* of this particular drug in man can be predicted from rodents. Since numerous physical and physiological parameters have been shown to be a function of body weight in animal models, as well as in humans, it follows logically that pharmacokinetic parameters should also be contingent on body weight. Boxenbaum [1982] derived an allometric model taking total body weight into account for clearance

$$Cl_i = a \cdot BW_i^b \qquad (3:353)$$

Taking the natural logarithm *ln* of both sides gives

$$\ln(Cl_i) = \ln(a) + b \cdot \ln(BW_i) \qquad (3:354)$$

When plotted on a log-log scale, the relationship is a straight line with intercept *ln(a)* and slope *b*. The physical nature of parameters *a* and *b* is different. We commonly say that *a* is drug-dependent and *b* is not. The latter, is rather, dependent on the type of kinetic or physiological variable being analyzed e.g., *Cl*, *V* or $t_{1/2}$.

The impact of two different exponents on the relationship between *Y* and body weight in Equations 3:353 and 3:354 is demonstrated graphically in Figure 3.100.

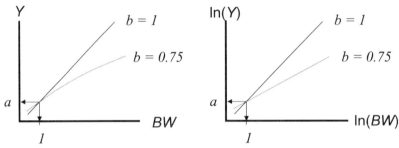

Figure 3.100 *The left hand figure demonstrates on a linear scale the relationship between a physiological variable Y and body weight for two different exponents acting on body weight. When b is equal to 1 (unity), a direct proportionality is present between Y and body weight. When b is equal to 0.75, a curvilinear relationship is present. Both relationships have the same value when body weight is equal to 1 (unity). The right hand side figure shows the same relationships plotted on a log-log scale, where ln denotes the natural logarithm (log_e).*

Schmidt-Nielsen [1984] wrote that it is unrealistic to expect that a regression line (or an equation) will tell precisely what the specific metabolic rate (or other variable in question) will be for an animal of a given body mass. Rather, what the equation tells is the expected mean value for a *typical* mammal of the given size. Real animals will always deviate more or less from this idealized norm as illustrated in Figure 3.101. Also, a good *correlation* does not imply a good *prediction* (see also Bonate and Howard [2000]).

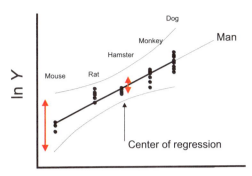

Figure 3.101 Plasma clearance(Y) versus body weight. Note how well clearance can be predicted from rat to man. However, be cautious in the use of allometric plots and techniques - as shown, the regression line gives a mean prediction of the values, around which there may be considerable variation as indicated by the dashed curves (95% confidence interval). See also exercise PK 28 on methadone.

Figure 3.102 shows how the exponent might differ depending on whether data are obtained within a species or among several species. The 0.67 exponent may well be a correct descriptor of how the metabolic rate varies within a species, although among several species, an exponent of 0.75 (3/4) emerges to be a more precise descriptor.

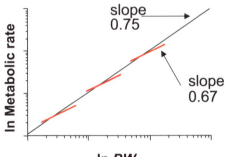

Figure 3.102 Allometric plot showing how the exponent may vary within and between species. Plot adapted from Schmidt-Nielsen [1984].

An allometric model taking total body weight into account has also been derived for volume of distribution. The exponent d of the volume term is often close to unity

$$V_i = c \cdot BW_i^d \tag{3:355}$$

or

$$\ln(V_i) = \ln(c) + d \cdot \ln(BW_i) \tag{3:356}$$

When plotted on a log-log scale, this is a straight line with intercept *ln(c)* and slope *d*. Substitution of the volume and clearance in a mono-exponential model by Equations 3:355 and 3:356 gives

$$C = \frac{D_{iv}}{V_i} \cdot e^{-\frac{Cl_i}{V_i} \cdot t} = \frac{D_{iv}}{c \cdot BW_i^d} \cdot e^{-\frac{a \cdot BW_i^{b-d}}{c} \cdot t} \tag{3:357}$$

Integrating Equation 3: 357 yields

$$AUC_0^\infty = \frac{D_{iv}}{V_i} \cdot \int_0^\infty e^{-\frac{Cl_i}{V_i} \cdot t} dt = \frac{D_{iv}}{c \cdot BW_i^d} \cdot \int_0^\infty e^{-\frac{a \cdot BW_i^{b-d}}{c} \cdot t} dt \qquad (3{:}358)$$

which after evaluation reduces to

$$AUC_0^\infty = \frac{1}{a} \qquad (3{:}359)$$

A plot of $C/(D_{iv}/BW^b)$ versus $t/(BW^b)$ for each species yields superimposable curves with $AUC_{0-\infty}$ equal to $1/a$. This is reasonable, since the exposure *(AUC)* is drug-dependent *(a)* and not dependent on the scaling factor *(b)*, which is generally drug-independent. The *b* parameter is, rather, energy-dependent.

Several authors (Boxenbaum [1982], McMahon and Bonner [1983], Schmidt-Nielsen [1984], Boxenbaum and DiLea [1995]) propose that the parameter of primary interest is the exponent parameter, *b* (or *d*), as it characterizes rate of increase of the dependent parameter as a function of the independent variable, *BW*

$$\% Y = 100 \cdot (2^b - 1) \qquad (3{:}360)$$

where *% Y* is the percent change in the dependent variable when the independent variable doubles, and *b* is the allometric exponent.

Boxenbaum and DiLea [1995] explained that when the exponent is less than 1, a 100% (two-fold) increase in body mass produces less than a 100% increase in the dependent variable. Clearances frequently fall into this category. Consequently, with larger species, mg/kg doses generally get smaller. When the exponent is 1, a constant ratio exists between variables. Blood volumes in terrestrial mammals scale with an exponent of 0.99 and tend to be about 8.5% of body mass, regardless of size. Skeletal mass in terrestrial mammals scales with an exponent of 1.09. Thus, larger mammals have proportionately more of their body mass dedicated to bone than smaller ones. For example, a shrew's skeleton is about 5% of body mass, whereas an elephant's is 27%. Boxenbaum [1996] summarized the impact of doubling the independent variable, body weight, on dependent variables with different exponents (Table 3.11).

Table 3.11 Impact of a doubling of body weight on physiological variables with different exponents. Adapted from Boxenbaum [1996]

Exponent	Factor for change in dependent variable as BW doubles	Examples of dependent variables
0.0	0.0	Hematocrit
0.25[a]	1.19	Lifespan, disposition half-life, breath time, heart beat time
0.67	1.59	Body surface area
0.75	1.68	Metabolic rate, clearance
1.0	2.0	Blood volume, volume of distribution
1.08	2.11	Skeletal mass

a   corresponds to an exponent of 0.28 in Equation 3:361-363.

3.8.3   Time scales

Physiological time is defined as a species-dependent unit of chronological time required for the completion of a species-independent physiological event. For example, the life expectancy is 14 years in the dog and 98 years in man. This means that the dog consumes 7.14% of its life every year, whereas a human consumes 7.14% of his life every 7 years. Thus, one year for the dog and 7 years for man correspond to 7.14% of the lifetime of these species (Boxenbaum [1982]). Boxenbaum concluded that one year in the dog and seven years in the man are equivalent physiological times necessary to produce a species-independent physiological event, i.e., living 7.14% of a lifetime.

Other good examples of physiological times are breath time and heart beat time. These differ considerably between mammalian species as follows

$$Breath\ time = 0.169 \cdot BW^{0.28} \tag{3:361}$$

$$Heart\ beat\ time = 0.0428 \cdot BW^{0.28} \tag{3:362}$$

$$\frac{Breath\ time}{Heart\ beat\ time} = \frac{0.169 \cdot BW^{0.28}}{0.0428 \cdot BW^{0.28}} \approx 4 \tag{3:363}$$

Dividing breath time by heart beat time gives a value of 4.0. In other words, all mammals have 4 heart beats for each respiratory cycle. Lifetime (or lifespan) scales with a factor of 0.28. This indicates that most mammals have nearly the same number of breaths per lifetime and the same number of heart beats per lifetime. This relates also to pharmacokinetics. Mammals tend to clear drugs at a similar pace, measured by their own internal clocks (Boxenbaum [1982]). The relationship between turnover time (mean residence time) and terminal half-life of a drug is

$$t_t = \frac{1}{\ln(2)} \cdot t_{1/2z} \cdot \frac{V_{ss}}{V_z} \tag{3:364}$$

where V_{ss}/V_z is supposed to be a constant ratio, close to unity. For different species, this is therefore

$$t_{1/2} = \ln(2) \cdot \frac{V}{Cl} = \ln(2) \cdot \frac{c \cdot BW_i^{d}}{a \cdot BW_i^{b}} = \text{constant} \cdot BW^{0.25} \qquad (3:365)$$

where BW^{d-b} is approximately $BW^{0.25}$ since d (exponent for volume) is close to 1 and b (exponent for clearance) is approximately 0.75.

Consider a drug given as an intravenous bolus dose to the rat, monkey, dog, and man. The time from each species may be converted into human time *(t_{human})* using the following equations

Equivalent biological time (i.e., plasma half-life)

$$t_{human} = t_{human} \cdot \left[\frac{BW_{human}}{BW_{animal}} \right]^{0.25} \qquad (3:366)$$

Kallynochron which is defined as a unit of pharmacokinetic time during which the different species clear the same volume of plasma per kilogram of body weight, is

$$t_{human} = t_{human} \cdot \left[\frac{BW_{human}}{BW_{animal}} \right]^{1-b} \qquad (3:367)$$

Apolysichron is defined as the unit of pharmacokinetic time during which animal species eliminate the same fraction of drug and clear the same volume of plasma per BW^d of body weight. It is defined as

$$t_{human} = t_{human} \cdot \left[\frac{BW_{human}}{BW_{animal}} \right]^{d-b} \qquad (3:368)$$

In chronological time, the *apolysichron* is equivalent to BW^{d-b} units.

Equivalent *biological time, kallynochron* and *apolysichron* are equal when b equals 0.75 and d equals unity (1). The rates of metabolism of many compounds in man are slower than in other species. This slower metabolism, can be one of several reasons why humans live longer than other animals. This longevity depends on various factors including brain weight. The maximum lifespan potential (*MLP*) is the maximum documented longevity for a species. Sacher [1959] fitted the following equation to data of 63 mammalian species

$$MLP = 10.839 \cdot W^{0.636} \cdot BW^{-0.225} \qquad (3:369)$$

BW and *W* are the body weight and brain weight, respectively. Using the equation above, the ratio of *MLP_{man}* to *MLP_{dog}* was a factor of 4.7 (Table 3.12, Boxenbaum [1982]). Comparing *MLP_{man}* to *MLP_{rat}* the factor is 20. Taking only *MLP* into account, a concentration of unity in the rat during 1 h corresponds to a 20 h period in man at the same concentration level, even though the absolute *AUC* will be 20 times greater in man.

Table 3.12 Maximum lifetime potential of different species

Species	Body Weight (BW)(g)	Brain Weight (W)(g)	% W/BW	MLP (years)
Mouse	23	0.334	1.45	2.7
Rat	250	1.88	0.75	4.7
Dog	14200	75.4	0.53	19.7
Monkey	4700	62	1.32	22.3
Man	70000	1530	2.19	93.4

3.8.4 Estimation of parameters

It has been shown that each species has a different influence on the predicted kinetic parameters in humans. A species close to the center of the regression will have less influence on the predicted curve because its data point(s) lie close to the center of the regression (Gabrielsson and Weiner [1994] and Lave *et al* [1995]), as shown in Figure 3.103 below.

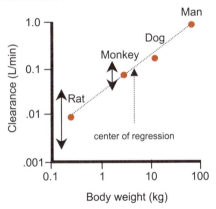

Figure 3.103 Log-log plot of plasma clearance versus body weight. Note that rat data will have a higher leverage than e.g., data from the monkey. The rat will therefore have a greater impact upon the regression parameters (Gabrielsson and Weiner [1994]).

These problems can be largely overcome by using pharmacokinetic time units, where each animal species potentially has a similar impact on the prediction for man (Dedrick [1974], Gabrielsson and Weiner [1994], Lave *et al* [1995]). To do this, the influence of each animal species is related to the number of samples collected to describe the kinetics, rather than its body weight. As protein binding differences are known to affect interspecies comparisons and their predictability for man, a better understanding of the protein binding and intrinsic clearance processes across species could help to elucidate the observed differences between kinetics in man and other species.

An additional explanation for the apparent differences between kinetic parameters in animals and man might be that comparisons are based on different sources of information in which different doses, sampling schedules, analytical methods, and routes of administration have been used.

3.8.5   The elementary Dedrick plot

Plasma concentration-time data of methadone were obtained from mouse, rat and man after receiving an intravenous bolus of 25, 500, and 100,000 μg, respectively (Figure 3.104). The corresponding body weights were 23 g (mouse), 250 g (rat) and 70 kg (man), and cover a 3000-fold weight range (Gabrielsson and Weiner [1994]). We assume instantaneous distribution upon administration. A plot of dose and body weight normalized (scaled) plasma concentrations *(C/D/BW)* versus body weight normalized times *(t/BW^{1-b}*, which is time normalized or scaled by clearance) shows a straight line and allows the superposition of the curves (Figure 3.104). This is called *the elementary Dedrick plot*. The *X* axis represents the *kallynochron,* which is defined as a unit of pharmacokinetic time during which the different species clear the same volume of plasma per kilogram of body weight.

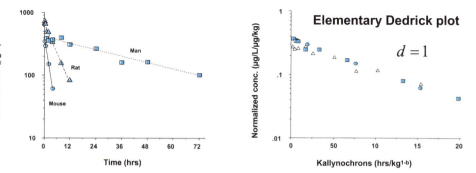

Figure 3.104 Observed (symbols) and predicted (lines) plasma concentrations in mouse, rat, and man following an intravenous bolus dose of methadone (left figure). The right figure shows concentrations normalized for dose and body weight, versus time normalized for clearance.

We will model the disposition of methadone in the sited example by means of the following allometric expressions

$$Cl_i = a \cdot BW_i^{b} \tag{3:370}$$

$$V_i = c \cdot BW_i^{d} \tag{3:371}$$

where *i* represents mouse, rat and man, respectively. Equations 3:264 and 3:265 are then fed into the disposition function for each species.

$$\frac{dC}{dt} = -K \cdot C = -\frac{Cl}{V} \cdot C \tag{3:372}$$

It will take the following form for the mouse

$$\frac{dC_{mouse}}{dt} = -\frac{a \cdot 23^{b}}{c \cdot 23^{d}} \cdot C \tag{3:373}$$

for the rat

$$\frac{dC_{rat}}{dt} = -\frac{a \cdot 250^b}{c \cdot 250^d} \cdot C \tag{3:374}$$

and for man

$$\frac{dC_{man}}{dt} = -\frac{a \cdot 70000^b}{c \cdot 70000^d} \cdot C \tag{3:375}$$

As can be seen, Cl and V are not estimated *per se* but, rather, the allometric parameters a, b, c, and d are estimated. Of course, the kinetics for each species might also be specified with a mono-exponential model in this case, such as

$$C = \frac{D_{iv}}{V_i} \cdot e^{-\frac{Cl_i}{V_i} \cdot t} = \frac{D_{iv}}{c \cdot BW_i^d} \cdot e^{-\frac{a \cdot BW_i^{b-d}}{c} \cdot t} \tag{3:376}$$

where Cl and V are specified as above. Cl_i and AUC_i are estimated as functions of the parameters for each species, together with the corresponding half-lives (Table 3.13).

Table 3.13   Comparison of parameters.

Species	Cl	V	AUC	$t_{1/2}$
Mouse	$a \cdot 23^b$	$c \cdot 23^d$	$\dfrac{Dose}{a \cdot 23^b}$	$\dfrac{\ln(2) \cdot c \cdot 23^d}{a \cdot 23^b}$
Rat	$a \cdot 250^b$	$c \cdot 250^d$	$\dfrac{Dose}{a \cdot 250^b}$	$\dfrac{\ln(2) \cdot c \cdot 250^d}{a \cdot 250^b}$
Man	$a \cdot 70000^b$	$c \cdot 70000^d$	$\dfrac{Dose}{a \cdot 70000^b}$	$\dfrac{\ln(2) \cdot c \cdot 70000^d}{a \cdot 70000^b}$

By including Cl and V as functions of body weight into Equation 3:376 we obtain

$$C = \frac{D_{iv}}{V_i} \cdot e^{-\frac{Cl_i}{V_i} \cdot t} = \frac{D_{iv}}{c \cdot BW_i^d} \cdot e^{-\frac{a \cdot BW_i^{b-d}}{c} \cdot t} \tag{3:377}$$

This demonstrates the results from a simultaneous fit of an allometric model (Equation 3:377) to concentration-time data obtained from three different species. The terminal half-lives were found to be 1.5, 3.9 and 35 hours for mouse, rat, and man, respectively. Based on the plasma-to-blood concentration ratio and the hepatic blood flows for each animal this drug is classified as a low extraction drug with a hepatic extraction of less than 10%.

The area under the curve, *AUC*, is equal to *1/a*, where *a* is the allometric coefficient of clearance. Since $a = 0.319$, *AUC* becomes 3.13 which is consistent with 3.04 calculated from the intercept of the ordinate (≈ 0.35) divided by the slope of the curve (≈ 0.119).

One has to be cautious when interpreting this type of data unless data are obtained from two or more dose levels for each species, multiple dose (or steady state) administration measures are done, or model misspecification is ruled out.

3.8.6 The complex Dedrick plot

We assume in the elementary Dedrick plot that *V* is proportional to body weight (i.e., *d*=1). When the volume of distribution is not proportional to body weight (*d* >1), the curves can no longer be superimposed on an elementary Dedrick plot. In the case of multi-exponential disposition functions, superimposable curves will only occur if V_c, V_z, and V_{ss} individually represent the same fraction of body weight between species (Boxenbaum [1982]). On the *Y* axis, plasma concentration is divided by the ratio *Dose/BW^d* and on the *X* axis, time is divided by BW^{d-b}.

The generated curve of *C/D/BW^d* versus t/BW^{d-b} is called the complex Dedrick plot and enables multicompartment kinetics to be superimposed. The *X* axis represents the *apolysichron*, equal to

$$time\ scale = \frac{t}{BW^{d-b}} \qquad (3:378)$$

It is defined as the unit of pharmacokinetic time during which animal species eliminate the same fraction of drug and clear the same volume of plasma per *BW^d* of body weight. In chronological time, the *apolysichron* is equivalent to BW^{d-b} units.

The area under the curve in the complex Dedrick plot, *AUC*, is still proportional to *1/a*, where *a* is the allometric coefficient of clearance. Plasma concentration-time data were simulated for the mouse, rat, monkey, dog and man who received an intravenous bolus of 10, 125, 200, 6000, and 12000 µg, respectively of compound X (Figure 3.106).

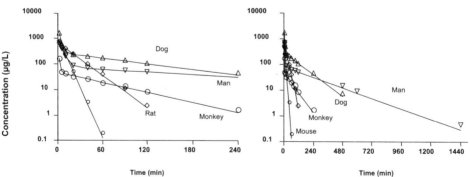

Figure 3.106 *Observed and predicted data for each individual species plotted versus time. Left hand figure shows time interval 0-240 min and right hand figure 0-1500 min.*

The corresponding body weights were 20 g (mouse), 250 g (rat), 3.5 kg (monkey), 14 (dog), and 70 kg (man) which cover a 3500-fold weight range. We will model the disposition of X by means of the following allometric expressions

$$Cl_i = a \cdot BW_i^b \tag{3:379}$$

$$V_i = c \cdot BW_i^d \tag{3:380}$$

$$Cl_{di} = g \cdot BW_i^b \tag{3:381}$$

$$V_{ti} = e \cdot BW_i^d \tag{3:382}$$

where i represent mouse, rat, monkey, dog, and man, respectively. Note that the inter-compartmental distribution parameter Cl_d, is scaled with parameters g and b, and the extravascular tissue volume is scaled with parameters e and d. The reason is that the inter-compartmental distribution parameter is similar to clearance and blood flows with respect to the exponent b, whereas the coefficient g is assumed to be different for different parameters. The same reasoning is assumed to be valid for the extravascular tissue volume, which scales like a volume term with exponent d similar to the central volume.

Since there are five species with two functions each, the total system contains ten differential equations (two per species) or five two-exponential models (one per species). Drug concentration in the central compartment is given by

$$V_c \frac{dC}{dt} = In - Cl \cdot C - Cl_d \cdot C + Cl_d \cdot C_t \tag{3:383}$$

where In is the bolus input function. The kinetics of drug in the peripheral compartment

$$V_t \frac{dC_t}{dt} = Cl_d \cdot C - Cl_d \cdot C_t \tag{3:384}$$

Kinetics in plasma for intravenous *bolus input* is defined as a sum of exponentials

$$C = \frac{D_{iv}}{V_c} \cdot \left\{ \frac{(k_{21} - \alpha) \cdot e^{-\alpha \cdot t}}{\beta - \alpha} + \frac{(k_{21} - \beta) \cdot e^{-\beta \cdot t}}{\alpha - \beta} \right\} \tag{3:385}$$

The intercompartmental diffusion parameter, which is equal in both directions and the volume of the central compartment, can be used for calculations of the micro-constants

$$k_{12} = \frac{Cl_d}{V_c} \tag{3:386}$$

$$k_{21} = \frac{Cl_d}{V_t} \qquad\qquad\qquad (3{:}387)$$

where k_{12} and k_{21} are the (fractional) intercompartmental rate constants which determine the rate of drug movement from the central to the peripheral and the peripheral to the central compartment.

$$k_{10} = \frac{Cl}{V_c} \qquad\qquad\qquad (3{:}388)$$

$$\alpha = \frac{k_{21} \cdot k_{10}}{\beta} \qquad\qquad\qquad (3{:}389)$$

$$\beta = \frac{1}{2}\left[k_{12} + k_{21} + k_{10} - \sqrt{(k_{12} + k_{21} + k_{10})^2 - 4k_{21} \cdot k_{10}} \right] \qquad (3{:}390)$$

The final parameter estimates of the allometric parameters were obtained by fitting a bi-exponential expression to the concentration-time data of each species. The macro- and micro-constants in Equations 3:388 to 3:390 were defined in terms of Cl, V_c, Cl_d and V_t. The latter four parameters were then expressed as allometric functions for each species, taking body weight into account. The resulting parameter values a-g were estimated to be 0.021, 0.74, 0.076, 1.18, 0.56 and 0.075. Exponent b was 0.74 and d was 1.18. A complex Dedrick plot including observed data of all species is shown below (Figure 3.107). One can observe a distinct bi-exponential behavior with perfectly superimposed data.

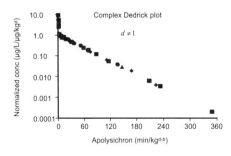

Figure 3.107 Complex Dedrick plot of dose and body weight normalized concentration-time data plotted for each individual species. Since data were simulated with a very low added error, they superimpose.

3.8.7   Integration of concepts

By means of information about clearance and dose in one species, one may be able to predict the appropriate exposure or dose level in another animal or in man. The body weight rule offers the investigator a very powerful tool with which to obtain general guidelines for making reasonable interspecies predictions. Furthermore, pre-clinical evaluation and analysis may reduce the frequency and severity of unexpected toxic events during single-dose phase I testing, because such data enable predictions to be made about dose and exposure. For example, if we want to establish similar exposure AUC in the laboratory animal and man

$$AUC_{rat} = AUC_{man} = \frac{Dose_{man}}{Cl_{man}} = \frac{Dose_{rat}}{Cl_{rat}}$$ (3:391)

$$\frac{Dose_{man}}{a \cdot BW_{man}^{b}} = \frac{Dose_{rat}}{a \cdot BW_{rat}^{b}}$$ (3:392)

$$Cl_{i} = a \cdot BW_{i}^{b}$$ (3:393)

$$Cl_{man} = Cl_{rat} \cdot \left[\frac{BW_{man}}{BW_{rat}}\right]^{b}$$ (3:394)

When Equation 3:392 to 3:394 are combined, we obtain

$$Dose_{man} = Dose_{dog} \cdot \left[\frac{BW_{man}}{BW_{dog}}\right]^{b}$$ (3:395)

under the assumption that a representative animal species is used and scaled appropriately with respect to elimination and sensitivity of drug. Boxenbaum and DiLea [1996] provide an excellent review of these concepts.

We firmly believe that the allometric approach is far too often underestimated in its potential usefulness when predicting e.g., the time course of drug, enzyme or protein in man, or when estimating of an appropriate dose and duration of exposure in man based on data from animals. The scaling does not only relate to first-order processes but also to zero-order input/output processes or other turnover processes such as the lifespan of the red blood cell as depicted in Figure 3.108 below.

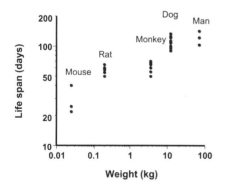

Figure 3.108 Lifespan of the red blood cell versus body weight for mouse, rat, monkey, dog and man.

Both erythrocytes and hemoglobin have a mean *in vivo* lifespan of 120 days in man. The term half-life is not applicable here because it relates to a first-order process, whereas the lifespan of the red blood cell does not follow a first-order process.

Tables 3.14 and 3.15 contain a compilation of organ weights and perfusion rates for a number of commonly used laboratory animals such as the mouse, gerbil, rat, marmoset, guinea pig, rabbit, rhesus monkey, dog, sheep and man.

Table 3.14 Organ weights (g) and CSF volume (mL) in mouse, gerbil, hamster, rat, marmoset, guinea pig, rabbit, rhesus monkey, beagle dog, sheep and man

	Mouse (0.02kg)	Gerbil (0.067kg)	Hamster (0.15kg)	Rat (0.25kg)	Marmoset (0.3kg)	Guinea pig (0.85kg)	Rabbit (2.5kg)	Rhesus monkey (5kg)	Dog (10kg)	Sheep (50kg)	Human (70kg)
Brain	0.36	0.98	1.4	1.8	2.2	4.4	14	90	80	400	1500
Liver	1.75	2.1	6.9	10.0	12.0	38	77	150	320	1600	1500
Kidneys	0.32	0.55	2.0	2.0	2.4	8.8	13	25	50	250	310
Heart	0.08	0.25	0.6	1.0	2.4	3.6	5	18.5	80	400	330
Spleen	0.1	0.049	0.54	0.75	0.5	2.1	1	8	25	125	180
Adrenals	0.004	0.038	-	0.05	0.096	0.90	0.5	12	1	10	14
Lung	0.12	0.32	0.74	1.5	1.8	10.1	18	33	100	1000	1000
Muscle	10.	27.	68.7	120	144	320*	1350	2500	4600	27700	35000
GI tract	1.5	2.8	19.2	6.3	7.6	43	120	230	190	3950	2100
Fat	-	4.6	17.	17.	20	-	-	-	-	-	10000
Marrow	0.6	-	-	-	-	-	40	135	110	-	1400
Skin	-	-	-	-	-	167	-	-	-	-	-
Bone	-	-	-	-	-	-	-	-	1180	6930	10000
CSF mL	0.035	0.04	0.15	0.15	0.18	-	2.3	-	-	-	100

Table 3.15 Organ blood flows, urine flow, bile flow, CSF flow and GFR in mouse, gerbil, hamster, rat, marmoset, guinea pig, rabbit, rhesus monkey, beagle dog, sheep and man

Blood flows (mL/min)	Mouse (0.02kg)	Gerbil (0.067kg)	Hamster (0.15kg)	Rat (0.25kg)	Marmoset (0.3kg)	Guinea pig (0.85kg)	Rabbit (2.5kg)	Rhesus monkey (5kg)	Dog (10kg)	Sheep (50kg)	Human (70kg)
Brain	0.46	1.13	1.45	1.3	1.5	4.9	-	72	45	150	700
Liver	1.8	5.6	6.5	13.8	15.8	46	177	218	309	1033	1450
Kidneys	1.3	2.82	6.2	9.2	10.5	36	80	138	216	722	1240
Heart	0.28	1.12	1.4	3.9	4.47	9.9	16	60	54	181	240
Spleen	0.09	0.07	0.25	0.63	0.72	4.6	9	21	25	84	77
Gut	1.5	3.58	5.3	7.5	8.6	39	111	125	216	722	1100
Muscle	0.91	0.1	4.5	7.5	8.6	-	155	90	250	836	750
Marrow	0.37	-	-	-	-	-	24	51	37	-	270
Adipose	-	-	0.77	0.4	0.46	-	32	20	35	117	260
Skin	0.41	-	-	5.8	6.6	29	-	54	100	334	300
Hepatic artery	0.35	0.30	1.2	4.0	4.6	7.4	37	51	79	264	300
Portal vein	1.45	5.3	5.3	9.8	11.2	-	140	167	230	769	1150
Cardiac output	8.0	-	40	74.0	85	248	530	1086	1200	4010	5600
Urine flow (mL/day)	1.0	-	-	50.0	57	<100	150	375	300	1000	1400
Bile flow (mL/day)	2.0	-	-	22.5	25.8	-	300	125	120	400	350
CSF flow (mL/h)	0.018	-	-	0.18	0.21	-	0.6	2.5	-	-	21
GFR (mL/min)	0.28	-	-	1.31	1.5	-	7.8	10.4	61.3	205	125

Peeters *et al* [1980], Matsumoto M *et al*. [1982], Gerlowski and Jain [1983], Welling and Tse [1988], Davies and Morris [1993], Davson *et al* [1987]. Marmoset data were scaled from rat.

3.8.8 Allometric scaling of turnover parameters

Most rate-processes like blood flows, clearances, secretion-rates, and synthesis rates require energy and scale generally proportionally to body weight raised to the power b of 0.6-0.8. Turnover rate R_{in}, becomes

$$R_{in} = a \cdot BW^b \qquad (3{:}396)$$

The amount of a substance in the organism at steady state A_{ss} scales proportionally with body weight according to

$$A_{ss} = c \cdot BW^{\sim 1} \qquad (3{:}397)$$

The fractional turnover rate k_{out} scales like any first-order rate constant or frequency proportionally to body weight raised to the power of ~ -0.25

$$k_{out} = d \cdot BW^{-0.25} \qquad (3{:}398)$$

The turnover time is derived as the ratio of A_{ss} to R_{in} and scales proportionally with body weight according to

$$Turnover\text{-}time = \frac{A_{ss}}{R_{in}} = \frac{c \cdot BW^1}{a \cdot BW^{0.75}} = e \cdot BW^{0.25} \qquad (3{:}399)$$

and rearranged for half-life gives

$$t_{1/2kout} = \frac{\ln(2)}{k_{out}} = \ln(2) \cdot e \cdot BW^{0.25} \qquad (3{:}400)$$

A general expression for half-life of response is then written as

$$t_{1/2response} = \text{constant} \cdot BW^{0.25} \qquad (3{:}401)$$

The simulated time course of response for rat and man is illustrated in Figure 3.109. We assume that potency and intrinsic activity are the same for the drug in the two species and the only difference is due to $t_{1/2response}$ ($t_{1/2kout}$).

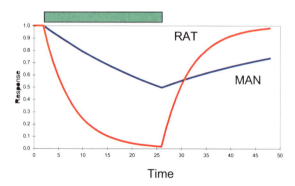

Figure 3.109 *Response-time*
course in rat (red line) and man
(blue line) follwing the same
period of exposure (horizontal
green bar. Time is arbitrary
units.

If the half-life of response $ln(2)/k_{out}$ is proportional to the turnover time of biochemicals causing the effect, then the principles of allometric scaling are applicable to the pharmacodynamic response.

3.8.9   General conclusions of exposure and scaling

Fundamental is that toxicological responses should be correlated to drug exposure in target organs or a biomarker for exposure (e.g., plasma concentrations). These conditions can be achieved if controlled drug input (e.g., constant intravenous or intraduodenal infusion) and drug monitoring are applied. Advantage should also be taken of pharmacokinetic data in the experimental design of e.g., toxicokinetic (exposure) studies, and prediction of drug (or metabolite) disposition in tissues may be achieved by means of *physiologically based pharmacokinetic PBPK* models (Gabrielsson and Larsson [1985]). The advantage of *PBPK* models is that the complete concentration-time course of parent compound and/or metabolite(s) can be visualized in organs or tissues such as the brain or fetus.

CHAPTER 4 - Pharmacodynamic concepts

Objectives
◆ **To give a general introduction to pharmacodynamic concepts**
◆ **To add insight into experimental design of pharmacodynamic studies**
◆ **To consider a wide range of pharmacodynamic systems**

4.1   Background and Definitions

Pharmacodynamics can be defined as the study of the biological effects of drugs and their mechanisms of action. The objectives of the analysis of drug action are to delineate the chemical and/or physical interactions between a drug and the target cell and to characterize the full sequence and scope of actions of each drug (Ross [1996], Goodman and Gilman [1996]). This chapter provides an introduction to the concept of receptors, drug-receptor binding and pharmacodynamic models. Models describing the relationship between the concentration of drug in blood or plasma and drug receptors provide clinically useful information regarding baseline and maximal drug effects (sensitivity aspects) and the change in the observed effect with time (temporal aspects) (Figure 4.1).

Pharmacodynamics

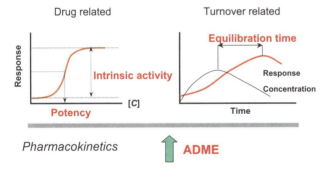

Figure 4.1 The pharmacodynamic characteristics of a drug: intrinsic activity and potency (sensitivity aspects) and time to equilibrium (temporal aspects). The drug kinetics (absorption, distribution, metabolism and excretion - ADME) is considered to 'drive' the dynamics.

We begin our discussion of pharmacodynamic concepts by including a small number of definitions which are crucial to the concepts considered in the chapter as a whole. This is followed by an overview of the law of mass-action (section 4.2) and of receptor binding models for saturation and displacement studies (section 4.3). After this, we move on to the principles of pharmacodynamic modeling (sections 4.4-4.12).

Although we base our reasoning on the law of mass-action and on receptor occupancy theory, we also mention other receptor interaction models (e.g., the model by

Paton [1961]). We present a number of more commonly used relationships that were obtained by Levy [1966], Jusko [1971], Rang and Dale [1987], Matthews [1993], Jusko [1994], Levy [1994], and Ross [1995]. For a more thorough review of receptor binding models and related areas, see Black and Leff [1980], Jenkinson *et al* [1995], Kenakin [1993, 1997] and van der Graaf [1996].

Definitions: There are two classes of response. One is called a *graded* or *continuous* response, and includes parameters such as blood pressure, heart rate or acetylcholin-esterase *AChE* activity. In such a response, one can measure the changes as a continuous function of dose, concentration, or time. The other class of response is a *quantal* or *dichotomous* response. Epileptic seizures, cancer, and death exemplify this class. In this class, there is either a response or no response, but nothing in between. Analysis of these two classes of response requires different approaches.

The following definitions are taken from Jenkinson *et al* [1995]. *Receptors* are macromolecules that are concerned directly and specifically in chemical signaling between and within cells. An *agonist* is a ligand that binds to receptors and thereby alters (stabilizes) the proportion of them that are in an active form, resulting in a biological response. Conventional agonists increase this proportion, whereas inverse agonists reduce it. An *antagonist* is a compound that reduces the action of another compound, generally an agonist.

Potency is an expression of the activity of a compound, either in terms of the concentration or amount needed to produce a defined effect, or, less acceptably, with regard to the maximal effect attainable. EC_{50} is the (preferably molar) concentration of an agonist that produces 50% of the maximal possible effect. IC_{50} is the (preferably molar) concentration of an antagonist that reduces a specified response to 50% of its former value.

Desensitization or *tachyphylaxis* (which contrasts *anaphylaxis*) refers to the spontaneous decline in a response during continuous application of an agonist, or during repeated applications of doses. It is recommended that the term desensitization be used when tachyphylaxis is considered to involve the receptor itself, or to be a direct consequence of receptor activation.

Efficacy is a concept originally introduced by Stephenson [1956] to express the degree to which different agonists produce varying responses, even when occupying the same proportion of receptors. In some tissues, agonists of high efficacy can produce a maximal effect, even when only a small fraction of the receptors is occupied. The tissue is said to possess *spare receptors* when, for a given level of response, there is a large receptor reserve for the action of the agonist (Jenkinson *et al* [1995]).

4.2   Law of Mass-action

Different proteins are targets for drug binding in the body. Some of these targets are *enzymes*, *transporters*, *ion channels* and *receptors*. Receptors can be regarded as the sensing element in a system of chemical communications that co-ordinate cellular function (Rang and Dale [1991]). Although several receptor models can be defined (Paton [1961]), the most applicable is the receptor-occupancy model based on the law of mass-action. Drug binding to the receptor is usually reversible, as shown in Figure 4.2

Inactive form   *Active form*

$$[Drug] \; + \; [Receptor] \underset{k_{-1}}{\overset{k_1}{\rightleftharpoons}} [Drug\text{-}Receptor] \longrightarrow E$$

Figure 4.2 *Schematic illustration of drug, receptor, and drug-receptor complex equilibrium. The transduction step is when the drug-receptor complex causes an effect E. [Drug] corresponds to [C] in the text.*

Let R represent the concentration of available receptors, C the drug concentration in the biophase ([*Drug*]), and RC the concentration of the drug-receptor complex ([*Drug-Receptor*]). It is further assumed that response (effect E) is proportional to the concentration of receptors that are occupied at a particular time. There are, however, exceptions where the full effect may be established with only a portion of the receptors being occupied. In this case there would be *spare receptors*.

At equilibrium, the rate of formation (rate of association governed by the association rate constant k_1 ($M^{-1}s^{-1}$) where M denotes the molar concentration) of the drug-receptor complex is equal to its rate of degradation (rate of dissociation governed by the association rate constant k_{-1} s^{-1})

$$\frac{d[RC]}{dt} = k_1 \cdot [R] \cdot [C] - k_{-1} \cdot [RC] \tag{4:1}$$

At equilibrium

$$\frac{d[RC]}{dt} = 0 \tag{4:2}$$

$$k_1 \cdot [C] \cdot [R] = k_{-1} \cdot [RC] \tag{4:3}$$

Rearrangement yields

$$\frac{[C] \cdot [R]}{[RC]} = \frac{k_{-1}}{k_1} = K_d \tag{4:4}$$

K_d is the equilibrium dissociation constant of the receptor-drug complex (actually the equilibrium constant for the binding reaction, $K_d = k_{-1}/k_1$). The smaller the K_d, the higher is the drug-receptor affinity. In other words, the smaller the K_d, the more the equilibrium is *pushed* to the right to the active form of the receptor. Assume that the total concentration of receptors is *[R$_t$]*, which equals *[RC]* + *[R]*. If we then solve for *[R]*, we obtain the following expression

$$\frac{[C] \cdot ([R_t] - [RC])}{[RC]} = \frac{k_{-1}}{k_1} = K_d \tag{4:5}$$

Equation 4:5 can be rearranged to give

$$\frac{[RC]}{[R_t]} = \frac{[C]}{[C] + K_d} \tag{4:6}$$

Ariens[1] [1954] introduced the concept that response is directly proportional to the fractional occupancy of the total receptor population. In other words, the response E, is proportional to the concentration of drug-receptor complex. The proportionality constant was called *intrinsic activity*, $\varepsilon = [RC]/E$. The maximal drug-induced response E_{max}, occurs when all receptors are occupied, so that $E_{max} = \alpha [R_t]$, where α is efficacy.

$$\alpha = \frac{[RC]}{E} \tag{4:7}$$

$$E_{max} = \alpha \cdot [R_t] \tag{4:8}$$

Substitution of these values into Equation 4:6 yields

$$\frac{E}{E_{max}} = \frac{[C]}{[C] + K_d} \tag{4:9}$$

and rearranged yields

$$Occupancy \ or \ Stimulus = \frac{E}{E_{max}} = \frac{[C]}{[C] + K_d} \tag{4:10}$$

$$E = \frac{E_{max}[C]}{[C] + K_d} \tag{4:11}$$

Stephenson [1956] described *efficacy* as the *strength* of a single drug-receptor complex in evoking a response of the tissue. Subsequently, it was appreciated that characteristics of the tissue (e.g., the number of receptors that it possesses and the nature of the coupling between the receptor and the response) and of the drug-receptor complex were important, and the concept of *intrinsic efficacy* was developed (see Kenakin [1993, 1997], Rang and Dale [1987], Leff *et al* [1990]). The relationship

---

[1]In order to produce an effect, the drug has to satisfy at least two conditions. There must be an *affinity* between the drug and the specific receptors, in other words a pharmacon-receptor complex has to be formed, and this complex must have the properties necessary to intervene in the biochemical or biophysical processes in such a way that an effect is produced - *efficacy*. The contribution to the effect per unit of pharmacon-receptor complex is called *intrinsic activity*.

between occupancy and response E can thus be represented

$$E = \frac{\varepsilon \cdot N_{tot}[C]}{[C] + K_d} \qquad (4:12)$$

ε, N_{tot}, K_d, and C are the intrinsic efficacy, total number of receptors, concentration of 50% of maximal response, and concentration of drug at the receptor site, respectively. Equation 4:12 represents the *transduction function*, which describes the characteristics of the responding system (Rang and Dale [1987]). Transduction basically means that a chemical stimulus (binding between ligand and receptor) is converted into some physical action such as lowering of blood pressure. In the following discussion, we use a form of Equation 4:12 that is shown in Equation 4:11. Thus, E_{max} is the notation for εN_{tot} and EC_{50} for K_d.

Because occupancy is often not directly related to response, and signal amplification occurs between receptor occupancy, transducer, effector and second messenger activation, and ultimately response, concentration-response curves often fall to the left of the receptor-occupancy profiles as seen in Figure 4.3 (Ross [1995]).

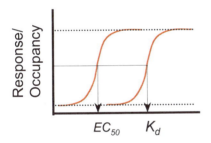

Figure 4.3 *Stimulus-response, and relationship between K_d and EC_{50}. Response represents both binding to the receptor (K_d) and functional response (EC_{50}) in this case.*

It can be shown mathematically that the sensitivity of the total response of two or more hyperbolic processes will be greater than the sensitivity of any one of them (Kenakin, [1996]). Figure 4.4 demonstrates the cascade starting with drug binding to the receptor that produces the initial stimulus. This then feeds into the cascade. The activation of a small fraction of the receptor population can result in a strong physiological signal (functional response) (Kenakin, [1996]).

$$Stimulus_i = \frac{[C]}{K_d + [C]} \qquad (4:13a)$$

$$Response = \frac{Stimulus_i}{\beta + Stimulus_i} \qquad (4:13b)$$

The β parameter represents the stimulus that causes 50% of maximal response (see also Section 4.13 and PD26).

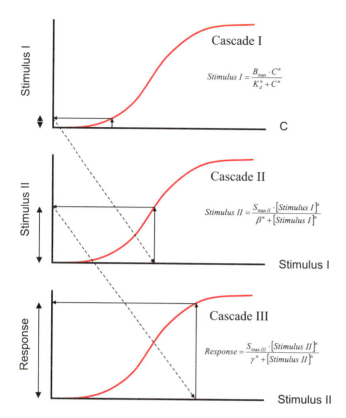

Figure 4.4 *The cascade effect. Drug stimulates only a small portion of the available receptor population, which generates Stimulus I. That stimulus then enters the cascade and causes a Stimulus II, which then generates a functional response. A small fraction of the originally occupied receptors (e.g., 10%) will eventually generate an almost complete response (e.g., 90%) when the cascade process has fully developed. B_{max} and K_d refer to maximum binding capacity and affinity between ligand and receptor, respectively. β and γ denote the half-maximal stimuli for stimulus I and II, respectively. The n parameter is the sigmoidicity factor for the three systems. It may not have the same value in the three systems. Adapted from Kenakin [1997].*

The *spare receptor* concept means that not all receptors need to be occupied in order for a ligand to elicit full maximum effect. *Spare receptors* are common in pharmacology and, as a result, a certain number of receptors can be lost without diminution of the maximal observable response. For curve No. 1 in Figure 4.5, there is a large number of *spare receptors* as compared to curve No. 3.

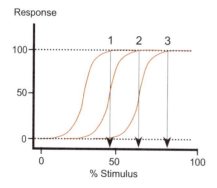

Figure 4.5 *Stimulus-response curves for three situations where there are various degrees of spare receptors. Situation 1 has most spare receptors.*

In order to obtain full effect in the same effector system, only a limited proportion of receptors needs to be occupied (e.g., 45%) for curve No. 1, whereas for curve No.3 about 85% of the receptors need to be occupied.

4.3   Receptor Binding Models

4.3.1   Saturation studies

In pharmacological binding studies, the quantity of radioactive ligand that is bound to protein is measured. It is assumed in these types of studies that the ligand binds to a specific population of receptors. However, binding also occurs at other sites. The total number of non-receptor binding sites is collectively referred to as non-specific binding (*NSB*) (Kenakin [1997]).

Two types of binding therefore exist, namely specific binding and *NSB* (Figure 4.6). The sum of the two is called total binding, and can be described by Equation 4:14, where B_{max} is maximum ligand binding capacity.

$$Total\ binding = \frac{B_{max}[C]^n}{[C]^n + K_d^n} + a \cdot [C] \qquad (4:14)$$

It is important to note that total binding is seldom analyzed but rather the specific binding which is total-*NSB*.

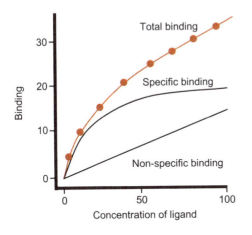

Figure 4.6 Combined specific and non-specific binding versus free ligand concentration on a linear scale

Equation 4:14 assumes that the specific binding is saturable and that *NSB* is non-saturable with a linear function of ligand concentration. Under these circumstances, a curve representing total binding of ligand to membrane preparation can be dissected into the receptor and non-specific site binding portions (Kenakin [1997]). A simultaneous curve-fitting of the total binding and *NSB* to Equation 4:14 will yield estimates of B_{max}, n, K_d and a. We assume for simplicity that the *NSB* can be described by a linear function within a certain concentration range. This may not always be so. *NSB* can also be described by means of a saturable expression if a larger concentration range is covered.

4.3.2   Displacement studies

Displacement experiments are also performed to measure ligand affinity (Figure 4.7). In displacement experiments, a fixed quantity of radioactive ligand is equilibrated with the receptor preparation and then a range of concentrations of another ligand (which binds to the same receptor) is added to the medium. As the non-radioactive ligand displaces the radioactive one, the radioactivity associated with the receptor is reduced (Kenakin [1993]).

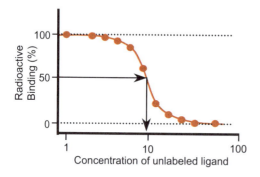

Figure 4.7 Binding of radioactive ligand versus displacer (non-radioactive ligand) concentration on a semi-logarithmic scale. The IC_{50} value is indicated with a vertical arrow pointing at the concentration axis.

This relationship is expressed mathematically as follows

$$\% \, Binding = Baseline - \frac{B_{max}[C]}{[C] + IC_{50}} \qquad (4{:}15)$$

where

$$IC_{50} = K_i \cdot \left(1 + \frac{[L]}{K_d}\right) \qquad (4{:}16)$$

Equation 4:16 is the Cheng-Prusoff equation, and rearranged gives

$$K_i = \frac{IC_{50}}{1 + \dfrac{[L]}{K_d}} \qquad (4{:}17)$$

In displacement experiments, *[C]* is the concentration of the unlabelled ligand, *[L]* is the concentration of the radioligand and K_d is the binding constant of the ligand (Cheng and Prusoff [1973], Barlow [1995]). Note that K_i is used as notation for the displacer concentration responsible for displacing 50% of the ligand.

The non-radioactive ligand competes with the radioactive ligand *[A*]* according to the equation for competitiveness below

$$[A^*R] = \frac{R_t[A^*]}{[A^*] + K_a\{1 + [B]K_b\}} \qquad (4{:}18)$$

The single- and two-site models have been used to describe binding of an experimental compound to certain structures of the brain. As the concentration of unlabelled ligand increases, it displaces radioligand from the receptor. Figure 4.8 displays observed displacement data, as well as predicted data for both a single-site and two-site binding model. The solid line represents the two-site binding model and the dashed line represents the single-site binding model. Observe the systematic deviation between experimental data (symbols) and the single-site-binding model at higher ligand concentrations. This systematic deviation is not apparent when we apply the two-site binding model.

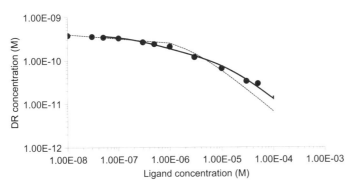

Figure 4.8 *Labeled drug-receptor binding complex [DR] versus unlabeled ligand concentration. The experimental data are shown as closed circles. Solid line is two-site model and dashed line is single-site model.*

The single- and two-site models are described by Equations 4:19 to 4:21 below

$$[A_s] = \frac{[A]}{1 + K_b[C]} \tag{4:19}$$

One-site binding model

$$[RC] = R_t \frac{1 - K_i[A_s]}{1 + K_i[A_s]} \tag{4:20}$$

Two-site binding model

$$[RC] = R_t \left\{ 1 - \frac{F_1 K_1[A_s]}{1 + K_1[A_s]} + \frac{(1 - F_1)K_2[A_s]}{1 + K_2[A_s]} \right\} \tag{4:21}$$

The unlabeled ligand ($C = Drug$) displaces the radioligand [A] from the receptor R, the shifted ligand concentration is A_s, affinity constant K_a, drug concentration C, receptor concentration R_t, fraction of binding to site one F_1, affinity constant for site one K_1 and affinity constant for site two K_2. R_t and K_a are held constant, and K_1, K_2, and F_1 are the model parameters to be estimated.

While saturation and displacement studies are the commonly performed ligand-binding experiments. Alternative protocols, such as 'mixed'-type protocols and multiligand experiments, might have a superior resolving power compared to equilibrium experiments (for review see Rovati [1999]). A careful choice of the design, and computer-driven analysis of the data may result in dramatic improvements in parameter precision over conventional approaches.

4.4   Basic Pharmacodynamic Models

4.4.1   The linear effect concentration model

The simplest relationship between concentration C and a pharmacological effect E can be described by the linear function

$$E = S \cdot C$$

(4:22)

where S is the slope parameter. The linear model assumes linear and direct proportionality between drug concentration and effect. However, it rarely applies to actual cases. If the measured effect has some baseline value E_0 (e.g., including behavior of disease model) when drug is absent then the model may be expressed as

$$E = E_0 + S \cdot C$$

(4:23)

The parameters of this model, S and E_0, may easily be estimated by linear regression as shown in Figure 4.9.

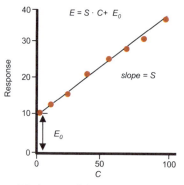

Figure 4.9 *Observed (filled circles) and predicted (line) effect E versus concentration C. Note that a straight line with a slope S and an intercept E_0 could well approximate the effect.*

If the precision of the baseline effect E_0 is much greater than that of the pharmacological effect E, then S can be estimated using

$$(E - E_0) = S \cdot C$$

(4:24)

However, these models are only valid in the *linear* range of the effect-concentration curve. One should be careful with extrapolations beyond the observational range.

4.4.2 The log-linear effect concentration model

If the effect is measured over a large concentration range, the relationship between effect and concentration may appear curvilinear. It may therefore be convenient to plot log-concentration *versus* effect as this will appear linear within a certain effect-range, often between 20–80% of maximal effect. This log-transformation of the concentration axis enables graphical estimation of the *slope (m)* of the apparently linear segment of the curve

$$E = m \cdot \ln(C + C_0)$$

$$\text{(4:25)}$$

$$E_0 = m \cdot \ln(C_0)$$

$$\text{(4:26)}$$

where E, m, and C are already defined. We advocate using Equation 4:25 rather than $E = m \cdot ln(C) + b$, since $ln(C)$ becomes negative for C less than 1 and the sum of $m \cdot ln(C) + b$ is less than 0. In Figure 4.10 we have depicted the effect *versus* log concentration with a baseline value, E_0, and C_0 denotes the concentration of the endogenous agonist. Note that there is no maximum effect within the observed concentration-effect range.

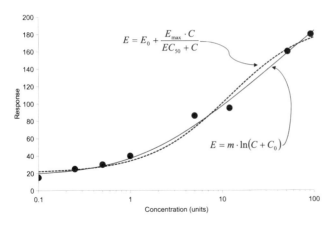

$$E = E_0 + \frac{E_{max} \cdot C}{EC_{50} + C}$$

$$E = m \cdot \ln(C + C_0)$$

Figure 4.10 Observed (●) and predicted (lines) effect versus log-concentration. Note that the effect increases exponentially, but no plateau is reached. Equation 4:25, which includes a baseline value, is probably a good candidate for this dataset. Equation 4:28 is also included but levels off at concentrations above 100 units. See PD27 for details.

Equation 4:25 is valid for the lower region of the E_{max} concentration-effect relationship in section 4.4.3.

The pharmacological effect cannot be estimated when the concentration is zero because of the logarithmic function, unless one adds the C_0 parameter. There is also no maximal effect. This limitation violates some widely accepted principles that competitive and non-competitive inhibitors are often differentiated based on their effects on maximum response. In spite of this, the model has been successfully applied in a number of instances, although it lacks biological meaning.

4.4.3 The ordinary E_{max} model

The measure of the tendency of the ligand and receptor to stay together or to have affinity for one another is also called the *potency* of the ligand (expressed as K_d in receptor binding studies). For functional responses that we wish to correlate to plasma or tissue measurements, we often denote this parameter as EC_{50}. If two compounds were to be compared with respect to potency, the one with the lowest EC_{50} value would have the highest potency.

$$E = \frac{E_{max} \cdot C}{EC_{50} + C}$$
(4:27)

The relationship between concentration and effect is shown in Figures 4.11 on a linear concentration scale (left) and logarithmic concentration scale (right). By using a logarithmic transformation of concentrations, the concentration at EC_{50} is easier to determine. Potential low baseline values are also more easily identifiable and are easier to estimate graphically. Figure 4.11 represents an agonistic effect.

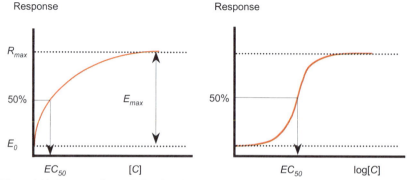

Figure 4.11 *Agonistic effect including baseline versus concentration at steady state. Note that EC_{50} is based on the interval between baseline, E_0 and the maximal response, where $R_{max} = E_0 + E_{max}$. EC_{50} is not derived from the interval between R_{max} and zero.*

The maximum observed effect R_{max} coincides with the maximum drug-induced effect E_{max}, and R_{max} is the sum of $E_{max} + E_0$ (Figure 4.11). There are various forms of this function for agonistic and antagonistic (inhibitory) effects that include a baseline effect, E_0. Figure 4.11 displays the baseline effect relationship, which is given mathematically in Equation 4:28.

$$E = E_0 + \frac{E_{max} C}{EC_{50} + C}$$
(4:28)

or alternatively

$$E = \frac{E_{max} \left[C + C_0 \right]}{EC_{50} + \left[C + C_0 \right]}$$
(4:29)

C_0 in Equation 4:29 is the concentration that would be required to generate a baseline effect such that E includes the baseline effect (Colburn [1986]). In a real situation C_0 could be the baseline concentration of an endogenous drug.

In Equation 4:28 the drug has an agonistic effect which increases from a baseline value E_0. The relationship between concentration and effect for an antagonistic effect including a baseline value is shown in Equation 4:30 and Figure 4.12.

$$E = E_0 - \frac{I_{max}C}{IC_{50} + C}$$

(4:30)

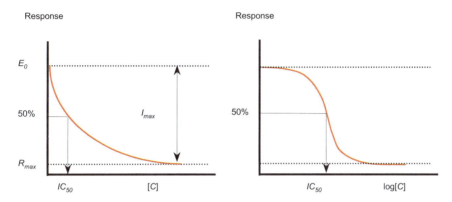

Figure 4.12 _Antagonistic effect including baseline versus concentration at steady state. Note that IC_{50} is based on the interval between the baseline, E_0 and the maximal response, R_{max} or E_0–I_{max}. IC_{50} is not derived from the interval between E_0 and zero, but between E_0 and R_{max}._

We propose that you discriminate between maximum drug induced effect E_{max} and the maximum observed effect R_{max}. Since the potency EC_{50} or IC_{50} is pharmacologically related to E_{max}, other parameterizations should be carefully scrutinized.

4.4.4 The sigmoid E_{max} model

By the addition of a single parameter n (Equation 4:31) to the *ordinary E_{max}* model (Equation 4:28) it is possible to modify the steepness or curvature of the response-concentration curve, thus accounting for curves that are shallower or steeper. The more general or basic for of the equation that describes the relationship shown by different curves in Figure 4.13 is referred to as the Hill equation or *sigmoid E_{max}* model.

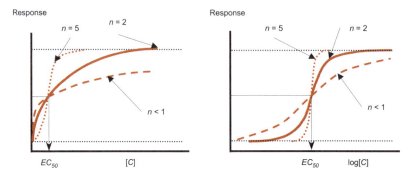

Figure 4.13 *Schematic presentation of the sigmoid E_{max} model for three different values of the exponent n (n<1, n=2 and n=5) on a linear scale (left) and semi-logarithmic scale (right). When the exponent n increases, the 'steepness' of the tangent to the curve at EC_{50} will increase. Note that EC_{50} is the same for the three curves.*

As can be seen, the concentration and EC_{50} are raised to the power of *n*. Thus, in the ordinary E_{max} model (Equation 4.27), *n* is equal to 1.

$$E = \frac{E_{max}C^n}{EC_{50}^n + C^n}$$
(4:31)

This additional parameter, also called the sigmoidicity factor *n*, was first introduced into pharmacokinetics by Wagner [1968] who used the above relationship. The equation is known as the Hill equation, and had been used by Hill [1910] to describe the relationship for oxygen-hemoglobin association. The sigmoidicity parameter does not necessarily have a direct biological interpretation and should be viewed as an extension of the original E_{max} model to account for the curvature. It provides a further degree of flexibility in the sensitivity of the response-concentration relationship. The larger the value of the exponent, the more curvature (steeper) is the line around the EC_{50} value. The solid line in Figure 4.13 has an exponent of 2 whereas the dotted line has an exponent of 5. Both functions have the same EC_{50} value.

A high exponent results in an all-or-nothing effect, and thus, within a narrow concentration range, the observed effect goes from all to nothing or *vice versa*. An exponent less than unity (e.g., 0.4) could be an indication of active metabolites and/or multiple receptor sites. Note that the potency and intrinsic activity are the same for curves with different exponents. In other words, the only difference is the exponent and, as a result, the steepness of the curve.

The *inhibitory sigmoid E_{max}* model is functionally described by Equation 4:32, and depicted in Figure 4.14.

$$E = E_0 - \frac{I_{max}C^n}{IC_{50}^n + C^n}$$
(4:32)

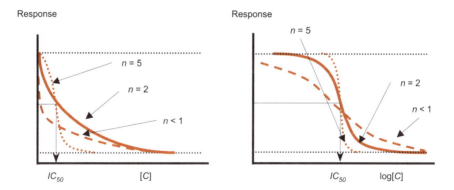

Figure 4.14 *The inhibitory sigmoid E_{max} model, using three different exponents, n, on a linear concentration scale (left) and a semi-logarithmic concentration scale (right).*

As for the *excitatory sigmoid E_{max}* model (Equation 4:33), a change in exponent only affects the steepness of curvature.

$$E = E_0 + \frac{E_{max}C^n}{EC_{50}^n + C^n}$$ (4:33)

Figure 4.15 shows the effect *versus* log-concentration relationship for the excitatory model. This plot can be used to derive initial estimates of parameters in the *sigmoid E_{max}* model. By plotting effect *versus* concentration data on a log-concentration scale it is usually easier to locate the EC_{50} value. The slope *m* of the center of this curve around the EC_{50} value can be used for estimation of the sigmoidicity factor or Hill coefficient *n*. The curve in Figure 4.15 is symmetrical around the EC_{50}.

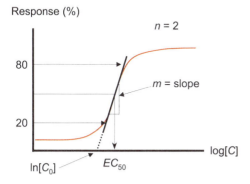

Figure 4.15 *Response versus log-concentration. The slope of the curve m, in the region of 20-80% of the maximal effect is actually the tangent of the function at a concentration equal to EC_{50} (Levy [1995]).*

When plotted against the logarithm of the concentration, the effect appears *linear* in the 20 - 80% interval of maximal effect. Levy [1995] derived the slope at a concentration equal to EC_{50} according to

$$E = \frac{E_{max}e^{n\ln(C)}}{EC_{50}^n + e^{n\cdot\ln(C)}} \tag{4:34}$$

The derivative of E with respect to lnC is then

$$\frac{dE}{d\ln(C)} = \frac{n \cdot E_{max}EC_{50}^n C^n}{\left\{EC_{50}^n + C^n\right\}^2} \tag{4:35}$$

The slope at $C = EC_{50}$ is then

$$m = \frac{dE}{d\ln(C)} = \frac{n \cdot E_{max}EC_{50}^{2n}}{\left[2 \cdot EC_{50}^n\right]^2} \tag{4:36}$$

and rewritten becomes

$$m = \frac{n \cdot E_{max}}{4} \tag{4:37}$$

The m parameter can be obtained from the tangent to the response *versus* log-concentration at EC_{50} according to

$$m = \frac{E_2 - E_1}{\ln(C_2) - \ln(C_1)} = \frac{n \cdot E_{max}}{4} \tag{4:38}$$

n is then solved as

$$n = \frac{m}{E_{max}} \tag{4:39}$$

The extrapolated line intercepts the concentration axis at

$$Intercept = \ln(C_0) = \ln(EC_{50}) - \frac{2}{n} \tag{4:40}$$

$$C_0 = EC_{50} \cdot e^{-2/n} \tag{4:41}$$

Equations 4:38 and 4:39 are useful in order to calculate the initial estimate of n in the *sigmoid E_{max}* model (Levy [1995]). Most nonlinear regression software is reasonably good at estimating n, provided the data contains little noise. Figure 4.16 shows observed and predicted effect-concentration data using the *ordinary inhibitory I_{max}* model ($n = 1$) and the *sigmoid inhibitory I_{max}* model ($n > 1$). By adding the n parameter, the model is given more flexibility and follows the observed data better. A parameter for the baseline value E_0 was used in both models.

$$E = E_0 - \frac{I_{max}C^n}{IC_{50}^n + C^n} \tag{4:42}$$

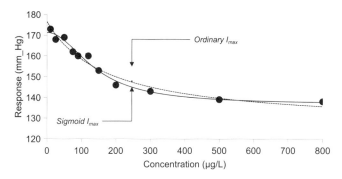

Figure 4.16 Response versus concentration. The fitted lines represent the ordinary I_{max} (dashed) model and the sigmoid I_{max} model (solid).

4.4.5 Composite E_{max} models

The situation of combined drug action arises either when two or more active compounds exert an effect on a certain biological system or when a single drug acts simultaneously at two different receptors. Several authors (Ariens [1964], Gero [1971], Kenakin [1993], Paalzow and Edlund [1979], Ebling *et al* [1991], and Lundström *et al* [1992]) have described this approach to modeling.

The relationship between an effect and plasma concentration of a new experimental compound was studied in rats, and analyzed by Lundström *et al* [1992]. Data in Figure 4.17 are simulated by means of the parameter estimates presented by the authors (Equation 4:43) with a constant coefficient of variation error model.

$$E = E_0 - \frac{I_{max1}C^{n1}}{IC_{51}^{n1} + C^{n1}} - \frac{I_{max2}C^{n2}}{IC_{52}^{n2} + C^{n2}} \qquad (4:43)$$

The IC_{50} values were estimated to be 1.8 and 23 and well separated; I_{max1} and I_{max2} were 4 and 3.2, respectively, and the corresponding exponents were 1.4 and 4.7. Presumably, the first phase (A) corresponded to the predominantly autoreceptor-mediated effects, whereas the later phase (B) reflected a postsynaptic receptor blockade (Lundström *et al* [1992]).

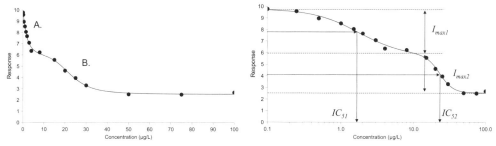

Figure 4.17 The observed effect versus steady state concentration of a new compound. Left hand figure is on a Cartesian scale and right figure is the same data on a semi-logarithmic scale. The initial phase A is represented by the I_{max1} term and the later phase B by the I_{max2} term in Equation 4:43 (Lundström et al [1992]).

Equation 4:44, depicted in Figure 4.18 (simulated data), is sometimes referred to as a *bell-shaped* or *u-shaped* pharmacodynamic effect relationship. Such relationships are common in pharmacology.

$$E = \frac{E_{max}C^{n_1}}{EC^{n_1} + C^{n_1}} - \frac{I_{max}C^{n_2}}{IC^{n_2} + C^{n_2}} \tag{4:44}$$

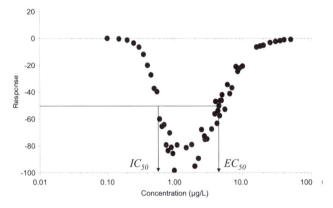

Figure 4.18 *Observed and predicted effect at steady state versus concentration of a new experimental agent. The relationship is clearly u-shaped. I_{max} is equal to E_{max}.*

This model implicitly relates the observed effect to a plasma or effect compartment concentration. E_{max} and I_{max} are the maximum opposite effects, EC_{50} and IC_{50} correspond to the potencies, and n_1 and n_2 correspond to the sigmoidicity factors. A schematic diagram of the data in Figure 4.18 and of the composite I_{max}/E_{max} model is shown in Figure 4.19.

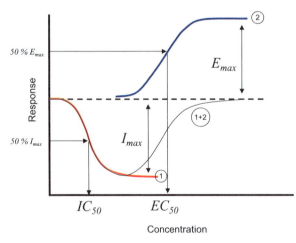

Figure 4.19 *Schematic illustration of the composite I_{max} model. The red line (No. 1) is characterized by means of I_{max} and IC_{50} and the blue line (No. 2) by means of E_{max} and EC_{50}. The composite model (No. 1+2) is represented by the black line.*

Unless effect data exhibit low variability and are well spaced with respect to the EC_{50} values, there are inherent deficiencies in this model with opposing expressions for the effect (Ebling *et al* [1981]). In other words, the utility of the model depends on the degree of separation of the EC_{50} and IC_{50} values. At high concentrations (where $C >>$

EC_{50} and IC_{50})

$$E \approx E_{max} - I_{max} \qquad\qquad (4:45)$$

At low concentrations (where $C << EC_{50}$ and IC_{50})

$$E \approx \frac{E_{max} C^{n_1}}{EC_{50}^{n_1}} - \frac{I_{max} C^{n_2}}{IC_{50}^{n_2}} \qquad\qquad (4:46)$$

At intermediate concentrations (where $EC_{50} << C << IC_{50}$)

$$E \approx E_{max} - \frac{I_{max} C^{n_2}}{IC_{50}^{n_2}} \qquad\qquad (4:47)$$

Two or more parameters (often the I_{max} parameters) will generally be highly correlated since they appear as a sum. Alternatively, more empirical models may be utilized to describe data, but they often lack physiological significance (Ebling *et al* [1991]). Therefore, we may conclude that the sum of the E_{max} function is a flexible system to summarize data for a situation where we know the underlying receptor model (Lundström *et al* [1992]) and for more empirical situations.

4.5   Interaction Models

4.5.1   Competitive antagonism

Remember from the previous discussion (see section 4.1) that an agonist stabilizes the active form of a receptor. A simple competitive antagonist is a ligand that binds to the same site on a receptor as the agonist, but is incapable of stimulating the receptor. Simple competitive antagonists have zero intrinsic activity (Kenakin [1993], Matthews [1993]). Drug A may reduce the effectiveness of C according to the *ordinary E_{max}* type model. The A and EA_{50} represent ligand A concentration and potency, respectively.

$$E = \frac{E_{max} C}{\left[1 + \dfrac{A}{EA}\right] \cdot EC + C} \qquad\qquad (4:48)$$

From this relationship one may appreciate the impact of different ratios of A/EA_{50} on the effect of C. Competitive antagonists do not alter E_{max} nor EC_{50}, although they change the location of the inflection point of the curve to $[1 + EA_{50}] \cdot EC_{50}$. The value of EC_{50} is the same in Equation 4:48 as it is in Equation 4:27 (Matthews [1993]). However, a larger concentration of drug is required to give the same effect in the presence of A (Figure 4.20).

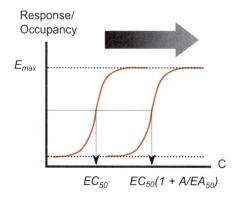

Figure 4.20 *Competitive antagonists shift the original effect-concentration curve to the right, without altering the E_{max} or EC_{50}.*

4.5.2   Noncompetitive antagonism

Drug A may reduce E_{max} of drug C according to the following variation of the E_{max} model.

$$E = \frac{E_{max} \cdot EA_{50} \cdot C}{\left[A + \dfrac{C}{EA_{50}} \right] \cdot \left[EC_{50} + C \right]} \qquad (4:49)$$

From this relationship, one may appreciate the impact of different values of A on the effect caused by C. An example of such antagonism is shown in Figure 4.21.

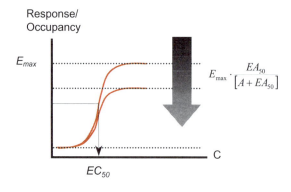

Figure 4.21 *Noncompetitive antagonism shifts the observable maximum effect.*

4.5.3   General empirical dynamic model for two drugs

A general model, incorporating both competitive and noncompetitive interaction, and allowing for synergism/antagonism can be obtained by modifying the general form common to both types of interaction models (Rang and Dale [1987], Sheiner [1990], Kenakin [1993, 1997], and Matthews [1993]).

$$E = E_{max} \cdot \left[\frac{\dfrac{C}{EC_{50}} + \alpha \cdot \dfrac{A}{EA_{50}} + \beta \cdot \dfrac{C}{EC_{50}} \cdot \dfrac{A}{EA_{50}}}{1 + \dfrac{C}{EC_{50}} + \delta \cdot \dfrac{A}{EA_{50}} + \gamma \cdot \dfrac{C}{EC_{50}} \cdot \dfrac{A}{EA_{50}}} \right] \qquad (4:50)$$

The restrictions are that $0 \le \alpha$, $\beta \le 1$, and $\beta = 0$ if $\delta = 0$; and $\beta \ge 0$ if $\delta > 0$ (Table 4.1).

Table 4.1 Summary of general empirical dynamic model

β	δ	Model
$1 + \alpha$	1	Two separate E_{max} models, effects add Max effect $= E_{max}\,(1+\alpha)$
1	1	Noncompetitive interaction
0	0	Competitive interaction
$0<\beta<1$	β	Competitive and noncompetitive
$>\delta$	>0	Synergism (Max effect $= E_{max}\,(\beta/\delta)$)
$<\delta$	>0	Antagonism (Max effect $= E_{max}\,(\beta/\delta)$)

4.5.4 Enantiomer interaction models

To characterize the competitive pharmacodynamic interaction between two agents (e.g., racemic mixture of R(-)- and S(+)-ketamine) acting at the same receptor site, an inhibitory sigmoid I_{max} model was proposed, yielding estimates of I_{max} and IC_{50} for the racemic mixture (Equation 4:51) and enantiomeric forms (Equation 4:52) of ketamine (Schüttler *et al* [1987], Drayer [1986]). The model includes a baseline effect E_0 in addition to I_{max1}, IC_{51}, and n_1 for the S(+)-form and I_{max2}, IC_{52}, and n_2 representing the R(-)-form of ketamine.

$$E = E_0 - \left[\frac{I_{max\,1} \cdot \left[\dfrac{C}{IC_{51}} \right]^{n_1} + I_{max\,2} \cdot \left[\dfrac{C}{IC_{52}} \right]^{n_2}}{1 + \left[\dfrac{C}{IC_{51}} \right]^{n_1} + \left[\dfrac{C}{IC_{52}} \right]^{n_2}} \right] \qquad (4:51)$$

The effect *versus* enantiomer concentration profile is given by

$$Effect\ (R(\text{-})\text{-}\ or\ S(+)\text{-}\ ketamine) = E_0 - \frac{I_{max}C^n}{IC_{50}^n + C^n} \qquad (4:52)$$

Enantiomer data were fit by means of Equation 4:52 including one function for S(+)-ketamine data and a separate function for R(-)-ketamine data (Figure 4.22, right). For a more thorough analysis, see Schüttler *et al* [1987].

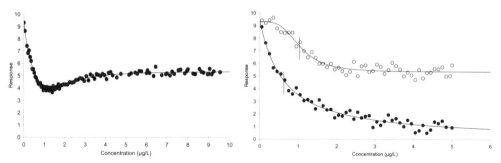

Figure 4.22 *Observed and predicted effects of a racemic mixture of ketamine versus concentration (left). The predicted curve was generated by Equation 4:51. Observed and predicted effects of the individual enantiomers versus concentration (right). Open circles represent R(-)-ketamine and solid circles S(+)-ketamine (right). Vertical lines in the graph indicate the respective IC_{50} values. The predicted curves were generated by Equation 4:52. Parameter values were obtained from Schüttler et al [1987]. Effect-concentration data were then generated with a 10% constant CV.*

It might be appreciated from comparison of the two diagrams in Figure 4.22 that the R(-)-ketamine enantiomer *dilutes* the effect that S(+)-ketamine has on the receptor. The intrinsic activity and potency of R(-)- ketamine are lower than for S(+)-ketamine. The initial decline in the effect (Figure 4.22, left) is mainly due to S(+)-ketamine with its higher potency and larger intrinsic activity. When the concentration of the racemate increases the initial drop gradually levels off due the diluting effect from R(-)-ketamine. This has a smaller intrinsic activity but will compete with the S(+)-form for the receptor binding sites. See also exercise PD16 for implementation of models.

4.5.5   Additional sigmoidal models

Several additional mathematical functions have been proposed for modeling sigmoidal curves. Some of these are, in addition to the *Hill* function, the *Gompertz*, the *logistic*, the *Richards*, the *Morgan-Mercer-Flodin* and the *Weibull* functions.
The *Gompertz* function takes the form

$$Y = \alpha \cdot e^{-e^{\beta - \gamma \cdot x}} \tag{4:53}$$

The *logistic* function forms a whole class of mathematical functions (Equations 4:54 to 4:57). *X* is the independent variable such as concentration and *Y* is the effect.

$$Y = \frac{\alpha}{1 + e^{\beta - \gamma \cdot x}} \tag{4:54}$$

$$Y = \frac{1}{\alpha + e^{\beta - \gamma \cdot x}} \tag{4:55}$$

$$Y = \frac{\alpha}{1 + \beta \cdot e^{-\gamma \cdot x}} \tag{4:56}$$

$$Y = \frac{1}{\alpha + \beta \cdot e^{-\gamma \cdot x}} \tag{4:57}$$

The *Richards* function

$$Y = \frac{\alpha}{\left[1 + e^{\beta - \gamma \cdot x}\right]^{\frac{1}{\delta}}} \tag{4:58}$$

The *Morgan-Mercer-Flodin* function

$$Y = \frac{\beta \cdot \gamma + \alpha \cdot x^{\delta}}{\gamma + x^{\delta}} \tag{4:59}$$

The *Weibull* function

$$Y = \alpha - \beta \cdot e^{-\gamma \cdot x^{\delta}} \tag{4:60}$$

All these functions are analyzed in exercise PD8, where also the asymmetrical characteristics of the *Weibull* function is further explored. See also van der Graaf [1999] for review. The basic logistic function is examined in detail in PD20 and PD21.

4.5.6   Kinetics of pharmacological response

The previous pharmacodynamics sections have elaborated on *concentration-response models* when equilibrium is established between these two. However, the characterization of the pharmacodynamics of a drug is often done at non-equilibrium, in which the model must account for the effect of *time*. We therefore have to add *time* as a variable in our non-steady state analysis. Figure 4.23 is aimed to demonstrate the interrelationship between *concentration* and *response*, *kinetics* (non-steady state) and *response* and the resulting *response-time course* when time has been taken into account.

The primary objective in pharmacodynamics is to characterize drug specific factors (e.g., EC_{50}, E_{max}) and system specific factors (e.g., k_{in}, k_{out}) that govern the 'on/off' response-time courses. The effects of a drug depends usually on the drug-receptor interaction (primary effect) and on the responses of the physiological system and its control mechanisms to the disturbance caused by the primary effect (Struyker-Boudier *et al* [1990]). Drug actions trigger the onset of various kinds of physiological responses, which have their own time-scales of operation (e.g., baroreceptor activation occurs in seconds whereas baroreceptor adaptation occurs in hours or days; vascular remodeling may take weeks or months). In ignoring 'off' responses, which are often markedly asymmetric to 'on' responses, one miss a powerful constraint on *PD* modeling, which is that the model has to be able to simulate both the 'on' and the 'off responses (personal communication with Dr. J. Urquhart [2000]).

Dynamics

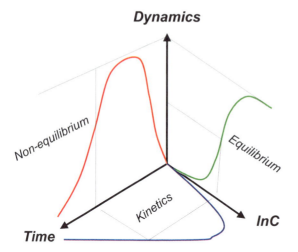

Figure 4.23 Schematic illustration of the inter-relationship between concentration and response (assuming equilibrium, green curve), concentration and time (kinetics, semi-logarithmic scale, blue curve) and response and time (non-steady state dynamics, red curve). The half-maximal response is shown in all three relationships and how it is related to concentration and time.

Consider the disposition of a drug obeying one-compartment kinetics after an intravenous bolus dose (Equations 4:61 and 4:62 and Figure 4.24) (Levy [1966]).

$$C = C_0 \cdot e^{-K \cdot t} \tag{4:61}$$

$$\ln(C) = \ln(C_0) - K \cdot t \tag{4:62}$$

Bolus input

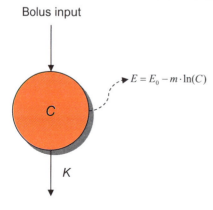

$E = E_0 - m \cdot \ln(C)$

Figure 4.24 Schematic representation of a one-compartment model (Equation 4:61) and its relationship to the response.

A common relationship between a response and the plasma concentration is empirically modeled as a linear regression of *ln(C) versus E*.

$$E = m \cdot \ln(C) \tag{4:63}$$

$$E_0 = m \cdot \ln(C_0) \tag{4:64}$$

Equations 4:61 to 4:64 combined yield

$$E = E_0 - m \cdot K \cdot t \tag{4:65}$$

which states that the decline of response is zero-order in spite of the fact that decline of drug in plasma may be first-order. The zero-order decline in response is, however, only valid within a certain concentration range. The duration of response t_d above a certain baseline value (either response or concentration) takes the following form

$$t_d = m \cdot K \cdot \left[\ln(A) - \ln(A_{min}) \right] \tag{4:66}$$

where A is the amount of drug in the system and A_{min} is the minimum effective amount of drug. Equation 4:66 demonstrates that a doubling of the dose increases the duration of the effect according to Equation 4:67.

$$\Delta t_d = m \cdot K \cdot \ln(2) \tag{4:67}$$

The area under the effect *versus* time curve (AUC_E) for a one-compartment drug can be derived according to

$$E = \frac{E_{max} \cdot \dfrac{D}{V} \cdot e^{-K \cdot t}}{EC_{50} + \dfrac{D}{V} \cdot e^{-K \cdot t}} \tag{4:68}$$

which then integrated gives AUC_E

$$AUC_E = \int_0^\infty E dt = \frac{E_{max}}{K} \cdot \ln\left(1 + \frac{C_0}{EC_{50}}\right) = \frac{E_{max}}{K} \cdot \ln\left(1 + \frac{D}{EC_{50} \cdot V}\right) \tag{4:69}$$

Equation 4:69 demonstrates that a doubling of the dose increases the AUC_E by a factor of the logarithm of that increase and does not double the AUC_E.

When the concentration in plasma follows one-compartment first-order kinetics and the relationship between concentration and response is described by the *Hill* function in Figure 4.25, we get a concentration and effect time course as shown in Figure 4.26.

Bolus input

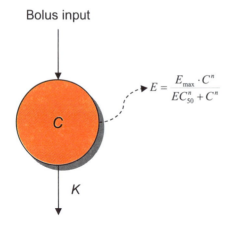

$$E = \frac{E_{max} \cdot C^n}{EC_{50}^n + C^n}$$

Figure 4.25 *A one-compartment model (Equation 4:61) and its relationship to the response. We assume that there is an instantaneous equilibrium between concentration and response. This type of model is also called a 'direct' response model in contrast to the 'indirect' response model (turnover response model) that we discuss in section 4.6.*

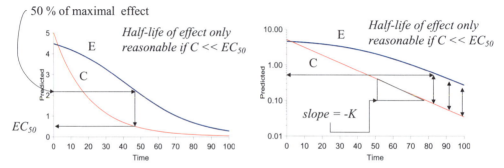

Figure 4.26 *Concentration (C) and effect (E) versus time on a linear scale (left) and semi-logarithmic scale (right). Concentration-time and effect-time have arbitrary units. Note that the effect declines 'linearly' on a semi-logarithmic scale when the concentration falls below EC_{50}. The half-life of the effect is only reasonable when concentration is less than EC_{50}.*

When you have done your experiment, look at your data in various ways such as *concentration-time, effect-time,* and *effect-concentration* plots. Ask yourself questions such as *Have I reached maximum effect? Can I obtain an estimate of the concentration at half-maximal effect? Do I have information about the kinetics? What model do I want to propose?* Look at your data and perform a preliminary analysis by hand. Then derive models, fit them to data and observe the consistency or lack of consistency between fitted and observed data. Ask yourself *What decisions should I make?* Scrutinize the design of the study - *Which pieces of information do I want to collect?* You might be forced to increase the dose or characterize the dynamics with a simpler model. Logical thinking combined with appropriate study design and expertise in modeling should drive your scientific approach.

4.6 Indirect Response (turnover) Model

4.6.1 Background

Currently there are two conceptually different approaches to relate kinetics to dynamics, namely *biological/mechanistic* and *empirical*. This section will demonstrate the *biological* or *mechanistic* approach and how a turnover model (also called the *indirect response model*) is constructed, how to analyze the behavior of the turnover model, and what the limiting values of the model are. This class of model is typically based on sound biological principles with prior knowledge about the mechanisms of drug action. Hence, the variables and parameters have a physiological meaning and can often be related to *in vitro* or other physiological data. Therefore, this approach gives us a better understanding of both intra- and interindividual variability in drug response. The *empirical* approach will be discussed in section 4.7 on effect-compartment (also called *distribution* or *link*) models.

When a pharmacological effect is seen immediately such that equilibrium is established rapidly between plasma concentration and response, we say that the response is directly related to the drug concentration. In such cases, a pharmacodynamic model such as a *linear* model or a *sigmoid* E_{max} model is applied to characterize the relationship between drug concentrations and response. When the pharmacological response takes time to develope, and the observed response is not apparently related to plasma concentration of the drug, a *link model* (also called the effect-compartment model) has usually been applied to relate the pharmacokinetics of the drug to its pharmacodynamics.

Following dosing, there is a build up of the drug response which is governed by the various inhibitory and/or stimulatory factors controlling this response (Ackerman *et al* [1964], Nagashima *et al* [1969], Jusko [1971, 1990], Dayneka *et al* [1993], Jusko [1995]). The length of the time frame of the stimulus and the resolution of data very often determine the category (*direct or indirect*) in which to place the observed response. However, as we will discuss later (Section 4.6.3), we consider all responses to be *indirect* in nature. Figure 4.27 aims to depict the practical difference between a *direct* and an *indirect* response.

Instantaneous equilibrium - 'Direct' response

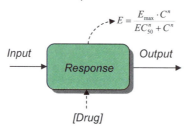

$$E = \frac{E_{max} \cdot C^n}{EC_{50}^n + C^n}$$

Input   Output

Response

[Drug]

Delayed equilibrium - 'Indirect' response

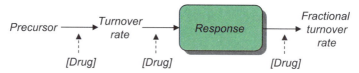

Precursor $\longrightarrow$ Turnover rate $\longrightarrow$ Response $\longrightarrow$ Fractional turnover rate

[Drug]   [Drug]   [Drug]

Figure 4.27 *A schematic presentation of the difference between the direct and indirect response models. [Drug] indicates sites of potential inhibition or stimulation of drug. In the direct response model, drug affects the response (E) without any time delay other than the time it takes for drug to reach the sampled compartment. In the indirect response model, drug acts on build-up or loss of response, but not on the response level as such. See also section 3.5, Turnover models.*

4.6.2   Indirect response models – Reversible mechanisms

Jusko and co-workers have elegantly documented the behavior and extended the use of indirect response (turnover) models to include drug systems (Dayneka *et al* [1993], Nagashima *et al* [1969]). The models are based on the effects of drugs (i.e., inhibition or stimulation) on factors that control the production or the dissipation of the drug response. The *indirect* response model has implications that go beyond distribution models (section 4.7) and may be useful in the interpretation and prediction of pharmacodynamic events that are not necessarily due to distributional effects, but rather affect onset and loss of response.

The measured response R to a drug may be due to factors controlling the turnover rate k_{in} (input or production), or the fractional turnover rate k_{out} (loss) of the response. Both of these may be either inhibited or stimulated. The rate of change of the response over time with no drug present can be described by

$$\frac{dR}{dt} = k_{in} - k_{out} \cdot R \qquad\qquad (4{:}70)$$

At equilibrium

$$\frac{dR}{dt} = k_{in} - k_{out} \cdot R = 0 \qquad\qquad (4{:}71)$$

$$R_0 = \frac{k_{in}}{k_{out}} \qquad\qquad (4{:}72)$$

where R_0 is the baseline value of the response.

The k_{in} parameter represents the zero-order constant for production of response and k_{out} defines the first-order rate constant for loss of the response. The response variable R may be a directly measured entity or an observed response. In Table 4.2, Model 1 and Model 2 (Table 4.2) represent inhibitory processes that operate according to the following inhibitory function, $I(t)$

$$I(t) = 1 - \frac{I_{max} \cdot C^n}{IC_{50}^n + C^n} \tag{4:73}$$

C is the plasma concentration of drug, IC_{50} is the drug concentration which produces 50% of maximum inhibition achieved at the effect site, and $0 < I_{max} \leq 1$. A kinetic model is often used to compute C as a smooth function. Processes that stimulate the factors controlling drug response operate according to the stimulation function $S(t)$

$$S(t) = 1 + \frac{E_{max} \cdot C^n}{EC_{50}^n + C^n} \tag{4:74}$$

where E_{max} is the maximum effect attributed to the drug and EC_{50} the drug concentration producing 50% of the maximum stimulation achieved at the effect site. There are four indirect response models, as shown in Table 4.2.

Table 4.   Overview of the Indirect Response Models

Model 1: Inhibition of build-up	$\dfrac{dR}{dt} = k_{in} \cdot \left[1 - \dfrac{I_{max} \cdot C^n}{IC_{50}^n + C^n} \right] - k_{out} \cdot R$
Model 2: Inhibition of loss	$\dfrac{dR}{dt} = k_{in} - k_{out} \cdot \left[1 - \dfrac{I_{max} \cdot C^n}{IC_{50}^n + C^n} \right] \cdot R$
Model 3: Stimulation of build-up	$\dfrac{dR}{dt} = k_{in} \cdot \left[1 + \dfrac{E_{max} \cdot C^n}{EC_{50}^n + C^n} \right] - k_{out} \cdot R$
Model 4: Stimulation of loss	$\dfrac{dR}{dt} = k_{in} - k_{out} \cdot \left[1 + \dfrac{E_{max} \cdot C^n}{EC_{50}^n + C^n} \right] \cdot R$

Adapted from Dayneka *et al* [1993] to also include the I_{max} parameter to account for fractional inhibition.

Model 1 (Figure 4.28) describes drug response that results from inhibition of the input or production of the response (i.e., inhibition of the build-up of response). In this model, the rate of change of response is described by

$$\frac{dR}{dt} = k_{in} \cdot I(t) - k_{out} \cdot R \tag{4:75}$$

The time to pharmacodynamic steady state, assuming a constant plasma concentration C_{ss}, is governed by k_{out}. The level of response at pharmacodynamic steady state R_{ss} is independent of time but dependent on the actual plasma drug concentration C_{ss} which gives

$$\frac{dR}{dt} = k_{in} \cdot I(C_{ss}) - k_{out} \cdot R_{ss} = 0 \tag{4:76}$$

This implies

$$R_{ss} = \frac{k_{in}}{k_{out}} \cdot I(C_{ss}) = \frac{k_{in}}{k_{out}} \cdot \left[1 - \frac{I_{max} C_{ss}^n}{IC_{50}^n + C_{ss}^n} \right] \tag{4:77}$$

and

$$R_{min} = \frac{k_{in}}{k_{out}} \cdot \left[1 - I_{max} \right] \tag{4:78}$$

where R_{min} is the minimum steady state response. The difference between R_{ss} (R_{min}) and R_0 is given by Equation 4:79

$$\Delta R = R_0 - R_{min} = \frac{k_{in}}{k_{out}} \cdot \left[1 - 1 + I_{max} \right] = \frac{k_{in}}{k_{out}} \cdot I_{max} \tag{4:79}$$

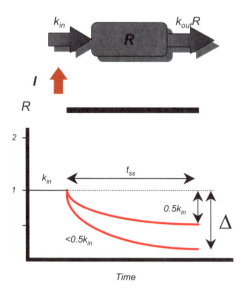

Figure 4.28 *Response versus time profile for Model 1 during a constant rate input. The bar represents the drug exposure period. Vertical arrows show that a change in dose affects the level of inhibition I, but not the time to dynamic steady state t_{ss}. Equation 4:79 is represented by Δ in the plot.*

Figure 4.29 displays an example of Model 1, illustrating a semi-logarithmic plot of the time course of response to warfarin *versus* time, with total blockage of input rate being inferred. See also exercise PD4 for further discussion on the action of warfarin.

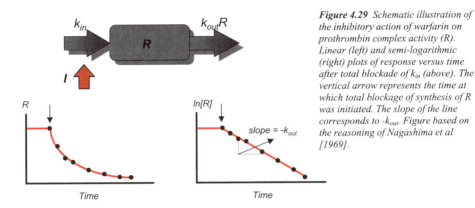

Figure 4.29 Schematic illustration of the inhibitory action of warfarin on prothrombin complex activity (R). Linear (left) and semi-logarithmic (right) plots of response versus time after total blockade of k_{in} (above). The vertical arrow represents the time at which total blockage of synthesis of R was initiated. The slope of the line corresponds to $-k_{out}$. Figure based on the reasoning of Nagashima et al [1969].

Model 2 (Table 4.2, Figure 4.30) describes drug response that results from inhibition of the factors controlling the loss (fractional turnover rate) of response. Thus, the response level builds up. It is described by the following relationship

$$\frac{dR}{dt} = k_{in} - k_{out} \cdot I(t) \cdot R \qquad (4:80)$$

The time to pharmacodynamic steady state, assuming a constant plasma concentration C_{ss}, is governed by k_{out} as well as the actual drug concentration and grade of inhibition $I(C_{ss})$. The level of response at pharmacodynamic steady state R_{ss} is independent of time but dependent on the actual plasma drug concentration C_{ss}. Thus

$$\frac{dR}{dt} = k_{in} - k_{out} \cdot I(C_{ss}) \cdot R_{ss} = 0 \qquad (4:81)$$

Rearranged, this yields

$$R_{ss} = \frac{k_{in}}{k_{out}} \cdot \frac{1}{I(C_{ss})} = \frac{k_{in}}{k_{out}} \cdot \frac{1}{1 - \dfrac{I_{max}C_{ss}^n}{IC^n + C_{ss}^n}} \qquad (4:82)$$

and

$$R_{max} = \frac{k_{in}}{k_{out}} \cdot \frac{1}{1 - I_{max}} \qquad (4:83)$$

The difference between R_{ss} (R_{max}) and R_0 is given by Equation 4:84

$$\Delta R = R_{max} - R_0 = \frac{k_{in}}{k_{out}} \cdot \left[\frac{1}{1 - I_{max}} - 1 \right] = \frac{k_{in}}{k_{out}} \cdot \frac{I_{max}}{1 - I_{max}} \tag{4:84}$$

where R_{max} is the maximum steady state response.

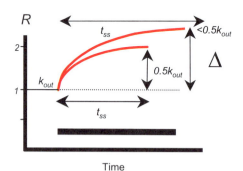

Figure 4.30 Response versus time profile for Model 2 during a constant rate input. The horizontal bar represents the drug exposure period. It can be seen that a change in dose affects both the level of response and the time to steady state t_{ss}. Δ denotes the changes from baseline and is mathematically expressed in Equation 4:84.

Model 3 (Table 4.2, Figure 4.31) represents drug response originating from stimulation of the input or production of the response (i.e., stimulation of build-up of response), and is described as follows

$$\frac{dR}{dt} = k_{in} \cdot S(t) - k_{out} \cdot R \tag{4:85}$$

The time to dynamic steady state, assuming a constant plasma concentration C_{ss}, is governed by k_{out}. The level of response at dynamic steady state is then independent of time but dependent on the actual plasma drug concentration C_{ss} that gives

$$\frac{dR}{dt} = k_{in} \cdot S(C_{ss}) - k_{out} \cdot R_{ss} = 0 \tag{4:86}$$

and

$$R_{ss} = \frac{k_{in}}{k_{out}} \cdot S(t) = \frac{k_{in}}{k_{out}} \cdot \left[1 + \frac{E_{max} C_{ss}^n}{EC_{50}^n + C_{ss}^n} \right] \tag{4:87}$$

and

$$R = \frac{k_{in}}{k_{out}} \cdot \left[1 + E_{max} \right] \tag{4:88}$$

The difference between R_{ss} (R_{max}) and R_0 is given by Equation 4:89

$$\Delta R = R_{max} - R_0 = \frac{k_{in}}{k_{out}} \cdot E_{max} \tag{4:89}$$

where R_{max} is the maximum steady state response.

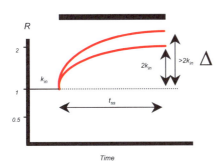

Figure 4.31 *Response versus time profile for Model 3 during a constant rate input. The bar represents the drug exposure period. Vertical arrows show that a change in dose affects the level of stimulation S but not the time to dynamic steady state t_{ss}. Δ denotes Equation 4:89.*

Model 4 (Table 4.2, Figure 4.32) describes drug response that results from stimulation of the factors controlling the dissipation of the response variable. The rate of change of response is described by

$$\frac{dR}{dt} = k_{in} - k_{out} \cdot S(t) \cdot R \tag{4:90}$$

The time to dynamic steady state, assuming a constant plasma concentration C_{ss}, is governed by k_{out} as well as the actual drug concentration and grade of inhibition $S(C_{ss})$.

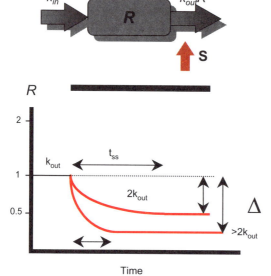

Figure 4.32 *Response versus time profile for Model 4 during a constant rate input. The bar represents the drug exposure period. A change in dose will affects the level of stimulation of loss of response and the time to dynamic steady state t_{ss}. Δ denotes the change from baseline described by Equation 4:94.*

The level of response at dynamic steady state R_{ss} is then independent of time but dependent on the actual plasma drug concentration C_{ss} as follows

$$\frac{dR}{dt} = k_{in} - k_{out} \cdot S(C_{ss}) \cdot R_{ss} = 0 \tag{4:91}$$

Hence

$$R_{ss} = \frac{k_{in}}{k_{out}} \cdot \frac{1}{S(C_{ss})} = \frac{k_{in}}{k_{out}} \cdot \frac{1}{1 + \dfrac{E_{max} C_{ss}^n}{EC_{50}^n + C_{ss}^n}} \tag{4:92}$$

The minimum value of response R_{min} is given by

$$R_{min} = \frac{k_{in}}{k_{out}} \cdot \frac{1}{1 + E_{max}} \tag{4:93}$$

The difference between R_0 and R_{ss} (R_{min}) is

$$\Delta R = R_0 - R_{min} = \frac{k_{in}}{k_{out}} \cdot \frac{E_{max}}{1 + E_{max}} \tag{4:94}$$

As stationarity (time constant parameters) is assumed for all models, the response variable R begins at a predetermined baseline value R_0, changes with time following drug administration, and eventually returns to R_0 once the stimulus has been removed.

The basic indirect response models yield both common and differential characteristics and may serve to classify appropriately collected dose response-data.

4.6.3 Summary of model characteristics

Using Model 3 (Table 4.2) as an example, the baseline, steady state, and maximum response for the ordinary E_{max} model and the indirect response model may be compared as shown in Figure 4.33.

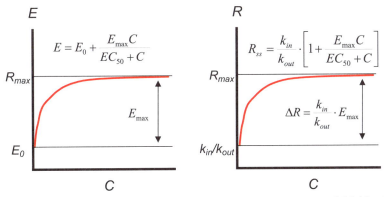

Figure 4.33 *Response versus steady state concentration for the ordinary E_{max} model (left hand figure) and the indirect response Model 3 (right hand figure), both with baseline effect. Note that E_{max} has different meanings in the two models. In the left hand figure, E_{max} is the absolute distance $R_{max} - E_0$, whereas in the right hand figure, E_{max} is a multiplier of k_{in}/k_{out}.*

Table 4.3 summarizes the inhibition ($I(t)$) or stimulation ($S(t)$) function $H(t)$, and its impact on the time to dynamic steady state t_{ss}, response level at steady state R_{ss} and maximum or minimum response R_{max} or R_{min}, respectively.

Table 4.3 Summary table of the impact of $H(t)$ on t_{ss} and R

Model	H(t)	t_{ss} determined by	Effect on $t_{1/2kout}$	Effect on t_{ss}	Effect on R_{ss}	R_{max} or R_{min}	ΔR	Limit of R	Return to R_0
1	Inhibition of k_{in}	k_{out}	$\Leftrightarrow$	$\Leftrightarrow$	$\Downarrow$	$R_{min} = \dfrac{k_{in}}{k_{out}} \cdot [1 - I_{max}]$	$\dfrac{k_{in}}{k_{out}} \cdot I_{max}$	0	$\Leftrightarrow$
2	Inhibition of k_{out}	$k_{out} \cdot I(C)$	$\Uparrow$	$\Uparrow$	$\Uparrow$	$R_{max} = \dfrac{k_{in}}{k_{out}} \cdot \dfrac{1}{1 - I_{ma}}$	$\dfrac{k_{in}}{k_{out}} \cdot \dfrac{I_{max}}{1 - I_{max}}$	∞	$\Downarrow$
3	Stimulation of k_{in}	k_{out}	$\Leftrightarrow$	$\Leftrightarrow$	$\Uparrow$	$R_{max} = \dfrac{k_{in}}{k_{out}} \cdot [1 + E_{max}]$	$\dfrac{k_{in}}{k_{out}} \cdot E_{max}$	∞	$\Leftrightarrow$
4	Stimulation of k_{out}	$k_{out} \cdot S(C)$	$\Downarrow$	$\Downarrow$	$\Downarrow$	$R_{min} = \dfrac{k_{in}}{k_{out}} \cdot \dfrac{1}{1 + E_{ma}}$	$\dfrac{k_{in}}{k_{out}} \cdot \dfrac{E_{max}}{1 + E_{max}}$	0	$\Uparrow$

$\Uparrow$, $\Downarrow$ and $\Leftrightarrow$ denote increase, decrease and no change, respectively.

Table 4.4 summarizes some functional forms of the inhibition and stimulation functions and their limits.

Table 4.4   Summary table of various forms of $H(t)$

$H(t)$	$H(t)$	Comments	Limits
Stimulation	$1 + \dfrac{E_{max} C}{EC_{50} + C}$	Full sigmoidicity	$1 \leq H(t) \leq E_{max}$
Stimulation	$1 + \left[\dfrac{C}{EC_{50}}\right]^n$	Infinite stimulation	$1 \leq H(t) \leq \infty$
Inhibition	$1 - \dfrac{I_{max} C^n}{IC_{50} + C^n}$	Partial inhibition	$1 \leq H(t) \leq 1 - I_{max}$
Inhibition	$1 - \dfrac{C^n}{IC_{50} + C^n}$	Full inhibition	$0 \leq H(t) \leq 1$
Inhibition	$\dfrac{1}{1 + \dfrac{I_{max} C^n}{IC_{50} + C^n}}$	Full sigmoidicity	$0 \leq H(t) \leq 1$
Stimulation	$1 + s \cdot C$	Linear	$1 \leq H(t) \leq \infty$
Stimulation	$1 + \ln\left[C^m + 1\right]$	Logarithmic	$1 \leq H(t) \leq \infty$

Krzyzyanski and Jusko [1998] derived a number of useful expressions for estimation of initial parameter estimates of the indirect response (turnover) class of models (Equations 4:95 to 4:97).

$$\frac{(R_{ss1} - R_{ss2}) \cdot C_{ss1} \cdot C_{ss2}}{(R_0 - R_{ss1}) \cdot C_{ss2} - (R_0 - R_{ss2}) \cdot C_{ss1}} = \begin{cases} IC_{50} \text{ Models I and II} \\ EC_{50} \text{ Models III and IV} \end{cases} \tag{4:95}$$

$$\frac{(R_0 - R_{ss1}) \cdot (R_0 - R_{ss2}) \cdot (C_{ss2} - C_{ss1})}{(R_0 - R_{ss1}) \cdot R_0 \cdot C_{ss2} - (R_0 - R_{ss2}) \cdot R_0 \cdot C_{ss1}} = \begin{cases} I_{max} \text{ Model I} \\ -E_{max} \text{ Model III} \end{cases} \tag{4:96}$$

$$\frac{(R_0 - R_{ss1}) \cdot (R_0 - R_{ss2}) \cdot (C_{ss2} - C_{ss1})}{(R_{ss1} - R_0) \cdot R_{ss2} \cdot C_{ss2} - (R_{ss2} - R_0) \cdot R_{ss1} \cdot C_{ss1}} = \begin{cases} I_{max} \text{ Model II} \\ -E_{max} \text{ Model IV} \end{cases} \tag{4:97}$$

Figure 4.34 summarizes schematically the behavior of the four classes of indirect response (turnover) models we have discussed earlier. It can be seen that factors that act on the turnover rate only affect the actual steady state level (plots A and C). This is analogous with changing the infusion rate of a drug that follows first-order kinetics (i.e., there is a change in steady state concentration but not in time to steady state). On the other hand, factors that act on the disposition parameter (fractional turnover rate and thereby its half-life) change the steady state level of response as well as the time to steady state t_{ss} (plots B and D). This is analogous with changing the clearance of the compound that is being infused. It will be remembered that clearance determines both the steady state concentration and the time to steady state via the half-life of the drug.

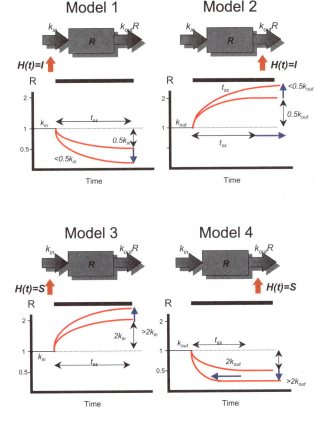

Figure 4.34 *Impact of dose on the time to dynamic steady state t_{ss} and the actual level at steady state R_{ss}. When drug is acting through inhibition (Model 1) or stimulation (Model 3) on k_{in}, only R at steady state is affected. When drug inhibits (Model 2) or stimulates (Model 4) the loss of response k_{out}, both R_{ss} and t_{ss} are affected. The red arrows below the turnover models indicate increase in drug dose (concentration) level. The effect of an increase in drug dose is mirrored by the blue arrows in the graphs which indicate the direction of the corresponding change in R_{ss} and t_{ss}. The horizontal black bars above the simulated curves represent the length of infusion.*

The prolonged approach to the new dynamic steady state with an increase in dose (Model 2, Table 4.2) is due to the increased half-life of loss of response. One has the opposite situation when drug stimulates the loss of response (Model 4, Table 4.2). Then, the half-life is shortened and the time to the new steady state is shorter. In section 5.4 on how to obtain initial parameter estimates, we discuss how to derive initial estimates for k_{in}, k_{out}, E_{max}/I_{max}, EC_{50}/IC_{50}, and n.

One may also speculate about the possibility of a combination of drug distribution into the biophase (effect-compartment) and indirect response (Figure 4.35). However, we have not seen data that supports such estimation of the model parameters with reasonable precision and consistency between observed and predicted data.

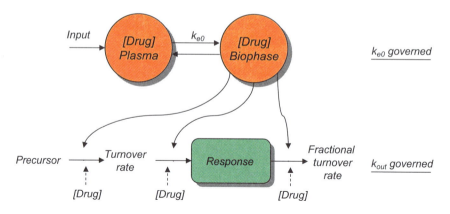

Figure 4.35 *The theoretical combination of distribution (effect-compartment) and indirect response models.*

   As pointed out earlier, we also believe that the majority of functional responses (observed effects) are indirect in nature. It takes time for a drug to provoke a stimulus that starts a reaction (synthesis, catabolism etc.) that eventually causes a measurable functional response. Some reactions occur so rapidly that they appear to be direct effects, but it is usually a matter of time and resolution of the data that determine the category one selects.

   When the effect-compartment model (see section 4.7.6 for the full example) is fit to data generated by Models 1 - 4 above (Table 4.2), dose dependency of estimated parameters (in the link/effect-compartment models) creates a biologically implausible situation where sensitivity IC_{50}/EC_{50}, capacity E_{max}, and n for a system change with dose. Therefore, the distribution (effect-compartment) model, while seemingly capable of characterizing the data, has severe limitations when applied to fitting data described by an indirect response. For further information about the above mentioned indirect dynamic response models, see Dayneka *et al* [1993] and Levy [1995].

4.6.4   Indirect response models – Irreversible mechanisms

We have in the previous sections (4.6.1 to 4.6.3) discussed *reversible* drug effects on the production or loss of response. In the following section *irreversible* drug action is elaborated. Figure 4.36 demonstrates the response-time courses of an irreversible system at two dose levels. See PD13 for further information about data, model and analysis.

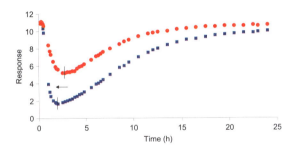

Figure 4.36 *Observed response-time data obtained from a study where oral (400 mg dose ● and 1600 mg ■) doses were administered to a single subject at two different occasions separated by a week. Note the shift to the left in the trough value (t_{max} decreases) with an increase in dose. The irreversible drug action is on the loss of response.*

Let Equation 4:98 express the turnover of response R of e.g., bacterial growth ($\mu \cdot R$) in the presence of bacterial killing ($-K \cdot D \cdot R$)

$$\frac{dR}{dt} = \mu \cdot R - K \cdot D \cdot R \tag{4:98}$$

K denotes the rate constant of bacterial killing and D is a function of drug concentration (Jusko [1971]). The μ term represents the availability of growth medium according to Equation 4:99.

$$\mu = \mu \ \frac{[S]}{K + [S]} \tag{4:99}$$

At high substrate concentrations $[S] \gg K$, Equation 4:99 can be simplified as

$$\mu = \mu_{max} \cdot \frac{[S]}{K \ll [S]} \approx \mu_{max} \cdot \frac{[S]}{[S]} = \mu_{max} \tag{4:100}$$

and becomes a constant first-order term μ_{max}. At low substrate concentrations, the expression of Equation 4:99 is valid.

$$\mu = \mu_{max} \cdot \frac{[S]}{K \gg [S]} \approx \mu_{max} \cdot \frac{[S]}{K} \approx \mu' \cdot [S] \tag{4:101}$$

For further reading on irreversible response modeling we refer the reader to PD13 and PD23.

4.7   Effect-Compartment Models

4.7.1   Background

The temporal displacement with respect to the effect and concentration may be due to an active metabolite, arterial/venous differences, drug tissue distribution phenomena and/or sensitization (e.g., up-regulation of the number of receptors). The dynamics of a drug can be studied in the absence of steady state conditions through the application of an effect-compartment model, linking a kinetic model and a dynamic model (e.g., EC_{50}, E_{max}). The latter relates the concentration of the drug in the biophase or effect-compartment C_e to a response (Figure 4.37). The effect-compartment (also called link-model) model can be viewed as a first-order distribution model that relates concentration in plasma and the biophase by means of a first-order constant. After drug is bound to the receptor (stimulus) a cascade of events may take place, eventually resulting in a biochemical or functional response.

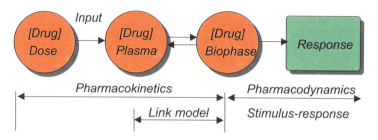

Figure 4.37 *The relationship between drug kinetics, the effect-compartment (link) model and drug dynamics.*

The effect-compartment model allows estimation of the *in vivo* pharmacodynamic relationship from non-steady state effect *versus* time and plasma concentration *versus* time. The model is valid for situations where there is an apparent temporal displacement between concentration and response (Segre [1950], Wagner *et al* [1968], Dahlström *et al* [1978], Sheiner *et al* [1979]). This time delay is often called hysteresis. Hysteresis may be counter clockwise (Figure 4.38) or clockwise.

If there is a time delay between the observed pharmacological (or toxicological) effect and plasma concentration, a plot of response *versus* concentration will demonstrate a hysteresis loop. Figure 4.38 (top) shows schematically the concentration and response-time profiles following an extravascular dose for a single valued system (left) and a delayed response (right). The bottom of Figure 4.38 shows the response *versus* concentration profile for these same two systems.

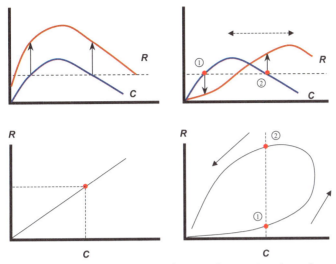

Figure 4.38 *Plasma concentration C and response R versus time t (upper figures) of a single valued system (upper left) and a system with delayed response (upper right). Effect versus plasma concentration (bottom figures) for the single valued system (bottom left) and delayed response system (bottom right). The arrows in the bottom right graph indicate the time course of response. This example of hysteresis may be caused by several factors, including an equilibrium delay between the sampling and effect-compartments. In the right hand figures, ① corresponds to early time points and ② to late time points.*

Figure 4.39 shows schematically the relationship between plasma and effect-compartment kinetics and response. The goal of fitting the effect-compartment model to response-time data is to obtain an estimate of k_{e0} causing the time delay between C and the effect-compartment concentration C_e and the pharmacodynamics (E_{max}, EC_{50} and n).

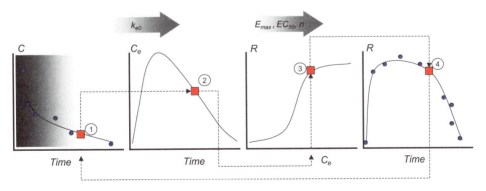

Figure 4.39 *Schematic diagram of the effect-compartment model. The drug is given as an intravenous bolus dose. The concentration at time point ① in plasma corresponds to the effect-compartment concentration at ②. The k_{e0} parameter will enable us to generate the C_e versus time relationship. The k_{e0} parameter gives us the time delay between C and C_e. The C_e at ② will generate a response corresponding to ③ provided response R versus time data are available (④) for estimation of E_{max}, EC_{50} and n.*

4.7.2   One-compartment models

A short summary of the equations for plasma and the hypothetical biophase concentration are derived for the one-compartment plasma disposition model.

$$A_p = D \cdot e^{-K \cdot t} \tag{4:102}$$

The rate of change of amount of drug A_e in the hypothetical effect-compartment can be expressed as

$$\frac{dA_e}{dt} = k_{1e} \cdot A_p - k_{eo} \cdot A_e \tag{4:103}$$

A_p is the amount of drug in the central (e.g., plasma) compartment of the pharmacokinetic model, linked to the effect-compartment, with rate constants k_{1e} and k_{e0}. The corresponding expression for the amount of drug in the effect-compartment, for a one-compartment model with bolus input of dose D is

$$A_e = \frac{k_{1e} \cdot D}{k_{e0} - K} \cdot \left[e^{-K \cdot t} - e^{-k_{e0} \cdot t} \right] \tag{4:104}$$

where K is the elimination rate constant. The concentration of drug in the effect-compartment C_e is obtained by dividing A_e by the effect-compartment volume V_e

$$C_e = \frac{k_{1e} \cdot D}{V_e \cdot (k_{e0} - K)} \cdot \left[e^{-K \cdot t} - e^{-k_{e0} \cdot t} \right] \tag{4:105}$$

At equilibrium, the rates of drug transfer between the central and effect-compartments are equal

$$k_{1e} \cdot A_p = k_{e0} \cdot A_e \tag{4:106}$$

or

$$k_{1e} \cdot V_c \cdot C = k_{e0} \cdot V_e \cdot C_e \tag{4:107}$$

V_c is the central compartment volume and C the plasma concentration. If the partition coefficient K_p equals C_e/C at equilibrium, then we can rearrange Equation 4:107 to get

$$V_e = \frac{k_{1e} \cdot V_c}{K_p \cdot k_{e0}} \tag{4:108}$$

Substituting for V_e in Equation 4:108 (i.e., assuming $k_{1e} = k_{e0}$ at equilibrium) yields

$$C_e = \frac{k_{1e} \cdot D \cdot K_p}{V_c \cdot (k_{e0} - K)} \cdot \left[e^{-K \cdot t} - e^{-k_{e0} \cdot t} \right] \tag{4:109}$$

At equilibrium, C will be equal to C_e/K_p by definition, which gives

$$C_e = \frac{k_{e0} \cdot D}{V_c \cdot (k_{e0} - K)} \cdot \left[e^{-K \cdot t} - e^{-k_{e0} \cdot t} \right] \qquad (4{:}110)$$

Table 4.5 contains a summary of one-compartment plasma equations for bolus, first-order and zero-order input and their corresponding C_e functions.

Table 4.5 Comparison of one-compartment link models.

Input	Plasma equation ($C =$)	Effect-compartment equation ($C_e =$)*
Bolus	$\dfrac{D_{iv}}{V} \cdot e^{-K \cdot t}$	$\dfrac{k_{1e} \cdot D_{iv}}{V \cdot (k_{e0} - K)} \cdot \left[e^{-K \cdot t} - e^{-k_{e0} \cdot t} \right]$
First-order	$\dfrac{K_a F D_{po}}{V \cdot (K_a - K)} \cdot \left[e^{-K \cdot t} - e^{-K_a \cdot t} \right]$	$\dfrac{k_{1e} \cdot K_a F D_{po}}{V} \cdot \left[\dfrac{e^{-K_a \cdot t}}{(K - K_a)(k_{e0} - K_a)} + \right.$ $\left. \dfrac{e^{-K \cdot t}}{(K_a - K) \cdot (k_{e0} - K)} + \dfrac{e^{-k_{e0} \cdot t}}{(K_a - k_{e0}) \cdot (K - k_{e0})} \right]$
Zero-order	$\dfrac{R_{in}}{K \cdot V} \cdot \left[1 - e^{-K \cdot T_{inf}} \right] e^{-K \cdot t'}$	$\dfrac{k_{1e} \cdot R_{in}}{K \cdot V \cdot (k_{e0} - K)} \cdot \left[1 - e^{-K \cdot T_{inf}} \right] \cdot e^{-K \cdot t'} +$ $\dfrac{k_{1e} \cdot R_{in}}{k_{e0} \cdot V \cdot (K - k_{e0})} \cdot \left[1 - e^{-k_{e0} \cdot T_{inf}} \right] \cdot e^{-k_{e0} \cdot t'}$

*Corrected with Equation 4:106.

Figure 4.40 shows three simulations of the concentration-time course in the effect-compartment C_e for different k_{e0} values and a constant plasma concentration. As you can see, C_e (solid lines) will approach the steady state with different rates depending on how large the k_{e0} value is. Also, at equilibrium, C_e will equal the plasma concentration provided K_p is equal to unity.

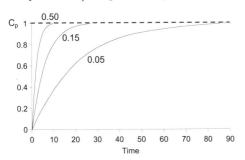

Figure 4.40 *The behavior of C_e (solid black lines) with changing k_{e0} values and a constant plasma concentration (C_p, dashed blue line). Note that C_e and C_p are equal at steady state when K_p is equal to unity.*

4.7.3 Two-compartment models

The equations for plasma and the hypothetical biophase concentration are given below for the two-compartment plasma disposition model. Concentrations in plasma for intravenous *bolus input*

$$C = \frac{D_{iv}}{V_c} \cdot \left\{ \frac{k_{21} - \alpha}{\beta - \alpha} \cdot e^{-\alpha \cdot t} + \frac{k_{21} - \beta}{\alpha - \beta} \cdot e^{-\beta \cdot t} \right\} \qquad (4:111)$$

Concentrations in plasma for *first-order input*

$$C = \frac{K_a F D_{po}}{V_c} \left\{ \frac{(k_{21} - \alpha) \cdot e^{-\alpha \cdot t}}{(K_a - \alpha) \cdot (\beta - \alpha)} + \frac{(k_{21} - \beta) \cdot e^{-\beta \cdot t}}{(K_a - \beta) \cdot (\alpha - \beta)} + \frac{(k_{21} - K_a) \cdot e^{-K_a \cdot t}}{(\alpha - K_a) \cdot (\beta - K_a)} \right\} \qquad (4:112)$$

Concentrations in plasma for *zero-order input* (constant intravenous infusion)

$$C = \frac{R_0}{V_c} \left\{ \frac{(k_{21} - \alpha) \cdot (1 - e^{-\alpha \cdot T_{inf}}) \cdot e^{-\alpha \cdot (t - T_{inf})}}{\alpha \cdot (\beta - \alpha)} + \frac{(k_{21} - \beta) \cdot (1 - e^{-\beta \cdot T_{inf}}) \cdot e^{-\beta \cdot (t - T_{inf})}}{\beta \cdot (\alpha - \beta)} \right\} \qquad (4:113)$$

where T_{inf} is the duration of infusion and $t - T_{inf}$, time after end-infusion.

The concentration of the drug in the effect-compartment for *intravenous bolus input* is

$$C_e = \frac{k_{1e} \cdot D_{iv}}{V_c} \cdot \left[\frac{(k_{21} - \alpha) \cdot e^{-\alpha \cdot t}}{(k_{e0} - \alpha) \cdot (\beta - \alpha)} + \frac{(k_{21} - \beta) \cdot e^{-\beta \cdot t}}{(k_{e0} - \beta) \cdot (\alpha - \beta)} + \frac{(k_{21} - k_{e0}) \cdot e^{-k_{e0} \cdot t}}{(\alpha - k_{e0}) \cdot (\beta - k_{e0})} \right] \qquad (4:114)$$

The concentration of the drug in the effect-compartment for *first-order input* is

$$C_e = \frac{k_{1e} \cdot K_a F D_{po}}{V_c} \cdot \left[\frac{(k_{21} - K_a) \cdot e^{-K_a \cdot t}}{(\alpha - K_a) \cdot (\beta - K_a) \cdot (k_{e0} - K_a)} + \frac{(k_{21} - \alpha) \cdot e^{-\alpha \cdot t}}{(K_a - \alpha)(\beta - \alpha) \cdot (k_{e0} - \alpha)} + \right.$$

$$\left. + \frac{(k_{21} - \beta) \cdot e^{-\beta \cdot t}}{(K_a - \beta) \cdot (\alpha - \beta) \cdot (k_{e0} - \beta)} + \frac{(k_{21} - k_{e0}) \cdot e^{-k_{e0} \cdot t}}{(\alpha - k_{e0}) \cdot (\beta - k_{e0}) \cdot (K_a - k_{e0})} \right] \qquad (4:115)$$

The concentration of the drug in the effect-compartment for *zero-order input* is

$$C_e = \frac{k_{1e} \cdot R_0}{V_c} \cdot \left[\frac{(k_{21} - \alpha) \cdot (1 - e^{-\alpha \cdot T_{inf}}) \cdot e^{-\alpha \cdot (t - T_{inf})}}{\alpha \cdot (\beta - \alpha) \cdot (k_{e0} - \alpha)} + \frac{(k_{21} - \beta) \cdot (1 - e^{-\beta \cdot T_{inf}}) \cdot e^{-\beta \cdot (t - T_{inf})}}{\beta \cdot (\alpha - \beta) \cdot (k_{e0} - \beta)} + \right.$$

$$\left. + \frac{(k_{21} - k_{e0}) \cdot (1 - e^{-k_{e0} \cdot T_{inf}}) \cdot e^{-k_{e0} \cdot (t - T_{inf})}}{k_{e0} \cdot (\alpha - k_{e0}) \cdot (\beta - k_{e0})} \right] \qquad (4.116)$$

The effect-compartment concentrations above obtained from Colburn [1981].

4.7.4   Integration of time into the Hill equation

The parameters to be estimated by means of the effect-compartment model are E_{max}, EC_{50}, k_{e0}, and n of the Hill equation. That is achieved iteratively after having substituted the aforementioned expressions for C_e in the effect equation

$$E = \frac{E_{max} \cdot \left(k_{e0} \cdot C_e / k_{1e}\right)^n}{EC_{50}^n + \left(k_{e0} \cdot C_e / k_{1e}\right)^n} \tag{4:117}$$

Figure 4.41 shows the effect-time course after a bolus dose for a system where there is a delay between plasma kinetics and drug response.

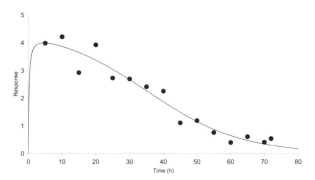

Figure 4.41 *Observed and predicted response versus time data. The effect-compartment model that relates the kinetics in plasma to the kinetics of the drug in the effect-compartment may be used together with the effect equation (e.g., ordinary E_{max} model) for estimation of E_{max}, EC_{50}, and k_{e0}.*

Colburn compiled a substantial number of effect-compartment models for the one-, two-, and three-compartment models with bolus, first-order and zero-order input (see Colburn [1981]).

4.7.5   Alternative parameterizations

If we assume that there is no partitioning (no concentration differences at steady state) of drug between plasma and the effect-compartment. We can then view the effect-compartment concentration by means of convolution (* denotes the convolution operator. See also Section 5.4.2) of the plasma equation with the first-order rate constant k_{e0} which gives

$$C_e = k_{e0} * C^0 \cdot e^{-K \cdot t} \tag{4:118}$$

This (Equation 4:118) assumes that drug is given as a bolus and follows one-compartment disposition kinetics, and that rate in and rate out of the effect-compartment is governed by the first-order rate constant k_{e0}. The effect-compartment concentration then becomes

$$\frac{dC_e}{dt} = k_{e0} \cdot C - k_{e0} \cdot C_e \tag{4:119}$$

$$C_e = \frac{k_{e0} \cdot D}{V_c \cdot \left(k_{e0} - K\right)} \cdot \left[e^{-K \cdot t} - e^{-k_{e0} \cdot t}\right] \tag{4:120}$$

If the plasma concentration could be established instantaneously, the effect-compartment concentration would approach the plasma concentration with a rate determined by k_{e0}. The larger the k_{e0} the shorter the time to equilibrium in the biophase. $t_{1/2ke0}$, which is $\ln(2/k_{e0})$, will then determine the time to dynamic equilibrium.

4.7.6   Problems and pitfalls

Using Model 3 (Table 4.2 section 4.6.2), we generated three data sets for the effect-time profile for each of three dose levels. A nonlinear drug function $H(t)$ (Table 4.4) was used. Three per cent error (3% CV) was added to the data. These data sets were then fit simultaneously and individually by means of the effect-compartment model. When fitting all three data sets simultaneously, the predicted curves peaked at the same time point (implicit in the distribution model) and systematic deviations were seen between observed and predicted response (Figure 4.42).

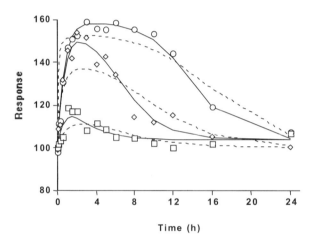

Figure 4.42 *Observed (symbols) and predicted effects (dashed and solid lines). Solid lines represent predicted values of the indirect response model and dashed lines the effect-compartment model. Note how the effect-compartment model peaks at the same time independent of dose, whereas the indirect model peaks later with increasing doses.*

When fitting the data sets separately, three different sets of final parameter estimates of implausible EC_{50} values were obtained. The final parameter values are summarized in Table 4.6.

Table 4.6 Comparison of parameter estimates. The values at the bottom of the table were used in the indirect response (*IRP*, turnover) model to generate the data

Fitting	k_{e0}	E_{max}	EC_{50}	n
Simultanously	0.0075	55.8	0.206	0.91
Dose 1	0.0166	447	0.681	1.33
Dose 10	0.0096	437	4.85	0.88
Dose 100	0.0031	68.1	0.941	1.62
Original *IRP*	k_{in} 2.5, k_{out} 0.024	0.532	0.164	1.47

IRP denotes the indirect response model.

Note that experimental data peaked later with increasing doses. For a more thorough comparison between the two classes of models, see Dayneka *et al* [1994].

When the traditional effect-compartment model (see Gabrielsson and Weiner [1997] for the full example) is fit to data generated by Models 1 - 4 in sections 4.6.2, dose dependency of estimated effect-compartment model parameters creates a biologically implausible situation where sensitivity IC_{50}/EC_{50}, the capacity E_{max}, and the n value for a system change with dose. Therefore, the distribution (effect-compartment or link) model, while seemingly capable of characterising the data, has severe limitations when applied to fitting data described by an indirect response. For further information about the above-mentioned indirect dynamic response models, see Dayneka *et al* [1993] and Levy [1995]. See PK6 for a discussion about the underlying factors to peak-shifts.

4.8   Dose-Response-Time Models

4.8.1   Background

For dosage forms that deliver drug directly to the blood for systemic distribution, the measurement of drug concentrations in an accessible biological fluid (e.g., blood or urine) as a function of time is preferred for bioequivalence testing. However, for oral drug products that are not well absorbed or for local drug deliver (e.g., topical dermatological or inhalation products or drugs instilled into the eye), quantification of drug concentrations in biological fluids may not be appropriate or have sufficient accuracy, precision or sensitivity to allow for bioequivalence assessment (Lesko and Williams [1994]).

Response *versus* time data may contain information about a drug's biophase kinetics and dynamics (biophase may include e.g., total or free plasma or blood concentration, tissue concentration, CSF, urine, blood cell concentration). For a recent review see Gabrielsson *et al* [2000]. Thus, the question is, can any useful information about a drug's kinetic/dynamic relationship be obtained when only effect *versus* time data are available?

Figure 4.43 dissects the problem of what pure drug-related and response turnover issues are, and how we can discriminate between the two. The drug kinetics A, is a function of K, K_a and F, and the turnover of response is a function of k_{in} and k_{out}. The pharmacodynamic determinants are then potency ED_{50}/ID_{50} and efficacy E_{max}/I_{max}. By means of optimal (or near optimal) design it would be possible to dissect the response time course and to determine the aforementioned parameters with good precision.

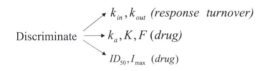

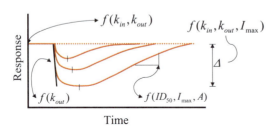

Figure 4.43 *Response turnover, biophase kinetics and pharmacological factors that all contribute to the response-time profile. The bottom part shows what determines the various parts of the response-time profile (Gabrielsson [1996]).*

There are situations where clinical endpoints (biomarkers) are more relevant to bioequivalence assessment than pharmacokinetic measures for certain drugs. For review see Smolen [1971], Levy [1969, 1993], Lesko and William [1994], Colburn [1995], Gabrielsson *et al* [2000].

4.8.2   Miotic data

Figure 4.44 shows a schematical diagram of the eye, with the pharmacological action of a prostaglandin drug indicated by the heavy downgoing arrow (Gabrielsson and Weiner [1997], Gabrielsson *et al* [2000]).

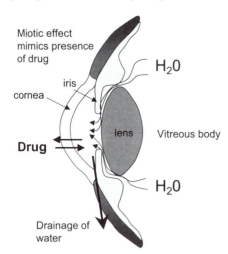

Figure 4.44 *The prostaglandin is instilled into the conjuctival cavity and passes across the cornea. The step across the cornea is assumed to be the rate-limiting step. Some of the drug may be degraded during this passage. Drug acts by increasing the outflow of water. The more drug present, the larger the outflow of water is assumed to be. The miotic effect is a surrogate marker for the presence of drug in the eye and indirectly the effect on lowering water pressure (Gabrielsson [1996]).*

A combined parametric model was fit simultaneously to the response-time data collected from three dose levels (Figure 4.45). The biophase kinetics of a new prostaglandin was characterized by a first-order input-output model including a lag-time (Figure 4.45 and Equation 4:121). The pharmacodynamic component comprised an inhibitory sigmoid E_{max} model Equation 4:122, incorporating the kinetic model variable A_{ev}. A_{ev} is the amount of drug in the biophase at the receptor and is seldom identical to plasma concentration. Observed and predicted response *versus* time data are shown in Figure 4.45 (top left).

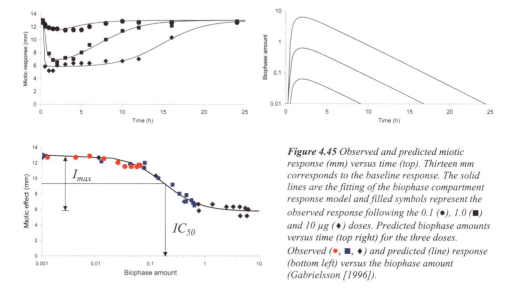

Figure 4.45 *Observed and predicted miotic response (mm) versus time (top). Thirteen mm corresponds to the baseline response. The solid lines are the fitting of the biophase compartment response model and filled symbols represent the observed response following the 0.1 (●), 1.0 (■) and 10 μg (◆) doses. Predicted biophase amounts versus time (top right) for the three doses. Observed (●, ■, ◆) and predicted (line) response (bottom left) versus the biophase amount (Gabrielsson [1996]).*

Equation 4:121 describes the time profile for biophase amount A_{ev} of the drug after extravascular (*ev*) dosing, expressed as a fraction of the applied dose (i.e. D_{ev} - where the doses were 0.1, 1.0 and 10 μg), where the ratio of the volume of the biophase compartment V to the biophase availability of the applied dose F_{eff} is set to unity.

$$A_{ev} = \frac{k_a F D_{ev}}{k_a - k} \cdot \left[e^{-k_a(t - t_{lag})} - e^{-k(t - t_{lag})} \right] \qquad (4:121)$$

Equation 4:122 predicts that, with increases in A_{ev}, the effect E (i.e. diameter of the pupil) will decrease from pre-dose baseline values E_0, following a sigmoid E_{max} relationship.

$$E = E_0 - \frac{I_{max} \cdot A_{ev}^n}{ID_{50}^n + A_{ev}^n} \qquad (4:122)$$

Simulated pupil diameter (effect) *versus* hypothetical biophase amount A_{ev} is shown in the bottom left graph of Figure 4.45. This relationship is also commonly called the transduction function. E_{max} is the drug induced maximal constriction of the iris. EC_{50} is the hypothetical biophase concentration at 50 % of E_{max} and n the sigmoidicity parameter. The parameters K_a, t_{lag}, and K are respectively, the first-order input rate constant, lag-time during input, and first-order disposition rate constant (assuming K_a >> K for the drug with respect to the biophase). Therefore, based upon simultaneous fitting of Equations 4:121 and 4:122 to effect *versus* time data, the underlying temporal pattern, e.g., K_a, t_{lag}, and K of the kinetics in the biophase, can be estimated and is presented in Table 4.7.

Table 4.7   Parameter estimates and their corresponding $\pm$ CV % of the miotic dose-response-time data model. n is the exponent in Equation 4:122.

	$t_{1/2}Ka$ (min)	$t_{1/2}K$ (min)	t_{lag} (min)	I_{max} (mm)	ID_{50} (μg)	n
Mean $\pm$ CV %	35.4 ± 27	139 ± 6.8	25.7 ± 5.6	7.30 ± 2.0	0.175 ± 12	1.29 ± 7.5

The assumption that there is only one effect observation for any single A_{ev} (single valued response) is important if the true underlying time-course of the drug at the receptor (biophase) is to be modeled according to Equation 4:122. However, for this to be valid it must be assumed that the measured effect is in instantaneous equilibrium with A_{ev}, and that the F_{eff}, and E_0 parameters display stationarity (i.e., remain time-invariant). In addition, only one pharmacologically active species may be present i.e., pharmacologically active drug may not concurrently be present with metabolites that may act as agonists/antagonists with respect to the measured effect.

4.8.3   Acetylcholinesterase turnover

In this example, the aim is to elaborate on a dose-response-time model based on acetylcholinesterase *AChE* activity in red blood cells rather than to use the plasma concentration-driven effect model after administration of the drug. Blood *AChE* (surrogate marker) is thought to reflect activity of brain *AChE*. Brain *AChE*, which is the main form of cholinesterase in the mammalian brain, is thought to be related to cognitive effects associated with changes in *ACh* levels.

A turnover model was fitted to *AChE* time data (Taylor *et al* [1994]). The level of *AChE* activity was proportional to turnover rate k_{in} and inversely proportional to fractional turnover rate k_{out}. Figure 4.46 illustrates a tentative model for the drug action on *AChE*. Drug binds reversibly to active enzyme R_a and forms R_i, which is the inactive form of the enzyme. The enzyme is then irreversibly eliminated via k_{out} or is regenerated via k_2.

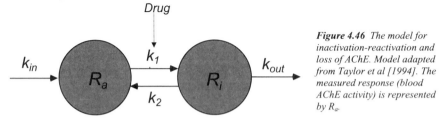

Figure 4.46   The model for inactivation-reactivation and loss of AChE. Model adapted from Taylor et al [1994]. The measured response (blood AChE activity) is represented by R_a.

The rate of change of the response over time can be described by

$$\frac{dR}{dt} = k_{in} - k_1 \cdot R \cdot H(t) + k_2 \cdot R_i \qquad (4:123)$$

$$\frac{dR_i}{dt} = k_1 \cdot R \cdot H(t) - k_2 \cdot R_i - k_{out} \cdot R_i \qquad (4:124)$$

The turnover parameters k_{in} and k_{out} are the zero-order turnover rate and the first-order fractional turnover rate, respectively. $H(t)$ is the driving function of drug on the response. The parameters of the input function (the rate constant K for loss of orally dosed drug D_{po} and potency ED_{50}) were estimated simultaneously with k_{in} and k_{out}. The total (parent drug and potentially active metabolites) amount of drug in the biophase A_B is given by Equation 4:125

$$A_B = D_{po} \cdot e^{-K \cdot t} \qquad (4:125)$$

$$H(t) = 1 + \left(\frac{A_B}{ED_{50}} \right)^n \qquad (4:126)$$

The level of response at pharmacodynamic steady state R_{ss} is then assumed to be time-independent, and can be expressed as follows

$$R_{ss} = \frac{k_{in}}{k_{out}} \cdot \frac{k_2 + k_{out}}{k_1 \left[1 + \dfrac{A_B}{ED_{50}} \right]} \qquad (4:127)$$

Observed and model predicted data proved to be consistent, as can be seen in Figure 4.47. The left hand plot shows data from acute (one week) exposure to the drug and the right hand plot shows data from extended (two months) exposure to drug.

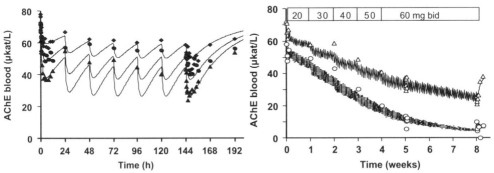

Figure 4.47 The mean AchE activity versus time relationship at the fixed 40, 60 and 80 mg daily dose levels are shown during a one week treatment period (left). The individual AChE activity versus time relationship of two patients are shown during a two month treatment period (right) with escalating daily doses of 40 to 120 mg. Each of the first doses was given for a week and the highest (120 mg) was given from weeks five to eight (Gabrielsson [1996, 1998]).

The response at steady state was determined by Equation 4:127, where ID_{50} corresponds to 50 % of the baseline value. Figure 4.48 (right) shows a simulation of the response during and after dosing. Note the slow recovery period. This type of analysis lends itself for experimental design utilizing a more optimal dose and sampling schedule in future studies/protocols. The model predicts 100 % inhibition at an infinite dose. Figure 4.48 (left) also displays the response *versus* time data at three different dose levels. As can be seen, the model also predicts that the time to steady state is reduced with increasing doses.

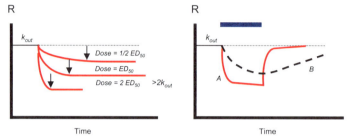

Figure 4.48 *Decrease in time to steady state as a function of dose. Arrows indicate an approximate steady state level at three dosing rates (left). Effect versus time in two different subjects (A and B), each given one of two dose levels during and after dosing (right). The horizontal bar in the right hand figure indicates the dosing period (Gabrielsson [1996]).*

4.8.4   Antinociception

In the following example, an analgesic was given via the intravenous and subcutaneous routes, and *time* taken for the animals to respond to a noxious stimulus (i.e., response) was determined at different times after administration. Response-time data were obtained from each dose level and route after a single dose (Figure 4.49). Figure 4.50 shows the response versus log-dose relationship (Gabrielsson and Weiner [1997], Gabrielsson *et al* [2000]).

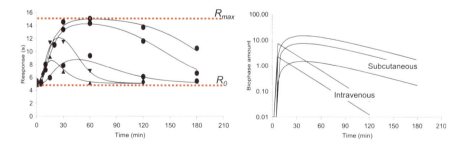

Figure 4.49 *The left plot contains observed and predicted responses versus time (top) following intravenous (▲▼) and subcutaneous administration (●). The horizontal dashed lines indicate the baseline value (R_0) and maximum allowed effect (R_{max}). The upper right plot shows predicted biophase amounts versus time (center). Note the flip-flop behavior after subcutaneous dosing.*

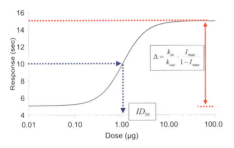

Figure 4.50 Predicted response versus the dose (Gabrielsson [1998]). The delta (Δ) response and ID_{50} values are also shown.

The underlying kinetic model is shown in Equations 4:128 and 4:129. A first-order input/output is assumed for extravascular dosing. The kinetic model then drives the inhibition function of the dynamic model. The dynamic behavior is supposed to be described by an indirect response model, which is given by Equation 4:131. The amount of drug in the biophase after an intravenous dose can be modeled with mono-exponential decline

$$A_s = A_i \cdot e^{-a_i \cdot t} \tag{4:128}$$

The amount of drug in the biophase A_{ev} after an extravascular dose can be modeled with first-order input/output kinetics

$$A_{ev} = \frac{k_a F D_{ev}}{k_a - k} \cdot \left[e^{-k_a (t - t_{lag})} - e^{-k(t - t_{lag})} \right] \tag{4:129}$$

The inhibitory effect of the drug is

$$H(t) = 1 - \frac{I_{max} \cdot A}{ID_{50} + A} \tag{4:130}$$

The turnover of response

$$\frac{dR}{dt} = k_{in} - k_{out} \cdot R \cdot H(t) \tag{4:131}$$

Following subcutaneous (sc) administration, the half-life of absorption was about 36 min and the half-life for elimination was 15 minutes. ID_{50} was 1.1 μg and the predicted baseline value was 5.1 seconds. The estimated half-life of response was about 3 min.

The predicted subcutaneous bioavailability (66%), by means of the proposed dynamic model, was consistent with that based on plasma concentration-time. See also PD24 for further information about this particular dataset.

4.8.5   Body temperature

The primary goal of out body temperature modeling exercise is to demonstrate a dose-response relationship of the effect of *8-OH-DPAT* on the hypothermic response, from which one can select an appropriate dose level for subsequent safety studies (Deveney *et al* [1999]). *8-OH-DPAT* lowers the turnover rate of 5-hydroxytyramine *5-HT*, an important mediator of temperature raising mechanisms. Thus, the net (indirect) effect of blocking the synthesis rate of *5-HT* is decreased body temperature.

Five groups of rats received a subcutaneous dose of 0, 0.03, 0.125, 0.50 or 2 mg *8-OH-DPAT* in a dose-ranging study (Deveney *et al* [1999]). The decrease in body temperature from the baseline was obtained and is plotted in Figure 4.51 as mean values against time (Gabrielsson [1996], Gabrielsson *et al* [2000]).

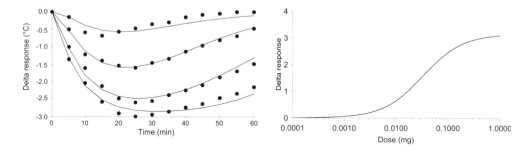

Figure 4.51 *Observed and predicted delta response-time data after four subcutaneous doses (0.03, 0.125, 0.5 and 2 mg) of 8-OH-DPAT (left). Simulated delta-response versus dose profile (right). The delta-response corresponds to absolute change in body temperature from the baseline value. Maximum reduction of body temperature, delta (Δ), is equal to 3 °C. Data were adapted from Deveney et al [1999].*

The selected dose-driven model is shown in Figure 4.43 (Model 3 in Table 4.2, see Gabrielsson *et al* [2000] for a more indepth discussion of the model). We assume that we can *approximate* the biophase compartment kinetics by a first-order input/output model, where the input-rate constant is equal to the output-rate constant (k') in order to reduce the number of model parameters. With these assumptions, the biophase amount A_{ev} after extravascular dosing of drug becomes

$$A_{ev} = Dose_i \cdot k' \cdot t \cdot e^{-k' \cdot t} \tag{4:132}$$

Note that Equation 4:132 will never exhibit a straight terminal slope on a semi-logarithmic scale since k' is multiplied by time (t). Equation 4:132 was then fed into the inhibitory function (Equation 4:133) acting on production of response (temperature). The inhibitory effect of drug is

$$H(t) = 1 - \frac{I_{max} \cdot A_{ev}}{ID_{50} + A_{ev}} \tag{4:133}$$

Since Δ-temperature (change from baseline) is modeled, we wrote the equation to be fitted to data as

$$\Delta\text{-}Temp = k_{in} \cdot H(t) - k_{out} \cdot R - R_0 \tag{4:134}$$

R_0 is the baseline response defined as k_{in}/k_{out}. Because k_{in} only shifts the response horizontally, k_{in} can be set to any constant value. The function of the response at steady state is

$$R_{ss} = R_0 \cdot H(t) = \frac{k_{in}}{k_{out}} \cdot \left[1 - \frac{I_{max} A^n}{ID_{50}^n + A^n} \right] \qquad (4{:}135)$$

This relationship is the dose-response function for the studied compound with respect to body temperature in the rat. The function starts at a value equal to k_{in}/k_{out} and decreases with increasing values of A (or *Dose*). The maximum obtained inhibition (Δ_{max}), when $A >> ID_{50}$, becomes

$$\Delta_{max} = R_0 - R_0 \cdot H(t) = \frac{k_{in}}{k_{out}} - \frac{k_{in}}{k_{out}} \cdot \left[1 - \frac{I_{max} A_\infty^n}{ID_{50}^n + A_\infty^n} \right] = \frac{k_{in}}{k_{out}} \cdot I_{max} \qquad (4{:}136)$$

The maximum obtained effect of drug was estimated to be 3°C ($k_{in} I_{max}/k_{out}$) and the potency (ID_{50}) was about 33 µg. Figure 4.51 (left) contains observed and predicted body temperature. The trough value is shifted to the right with increasing dose. Figure 4.51 (right) plots the change (decrease) in temperature from the baseline value against dose. The half-life of k' was about 8 min.

4.8.6   Conclusions about dose-response-time data modeling

The underlying principles of modeling dose-response-time data and duration of effect have been laid out in a number of articles (Levy [1964, 1969], Jusko [1971], Smolen [1971, 1976], Gabrielsson *et al* [2000]). The kinetic part of the model is a function of dose, route and some arbitrary parameters, and serves as the input function. This part is *deconvolved* by means of regression. A number of questions need to be considered. Does the drug have an inhibitory or stimulating effect? What is the mechanism of action? How does the effect-time course change with dose? Are multiple dose data available? Does the effect reach a plateau with higher doses? Does the response peak time change with dose? It is important to realize that information about the mechanism is central to designing a model.

The steps involved in dose-response-time data modeling are summarized below but are further evaluated in Gabrielsson *et al* [2000]. *Step one* includes the anticipated kinetic model; in other words, how the drug enters, and is eliminated from, the biophase. *Step two* is the construction and implementation of the dynamic model, and requires that initial estimates for the parameters are obtained. This is the most difficult part of the process, and we recommend that you obtain data from different routes and dose levels. *Step three* involves simulation exercises with the complete model. Does the model mimic the observed data? Are the initial parameter estimates reasonable? *Step four* involves fitting the model to the data. This should ideally involve fitting combinations of dynamic data from different routes of administration (e.g., intravenous and extravascular) in order to estimate the biophase availability. *Step five* involves construction of the dose-response profiles.

It has been assumed that dose-response-time data analysis might be expected to fail unless there is linear kinetics and dynamics, time constant parameters, no tolerance/adaptation/rebound, and no active metabolites.

We believe, however, that the drug effect can be indirect (*AChE* inhibition and antinociception) or exhibit hysteresis with the drug levels in the biophase (*AChE*-inhibition, body temperature), or display nonlinear kinetics (AChE inhibition). We do not rule out that dichotomous response data can also be modeled, provided the data are of good quality. We also think that a good design of dose-response-time data including two or more dose levels and repeated dosing could identify nonlinearities such as capacity (Michaelis-Menten elimination kinetics) and time-dependencies (tolerance), although the complexity would increase. Crucial determinants of the success of modeling dose-response-time data are dose selection, repeated dosing, and, to some extent, different input rates and routes. Smolen proposed that the use of pharmacological data for bioequivalence studies etc. should always be considered irrespective of whether or not pharmacokinetic data are available, and the decision to use either or both types of data should depend upon their relative sensitivity, precision, convenience, and economy (Smolen [1976]). For similar recent work, see Verotta and Sheiner [1991], Warwick *et al* [1996].

4.9   Tolerance and Rebound Models

4.9.1   Background

So far, we have discussed various models of drug action with and without a time delay between kinetics and dynamics. However, we have not taken into account what will occur when the attained response is counterbalanced by some endogenous mechanism. Exposure of drugs to cells and tissues over an extended period of time generally causes progressive reduction in the response to the drug. This phenomenon is commonly termed desensitization, tachyphylaxis (Goodman and Gilman [1990]) or *tolerance*. Tolerance phenomena may occur as a result of down regulation of receptors, or depletion of co-factors, precursors, endogenous compounds, or messengers (Jusko [1996]). The pattern of tolerance varies according to the extent to which these different systems are modified. In some cases, only the signal from the stimulated receptor is altered. This is denoted *homologous* desensitization. In *heterologous* desensitization, the desensitization is due to post-receptor events. Tachyphylaxis (contrasts *anaphylaxis*) generally refers to the acute phenomenon when a pool of endogenous substance is depleted (e.g., repeated administration of a stimulant of histamine release depletes histamine in the mast cells). A related phenomenon to tolerance is the so-called *rebound*, when a system after cessation of stimulation exhibits a response which is less than the baseline response. Rebound may also occur in the opposite direction, in that inhibition of a system causes an excessive response above baseline upon cessation of inhibition. Figure 4.52 illustrates schematically the time course of response before, during and after drug exposure for a stimulatory drug effect (left) and an inhibitory drug effect (right).

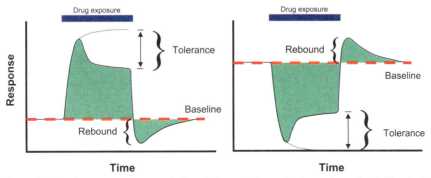

Figure 4.52 The time course of response before, during and after a stimulatory drug effect (left) and after an inhibitory drug effect (right). The dashed red line corresponds to the baseline response and the blue line to the drug exposure period. The fine black line represents the time course of response in a system with no tolerance development. The vertical arrow indicates the lack of full response due to tolerance development.

The easiest way to observe tolerance is by means of repeated doses or continuous controlled input. The dose size, frequency of dosing, and duration are kinetic determinants of the rate and extent to which tolerance and rebound develop. The ability to discriminate between effect and rebound is partly determined by the kinetics of drug. A drug with slow kinetics, such as a controlled release dosage form, might still be able to develop tolerance, although very little rebound may be seen.

In the following section we will lay out the concepts of the *feedback* and *pool* models and demonstrate some of their differences (Figure 4.53). The models presented provide a simplistic view of how the system will behave when opposing processes are operating simultaneously, and will enable us to predict the time course of drug action (R) as well as the hypothetical modulator (M) or the pool.

The feedback model originates from the physiological examples such as, glucose-insulin regulation (Ackerman *et al* [1964], Ceresa *et al* [1968] and Segre *et al* [1970], Cobelli *et al* [1982]) and regulation of follicle stimulating hormone and estradiol (Rang and Dale [1992]). This model may also be used to model the plasticity of enzyme induction. This is further discussed in section 3.6.3.

Amongst other things, the pool or precursor model has been used to model prolactin balance (Sharma *et al* [1995]). It can conceptually be compared to tachyphylactic systems, although its flexibility reaches beyond that. Figure 4.53 displays schematically the concepts of the negative feedback model (left) and pool model (right).

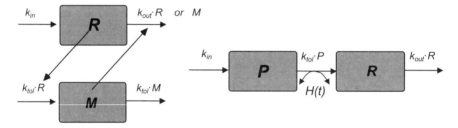

Figure 4.53 *The negative feedback control model (left) and the pool model (right). R represents the drug effect, M the effect of the endogenous modulator, P the precursor and k_{out} and k_{tol} the first-order rate constants of R and M (P). H(t) indicates where drug has to act in the pool model in order to display tolerance/rebound in the R-compartment.*

Tolerance is often manifested by a clockwise response *versus* concentration relationship. However, this may not always be so. Anti-clockwise response *versus* concentration may also occur, provided the rate of presentation of compound into the system is more rapid than the rate of distribution or tolerance development.

4.9.2   Feedback model - 1

Finkelstein and Carson [1985] described the form of control shown in Figure 4.53 (left), which is known as control by feedback. The information concerning the state of the system to be controlled, in this case R, is *fed back* so as to alter variables affecting the system state, in this case the outflow rate, and thus the system state. The feedback is said to be negative because a transient rise of R causes, through the resultant increase in outflow, a reduction of R. It is even possible to come up with a function for non-stationarity in k_{in}, such as the circadian rhythm.

Figure 4.54 shows a schematic response-time curve before, during and after drug exposure to a drug that induces both tolerance and rebound. The response-time curve includes the *overshoot* immediately after drug stimulus, the *shoulder* during drug stimulus and the *rebound* after cessation of drug stimulus. Note also that the AUC_E is asymmetric with AUC_R.

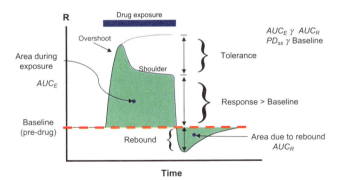

Figure 4.54 *Schematic illustration of tolerance and rebound during the response-time course of a drug (feedback model). Drug exposure is indicated by the horizontal bar.*

Figure 4.54 showed schematically the principal parts of a system that exhibited feedback, tolerance and rebound. Figure 4.55 shows literature data of *ACTH/Cortisol* dynamics. Note the baseline, overshoot, shoulder, rebound and fluctuation about the baseline. Also note the asymmetric shapes of the response above and below baseline. A more complete analysis of this system is given in exercise PD11.

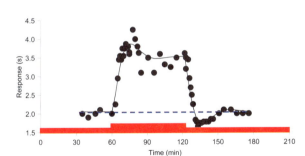

Figure 4.55 *Time course of observed (●) and predicted (solid line) cortisol secretion rate following a stepwise increase (red bars) in ACTH concentration from 1 to 2 µU/mL, and then following a stepwise decrease back to 1 µU/mL. The dashed blue line indicates baseline. Note the fluctuation around the baseline between 120 and 180 min. Data scanned and adapted from Urquhart and Li [1969].*

Let us assume that the activation and loss of response R, can be modeled by the following relationship

$$\frac{dR}{dt} = k_{in} \cdot H(t) - k_{out} \cdot M \qquad (4{:}137)$$

$H(t)$, which represents the presence of a drug, stimulates the production of R which is then counterbalanced by means of the endogenous modulator that we denote as M. Note that the loss of R is indirectly governed by means of M. M is governed by R and the rate constant for development of tolerance k_{tol}, and can be written as

$$\frac{dM}{dt} = k_{tol} \cdot R - k_{tol} \cdot M \qquad (4{:}138)$$

The k_{tol} parameter was selected to govern both production and elimination of M in this particular example. However, this may not always be the case where data contain more information about the different rate processes.

When R increases, the production of modulator M is stimulated and then increases. The increase in M counterbalances R by increasing the rate of loss of R. For simplicity, we assume that rate in and out of M are governed by the rate constant for tolerance k_{tol}.

Let us assume that a compound stimulates $(H(t))$ a loss of response in the following manner. The time to pharmacodynamic steady state, assuming a constant plasma concentration C_{ss}, is governed by k_{tol} as well as by the actual drug concentration and grade of inhibition $H(t)$. M is directly proportional to R at steady state. When no drug is present $H(t)$ is equal to unity. Thus, both no drug input differential equations are equal to zero at equilibrium. Also, M is equal to R, because k_{tol} determines both rate of build-up and loss of M. It may not necessarily be so, but in the analysis of our present data set, there is not enough information in the data to apply two different rate constants. Therefore

$$\frac{dR}{dt} = k_{in} \cdot H(t) - k_{out} \cdot M = 0 \tag{4:139}$$

$$\frac{dM}{dt} = k_{out} \cdot R - k_{out} \cdot M = 0 \tag{4:140}$$

$$k_{out} \cdot R - k_{out} \cdot M = 0 \tag{4:141}$$

$$k_{out} \cdot R = k_{out} \cdot M \tag{4:142}$$

Since $R = M$ (with or without drug) at equilibrium, one could then solve for R.

$$k_{in} \cdot H(t) - k_{out} \cdot R = 0 \tag{4:143}$$

The baseline value (when $H(t)$ is 1) of response R_0 is determined by

$$R_0 = \frac{k_{in}}{k_{out}} \tag{4:144}$$

The steady state value of response R_{ss} is determined by

$$R_0 = \frac{k_{in}}{k_{out}} \cdot H(t) \tag{4:145}$$

The impact of drug stimulus can then take the form of an E_{max} model.

$$R_{ss} = \frac{k_{in}}{k_{out}} \cdot \left[1 + \frac{E_{max} \cdot C_{ss}}{EC_{50} + C_{ss}} \right] \tag{4:146}$$

This gives the maximum response R_{max} if $C_{ss} \rightarrow \infty$

$$R_{max} = \frac{k_{in}}{k_{out}} \cdot \left[1 + E_{max} \right] \tag{4:147}$$

A short description of the behavior of Equations 4:137 and 4:138, describing the model in Figure 4.54, is needed. Let us look at the equations again. The half-life of response with no drug present becomes

$$t_{1/2kout} = \frac{\ln(2)}{k_{out}} = \frac{\ln(2)}{\dfrac{k_{in}}{R_0}} = \frac{\ln(2) \cdot R_0}{k_{in}} \tag{4:148}$$

The half-life of response with drug present becomes

$$t_{1/2kout} = \frac{\ln(2)}{k_{out}} = \frac{\ln(2) \cdot R_{ss}}{k_{in}} \cdot \frac{1}{H(t)} \tag{4:149}$$

Equation 4:149 states that the larger the stimulus the shorter is the half-life.

4.9.3   Feedback model - 2

Let us assume that the activation and loss of response R, can be modeled by the following relationship

$$\frac{dR}{dt} = k_{in} \cdot H(t) - k_{out} \cdot R \cdot \left[1 + \frac{M}{M_{50}} \right] \tag{4:150}$$

H(t) represents the kinetics of a drug that stimulates the production of R which is then counterbalanced by means of the endogenous modulator that we denote as M. The loss of response is not only governed by the actual level of response R, but also by means of M. Let us also assume that M_{50} (i.e., 50% of maximal effect of the modulator) is equal to unity for simplicity. This reduces Equation 4:150 to

$$\frac{dR}{dt} = k_{in} \cdot H(t) - k_{out} \cdot R \cdot (1 + M) \tag{4:151}$$

M is governed by R and the rate constant k_{tol} and can be written as

$$\frac{dM}{dt} = k_{tol} \cdot R - k_{tol} \cdot M \tag{4:152}$$

At steady state Equation 4:151 is equal to zero

$$\frac{dR}{dt} = k_{in} \cdot H(t) - k_{out} \cdot R \cdot (1 + M) = 0 \qquad (4:153)$$

The same holds true for Equation 4:151

$$\frac{dM}{dt} = k_{tol} \cdot R - k_{tol} \cdot M = 0 \qquad (4:154)$$

Equation 4:154 rearranged becomes

$$k_{tol} \cdot R = k_{tol} \cdot M \qquad (4:155)$$

and, since at equilibrium $R = M$ (with or without drug), one could then solve for R.

$$k_{in} \cdot H(t) - k_{out} \cdot R \cdot (1 + R) = 0 \qquad (4:156)$$

$$R^2 + R - \frac{k_{in}}{k_{out}} \cdot H(t) = 0 \qquad (4:157)$$

which takes the form of a second order polynomial. In this particular case

$$x^2 + x - a = 0 \qquad (4:158)$$

$$x_{1,2} = -\frac{1}{2} \pm \sqrt{\frac{1}{4} + a} \qquad (4:159)$$

The baseline condition (R_0) with no drug present (no stimulus, i.e., $H(t)$ is equal to unity) becomes

$$R_0 = -\frac{1}{2} + \sqrt{\frac{1}{4} + \frac{k_{in}}{k_{out}}} \qquad (4:160)$$

The steady state condition with drug present (R_{ss}) becomes

$$R_{ss} = -\frac{1}{2} + \sqrt{\frac{1}{4} + \frac{k_{in}}{k_{out}} \cdot H(t)} \qquad (4:161)$$

The drug stimulus can then take the form of an E_{max} model.

$$R_{ss} = -\frac{1}{2} + \sqrt{\frac{1}{4} + \frac{k_{in}}{k_{out}} \cdot \left[1 + \frac{E_{max} \cdot C_{ss}}{EC_{50} + C_{ss}}\right]} \qquad (4:162)$$

which gives the maximum response R_{max} as $C_{ss} \to \infty$

$$R_{max} = -\frac{1}{2} + \sqrt{\frac{1}{4} + \frac{k_{in}}{k_{out}} \cdot [1 + E_{max}]} \qquad (4:163)$$

4.9.4   A simulation with feedback model - 2

Figure 4.56 depicts simulations of the behavior of R and M in the feedback model including the *baseline effect, overshoot, dynamic steady state* in the presence of drug, and *the rebound time course* after removal of drug. The horizontal blue bar in Figure 4.56 illustrates the time course of drug action, $H(t)$. For simplicity, we have assumed a square wave of stimulation (with an instantaneous onset and loss of drug action) in order to observe the impact on R and M.

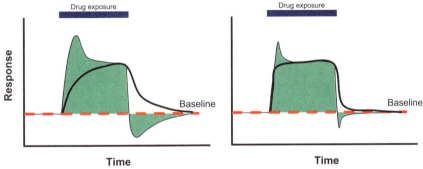

Figure 4.56 *Time course of R and M when $k_{tol} << k_{out}$ (left) and $k_{tol} >> k_{out}$ (right). In both cases, the kinetics of drug upon input and loss is assumed to be very rapid. Green area represents the time course of response R and thick black line the modulator M. The dashed red line is the baseline level R_0.*

At the baseline, R and M balance each other. When drug stimulates the production or build-up of response, R increases from its baseline value R_0. An *overshoot* (Figures 4.54 and 4.56) is seen for some responses if the drug is rapidly introduced. Since R increases, M will also increase, but the level of M changes at another rate (k_{tol}) than does R (k_{in}, k_{out}). M starts acting on the increasing level of R, since they are no longer at equilibrium with each other. R will peak when the product of production of response and the drug function ($k_{in} \cdot H(t)$) is equal to the loss of response ($k_{out} \cdot R(1 + M/M_{50})$) and then decay back to R_0 when drug is gone and equilibrium has been re-established. Note that the response has leveled out at a plateau (dynamic steady state) that differs from the baseline value. Only three parameters will determine the basic parts of the curve, namely k_{in}, k_{out} and k_{tol}. The shape of the response-time profile in Figure 4.56 can differ vastly depending on the type of stimulation/inhibition (namely, route, rate, and duration of drug exposure).

If the stimulant effect of drug is removed more rapidly than the kinetics (k_{tol}) of the moderator M, e.g., when drug wash-out is fast, the level of M is high relative to R. In this case, the loss of R will be rapid since M acts upon the rate of elimination of R. Before M and R balance each other a level of $R < R_0$ may be reached for a period of time. This sub-baseline state is called the *rebound*. Its depth or intensity and duration characterize the rebound. Rebound is seen after extensive exposure to a number of drugs (e.g., cocaine and amphetamines).

With a controlled release formulation (e.g., the transdermal patch) substantial tolerance may develop over time, whereas very little of the rebound effect may be seen since the elimination of drug (K for kinetics) is slower than development of tolerance ($k_{tol} > K$). The duration of the effect might be different from the duration of the rebound as indicated in Figure 4.55 (right).

Note that the area between baseline and response AUC_E is not symmetric with the area between rebound and baseline AUC_R. We have found this to be the major difference between the feedback and precursor models. The intensities of the effect at steady state and rebound might also differ. For a discussion of the pool model of tolerance and rebound, see sections 4.9.5 and 4.9.6. In addition, examples PD9 and PD10 in the applications section contain a comparison of the behavior of these models.

A good example of the rebound effect in clinical practice is the use of benzodiazepines (*BZP*). If a *BZP* is being used over a long period of time and is then suddenly removed, the patient may experience a period of exaggerated anxiety and even a state of psychosis. During *BZP* treatment, the patient adapts to the blockage of the receptor. Thus, when *BZP* is suddenly removed and patient is *starved* of stimulus, the overcapacity causes an excessive negative effect, resulting in the rebound effect.

4.9.5   Pool model – Unidirectional flow

One may view the basic pool model as a catenerary system (Figure 4.57, A). In its simplest form this means that the flux through the pool equals the flux through the response compartment. Figure 4.57 (A) demonstrates the physiological pool or precursor model that has been proposed for compounds like prolactin (Sharma *et al* [1995]). A pool separate from the response is produced and lost by means of k_{in} and k_{tol}. The mass from the pool is then input to the response compartment. The pool may then be activated by means of stimulation or inhibition by the drug, *H(t)*. The parameters are k_{in} and k_{tol}, the zero-order input and first-order exit rate constant from the pool compartment, respectively.

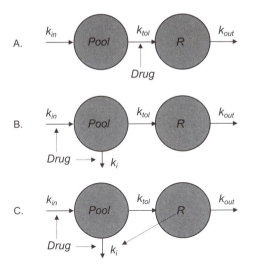

Figure 4.57 *Schematic illustration of the pool model. The basic pool model is depicted as model A. Models B and C are two derivatives of the basic model that have applications in pharmacology and physiology. The dashed arrows denote the site of action of drug (H(t)), which is either stimulation or inhibition of the process. The k_i parameter may denote any of the other first-order rate constants or a different parameter.*

If a certain fraction of the pool size is pushed into the response compartment, resulting in an area above the baseline response AUC_E, a similar fraction will be needed to refill the pool before the original equilibrium can be re-established (Figure 4.58). This refill results in an area below the baseline response AUC_R, such that AUC_R is equal to AUC_E. This is an intrinsic characteristic of the basic pool model, which makes it different from the ordinary feedback model. The pool model can, however, be adjusted to accommodate a steady state response (fractional tolerance development) that differs from the original baseline value.

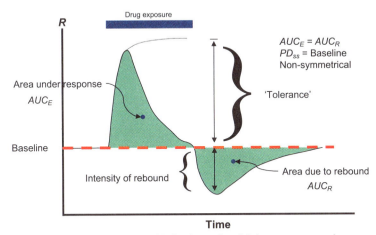

Figure 4.58 *The response-time profile for the pool model during a constant drug exposure (horizontal blue bar). Note that the response returns to its baseline value during drug exposure and that AUC_E is equal to AUC_R. PD_{ss} is the steady state response level.*

We have also elaborated on two versions of the pool model (Models B and C in Figure 4.57). Figure 4.57 (B) demonstrates the situation where a pool separate from the response is produced and lost by means of k_{in}, k_i, and k_{tol}. The level in the pool compartment drives the response compartment by means of the product of $k_{tol} P$ where P is the pool. The pool may thus be activated by means of stimulation or inhibition by the drug on k_{in} and k_{tol}. This model therefore includes one additional parameter, and hence requires more experimentation. There may be problems in identifying parameters in this model unless the biochemical processes involved are well understood. Model C in Figure 4.57 is an additional permutation of A including feedback action from the response R compartment.

Let us consider the basic form of the pool model (Model A in Figure 4.57). It is assumed that a precursor to R is synthesized or secreted into a pool compartment and lost from that compartment by a rate proportional to the pool size

$$\frac{dP}{dt} = k_{in} - k_{tol} \cdot P \tag{4:164}$$

This loss from the pool compartment is, at the same time, the input into the response compartment, R.

$$\frac{dR}{dt} = k_{tol} \cdot P - k_{out} \cdot R \tag{4:165}$$

Therefore, while the pool governs the kinetics of R, the reverse is not true in this unidirectional flow model. The baseline condition for pool size is

$$P_0 = \frac{k_{in}}{k_{tol}} \tag{4:166}$$

and the baseline response R_0 is

$$R_0 = \frac{k_{in}}{k_{out}} \tag{4:167}$$

Note that both the level (size) of P_0 and R_0 are proportional to k_{in} at baseline, and are inversely proportional to their respective exit rate constants (k_{tol} and k_{out}, respectively). Hence, the area under the response-time curve is equal to the area under the rebound-time curve, unless some of the pool circumvents the response compartment. The pool size can be modified by changes in the build-up or loss by means of $H(t)$. A change in loss is shown below

$$\frac{dP}{dt} = k_{in} - k_{tol} \cdot P \cdot H(t) \tag{4:168}$$

where $H(t)$ represents the kinetics of a drug that either stimulates or inhibits the loss of P (Sharma *et al* [1995]). The flux into R is therefore

$$\frac{dR}{dt} = k_{tol} \cdot P \cdot H(t) - k_{out} \cdot R \qquad (4{:}169)$$

The adapted pool level at steady state P_{ss} is

$$P_{ss} = \frac{k_{in}}{k_{tol}} \cdot \frac{1}{H(C_{ss})} \qquad (4{:}170)$$

where $H(C_{ss})$ is the drug function for inhibition or stimulation at constant drug exposure (C_{ss}). The steady state condition for the response R_{ss} is

$$R_{ss} = \frac{P_{ss} \cdot k_{tol}}{k_{out}} \cdot H(C_{ss}) = \frac{P_{ss} \cdot k_{tol}}{k_{out}} \cdot \frac{k_{in}}{k_{tol} \cdot P_{ss}} = \frac{k_{in}}{k_{out}} \qquad (4{:}171)$$

The ratio of P_{ss} to R_{ss} gives

$$\frac{P_{ss}}{R_{ss}} = \frac{\dfrac{k_{in}}{k_{tol}} \cdot \dfrac{1}{H(C_{ss})}}{\dfrac{k_{in}}{k_{out}}} = \frac{k_{out}}{k_{tol}} \cdot \frac{1}{H(C_{ss})} \qquad (4{:}172)$$

This ratio gives the 'amount' of response in P compartment relative to that in the R compartment. In other words, by taking the ratio of the first-order rate constants one may estimate the total capacity of the system with respect to response.

The impact of $H(t)$ on P but not on R at steady state is obvious from Equations 4:170 and 4:171. This means that the basic pool model will predict a return of response to its baseline value during constant stimulation or inhibition (Figure 4.58).

4.9.6 Pool model – Bidirectional flow

The pool model may be extended to a more generic system that captures the turnover of enzymes and other endogenous compounds such as is depicted in Figure 4.59. The main difference between this model and the basic pool model that we discussed in the previous section is the reversible (i.e., bidirectional) process between *Pool* and R_i. This is taken into account by the rate constants k_1 and k_2. We will exemplify this system by means of acetycholinesterase (*AChE*) activity.

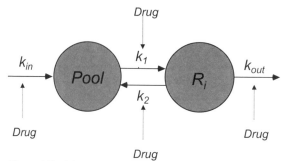

Figure 4.59 *Schematic illustration of the bidirectional pool model. The various potential sites of drug action are also indicated by means of arrows.*

AChE activity R_a is measured in red blood cells. Figure 4.60 illustrates how the substrate (drug or acetylcholine) binds to the free active enzyme R_a and shifts the equilibrium to the inactive or substrate-bound pool R_i. As long as the substrate is bound to *AChE*, the enzyme is not available for binding additional substrate molecules because it is in the substrate-bound pool R_i. When the substrate has been irreversibly degraded, the enzyme can either be lost via k_{out} or reactivated via k_2.

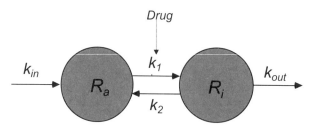

Figure 4.60 *Schematic diagram of the pool model for turnover of AChE-activity in blood. The k_{in} parameter is the zero-order production of enzyme. R_a is the active (unbound) enzyme and R_i the inactive (substrate-bound) enzyme. The k_2 and k_{out} parameters are first-order rate constants and k_1 is a second-order rate constant. The AChE activity in blood is represented by R_a.*

The turnover of free or unbound *AChE* is determined by Equation 4:173

$$\frac{dR_a}{dt} = k_{in} - k_1 \cdot R_a \cdot H(t) + k_2 \cdot R_i \qquad (4:173)$$

and the turnover of the substrate-bound form of the enzyme then becomes

$$\frac{dR_i}{dt} = k_1 \cdot R_a \cdot H(t) - k_2 \cdot R_i - k_{out} \cdot R_i \qquad (4:174)$$

The steady state level of *AChE*-activity is estimated to be

$$R_{ass} = \frac{k_{in}}{k_{out}} \cdot \frac{k_2 + k_{out}}{k_1 \cdot H(t)} = \frac{k_{in}}{k_{out}} \cdot \frac{K_m}{H(t)} \qquad (4:175)$$

where $H(t)$ is the inhibition or stimulation function. The ratio of $(k_2 + k_{out})/k_1$ corresponds to the Michaelis-Menten constant K_m in a traditional enzyme kinetics system. In other words, a subject with a large active pool R_{ass} of free enzyme at steady state relative to the bound pool will also have a large K_m value, since R_{ass}/R_{iss} is K_m. The product of k_{out} and total enzyme in the system R_a+R_i corresponds to V_{max}, and the inactivated enzyme at steady state R_{iss} is then

$$R_{iss} = \frac{k_{in}}{k_{out}} \tag{4:176}$$

The response at the ED_{50} dose R_{ED50} can be simplified to

$$R_{ED_{50}} = \frac{1}{2} \cdot R_0 \tag{4:177}$$

where R_0 is the baseline value of *AChE* without drug in the system. The production of the enzyme is assumed to occur by a zero-order process k_{in}. The equilibrium between free enzyme and the substrate-enzyme complex is governed by k_1 and k_2. The k_{out} parameter determines the first-order irreversible loss of enzyme. Estimates of k_{in}, k_{out}, k_1, k_2, ED_{50} and K can be obtained by fitting Equations 4:173-4:174 to data.

It is interesting to note from the steady state conditions of R_a and R_i that R_a but not R_i is dependent on $H(t)$. During constant exposure to drug, R_i returns to the baseline value. The ratio of R_{iss} to R_{ass} is then given by

$$\frac{R_{iss}}{R_{ass}} = \frac{\dfrac{k_{in}}{k_{out}} \cdot \dfrac{k_2 + k_{out}}{k_1}}{\dfrac{k_{in}}{k_{out}}} = \frac{k_2 + k_{out}}{k_1} \tag{4:178}$$

Hence, when k_{out} is much greater than k_2, the bidirectional flow model reverts to the pool model according to Equations 4:179 and 4:180.

$$\frac{R_{iss}}{R_{ass}} = \frac{k_{out}}{k_1} \tag{4:179}$$

$$\frac{P_{ss}}{R_{ss}} = \frac{\dfrac{k_{in}}{k_{tol}}}{\dfrac{k_{in}}{k_{out}}} = \frac{k_{out}}{k_{tol}} \tag{4:180}$$

Both the bidirectional and the unidirectional (basic) pool models display a conservative behavior with respect to R_i and R if $H(t)$ acts upon k_1 or k_{tol}. They both return to their baseline values and their area above (or below if inhibition) the baseline value during stimulation, is equal to the area of the rebound curve. Furthermore, t_{max} value for the effect is less than t_{max} for the rebound and the height of the peak effect is greater than that of the rebound. From this, one can conclude that the effect and rebound curves are not symmetrical.

4.9.7   Comparisons with other models

If the response R is proportional to the E_{max} relationship when no counter-action or moderator is present, it is proportional to the square root of the same relationship when tolerance is added, provided the feedback model is used. Figure 4.61 schematically demonstrates the characteristics of the baseline value k_{in}/k_{out}, maximum observed response R_{max}, and determinants of the span between R_{max} and baseline. If feedback mechanisms are overlaid on a non-tolerance system, the baseline value and the maximum observed response at equilibrium will be lowered relative to the normal state.

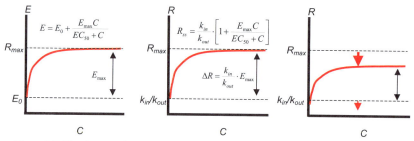

Figure 4.61 *The pharmacodynamic differences between the ordinary E_{max} model without tolerance (left), the indirect response model with stimulation of k_{in} but no tolerance (middle) and the indirect response model with tolerance (right). Note the shift in (or decrease in) R_{max} and k_{in}/k_{out} with tolerance (right), compared with the no tolerance situations (left and middle).*

If one studies a plot of effect *versus* concentration after each of several doses, tolerance might manifest itself as either an apparent increase in EC_{50} and/or a decrease in the E_{max} value. Figure 4.62 displays one such situation where EC_{50} has apparently increased and E_{max} decreased.

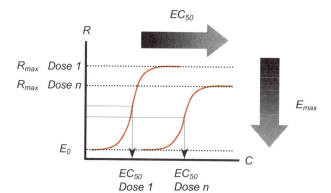

Figure 4.62 *An apparent shift in the concentration-response curve between the first and the n:th dose of a drug. The apparent E_{max} has decreased and EC_{50} increased. Unless the number of receptors or the binding capacity of the tissue has changed (decreased) or the affinity between ligand and receptor has changed (decreased), these changes are not real. Rather, they are apparent, confounded by feedback or other mechanisms.*

However, what one is observing may only be an apparent change in EC_{50} and/or E_{max}, as other processes (such as an endogenous modulator M) may be balancing the response. While this appears as a change in potency or intrinsic activity when data are analyzed with standard methods/models, the system could be modeled accurately by means of the feedback model with a time constant EC_{50}. This exemplifies the importance of mechanistic information for appropriate modeling, as shifts in EC_{50} and/or E_{max} may be misinterpreted unless one clarifies whether they are real or apparent (see Figures 4.60 and 4.61).

There are very practical but less mechanistic ways to model changes in the parameters. Colburn and Eldon [1994] reviewed these for the E_{max} model. EC_{50}, and K were obtained by fitting Equations 4:181-4:182 to data.

$$E = \frac{E_{max} \cdot C}{EC_{50} + C} \tag{4:181}$$

$$EC_{50} = EC_{50} \cdot \left[1 + Q \cdot (1 - e^{-K \cdot t})\right] \tag{4:182}$$

where $0 < Q < K$. Q and K are arbitrary parameters governing the change in EC_{50}.

This is a flexible approach to smoothing data without unveiling or tackling the underlying mechanisms. Unfortunately, however, these functions do not include the return of EC_{50} to baseline, nor do they capture the rebound effect. They should preferably be used for single dose administration.

In order to capture the underlying model for tolerance, we recommend that a more comprehensive approach be adopted, based on several dose levels and multiple dosing. Ideally one would study the effect after various routes of administration, because it is possible to screen for the presence of active metabolites without the necessity of having the metabolites.

We have listed the steady state response expressions for 4 different classes of models below. These provide a comparison of the parameterizations.

♦ **Ordinary E_{max} model** $$E = E_0 + \frac{E_{max} \cdot C_{ss}}{EC_{50} + C_{ss}}$$

♦ **Indirect response model 3** $$R_{ss} = \frac{k_{in}}{k_{out}} \cdot \left(1 + \frac{E_{max} \cdot C_{ss}}{EC_{50} + C_{ss}}\right)$$

♦ **Feedback model 1** $$R_{ss} = \frac{k_{in}}{k_{out}} \cdot \left(1 + \frac{E_{max} \cdot C_{ss}}{EC_{50} + C_{ss}}\right)$$

♦ **Feedback model 2** $$R_{ss} = -\frac{1}{2} + \sqrt{\frac{1}{4} + \frac{k_{in}}{k_{out}} \cdot \left(1 + \frac{E_{max} \cdot C_{ss}}{EC_{50} + C_{ss}}\right)}$$

♦ *Unidirectional pool model* $$R_{ss} = \frac{k_{in}}{k_{out}}$$

The maximum/minimum predicted response R_{max}/R_{min} is summarized below.

♦ *Ordinary E_{max} model* $$R_{max} = E_0 + E_{max}$$

♦ *Indirect response model 3* $$R_{max} = \frac{k_{in}}{k_{out}} \cdot \left(1 + E_{max}\right)$$

♦ *Feedback model* $$R_{max} = -\frac{1}{2} + \sqrt{\frac{1}{4} + \frac{k_{in}}{k_{out}} \cdot \left(1 + E_{max}\right)}$$

♦ *Unidirectional pool model* $$R_{max} = \frac{k_{in}}{k_{out}} \cdot \frac{1}{H(t)}$$

4.9.8   Modeling of generic response-time data

A new anesthetic agent was given as a single intravenous infusion and plasma concentrations and EEG spectra were obtained (Figure 4.63). The observed and model predicted response (mean EEG frequency) is plotted *versus* time in Figure 4.63 (left) and observed response *versus* plasma concentration (right). Data, model and results are thoroughly discussed in exercise PD9. This dataset exemplifies the use of the feedback class of models when we have pre-, peri- and post-dosing response data.

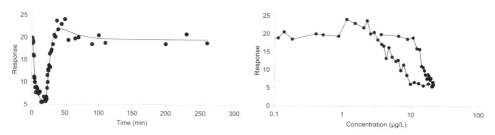

Figure 4.63 *Observed and model-predicted response-time (left) and observed response-concentration (right) data. Note the slight rebound (peak) in the response-time data and the hysteresis in the response-concentration data. The length of infusion was 30 minutes.*

4.9.9   Modeling of anesthetic response-time data

When two consecutive intravenous infusions of an anesthetic are given, adaptation may be observed. Figure 4.64 shows such adaptation.

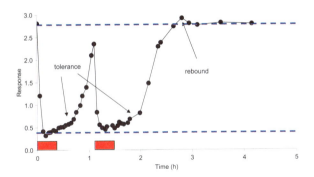

Figure 4.64 The observed (●) response versus time during and after anesthetic infusion, and the baseline, tolerance, and rebound. The two horizontal bars represent the two periods of intravenous infusion.

A feedback model, taking negative feedback into account, was successfully fit to the data. Note the rapidly induced tolerance as indicated by the arrows in Figure 4.64. This results from desensitization that is already acting during the infusion regimens. Note also that the response trough is not the same during the second infusion regimen. Finally, there is a slight rebound after cessation of the second infusion regimen. Final parameter estimates are given in Table 4.8. Note the rate constant that governs the tolerance development k_{tol} is larger than the rate constant for the decline of response k_{out}. This means that one will be able to observe tolerance provided the experimental design is appropriate.

Table 4.8　Model parameters

Parameter	Estimates (mean ± CV %)
k_{in} *(units/h)*	30.0 ± 5
k_{out} *(h^{-1})*	2.89 ± 6
EC_{50} or IC_{50}	348 ± 1
n	7.36 ± 7
k_{tol} *(h^{-1})*	4.21 ± 10
E_{max}	9.75 ± 4

4.9.10　Modeling of cocaine response-time data

Data are taken from Holford *et al* [1990] where an analysis was done on data on euphoria, heart rate and blood pressure obtained from humans. Below we discuss how the model of euphoria was constructed. Table 4.9 contains the final results.

Cocaine was assumed to act according to the scheme of a synapse shown in Figure 4.65.

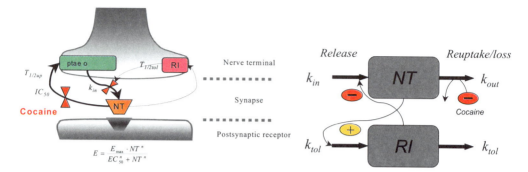

Figure 4.65 *Schematic illustration of the synapse and the turnover of neurotransmitter NT, release inhibitor RI and the action of cocaine on the reuptake of NT (left). The right hand figure shows a flow diagram of turnover of NT and RI and the indirect action of cocaine on response.*

NT denotes the neurotransmitter responsible for the euphoric effect when it binds to the postsynaptic receptor. NT is stored in presynaptic vesicles (green pool). The release inhibitor RI governs the release of NT from its presynaptic pool. When NT is released into the synapse, it can either bind to the postsynaptic receptor or be taken up by the presynaptic pool. Cocaine is known to block the reuptake of neurotransmitter. Released NT in the synapse will positively affect RI, in that the more NT in the synapse, the more RI is produced. The latter then blocks the release of NT. This is called negative feedback.

The turnover of neurotransmitter NT is given by

$$\frac{dNT}{dt} = \frac{k_{in}}{RI} - \frac{k_{out} \cdot NT}{H(t)} \tag{4:183}$$

The turnover of release inhibitor RI is given by

$$\frac{dRI}{dt} = k_{tol} \cdot NT - k_{tol} \cdot RI \tag{4:184}$$

The relationship between the NT concentration in the synapse and the response E is given by

$$E = \frac{E_{max} \cdot NT^n}{EC_0^n + NT^n} \tag{4:185}$$

The rate of change in the turnover of NT is equal to zero at steady state

$$\frac{dNT}{dt} = \frac{k_{in}}{RI_0} - k_{out} \cdot NT_0 = 0 \tag{4:186}$$

NT_0 then becomes after rearrangement

$$NT_0 = \frac{k_{in}}{k_{out}} \cdot \frac{1}{RI_0} \qquad (4{:}187)$$

The rate of change in the turnover of *RI* is equal to zero at steady state

$$\frac{dRI}{dt} = k_{tol} \cdot NT_0 - k_{tol} \cdot RI_0 = 0 \qquad (4{:}188)$$

which gives

$$k_{tol} \cdot NT_0 = k_{tol} \cdot RI_0 \qquad (4{:}189)$$

and after canceling common terms

$$NT_0 = RI_0 \qquad (4{:}190)$$

RI_0 reinserted into Equation 4:187 gives

$$NT_0 = \frac{k_{in}}{k_{out}} \cdot \frac{1}{NT_0} \qquad (4{:}191)$$

which may be rearranged to give

$$NT_0^2 = \frac{k_{in}}{k_{out}} \qquad (4{:}192)$$

$$NT_0 = \sqrt{\frac{k_{in}}{k_{out}}} \qquad (4{:}193)$$

When the system is exposed to cocaine, the steady state level of *NT* is given by

$$NT_{ss} = \sqrt{\frac{k_{in}}{k_{out}}} \cdot H(t) \qquad (4{:}194)$$

Table 4.9 Parameter estimates from Holford *et al* [1990].

Response	E_{max}	EC_{50}	E_0	n	$t_{1/2up}$ sec	I_{50}	$t_{1/2tol}$ min
Euphoria, 0-100	100	127	0	1.4	<1	8.4	230
HR, bpm	56	43	70	3.3	<1	5.4	29
BP, mm Hg	43	115	115	2.4	<1	8.4	69

$t_{1/2up}$ and $t_{1/2tol}$ are the inverses of the respective rate constants multiplied by ln(2).

4.9.11   General thoughts about tolerance /dependence models

Collier [1964] proposed that *tolerance* to a drug might arise from a decrease in the number of receptors or from an increase in silent receptors. Although, these changes are inadequate to explain *physical dependence*. *Dependence* does not necessarily accompany tolerance, and the impact on a dependent system (e.g., cells) of withdrawing a drug tend to be the opposite to extended treatment with the drug. Table 4.10 summarizes the genesis of *tolerance* and *dependence*.

Table 4.10 Possible ways in which tolerance and dependence may arise (adapted from Collier [1964])

	Cause	Proposed model
Tolerance	Drug decrease in the number of pharmacological receptors	e.g., model Equations 4:137-4:138
Tolerance	Drug increases in the number of its silent receptors	
Dependence	Drug reduces the supply of an excitatory transmitter, causing the pharmacological receptors for the transmitter to increase in number. After withdrawal of the drug, the supply of transmitters rises to its normal level faster than the surplus receptors are removed.	
Dependence	Drug increases the supply of an inhibitory transmitter, causing the pharmacological receptors for the transmitter to decrease in number. After withdrawal of the drug, the supply of transmitters falls to its normal level faster than the surplus receptors are restored.	
Dependence	Drug occupies the receptors for an endogenous excitatory transmitter, causing the pharmacological receptors for the transmitter to increase in number. After withdrawal of drug, its molecules leave the receptors for the transmitter faster than the surplus receptors are removed.	

Models proposed by JG.

Physical dependence is always accompanied by *tolerance*, whereas *tolerance* need not be accompanied by *dependence* (Collier [1964]). The *feedback* model is a good example of both *tolerance* and *physical dependence*, including both *tolerance* (adaptation) and *withdrawal/rebound*. The *non-conservatory pool* model displays only *tolerance* but no *withdrawal/rebound*. Many, but not all, drugs induce *tolerance*. Only a few drugs induce *physical dependence*.

Figure 4.66 summarizes two classes (*feedback* and *pool/precursor*) of models that we have used in our discussion of tolerance and rebound. We have also split these classes into subclasses (*feedback models* 1 and 2 and *uni-* and *bidirectional pool* models).

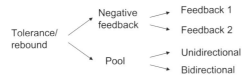

Figure 4.66 *Summary of the two classes of tolerance models we have discussed in section 4.9. Note also the subclasses in terms of at least two different feedback models and the unidirectional and bidirectional pool models.*

4.10   Logistic Response Models

If the pharmacological effect is difficult to grade, it may be useful to determine the probability of achieving such an effect as a function of e.g., plasma concentration. In such a case we may apply a logistic regression model. What distinguishes a logistic regression model from a general nonlinear regression model is that the dependent variable in logistic regression is binary or dichotomous (usually coded as *0* or *1*). This difference between logistic and nonlinear (linear) regression is reflected both in the choice of a parametric model and in the underlying assumptions.

The first difference between logistic and nonlinear regression concerns the nature of the relationship between the outcome (dependent variable) and the independent variable (Hosmer and Lemeshow [1989]).

A logistic regression model may be used to analyze the impact of a stimulus on a dichotomous response. In this example, *C* is the plasma concentration of the drug and is assumed to be a continuous variable. Figure 4.67 shows the observed and predicted probability of the response.

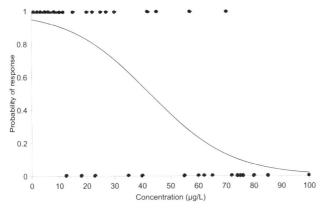

Figure 4.67 *Observed (●) and predicted (line) probability of response versus plasma concentration of a drug in the logistic regression model (Equation 4:195).*

Note that at low concentrations, there is a high probability of response (which is consistent with data predominantly coded as 1), while at high concentrations the converse is true. A logistic regression model is of the form

$$\pi(C) = \frac{e}{e+1} \tag{4:195}$$

$$L = \text{logit}(\pi) = \log_e \left[\frac{e^L}{e^L+1} \right] \tag{4:196}$$

where the general (multiple linear regression) functional form of the logit (L) is

$$L = \sum \theta_i \cdot C \tag{4:197}$$

For this particular example

$$L = \theta_1 \cdot S_1 - \theta_2 \cdot C \tag{4:198}$$

and $\pi(C)$ is the probability of response to a certain stimulus (in this case drug concentration, C). L is also called the logit, and has several desirable properties of the linear model. α_1 is the parameter expressing the logit of the probability that a certain stimulus (S_i which is either 1 or 0) yields a specific response. The probability of *no* response $Q(C)$ to a stimulus is

$$Q(C) = 1 - \pi(C) = 1 - \frac{e^L}{e^L+1} = \frac{1}{e^L+1} \tag{4:199}$$

The odds ratio (relative risk) of a response to no response is given by

$$\text{odds ratio} = \frac{\pi(C)}{Q(C)} = \frac{\dfrac{e^L}{e^L+1}}{\dfrac{1}{e^L+1}} = e^L \tag{4:200}$$

This can also be expressed as the log(odds ratio)

$$\ln(\text{odds ratio}) = L \tag{4:201}$$

The second difference between general nonlinear regression and logistic regression concerns the variance. For any given sample size N, the variance of the binomial distribution is greatest when $\pi(C) = 0.5$ and least when $\pi(C) = 0$ or $Q(C) = 1$. The mean variance for $\pi(C)$ is $\pi(C) \cdot Q(C)$ and consequently the weight WT becomes

$$WT = \frac{1}{\pi(C) \cdot Q(C)} = \frac{1}{\dfrac{e^L}{e^L+1} \cdot \dfrac{1}{e^L+1}} = \frac{\left[e^L+1\right]^2}{e^L} \tag{4:202}$$

Below, a logistic regression model was expanded to analyze the impact of four different stimuli, such as tetanus, trapezius squeeze, laryngoscopy and intubation, on three different dichotomous responses such as movement, heart rate, and blood pressure (Figure 4.68) (Hung *et al* [1990]).

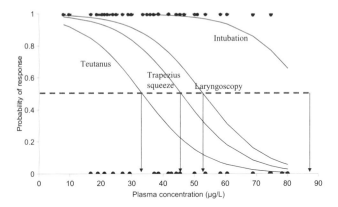

Figure 4.68 Observed (●) and predicted (line) probability of response versus plasma concentration of a drug using logistic regression (teutanus, trapezius squeeze, laryngoscopy and intubation). The vertical arrows give the respective concentrations of 50% probability of a response. Data adapted from Hung et al [1990].

Plasma concentration C is the independent variable. The logit L is linear in its parameters and ranges from $-\infty$ to $+\infty$ depending on θ_5 and C. The θ_i parameter expresses the logit of the probability that a certain stimulus (S_i which is either 1 or 0) yields a specific response such as movement, change in blood pressure or change in heart rate change). S_1 corresponds to tetanus, S_2 to trapezius squeeze, S_3 to laryngoscopy and S_4 to intubation. The model includes five parameters, one for each stimulus corresponding to a specific response (movement) plus an additional parameter relating C to the probability of movement. The model has the general form of Equation 4:195 where the logit in this particular case is defined as

$$L = \theta_1 \cdot S_1 + \theta_2 \cdot S_2 + \theta_3 \cdot S_3 + \theta_4 \cdot S_4 - \theta_5 \cdot C \qquad (4:203)$$

Equations 4:199 and 4:200 give probability of no response and odds ratio.

4.11   Oscillating Processes

Oscillating functions have previously been applied to a variety of phenomena such as diurnal changes in body temperature, immunological responses and turnover of nicotine (Moore-Ede *et al* [1982], Jusko [1992], Haefner [1996], Gries *et al* [1996], Lemmer [1997], Chakrobarty *et al* [1999]). Below we show two examples of such functions.

A new device for pulsatile delivery is being tested on rats. The microdevice can be pre-programmed with respect to rate (flow), frequency, and amplitude. This type of device has potential for, amongst other things, delivery of hormone or other compounds (such as nicotine or cocaine) when the aim is to obtain narrow plasma concentration peaks with a desired pre-programmed frequency. Figure 4.69 shows observed and predicted data during a six hour observational period using the device in the rat.

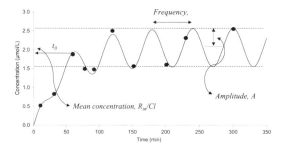

Figure 4.69 Comparisons of observed and predicted concentration-time data obtained from an oscillating input-function. The t_0 parameter is indicated by the left hand horizontal arrow and the amplitude (A) by the double arrow. Mean input is given by Equation 4:204.

The following expression was used to mimic the input

$$Input = mean + A \cdot \cos[\omega \cdot (t - t_0)] \qquad (4:204)$$

where *mean* is the average input rate, A the amplitude and t_0 the shifted peak (60 min). The ω parameter scales the frequency of oscillations of the function to the physical frequency $(2\pi/60)$, i.e., the cycle is 60 min and $\pi \approx 3.14$. The implemented model is written as

$$Input = R_{in} \cdot (1 + A \cdot \cos[\omega \cdot (t - t_0)]) \qquad (4:205)$$

A will be the fraction of the mean rate of input R_{in} that corresponds to the amplitude. The differential equation model for the pharmacokinetic system employs volume and clearance as model parameters

$$V \frac{dC}{dt} = Input - Cl \cdot C \qquad (4:206)$$

Results from fitting Equation 4:206 to the data are shown in Figure 4.68. The amplitude then becomes about 25 % of the mean rate of input R_{in} which is indicated in Figure 4.68. Note also the 60 min time shift t_0.

Jusko and co-workers [1993] used a circadian input process for the turnover rate of cortisol according to

$$k_{in} = R_m + R_b \cdot \cos[\omega \cdot (t - t_0)] \qquad (4:207)$$

R_b is the amplitude and R_m the mean turnover rate. The ω parameter is $(2\pi/24)$ and t_0 is zero (0). The differential equation model for the pharmacodynamic system then employs Equation 4:207 as input according to

$$\frac{dR}{dt} = k_{in} \cdot H(t) - k_{out} \cdot R \qquad (4:208)$$

where $H(t)$ is the inhibition function.

We obtained baseline, placebo and drug-effect data for body temperature in the rat over 24 h. Data were filtered, and the resulting observational data used for modeling are shown in Figure 4.70, including the baseline oscillation within a 24 h period and the effect of treatment with placebo or 100 µg drug.

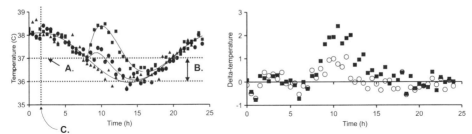

Figure 4.70 *Observed (symbols) and predicted responses (solid lines) including the baseline (▲), the placebo dose (●) and test dose (■) (left). The mean body temperature (A), amplitude (B) and peak shift (C) are superimposed on observed and model-predicted data. Baseline data were then subtracted from test dose (dose 100 µg) and placebo data (right). The resulting response-time curves are pure drug (■) and placebo (○) responses when the diurnal variation has been subtracted. Note that there are still oscillations around the baseline (horizontal black line).*

The rate constants of the first-order one-compartment model are 1.6 and 0.7 h^{-1}, and the volume is approximately 1 L/kg. A is a fraction of the amplitude and k_{inm} is the mean turnover rate. ω is $2\pi/24$ and t_0 is zero (Equation 4:209). The k_{inm} parameter is either estimated by taking the mean value (37°C) of the baseline data as a constant during the regression, or is included as a model parameter.
We then applied the circadian input process for the turnover rate

$$k_{in} = k_{inm} + A \cdot k_{inm} \cdot \cos[\omega \cdot (t - t_0)] \qquad (4:209)$$

The differential equation model for the pharmacodynamic system then employs Equation 4:209 as input according to

$$\frac{dR}{dt} = k_{in} - k_{out} \cdot H(t) \cdot R \qquad (4:210)$$

Experimental and predicted data were consistent and the parameters had high precision. Figure 4.70 (left) shows superimposed observed and model (Equation 4:210) predicted data. Figure 4.70 (right) shows the baseline-subtracted placebo and drug-treated data.

Figure 4.71 shows an asymmetric response-time curve (upper curve) within a 24 h period, due to the non-stationary turnover rate (lower curve). See also Figure 3.63 in Section 3.5.8 for a discussion on variable baselines.

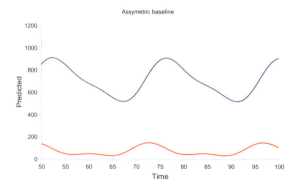

Figure 4.71 *Schematic illustration of an asymmetric response-time course (upper blue curve) due to the non-stationary turnover rate (lower red curve).*

To do a thorough analysis of response-time data, we highly recommend that the analyst use multiple doses and dose levels. However, it is even more important to study the baseline variation for an extended period of time. If there is a pronounced circadian variation, this should be documented before, between and maybe after dosing. For an in-depth analysis of asymmetrical and symmetrical circadian baseline variations, we recommend Chakrobarty *et al* [1999].

4.12   Synergistic Effects Modeled by Turnover Functions

A special situation occurs when two compounds simultaneously act on different production/loss sites of the turnover model (Figure 4.72).

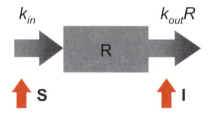

Figure 4.72 *Schematic illustration of the turnover model with two drugs acting on stimulation of the production of response and inhibition of the loss of response, respectively.*

Consider the turnover function below

$$\frac{dR}{dt} = k_{in} \cdot S(A) - k_{out} \cdot I(B) \cdot R \qquad (4{:}211)$$

where drug *A* stimulates *S(A)* the production of response *R* and drug *B* inhibits *I(B)* the loss of *R*. At steady state the relationship between response R_{ss} and drug exposure becomes

$$R_{ss} = \frac{k_{in}}{k_{out}} \cdot \frac{S(A)}{I(B)} \qquad (4{:}212)$$

If the apparent stimulatory effect of drug *A* is e.g., 4, and the apparent inhibitory effect of *B* is 4 (i.e., *I(B)* is equal to 0.25, and the ratio of one over *I(B)* becomes 4 which multiplies the baseline), then the combined effect of *A* and *B* is not 8 (4+4) but rather 16 (4/0.25). This is illustrated in Figure 4.73.

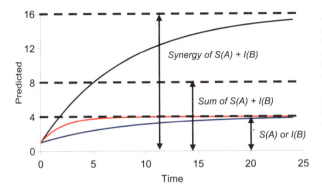

Figure 4.73 Schematic illustration of the synergistic effects when drug A and drug B are combined. Their individual impact on the response is 4 and one would believe that their combined impact would then be 4+4 (8). However, since they act on different 'sides' of the turnover system they form a ratio (multiplicative effect) rather than a sum (additive effect). This ratio becomes 16, and is equal to their combined effect.

Figure 4.74 shows that the time to the new steady state is determined by the time it takes for the slowest process to occur (in this case the inhibitory process).

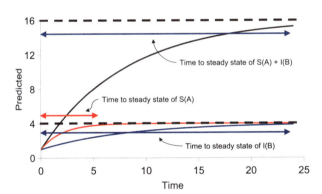

Figure 4.74 Schematic illustration of the time to steady state when two different drugs have the same pharmacological action (e.g., blood pressure lowering), but act via different mechanisms. S(A) (red curve) acts via stimulation of the production of response. I(B) (red curve) acts via inhibition of loss of response. Both mechanisms acting simultaneously, result in a positive (increasing) response (black curve).

This generic reasoning is also applicable to induction/inhibition of enzymes, and may explain why certain compounds exhibit metabolic interactions (with synergistic potential) when combined together.

4.13   Synergistic Effects Modeled by Hyperbolic Functions

Synergistic or cascade effects may be predicted from a nested system of hyperbolic functions (see Figure 4.4 for a schematic illustration of nested hyperbolic functions). Equations 4:213 and 4:214 are nested since the former drives the latter.

$$Stimulus\ I = \frac{B_{max} \cdot C^n}{K_d^n + C^n} \tag{4:213}$$

which then enters

$$Response = R = \frac{Stim_{max} \cdot Stimulus\ I^n}{beta^n + Stimulus\ I^n} \qquad (4:214)$$

This system (Equations 4:213 and 4:214) is fit to the experimental data shown in Figure 4.75 and the resulting parameter estimates are shown in the figure text. The model parameters were n_1, K_d, n_2 and *beta*, but B_{max} and $Stim_{max}$ were constants, since *Stimulus I* and *Response* were normalized to 100 per cent, respectively.

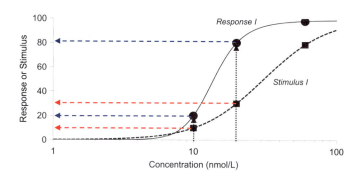

Figure 4.75 Response/Stimulus I versus drug concentration. By doubling the drug concentration from 10 to 20 nmol/L Stimulus I increases from 10 to 30 units (3-fold increase) and Response from 20 to 80 units (4-fold increase). K_d and n_1 were 30.9 ± 0.17 % (mean ± CV %) nmol/L and 1.95 ± 0.27, respectively. Beta and n_2 were 17.3 ± 0.43 nmol/L and 2.52 ± 0.44, respectively.

Another way of viewing this synergistic system is to study Figure 4.76 where the transduction function for ligand concentration-stimulus is separated from stimulus-response.

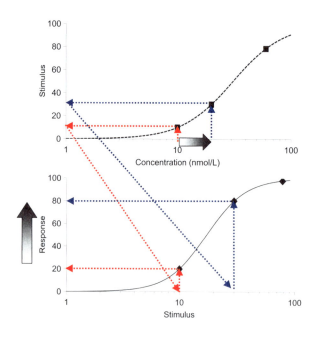

Figure 4.76 Ligand-concentration versus Stimulus (upper figure) and Stimulus versus Response (bottom figure). The doubling of the ligand concentration from 10 to 20 nmol/L(2-fold increase shown by the gray horizontal arrow) increases the response from 20 to 40 units (4-fold increase shown by the gray vertical arrow). Solid symbols represent measured data of ligand concentration-stimulus (■) and stimulus-response (●). Solid lines are model predicted. The 10 nmol/l ligand concentration-response relationship is shown by the dotted red line. The 20 nmol/L ligand concentration-response relationship is shown by the dotted blue line.

The magnitude of amplification throughout the cascade is, of course, very much dependent on the individual $Stim_{max}$ values (Equation 4:214). The relative magnitude of two different $Stim_{max}$ values may differ substantially. See also Kenakin [1997] and PD26 for a discussion on cascade systems modeled by hyperbolic functions.

CHAPTER 5 – Modeling strategies

Objectives

◆ **To review exploratory data analysis**

◆ **To set up a framework for model specification**

◆ **To select a minimization algorithm**

◆ **To demonstrate different ways of obtaining initial estimates**

◆ **To review the iteration process**

◆ **To assess the goodness-of-fit criteria**

◆ **To add insight into the statistical tools for model discrimination**

◆ **To add insight into experimental design**

5.1   Background

Modeling requires a solid theoretical background, intuition, patience, and experience. With practice, you will develop skill at selecting appropriate models and weighting schemes, and become adept at troubleshooting difficult model fitting problems. In this chapter, we provide some guidelines for model selection and weighting schemes based on graphical methods, residual analyses, and other diagnostic tools, such as the *Akaike* and *Schwarz* information criteria. We also include a checklist that can be used when assessing the quality of the fit of a model to a set of data, and consider ways of dealing with data sets involving several subjects. This chapter explains some of the basics of *going from data to model* (Figure 5.1).

Before designing a full experiment, one should use existing prior data/information regarding turnover/dynamics/kinetics. Based on these data, make predictions – in other words, simulate! The reason for this is that a plot of the concentration-time course of drug will be particularly useful when selecting the design points for sampling (i.e., time points (kinetics) or plasma concentrations (dynamics)) using the tentative parameter estimates and the proposed model. Plot e.g., the partial derivatives of the dependent function (such as plasma concentration or pharmacological response) with respect to each of the model parameters (Figure 5.1) and observe for extreme value (maxima and/or minima).

Having done all of this, one is ready to carry out the experiment in order to test the tentative model (see Endrenyi [1981] for the theoretical aspects, and PK1 to PK45 and PD1 to PD26 for practical aspects of experimental design) and to collect the data. The analytical assay, the number of animals, the total blood volume and the number of samples needed per animal have to be considered.

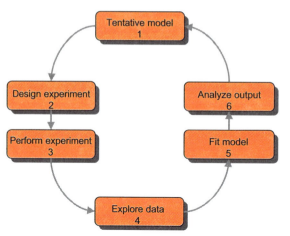

Figure 5.1 From tentative model to plot of data. Start out with a notion about the model (1) and an experiment (2) measuring e.g., effect at various concentrations. You obtain your effect-concentration data (3) and analyze the data graphically (4). Step number 4 is also called exploratory data analysis. Fit the models (5) and compare the fits (6). Then plan for a new experiment (1) to get more accurate and precise parameters by means of simulations.

5.2   Plot and Explore Data

5.2.1   Understand your experimental data better

Going from data to insight requires practice. Before one can start practicing, a certain amount of knowledge is needed. Knowledge means structuring new information and then applying previous knowledge. To help structure the information, we have supplied a framework of questions and guidelines that are meant to serve as a first-aid toolbox. The inexperienced student will benefit from applying these rules until more experience has been gained. The only way of becoming a good modeler is to practice on real problems over and over again – in other words, the only way to escape a problem is to solve it. Do not rush into immediate use of equations. Rather, let the data lead you to the type of model and/or equation that is best.

The first step in *exploratory analysis* is the most important one, and requires making a plot of the data. In other words, describe the behavior of the observations in terms of a concentration- or response-time curve, or a response-concentration curve and then ask yourself questions. Do the data contain experimental error and to what magnitude? Is the behavior linear, nonlinear and/or mixed? Does the extrapolated area contribute to a large or small portion of the curve?

What are C_{max}, t_{max}, $t_{1/2}$ and so on? Does the plasma concentration-time curve decline in a mono- or multi-exponential manner, or does it stay flat during a large portion of the dosing interval? Are dose data single and/or multiple, and what do they show? What is the accumulation (index)? Are there single or multiple peaks or other forms of unconventional behavior in the extravascular data? Could one anticipate solubility problems with the drug in the administered solution, gastrointestinal tract or plasma? Is the solubility *pH*-dependent? What about co-solvent, caking, sorption to injection sets, stability *et cetera*?

If the data are dose-normalized, do the curves superimpose? Do the dose-normalized areas superimpose? Are the dose-normalized areas equal or do they increase/decrease with dose? Do the dose-normalized curves change in shape? Is there a peak shift with increasing doses (concentrations)? Is there a baseline concentration value? Is the drug an endogenous compound? If so, does the baseline value vary over time and what is the behavior of that variation? Is the baseline constant or circadian, or is it feedback regulated? If circadian, is it symmetric?

Is total blockade of the response observed? Are there large inter-individual differences with respect to the plateau, inflexion point or curvature? Are the data single-valued or do they contain hysteresis? Have the dynamics/kinetics been characterized? Do the data contain information about free concentrations and plasma protein binding? Are there kinetic/dynamic data from two or more species? Do the data scale?

In other words, use a question and answer technique to help in making overall interpretations of the data and the model.

Characterize the kinetics/turnover of the compound. In other words, determine how long it takes to reach steady state and the routes of elimination (biliary excretion, metabolism, renal excretion, or expired air).

Model selection would preferably begin with an understanding of the underlying kinetics or the pharmacological effect. Again, this means questions to be answered. Is the observed response a result of inhibition or stimulation? Is the response a direct or indirect action, or even a sequence of events? Does it include a baseline value and a maximum? Is anything known about the physiology of the effect? Is the kinetics linear or nonlinear? Note that if there is a plasma profile to be followed, one usually knows something about the half-life or clearance of the drug or of related compounds.

We believe that exploratory analysis is often taken too lightly, as predetermined expectations influence the experimentalist. This carries with it the risk of inadequate and/or inappropriate interpretation of the data, which may have dire consequences. For this reason it is important that plotting and exploring the data are not approached casually. For additional reading on practical, statistical and philosophical considerations when fitting kinetic and dynamic models, see Endrenyi [1981], Boxenbaum [1992], Jusko [1992] and Heafner [1996].

5.2.2   Pooling of data

Assume steps 1-3 are now complete (tentative model, experimental design, collection of data – Figure 5.1). It is therefore time to explore and prepare the experimental data. Start by inspecting the data from each subject and by pooling data from several subjects. Pooling of data becomes necessary when the data density from individuals is not sufficient for analysis, or the data are highly variable. One can either average (e.g., the concentrations at each time point) and fit a model to the mean data, or one can fit a model to all individual data observations simultaneously. These two approaches are commonly called the *Naive Averaged Data* (*NAD*) and *Naive Pooled Data* (*NPD*) approaches, respectively. They give (theoretically) identical estimates provided the number of observations at each time point is the same.

We usually start by making a *spaghetti-plot* of the results from all subjects (Figure 5.2). This approach shows the overall tendency of the response. From this plot, it is easy to select individuals or groups of individuals who deviate from the central tendency with respect to half-life, absorption, clearance and so on. Pooling runs the risk of masking individual behavior, although it may still function as a general guide. Our view is that a spaghetti plot gives a far better picture of the variability than error bars attached to a mean curve.

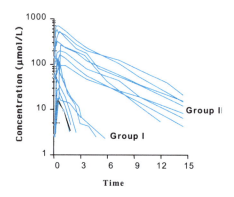

Figure 5.2 Spaghetti plot of plasma concentrations of a new compound in two different populations. Data were obtained within a 24-hour dosing interval at steady state. It is clear that the subjects in these two populations fall into two groups (I and II).

From the plot shown in Figure 5.2, one sees that subjects tend to fall into one of two groups. One may speculate that these two groups have different volume terms V/F because C_{max} differs, and we know that dose divided by C_{max} gives a hint of the volume size. The bioavailability may also differ due to different areas under the curve. Let us assume, however, that the volume terms are the same. Then the oral clearance (Cl/F or Cl_o) for group I is larger than for group II. This explains why the area is smaller and the half-life shorter for group I.

Even though there is substantial variation within each of the two groups in Figure 5.2, one can observe two distinct tendencies, one with rapid and the other with slower kinetics. From these concentration-time curves, a common time-interval for assessment of the terminal half-life for each group can be selected.

It is also recommended to *dose-normalize* plasma concentrations and plot for example, the average concentration (C_{ave}) or Cl_o versus dose if more than one dose level is used. This manipulation picks out nonlinear tendencies in the kinetics. If data from multiple dosing were available, we would recommend plotting the pertinent pharmacokinetic estimates (e.g., C_{ave} or Cl_o) versus time, in order to study time-dependent changes. Similar exploratory analysis could also be undertaken for metabolite data and other covariates such as age, body weight, gender, and so on. Holford [1995] recommended the more physiologically correct use of clearance rather than AUC in clinical summaries. Since clearance or clearance over bioavailability (Cl_o) is easily calculated from AUC and dose, there is no intrinsic merit in AUC. Cl or Cl_o, on the other hand, can immediately be interpreted in a physiological context. For more details, see Section 3.4.

5.2.3   Transformation for exploration

It is also important to consider if raw data should be used, or if they should be transformed for exploratory purposes. It is worth pointing out here that transformed data alter the distribution of the data and can conceal baseline variability.

The six plots in Figure 5.3 illustrate how three different models, the ordinary E_{max} model, the ordinary E_{max} model with baseline, and the sigmoid E_{max} model with baseline effect, mimic the same set of data. Note how the linear plots hide the lack of fit for the first two models, while the effect *versus* log-concentration plot clearly demonstrates the systematic deviations between observed and predicted data. The more flexible sigmoid E_{max} model with a baseline parameter successfully described the overall trend in these data.

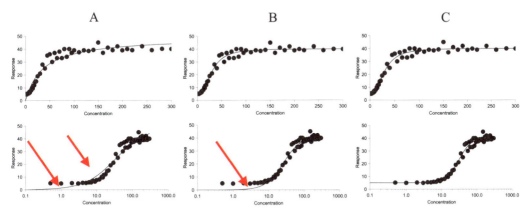

Figure 5.3 *Observed (●) and predicted (curve) effect-concentration data using the ordinary E_{max} model (A), the ordinary E_{max} model with baseline (B) and the sigmoid E_{max} model with baseline (C). Data are plotted with log-transformed concentrations (bottom plots) and untransformed concentrations (upper plots). Observe the lack of fit of A and B indicated by red arrows.*

When data have been plotted, look at them and ask: Has E_{max} been reached or not? Can you obtain good (initial) estimates of EC_{50}? In the plots in Figure 5.3, it is easier to estimate EC_{50} from the plot on a log-scale. This scale also gives the slope of the tangent at the EC_{50} value. Note that you should be able to perform the analysis by hand, and always remember that you (not the program) are in charge of the analysis. In other words, let the data lead you to the type of model and/or equation that is appropriate.

Generally, as most of the commonly employed kinetic models involve exponential terms, it is preferable to plot the data on a semi-log scale, using a logarithmic scale for the concentration axis and an arithmetic scale for the time axis. The plot should be examined to determine if the log-scaled concentrations decline in a linear fashion and, if so, how many linear segments are involved. This concept is often described as mono-, bi- or three-exponential decay and corresponds to a one-, two- or three-compartment kinetic model, respectively. Figure 5.4 shows observed concentrations exhibiting multi-exponential decay following administration of a drug as an intravenous bolus, along with results from fitting a one-, two- or three-compartment model to data.

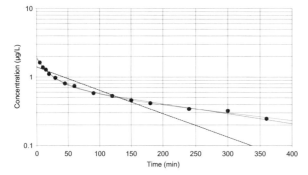

Figure 5.4 *Observed (●) and predicted (curves) concentrations of a one-, two- and three-exponential model.*

If a drug was administered intravenously, the one-, two- and three-compartment models would include one, two or three exponential terms, respectively. However, if it was administered extravascularly, the models would include two, three or four exponential terms, respectively, as an additional term would be needed to account for the absorption of the drug. Figure 5.5 displays the plasma kinetics following extravascular administration of a two-compartment drug. Note the three distinct phases; the upswing, the initial drop, and the terminal decline.

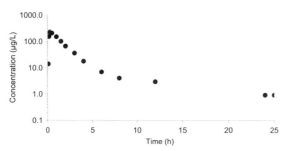

Figure 5.5 *Concentration-time course of a two-compartment drug following first-order input. Note the early peak, the rapid initial decline between 1 and 5 h, and the slow terminal phase after 10 h.*

If data are available for several subjects, plots may suggest different models should be utilized for different subjects. For example, a one-compartment model may seem most appropriate for some subjects while a two-compartment model may seem most appropriate for others. There is no universally accepted method for dealing with this situation. Some scientists would report the results of fitting different models to the two groups of subjects, and others would fit the more complex model to all of the subjects. The latter approach assumes that all subjects were randomly selected from the same population. If this method is employed, it is generally preferable to summarize the primary pharmacokinetic parameter values (across subjects) using weighted means (with individual weights equal to the reciprocal of the square of the standard error) or nonparametric methods. It may also be possible to avoid some of the difficulties in the choice of the appropriate model by calculating the primary kinetic parameters such as *Cl* and *V*, and making an overall summary of the derived values for *Cl* and *V* for the whole study population.

The choice of a particular model generally depends on the aim of the study. In our experience, the precision of clearance is usually more or less independent of the number of exponentials.

If a kinetic model is to be used as the input to a dynamic model, we would recommend that the kinetic data be smoothed as much as possible. This means it may be necessary to use more exponentials than there is information for in the data so as to obtain precise and accurate estimates. This does not matter, as long as the kinetic model mimics the data correctly and one is not predicting responses outside the range of observed concentrations or doses.

Although it is usually inappropriate to fit transformed data, it is often helpful to display transformed data. In fact, many times it is easier to visually explore and interpret transformed than untransformed shapes. Whether one should transform or not for fitting purposes depends on the underlying error distribution.

5.2.4 Transformation for fitting

Prior to the era of personal computers, nonlinear regression was not readily available to most scientists. Instead they transformed their data to force curvilinear relationships into straight lines that were then analyzed by means of linear regression. Scatchard and Lineweaver-Burke plots of binding data and log-linear plots of kinetic data are examples of such linearization. Today, these methods should rarely be used for final regression analysis of data, though they are suitable to obtain initial estimates and for exploratory data analysis. Linear regression assumes that the variability among replicate concentration values follows Gaussian distributions, with standard deviations that do not depend on time (the independent variable t). This assumption is less valid for transformed data. Linear regression also assumes that concentration and time are measured independently. With the Scatchard plot, the X and Y values are intertwined during the transformation, invalidating linear regression assumptions of normality.

Traditionally, graphical methods have been applied for estimation of K_m and V_{max} (see Leatherbarrow [1990] and Rovati [1998] for a review). However, the ease of linear regression analysis makes it very tempting to apply this method to situations where it is not strictly appropriate. The most common misuse is to take a nonlinear equation, rearrange it to a linear form (e.g., double reciprocal plot) and then perform linear regression to fit the data. Such manipulations are very common, and include examples such as the Scatchard plot of binding data and the Lineweaver-Burke plot of enzyme kinetics. However, the rearrangements that are performed also rearrange the error distribution, thus invalidating the assumptions behind the linear regression technique. Simple linear regression should not be used to analyze such data, except perhaps to obtain initial estimates for the more appropriate nonlinear regression analysis. The degree of inaccuracy introduced by such analysis depends upon the rearrangements made and how much error is present in the original data. The effect that rearrangement of data has on error structure is illustrated in Figure 5.6. Here, enzyme kinetic data and a common transformation are shown. The raw data (A) are assumed to contain errors in rate that are the same for each data point – in other words, there is a constant absolute error. The model is, however, nonlinear in its parameters V_{max} and K_m.

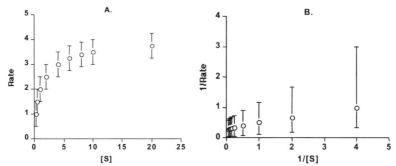

Figure 5.6 *(A.) Rate (amount eliminated per unit of time) versus substrate concentration [S] (amount per volume) for a typical capacity-limited system. Bars represent the standard deviation with a constant absolute variance. (B.) A Lineweaver-Burke plot of the inverse of rate versus the inverse of the substrate concentration for a typical capacity limited system. Bars represent the inverse of standard deviation (with a constant variance). Note the distortion of the error distribution in the Lineweaver-Burke plot.*

As can be seen from Figure 5.6 (B), the Lineweaver-Burke double reciprocal plot distorts the error distribution considerably. Simple linear regression is therefore inappropriate to analyze such data, although the plot is still very useful when visualizing the results.

With some equations, it is not possible to fit a function of X to the data by calculating the partial differential equations of the sum of residuals and setting them equal to zero. This is because one obtains equations which cannot be solved directly. An example of this is the ordinary E_{max} model

$$E = \frac{E_{max} \cdot C}{EC_{50} + C}$$
(5:1)

This can be rearranged so as to obtain the equation for a straight line (*Lineweaver-Burke linearization*)

$$\frac{1}{E} = \frac{1}{E_{max}} + \frac{EC_{50}}{E_{max}} \cdot \frac{1}{C}$$
(5:2)

A plot of *1/E versus 1/C* gives a straight line where *$1/E_{max}$* is the intercept and *EC_{50}/E_{max}* is the slope. However, remember that the transformation distorts the error distribution.

The problems described above are easily avoided by analyzing the original data using nonlinear regression. Although nonlinear regression is conceptually the same as linear regression, it fits the best curve (rather than the best line) to the data through an iterative procedure. Consequently, there is no need to rearrange the original experimental results prior to analysis, and therefore there is no distortion of the original error distribution (Leatherbarrow [1990]).

5.2.5   Normalizing data

Thus far, we have assumed that the drug concentrations have declined in a linear fashion. What if this were not the case, such as in Figure 5.7. In this example, data were obtained from two healthy young individuals who had received a different dose of a new compound by intravenous bolus. Their body size (weight) and age were very similar. The measured plasma concentrations (Figure 5.7, left) included both free and bound drug. We assume that the major source of variability is the Michaelis-Menten constant K_m rather than the maximum metabolic rate V_{max} or the volume of distribution V. We also assume that distribution occurs rapidly upon administration and that V is directly proportional to body weight. The proposed system is a one-compartment model with Michaelis-Menten elimination kinetics.

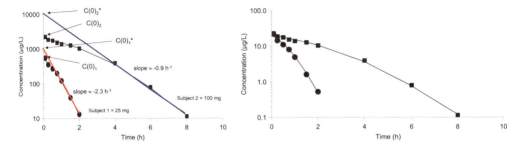

Figure 5.7 *The observed plasma concentration-time profiles of a compound following two different intravenous doses to two subjects. Left panel is actual data and right panel is dose-normalized data for the high dose. Dose normalized C^0 values indicate similar volumes of distribution. Different slopes at low concentrations indicate different K_m values.*

Overlaying the predicted data on the observed data provides guidance as to the next step. Since the higher dose was 100 and the lower 25, we divided the concentrations for the higher dose by a factor of 4.

The dose-normalized concentrations (Figure 5.7, right) suggest that the volume of distribution estimated from $Dose/C^0$ is the same for the two subjects, whereas clearance differs. Since clearance obeys a capacity-limited process, we may propose a Michaelis-Menten type of clearance model. There are two parameters that explain the Michaelis-Menten behavior, namely V_{max} and K_m. Since data contain total drug concentrations, we would suggest that V_{max} is the same, but K_m differs between the two subjects. K_m also includes protein binding f_u. If protein binding differs (i.e., is concentration dependent), the K_m values will also differ. This causes the terminal fall-off of the two subjects to be non-parallel. Parallel decline would suggest similar K_m values. See also Gibaldi and Perrier (1982) and our applications section on capacity-limited systems (PK17-PK20) for a more thorough description of estimation of V_{max} and K_m by means of graphical methods.

Modeling these types of data is generally difficult and often requires data from more than one dose level. It is difficult not only because the source of the nonlinearity has to be determined, but also because models involving nonlinear kinetics can generally only be described by differential equations. More than one dose level may be required for two reasons: (1) to confirm that the kinetics are nonlinear, one needs to determine if concentrations (and AUC) change in proportion to changes in dose and (2) to ensure that the model parameters are estimated with a high degree of precision and a low correlation. We highly recommend dose-normalization of concentrations or areas to see whether they superimpose.

Note that even if the concentrations tend to decline in a linear fashion, the underlying kinetics could still be nonlinear. This is because data from a single dose of a drug whose kinetics is nonlinear can often be well approximated by linear kinetics. To see this, use a program to simulate a model with nonlinear kinetics and output the predicted values. Then treat the predicted values as data, and fit them to a linear compartmental model. You will often find that there is a very good fit to the data, which is precisely why a dose ranging study should be run very early in the development program.

Another reason why a model selected from fitting single dose data should be tested for adequacy by studying the transition from single to multiple doses, is that predicted and observed effects will systematically diverge when multiple doses are administered if an incorrect model has been chosen (Colburn [1981]).

5.3   How Complicated a Model?

5.3.1   How many parameters?

Another area that receives inadequate attention is that of parameter estimability. In this case the question is how complicated a model can be fit to the data? The number of parameters (NP) which can be calculated for a given model is dependent on the number of exponentials (EX) visible in the plasma concentration-time profile, the number of elimination or excretory pathways (PE) suitably measured, the number of tissue spaces or binding proteins (TS) analyzed, and the number of visible nonlinear features (NL) in the data. This relationship is given by

$$NP = 2 \cdot EX + PE + 2 \cdot TS + NL \qquad (5:3)$$

Equation 5:3 is applicable if sufficient and accurate data are obtained (Jusko [1986]). For example, if we have data that decline in a bi-exponential fashion (that is, with an α- and a β-phase), then it is possible to estimate $2EX = 4$ parameters (i.e., A, α, B and β). If we also have measured drug excretion in urine we could estimate renal clearance or fraction of dose excreted via the urine. If we, in addition to the bi-exponential decline in plasma ($2EX$), have a nonlinear feature (NL) and urinary data (PE), we might be able to estimate six parameters.

Taking the E_{max} model as an example of a model that behaves in a particularly nonlinear fashion we will shortly demonstrate the topic of estimability (Section 5.3.5). The parameters may be theoretically identifiable. That is, it may be theoretically possible to discriminate between the parameters if a correct experiment is performed. However, it may still be impossible to discriminate between the parameters because of experimental noise or inappropriate study design.

5.3.2   How do we specify the model?

Assume that we have a first-order one-compartment system (mono-exponential decline) as depicted in Figure 5.8. How do we then specify the model mathematically?

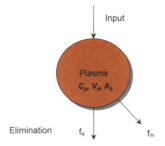

Input

Plasma
C_p, V_d, A_b

Elimination f_e f_m

Figure 5.8 Schematic representation of the one-compartment model with first-order elimination. C_p is plasma concentration, V_d is the volume and A_b the amount of drug in the body. The f_e and f_m parameters are the fractions of the dose excreted via the kidneys and being metabolized by the liver, respectively.

The simplest way to mathematically describe such a system is by means of a first-order differential equation. Figure 5.9 demonstrates the concentration-time course of a one-compartment system in plasma on linear and semi-logarithmic scales.

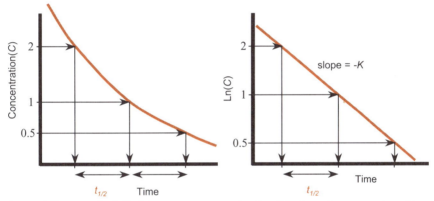

Figure 5.9 Cartesian plot (left) of plasma concentration-time data following an intravenous dose of drug. Semi-logarithmic plot (right) of the same data. The data show mono-exponential decline as described by Equations 5:4 and 5:5.

The rate of change of the plasma concentration, *dC/dt,* is directly proportional to the plasma concentration by means of the first-order elimination-rate constant. That is,

$$\frac{dC}{dt} = -K \cdot C = \frac{Cl}{V}C \qquad (5:4)$$

This equation describes the behavior as a differential equation. The integrated solution of Equation 5:4 becomes

$$C = C^0 \cdot e^{-K \cdot t} = \frac{D_{iv}}{V} \cdot e^{-\frac{Cl}{V}t} \qquad (5:5)$$

The integrated function calculates only the concentration at each observed time point for each iteration. If we have 3 observations, there will be only 3 function evaluations for each iteration (i.e., one per time point). However, the differential equation needs to be integrated throughout the whole observation period from time zero (starting value) to the last observation. This complete function evaluation, which also includes intermediate time points, makes fitting the differential equation model much slower than the integrated function. The difference between the two approaches is shown in Figure 5.10 below.

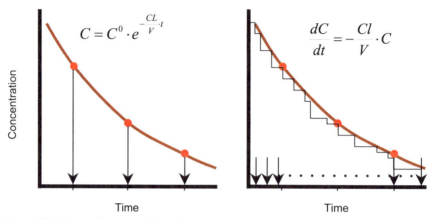

Figure 5.10 *Schematic illustration of the difference between the integrated model (left) and the differential equation model (right). The integrated model requires only three function evaluations (one at each observed time point) whereas the differential equation model performs one function evaluation for each integration step.*

It can be seen that either a first-order differential equation and/or a mono-exponential equation may represent such a system. However, this is still a nonlinear regression model with respect to *Cl*, *V* and *K*, and a linear model with respect to kinetics. If the elimination-rate is saturable, as in Equation 5:6,

$$V\frac{dC}{dt} = -\frac{V_{max} \cdot C}{K_m + C} \qquad (5:6)$$

we cannot integrate the equation to get a simple analytical solution. However, the plasma concentration-time profile is available by means of numerical integration. Such a system is then best represented as a differential equation. The rate of metabolic capacity (*rate*) can be written as a function of plasma concentration as follows

$$rate = \frac{V_{max} \cdot C}{K_m + C} \qquad (5:7)$$

From this relationship, the metabolic capacity can be obtained at different plasma concentrations, though concentration *versus* time is not available.

5.3.3 Combining several sources of data for modeling

When fitting a model, we strenuously advise the use of several sources of data (e.g., plasma and urine) simultaneously. Examples of the usefulness of this approach are shown in Table 5.1

Table 5.1   Sources of data, parameters and examples of simultaneous modeling

Source of data	Parameter(s)	System(s)	Example(s)
Plasma, urine	f_e, Cl_R, F	Excretion, metabolism	PK5, PK6, PK16
Intravenous and extra-vascular dosing	F	Absorption and disposition kinetics	PK10
Drug, metabolite	f_m, Cl_m,	Ideally, intravenous dosing of metabolite gives disposition kinetics	PK19, PK45
Several species	*Allometric parameters (a, b, c and d)*	*In vivo/in vivo* scaling	PK28, PK29
Several doses	V_{max}, K_m	Capacity dependencies	PK17-PK20
Several inhibitory concentrations	V_{max}, K_m, K_I	Inhibition studies	PK44
Repeated doses	k_d	Time dependencies, induction/inhibition	PK21, PK22
Several doses, response	k_{in}, k_{out}, EC_{50}/IC_{50}, E_{max}/I_{max}	Turnover *versus* distribution related response, logistic regression, dose-response-time analysis	PD6, PD7, PD21 to PD25
Several tissues	τ	Operational model	PD2
Repeated exposure	k_{tol}	Time dependencies, tolerance/sensitization	PD10, PD11
Several subjects		Intra *versus* inter-individual differences	Not covered in this book

As a bonus, new information will be made available using this approach, and one will obtain more accurate and precise parameter estimates. Different sources of data also complement each other. Thus, when data are lacking or of poor quality with respect to one or several parameters in one source of data, this information may be available in the other source. There are several examples in later sections of simultaneous modeling of different sources of data (see Table 5.1 for details).

Some parameters are often highly correlated, such as V_{max} and K_m, so that the estimate of one almost governs the estimate of the other. Estimates of these parameters are also often skewed to the right, with many of the estimates being many times larger than the true parameter value. They also have high variances, in spite of the fact that the variances computed from individual data sets are underestimated. Also, there is a high correlation between the estimates of V_{max} and K_m with their respective standard deviations. Thus, the estimated precision of V_{max} and K_m, given by regression programs, does not give a true picture of the uncertainty in these estimates (Metzler and Tong [1981]). To avoid some of these problems, we advocate the modeler to fit data obtained from two or more dose levels simultaneously, or data covering a large enough concentration range (PK17-PK20). See also Kakkar *et al* [2000] on minimal experimental design of enzyme kinetics.

5.3.4   Parameter identifiability

Before actually proceeding with the analysis to estimate the unknown parameters of a model from experimental data, it is necessary to determine whether useful estimates can be obtained. That is, assuming ideal data from the experiment, is it theoretically possible to determine unique estimates of all unknown model parameters? This question is often referred to as that of theoretical or *a priori* identifiability (Finkelstein and Carson [1985]). The mathematical property of a model concerning the uniqueness of a set of one or more parameter values is often called *identifiability*.

For model parameters to be identifiable, they must have a mathematically unique influence on the observed model response. Identifiability is dependent on the model and its parameterization. Identifiability indicates whether *unique* parameter values can be estimated from a *noise free* model response (Dutta *et al* [1996]).

Details of available methods of testing for identifiability can be found elsewhere (Cobelli and DiStefano [1980], Carson *et al* [1983], Godfrey [1983] and Dutta *et al* [1996]). Having carried out this identifiability analysis of a model and given ideal test data, all the parameters can be estimated uniquely. Such a model is said to be *uniquely identifiable*.

Another possibility is that although some parameters can be estimated uniquely, there are others for which a finite number of feasible numbers would be compatible with the given ideal data. Assume that we have a first-order one-compartment system that can be modeled by

$$C = A \cdot e^{-K \cdot t} - A \cdot e^{-K_a \cdot t} \qquad\qquad (5{:}8)$$

For this model, K_a and K are not identifiable. Hence if K_a and K are close to each other, instability of the sum of squares function will result. There are also two solutions to the problem in that the residual sum of squares has at least two minima. This situation is commonly known as the *flip-flop* situation. The only way of discriminating between the two rate constants is by giving an intravenous bolus dose and observing the disposition kinetics of the drug to obtain the true value of K.

Another situation where one cannot estimate the volume and rate constants uniquely, unless there are data from both compartments, is the two-compartment model depicted in Figure 5.11 which contains five parameters. Plasma data of a bi-exponential system do not contain information about five parameters. Rather, a bi-exponential system is fully described by means of only four parameters.

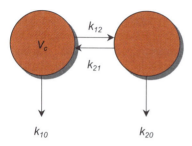

Figure 5.11 *Schematic diagram of a two-compartment model with two routes of elimination from both compartments.*

5.3.5 Parameter estimability

The ability to estimate the parameters of a model accurately and precisely depends on the noise level in the data, the study design, the model and the algorithm. Ill-conditioned problems of various kinds can be inherent in the model itself and poor experimental designs can make matters worse. However, even with good designs and sufficient experimental data it may not be possible to eliminate severe ill-conditioning. Very accurate experimental data can still give rise to extremely imprecise and highly correlated estimates. This may occur even though the model is correct and fits the data well (Metzler [1986], Seber and Wild [1989], Ebling [1995], Dutta *et al* [1996]).

Unlike identifiability, which can be considered a mathematical property of the model and design points, *estimability* is a statistical property that requires a statement concerning the degree of uncertainty that the modeler is willing to accept. For example, a parameter that is estimated within ± 50% might be considered estimable for one application, but not for others. Although a problem may be ill-conditioned, certain parameters or data descriptors may be highly estimable. Confidence intervals reported by nonlinear regression programs are not necessarily reliable indicators of parameter estimability. This and other issues of experimental design and nonlinear regression are summarized in excellent fashion by Dutta *et al* [1996]. A classic example of reduced estimability can be seen for two or more parameters sometimes referred to as the *u-shaped* pharmacodynamic model (Figure 5.12).

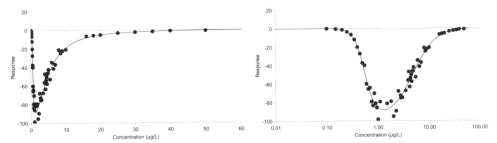

Figure 5.12 *Observed and predicted effect-concentration profiles of a new experimental agent. The profiles follow a distinct u-shaped relationship. The left hand plot displays the data on a linear concentration scale, and the right hand plot on a semi-logarithmic scale.*

The model used in Figure 5.12 is

$$E = \frac{E_{max} C^{n_1}}{EC_{50}^{n_1} + C^{n_1}} - \frac{I_{max} C^{n_2}}{IC_{50}^{n_2} + C^{n_2}} \tag{5:9}$$

This model implicitly relates the observed effect to a plasma or biophase concentration. E_{max} and I_{max} are the maximum opposite effects, and EC_{50} and IC_{50} correspond to their respective potencies. The n_1 and n_2 parameters correspond to the sigmoidicity factors. There are inherent deficiencies in this model unless effect data exhibit low variability and are well-spaced with regard to the EC_{50}/IC_{50} values. In other words, the utility of the model depends on the degree of separation of the EC_{50} values. At high concentrations (where $C \gg EC_{50}$ and IC_{50})

$$E \approx E_{max} - I_{max} \tag{5:10}$$

At low concentrations (where $C \ll EC_{50}$ and IC_{50})

$$E \approx \frac{E_{max}C^{n_1}}{EC_{50}^{n_1}} - \frac{I_{max}C^{n_2}}{IC_{50}^{n_2}} \tag{5:11}$$

At intermediate concentrations (where $EC_{50} \ll C \ll IC_{50}$)

$$E \approx E_{max} - \frac{I_{max}C^{n_2}}{IC_{50}^{n_2}} \tag{5:12}$$

Unfortunately, two or more parameters (often the E_{max} parameters) are often highly correlated because they only appear as a difference. This would therefore render the model inappropriate.

More empirical models may be utilized to describe such data, but they often lack physiological significance (Ebling *et al* [1991]).

5.4   Obtaining Initial Estimates

So far we have discussed data collection and the graphical exploratory aspects of modeling, as well as the number of potential parameters there will be in a model. Here, we elaborate on the initial parameter estimates and discuss why these are so important (Step 4 in Figure 5.13).

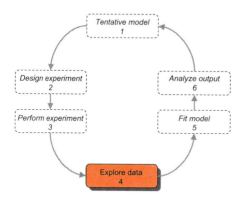

Figure 5.13 From tentative model to plot of data. Exploratory data analysis (step 4) involves calculating initial parameter estimates from the experimental data obtained in the experiment that had been designed around the proposed model (steps 1-3). This important step will then allow appropriate analysis of the results after the data have been fit to the model (steps 5 and 6), and, enable a new experiment to be planned to obtain better estimates (i.e., the cycle is repeated.)

Invest time in calculating initial parameter estimates. These initial estimates are more important when your data have a lot of scatter, do not span a large enough range of *X*-values to define a full curve, or do not really fit the model. In fact, our own experience with the Hill equation shows that, if the initial estimates are poor, convergence to the wrong final values can easily occur. There is no golden rule for obtaining initial estimates. One uses whatever information is available. However, good starting values will often allow an iterative process to converge to a solution much faster than would otherwise be possible. Also, if multiple minima exist or if there are several local minima in addition to an absolute minimum, poor starting values may result in convergence to an unwanted point in the sum of squares surface. This unwanted point may have parameter values that are physiologically impossible or that do not provide the true minimum value of the objective function (Draper and Smith [1998]).

Graphical methods and linear regression (section 5.4.1), *non-compartmental analysis* (section 3.7) or *convolution-deconvolution methods* (section 5.4.2) can obtain initial parameter estimates.

Build a knowledge base around the compound as early as possible and state the objectives for the regression exercise. Modeling can be an endless journey if there is not a clearly defined goal from the outset.

5.4.1   Graphical methods and linear regression

5.4.1.1   Kinetic data

In our first example (Figure 5.14), we have plotted the concentration-time course of drug in plasma of two different subjects after an identical intravenous bolus dose.

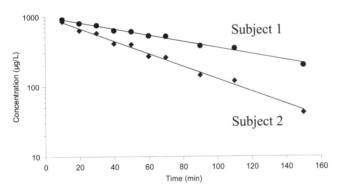

Figure 5.14 Semi-logarithmic plot of observed (Subject 1 •, Subject 2 ◆) and predicted (lines) concentration-time data of drug from two subjects following a 10 mg intravenous bolus dose.

Because all points seem to fall on the linear portion of the curve, we use the highest and lowest concentration values for the calculation of K. In this exercise, K is calculated from the slope of the plasma concentration-time curve by means of log-linear regression

$$slope = \frac{y_2 - y_1}{x_2 - x_1} = -K = \frac{\ln(920) - \ln(200)}{10 - 150} = 0.011\,\text{min}^{-1} \qquad (5\!:\!13)$$

and the half-life becomes

$$t_{1/2} = \frac{\ln 2}{K} = \frac{0.693}{0.011} = 63\,\text{min} \qquad (5\!:\!14)$$

The area under the curve AUC for subject 1, is estimated as

$$AUC_0^\infty = \int_0^\infty C^0 e^{-K \cdot t} dt = C^0 \left[\frac{e^{-K \cdot \infty}}{-K} - \frac{e^{-K \cdot 0}}{-K} \right] = \frac{C^0}{K} = \frac{1000}{0.011} \approx 91000 \, \mu g \cdot \min / L \quad (5:15)$$

Since it is obvious from the plot that subject 1 has a greater area than subject 2, one knows that clearance Cl for subject 1 will be less than clearance for subject 2. For subject 1 plasma clearance is

$$Cl = \frac{D_{iv}}{AUC_0^\infty} = \frac{D_{iv}}{\left[\frac{C^0}{K} \right]} = \frac{10000}{91000} = 0.11 \, L/\min \qquad (5:16)$$

The volume of distribution for both subjects is

$$V = \frac{D_{iv}}{C^0} = \frac{10000}{1000} = 10 \, L \qquad (5:17)$$

Due to the simple dataset and model, one can use the values of Cl, V and K from subject 1 as initial parameter estimates for subject 2.

In the following example, we model bi-exponential decline by means of Equation 5:18. The primary model parameters in Equation 5:18 are A, α, B and β.

$$C = A \cdot e^{-\alpha \cdot t} + B \cdot e^{-\beta \cdot t} \qquad (5:18)$$

These can be estimated graphically, where α and β correspond to the initial and terminal slope factors, respectively, and A and B to the intercepts. The concentration time course of Equation 5:18 is displayed in Figure 5.15.

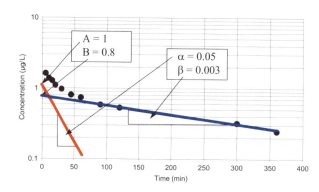

Figure 5.15 Bi-exponential behavior of a drug given as a bolus dose. $A \approx 1.0 \, \mu g/L$ and $B \approx 0.8 \, \mu g/L$. $\alpha \approx 0.05 \, min^{-1}$ and $\beta \approx 0.003 \, min^{-1}$. Note that A intercepts the abscissa below the observed values.

The B and β parameters are determined for a log-linear regression of the terminal phase. B is obtained from the intercept with the Y axis by back-extrapolation of the terminal slope and β is the actual slope. The A and α parameters are obtained by curve stripping the back-extrapolated phase from the initial phase, and then fitting the residual line by log-linear regression. A is the intercept of the Y axis of the residual line and α is the corresponding slope. By means of the intercepts (A and B), slopes (α and β) and dose, one can derive estimates of primary kinetic parameters such as Cl and V_{ss}.

5.4.1.2 Dynamic steady state data

Initial parameter estimates such as EC_{50} and E_{max} can usually be obtained graphically from inspection of the effect-concentration plot of untransformed (lin-lin plot) or transformed (semi-log plot) data (see Figure 5.16).

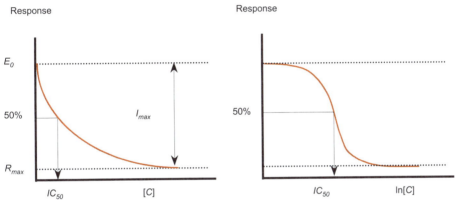

Figure 5.16 *Response versus steady-state concentration for an antagonistic compound. Note how the shape of the curve is affected by the change in the concentration scale. The semi-logarithmic scale (right) displays the symmetrical shape of the curve. It also expands the section below IC_{50}, which makes it easier to obtain an initial estimate of IC_{50}.*

When E is plotted against the natural logarithm of the plasma concentration $ln(C)$, both the hyperbolic and logistic relations are *S-shaped* and may be symmetrical about $ln(EC_{50})$. The human eye can judge this quite well, though only if data encompass a wide range of concentrations (Barlow [1991]). There is more difficulty in judging the exponent n, by eye because the steepness of the line depends on the choice of the concentration scale. See also PD2 for asymmetric models.

5.4.1.3 Dynamic non-steady state data

Linear regression may be used in many cases. For example, in the indirect response model for prothrombine complex activity, it can be used to obtain starting values for k_{out} (Figure 5.17). The k_{out} parameter is determined from the downswing of the curve. The easiest way of measuring k_{out} is by giving a large dose of warfarin that totally blocks the synthesis k_{in} of the prothrombin complex. The prothrombin complex activity then declines in a first-order manner (straight line in a semi-log plot of prothrombin complex activity *versus* time) until the blockade by warfarin of k_{in} has vanished.

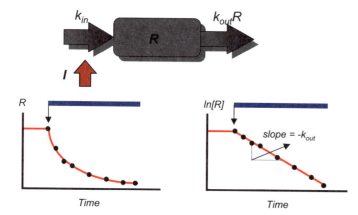

Figure 5.17 *The indirect response model. The red arrow symbolizes the inhibitory action of warfarin on the synthesis rate of the clotting factor (Nagashima et al [1969]). The figures illustrate the response-time course upon total inhibition of synthesis of prothrombin complex on a linear (left) and semi-logarithmic (right) scale.*

Experimental data of prothrombin complex activity (*PCA*) are plotted in Figure 5.18. From the slope of the downswing of the curve, log-linear regression gives k_{out}.

$$k_{in} = P_0 \cdot k_{out}$$

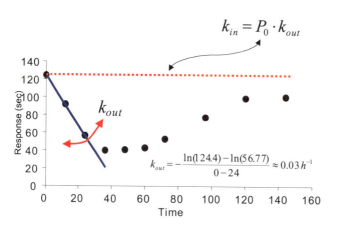

Figure 5.18 *Observed (●) and predicted (line) PCA time course following the administration of an intravenous bolus dose of warfarin. Data kindly supplied by Dr. Leon Aarons, Manchester.*

$$slope = \frac{\ln(R_2) - \ln(R_1)}{t_2 - t_1} = -k_{out} = -\frac{\ln(124) - \ln(56.77)}{24 - 0} = 0.03\,\mathrm{h}^{-1} \qquad (5:19)$$

From the intercept of the effect axis, the baseline (120 sec) value can be obtained. This baseline value P_0 is the ratio of k_{in}/k_{out} (Models 1 and 4 in section 4.6). From the intercept and slope, k_{in} can then be calculated (3.6 sec·h^{-1}).

When both kinetic and dynamic data exist, we recommend three types of plots: *concentration versus time, effect versus time* and *effect versus concentration*. It is also wise to transform the concentration axis, and plot effect *versus* log-concentration (see also Figure 5.3). In the example presented in Figures 5.19 and 5.20, we include concentration-time and effect-time data. We have also selected two out of four indirect response models. From the plots, initial estimates are derived for the model.

Figure 5.19 shows the kinetics of a new compound after two rapid infusions, with a rapid increase and washout of the plasma concentration.

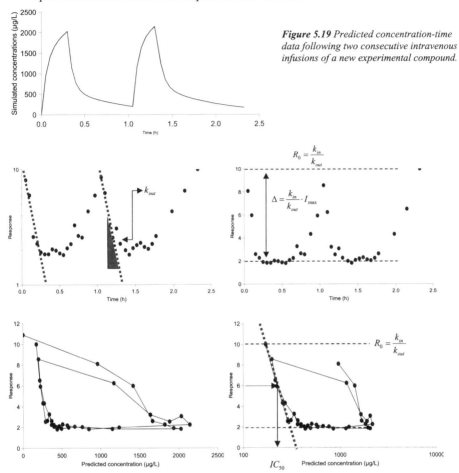

Figure 5.19 *Predicted concentration-time data following two consecutive intravenous infusions of a new experimental compound.*

Figure 5.20 *Plots of log-response versus time (upper left), response versus time (upper right), response concentration (lower left) and response versus log concentration (lower right) with the baseline value and slope at IC_{50}. Note the baseline value R_0 and the slope at IC_{50}. Data were obtained following two consecutive intravenous infusions of a new experimental compound. The vertical arrow indicates the IC_{50} value. The predicted concentration-time data are shown in Figure 5.21.*

From the response *versus* time graphs (Figure 5.20), note the steep decline in response and the flat portions at maximal response. That is, the response levels off at a constant fraction of the baseline value. This means it will not be possible to increase the response by increasing the concentration or dose. The response *versus* concentration curves (Figure 5.20) show the maximum response is achieved at a concentration of about 500 µg/L. Because the response *versus* concentration hysteresis loops of both infusions are superimposed, there was no apparent tolerance development with the dosage regimen used. Models 1 (Equation 5:20) and 4 (Equation 5:21) refer to the indirect response models that capture the decline of response.

Model 1

$$\frac{dR}{dt} = k_{in} \cdot \left[1 - \frac{I_{max}C_{ss}^n}{IC_{50}^n + C_{ss}^n}\right] - k_{out} \cdot R \qquad (5:20)$$

Model 4

$$\frac{dR}{dt} = k_{in} - k_{out} \cdot \left[1 - \frac{E_{max}C_{ss}^n}{EC_{50}^n + C_{ss}^n}\right] \cdot R \qquad (5:21)$$

The following scheme shows the derivation of response at baseline and steady state and the initial estimates of EC_{50} and the exponent n using Model 4 (Equation 5:21). At time 0, $C = 0$ and $dR/dt = 0$ since we are at response steady state. R_0 is obtained by setting

$$\frac{dR}{dt} = k_{in} - k_{out} \cdot R_0 = 0 \qquad (5:22)$$

and rearranged gives

$$R_0 = \frac{k_{in}}{k_{out}} \approx 10 \qquad (5:23)$$

This is the baseline value. This expression can be rewritten as

$$k_{in} = k_{out} \cdot 10 \qquad (5:24)$$

As C goes to infinity, $dR/dt = 0$, since we are at a drug-induced response steady state and R_{ss}. R_{ss} is obtained from

$$\frac{dR}{dt} = 0 = k_{in} - k_{out} \cdot R_{ss} \cdot (1 + E_{max}) \qquad (5:25)$$

which implies that

$$R_{ss} = \frac{k_{in}}{k_{out}} \cdot \frac{1}{1 + E_{max}} \approx 2 \qquad (5:26)$$

as stimulation is assumed to act upon loss of response. Substituting Equation 5:23 into Equation 5:26 and rearranged gives E_{max}

$$E_{max} = \frac{R_0}{R_{ss}} - 1 \approx \frac{10}{2} - 1 = 4 \qquad (5:27)$$

The k_{out} parameter is determined from the downswing of the response-time curve as

$$slope = \frac{\ln(R_2) - \ln(R_1)}{t_2 - t_1} = -k_{out} = -\frac{\ln(2.5) - \ln(7.5)}{0.25 - 0} \approx 4\,h^{-1} \qquad (5:28)$$

and k_{in} is obtained from Equation 5.24 above as

$$k_{in} = k_{out} \cdot R_0 = 4 \cdot 10 = 40 \qquad (5{:}29)$$

The exponent n is most easily derived from m and E_{max} (Levy [1995]). The m parameter is the slope of the response *versus* log-concentration curve.

$$m = \frac{E_2 - E_1}{\ln(C_2) - \ln(C_1)} = \frac{n \cdot E_{max}}{4} \Rightarrow n > 5 \qquad (5{:}30)$$

E_1 and E_2 are obtained from Figure 5.20 (lower right). IC_{50} is most readily read from graph of response *versus* log concentration. A rough estimate of IC_{50} is obtained from the upswing of the curve, which is indicated with arrows. I_{max} is most readily read from graph of response *versus* time. The flat portion of the curve is about 25% of the baseline effect, which means that a reasonable value of I_{max} is $1-0.25 = 0.75$. The final fits are superimposed on observed data in Figure 5.21.

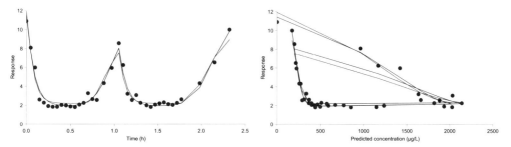

Figure 5.21 *Response versus time of observed (●) and predicted (lines) data (left). Response versus predicted plasma concentration of observed and predicted data (right).*

Table 5.2 compares the final parameter estimates and their precision.

Table 5.2 Parameter estimates and initial estimates

Parameter	Model 1 mean ± CV%	Model 4 mean ± CV%	Initial estimates
k_{in}	143 ± 10	34.2 ± 13	30-40
k_{out}	12.5 ± 9	2.87 ± 13	6-10
E_{max}	–	4.41 ± 10	4–6
I_{max}	0.83 ± 1.5	–	0.8
EC_{50}	–	312 ± 4.2	250-350
IC_{50}	244 ± 2.6	–	250
n	7.0 ± 13	19 ± 75	>10
+/–	14	14	
WRSS/AIC	7.3/83	12/102	

WRSS, +/- and *AIC* denotes the weighted residual sum of squares (objective function), the changes of the sign of the residuals and the Akaike Information Criterion, respectively, in Table 5.2. The final parameter estimates are surprisingly close to the majority of initial parameter estimates.

In this example, the drug blocked the effect only partially, although a maximum response was still reached. The model of choice is the *inhibitory* model (i.e., Model 1) because the parameter precision was better and the weighted residual sum of squares (*WRSS*) was lower for that model. The next step would be to compare the fits of both models on more than one subject.

5.4.1.4 Dynamic repeated dose data

The next example involves data that were collected on a new anxioyltic compound during a phase II study in psychiatric patients. The goal was to generate pivotal information that could serve as a basis for a large multi-center clinical trial design. The investigators aimed to establish the turnover characteristics (k_{in}, k_{out}) of a biomarker, in addition to drug specific information (IC_{50}). The k_{in} and k_{out} parameters represent the *turnover rate* and *fractional turnover rate* of response, respectively. The parameters could be compared to the infusion rate and elimination rate constant, respectively, in a pharmacokinetic model. For further reading on turnover models (indirect response models) and their parameterization, see section 4.6)

The pharmacodynamic results of the compound after repeated dosing are illustrated in Figure 5.22. Observe how the slopes of the downswing after the first and fourth doses fall in parallel.

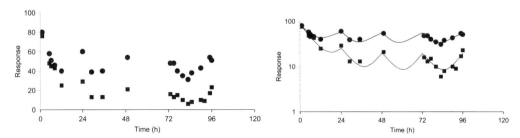

Figure 5.22 *Observed response-time data following administration of oral doses of 5 (●) and 25 (■) mg of an anxiolytic drug given once daily for four days. The left hand figure contains data plotted on a linear scale and the right hand figure is a semi-logarithmic plot of the data and includes the predicted relationship. Data taken from Gabrielsson and Weiner [1998].*

The pharmacokinetic model of the drug in plasma takes the following form

$$C = \frac{K_a D_{po}}{(V/F) \cdot (K_a - K)} \left[e^{-K \cdot t} - e^{-K_a \cdot t} \right]$$ (5:31)

The compound is known to fully block the response at high concentrations *in vitro*. Therefore, it is reasonable to assume that the inhibitory effect of the drug on production of response $I(C)$ can take the form

$$I(C) = 1 - \frac{C_{ss}^n}{IC_{50}^n + C_{ss}^n} \tag{5:32}$$

The turnover of response is described by means of a zero-order production and first-order loss equation according to

$$\frac{dR}{dt} = k_{in} \cdot I(t) - k_{out} \cdot R \tag{5:33}$$

The drug acts by means of blocking the production of response. By setting this equation equal to zero

$$\frac{dR}{dt} = k_{in} \cdot I(C_{ss}) - k_{out} \cdot R_{ss} = 0 \tag{5:34}$$

and then solving for R, one obtains the response level at steady state R_{ss} by means of

$$R_{ss} = \frac{k_{in}}{k_{out}} \cdot I(t) = \frac{k_{in}}{k_{out}} \cdot \left[1 - \frac{C_{ss}^n}{IC_{50}^n + C_{ss}^n} \right] \tag{5:35}$$

Depending on the steady state drug concentration (C_{ss}), R_{ss} varies between the baseline value (k_{in}/k_{out}) and zero. The latter value corresponds to high concentrations of drug.

$$R_{min} = 0 \tag{5:36}$$

The baseline response is given by Equation 5:37

$$R_0 = \frac{k_{in}}{k_{out}} \tag{5:37}$$

The initial parameter estimates should be derived prior to implementing the model parameterized with k_{in}, k_{out}, IC_{50} and n. The response will obey a first-order decline (k_{out}) when the input of response (k_{in}) is blocked by the drug ($I(C)$). The dashed line in Figure 5.23 illustrates this behavior, where response-time data are plotted on a semi-logarithmic scale. The slope of the initial decrease corresponds to k_{out}.

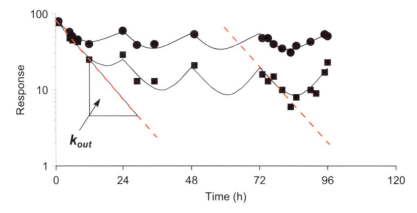

Figure 5.23 *Semi-logarithmic plot of observed (symbols) and predicted (lines) response-time data following administration of oral doses of 5 (●) and 25 (■) mg of an anxiolytic drug given once daily for four days. The red dashed lines indicate the slope of the downswing of the response-time curves, from which the fractional rate constant k_{out} is obtained. The slope is $-k_{out}$. The units of the response variable are intentionally omitted.*

Equation 5:38 shows an initial estimate of the slope based on experimental data.

$$k_{out} = \frac{\ln(R_1/R_2)}{t_1 - t_2} = \frac{\ln(80/48)}{0-8} \approx 0.06\,\text{h}^{-1} \tag{5:38}$$

k_{in} was obtained by rearrangement of Equation 5:37

$$k_{in} = k_{out} \cdot R_0 = 0.06 \cdot 80 = 4.8\ units \tag{5:39}$$

IC_{50} is the plasma concentration at half-maximal response. The closest one gets to equilibrium between plasma concentration and response is during the upswing of the response-time curve, and not during the downswing of the curve where response lags behind concentration. The response is approximately 50% of maximal response at about 86 h (steady-state), and the corresponding plasma concentration is then obtained by setting t equal to 86 h in Equation 5:31, which gives approximately 0.25 μg/L.

Figure 5.24 demonstrates how to translate the terminal response data at 50% (which is approximately 40 units) of maximal response (approximately 80 units at baseline) into plasma concentrations (IC_{50}) at the same time point. Both response (observed) and concentration (simulated) data are plotted on a semi-logarithmic scale.

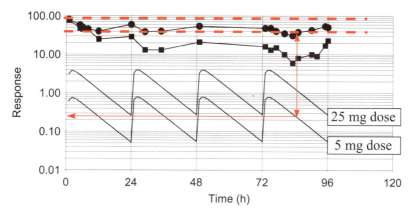

Figure 5.24 *Semi-logarithmic plot of observed (symbols) and predicted (lines) response-time data following administration of oral doses of 5 (●) and 25 (■) mg of an anxiolytic drug given once daily for four days. Response-time relationships for observed response (upper pair of curves) and simulated plasma concentrations (lower pair of curves) following repeated dosing.*

K_a, K and V/F were estimated by means of fitting Equation 5:31 to plasma concentration-time data. These parameters were equal to 1.1 h^{-1}, 0.128 h^{-1} and 5.0 L/kg, respectively, and should be treated as constants during regression of the response-time data. In other words, Equation 5:31 serves as input (with fixed kinetic parameters) to the pharmacodynamics during the regression of the response-time data.

The fits of the two response-time profiles are shown in Figure 5.23. It is obvious from these two curves that the peak of the response (trough value) was shifted to the right with an increase in dose. This can be seen after the first dose as well as at steady state. The fluctuation in response within a dosing interval was less pronounced at the higher dose level.

The predicted baseline value (k_{in}/k_{out}) was about 80 units and the potency of the drug 0.25 μg/L. Half-life of response was 6.9 hours as compared to the plasma half-life of the drug, which was 5.4 hours. On the basis of these results, it was predicted that 15-25 mg per day would probably be an adequate dose range in the planned clinical trial as the drug is non-toxic at that dose level.

5.4.2 Convolution and deconvolution analysis

5.4.2.1 Background

Our presentation of the convolution and deconvolution concept is based partly on the application of a software package (*QWERT*, [1993]) that is designed for numerical convolution/deconvolution. Langenbücher [1982] covers the theory of this particular algorithm in more depth. For a general review of different deconvolution techniques see Madden *et al* [1996].

Deconvolution is primarily used in kinetics to obtain the *input function* and the *integral of the input function*. The input function reveals information about multiple peaks, the rate of absorption and so on. The integral of the input function gives the bioavailability, also called the extent of absorption.

Consider the plasma concentration curve of a compound after an intravenous bolus dose. From the moment a compound enters the blood stream, plasma concentration represents a mixture of distribution, redistribution and elimination processes, or in other words the disposition. The input function after a bolus dose can be described as a *spike* (Figure 5.25).

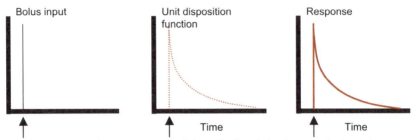

Figure 5.25 *The input function versus time (left panel) after a bolus dose convolved with the unit disposition function or weighting function versus time after a bolus input (middle panel) gives the response function (right panel). Response denotes the concentration-time course.*

Consider now the plasma concentration curve after an oral dose of the same compound. This curve describes the time course of the drug entering plasma *and* what happens to the compound afterwards. The input from the gut compartment into blood is shown in Figure 5.26 as a first–order loss from the gut.

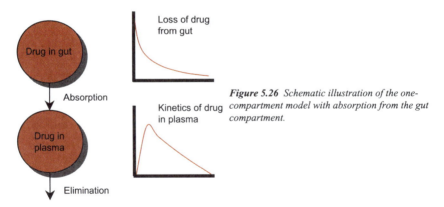

Figure 5.26 *Schematic illustration of the one-compartment model with absorption from the gut compartment.*

Conceptually, we have two phases: *absorption* and *disposition*. Absorption covers all processes that drive the compound into the body. Disposition is a combination of distribution and elimination processes with no interference from any absorption process. The decline in plasma after a bolus dose mirrors disposition. If a drug is given extravascularly, we have movement of drug into the blood stream in addition to disposition processes.

As a first attempt to describe the absorption phase, assume the stomach to be a container from which a drug is injected into the blood system as a number of *intravenous bolus doses.* Suppose we know the magnitude of these *doses* and when they are *injected.* Then, the blood concentration at a given time can be predicted by calculating and summing the contribution from each *dose.* We must also know the disposition curve from each of the *doses.* However, if the *superposition principle applies* and if *the presence of the drug does not affect its kinetics,* then each of the disposition curves should look like a disposition curve from an intravenous bolus dose.

That is, if we know the disposition curve from a *unit* intravenous (bolus) dose, which we will call the *weight function* or *unit disposition function* and denote *w(t)*, and the rate at which the drug is put into the plasma or the *input rate function i(t),* then we can calculate the plasma disposition curve from that input. We call this the *response function, r(t).*

This kind of reasoning is by no means restricted to intravenous bolus or extravascular administration. It is applicable to any problem that can be formulated as a function describing how the drug enters a *system* (input rate function), a function describing the disposition (weight function) after a unit dose and a function describing the input and disposition together (the response function). The natural response of a system is its response to a change from equilibrium in the absence of further inputs. This response function, called the unit input response, is the characteristic behavior of the system.

Figure 5.27 displays three kinds of input functions and their resulting response functions provided the unit input disposition function in the lower left curve is used. A spike, a square wave, and exponential decay pattern are expected after a bolus, zero-order and first-order input, respectively. These are therefore what we would expect or should look for as a resulting input function, if bolus, infusion, and first-order absorption data in plasma are deconvolved with a unit disposition function.

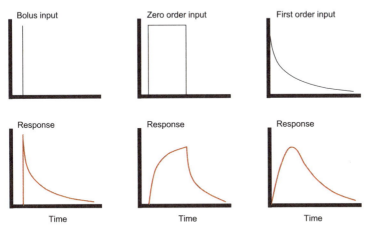

Figure 5.27 *Input functions versus time (upper figures) and their corresponding concentration (denoted response) versus time curves (bottom figures), assuming the weighting function to be a unit disposition function shown in Figure 5.25.*

5.4.2.2   Theory

We have exemplified the input function as a set of *bolus doses* and the corresponding

Gabrielsson & Weiner

times they were *injected*. Taking a particular *dose*, dose k, the amount $a(x_k)$ is *injected* at time x_k. Assume the dose is injected over a short time Δx_t at a constant rate, $i(x_k)$. Then the amount can be written as: $a(x_k) = \Delta x_t \, i(x_k)$. The predicted plasma concentration at time t is then the sum of n injected doses

$$C = r(t) = \sum_{k}^{n} i(x_k) \cdot w(t - x_k) \Delta t \tag{5:40}$$

The term $t{-}x_k$ comes from the fact that the contribution from dose k is the time elapsed since this particular dose.

So far we have only derived the ordinary formula for multiple dosing. Suppose now that the *injections* are given all the time, that is the compound is continuously put into the blood stream. It can then be seen that the equation above can be written as

$$C = r(t) = \int_{0}^{t} i(x) \cdot w(t - x) dt \tag{5:41}$$

The integral of this form is known as the convolution integral. It is often written as

$$C = r(t) = i(t) * w(t) \tag{5:42}$$

where $*$ is the convolution operator. In the opposite case, which is called deconvolution, we know $r(t)$ and one of the other functions, e.g., $w(t)$, and we want to calculate the third, in this case $i(t)$. There is no standard notation for this but Langenbücher [1983] suggests the notation $//$ as the deconvolution operator. The response function, $r(t)$, deconvolved with the weight function, $w(t)$, gives

$$i(t) = r(t) // w(t) \tag{5:43}$$

and the response function deconvolved with the input function, $i(t)$, gives the weight function

$$w(t) = r(t) // i(t) \tag{5:44}$$

There are two approaches to solve a convolution/deconvolution problem. One is to model the functions mathematically and solve the convolution integral for the unknown function. This approach has the limitation that it requires a mathematical model for the kinetics. Furthermore, the integral may be impossible to solve for deconvolution problems.

The other is by means of some numerical method that convolutes or deconvolutes the actual data. *QWERT* uses the latter approach in a scheme devised by Langenbücher [1983]. The idea is as follows. Assume that measurements are sampled with equal time interval $\Delta x_t = T$ for all k. Then the convolution integral can be approximated with the sum

$$r(t) = r(x_n) = \left[r(t)/T - \sum_{k=1}^{n} i(x_k) \cdot w(x_{n-k+1}) \right] / i(x_i) \tag{5:45}$$

where $x_k = k\,T$ and $x_n = 1$. This sum can be reversed into the forms

$$w(t) = w(x_n) = \left[r(t)/T - \sum_{k=1}^{n} i(x_k) \cdot w(x_{n-k+1}) \right] / i(x_i) \qquad (5:46)$$

$$i(t) = i(x_n) = \left[r(t)/T - \sum_{k=1}^{n} i(x_k) \cdot w(x_{n-k+1}) \right] / w(x_i) \qquad (5:47)$$

The value $r(x_k)$ is taken as the response at time x_k while the values $i(x_k)$ and $w(x_k)$ are the average input rate and the average weight between the times x_{k-1} and x_k respectively. Note that the input must not be zero at the first time interval when deconvoluting for weight, since that would cause a division by zero. By analogy, the weight must not be zero at the first time interval when deconvoluting for input. In contrast to convolution, deconvolution may be subject to numerical instability. Thus, the time step T must be chosen with care.

If the data sets are not sampled at equal time intervals, which they seldom are, *QWERT* calculates the intermediate values by linear interpolation. *QWERT* uses *input, weight, and response*. Given any two of these, the third can be estimated. The weight is by definition measurements of the unit input response. There are usually concentration units for the weight function. The response corresponds to measurements in plasma after the dose has been given by some other route. This is usually the plasma concentration-time curve after say extravascular administration. The units are the same as for the weighting function. The input function, however, is the fractional input rate with which the drug enters the system and has units per time (t^{-1}).

Let the weight be the concentration-time profile following an intravenous bolus dose and the response be the concentration-time profile after an oral dose (Figure 5.28). Then the input will be the fractional absorption rate. Thus, given input and weight data, an estimate of response is obtained by convolution. In contrast, given response and weight, the input is obtained by deconvolution. Given response and input, weight is estimated by deconvolution.

There are two conditions that must be met. Firstly, the process must be linear, which means that the superposition principle applies. Secondly, the process must be time-invariant, which basically means that the response to a certain dose does not depend on when the dose is given.

In the example shown in Figure 5.28, numerical deconvolution of oral data C_{po} with intravenous data C_{iv} revealed a 15 min lag-time until absorption started and a bioavailability or cumulative input of about 50% after dose correction. The resulting unit input function was plotted on a semi-logarithmic scale (Figure 5.28 (left)). Note that a linear behavior on this scale corresponds to a first-order input process.

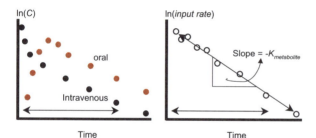

Figure 5.28 *Observed (filled symbols) and predicted (solid lines) concentration time data of drug after an intravenous bolus (left) and an oral dose, respectively. Log-input rate versus time (right) obtained after deconvolution of oral data by intravenous data. The slope of the log-input rate gives the absorption rate constant.*

In contrast to convolution, deconvolution can be subject to numerical instability such as sudden oscillations in the input function. Thus, if errors occur in the earlier estimates, as a result of noise in data for example, these will have a cumulative effect upon subsequent estimates (Finkelstein and Carson [1985]). Nevertheless, although the input function may be ill-conditioned, it can still provide reasonable estimates of the total input and bioavailability.

5.4.2.3   Oral solution *versus* tablet

In the example shown in Figure 5.29, a compound is given as an oral solution, and the resulting plasma concentration curve is assigned as the weight function. The drug is then given as a tablet and the resulting plasma concentration is assigned as the response function. We then deconvolute the plasma concentration of the tablet, the response function, with the plasma concentration-time curve obtained from the solution experiment (i.e., the weight function). The input function will then describe how the tablet dissolves.

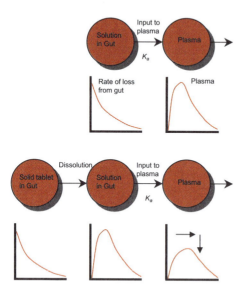

Figure 5.29 *Schematic illustration of different processes when a drug is given as an oral solution (upper figure) or a tablet (bottom figure). The two upper graphs demonstrate the drug kinetics in gut and plasma. The lower three graphs demonstrate the kinetics of amount of drug in tablet form, solution kinetics of drug in gut, and plasma kinetics of drug after being absorbed. The arrows in the lower right plot demonstrate how the concentration peak in plasma is lower and somewhat delayed in comparison to the concentration of drug in the gut.*

5.4.2.4   Transdermal input

Figure 5.30 shows observed plasma concentrations of nicotine after application of a nicotine patch for 16 h. The shaded bars indicate the rapid (burst) and extended input functions of nicotine from the patch. Note that the burst input lasted for about 6 h and the maintenance input for about 18 h.

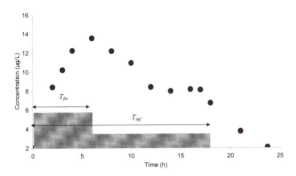

Figure 5.30 Observed concentration-time data after the application of a nicotine patch. The shaded area corresponds to the initial burst over approximately six hours overlapped with the maintenance release over 17-18 h. The length of the rapid (T_{fst}) and prolonged (T_{inf}) input times will be estimated in parallel to the relative amount released during the two release periods.

Figure 5.31 demonstrates schematically the behavior of the input, weighting, and response functions in a system like the patch. The input function, shown as a rapid infusion overlapped by a slow maintenance infusion (left hand figure), convoluted with the unit disposition function of nicotine (middle figure) gives the response function (right hand figure) of nicotine in plasma. Thus, one may approximate the release of nicotine from the patch as two superimposed *square waves* originating from the initial burst input (high and short) and the maintenance input (low and extended).

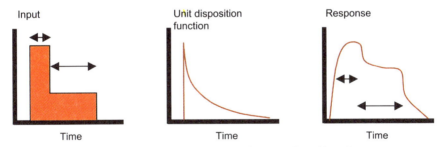

Figure 5.31 The input function, shown as a rapid bolus infusion overlapped by a slow maintenance infusion (left), convoluted with the unit disposition function (middle) gives the response function (right). The arrows indicate the length of the rapid and slow inputs.

5.4.2.5   Drug and metabolite data

One may also use data on a parent compound and metabolite to derive the unit disposition function of the metabolite. By deconvolution of the metabolite concentration-time profile with drug concentration-time data, it is possible to estimate the disposition function of the metabolite. Figures 5.32 show this schematically.

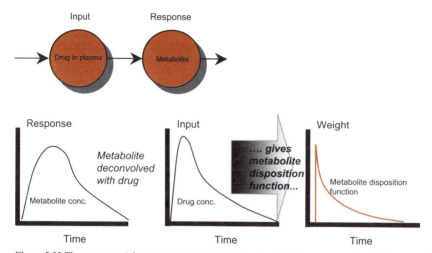

Figure 5.32 *The processes (plasma concentration-time curve) of the drug system suit as the input to the metabolite system (metabolite plasma concentration-time curve also called the response function). Note that response function in the sense of convolution/deconvolution does not relate to the pharmacological effects. Response (metabolite concentration) deconvolved with input (drug concentration) gives the unit disposition function of the metabolite.*

Figure 5.33 shows this method of deconvolution for the parent compound and its metabolite. As was discussed earlier, all processes must be linear.

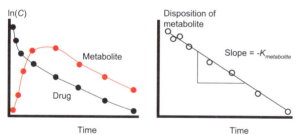

Figure 5.33 *Observed (filled symbols) and predicted (solid lines) concentration-time data of drug and metabolite. The two doses were 80 and 480 μmol/kg, respectively. The disposition function of the metabolite was obtained by deconvolution of metabolite data by means of the drug data. The slope of the log-disposition function corresponds to the elimination rate constant of metabolite – $K_{metabolite}$.*

The resulting unit disposition function (weighting function) for the drug is shown in the left hand panel of Figure 5.33. The slope of the solid line is equivalent to the elimination rate constant of the metabolite. This plot shows clearly that the disposition function follows mono-exponential decline. The behavior of the estimated disposition function is an almost ideal first-order process. One often sees substantial fluctuation in the terminal time points due to noisy data. However, if the decline of the obtained disposition behaves linearly on a logarithmic scale, this implies that the elimination rate obeys first-order kinetics.

Numerical convolution and deconvolution by means of *QWERT* are independent methods we have found very useful for obtaining measures such as initial estimates, input rate functions, lag-time, bioavailability and metabolite disposition curves without compartment modeling (see also Nakashima and Benet [1988, 1989]). *QWERT* also allows generation of negative input curves. Consider Figure 5.34 which shows the flux of a drug (between gut and blood stream) given orally. In Figure 5.34 (A) we have all drug dissolved in the gut at time zero and no drug has entered the blood stream. In Figure 5.34 (B) the absorption of drug into the blood stream has started and is governed by simple passive diffusion across the gutwall. The gut-to-blood concentration gradient is initially very high. Figure 5.34 (C) demonstrates the exsorption of drug into the gut from the blood stream by means of efflux proteins. The concentration of drug is relatively low in both the blood stream and gut enabling a net flux of drug in the direction towards the gut. One may anticipate this situation for drugs administered by means of other routes than the oral route (e.g., via the intravenous, subcutaneous, pulmonary route). For review, see Arimori and Nakano [1998].

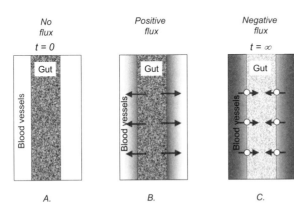

Figure 5.34 Schematic illustration of oral administration of drug. A shows drug dissolved in the gut content at time zero. B shows the net flux of drug by means of passive diffusion into the blood stream. C shows exsorption of drug into the gut by means of efflux proteins indicated by arrows with white circles.

Figure 5.34 (B) illustrates that the flux (primarily by passive diffusion) of drug is from gut towards the blood stream. In Figure 5.34 (C), efflux proteins pump the drug from the blood stream towards the gut (illustrated as arrows with white circles). We assume that the flux of drug from blood to gut occurs primarily by means of an active process.

5.4.3   When all else fails

Another useful technique when good initial estimates are not readily available is to make an educated guess as to plausible values of the model parameters and then to use these to simulate the model. By comparing the predicted values from the simulation with the actual data, one can assess how reasonable the initial estimates are for model fitting. We recommend that the user first do a simulation with user-defined models prior to fitting the model to data. This will ascertain if the models have been correctly implemented and capture any trends in the data.

WinNonlin will determine initial parameter estimates for all single dose models in the program library. Although some programs will not automatically determine initial estimates for other models (e.g., multiple dose models or user-defined models), other options are available. If the user specifies lower and upper boundaries for the model parameters (without specifying initial estimates), a modern program will perform a grid search using the boundaries to obtain initial estimates. However, this may take some time if the model involves a large number of parameters or if it can only be expressed as a system of differential equations. In these instances, the user should consider using initial estimates derived from the literature or from previous studies involving the drug. For example, if your model includes a clearance parameter, you have a rough idea about the physiological range that clearance can vary within (say 0 - 2 L/min for humans). You should then try to use this range when setting the parameter boundaries.

Once a model has been selected and implemented into the software package, and reasonable initial estimates have been obtained from graphical exploratory analysis, it is time to start the regression and fit the model to data. We will discuss the minimization procedure briefly in the next section.

5.5   Selection of the Minimization Algorithm

This section may interest users of the *WinNonlin* program. When fitting deterministic models such as the classical kinetic and dynamic models, we recommend using *Hartley's* modification to the *Gauss-Newton* algorithm with the *Levenberg-Marquardt* modification. This algorithm has been extensively tested over the years and, when used in conjunction with lower and upper boundaries (as implemented in *WinNonlin*), is one of the most powerful and robust algorithms available. The program reparameterizes the model when bounds are provided. Because this makes the model fitting process much more stable, we recommend that users always include parameter boundaries.

No one algorithm will work in every circumstance, however, and when the default method fails to converge, we recommend rerunning the problem using the *Nelder-Mead simplex* algorithm (*METHOD 1*). Unlike the default method, the simplex method does not require derivatives, and it will generally manage to converge when the default method has failed.

When fitting probabilistic models such as probit, logit or logistic regression models, the *Gauss-Newton* method without the *Levenberg-Marquardt* modification should be employed (*METHOD 3*) in conjunction with iterative reweighting. In addition to the *METHOD 3* command, the user should also specify:

> MEANSQUARE 1   CONVERGENCE 0   ITERATIONS 10

This combination of commands, when used in conjunction with iteratively reweighted least squares, will produce the maximum likelihood estimates that are recommended for these types of models.

It is important to use the correct weighting function. Two examples of fitting a logistic model can be found later in the applications section (PD20 and PD21). The *MEANSQUARE 1* command will set the estimated variance (σ^2) equal to one; *CONVERGENCE 0* will turn off the convergence checks; and *ITERATIONS 10* tells the program to stop the iteration process after 10 iterations. Note that for these types of models we are not attempting to minimize the (weighted) residual sum of squares. However, the weighted residual sum of squares should stabilize (remain constant) for the last few iterations. Only if this is the case have maximum likelihood estimates been determined.

The program will also estimate the partial derivatives with respect to the model parameters, and these can then be plotted when needed (see section 5.9.4).

5.6 Iterations

The next step in the analysis after the exploratory analysis is to fit the proposed model(s) to the data. This step is not synonymous with modeling, as modeling includes all steps in the carousel (Figure 5.35).

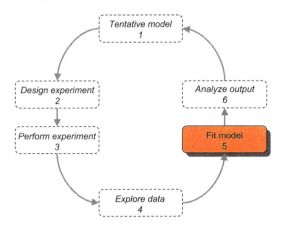

Figure 5.35 *The fifth step in the modeling carousel. This step covers the regression of the proposed model to the experimental data. It will be greatly improved by prior exploratory analysis of the data (step 4), thus enhancing the goodness-of-fit (step 6).*

Given a set of data and a proposed model, we want to estimate a set of parameters (e.g., *Cl, V, K*) such that the differences between the observed and predicted data are *small*. Nonlinear regression programs do so by adjusting the parameter values iteratively with the weighted residual sum of squares *WRSS* as the objective function, as shown in Figure 5.36.

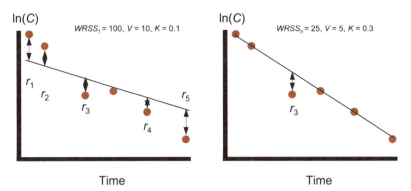

Figure 5.36 *Semi-logarithmic plot of observed (filled circles) and predicted (solid line) concentration versus time for the first iteration (left hand plot) and last iteration (right hand plot). Arrows indicate the residuals.*

Figure 5.37 demonstrates schematically three observations, their distributions and the predicted curve for the ith iteration. The predicted curve, denoted as *ln(C-hat)*, is assumed to coincide with the theoretical mean of each distribution, provided normal distribution of the data. The residuals are indicated as r_i ($r_1, r_2 \ldots$ and r_n).

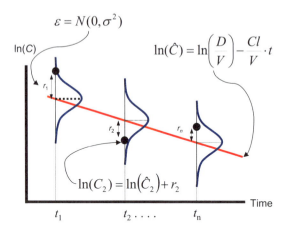

Figure 5.37 Semi-logarithmic plot of observed (filled circles) and predicted (red line) concentration-time course of iteration i. The blue lines represent the theoretical distribution of data. The slope of the line is Cl/V and the intercept with the concentration axis is ln(D/V). The residuals are denoted as r_i.

Figure 5.38 shows the *WRSS* plotted against various sets of parameter values. Observe that the *WRSS* is a three dimensional landscape for a two parameter model, and that there may be several minima. By choosing poor initial estimates of the parameters (starting point *A* in Figure 5.38), the likelihood of ending up in a local minimum increases, as indicated in the figure. Starting point *B* in Figure 5.38 is better, since the program moves towards the global minimum.

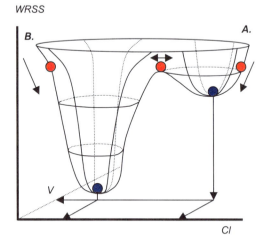

Figure 5.38 Schematic illustration of the parameter space for a one-compartment model parameterized with Cl and V. Depending on how well the initial parameter estimates are derived, you may end up in either a local minimum (right) or a global minimum (left). Note that the final estimate for V happens to be the same, whereas the Cl estimates are different.

There is no general method that is guaranteed to find the best global solution to a nonlinear least square problem. A minimum-seeking program will invariably succeed in finding a minimum; however, it may not be the global minimum.

The default search method in *WinNonlin*, is the *Gauss-Newton* method with the modifications by *Hartley* and *Levenberg*. The *Levenberg* modification is a *Marquardt-like* algorithm. Using this method, the program will present the weighted residual sum of squares (*WRSS*), parameter values, condition number and rank for each iteration. As you can see in the box below, *WRSS* decreases with each iteration until the program converges. You can also see how the parameter values of *V* and *K* change from one iteration to another.

```
   ITERATION WEIGHTED SS        VOLUME            K
      0          303389.        14.00        .2500E-01
      1          6893.38        10.68        .1768E-01
      2          5105.43        10.19        .1906E-01
      3          5101.95        10.18        .1905E-01
CONVERGENCE ACHIEVED, RELATIVE CHANGE IN WEIGHTED
 SUM OF SQUARES LESS THAN   .000100
```

The difference in parameter values from iteration to iteration usually decreases as the algorithm approaches the minimum of the sum of squares surface. The program converges when the relative change between $WRSS_{old}$ and $WRSS_{new}$ is less than a predetermined value (e.g., 0.0001 which is the default). Thus, the *relative change* is expressed as

$$relative\ change = \frac{\left|WRSS_{old} - WRSS_{new}\right|}{WRSS_{old}} \tag{5:48}$$

Although there are no guarantees, there are certain techniques that can be used to increase the likelihood of success in finding the *best fit* to nonlinear models. Firstly, no nonlinear least squares program will work well unless it is provided with approximations to the answer that are *close enough* so that this minimum may be found. The development of good initial estimates is accomplished primarily by using your own physical knowledge of the problem at hand and your best intuition and experience (see section 5.4). Generally, it is useful to do one or more simulations while varying parameter values in order to develop a feel for how changes in parameter values may affect the calculated curves. Secondly, we recommend you *always* use lower and upper bounds on the parameters. Placing bounds on the parameter space results in better numerical stability.

Once a minimum is found, how do we know that it represents the absolute minimum? Unfortunately, there is no assurance that it does! Again, though, there are several techniques that may be used to help to validate answers provided by the program. One simple approach is to run the program several times, using different starting estimates and different minimization algorithms, such as the *Nelder-Mead Simplex* algorithm (*METHOD 1*), and note whether a different minimum is found, possibly with a smaller sum of squares. If no other minima can be found, you can be fairly confident that the proposed minimum is the best answer. Refer back to section 2.4

for guidelines for the selection of weighting schemes.

The generally available statistics in nonlinear regression packages offer some help in choosing an appropriate weighting factor. In addition, however, the importance of visual inspection of the observed and predicted data cannot be overestimated. If predicted and observed data are not consistent the parameter values are of no value, no matter how precise they happened to be.

When we adopt models whose behavior very closely approximates the behavior of linear models, there will always be a unique minimum, and the speed of convergence of the unmodified *Gauss-Newton* method to that minimum will usually be very rapid. This follows on from the fact that, for a linear model, the sum of squares surface has a single minimum and the *Gauss-Newton* method will find that minimum in a single iteration from any set of initial parameter estimates. As the model becomes more and more nonlinear, convergence may take more iterations or may not even occur at all. The response surface may have more than one minimum, and convergence to the right minimum is never guaranteed (Ratowsky [1983]).

5.7   Assessing the Goodness-of-Fit

Assessment of the *goodness-of-fit* (the last step in the first round of the modeling carousel, Figure 5.39) is the heart of this book. *Goodness-of-fit* relates in part to the distribution of the data about the fitted curve. This ought to be random.

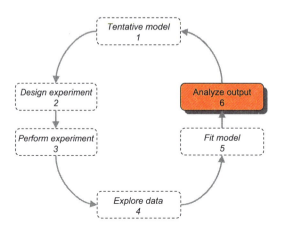

Figure 5.39 *Analyzing output from the regression and assessing the goodness-of-fit comprise the last step of the first round of the modeling carousel.*

5.7.1   Analyzing the residuals

The assessment of goodness-of-fit often tends to be subjective and it behoves the investigator to inspect a plot of the fitted curve superimposed on the observed data. In addition, we highly recommend the use of residual plots. A residual is the vertical difference between an observed concentration and the predicted concentration (Figure 5.40). If the residual ε is positive, it means that the point is above the curve. If the residual is negative, the point is below the curve.

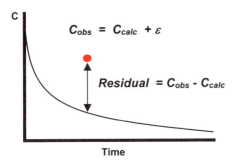

Figure 5.40 *The relationship between observed (C_{obs}) and predicted (C_{calc}) data and the residual ε.*

The residual is an error that the model cannot explain. Residual error has traditionally been viewed as though it has properties similar to those of assay error. This may not necessarily be so, since assay error is often only a minor part of the difference between observed and predicted concentrations. We generally assume that the errors (residuals) are independent, that they have a zero mean and a constant variance equal to sigma squared (σ^2), and that they follow a normal distribution. The last assumption is required for making *F*-tests.

$$\varepsilon = N(0,\sigma^2) \tag{5:49}$$

This error should be randomly distributed about the mean, i.e., about the model-predicted curve. With plots of observed and fitted curves *versus* the independent variable (usually time), the visual inspection can be misleading in certain regions of the curve. The distribution can be difficult to assess if the residuals are small relative to the calculated value. This is particularly so in Cartesian plots (lin-lin plots). An alternative presentation that highlights the distribution is to plot the residuals *versus* the independent variable or the calculated dependent variable. The principal ways of plotting the residuals are *overall*, *in time series* and *against the fitted* or *predicted value(s)*.

One should examine such plots to determine if the residuals tend to be randomly scattered within a lower and upper horizontal band. If the residuals do not appear to be randomly scattered, this would suggest that either the model or weighting scheme is incorrect.

The Figure 5.41 shows plots of weighted residuals (i.e., the square root of the weight times the residual) *versus* concentration, where the residuals are randomly distributed about the mean (0) but display a cone when weighted according to a constant absolute variance (left, not optimal weighting) or constant relative variance (right).

In the first case (constant absolute error), the weighted residual corresponds to the absolute residual, since the weight is equal to unity (1). In the second case (constant relative error) , the weighted residual corresponds to the relative residual, since the weights are equal to the reciprocal of the predicted value squared.

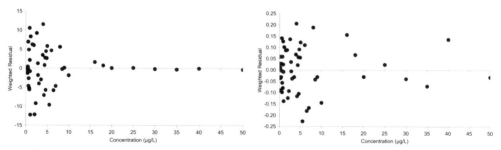

Figure 5.41 *Weighted (absolute) residual of response versus the predicted concentration using a constant absolute variance model (left). Weighted (relative) residual of response versus the predicted concentration using a constant relative variance model (right).*

One thing to look for is whether or not there are any *runs* (i.e., sequences of residuals having the same sign) in the residuals. Although there are statistical methods for testing for significance of the number of runs, these methods generally require large numbers of observations. For smaller data sets such as data commonly obtained in kinetic and dynamic studies, the user must make a judgement call.

The box below includes an example of residual analysis showing the independent variable *X* (e.g., time), the observed dependent variable *OBS Y* (e.g., observed concentration), the calculated dependent variable *CALC Y* (e.g., model-predicted concentration), the residuals *RES* (observed concentration minus calculated concentration), the weights *W*, standard deviation of the calculated dependent variable *SDYHAT* and the standardized residual *STAND RES*. For details of the standard deviation of the calculated dependent variable and the standardized residual, see Draper and Smith [1998].

X	OBS Y	CALC Y	RES	W	SDYHAT	STAND RES.
10.00	769.0	812.0	-43.03	1.000	19.47	-2.676
20.00	710.0	671.2	38.80	1.000	12.95	1.790
30.00	585.0	554.8	30.22	1.000	9.982	1.303
40.00	472.0	458.6	13.44	1.000	9.501	.5742
50.00	363.0	379.0	-16.03	1.000	9.947	-.6907
60.00	300.0	313.3	-13.29	1.000	10.40	-.5777
70.00	256.0	259.0	-2.956	1.000	10.58	-.1289
90.00	170.0	176.9	-6.920	1.000	10.11	-.2990
110.0	109.0	120.9	-11.87	1.000	8.944	-.5027
150.0	52.00	56.42	-4.420	1.000	6.112	-.1804

```
SUM OF WEIGHTED SQUARED RESIDUALS =   5101.94
S =  25.2536    WITH   8 DEGREES OF FREEDOM

CORRELATION (Y,YHAT) = .996
```

One should carefully examine the residuals to determine if there are any *outliers* (i.e., data values which were not fit well by the model) or *runs*. In the example above, the first row contains a residual that is somewhat larger than for the rest of the curve. The standardized residual is about 2.7 whereas the rest are less than 1.8. We can also observe three runs (-+++------) in the residuals. The first block (-) is line 1, the second (+++) includes lines 2-4, and the third (------) lines 5-10. Note that the model overestimated all the last six data values. This suggests either that a model with an additional exponential term should be fit to the data or that a weighting function should be employed. *WRSS* is estimated as

$$WRSS = \sum_{i=1}^{n} W_i (C_i - \hat{C}_i)^2 \qquad (5:50)$$

WRSS is also known as the objective function where the *hat* (^) denotes a predicted value. The mean standard deviation *S* is equal to the square root of *WRSS* divided by the degrees of freedom, $df = N_{obs} - N_{par}$. N_{obs} is the number of observations and N_{par} the number of parameters. Mean standard deviation then becomes

$$\sigma = \sqrt{\frac{WRSS}{N_{obs} - N_{par}}} \qquad (5:51)$$

5.7.2   Graphical presentation of residuals

In this section we demonstrate the types of residual plots that may be used as diagnostic tools. As a general rule of thumb, we want the residuals to be randomly distributed around the predicted curve. However, sometimes there are clusters of observations above or below the predicted curve. This is much easier to see in a residual plot. As pointed out above, clusters of consecutive positive or negative residuals are called a run, and they can be analyzed with respect to their randomness. The *runs test* calculates a P value, testing the null hypothesis that each point is equally likely to be above or below the curve. For further information about the *runs test*, see Draper and Smith [1998].

Remember from previous sections that residuals represent unexplained error, and that this is assumed to be normally distributed about zero with a variance of σ^2. A relative residual that displays a non-random pattern as in Figure 5.42 (upper left) indicates problems with the structural model (kinetic or dynamic). A relative residual that displays a cone shaped pattern as in Figure 5.42 (upper right), indicates problems with the error model (variance or weighting). A possible solution to the banana-shaped structure in Figure 5.42 (upper left) would be to add an additional exponential term. Instead of a mono-exponential model, use a bi- or three-exponential function. The residual plot in Figure 5.42 (upper right), which comes from the fit of a model to data obtained after a bolus dose, suggests that data should be weighted by the reciprocal concentration raised to power λ.

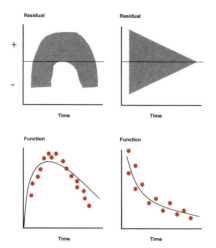

Figure 5.42 *Relative residual plotted versus time. In the left hand figure, there is a lack of fit due to use of the wrong structural model. In the right hand figure, the lack of fit is due to an incorrect error model (i.e., wrong weighting scheme).*

Figure 5.43 shows the ordinary and sigmoid E_{max} models fit to data and displayed on a linear and semi-logarithmic scale, respectively. There are obvious clusters of points above and below the predicted curve.

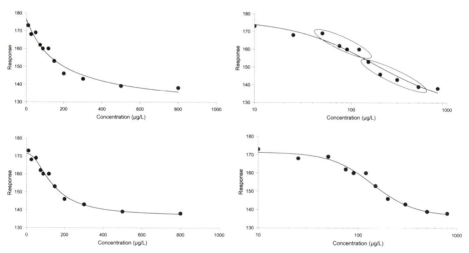

Figure 5.43 *Upper panel shows superimposed observed (●) and predicted (line, ordinary E_{max} model) response versus concentration (left) and effect versus log-concentration (right). Lower panel shows superimposed observed (●) and predicted (line, sigmoid E_{max} model) response versus concentration (left) and effect versus log-concentration (right). Systemic deviations are indicated for the ordinary E_{max} model. Note the lack of systematic deviations with the sigmoid E_{max} model.*

The same figure shows the function plots of the sigmoid E_{max} (extended) model (lower panels), while Figure 5.44 shows the resulting residual plots.

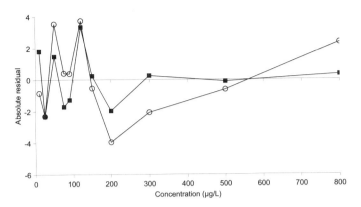

Figure 5.44 Absolute residual versus concentration. Note the clusters of positive and negative residuals. Open (○) symbols correspond to residuals of the ordinary E_{max} model and closed (●) symbols residuals of the sigmoid E_{max} model. See Figure 5.43 for the response-concentration data.

If the model tends to overestimate the larger concentrations while underestimating the smaller concentrations, this would either suggest re-running the fit while utilizing a weighting scheme that gives more emphasis to the smaller concentrations. Alternatively, one could add an additional exponential term. The weighting scheme can be accomplished by using weights inversely proportional to the observed value or predicted value, raised to the power of λ.

$$Weight = C_i^\lambda = \hat{C}_i^\lambda \tag{5:52}$$

The *WEIGHT* $-\lambda$ command is the reciprocal of the observed concentration C_i raised to λ. The *REWEIGHT* $-\lambda$ command is then the predicted concentration C_i raised to $-\lambda$. Figure 5.45 demonstrates the distribution of error when we have constant absolute error (left) and a constant relative error (right) in a semi-logarithmic plot (upper panel) and a Cartesian plot (lower panel).

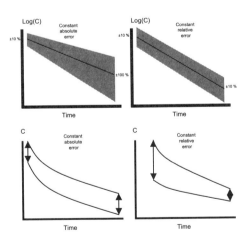

Figure 5.45 Semi-log plots of concentration versus time (upper panel). The shaded areas represent a constant absolute error distribution (upper left) and a constant relative error distribution (upper right). Linear plots of concentration versus time (lower panel). The lower left plot represents a constant absolute error distribution and the lower right plot a constant relative error distribution.

Let us look at Figure 5.46 in which we have plotted observed data (filled circles) and fits of the mono-, bi- and three-exponential models. The mono-exponential model clearly deviates from the terminal data points, which means it represents a poor fit.

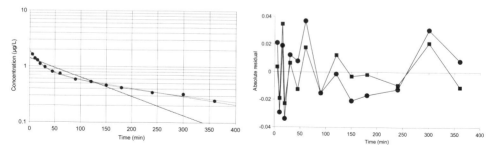

Figure 5.46 *Observed and predicted concentrations for the mono-, bi- and three-exponential models (left). The bi- and three-exponential models almost superimpose whereas the mono-exponential model shows a systematic deviation from observed data. The absolute residual versus time for the bi- (●) and three-exponential (■) models (right). This plot reveals that the three-exponential model fits the data better than the bi-exponential model. The three-exponential model displays more random scatter and less amplitude in the residuals.*

The quality of the fit of the bi- and three-exponential models is more difficult to judge from the function plot (Figure 5.46, left). Therefore, we have also plotted the residuals (Figure 5.46 right). The mono-exponential model is not shown in the residual plot since the poor fit is already shown in the function plot. One is able to discriminate between the bi- and three-exponential models in the residual graph, with the three-exponential model showing more of a random scatter. We have chosen to connect the absolute residuals with a line for each fit in order to discriminate more easily between the models. It is obvious from the function plot of the one-compartment model how poor the fit is to the data. That is, there is an obvious trend with time. As the concentration declines the residual decreases and then continuously increases with time to become more and more positive. The three-compartment model shows more runs and more random scatter about the predicted values. The frequency of changing *plus/minus* signs is higher for the three-exponential model (12) than it is for the bi-exponential model (7), indicative of fewer runs for the former model. The absolute magnitude of the residuals, the amplitude, is also smaller for the three-exponential model. Therefore, we would suggest that the three-exponential model fits the data better than the bi-exponential model with respect to the absolute residuals.

In the above example we have used the *absolute* residual to distinguish between two models. The *relative* residual is obtained by dividing the *absolute* residual by its corresponding predicted concentration. The advantage of using the relative residual rather than the absolute residual is that one observes more easily where one have the largest contributions to the residuals (deviations) on a relative scale.

If fitting an extravascular model, the first few residuals should be carefully examined to determine if a lag-time (start of absorption) is required. Deconvolution of oral data by means of intravenous data provides a means of independently assessing the shape of the input function (Figure 5.47). It is clear that the model without a lag-time had three runs, whereas the lag-time model had six runs. The latter indicates more random scatter and is therefore the preferred model.

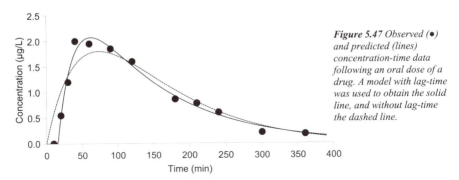

Figure 5.47 Observed (•) and predicted (lines) concentration-time data following an oral dose of a drug. A model with lag-time was used to obtain the solid line, and without lag-time the dashed line.

One should also check to see if the predicted values mimic the trends in the data. For example, do the predicted curve and the data peak at the same time? Figure 5.48 displays the observed and predicted concentrations after an oral dose of a multi-compartment drug. Run #1 corresponds to unweighted data and run #2 to data weighted according to a constant *CV* (proportional) error model. Observe the lack of fit of the terminal observations in run #1, which is obviously not the better model.

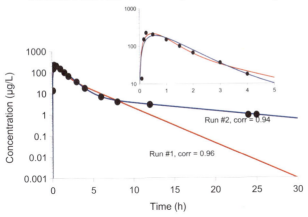

Figure 5.48 Predicted and observed concentration-time data following oral dosing. Run #2 (blue line) represents the Iterative Reweighted Least Squares (IRLS) fit and Run #1 (red line) the Ordinary Least Squares (OLS) fit using a model without lag-time. The inset shows the 0-5 h time range.

Statistical tests for *goodness-of-fit* are an important complement to visual inspection and allow the model to be accepted or rejected at a known level of confidence. For further discussion of residual analysis, the reader is referred to Draper and Smith [1998].

5.7.3   Parameter estimates - Accuracy

What are accuracy and precision? When can I trust my parameter estimates? What is a good parameter estimate? These are frequently asked questions about parameter estimates. Figure 5.49 shows the differences between accuracy and precision, using a dart board. The aim, of course, is to hit the centre of the dart board and thus gain the maximum score.

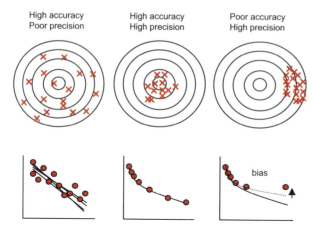

High accuracy   High accuracy   Poor accuracy
Poor precision   High precision   High precision

Figure 5.49 *Accuracy, precision and bias exemplified by dart boards (upper row). The corresponding experimental situations are shown as concentration-time plots below each dart board.*

bias

Figure 5.49 (left) shows a situation with reasonable accuracy because the mean of all observations corresponds to the center (mean) of the dartboard. The precision, however, is low, because there are *hits* all over the board. In the picture on the right, the hit-pattern is well away from the center of the board. This means that accuracy is poor. We can also say that the mean estimate is biased, where bias may be estimated as the true value minus the estimated value divided by the true value. Precision, on the other hand, is good as there is little variation between the observations. The middle panel shows the ideal picture, with the mean estimate coinciding with the bull's-eye and the hit having high precision.

Even though a fit is good or acceptable in terms of good correspondence between the observed and predicted values, the resulting parameter estimates may be of no value. Any further interpretation of the actual values then becomes meaningless.

Accuracy deals with the parameter estimate (*p-hat*) in relation to the true parameter value (*p*). If *p-hat* differs by a factor *x* from the true *p,* it is said to be biased. *P-hat* may be correct on average, but different individual estimates *bounce* around the true value of *p* by a factor of σ^2. The precision is said to be σ or $100\,\hat{\sigma}/\hat{p}$. The latter is called the coefficient of variation denoted *CV*%.

Remember also that the properties of the parameter estimates depend on the nonlinearity of the model, the number of observations and the values of t_i (design points), the distribution of ε_i (noise level), the choice of parameters, the objective function, the algorithm (and implementation), transformations of parameters, and initial estimates of parameters.

5.7.4   Parameter estimates - Precision

Estimated model parameters *p-hat* have little or no value unless they have a fair degree of precision. This can be assessed by computing each parameter's coefficient of variation (i.e., the parameter's standard error (*SE* denoted *SD* in Equation 5:53) divided by the corresponding parameter estimate).

$$CV\% = \frac{SD}{\hat{p}} \cdot 100 \qquad\qquad (5{:}53)$$

Note that a large parameter *CV* does not necessarily imply that the model is incorrect. It may be due to not having collected enough samples or not having collected samples at the appropriate times. When a model contains several parameters which differ markedly in magnitude (e.g., $V = 1000 \pm 30$ L, $CV = 3\%$ and $K = 0.01 \pm 0.005$ h^{-1}, $CV = 50\%$), the *CV* is a better way of expressing in which parameter the variability lies than is the *SE*. This is because *CV* is a relative measure, relative to the magnitude of *p-hat*, whereas *SE* is an absolute measure.

There are two types of confidence intervals for the model parameters. The *univariate* method is computed as the parameter estimate plus and minus a *t value* multiplied by the *SE* of the parameter. The *planar limits* are more difficult to compute, but take into account the correlation amongst the parameters. Planar limits are always wider than univariate limits. Again, relatively wide confidence intervals suggest that the parameter estimates may have little or no practical value.

The better model is generally based on a random scatter about the predicted curve, a smaller residual sum of squares and smaller 95% confidence limits for the model parameters (i.e., higher parameter precision). Testing for the robustness of the model is equally encouraged, since an obvious goal of modeling is to predict events or outcomes for a larger group of subjects. The more perturbations that can be explained by the model, the greater the predictability to a real system.

The statistical information provided by modern packages should provide at least the final estimate for each parameter, the asymptotic *SE* of each estimate (also sometimes called *SE*), the univariate and support plane 95% confidence ranges for each parameter, and the correlation matrix. Be aware that the minimum sum of squares is found with some degree of uncertainty due to the inevitable errors in experimental data. Experience has shown that, while there may be some uncertainty in the exact location of the minimum, the 95% confidence regions are frequently much better defined. If the experiment were repeated a large number of times, and the confidence interval computed for each run, approximately 95% of the confidence intervals would contain the *true* value.

The univariate and support plane confidence ranges represent boundaries of regions that have a given probability of containing the true value of the estimated parameter. The principal difference between confidence ranges is the assumption that the limits of the univariate confidence interval are calculated without consideration to values assumed by any other parameters of the model. In other words, correlation between any of the model parameters is ignored. In the support plane confidence limits, the correlation between two or more parameters is taken into account. A consequence of this correlation is that the support plane confidence limits have a larger range than the univariate confidence limits. The box below shows output from *WinNonlin* with respect to the parameter estimate, standard error (*SE* or *SD*) and the univariate and planar 95% confidence limits.

PARAMETER	ESTIMATE	STANDARD ERROR	95% CONFIDENCE LIMITS		
VOLUME	10.1789	.311081	9.46	10.89	UNIVARIATE
			9.19	11.16	PLANAR
K	.019048	.000878	.017	.0210	UNIVARIATE
			.016	.0218	PLANAR

5.7.5   Correlation between observed and predicted values

The correlation coefficient r between the observed concentrations C_i and average predicted values *C-bar* is calculated as

$$r = 1 - \frac{\sum(X_i - \overline{X}) \cdot (C_i - \overline{C})}{\sum(X_i - \overline{X})^2 \sum(C_i - \overline{C})^2} \qquad (5:54)$$

where X_i and *X-bar* are the observed and averaged time points, respectively, and C_i and *C-bar* the corresponding concentration points.

    The correlation coefficient between observed and predicted values is probably one of the most misused goodness-of-fit criterion. Hence, the correlation coefficient should be interpreted with caution. While models that provide a good fit will necessarily be associated with a relatively high r value, the converse in not true. That is, a high r value does not necessarily imply a good fit, particularly for large concentration ranges. This can be seen in Figure 5.50. In this example, the model systematically under estimated the last three observations in the data set using Ordinary Least Squares (*OLS*), yet the r value was high (0.96). However, in a second run the r value was lower (0.94) even though the overall fit was better. It is important therefore not to place too much value on a high r value as it is not necessarily indicative of the best model solution.

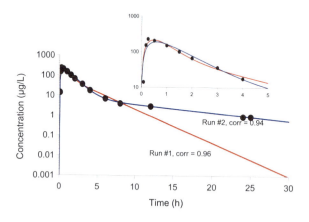

Figure 5.50 Observed (●) and predicted (lines) concentration-time data following oral dosing of a drug. Blue line represents the IRLS fit and red line the OLS fit. IRLS and OLS denote Iteratively Reweighted Least Squares and Ordinary Least Squares, respectively.

5.7.6   Correlation between parameters

The statistical dependence between two or more parameter estimates is called *parameter correlation.* Correlation is often used to indicate an apparent linear relationship between parameters. Perturbing a parameter may cause a near proportional change in other correlated parameters at the same time as nearly identical values of the nonlinear minimization objective function (*WRSS*) are maintained (Dutta *et al* [1996]). Two or more parameters can be positively or negatively correlated or not correlated at all. Figure 5.51 illustrates these three situations. The ideal (model) case is the one where the parameters are totally uncorrelated with each other.

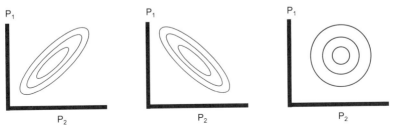

Figure 5.51 *Positively (left), negatively (middle), and uncorrelated (right) parameters. P_1 and P_2 are two hypothetical model parameters.*

The correlation matrix provides an estimate of how closely two or more parameters depend on each other. The matrix is ordered with parameter 1 through n reading from top to bottom and left to right. For example, if *VOLUME* and *K* were absolutely independent of each other, then the second element of the first column of the correlation matrix would be zero. Conversely, large off-diagonal elements of the correlation matrix are indicative of parameters that are very dependent on each other. For example, a large absolute value of say > 0.95 in the second element of the first column suggests that the *VOLUME* and *K* are highly correlated. Any small change in one of the parameters could be compensated for by making an appropriate adjustment in the other. High correlations between parameters suggest that there may be insufficient information in the concentration-time data in order to accurately and precisely determine both parameters. High correlation may also manifest itself in large standard deviations and 95% confidence intervals. The box below shows the correlation matrix for two parameters *VOLUME* and *K*.

If two parameters are highly correlated, one might expect that they would be estimated with poor precision, e.g., with high *CVs*. However, as indicated in Figure 5.52, this is not necessarily the case. The larger (outer) ellipsoid in the left hand plot shows a situation where we have highly correlated parameters with low precision - wide confidence intervals. The inner ellipsoid shows parameters with high correlation and high precision parameters - narrow confidence intervals. We would propose a study design including a wider dosage range for both cases (*inner* and *outer*) in order to discriminate between the parameters e.g., giving a high dose that covers a significant portion of the zero-order kinetics and a low dose that covers the first-order kinetics.

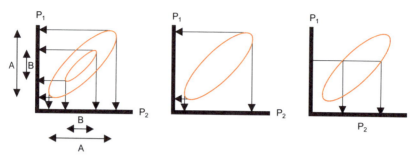

Figure 5.52 *The 95% confidence ellipsoid for P_1 and P_2. The left hand plot displays two systems with parameters that are equally highly correlated. The middle and right hand plots show the impact of setting one parameter (P_1) constant on the confidence limits of the remaining parameter (P_2). Setting one parameter (of two highly correlated parameters) to a constant value will generally reduce the confidence limits of the other parameter. A and B indicate the confidence limits for parameter P_1 and P_2. A and B on the Y axis may not necessarily be the same as A and B on the X axis.*

   In Figure 5.52 we have plotted the 95% confidence ellipsoids for two highly correlated parameters. Both parameters also have wide confidence intervals (*CI*) in the left figure as indicated by *A* and *B*. By fixing one parameter to a single value indicated on the *Y* axis in the graph, the uncertainty of the other parameter is affected as shown with a narrower *CI* (right *versus* middle figure).

   Sometimes an experiment is repeated many times with design *A* and *B*. Design *A* is good, with well-spaced time points, while design *B* is the opposite (i.e., poor, due to a very narrow range of selected time points). This is shown diagramatically in Figure 5.53 (upper panel). When fitting the models to the respective designs, the results would then be as shown in the lower panel. A low correlation between *V* and *K* (panel *C*) results if design *A* is applied, and high negative correlation if design *B* is applied. A small change in *K* would immediately be compensated for by a change in *V* in order to obtain similar residual sums of squares (case *B/D*).

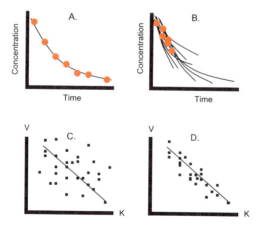

Figure 5.53 *Schematic illustration of good design with well-spaced time points (A) which results in poorly correlated parameters (C), and a poor design with narrow time interval (B) which results in highly correlated parameters (D). Each data point in C and D represents one experiment towards the subsequent regression. K and V are the elimination rate constant and the volume of distribution,*

Figure 5.54 illustrates how an incomplete design will affect the correlation between two parameters. In the lower panel, very little information is obtained about the I_{max} value because of the narrow concentration range. This results in a high correlation (0.95) between I_{max} and IC_{50} as indicated by the correlation matrix. In the upper panel, more data are collected at maximum response. As a result, the correlation between I_{max} and IC_{50} is reduced (0.65).

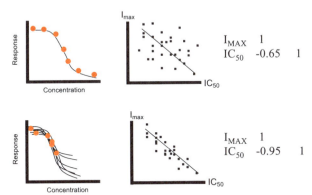

$$\begin{array}{ll} I_{MAX} & 1 \\ IC_{50} & -0.65 \quad 1 \end{array}$$

$$\begin{array}{ll} I_{MAX} & 1 \\ IC_{50} & -0.95 \quad 1 \end{array}$$

Figure 5.54 *Two designs of effect versus plasma concentration. The example at the bottom is the reduced design and provides very little information about I_{max}. The upper example is the full design which also includes effect measurements at the plateau. The correlation matrices of the two runs are inserted into the respective figures. Note how the second element of the first column decreases from -0.95 which is high due to poor design, to -0.65 with the improved design.*

The complications with high parameter correlations arise from both design issues (when to sample e.g., time, and the absolute range of the dependent variable, e.g., concentration) and an assay issue (or variability issue).

In the context of uncertainty and correlation between parameters, the function value may still be predicted with precision. A *data descriptor* is the concentration or response predicted from the estimated parameters (Dutta *et al* [1996]). Since data descriptors are calculated from estimated parameters, they also have an associated accuracy and precision. Ebling [1995] concluded that the data descriptors could sometimes be estimated with higher precision and accuracy than the poorly estimated, highly correlated parameters from which they are derived.

5.7.7   Condition number

The *eigenvalues* in themselves are of little or no value. However, the *condition number*, which is defined as the square root of the ratio of the largest to the smallest *eigenvalue*, is a measure of how *stable* the final solution is. There is no *magic* cut-off number available. However, for a simple two-parameter model, a *condition number* greater than 10^3 would generally suggest that one or more parameters cannot be estimated precisely. As a rule of thumb, the *condition number* should be less than or equal to $10^{N_{par}}$ (N_{par} is the number of parameters) for a well-defined model with respect to the information in the data. The box below shows a print-out of the *eigenvalues* and *condition number* from *WinNonlin*.

```
••• EIGENVALUES OF (Var - Cov) MATRIX •••

   NUMBER          EIGENVALUE
      1            2400E+10
      2            6590.

Condition number = 603.5
```

The *condition number* shown above (603.5) is different in size from the *condition number* presented after each iteration. This is because it was computed in the unconstrained parameter space.

5.8   Discrimination between Rival Models

In this section we cover a battery of statistical methods used for discrimination between rival models, namely the *F-test* and the *Akaike Information* and *Schwarz* criteria (respectively, *AIC* and *SC*).

The last step of the modeling carousel involves analysis of the output (Figure 5.55). One usually compares the fits of two or more models to the data, and uses the whole *battery of statistical tools* (i.e., not only the parameter estimates and their precision). With the *battery of statistical tools* we mean *goodness-of-fit criteria* (residual and function plots), *parameter estimates* and *their precision*, *Akaike Information* and *Schwarz citeria, condition number, correlation matrix, F-test*, and so on.

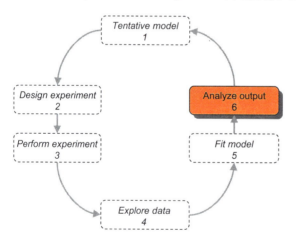

Figure 5.55 *Do a thorough analysis of the program output and compare outputs from fits of two or more competing models. The whole battery of statistical tools is used for model discrimination purposes.*

5.8.1   *F*-test

5.8.1.1   Background

When comparing hierarchical or nested models, the probability that additional parameters are without effect on the sum of squares is defined by an *F* distribution. A system of hierarchical models (also called full and reduced models) is used. If the reduced model is identical to the full model when one or more parameters are fixed to a specific value (usually zero), there is an F distribution. The mono-exponential model is the reduced form (B set to 0) of the bi-exponential model.

$$C = A \cdot e^{-\alpha \cdot t} + B \cdot e^{-\beta \cdot t} \qquad \textbf{(Full)}$$

$$C = A \cdot e^{-\alpha \cdot t} \qquad \textbf{(Reduced)}$$

The F-statistic is calculated according to Equation 5:55

$$F^* = \frac{\dfrac{|WRSS_1 - WRSS_2|}{|df_1 - df_2|}}{\dfrac{WRSS_2}{df_2}} \qquad (5{:}55)$$

$WRSS_1$ and $WRSS_2$ are the objective functions of the reduced and full models, respectively, and df_1 and df_2 are their corresponding degrees of freedom. Degrees of freedom are calculated as

$$df = N_{obs} - N_{par} \qquad (5{:}56)$$

N_{obs} is the number of observations and N_{par} is the number of parameters. If F^* is larger than F_{table} ($p = 0.05$, F_{table}; $\Delta df = |df_1 - df_2| = column$ and $df_2 = row$) it means that the full model is statistically superior to the reduced model. With mathematically distinct models, a function that is mechanistically correct may describe the data with fewer parameters than are required wforith a function that is mechanistically wrong. Such models are not hierarchical and the ratio does not have an F distribution.

5.8.1.2　　　The sigmoid E_{max} model *versus* the ordinary E_{max} model

Let us consider the ordinary E_{max} and the sigmoid E_{max} models and ask the question, Do they represent two nested models or not? The answer is yes. By fixing the exponent n to unity, the sigmoid E_{max} model collapses into the ordinary E_{max} model as shown below.

$$E = \frac{E_{max} C^n}{EC_{50}^n + C^n} \qquad \textbf{(Full)}$$

$$E = \frac{E_{max} C^1}{EC_{50}^1 + C^1} \qquad \textbf{(Reduced)}$$

One could fit both models to data and then test by means of the *F*-test (see section PD3) whether the sigmoidicity factor is significantly different from 1 or not. However, a simpler test that $n = 1$ can be obtained as follows. If the confidence intervals of the exponent (n) do not contain unity, n is said to be significantly different from 1.

5.8.1.3   The ordinary E_{max} model *versus* the linear response model

Two models that are not classified as nested models are the ordinary E_{max} model and the linear response (effect concentration) model, since no parameters in the ordinary E_{max} model can be fixed to obtain the linear response model unless $C << EC_{50}$.

$$E = \frac{E_{max} C^1}{EC_{50}^1 + C^1} \qquad \text{(Full)}$$

$$E = s \cdot C \qquad \text{(Not reduced)}$$

Remember that the *F*-test requires nested models and equal weighting (*OLS* or *WLS* denote the Ordinary Least Squares and Weighted Least Squares, respectively).

5.8.1.4   The hepatic distributed model *versus* the parallel-tube model

The parallel-tube model and the distributed model can be looked upon as reduced and full models, respectively.

$$Cl_H = Q_H \cdot \left[1 - e^{-\left\{ \frac{f_u \cdot Cl_{int}}{Q_H} + \frac{1}{2} \varepsilon^2 \left(\frac{f_u \cdot Cl_{int}}{Q_H} \right)^2 \right\}} \right] \qquad \text{(Full)}$$

$$Cl_H = Q_H \cdot \left[1 - e^{-\frac{f_u \cdot Cl_{int}}{Q_H}} \right] \qquad \text{(Reduced)}$$

The ε^2 parameter is the variance for each sinusoid in the whole liver. The distributed model reduces to the parallel-tube model when either $Q_H >> Cl_{int}$ or ε is very small.

5.8.2   Akaike and Schwarz criteria

The *Akaike Information Criterion (AIC)* and the *Schwarz Criterion (SC)* attempt to quantify the information content of a given set of parameter estimates by relating the *WRSS* to the number of parameters that were required to obtain the fit. When comparing two models with different numbers of parameters with the same weighting scheme, these criteria (*AIC* or *SC*) places the burden on the model with more parameters not only to have a lower *WRSS*, but also to quantify how much lower it must be for the model to be deemed more appropriate (Akaike [1978], Schwarz [1978], Landaw and DiStefano [1984]).

$$AIC = N_{obs} \cdot \ln(WRSS) + 2 \cdot N_{par} \qquad (5:57)$$

$$SC = N_{obs} \cdot \ln(WRSS) + N_{par} \cdot \ln(N_{obs}) \qquad (5:58)$$

Never judge the *goodness-of-fit* without a battery of statistical tools. One should be careful when interpreting the result of a fit solely from the *AIC* and *SC* values or the *F*-test. The *AIC* as defined above is dependent on the magnitude of data points as well as the number of observations. The most appropriate model is the one with the smallest value of *AIC*. How much smaller the value needs to be in order to be statistically significantly better cannot be determined because the distribution of these values is unknown. However, it is not known how much smaller the values have to be in order to be statistically significantly better, as the distributions of the *AIC* and *SC* values have not been tabulated. Note that while the model associated with the smallest value of the *Akaike (Schwarz)* criterion may give the best fit of those models compared, it still may not give an *adequate* fit (as determined by residual analysis and other means). Remember that the *Akaike* or *Schwarz* criteria do not require nested models but it is necessary to apply equal weighting of data (*OLS* or *WLS*).

5.9 Experimental Design Issues

5.9.1 Background
In this section on experimental design we propose a practical approach. This can be accomplished by means of *the delta function, the variance inflation factor (VIF), and the partial derivatives* of the predicted curve with respect to each of the model parameters (Godfrey [1983], Pharsight [1999]). For the reader who is interested in a more in depth theoretical discussion of optimal experimental design, see Endrenyi [1980].

5.9.2 Delta function
Assume that a pilot study has been performed in order to improve the sampling strategy for the main experiment. Let us also assume that we want to establish the relationship between the steady state concentration of a drug and its pharmacological effect. The relationship between concentration and effect can be described by some sort of E_{max} model. Our strategy now is to obtain the *optimal* design points to be able to discriminate between the two candidate models below. The box below shows two models that were fit to a data set. The full model uses four parameters (E_0, I_{max}, IC_{50} and n) and the reduced model uses three parameters (E_0, I_{max} and IC_{50}).

$$E = E_{0F} - \frac{I_{maxF}C^n}{IC_{50F}^n + C^n} \qquad \textbf{(Full)}$$

$$E = E_{0R} - \frac{I_{maxR}C^1}{IC_{50R}^1 + C^1} \qquad \textbf{(Reduced)}$$

We subtract the reduced (R) model from the full (F), and then square that difference to obtain

$$\Delta E = \left[\left(E_{0F} - \frac{I_{maxF}C^n}{IC_{50F}^n + C^n} \right) - \left(E_{0R} - \frac{I_{maxR}C^1}{IC_{50R}^1 + C^1} \right) \right]^2 \qquad (5:59)$$

This function is then simulated by implementing the parameter estimates from the fit of the sigmoid model and ordinary I_{max} models, respectively. Equation 5:59 (*Delta*) is then plotted against the plasma concentration according to Figure 5.56.

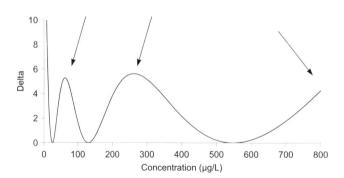

Figure 5.56 Delta function (Equation 5:59) versus plasma concentration. The three arrows indicate design points (concentration-response measurements) where the two functions in the box above differ mostly. These design points may be useful to include in future experiments.

Figure 5.56 indicates the design points to include in the experiment to optimally discriminate between the two models. The maximum difference between the models seems to be at approximately 0, 75, and 260 µg/L, and at the last observation. These are therefore the points to be included among the rest of the design points in a future experiment. Godfrey [1983] provides a good description of this approach.

5.9.3   Variance inflation factor

A pharmacodynamic study is to be conducted on a drug to be administered as a constant intravenous infusion and whose dynamics can be represented by a sigmoid E_{max} model. Ideally, samples need to be collected over a 0 - 800 µg/L concentration range (Figure 5.57). However, if samples are to be collected above 500 µg/L, then the study will jeopardize the safety of the subjects. That is, subjects would need to be supervised by an anesthetist. This would dramatically increase the cost of the study.

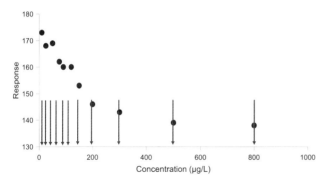

Figure 5.57 *Planned design points are indicated by arrows. The blue arrows correspond to data within the 0-500 µg/L concentration range and red arow corresponds to the extra design point at 800 µg/L.*

The question is, how much information are we losing by not collecting a sample at e.g., 800 µg/L? Or, to put the question another way, would the *value* of collecting a sample at say, 800 µg/L, exceed the cost required to collect the sample (e.g., the cost of extra staff)?

To address this question, we must define *value*. For this example, let us assume that *value* can be measured by an increase in the precision with which we can estimate the parameters. Thus, the question we are asking is, Would a sample collected at 800 µg/L produce enough of an increase in precision of the estimated parameters to justify the cost of its collection? To answer this question, we need to look at the variance inflation factor *VIF*.

The variance of a model parameter can be expressed as the product of the underlying variance of the residuals σ^2 and a multiplier, which we will call the *VIF*. For example, the variance of the mean of a sample is well known to be σ^2/n, where n is the sample size. In this instance, the *VIF* is therefore $1/n$. Note that *VIF* does not depend on the actual data which are measured, rather it is only dependent on the experimental design. Although the *VIFs* which we require are much more complicated to compute than that of a sample mean, *WinNonlin*, when using the *SIMULATION* option, automatically computes these values for model parameters, secondary parameters and the predicted values.

$$\text{var}(E_{max}) = \sigma^2 \cdot VIF \tag{5:60}$$

Remember from the Section 2.8, the variance of a parameter is calculated as

$$\text{var}(\hat{\beta}) = \sigma^2 \cdot (X'X)^{-1} \tag{5:61}$$

Similarly

$$VIF = (X'X)^{-1} \tag{5:62}$$

We will calculate the ratios of the *VIF* values of each set of parameters originating from the two different designs. By doing this, one is able to determine the *value* of adding an extra design point at 800 µg/L. Table 5.3 summarizes the *VIF*-values (including the *VIF* ratios) of model parameters already stated in paretheses for the 0-500 and 0-800 µg/L designs considered above.

Table 5.3   Data on *VIF* for model parameters

Parameter	Estimate	VIF 0-500	VIF 0-800	VIF ratio
I_{max}	34.7	4.955	2.488	1.99
IC_{50}	139.6	85.77	50.88	1.69
n	2.031	.060664	.042771	1.42
E_0	171.4	.71823	.67890	1.06

Table 5.3 suggests that the estimated variance for I_{max} would have halved by including a sample at 800 μg/L. In fact, according to the *VIF* ratios, the precision of I_{max}, IC_{50}, n and E_0 would have increased approximately 100, 70, 40 and 6%, respectively. A summary of the *VIF* values of the predicted concentrations, including their *VIF* ratios for the two designs, is given in Table 5.4. These data again suggest that variance would be reduced by including a sample at 800 μg/L. This reduction would be dramatic in the case of the 500 μg/L value.

Table 5.4   Data on *VIF* for the data descriptor

X	Calculated Y	VIF 0-500	VIF 0-800	VIF ratio
10.00	171.2	.6025	.5806	1.04
25.00	170.3	.3553	.3552	1.00
50.00	167.5	.2869	.2642	1.10
75.00	163.7	.3228	.3054	1.06
90.00	161.3	.2937	.2895	1.01
120.0	156.7	.2563	.2486	1.03
150.0	152.8	.3143	.2823	1.11
200.0	148.0	.3958	.3710	1.07
300.0	142.8	.3526	.3276	1.08
500.0	139.1	.8199	.3485	2.35
800.0	137.7	–	.6271	–

VIF will not give the *optimal* design but provides, rather, a relative measure for comparing different designs. Table 5.5 summarizes the final parameter precisions after having fit the *ordinary* and *sigmoid* I_{max} models to data from the two designs.

Table 5.5   Parameter precision

	Parameter precision CV%			
	0-500		0-800	
Parameter	Ordinary I_{max}	Sigmoid I_{max}	Ordinary I_{max}	Sigmoid I_{max}
E_{max}	13	15	8.0	9.6
EC_{50}	38	16	31	11
E_0	1.4	1.1	1.5	1.0
n	---	28	---	21
AIC	44	42	51	46

CV% is obtained as *100·SE/(p-hat)*. The content of Table 5.5 confirms what has already been seen in the simulations. In other words, by adding an extra effect measurement at a concentration of 800 µg/L, one can discriminate between the two models. From measurements within the 0 - 500 µg/L concentration range, one would probably not be able to make this discrimination. Thus, accepting the extra cost for medical personnel for sampling at 800 µg/L is probably acceptable since one will obtain a far better model for future predictions and study design.

5.9.4 Partial derivatives

The partial derivative of a predicted function with respect to each of its parameters is a useful measure when plotted against the independent variable. The partial derivative gives the sampling point where the maximum information about each particular parameter will be obtained. Take, for example, the bi-exponential model below

$$C = A \cdot e^{-\alpha \cdot t} + B \cdot e^{-\beta \cdot t} \qquad (5:63)$$

The partial derivatives of C with respect to the model parameters A and B then become

$$\frac{dC}{dA} = e^{-\alpha \cdot t} \qquad \frac{dC}{dB} = e^{-\beta \cdot t} \qquad \text{(5:64 a and b)}$$

and show a mono-exponential decay (Figure 5.58 left graph). The partial derivative of C with respect to the slope factors α and β are somewhat more complex since they include both an exponential decay with time and a factor that linearly increases with time. This function will therefore initially decrease (negative slope) and, when the exponential function increases more rapidly than the decrease from $-t$, approach zero (Figure 5.58 right graph).

$$\frac{dC}{d\alpha} = -t \cdot A \cdot e^{-\alpha \cdot t} \qquad \frac{dC}{d\beta} = -t \cdot B \cdot e^{-\beta \cdot t} \qquad \text{(5:65 a and b)}$$

$$\frac{d^2C}{dtd\alpha} = A \cdot t \cdot e^{-\alpha \cdot t} - A \cdot e^{-\alpha \cdot t} \qquad \frac{d^2C}{dtd\beta} = B \cdot t \cdot \alpha \cdot e^{-\beta \cdot t} - B \cdot e^{-\beta \cdot t} \qquad \text{(5:66 a and b)}$$

Set Equations 5:66 a and 5:66 b equal to zero and solve for t. This gives the optimal time points for sampling with respect to α and β.

$$t = \frac{1}{\alpha} \qquad \text{respectively} \quad t = \frac{1}{\beta} \qquad \text{(5:67 a and b)}$$

Ideally, one would then sample where the partial derivatives exhibit an extreme value such as the intercept for A and B (Figure 5.58, left) or minimum for α and β (Figure 5.58, right).

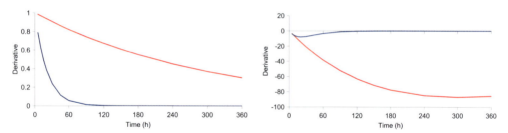

Figure 5.58 *Partial derivatives of concentration with respect to A (blue line) and B (red line) (left), and α (red line) and β (blue line) (right), versus time. A and B are the intercepts of Equation 5:63 and α and β are the slopes.*

Consider the *sigmoid* I_{max} function below.

$$E = E_0 - \frac{I_{max}C^n}{IC_{50}^n + C^n}$$

(5:68)

The partial derivatives of this function with respect to I_{max} and IC_{50} are, respectively

$$\frac{dE}{dI_{max}} = \frac{C^n}{IC_{50}^n + C^n}$$

(5:69)

$$\frac{dE}{dIC_{50}} = \frac{n \cdot I_{max} \cdot IC_{50}^{n-1} \cdot C^n}{\left[IC_{50}^n + C^n\right]^2}$$

(5:70)

The partial derivatives of the effect with respect to each parameter were obtained from the fit of the sigmoid I_{max} model to data of the 0-800 µg/L concentration range used in section 5.9.3. Most information about a certain parameter is obtained at its extreme value, in other words, when the derivative has a maximum or minimum value such as dE/dIC_{50}.

The first set of observations that contains information about the IC_{50} value should be centered at about half-maximal effect (Figure 5.59). The partial derivative of the effect with respect to IC_{50} peaks at the actual value of IC_{50}, namely 140 µg/L. It is interesting to note that maximum information about the sigmoidicity factor is obtained at approximately the 20 and 80% effect level, respectively. This interval is commonly said to be linear when effect is plotted against log-concentration. Maximum information about the IC_{50} is at the inflection point where the concentration equals IC_{50}. Maximum information about I_{max} is at the higher concentration levels or at the plateau of the effect, where sampling would give a good estimate of I_{max}. This is where the third set of samples would be taken.

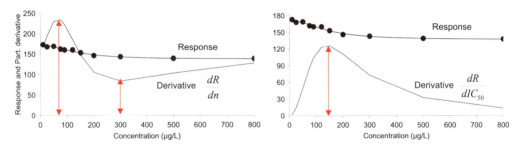

Figure 5.59 *Predicted effect (solid line with filled circles) versus steady state concentration superimposed upon the partial derivative with respect to the exponent (n) of the sigmoid I_{max} model (line without symbols) (left). Predicted effect (solid line with filled circles) versus steady state concentration superimposed upon the partial derivative with respect to IC_{50} of the sigmoid I_{max} model (line without symbols) (right). The red arrows indicate the extreme values (maximum or minimum) of the pratial derivatives and where on the concentration axis these values occur. dR/dn and dR/dIC$_{50}$ are the partial derivatives of response with respect to n and IC_{50}, respectively.*

Figure 5.60 (left) shows five design points with their respective information value. Observation one contains information about the baseline value. Observations 2 and 4 contain information about the sigmoidicity factor (exponent). It is interesting to note that these two points lie in the neighbourhood of the 20 and 80% of maximum response values. Observation number 3 gives the EC_{50} value and observation 5 gives E_{max}.

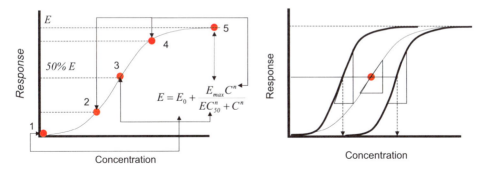

Figure 5.60 *The left hand plot shows schematically five important design points where one preferentially would like to sample in a single experiment. The dangers of pooling data (right). Two effect-concentration curves with different EC_{50} values but the same slope are pooled. The vertical placement of the new EC_{50} value is indicated on the curve by means of the red filled circle.*

Figure 5.60 (right) shows the danger of pooling data and then fitting the mean values. Observe that both the potency EC_{50} and the sigmoidicity n (commonly called the slope of the curve at EC_{50}) differ from their original individual values, and the mean curve bears no resemblance to the two individual curves.

A new compound was given to volunteers daily for a week. Response-time data were collected from day 1 to day 7. The model is based on drug effects (inhibition or stimulation) on the factors controlling either the input or the loss of drug response. The response was modeled as inhibition of production of response (Figure 5.61).

Figure 5.61 The simulated turnover model with inhibition I on the production of response as shown by the red arrow.

The partial derivatives of response with respect to k_{in}, k_{out} and IC_{50} were determined for the last dosing interval, using the following equation for modeling the turnover of response

$$\frac{dR}{dt} = k_{in} / H(t) - k_{out} \cdot R \qquad (5:71)$$

where *H(t)* is defined as

$$H(t) = 1 + \left[\frac{f(Dose, t, K)}{IC_{50}} \right]^n \qquad (5:72)$$

K is the kinetic rate constant for the input function. The predicted partial derivatives were based on the final parameter estimates and then plotted against time for the last dosing interval (day 7) (see Figure 5.62).

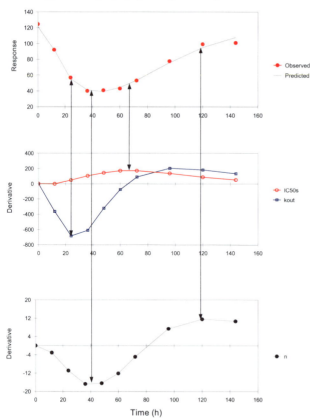

Figure 5.62 Partial derivatives of R with respect to k_{in}, IC_{50} and n versus time for an indirect response model (Equation 5:71) during the last dosing interval. The top figure shows observed and predicted response versus time data. The middle figure shows the absolute values of the partial derivatives of response with respect to k_{out} and IC_{50} versus time. The bottom figure shows the partial derivative of response with respect to n versus time. Arrows indicate where one would sample to obtain more information about the parameters in future studies.

The point of maximal response does not reveal very much about k_{in} and was therefore not included in the plot. The k_{in} parameter, as well as the infusion rate, determines the final level at steady state, but does not affect the time to steady state as does k_{out} or the elimination rate constant. In our example, the design points for k_{out} are two, namely very early and at about 100 hours post-dosing. The k_{out} parameter can be determined from the slope of the initial phase in the response-time curve. This may be useful information for the indirect response modeler since this class of model contains k_{in} and k_{out}.

We have demonstrated a graphical approach to experimental design by means of plots of the partial derivatives. No special assumptions about biological error have been made. Endrenyi's excellent text on design of experiments for estimating enzyme and pharmacokinetic parameters lays out optimal designs from a theoretical point of view, taking different error structures into account (Endrenyi [1981]).

5.10 Outliers

The importance of the graphical presentation of data cannot be over-emphasized. Take for example the data in Figure 5.63. Looking at these data in a table would hardly have shown two deviating points, namely A and B. As you can see, there are basically two types of outliers that one has to consider for the fitting procedure.

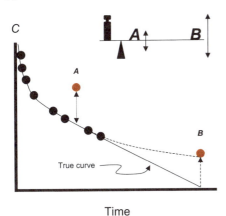

Figure 5.63 Concentration-time data with outliers A and B. Observation A does not have the same leverage as B, which is indicated by the inset. Observation B will tend to pull the fitted curve towards itself more strongly than observation A. A will result in reduced parameter precision whereas B will result in biased parameters (poor accuracy) with a high precision.

One type of outlier is represented by A and deviates vertically from the rest of the dataset. This outlier will not affect the parameter estimates to any greater extent but rather their precision, which will decrease. The other type of outlier is B, which deviates from the rest of the data set, particularly in the time dimension. This outlier has significant leverage and will therefore pull the parameter estimate of e.g., the terminal slope factor λ_z, in the direction of observation B so that the residual is reduced. One may also expect that the precision in λ_z will increase because the information about λ_z is substantial in that point.

Another situation that may arise is a mixture of both high leverage and extensive deviation in the Y-dimension (concentration dimension). This combination results in substantial influence on the parameter estimate of λ_z and its precision, which will be high. For more information on graphical regression analysis see Cook and Weisberg [1994].

5.11   A Checklist for Assessing Goodness-of-Fit

Boxenbaum [1992] has reviewed the philosophy of modeling. His text is recommended to everybody who has an ambition to become a modeler. In addition, we have compiled the following checklist to consider when assessing the adequacy of a model.

- Does the model have biological relevance?
- Does the fitted curve mimic trends in the data?
- Are the parameters estimated with adequate precision?
- Do the residuals show a lack of systematic deviation?
- Do the residual plots display a random scatter?

An answer of *no* to anyone of these questions would suggest that either the model or the weighting scheme is incorrect. Table 5.6 summarizes common problems, their symptoms and potential solutions.

Table 5.6   Common problems, their symptoms and potential solutions

Problem	Symptom	Solution
Much error in data	Poor precision	Improve design
Poorly selected data	Poor estimates	Improve design
Good fit, poor estimates	High correlation between parameters	Model too ambitious, data will not allow estimation, new design
Wrong structural model	Trend in residuals	New structural model
Wrong variance model	Trend in residuals	Improve variance model

Adapted from Aarons [1983].

Critically analyzing the residuals (residual *versus* independent variable, residual *versus* predicted value, runs etc.), checking parameter precision and accuracy, and verifying consistency between predicted and observed values, is often superior to statistical tests, such as the *F-test, Akaike* or *Schwarz Criterion* for model discrimination. In addition to residuals, parameter estimates, and precision, it is recommended to check the correlation between parameters and the condition number of the fit, where the latter, as a rule of thumb, should be less than or equal to 10^{Npar} (N_{par} is the number of parameters) for a well-defined model with respect to the information in the data. We strongly recommend that you *always use a battery of statistical tools* for determination of the goodness-of-fit and model validity. The model is good when:

- Predicted values mimic data and residuals show random scatter
- Parameter estimates are unbiased and have high precision
- Parameters have low correlation

For more difficult problems, the determination of best fit is a procedure that benefits greatly from experience, intuition and the application of scientific reason. A good answer requires good initial estimates. Furthermore, a strong software package should provide several means for improving on reasoned guesses, including the ability to do simulations, to display calculated curves graphically, and to refine estimates using the simplex search. Start the least square minimization with the best possible estimates and, if there are physical constraints on parameter values, specify the constraints to the program when entering initial estimates.

For complicated models, test the answer by running the problem again with perturbed parameter values; run the simplex algorithm to see if it finds another nearby minimum or wanders off in a different direction.

In chapter 5 we have reviewed the means for exploratory data analysis prior to specifying the actual model. We have tried to reinforce the idea of letting the data lead you to what type of model to use. This will be further emphasized thoughout the applications section.

We have demonstrated different ways of obtaining initial parameter estimates. We have looked at different minimization algorithms and the iteration process. Program output such as goodness-of-fit criteria and the *battery of statistical tools* are reviewed. Model discrimination and experimental design are also covered from a practical point of view.

Information on the Applications section

This section contains a collection of pharmacokinetic and pharmacodynamic datasets that we have encountered in our work. The analyses contain a more or less in-depth analysis of the data. In addition, we provide various degrees of program output. In general, we refer the reader to Chapters 3 and 4 for details on specific *PK/PD* concepts or Chapter 5 on modeling strategies. There is unavoidable some overlap, but that is intentional. According to our own experience some readers are more interested in the applications and the analysis, while others are more interested in the concepts. The first couple of exercises are explored more thoroughly with respect to data, output, and interpretation, whereas a few exercises present only the data and implementation of equations. Some datasets are generated from already published models and parameters. For the majority of exercises we have omitted, due to limited space, the iteration process, variance/covariance matrices, eigenvalues, and summary tables with observed and predicted values. We hope this still will not hamper the quality of the presentation. All models were proposed by JG or shared by generous colleagues, unless they were taken directly from the *WinNonlin* model library.

The enclosed diskette contains all command files (including commands, data and model). Due to the limited storage size, the output files had to be excluded.

Values of the initial parameter estimates are sometimes selected a bit off the more correct starting value. This is intentionally done to demonstrate that the program is efficient enough to finally reach the 'true' parameter value.

All calculations for this section were performed on a Compaq Deskpro with a Pentium II processor under MS Windows NT or Windows 95 using *WinNonlin* 2.0 or later versions.

PK1 - One-compartment intravenous bolus dosing

Objectives
◆ **To go from data to insight**
◆ **To analyze four intravenous bolus datasets with mono-exponential decline**
◆ **To obtain initial parameter estimates of *Cl*, *V* and *K***
◆ **To characterize the pharmacokinetics of drug in four individuals**
◆ **To apply an integrated (exponential) model**
◆ **To apply a differential equation model**
◆ **To discuss various aspects of experimental design**

Problem specification - 1

This exercise highlights how to *go from data to insight*, *obtain initial parameter estimates*, assess *goodness-of-fit*, and *interpret output*. For further reading about the one-compartment system analyzed in this exercise, see also Section 3.2.

Four volunteers were given an intravenous bolus dose of 10 mg of drug X. Assume a body weight of 70 kg. In the first part of the study, plasma samples were obtained from two females and the concentration-time data (Table 1.1) are plotted together with model-predicted data in Figure 1.1.

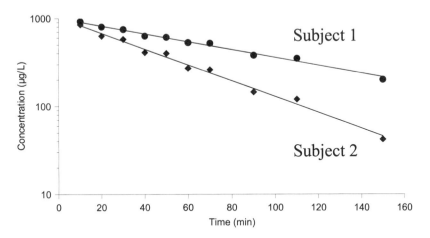

Figure 1.1 *Semi-logarithmic plot of observed and predicted concentration-time data from two female subjects (1 and 2) following a 10 mg intravenous bolus dose of drug X. Note the typical mono-exponential decline of concentration in both subjects and the smaller AUC (higher Cl) for subject 2. Intercepts with the concentration axis are the same, which indicates similar distribution volumes (V).*

Input the time and concentration data from Table 1.1 into *WinNonlin* and make a plot.

Look at the data and describe what you see. Then, analyze the disposition of drug, obtain initial parameter estimates, and apply a one-compartment model from the library

Table 1.1 Data from two females given an intravenous bolus of drug X

Time (min)	Concentration (µg/L)	
	Subject 1	Subject 2
10	920	850
20	800	630
30	750	580
40	630	410
50	610	400
60	530	270
70	520	260
90	380	145
110	350	120
150	200	42

Problem specification - 2

In the second part of the study, two male firemen were studied. Their drug plasma concentration-time data (Table 1.2) are displayed together with model-predicted data in Figure 1.2.

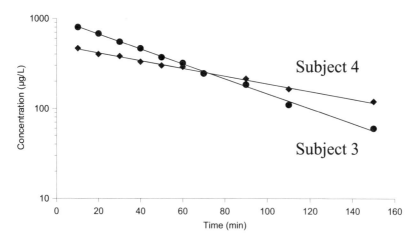

Figure 1.2 *Semi-logarithmic plot of observed and predicted concentration-time data of drug from male subjects (3 and 4) following a 10 mg intravenous bolus dose of drug X. Note the typical mono-exponential decline of concentration in both subjects and the lower intercept for subject 4 (larger V). It is difficult to graphically estimate which AUC is largest.*

Input the data from Table 1.2 into *WinNonlin* and make a plot. Look at the data and describe what you see. Analyze the disposition of drug, obtain initial parameter estimates, and apply a one-compartment user-specified model.

Table 1.2 Data from two males given an
intravenous bolus of drug X

Time (min)	Concentration (µg/L)	
	Subject 3	Subject 4
10	800	465
20	680	400
30	550	380
40	465	330
50	370	300
60	320	290
70	245	245
90	185	215
110	110	165
150	60	120

The four data sets in Tables 1.1 and 1.2 are typical one-compartment systems since they display mono-exponential decline in plasma after an intravenous bolus dose (see Figures 1.1 and 1.2, respectively).

Initial parameter estimates

Obtain the initial parameter estimates graphically for the one-compartment model (elimination rate constant and volume). Use the non-compartmental routines available in *WinNonlin*. Also write a user-specified model, either as an analytical solution (Equations 1:2) or as a differential equation (Equation 1:1 or 1:8), which is parameterized with clearance and volume. Figure 1.3 shows the necessary information you need to derive.

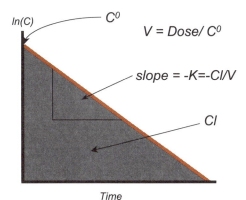

ln(C)

C^0

$V = Dose/ C^0$

slope = -K=-Cl/V

Cl

Time

Figure 1.3 Semi-logarithmic plot of the estimation of the principal parts of an intravenous bolus curve following mono-exponential decline. The negative slope is –K, clearance is Dose/AUC and V is C^0/V. K denotes the elimination rate constant, Cl clearance, V distribution volume and C^0 the concentration at time zero.

A first-order decline can be described by the following differential equation

$$\frac{dC}{dt} = -\frac{Cl}{V} \cdot C = -K \cdot C \qquad (1:1)$$

The analytical solution to Equation 1:1 becomes

$$C = C^0 e^{-K \cdot t} = \frac{Dose_{iv}}{V} e^{-K \cdot t} = \frac{Dose_{iv}}{V} e^{-\frac{Cl}{V} \cdot t} \qquad (1:2)$$

It is obvious from Figure 1.1 that the intercepts on the concentration axis are the same. This indicates that the two volumes of distribution are the same. However, the slope or elimination rate constant differs in these two subjects. For subject 1, the slope is more shallow and can be obtained from

$$slope = \frac{y_2 - y_1}{x_2 - x_1} = -K = \frac{\ln(920) - \ln(200)}{10 - 150} = -0.011 \, min^{-1} \qquad (1:3)$$

the half-life is

$$t_{1/2} = \frac{\ln 2}{K} = \frac{0.693}{0.011} = 63 \, min \qquad (1:4)$$

and the area under the curve *AUC* for subject 1, is estimated as

$$AUC_0^\infty = \int_0^\infty C^0 e^{-K \cdot t} dt = C^0 \left[\frac{e^{-K \cdot \infty}}{-K} - \frac{e^{-K \cdot 0}}{-K} \right] = \frac{C^0}{K} = \frac{1000}{0.011} \approx 91000 \, \mu g \cdot min / L \qquad (1:5)$$

Since it is obvious from the plot that subject 1 has a greater area than subject 2, then clearance for subject 1 will be less than clearance for subject 2. Plasma clearance for subject 1 is estimated as

$$Cl = \frac{D_{iv}}{AUC_0^\infty} = \frac{D_{iv}}{\left[\dfrac{C^0}{K} \right]} = \frac{10000}{91000} = 0.11 \, L / min \qquad (1:6)$$

The volume of distribution for both subjects is then

$$V = \frac{D_{iv}}{C^0} = \frac{10000}{1000} = 10 \, L \qquad (1:7)$$

The elimination rate constant, half-life, clearance and volume of distribution can be estimated in a similar manner for subject 2. Similar principles also apply to subjects 3 and 4, though of course both volume of distribution and slope differ in these two subjects.

In the solutions that follow, we have applied both a library rate constant model (*PK1.LIB*) and a user-specified clearance model (Equations 1:2 and 1:8). The library model uses volume and elimination rate constant as model parameters, and the clearance model uses clearance and volume.

Finally, you may also specify the model as a differential equation according to Equation 1:8.

$$\frac{dZ(1)}{dt} = -\frac{Cl}{V} \cdot Z(1) = -K \cdot Z(1) \qquad (1:8)$$

The derivative is written as $DZ(1)$ and the state variable C as $Z(1)$. The value of $Z(1)$ at time equal to zero is

$$Z(1) = \frac{Dose_{iv}}{V} \qquad (1:9)$$

The _WinNonlin_ code and output from fitting Equation 1:8 to the data of subject 1 are shown in solution 3, at the very end of the solutions section.

Interpretation of results and conclusions

We succeeded in fitting a one-compartment model to the data from each of the four subjects. The final parameter estimates for both the rate constant model and the clearance model are shown in Table 1.3. Table 1.4 shows the secondary parameters of _WinNonlin_ from fitting the library rate constant model to the data.

Table 1.3   Parameter estimates for the library rate constant model and user-specified clearance models

Subject	\multicolumn{3}{Run 1: Library model #1}			Run 2: Clearance model		
	Parameter	Estimate	CV%	Parameter	Estimate	CV%
1	K	0.010	4.2	Cl	0.10	2.8
2		0.021	5.5		0.20	3.3
3		0.019	1.8		0.19	1.1
4		0.010	3.5		0.20	2.4
1	V	9.98	2.1	V	9.98	2.1
2		9.82	3.8		9.81	3.8
3		10.2	1.2		10.2	1.2
4		19.9	1.7		19.95	1.7

Table 1.4   Pertinent pharmacokinetic parameters obtained as secondary parameters using the library rate constant model

Subject	Parameter	Estimate	CV%	Subject	Parameter	Estimate	CV%
1	Cl	0.10	2.8	1	V	9.98	2.1
2		0.20	3.3	2		9.82	3.8
3		0.19	1.1	3		10.2	1.2
4		0.20	2.4	4		19.9	1.7
1	AUC	97700	2.8	1	$t_{1/2}$	67.6	4.2
2		49300	3.3	2		33.6	5.5
3		51400	1.1	3		36.4	1.8
4		51100	2.4	4		70.6	3.5
1	AUMC	9520000	6.9	1	MRT	97.5	4.2
2		2390000	8.2	2		48.4	5.5
3		2700000	2.7	3		52.5	1.8
4		5200000	5.7	4		102	3.5

The parameter estimates were consistent between the two models and their precision was high. The clearance model had a lower condition number, which indicates that the clearance model is preferable to the rate constant model.

When a differential equation model (Equation 1:8) was used to fit the data of subject 1, the parameter estimates were the same as for the analytical model (mono-exponential model).

You have

☐ analyzed a typical one-compartment (mono-exponential) system)
☐ learned to look at data via different transfomations
☐ learned to derive the initial estimates
☐ characterized the disposition kinetics after single dose administration in four subjects

The next step in the characterization of kinetics would be to give multiple doses via the intended route (e.g., orally) of administration.

Solution 1-1 - Library model #1 for subject 1

```
MODEL 1
NOPAGE BREAKS
NVARIABLES 3
NPOINTS 300
XNUMBER 1
YNUMBER 2
NCONSTANTS 3
CONSTANTS 1,10000,0
METHOD 2  'Gauss-Newton (Levenberg and Hartley)
ITERATIONS 50
INITIAL 10,.023
LOWER BOUNDS .0001,.0001
UPPER BOUNDS 50,1
MISSING 'Missing'
NOBSERVATIONS 10
DATA 'WINNLIN.DAT'
BEGIN
```

PARAMETER	ESTIMATE	STANDARD ERROR	CV%	UNIVARIATE C.I.	
Volume	9.975866	.205180	2.06	9.502715	10.449016
K10	.010262	.000433	4.22	.009263	.011261

```
*** CORRELATION MATRIX OF THE ESTIMATES ***
PARAMETER  Volume     K10
Volume      1.00000
K10        -.806888          1.00000

 Condition_number=        801.8

   FUNCTION   1
```

X	OBSERVED Y	PREDICTED Y	RESIDUAL	WEIGHT	SE-PRED	STANDARDIZ RESIDUAL
10.00	920.0	904.7	15.35	1.000	15.60	.8896
20.00	800.0	816.4	-16.42	1.000	11.83	-.8201
30.00	750.0	736.8	13.20	1.000	9.325	.6197
40.00	630.0	664.9	-34.94	1.000	8.086	-1.602
50.00	610.0	600.1	9.915	1.000	7.895	.4532
60.00	530.0	541.6	-11.56	1.000	8.316	-.5322
70.00	520.0	488.7	31.26	1.000	8.958	1.456
90.00	380.0	398.1	-18.06	1.000	10.14	-.8626
110.0	350.0	324.2	25.80	1.000	10.81	1.253
150.0	200.0	215.1	-15.05	1.000	10.72	-.7292

```
CORRECTED SUM OF SQUARED OBSERVATIONS = 438490.
WEIGHTED CORRECTED SUM OF SQUARED OBSERVATIONS = 438490.
SUM OF SQUARED RESIDUALS =              4327.48
SUM OF WEIGHTED SQUARED RESIDUALS = 4327.48
S = 23.2580      WITH      8 DEGREES OF FREEDOM
CORRELATION (OBSERVED,PREDICTED) = .9951
```

```
AIC criteria =           87.72741
SC  criteria =           88.33258
```

SUMMARY OF ESTIMATED SECONDARY PARAMETERS

PARAMETER	ESTIMATE	STANDARD ERROR	CV%
AUC	97682.707082	2767.035242	2.83
K10-HL	67.545083	2.848644	4.22
Cmax	1002.419274	20.596793	2.05
Cl	.102372	.002903	2.84
AUMC	9518882.482577	654952.142582	6.88
MRT	97.446956	4.109724	4.22
Vss	9.975866	.205180	2.06

Solution 2-1 - User-specified clearance model for subject 1

```
MODEL
 COMMANDS
  NPARAMETERS 2
  PNAMES  'Cl', 'V'
END
    1: TEMPORARY
    2: T = X
    3: END
    4: FUNCTION 1
    5: F = (10000/V)*EXP(-(CL/V)*T)
    6: END
    7: EOM
NOPAGE BREAKS
NVARIABLES 3
NPOINTS 100
XNUMBER 1
YNUMBER 2
METHOD 2  'Gauss-Newton (Levenberg and Hartley)
ITERATIONS 50
INITIAL 1,10
LOWER BOUNDS .01,1
UPPER BOUNDS 5,50
MISSING 'Missing'
NOBSERVATIONS 10
DATA 'WINNLIN.DAT'
BEGIN
```

PARAMETER	ESTIMATE	STANDARD ERROR	CV%	UNIVARIATE C.I.	
CL	.102339	.002904	2.84	.095642	.109036
V	9.978294	.205186	2.06	9.505130	10.451457

```
*** CORRELATION MATRIX OF THE ESTIMATES ***
PARAMETER  CL          V
```

```
CL        1.00000
V        -.476741  1.00000


   Condition_number=        80.38


   FUNCTION   1
```

X	OBSERVED Y	PREDICTED Y	RESIDUAL	WEIGHT	SE-PRED	STANDARDIZ RESIDUAL
10.00	920.0	904.5	15.51	1.000	15.59	.8991
20.00	800.0	816.3	-16.32	1.000	11.83	-.8150
30.00	750.0	736.7	13.25	1.000	9.325	.6220
40.00	630.0	664.9	-34.93	1.000	8.086	-1.602
50.00	610.0	600.1	9.885	1.000	7.894	.4518
60.00	530.0	541.6	-11.62	1.000	8.315	-.5349
70.00	520.0	488.8	31.18	1.000	8.957	1.453
90.00	380.0	398.2	-18.17	1.000	10.14	-.8680
110.0	350.0	324.3	25.67	1.000	10.81	1.247
150.0	200.0	215.2	-15.19	1.000	10.73	-.7359

```
   CORRECTED SUM OF SQUARED OBSERVATIONS =  438490.
   WEIGHTED CORRECTED SUM OF SQUARED OBSERVATIONS =  438490.
   SUM OF SQUARED RESIDUALS =        4327.38
   SUM OF WEIGHTED SQUARED RESIDUALS =  4327.38
   S =  23.2577      WITH      8 DEGREES OF FREEDOM
   CORRELATION (OBSERVED,PREDICTED) =  .9951


  AIC criteria =        87.72717
  SC  criteria =        88.33234
```

Solutions 1-2 - Library model #1 for subject 2

```
MODEL 1
NOPAGE BREAKS
NVARIABLES 3
NPOINTS 300
XNUMBER 1
YNUMBER 2
NCONSTANTS 3
CONSTANTS 1,10000,0
METHOD 2  'Gauss-Newton (Levenberg and Hartley)
ITERATIONS 50
INITIAL 10,.023
LOWER BOUNDS .0001,.0001
UPPER BOUNDS 50,1
MISSING 'Missing'
NOBSERVATIONS 10
DATA 'WINNLIN.DAT'
BEGIN
```

PARAMETER	ESTIMATE	STANDARD ERROR	CV%	UNIVARIATE C.I.	
Volume	9.815776	.368829	3.76	8.965247	10.666305
K10	.020661	.001133	5.48	.018048	.023274

```
*** CORRELATION MATRIX OF THE ESTIMATES ***
PARAMETER  Volume    K10
Volume    1.00000
K10       -.811011        1.00000
```

Condition_number=   556.4

FUNCTION   1

X	OBSERVED Y	PREDICTED Y	RESIDUAL	WEIGHT	SE-PRED	STANDARDIZ RESIDUAL
10.00	850.0	828.6	21.40	1.000	24.12	1.127
20.00	630.0	673.9	-43.93	1.000	15.70	-1.665
30.00	580.0	548.1	31.87	1.000	12.19	1.131
40.00	410.0	445.8	-35.81	1.000	11.82	-1.264
50.00	400.0	362.6	37.41	1.000	12.39	1.332
60.00	270.0	294.9	-24.91	1.000	12.82	-.8929
70.00	260.0	239.9	20.14	1.000	12.84	.7222
90.00	145.0	158.7	-13.67	1.000	11.86	-.4828
110.0	120.0	105.0	15.04	1.000	10.14	.5189
150.0	42.00	45.93	-3.931	1.000	6.476	-.1310

CORRECTED SUM OF SQUARED OBSERVATIONS = 587404.
WEIGHTED CORRECTED SUM OF SQUARED OBSERVATIONS = 587404.
SUM OF SQUARED RESIDUALS =   7539.72
SUM OF WEIGHTED SQUARED RESIDUALS = 7539.72
S = 30.6996   WITH   8 DEGREES OF FREEDOM
CORRELATION (OBSERVED,PREDICTED) = .9936

AIC criteria =   93.27941
SC criteria =   93.88458

SUMMARY OF ESTIMATED SECONDARY PARAMETERS

PARAMETER	ESTIMATE	STANDARD ERROR	CV%
AUC	49307.912342	1616.609101	3.28
K10-HL	33.548006	1.838023	5.48
Cmax	1018.768151	38.242099	3.75
Cl	.202807	.006656	3.28
AUMC	2386480.394350	195859.049367	8.21
MRT	48.399542	2.651706	5.48
Vss	9.815776	.368829	3.76

Solution 2-2 – User-specified clearance model for subject 2

```
MODEL
 COMMANDS
  NPARAMETERS 2
  PNAMES  'Cl', 'V'
END
    1: TEMPORARY
    2: T = X
    3: END
    4: FUNCTION 1
    5: F = (10000/V)*EXP(-(CL/V)*T)
    6: END
    7: EOM
NOPAGE BREAKS
NVARIABLES 3
NPOINTS 100
XNUMBER 1
YNUMBER 2
METHOD 2   'Gauss-Newton (Levenberg and Hartley)
```

```
ITERATIONS 50
INITIAL 1,10
LOWER BOUNDS .01,1
UPPER BOUNDS 5,50
MISSING 'Missing'
NOBSERVATIONS 10
DATA 'WINNLIN.DAT'
BEGIN
```

PARAMETER	ESTIMATE	STANDARD ERROR	CV%	UNIVARIATE C.I.	
CL	.202791	.006654	3.28	.187446	.218136
V	9.812184	.368439	3.75	8.962553	10.661814

```
*** CORRELATION MATRIX OF THE ESTIMATES ***
PARAMETER  CL        V
CL         1.00000
V         -.211038  1.00000

 Condition_number=        56.65

FUNCTION   1
```

X	OBSERVED Y	PREDICTED Y	RESIDUAL	WEIGHT	SE-PRED	STANDARDIZ RESIDUAL
10.00	850.0	828.9	21.15	1.000	24.12	1.113
20.00	630.0	674.1	-44.09	1.000	15.70	-1.672
30.00	580.0	548.2	31.77	1.000	12.19	1.128
40.00	410.0	445.9	-35.87	1.000	11.82	-1.266
50.00	400.0	362.6	37.38	1.000	12.39	1.331
60.00	270.0	294.9	-24.91	1.000	12.82	-.8930
70.00	260.0	239.8	20.15	1.000	12.84	.7226
90.00	145.0	158.6	-13.64	1.000	11.86	-.4818
110.0	120.0	104.9	15.07	1.000	10.14	.5200
150.0	42.00	45.91	-3.907	1.000	6.478	-.1302

```
CORRECTED SUM OF SQUARED OBSERVATIONS =  587404.
WEIGHTED CORRECTED SUM OF SQUARED OBSERVATIONS =  587404.
SUM OF SQUARED RESIDUALS =           7539.87
SUM OF WEIGHTED SQUARED RESIDUALS =  7539.87
S =  30.6999     WITH      8 DEGREES OF FREEDOM
CORRELATION (OBSERVED,PREDICTED) =  .9936

AIC criteria =         93.27960
SC  criteria =         93.88477
```

Solution 1-3 - Library model #1 for subject 3

```
MODEL 1
NOPAGE BREAKS
NVARIABLES 3
NPOINTS 300
XNUMBER 1
YNUMBER 2
NCONSTANTS 3
CONSTANTS 1,10000,0
METHOD 2  'Gauss-Newton (Levenberg and Hartley)
ITERATIONS 50
INITIAL 10,.023
```

```
LOWER BOUNDS .0001,.0001
UPPER BOUNDS 50,1
MISSING 'Missing'
NOBSERVATIONS 10
DATA 'WINNLIN.DAT'
BEGIN
```

PARAMETER	ESTIMATE	STANDARD ERROR	CV%	UNIVARIATE C.I.	
Volume	10.223252	.121919	1.19	9.942102	10.504402
K10	.019041	.000343	1.80	.018250	.019831

```
*** CORRELATION MATRIX OF THE ESTIMATES ***
PARAMETER  Volume     K10
Volume     1.00000
K10        -.809616  1.00000

 Condition_number=      606.2

   FUNCTION   1
```

X	OBSERVED Y	PREDICTED Y	RESIDUAL	WEIGHT	SE-PRED	STANDARDIZ RESIDUAL
10.00	800.0	808.6	-8.572	1.000	7.567	-1.372
20.00	680.0	668.4	11.62	1.000	5.033	1.379
30.00	550.0	552.5	-2.502	1.000	3.879	-.2775
40.00	465.0	456.7	8.289	1.000	3.692	.9116
50.00	370.0	377.5	-7.528	1.000	3.865	-.8345
60.00	320.0	312.1	7.927	1.000	4.042	.8864
70.00	245.0	258.0	-12.97	1.000	4.112	-1.455
90.00	185.0	176.3	8.730	1.000	3.929	.9707
110.0	110.0	120.4	-10.45	1.000	3.477	-1.138
150.0	60.00	56.24	3.763	1.000	2.377	.3952

```
CORRECTED SUM OF SQUARED OBSERVATIONS =  537753.
WEIGHTED CORRECTED SUM OF SQUARED OBSERVATIONS =  537753.
SUM OF SQUARED RESIDUALS =          770.511
SUM OF WEIGHTED SQUARED RESIDUALS =  770.511
S =  9.81396     WITH     8 DEGREES OF FREEDOM
CORRELATION (OBSERVED,PREDICTED) =  .9993

AIC criteria =        70.47053
SC  criteria =        71.07570
```

SUMMARY OF ESTIMATED SECONDARY PARAMETERS

PARAMETER	ESTIMATE	STANDARD ERROR	CV%
AUC	51372.381022	558.788004	1.09
K10-HL	36.403591	.654429	1.80
Cmax	978.162324	11.653623	1.19
Cl	.194657	.002119	1.09
AUMC	2698040.466701	73399.182804	2.72
MRT	52.519280	.944141	1.80
Vss	10.223252	.121919	1.19

Solution 2-3 – User-specified clearance model for subject 3

```
MODEL
 COMMANDS
  NPARAMETERS 2
  PNAMES  'Cl', 'V'
END
   1: TEMPORARY
   2: T = X
   3: END
   4: FUNCTION 1
   5: F = (10000/V)*EXP(-(CL/V)*T)
   6: END
   7: EOM
NOPAGE BREAKS
NVARIABLES 3
NPOINTS 100
XNUMBER 1
YNUMBER 2
METHOD 2  'Gauss-Newton (Levenberg and Hartley)
ITERATIONS 50
INITIAL 1,10
LOWER BOUNDS .01,1
UPPER BOUNDS 5,50
MISSING 'Missing'
NOBSERVATIONS 10
DATA 'WINNLIN.DAT'
BEGIN
```

PARAMETER	ESTIMATE	STANDARD ERROR	CV%	UNIVARIATE C.I.	
CL	.194658	.002120	1.09	.189770	.199546
V	10.221835	.121839	1.19	9.940870	10.502800

```
*** CORRELATION MATRIX OF THE ESTIMATES ***
PARAMETER  CL         V
CL         1.00000
V         -.243517   1.00000

 Condition_number=       59.27

FUNCTION   1
```

X	OBSERVED Y	PREDICTED Y	RESIDUAL	WEIGHT	SE-PRED	STANDARDIZ RESIDUAL
10.00	800.0	808.7	-8.662	1.000	7.566	-1.386
20.00	680.0	668.4	11.56	1.000	5.033	1.372
30.00	550.0	552.5	-2.533	1.000	3.880	-.2810
40.00	465.0	456.7	8.276	1.000	3.692	.9101
50.00	370.0	377.5	-7.528	1.000	3.865	-.8346
60.00	320.0	312.1	7.935	1.000	4.042	.8872
70.00	245.0	258.0	-12.95	1.000	4.112	-1.454
90.00	185.0	176.3	8.749	1.000	3.930	.9728
110.0	110.0	120.4	-10.43	1.000	3.478	-1.136
150.0	60.00	56.22	3.778	1.000	2.378	.3968

```
CORRECTED SUM OF SQUARED OBSERVATIONS =  537753.
  WEIGHTED CORRECTED SUM OF SQUARED OBSERVATIONS =  537753.
  SUM OF SQUARED RESIDUALS =           770.520
  SUM OF WEIGHTED SQUARED RESIDUALS =  770.520
  S =  9.81402     WITH      8 DEGREES OF FREEDOM
```

```
CORRELATION (OBSERVED,PREDICTED) =  .9993
```

Solution 1-4 - Library model #1 for subject 4

```
MODEL 1
NOPAGE BREAKS
NVARIABLES 3
NPOINTS 300
XNUMBER 1
YNUMBER 2
NCONSTANTS 3
CONSTANTS 1,10000,0
METHOD 2  'Gauss-Newton (Levenberg and Hartley)
ITERATIONS 50
INITIAL 10,.023
LOWER BOUNDS .0001,.0001
UPPER BOUNDS 50,1
MISSING 'Missing'
NOBSERVATIONS 10
DATA 'WINNLIN.DAT'
BEGIN
```

PARAMETER	ESTIMATE	STANDARD ERROR	CV%	UNIVARIATE C.I.	
Volume	19.941712	.329403	1.65	19.182100	20.701323
K10	.009820	.000342	3.48	.009032	.010608

```
*** CORRELATION MATRIX OF THE ESTIMATES ***
PARAMETER  Volume     K10
Volume       1.00000
K10         -.807128        1.00000

Condition_number=       1633.
```

```
    FUNCTION   1
```

X	OBSERVED Y	PREDICTED Y	RESIDUAL	WEIGHT	SE-PRED	STANDARDIZ RESIDUAL
10.00	465.0	454.6	10.44	1.000	6.314	1.471
20.00	400.0	412.0	-12.04	1.000	4.822	-1.471
30.00	380.0	373.5	6.498	1.000	3.815	.7468
40.00	330.0	338.6	-8.568	1.000	3.301	-.9618
50.00	300.0	306.9	-6.900	1.000	3.205	-.7716
60.00	290.0	278.2	11.80	1.000	3.366	1.329
70.00	245.0	252.2	-7.175	1.000	3.629	-.8172
90.00	215.0	207.2	7.793	1.000	4.136	.9112
110.0	165.0	170.3	-5.259	1.000	4.449	-.6265
150.0	120.0	115.0	5.048	1.000	4.498	.6033

```
CORRECTED SUM OF SQUARED OBSERVATIONS =  104690.
WEIGHTED CORRECTED SUM OF SQUARED OBSERVATIONS =  104690.
SUM OF SQUARED RESIDUALS =          721.969
SUM OF WEIGHTED SQUARED RESIDUALS =  721.969
S =  9.49979     WITH     8 DEGREES OF FREEDOM
CORRELATION (OBSERVED,PREDICTED) =  .9966
```

```
AIC criteria =        69.81982
SC  criteria =        70.42499
```

SUMMARY OF ESTIMATED SECONDARY PARAMETERS

PARAMETER	ESTIMATE	STANDARD ERROR	CV%
AUC	51064.961063	1202.764970	2.36
K10-HL	70.584753	2.453769	3.48
Cmax	501.461469	8.275007	1.65
Cl	.195829	.004617	2.36
AUMC	5200061.042995	296464.048475	5.70
MRT	101.832273	3.540040	3.48
Vss	19.941712	.329403	1.65

Solution 2-4 – User-specified clearance model for subject 4

```
MODEL
 COMMANDS
  NPARAMETERS 2
  PNAMES  'Cl', 'V'
END
    1: TEMPORARY
    2: T = X
    3: END
    4: FUNCTION 1
    5: F = (10000/V)*EXP(-(CL/V)*T)
    6: END
    7: EOM
NOPAGE BREAKS
NVARIABLES 3
NPOINTS 100
XNUMBER 1
YNUMBER 2
METHOD 2  'Gauss-Newton (Levenberg and Hartley)
ITERATIONS 50
INITIAL 1,10
LOWER BOUNDS .01,1
UPPER BOUNDS 5,50
MISSING 'Missing'
NOBSERVATIONS 10
DATA 'WINNLIN.DAT'
BEGIN
```

PARAMETER	ESTIMATE	STANDARD ERROR	CV%	UNIVARIATE C.I.	
CL	.195739	.004620	2.36	.185085	.206393
V	19.949630	.329484	1.65	19.189832	20.709428

```
*** CORRELATION MATRIX OF THE ESTIMATES ***
PARAMETER  CL          V
CL        1.00000
V         -.491503   1.00000

 Condition_number=        81.89

FUNCTION   1
```

X	OBSERVED Y	PREDICTED Y	RESIDUAL	WEIGHT	SE-PRED	STANDARDIZ RESIDUAL
10.00	465.0	454.4	10.58	1.000	6.312	1.491

20.00	400.0	411.9	-11.95	1.000	4.822	-1.460
30.00	380.0	373.4	6.552	1.000	3.815	.7531
40.00	330.0	338.5	-8.547	1.000	3.301	-.9595
50.00	300.0	306.9	-6.907	1.000	3.205	-.7724
60.00	290.0	278.2	11.78	1.000	3.366	1.326
70.00	245.0	252.2	-7.223	1.000	3.629	-.8227
90.00	215.0	207.3	7.718	1.000	4.136	.9025
110.0	165.0	170.3	-5.348	1.000	4.450	-.6373
150.0	120.0	115.1	4.949	1.000	4.501	.5916

```
CORRECTED SUM OF SQUARED OBSERVATIONS =  104690.
WEIGHTED CORRECTED SUM OF SQUARED OBSERVATIONS =  104690.
SUM OF SQUARED RESIDUALS =          721.945
SUM OF WEIGHTED SQUARED RESIDUALS =  721.945
S = 9.49964     WITH      8 DEGREES OF FREEDOM
CORRELATION (OBSERVED,PREDICTED) = .9966

AIC criteria =         69.81948
SC  criteria =         70.42465
```

Solution 3-1 - Differential equation model for subject 1

```
MODEL
 COMM
  NFUN 1
  NDER 1
  NPAR 2
  NCON 1
  PNAM 'CL', 'V'
END
    1: TEMP
    2: DOSE = CON(1)
    3: END
    4: START
    5: Z(1)= DOSE/V
    6: END
    7: DIFF
    8: DZ(1)= -(CL/V)*Z(1)
    9: END
   10: FUNC 1
   11: F = Z(1)
   12: END
   13: EOM
NVARIABLES 2
NPOINTS 100
XNUMBER 1
YNUMBER 2
CONSTANTS 10000
METHOD 2  'Gauss-Newton (Levenberg and Hartley)
ITERATIONS 50
INITIAL .1,10
LOWER BOUNDS 0,0
UPPER BOUNDS .5,50
MISSING 'Missing'
NOBSERVATIONS 10
DATA 'WINNLIN.DAT'
BEGIN
```

PARAMETER	ESTIMATE	STANDARD	CV%	UNIVARIATE C.I.

		ERROR			
CL	.102307	.002904	2.84	.095610	.109004
V	9.979416	.205200	2.06	9.506219	10.452613

```
*** CORRELATION MATRIX OF THE ESTIMATES ***
PARAMETER          CL            V
CL              1.00000
V               -.476883       1.00000

   Condition_number=        80.39

     FUNCTION   1
```

X	OBSERVED Y	PREDICTED Y	RESIDUAL	WEIGHT	SE-PRED	STANDARDIZ RESIDUAL
10.00	920.0	904.4	15.58	1.000	15.59	.9027
20.00	800.0	816.3	-16.30	1.000	11.83	-.8139
30.00	750.0	736.8	13.24	1.000	9.325	.6214
40.00	630.0	665.0	-34.97	1.000	8.086	-1.604
50.00	610.0	600.2	9.823	1.000	7.894	.4490
60.00	530.0	541.7	-11.70	1.000	8.314	-.5385
70.00	520.0	488.9	31.08	1.000	8.957	1.448
90.00	380.0	398.3	-18.28	1.000	10.14	-.8733
110.0	350.0	324.4	25.55	1.000	10.81	1.241
150.0	200.0	215.3	-15.30	1.000	10.73	-.7416

```
CORRECTED SUM OF SQUARED OBSERVATIONS =  438490.
 WEIGHTED CORRECTED SUM OF SQUARED OBSERVATIONS =  438490.
 SUM OF SQUARED RESIDUALS =        4327.45
 SUM OF WEIGHTED SQUARED RESIDUALS =  4327.45
 S =  23.2579     WITH      8 DEGREES OF FREEDOM
 CORRELATION (OBSERVED,PREDICTED) =  .9951

AIC criteria =        87.72733
SC  criteria =        88.33250
```

PK2 – One-compartment oral dosing

Objectives

◆ **To analyze a dataset following oral administration**

◆ **To analyze the absorption kinetics**

◆ **To estimate K_a, K, *lag-time* and *V/F***

◆ **To obtain initial parameter estimates graphically and numerically**

◆ **To analyze the flip-flop situation**

◆ **To discuss experimental design**

Problem specification

This exercise demonstrates two approaches to modeling oral data by means of a first-order input model with and without a *lag-time*. For further reading about one-compartment models and extravascular dosing see Section 3.2.4.

Concentration-time data were obtained from a human volunteer following an oral dose of 100 µg of substance A. We start by plotting the concentration-time data (Figure 2.1 and program output) and then calculate initial *estimates* of the basal pharmacokinetic parameters such as K_a, K, t_{lag} and V/F. This could also be easily done by means of curve stripping. Assume a bioavailability of 100%.

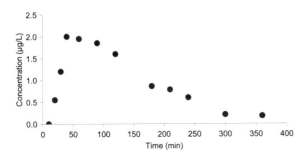

Figure 2.1 *Observed concentration-time data following a 100 µg oral dose of drug A.*

The explanation for a *lag-time* may be physiological and/or biopharmaceutical. A *lag-time* is often observed for controlled-release dosage forms when the disintegration and dissolution of the oral dosage form are the time-limiting steps (biopharmaceutical). The onset of absorption of the active compound may also be governed by the physiology/pharmacology (e.g., stomach emptying). For the present data, we will not make any interpretations of the underlying mechanism. The model equation takes the form

$$C = \frac{K_a FD_{po}}{V(K_a - K)} \left[e^{-K \cdot t} - e^{-K_a \cdot t} \right] \tag{2:1}$$

When we also include a lag-time for absorption, the expression becomes

$$C = \frac{K_a FD_{po}}{V(K_a - K)} \left[e^{-K \cdot (t - t_{lag})} - e^{-K_a \cdot (t - t_{lag})} \right]$$

(2:2)

Initial parameter estimates

Before we start the regression analysis, we obtain initial parameter estimates of K, K_a, V/F, and t_{lag} by means of non-compartmental analysis, method of residuals or curve stripping. K (min^{-1}) is the elimination rate constant. We assume that the elimination of drug is the rate-limiting step and not the absorption. K_a (min^{-1}), is the absorption rate constant and t_{lag} (min) is the *lag-time*. In practice it takes the disintegration of the tablet, the dissolution of the drug into gut fluid and the transport of drug across the gut lumen before any drug appears at the sampling site. See Gibaldi and Perrier [1981] and Section 3.3.5 for discussion of delayed absorption processes.

We start the analysis by estimating the rate constants of the absorption and elimination phase. The elimination rate constant is obtained from the slopes. The elimination rate constant is

$$K = -\frac{\ln(C_1) - \ln(C_2)}{t_1 - t_2} = -\frac{\ln(0.18/0.6)}{360 - 240} \approx 0.01 \, \text{min}^{-1}$$

The terminal slope is then back-extrapolated to the concentration axis (abscissa). By means of the method of residuals (Gibaldi and Perrier [1981]) we subtract the observed concentrations on the upswing of the curve (usually called the absorption phase) from the extrapolated line (see section 3.2.6 in *Pharmacokinetic Concepts*). This difference ($C_{residual} = C_{extrapolated} - C_{observed}$) is then plotted in the same figure and the data points are joined by a straight line (Figure 2.2).

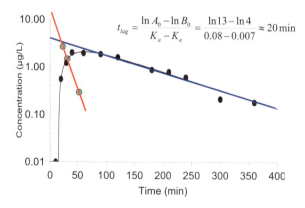

$$t_{lag} = \frac{\ln A_0 - \ln B_0}{K_a - K_e} = \frac{\ln 13 - \ln 4}{0.08 - 0.007} \approx 20 \, \text{min}$$

Figure 2.2 Schematic illustration of the graphical and numerical estimation of t_{lag}. The blue line is the terminal phase with a slope corresponding to $-K$ (also denoted K_e). The three green circles are the residual points and the red line is the regression line of the residual observations. The red line has a slope of $-K_a$. We assume that the latter is the absorption phase.

The slope of the new line gives the absorption rate constant K_a. You can also observe that the back-extrapolated terminal slope and the absorption slope intersect at about 15-20 min. This suggests that there is a *lag-time* before any drug is observed at the site of sampling.

Table 2.1 Tabulated formulas for estimation of initial values after extravascular dosing

Parameter	Calculation	Description
Elimination rate constant K (min^{-1})	$slope = -K = \dfrac{\ln C_1 - \ln C_2}{t_1 - t_2}$	(C_1, t_1) and (C_2, t_2) are obtained from the terminal portion of the curve
Absorption rate constant K_a (min^{-1})	$residual\ slope = -K_a = \dfrac{\ln R_1 - \ln R_2}{t_1 - t_2}$	(R_1, t_1) and (R_2, t_2) are obtained from the residual curve
V/F	$\dfrac{V}{F} \approx \dfrac{Dose}{C_{max}}$	This gives a rough estimate of the initial estimate of *V/F*
t_{lag} (min)	$t_{lag} = \dfrac{\ln A_0 - \ln B_0}{K_a - K}$	The time at the intercept of the extrapolated terminal phase and the residual curve. A_0 is the residual intercept with the Y axis and B_0 is the back-extrapolated intercept of the terminal slope.

$$
\begin{aligned}
V/F &= 30\ (\text{L}) && \leq Dose/C_{max} \approx 50 \\
K_a &= 0.1\ (\text{min}^{-1}) && \text{Curve stripping} \\
K &= 0.05\ (\text{min}^{-1}) && \text{Terminal slope } (0.01\ \text{min}^{-1}) \\
t_{lag} &= 20.0\ (\text{min}) && \text{Obtained by residual method (Table 2.1) or deconvolution}
\end{aligned}
$$

Note that the initial estimate of *V/F* (30) is slightly lower than $Dose/C_{max}$ (50). K (0.05 min^{-1}) was also a factor 5 from its more correct starting value of 0.01 (min^{-1}). The reason for this was to demonstrate the *WinNonlin* was capable of locating the correct final estimates.

Interpretation of results and conclusions

This exercise illustrates how to fit a first-order input-output model to single oral dose data. We also show how to improve the fit to these data by adding a *lag-time* to the model. Figure 2.3 displays the two model fits, superimposed on the observed data.

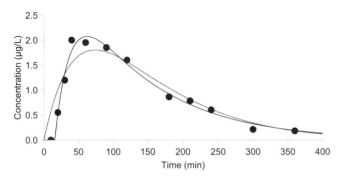

Figure 2.3 Observed (filled circles) and predicted (solid line lag-time model, dashed line model without lag-time) concentration-time data following a 100 µg oral dose of drug A.

In the first run, we use a model without a *lag-time* (Equation 2:1): Here, however, we cannot separate K_a (in the program code specified as k_{01}) from K (in program code specified as k_{10}), as both values are approximately 0.0132. The parameters also have high standard errors and are highly correlated (see correlation matrix in output). The condition number and trends in the residuals also indicate that the model does not provide a good fit to the data. The model markedly overestimates the first concentration-time point, and the predicted time for the maximum concentration (t_{max} 73 min) is 90% higher than the observed t_{max} (40 min). Note also that the program places a warning prior to parameter output for Solution I. This is the first indication that there is a pathological fit.

Equation 2:2 includes a *lag-time*, which markedly improves the fit. Not only can K_a and K be separated numerically, but their precision is also high and correlation low. There are no obvious trends in the residuals and the *lag-time* model captures both the first concentration-time point as well as t_{max}. The *Akaike (AIC)* and *Schwarz (SC) criteria* are also markedly reduced when using a *lag-time* model (Equation 2:2) relative to the fit using Equation 2:1.

Numerical deconvolution of oral data (C_{po}) with intravenous data (C_{iv}) from a separate experiment revealed a 15 min *lag-time* and a bioavailability (cumulated input *Cum*) of about 62% (Langenbücher [1982], *QWERT* [1993]). Output from *QWERT* is given in Table 2.2 and shown in Figure 2.4.

The slope of the terminal portion of intravenous data is about 0.01 min^{-1}, which can be calculated from the left hand panel in Figure 2.4. Looking at Figure 2.4 (right), there is no clear-cut indication of the type of input function (lower curve). The input function, plotted on a semi-log scale, should give a straight line if the absorption were purely a first-order process. We can, however, approximate our system with such a process and still obtain a good fit as shown in Figure 2.2.

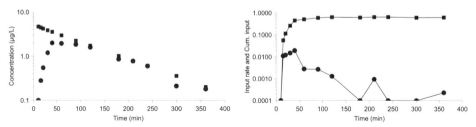

Figure 2.4 *Observed concentration versus time data (left) after intravenous (squares) and oral (circles) input. The corresponding input function (circles) and cumulated input (squares) (right) after having deconvoluted oral data with intravenous data.*

Table 2.2   Deconvolution data from *QWERT* [1993]

Time (min)	C_{po} (response)	C_{iv} (weight)	Input (min^{-1})	*Cum* Input
10	0.00	4.70	0.00000	0.000
15	0.28	4.48	0.01122	0.056
20	0.55	4.22	0.01168	0.114
30	1.20	3.87	0.01496	0.260
40	2.00	3.57	0.01932	0.450
60	1.95	2.97	0.00288	0.512
90	1.85	2.25	0.00272	0.593
120	1.60	1.74	0.00126	0.637
180	0.86	1.02	–0.00094	0.615
210	0.78	0.77	0.00095	0.645
240	0.60	0.61	–0.00011	0.646
300	0.21	0.36	–0.00092	0.609
360	0.18	0.20	0.00023	0.623

In light of the aforementioned information, we would select the *lag-time* model as our choice in this case. The information that we have obtained from the previous analysis is suitable as a pilot study. However, to fully characterize the kinetics of this drug, steady state data from 2-3 dose levels would be needed in conjunction with intravenous data.

Be aware of the *flip-flop* situation. The rate constants K_a and K are not uniquely identifiable unless one also administers an intravenous dose for estimation of the disposition rate constant (also called elimination rate constant for a one-compartment system) independently from K_a. In addition, be cautious when interpreting the estimate of V, because it also includes the bioavailability (i.e., it is really estimating V/F). Solution III contains output from a fit where the initial parameter estimates of K_a and K were reversed according to 0.01 and 0.1 min^{-1}. The final parameter estimates for K_a, K, V/F and t_{lag} were 0.0087 min^{-1}, 0.043 h^{-1}, 6.54 L and 16 min, respectively. Note also that the estimate of V/F (6.54 *versus* 30 L) changes in the *flip-flop* situation.

You have
☐ fit two different absorption models to data
☐ applied the *Akaike* and *Schwarz criteria*
☐ studied the flip-flop situation

The next step in the characterization of kinetics would be to give multiple doses (at several dose levels) via the intended route of administration.

Solution I – Oral model without lag-time

```
TITLE 1
ORAL ADM ONE-COMPARTMENT WITHOUT LAGTIME
MODEL 3
NVARIABLES 2
NPOINTS 500
XNUMBER 1
YNUMBER 2
NCONSTANTS 3
CONSTANTS 1,100,0
METHOD 2  'Gauss-Newton (Levenberg and Hartley)
ITERATIONS 50
INITIAL 30,.1,.05
```

```
LOWER BOUNDS 0,0,0
UPPER BOUNDS 90,10,.3
MISSING 'Missing'
NOBSERVATIONS 13
DATA 'WINNLIN.DAT'
BEGIN

***    WARNING    ***
VARIANCE  - COVARIANCE MATRIX IS NOT OF FULL RANK OR
IS ILL-CONDITIONED.  PARAMETER ESTIMATES AND THEIR
ASSOCIATED STANDARD ERRORS SHOULD BE INTERPRETED
WITH CAUTION.
```

PARAMETER	ESTIMATE	STANDARD ERROR	CV%	UNIVARIATE C.I.	
Volume/F	21.016885	6383.966353	30375.42	-14030.051983	14072.085753
K01	.013192	4.005115	30360.54	-8.802041	8.828424
K10	.013169	4.001516	30386.03	-8.794143	8.820481

```
*** CORRELATION MATRIX OF THE ESTIMATES ***
PARAMETER   Volume/F  K01        K10
Volume/F     1.00000
K01          1.00000      1.00000
K10         -1.00000     -1.00000     1.00000

 Condition_number=        .4401E+07


   FUNCTION   1
```

X	OBSERVED Y	PREDICTED Y	RESIDUAL	WEIGHT	SE-PRED	STANDARDIZ RESIDUAL
10.00	.0000	.5502	-.5502	1.000	.9481E-01	-1.796
15.00	.2800	.7726	-.4926	1.000	.1215	-1.660
20.00	.5500	.9645	-.4145	1.000	.1384	-1.433
30.00	1.200	1.268	-.6804E-01	1.000	.1519	-.2409
40.00	2.000	1.482	.5181	1.000	.1516	1.833
60.00	1.950	1.708	.2422	1.000	.1500	.8544
90.00	1.850	1.725	.1249	1.000	.1757	.4657
120.0	1.600	1.549	.5112E-01	1.000	.1889	.1972
180.0	.8600	1.054	-.1936	1.000	.1572	-.6925
210.0	.7800	.8277	-.4772E-01	1.000	.1503	-.1684
240.0	.6000	.6370	-.3701E-01	1.000	.1575	-.1325
300.0	.2100	.3611	-.1511	1.000	.1750	-.5623
360.0	.1800	.1965	-.1649E-01	1.000	.1667	-.0601

```
CORRECTED SUM OF SQUARED OBSERVATIONS =  6.20243
WEIGHTED CORRECTED SUM OF SQUARED OBSERVATIONS =  6.20243
SUM OF SQUARED RESIDUALS =          1.13124
SUM OF WEIGHTED SQUARED RESIDUALS =  1.13124
S =  .320687     WITH    11 DEGREES OF FREEDOM
CORRELATION (OBSERVED,PREDICTED) =  .9409

AIC criteria =        7.60314
SC  criteria =        9.29798

AUC (0 to last time) computed by trapezoidal rule = 330.875
```

PARAMETER	ESTIMATE	STANDARD	CV%

		ERROR	
AUC	361.310819	56.928520	15.76
K01-HL	52.543616	15936.589072	30330.21
K10-HL	52.635018	15977.712577	30355.67
CL/F	.276770	.043652	15.77
Tmax	75.870308	7.987822	10.53
Cmax	1.751921	.161865	9.24

Solution II – Oral model with lag-time

```
TITLE 1
ORAL ADM ONE-COMPARTMENT WITH LAGTIME
MODEL 4
NVARIABLES 2
NPOINTS 100
XNUMBER 1
YNUMBER 2
NCONSTANTS 3
CONSTANTS 1,100,0
METHOD 2   'Gauss-Newton (Levenberg and Hartley)
ITERATIONS 50
INITIAL 30,.1,.05,20
LOWER BOUNDS 0,0,0,0
UPPER BOUNDS 90,10,.3,50
MISSING 'Missing'
NOBSERVATIONS 13
DATA 'WINNLIN.DAT'
BEGIN
```

PARAMETER	ESTIMATE	STANDARD ERROR	CV%	UNIVARIATE C.I.	
Volume/F	32.063273	3.701984	11.55	23.688733	40.437813
K01	.042862	.011030	25.73	.017910	.067813
K10	.008794	.001365	15.53	.005705	.011883
Tlag	15.584100	1.783114	11.44	11.550382	19.617819

```
*** CORRELATION MATRIX OF THE ESTIMATES ***
```

PARAMETER	Volume/F	K01	K10	Tlag
Volume/F	1.00000			
K01	.86647	1.00000		
K10	-.92428	-.79660	1.00000	
Tlag	.41003	.67894	-.35007	1.00000

```
Condition_number=      7306.

   FUNCTION   1
```

X	OBSERVED Y	PREDICTED Y	RESIDUAL	WEIGHT	SE-PRED	STANDARDIZED RESIDUAL
10.00	.0000	.0000	.0000	1.000		
15.00	.2800	.0000	.2800	1.000		
20.00	.5500	.5272	.2282E-01	1.000	.1392	.5527
30.00	1.200	1.341	-.1414	1.000	.9242E-01	-1.263
40.00	2.000	1.788	.2122	1.000	.9986E-01	2.014
60.00	1.950	2.070	-.1204	1.000	.8691E-01	-1.036
90.00	1.850	1.878	-.2782E-01	1.000	.9583E-01	-.2551
120.0	1.600	1.522	.7815E-01	1.000	.8717E-01	.6732

180.0	.8600	.9208	-.6083E-01	1.000	.6841E-01	-.4751
210.0	.7800	.7090	.7102E-01	1.000	.7047E-01	.5595
240.0	.6000	.5450	.5496E-01	1.000	.7180E-01	.4356
300.0	.2100	.3217	-.1117	1.000	.6614E-01	-.8644
360.0	.1800	.1898	-.9817E-02	1.000	.5389E-01	-.0728

```
CORRECTED SUM OF SQUARED OBSERVATIONS =  6.20243
WEIGHTED CORRECTED SUM OF SQUARED OBSERVATIONS =  6.20243
SUM OF SQUARED RESIDUALS =             .189679
SUM OF WEIGHTED SQUARED RESIDUALS =   .189679
S =  .145174     WITH      9 DEGREES OF FREEDOM
CORRELATION (OBSERVED,PREDICTED) =  .9856
```

```
AIC criteria =          -13.61146
SC  criteria =          -11.35166
```

```
AUC (0 to last time) computed by trapezoidal rule =        330.875
```

PARAMETER	ESTIMATE	STANDARD ERROR	CV%
AUC	354.657234	23.231990	6.55
K01-HL	16.171730	4.157458	25.71
K10-HL	78.821036	12.226323	15.51
CL/F	.281962	.018489	6.56
Tmax	62.077235	4.433202	7.14
Cmax	2.072180	.086728	4.19

Solution III – Flip-flop situation

```
TITLE 1
ORAL ADM ONE-COMPARTMENT WITH LAGTIME
MODEL 4
NOPAGE BREAKS
NVARIABLES 2
NPOINTS 100
XNUMBER 1
YNUMBER 2
NCONSTANTS 3
CONSTANTS 1,100,0
METHOD 2  'Gauss-Newton (Levenberg and Hartley)
ITERATIONS 50
INITIAL 10,.01,.1,20
LOWER BOUNDS 0,0,0,0
UPPER BOUNDS 90,10,.3,50
MISSING 'Missing'
NOBSERVATIONS 12
DATA 'WINNLIN.DAT'
BEGIN
```

PARAMETER	ESTIMATE	STANDARD ERROR	CV%	UNIVARIATE C.I.	
Volume/F	6.541378	1.524345	23.30	3.026199	10.056557
K01	.008780	.001102	12.55	.006239	.011321
K10	.043079	.008982	20.85	.022366	.063791
Tlag	15.630340	1.437563	9.20	12.315281	18.945398

```
*** CORRELATION MATRIX OF THE ESTIMATES ***
PARAMETER  Volume/F  K01        K10        Tlag
Volume/F  1.00000
K01        .88088   1.00000
K10       -.97700   -.79520   1.00000
Tlag      -.63059   -.34793    .67762    1.00000

 Condition_number=      5415.

   FUNCTION   1
```

X	OBSERVED Y	PREDICTED Y	RESIDUAL	WEIGHT	SE-PRED	STANDARDI RESIDUAL
10.00	.0000	.0000	.0000	1.000		
20.00	.5500	.5242	.2581E-01	1.000	.1131	.7728
30.00	1.200	1.342	-.1423	1.000	.7519E-01	-1.566
40.00	2.000	1.790	.2101	1.000	.8116E-01	2.455
60.00	1.950	2.072	-.1221	1.000	.7060E-01	-1.292
90.00	1.850	1.878	-.2800E-01	1.000	.7788E-01	-.3162
120.0	1.600	1.522	.7841E-01	1.000	.7070E-01	.8307
180.0	.8600	.9209	-.6095E-01	1.000	.5555E-01	-.5859
210.0	.7800	.7093	.7069E-01	1.000	.5724E-01	.6855
240.0	.6000	.5455	.5450E-01	1.000	.5830E-01	.5316
300.0	.2100	.3222	-.1122	1.000	.5368E-01	-1.069
360.0	.1800	.1903	-.1028E-01	1.000	.4374E-01	-.0938

```
CORRECTED SUM OF SQUARED OBSERVATIONS =  5.74797
WEIGHTED CORRECTED SUM OF SQUARED OBSERVATIONS =  5.74797
SUM OF SQUARED RESIDUALS =          .111278
SUM OF WEIGHTED SQUARED RESIDUALS =   .111278
S =  .117940     WITH     8 DEGREES OF FREEDOM
CORRELATION (OBSERVED,PREDICTED) =  .9903

AIC criteria =       -18.34869
SC  criteria =       -16.40907
```

PARAMETER	ESTIMATE	STANDARD ERROR	CV%
AUC	354.868588	18.882115	5.32
K01-HL	78.943659	9.897071	12.54
K10-HL	16.090230	3.351420	20.83
CL/F	.281794	.015009	5.33
Tmax	62.003298	3.600099	5.81
Cmax	2.073697	.070462	3.40

PK3 – One-compartment first and zero-order input

Objectives

◆ **To analyze absorption kinetics**

◆ **To fit first- and zero-order absorption models**

◆ **To apply a one-compartment model in terms of K_a, K, t_{lag} and V/F**

◆ **To specify the model as an integrated solution**

◆ **To discuss experimental design**

Problem specification

This workshop problem highlights how to obtain *initial parameter estimates*, assess the *goodness-of-fit* and *interpret output*. For further reading on the one-compartment system with extravascular input, see also Section 3.2.4. The objectives of this exercise are to fit two types of absorption models to a dataset obtained after extravascular dosing with the compound. One is a first-order model including a lag-time (see also PK2), the other is a zero-order input model. After having fit both models to the data, we use the function and residual plots and the parameter precision and correlation between parameters, to select the model of choice.

A volunteer was given 20 mg orally of a highly polar drug. We will analyze the dataset with a one-compartment model including either first- or zero-order absorption. Data are shown in Figure 3.1 and in the program output.

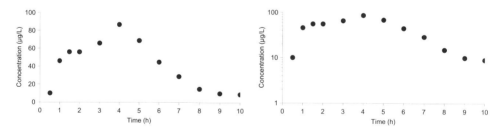

Figure 3.1 *Observed concentration-time data following an oral dose of 20 mg of drug X. The left hand figure uses a linear scale, and the right hand figure a semi-logarithmic scale. Note the delayed absorption with a maximum plasma concentration at 4 h.*

The absorption process from the gastrointestinal tract is complex and involves several processes including disintegration of the tablet, dissolution of the drug into the gastric juices, gastric emptying, diffusion across the gutwall, and so on. In general, absorption processes are assumed to occur by means of a first-order process, although there are exceptions. Under certain conditions it has been found that absorption of compounds is better described by zero-order kinetics. See also Section 3.3. The model of zero-order input is

$$C = \frac{R_0}{V \cdot K \cdot T_{abs}} \left[1 - e^{-K \cdot f(t)}\right] e^{-K \cdot (t - f(t))} \tag{3:1}$$

where

$$f(t) = \begin{cases} t & \text{if } t < T_{abs} \\ T_{abs} & \text{if } t \geq T_{abs} \end{cases}$$

Equation 3:1 is equivalent to a one-compartment continuous infusion model where the duration of the infusion T_{abs} is estimated as a parameter (Gibaldi and Perrier [1982]). The model of first-order input with lag-time is

$$C = \frac{K_a F D_{po}}{V(K_a - K)} \left[e^{-K \cdot (t - t_{lag})} - e^{-K_a \cdot (t - t_{lag})}\right] \tag{3:2}$$

Having previously built a simple first-order absorption model and parameterized it (see exercise PK2), you will now learn how to

☐ fit an alternative absorption model to your data
☐ see how that affects your output and how to interpret the output, and
☐ compare the results with that from a first-order absorption model

As you can see in Equation 3:1, the zero-order input model resembles the ordinary constant rate infusion model with wash-out (see section 3.2.2). However, clearance $V \cdot K/F$ in this example, is the oral clearance Cl_0 and can also be obtained from $Dose/AUC$ ($20 \cdot 10^3/320$). The dose is 20000 µg (or 20 mg) and AUC is about 320 µg·h/L, which is obtained by the linear trapezoidal method from 0-10 h. This gives

$$Cl_0 \approx \frac{20 \cdot 10^3}{320} = 67 \, L/h \tag{3:3}$$

Note that Cl_0 is Cl/F, which explains why Cl_0 can reach a very high value if e.g., F is low. Generally a clearance Cl value obtained from intravenous data and exceeding 90 L/h (hepatic blood flow) is considered to be high. In this example Cl_0, is about 67 L/h.

Initial parameter estimates

K_a	= 1.0 (h^{-1})	
K	= 0.5 (h^{-1})	
t_{lag}	= 0.5 (h)	
V/F	= 300 (L)	> $Dose/C_{max} \approx 200$
V/F	= 300 (L)	> $Dose/C_{max} \approx 200$
T_{abs}	= 1 (h)	Length of input time (or ≈ 4 from graph)
K	= 0.5 (h^{-1})	

Interpretation of results and conclusions

In this example we have fitted user-specified first- and zero-order absorption models to the data (Figure 3.2). An alternative would have been to use a first-order library model instead of the user-specified model.

We found that the zero-order model fitted the data best. However, that is not the same as saying that the zero-order absorption model is the correct absorption model for this drug. In order to draw that conclusion, we would need data from several doses (e.g., 200, 400, 800 . . .) given repeatedly. In addition, we would use intravenous dosing to see whether or not we have nonlinear kinetics in the disposition of drug. Our personal experience of modeling kinetic data is that when a drug with Michaelis-Menten-type hepatic elimination by the oral route results in a zero-order like pattern, the concentration declines as a straight line when plotted in a Cartesian diagram, while a semi-logarithmic plot generates a convex decline pattern. PK20 demonstrates such a system. The reason for this is that high portal concentration of drug entering the liver saturates the hepatic enzymes during the first-pass.

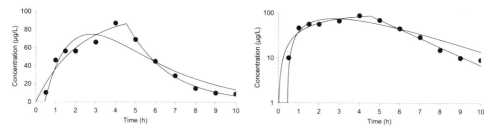

Figure 3.2 *Observed (filled circles) and predicted (solid lines) concentrations of the first- and zero-order absorption models following oral dosing of 20 mg of drug X. The zero-order model predicts the time course according to the discontinuous line at approximately 4 h. Note that the first-order model misses the peak concentration and displays systematic deviation between observed and predicted data.*

In order to make the final judgement about which model can best be used for multiple dose predictions we would still recommend that data be obtained from a multiple dose experiment. A dose range-finding study would also elucidate the absorption process better, i.e., whether it is a first- or zero-order input. Zero-order input is often energy-consuming and therefore saturable at higher concentrations. A true zero-order input system (i.e., of physiological or anatomical origin) is rare.

The estimated *mean standard deviation* (square root of *WRSS/df*) is lower in the zero-order input model compared to the first-order input. The superiority of the zero-order model is also seen for *AIC* (e.g. 76.2 *versus* 85.2). There is generally less correlation between the parameters in the zero-order model compared to the first-order absorption model, and the residuals indicate a more random scatter in the zero-order model fit. These criteria also favors the use of the zero-order input model.

You have
- [ ] analyzed a typical one-compartment (mono-exponential) system
- [ ] learned to derive the initial estimates
- [ ] explored first- and zero-order input models
- [ ] characterized the pharmacokinetics after single dose administration in four subjects
- [ ] selected a zero-order absorption model by means of residuals, predicted mean variance, *AIC* and the correlation matrix

The next step in the characterization of kinetics would be to give multiple doses via the intended route of administration.

For further information about absorption, single and multiple dose modeling, and discrimination between rival models, we recommend Endrenyi [1981] and Purves [1993]. Read also Chapters 3 and 5.

Solution I - First-order absorption

```
TITLE 1
SUBJECT C: 1st-order absorption
Model
 COMM
NCON 1
NFUN 1
NPARM 4
PNAMES 'ka', 'ke', 'Tlag', 'FV'
END
     1: TEMP
     2: DPO=CON(1)
     3: KA=P(1)
     4: KE=P(2)
     5: TLAG=P(3)
     6: FV=P(4)
     7: T=X
     8: END
     9: FUNC1
    10: IF (T LE TLAG) THEN
    11: F=0.0
    12: WT=0.0
    13: ELSE
    14: COEFF=(KA/FV)*DPO/(KA-KE)
    15: F = COEFF*(EXP(-KE*(T-TLAG))-EXP(-KA*(T-TLAG)))
    16: ENDIF
    17: END
    18: EOM
NVARIABLES 2
NPOINTS 500
XNUMBER 1
YNUMBER 2
CONSTANTS 20000
METHOD 2  'Gauss-Newton (Levenberg and Hartley)
ITERATIONS 50
INITIAL 1,.5,.5,300
LOWER BOUNDS .1,.01,.01,.001
UPPER BOUNDS 10,1.5,2,1000
MISSING 'Missing'
NOBSERVATIONS 12
DATA 'WINNLIN.DAT'
```

```
BEGIN

  PARAMETER  ESTIMATE      STANDARD        CV%           UNIVARIATE C.I.
                           ERROR
  KA          .432125     37.777122      8742.18     -88.897291      89.761541

  KE          .428275     37.331998      8716.83     -87.848581      88.705130

  TLAG        .440109       .434080        98.63       -.586334       1.466553

  FV        99.063184    8624.145262     8705.70   -20293.963566   20492.089933


  *** CORRELATION MATRIX OF THE ESTIMATES ***
  PARAMETER  KA          KE          TLAG        FV
  KA         1.00000
  KE         -.99999     1.00000
  TLAG        .73691     -.73528     1.00000
  FV          .99999     -.99999      .73564     1.00000

Condition_number=        .2081E+06

FUNCTION    1

     X          OBSERVED    PREDICTED    RESIDUAL    WEIGHT    SE-PRED    STANDARDIZ
                Y           Y                                             RESIDUAL
     .5000      10.00        5.092        4.908       .0000
    1.000       46.00       38.39         7.609      1.000     11.90       1.801
    1.500       56.00       58.61        -2.609      1.000      7.695      -.2606
    2.000       56.00       69.56       -13.56       1.000      8.516     -1.454
    3.000       66.00       74.25        -8.247      1.000      7.354      -.8032
    4.000       87.00       67.15        19.85       1.000      7.106      1.901
    5.000       69.00       55.94        13.06       1.000      7.050      1.246
    6.000       45.00       44.36          .6362     1.000      6.220       .5789E-01
    7.000       29.00       34.04        -5.043      1.000      5.620      -.4459
    8.000       15.00       25.52       -10.52       1.000      6.003      -.9465
    9.000       10.00       18.79        -8.791      1.000      6.869      -.8295
   10.00        9.000       13.65        -4.649      1.000      7.540      -.4589

  CORRECTED SUM OF SQUARED OBSERVATIONS =   7779.00
  WEIGHTED CORRECTED SUM OF SQUARED OBSERVATIONS =   6696.55
  SUM OF SQUARED RESIDUALS =            1116.42
  SUM OF WEIGHTED SQUARED RESIDUALS =  1116.42
  S =  12.6289      WITH       7 DEGREES OF FREEDOM
  CORRELATION (OBSERVED,PREDICTED) =   .7762

 AIC criteria =         85.19673
 SC  criteria =         86.78831
```

Solution II – Zero–order absorption

```
TITLE 1
 Zero-order input
Model
 COMM
NCON 1
NPARM 3
PNAMES 'VF','Tabs','Ke'
NFUN 1
END
    1: TEMP
    2: T = X
```

```
 3: DOSE = CON(1)
 4: VF = P(1)
 5: TABS = P(2)
 6: KE = P(3)
 7: FINF = DOSE/TABS
 8: END
 9: FUNC1
10: F = 0.0
11: IF T <= TABS THEN
12: F =(FINF/(KE*VF))*(1. - DEXP(-KE*T))
13: ENDIF
14: IF T > TABS THEN
15: F=(FINF/(KE*VF))*(1.-DEXP(-KE*TABS))*DEXP(-KE*(T-TABS))
16: ENDIF
17: END
18: EOM
NVARIABLES 2
NPOINTS 500
XNUMBER 1
YNUMBER 2
CONSTANTS 20000
METHOD 2   'Gauss-Newton (Levenberg and Hartley)
ITERATIONS 50
INITIAL 300,1,.5
LOWER BOUNDS 1,.1,.01
UPPER BOUNDS 1000,5,2
MISSING 'Missing'
NOBSERVATIONS 12
DATA 'WINNLIN.DAT'
BEGIN
```

PARAMETER	ESTIMATE	STANDARD ERROR	CV%	UNIVARIATE C.I.	
VF	96.262443	11.030301	11.46	71.309956	121.214930
TABS	4.540281	.240700	5.30	3.995775	5.084787
KE	.465660	.063139	13.56	.322828	.608491

```
*** CORRELATION MATRIX OF THE ESTIMATES ***
PARAMETER  VF         TABS       KE
VF         1.00000
TABS       -.78041    1.00000
KE         -.94145    .76471     1.00000

Condition_number=      523.7

   FUNCTION   1
```

X	OBSERVED Y	PREDICTED Y	RESIDUAL	WEIGHT	SE-PRED	STANDARDIZED RESIDUAL
.5000	10.00	20.41	-10.41	1.000	1.390	-1.719
1.000	46.00	36.58	9.416	1.000	2.126	1.613
1.500	56.00	49.40	6.603	1.000	2.502	1.161
2.000	56.00	59.55	-3.548	1.000	2.736	-.6360
3.000	66.00	73.96	-7.964	1.000	3.272	-1.508
4.000	87.00	83.01	3.987	1.000	4.066	.8485
5.000	69.00	69.75	-.7550	1.000	5.578	-.2756
6.000	45.00	43.79	1.213	1.000	3.211	.2281
7.000	29.00	27.49	1.514	1.000	3.050	.2797

```
 8.000      15.00      17.25      -2.253      1.000      2.845      -.4079
 9.000      10.00      10.83      -.8303      1.000      2.421      -.1451
 10.00      9.000      6.798       2.202      1.000      1.931       .3727
```

```
CORRECTED SUM OF SQUARED OBSERVATIONS =  7779.00
WEIGHTED CORRECTED SUM OF SQUARED OBSERVATIONS =  7779.00
SUM OF SQUARED RESIDUALS =          347.532
SUM OF WEIGHTED SQUARED RESIDUALS =  347.532
S =   6.21407     WITH      9 DEGREES OF FREEDOM
CORRELATION (OBSERVED,PREDICTED) =  .9774
```

```
AIC criteria =         76.21028
SC  criteria =         77.66500
```

PK4 – One-compartment repeated oral dosing I

Objectives

◆ **To simultaneously fit single and multiple (steady state) dose data**

◆ **To analyze four different absorption/disposition models**

◆ **To discuss ways of discrimination between competing models**

◆ **To discuss various aspects of correlation, accuracy, and precision**

◆ **To discuss various aspects of experimental design**

Problem specification

This exercise highlights how to obtain *initial parameter estimates*, assess the *goodness-of-fit* and *interpret output*. The primary aims of this exercise are to

☐ simultaneously fit single and multiple dose data
☐ analyze different absorption models with and without lag-time
☐ discuss ways of discriminating between competing models

For further reading on the one-compartment system with extravascular input, see also Section 3.2.4.

An oral dose of 352.3 µg of drug A was given at each of 0, 24, 48, 72, 96, 120, 144, 168, 192 and 216 h. We have obtained the plasma concentration-time profile after the first dose and within a 24 h dosing interval at steady state. A typical plasma concentration-time plot after the first oral administration is shown in Figure 4.1. Experimental data are shown in the program output. A schematic representation of the first-order input-output model is shown in Figure 4.2.

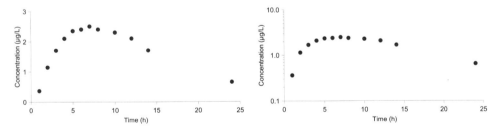

Figure 4.1 *Cartesian plot (left) and semi-logarithmic plot (right) of plasma concentration-time data following an oral dose of 352.3 µg of drug A. Data from the first dosing interval are plotted. Note the lag-time and the late C_{max}.*

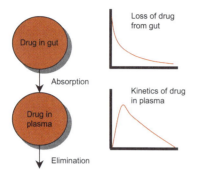

Loss of drug
from gut

Figure 4.2 *Schematic illustration of the one-compartment model with absorption from the gut compartment.*

Kinetics of drug
in plasma

We will fit a one-compartment model (with and without a lag-time) from the model library. We assume complete absorption (100% bioavailability). The differential equation for a drug that obeys one-compartment disposition characteristics, administered by the oral route, takes the following form

$$\frac{dC}{dt} = \frac{K_a FD_{po} e^{-K_a \cdot t}}{V} - K \cdot C \tag{4:1}$$

K_a denotes the first-order absorption rate constant and F denotes the bioavailability of the drug when administered orally. The first term $K_a FD_{po} e^{-K_a t}$ is equivalent to the input rate of the drug into the plasma compartment from the gut compartment. An alternative representation of the model is to specify the gut compartment with a differential equation according to

$$\frac{dA_g}{dt} = -K_a \cdot A_g \tag{4:2}$$

This is the rate of elimination of the drug from the gut. $A_g(0) = Dose$. We assume here that the drug is taken up by the gut wall with a first-order rate constant equal to K_a and that the bioavailability F governs the rate of the drug entering plasma. Solution of the differential equation yields

$$C = \frac{K_a FD_{po}}{V(K_a - K)} \left[e^{-K \cdot t} - e^{-K_a \cdot t} \right] \tag{4:3}$$

The slope of the linear portion of the terminal phase of the curve when plotted on a semi-loge scale is equal to $-K$, assuming that $K_a \gg K$. The intercept of the concentration axis is equal to C^0. The slope of the linear residual portion of the initial phase of the curve is equal to $-K_a$. See also section 5.4 on obtaining initial estimates in Chapter 5. K_a and K may be determined by the method of residuals. However, unlike the situation found after intravenous administration, C^0 extrapolated from the terminal portion of the curve, does not correspond to the term $D/V \cdot C^0$. Rather, it is a complex function of the amount of drug eliminated during the absorption phase, the time required for absorption, the dose, and the volume of distribution (Purves [1993]).

The time course of concentration in plasma of certain compounds suggests a lag-time between oral administration and the apparent onset of absorption. This is because of the time taken for dissolution of the drug particles and the absorption of the dissolved drug into the systemic circulation. These processes must occur before the concentration of drug becomes measurable in plasma. When a lag-time exists, the relationship shown in Equation 4:3 becomes

$$C = \frac{K_a F D_{po}}{V(K_a - K)} \left[e^{-K \cdot (t - t_{lag})} - e^{-K_a \cdot (t - t_{lag})} \right]$$ (4:4)

A difficulty sometimes encountered in least-squares fitting of a one-compartment model with first-order absorption is that estimated values of the rate constants of absorption and elimination are almost identical (K_a and K, respectively). This anomaly is explained by the existence of a class of datasets for which least-squares estimates of the rate constants are complex quantities. Such datasets may arise either from an unfortunate combination of random (e.g., assay) errors in the concentration values if K_a and K are sufficiently similar in magnitude, or from delayed absorption. The usual recommendation in such cases, given without discussion of the cause, has been to use the special equation

$$C = \frac{K' F D_{po}}{V} \cdot t \cdot e^{-K' \cdot t}$$ (4:5)

This is easily derived from Equation 4:1 as the limiting form with $K_a = K = K'$. We will demonstrate the applicability of this model (Equation 4:5) in this example. A lag-time may also be incorporated into Equation 4:5 and subsequently estimated as an additional parameter.

$$C = \frac{K' F D_{po}}{V} \cdot (t - t_{lag}) \cdot e^{-K' \cdot (t - t_{lag})}$$ (4:6)

When an oral multiple dosing regimen is initiated, plasma concentrations will increase, reach a maximum and then decrease. Generally, a second dose will be administered before the first dose is completely eliminated. Consequently, plasma concentrations resulting from the second dose are higher than those from the first dose. This accumulation continues to occur until steady state is reached. The extent to which a compound accumulates relative to the first dose can be quantified by an accumulation factor R, which is dependent on the relative magnitude of the dosing interval τ and the half-life of the drug (Gibaldi and Perrier [1982])

$$R = \frac{1}{1 - e^{-\varepsilon \cdot \tau}}$$ (4:7)

where ε is $1/t_{1/2}$. The smaller the ratio $\tau/t_{1/2}$ the greater will be the extent of accumulation.

When $\tau = t_{1/2}$ the average concentration at steady state will be about twice the average concentration after the first dose (Gibaldi and Perrier [1982]). It is also important to note that the time to 90% of steady state is only determined by the half-life, whereas the dosing rate and clearance determine the average steady state concentration C_{ss} as follows

$$C_{ss} = \frac{FD_{po}}{\tau} \cdot \frac{1}{Cl} = \frac{AUC}{\tau} \qquad (4:8)$$

AUC is either $AUC_{0-\infty}$ after a single dose or $AUC_{0-\tau}$ within a dosing interval.

We will fit Models 3 (Equation 4:3), 4 (Equation 4:4), 5 (Equation 4:5) and 6 (Equation 4:6) of the model library in *WinNonlin*. The model parameters will be K_a, K, V/F and t_{lag}. The reason for the ratio V/F is the inability to identify separately F and V. This is an inherent limitation of the model, since neither F nor V appears elsewhere in the model. Without information following an intravenous dose, unique values for Cl, F and V cannot be determined.

Initial parameter estimates

$$\begin{array}{lll} V/F & = 70 \text{ (L)} & < Dose/C_{max} = 352/2.8 \\ K_a & = 0.25 \text{ (h}^{-1}) \\ K & = 0.1 \text{ (h}^{-1}) \\ t_{lag} & = 1 \text{ (h)} \end{array}$$

Figure 4.3 demonstrates a graphical and a numerical method of estimating the lag-time parameter. K_a and K are obtained from the residual and terminal slopes, respectively.

$$t_{lag} = \frac{\ln A_0 - \ln B_0}{K_a - K_e} = \frac{\ln 11 - \ln 7}{0.7 - 0.09} \approx 0.7h$$

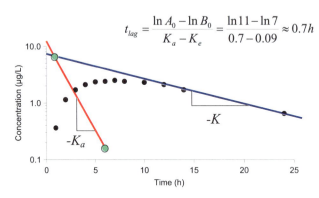

Figure 4.3 Schematic illustration of the estimation of lag-time. A_0 and B_0 are the back-extrapolated intercepts of the blue and red lines, respectively, on the concentration axis. The slopes of the blue and red lines are $-K$ (also denoted K_e) and $-K_a$.

Interpretation of results and conclusions

Introducing errors while making measurements is almost unavoidable. Therefore, it is essential to ensure that they are relatively small in magnitude and random in nature. If errors are large, parameter estimates will be imprecise, though not necessarily biased. If the error is non-random, the parameter estimates can be inaccurate and biased. The uncertainties associated with parameter estimates can be computed and expressed as either a standard deviation, coefficient of variation, and/or confidence interval.

If the precision of the parameters is poorer than expected, the parameters are said to be poorly determined. This situation is often the result of a high correlation between two or more parameters. A high correlation often arises when the number of parameters is high relative to the number of observations, or when two or more parameters (e.g., the rate constants) are very close in magnitude (e.g., the absorption rate constant approaches the elimination rate constant), or for Michaelis-Menten type systems. A poor experi-mental design (measurement times) can also contribute to poor precision.

In this particular example, high correlation is observed between several of the parameters even though their precision is high. The high correlation may be due to the fact that the estimated absorption and elimination rate constants are close: $K_a = 0.198$ h^{-1} and $K = 0.123$ h^{-1}, and their confidence intervals overlap. A lag-time of 0.7 h is also consistent with the fact that the drug was administered as a controlled-release formulation. Figure 4.4 shows the concentration-time profile of the observed and predicted data after the first oral dose.

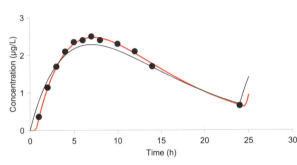

Figure 4.4 Observed and predicted concentration-time data after the first dose. The red line is identical to the fits of the two lag-time models (Equations 4:4 and 4:6) and the black line is identical to the fits of the two models without lag-time (Equation 4:3 and 4:5).

In order to help you interpret the results, a summary table is included of the pharmacokinetic parameters and goodness-of-fit criteria (Table 4.1). Dr. Arvidsson, AstraZeneca, Sweden, kindly supplied the plasma concentration-time data.

Table 4.1 Model comparisons (mean ± CV%)

	Model No. 3	Model No. 4	Model No. 5	Model No. 6
Equation	4:3	4:4	4:5	4:6
V/F	56.6 ± 3188	64.9 ± 12.2	56.3 ± 3.28	52.3 ± 1.22
K_a	0.142 ± 3182	0.196 ± 15.2	-	-
K	0.143 ± 3187	0.123 ± 12.3	-	-
K'	-	-	0.143 ± 3.99	0.154 ± 1.34
t_{lag}	-	0.697 ± 5.7	-	0.675 ± 5.17
Condition No.	$4 \cdot 10^{+5}$	5240	386	376
WRSS	0.907	0.107	0.903	0.108
+/–	8	17	6	19
AIC	3.46	-50.2	1.35	-51.7
No. iter.	7	3	3	4

Model 3 (Equation 4:3), without a lag-time, was initially fit to data but was not considered to be an acceptable choice. K_a and K are almost identical, which suggests that either a lag-time and/or another structure of the model are needed. Model 4 (Equation

4:4) was superior to Model 3. Model 5 (Equation 4:5) showed obvious trends in the residuals and was therefore discarded. Model 6 (Equation 4:6) was superior to Models 4 and 5, having higher parameter precision, improved condition number and increased randomness in the residuals. Note that there is very little accumulation between the first and n^{th} dose (see experimental data in program output). In conclusion, the presented example and the summary table above illustrate an approach that may be used to discriminate between models. We will therefore select Model 6 as our choice in this example.

You have
- [ ] analyzed a typical one-compartment (mono-exponential) system with first-order absorption
- [ ] learned to derive the initial estimates
- [ ] established a scheme for model discrimination (Table 4.1)
- [ ] characterized the pharmacokinetics after multiple dose administration in one subject

The next step in the characterization of kinetics would be to give an intravenous dose to study the disposition kinetics and the absolute bioavailability.

Solution I - Model 3 (Equation 4:3)

```
TITLE 1
Repeated administration of Super-High (SR)
MODEL 3
NVARIABLES 2
NPOINTS 1000
XNUMBER 1
YNUMBER 2
NCONSTANTS 21
CONSTANTS 10,352.3,0,352.3,24,352.3, 48,352.3,72,352.3,96,352.3,120, &
         352.3,144,352.3,168,352.3,192,352.3,216
METHOD 2  'Gauss-Newton (Levenberg and Hartley)
REWEIGHT -1
ITERATIONS 50
INITIAL 70,.25,.1
LOWER BOUNDS 0,0,0
UPPER BOUNDS 100,2,2
MISSING 'Missing'
NOBSERVATIONS 26
DATA 'WINNLIN.DAT'
BEGIN
```

PARAMETER	ESTIMATE	STANDARD ERROR	CV%	UNIVARIATE C.I.	
Volume/F	56.608984	1804.744105	3188.09	-3676.758612	3789.976580
K01	.142387	4.531092	3182.23	-9.230814	9.515589
K10	.143118	4.562359	3187.83	-9.294765	9.581001

```
*** CORRELATION MATRIX OF THE ESTIMATES ***
 PARAMETER  Volume/F   K01        K10
 Volume/F   1.00000
 K01         .999998  1.00000
 K10        -.999999  -.99999   1.00000
```

Condition_number=   .3588E+06
    FUNCTION   1

X	OBSERVED Y	PREDICTED Y	RESIDUAL	WEIGHT	SE-PRED	STANDARDIZED RESIDUAL
1.000	.3622	.7682	-.4060	1.302	.5716E-01	-2.471
2.000	1.140	1.332	-.1921	.7508	.8347E-01	-.9005
3.000	1.700	1.732	-.3233E-01	.5774	.9186E-01	-.1322
4.000	2.100	2.002	.9750E-01	.4995	.9190E-01	.3674
5.000	2.350	2.170	.1799	.4610	.9030E-01	.6469
6.000	2.400	2.258	.1423	.4431	.9078E-01	.5010
7.000	2.500	2.284	.2164	.4381	.9412E-01	.7601
8.000	2.400	2.263	.1374	.4422	.9918E-01	.4879
10.00	2.300	2.126	.1742	.4707	.1084	.6493
12.00	2.100	1.917	.1826	.5220	.1111	.7265
14.00	1.700	1.681	.1859E-01	.5953	.1063	.0793
24.00	.6500	.6915	-.4150E-01	1.449	.5954E-01	-.2699
216.0	.7400	.7388	.1235E-02	1.356	.7723E-01	.0081
216.5	.7000	1.114	-.4143	.8985	.7697E-01	-2.127
217.0	.7400	1.435	-.6946	.6977	.8491E-01	-3.130
218.0	1.800	1.932	-.1321	.5179	.9793E-01	-.5125
219.0	2.100	2.272	-.1720	.4404	.9940E-01	-.6096
220.0	2.600	2.487	.1128	.4023	.9307E-01	.3777
221.0	2.600	2.605	-.4910E-02	.3841	.8436E-01	-.0158
222.0	2.800	2.647	.1527	.3780	.7756E-01	.4872
223.0	2.700	2.632	.6767E-01	.3802	.7493E-01	.2162
224.0	2.800	2.574	.2255	.3887	.7633E-01	.7297
226.0	2.400	2.375	.2547E-01	.4215	.8429E-01	.0867
228.0	2.300	2.115	.1849	.4733	.9030E-01	.6748
230.0	1.900	1.838	.6195E-01	.5447	.9133E-01	.2449
240.0	.8000	.7388	.6123E-01	1.356	.7723E-01	.4030

CORRECTED SUM OF SQUARED OBSERVATIONS = 14.9088
WEIGHTED CORRECTED SUM OF SQUARED OBSERVATIONS = 11.1832
SUM OF SQUARED RESIDUALS =   1.23136
SUM OF WEIGHTED SQUARED RESIDUALS = .906757
S = .198555   WITH   23 DEGREES OF FREEDOM
CORRELATION (OBSERVED,PREDICTED) = .9696

AIC criteria =   3.45509
SC criteria =   7.22938

SUMMARY OF ESTIMATED SECONDARY PARAMETERS

PARAMETER	ESTIMATE	STANDARD ERROR	CV%
AUC	43.484329	1.589315	3.65
K01-HL	4.868037	154.757269	3179.05
K10-HL	4.843184	154.238157	3184.64
CL/F	8.101769	.296409	3.66
Tmax	7.005134	.481143	6.87
Cmax	2.283602	.118823	5.20

Solution II - Model 4 (Equation 4:4)

TITLE 1
Repeated administration of Super-High (SR)
MODEL 4
NOPAGE BREAKS
NVARIABLES 2
NPOINTS 1000

```
XNUMBER 1
YNUMBER 2
NCONSTANTS 21
CONSTANTS 10,352.3,0,352.3,24,352.3,48,352.3,72,352.3,96,352.3,120, &
         352.3,144,352.3,168,352.3,192,352.3,216
METHOD 2  'Gauss-Newton (Levenberg and Hartley)
REWEIGHT -1
ITERATIONS 50
INITIAL 70,.25,.1,1
LOWER BOUNDS 0,0,0,0
UPPER BOUNDS 100,2,2,2
MISSING 'Missing'
NOBSERVATIONS 26
DATA 'WINNLIN.DAT'
BEGIN
```

PARAMETER	ESTIMATE	STANDARD ERROR	CV%	UNIVARIATE C.I.	
Volume/F	64.945842	7.934188	12.22	48.491474	81.40021
K01	.196446	.029808	15.17	.134628	.258264
K10	.123751	.015216	12.30	.092196	.155306
Tlag	.696900	.039728	5.70	.614509	.779291

```
*** CORRELATION MATRIX OF THE ESTIMATES ***
PARAMETER  Volume/F  K01        K10        Tlag
Volume/F   1.00000
K01         .985490   1.00000
K10        -.994961   -.979395   1.00000
Tlag        .491011    .545177   -.493006   1.00000

Condition_number=        5235.

   FUNCTION   1
```

X	OBSERVED Y	PREDICTED Y	RESIDUAL	WEIGHT	SE-PRED	STANDARDIZED RESIDUAL
1.000	.3622	.3077	.5450E-01	3.257	.3297E-01	2.723
2.000	1.140	1.128	.1244E-01	.8872	.2754E-01	.1814
3.000	1.700	1.699	.6228E-03	.5886	.3318E-01	.00737
4.000	2.100	2.079	.2076E-01	.4810	.3484E-01	.2206
5.000	2.350	2.312	.3814E-01	.4326	.3395E-01	.3806
6.000	2.400	2.433	-.3274E-01	.4111	.3315E-01	-.3167
7.000	2.500	2.470	.3000E-01	.4049	.3378E-01	.2883
8.000	2.400	2.446	-.4577E-01	.4089	.3570E-01	-.4451
10.00	2.300	2.278	.2158E-01	.4389	.4005E-01	.2222
12.00	2.100	2.028	.7209E-01	.4931	.4150E-01	.8010
14.00	1.700	1.751	-.5141E-01	.5709	.3940E-01	-.6175
24.00	.6500	.6691	-.1911E-01	1.494	.1941E-01	-.3570
216.0	.7400	.7121	.2792E-01	1.404	.2513E-01	.5260
216.5	.7000	.6745	.2555E-01	1.483	.2507E-01	.4973
217.0	.7400	.9463	-.2063	1.057	.3667E-01	-3.627
218.0	1.800	1.700	.1004	.5885	.3362E-01	1.192
219.0	2.100	2.211	-.1112	.4523	.3866E-01	-1.158
220.0	2.600	2.537	.6332E-01	.3942	.3898E-01	.6102
221.0	2.600	2.720	-.1203	.3676	.3594E-01	-1.104
222.0	2.800	2.797	.2792E-02	.3575	.3216E-01	.0249
223.0	2.700	2.795	-.9494E-01	.3578	.2959E-01	-.8437
224.0	2.800	2.735	.6473E-01	.3656	.2898E-01	.5810

226.0	2.400	2.508	-.1078	.3987	.3128E-01	-1.020
228.0	2.300	2.209	.9077E-01	.4526	.3328E-01	.9267
230.0	1.900	1.894	.5501E-02	.5278	.3295E-01	.0611
240.0	.8000	.7121	.8792E-01	1.404	.2513E-01	1.656

```
CORRECTED SUM OF SQUARED OBSERVATIONS =  14.9088
WEIGHTED CORRECTED SUM OF SQUARED OBSERVATIONS =  13.6177
SUM OF SQUARED RESIDUALS =            .143507
SUM OF WEIGHTED SQUARED RESIDUALS =  .106624
S =  .696173E-01 WITH    22 DEGREES OF FREEDOM
CORRELATION (OBSERVED,PREDICTED) =  .9953


AIC criteria =        -50.19950
SC  criteria =        -45.16712


AUC (0 to last time) computed by trapezoidal rule =        216.532
```

SUMMARY OF ESTIMATED SECONDARY PARAMETERS

PARAMETER	ESTIMATE	STANDARD ERROR	CV%
AUC	43.834137	.539881	1.23
K01-HL	3.528439	.534864	15.16
K10-HL	5.601143	.687998	12.28
CL/F	8.037115	.099088	1.23
Tmax	7.053819	.161291	2.29
Cmax	2.470089	.033773	1.37

Solution III - Model 5 (Equation 4:5)

```
TITLE 1
Repeated administration of Super-High (SR)
MODEL 5
NOPAGE BREAKS
NVARIABLES 2
NPOINTS 100
XNUMBER 1
YNUMBER 2
NCONSTANTS 21
CONSTANTS 10,352.3,0,352.3,24,352.3, 48,352.3,72,352.3,96,352.3,120, &
         352.3,144,352.3,168,352.3,192,352.3,216
METHOD 2  'Gauss-Newton (Levenberg and Hartley)
REWEIGHT -1
ITERATIONS 50
INITIAL 70,.25
LOWER BOUNDS 0,0
UPPER BOUNDS 100,2
MISSING 'Missing'
NOBSERVATIONS 26
DATA 'WINNLIN.DAT'
BEGIN
```

PARAMETER	ESTIMATE	STANDARD ERROR	CV%	UNIVARIATE C.I.	
Volume/F	56.319982	1.849358	3.28	52.503125	60.136839
K	.142629	.005690	3.99	.130885	.154373

```
*** CORRELATION MATRIX OF THE ESTIMATES ***
 PARAMETER  Volume/F  K
 Volume/F     1.00000
 K          -.537691         1.00000
```

```
Condition_number=        385.5

FUNCTION   1
```

X	OBSERVED Y	PREDICTED Y	RESIDUAL	WEIGHT	SE-PRED	STANDARDIZED RESIDUAL
1.000	.3622	.7736	-.4114	1.292	.4547E-01	-2.501
2.000	1.140	1.342	-.2015	.7449	.7220E-01	-.9470
3.000	1.700	1.745	-.4482E-01	.5728	.8554E-01	-.1855
4.000	2.100	2.017	.8281E-01	.4955	.8964E-01	.3179
5.000	2.350	2.186	.1637	.4573	.8776E-01	.5994
6.000	2.400	2.275	.1252	.4395	.8242E-01	.4459
7.000	2.500	2.301	.1988	.4346	.7558E-01	.6991
8.000	2.400	2.280	.1196	.4386	.6878E-01	.4203
10.00	2.300	2.143	.1570	.4668	.5930E-01	.5655
12.00	2.100	1.933	.1666	.5176	.5738E-01	.6325
14.00	1.700	1.696	.4164E-02	.5903	.6007E-01	.0169
24.00	.6500	.6983	-.4829E-01	1.436	.5841E-01	-.3199
216.0	.7400	.7462	-.6165E-02	1.344	.6891E-01	-.0404
216.5	.7000	1.124	-.4242	.8907	.5200E-01	-2.134
217.0	.7400	1.447	-.7067	.6918	.4610E-01	-3.092
218.0	1.800	1.948	-.1477	.5136	.5392E-01	-.5571
219.0	2.100	2.290	-.1901	.4368	.6479E-01	-.6640
220.0	2.600	2.507	.9306E-01	.3990	.7122E-01	.3116
221.0	2.600	2.626	-.2571E-01	.3810	.7375E-01	-.0841
222.0	2.800	2.669	.1314	.3749	.7403E-01	.4266
223.0	2.700	2.654	.4628E-01	.3770	.7352E-01	.1507
224.0	2.800	2.596	.2044	.3855	.7323E-01	.6730
226.0	2.400	2.394	.5501E-02	.4180	.7480E-01	.0189
228.0	2.300	2.133	.1667	.4692	.7846E-01	.6127
230.0	1.900	1.854	.4569E-01	.5400	.8186E-01	.1821
240.0	.8000	.7462	.5384E-01	1.344	.6891E-01	.3531

```
CORRECTED SUM OF SQUARED OBSERVATIONS = 14.9088
WEIGHTED CORRECTED SUM OF SQUARED OBSERVATIONS = 11.0891
SUM OF SQUARED RESIDUALS =         1.21035
SUM OF WEIGHTED SQUARED RESIDUALS = .903165
S = .193989      WITH     24 DEGREES OF FREEDOM
CORRELATION (OBSERVED,PREDICTED) = .9696

AIC criteria =        1.35190
SC  criteria =        3.86809

AUC (0 to last time) computed by trapezoidal rule =        216.532
```

SUMMARY OF ESTIMATED SECONDARY PARAMETERS

PARAMETER	ESTIMATE	STANDARD ERROR	CV%
AUC	43.857314	1.555880	3.55
K-HL	4.859788	.193685	3.99
CL/F	8.032868	.285259	3.55
Tmax	7.011193	.279428	3.99
Cmax	2.301207	.075488	3.28

Solution IV - Model 6 (Equation 4:6)

```
TITLE 1
Repeated administration of Super-High (SR)
MODEL 6
NOPAGE BREAKS
```

```
NVARIABLES 2
NPOINTS 1000
XNUMBER 1
YNUMBER 2
NCONSTANTS 21
CONSTANTS 10,352.3,0,352.3,24,352.3,48,352.3,72,352.3,96,352.3,120, &
         352.3,144,352.3,168,352.3,192,352.3,216
METHOD 2   'Gauss-Newton (Levenberg and Hartley)
REWEIGHT -1
ITERATIONS 50
INITIAL 70,.25,1
LOWER BOUNDS 0,0,0
UPPER BOUNDS 100,2,2
MISSING 'Missing'
NOBSERVATIONS 26
DATA 'WINNLIN.DAT'
BEGIN
```

PARAMETER	ESTIMATE	STANDARD ERROR	CV%	UNIVARIATE C.I.	
Volume/F	52.251983	.638093	1.22	50.931996	53.571969
K	.153829	.002060	1.34	.149568	.158091
Tlag	.674630	.034861	5.17	.602514	.746745

```
*** CORRELATION MATRIX OF THE ESTIMATES ***
PARAMETER  Volume/F   K          Tlag
Volume/F    1.00000
K           -.555541   1.00000
Tlag        -.363767   .302522    1.00000

Condition_number=      375.8

    FUNCTION   1
```

X	OBSERVED Y	PREDICTED Y	RESIDUAL	WEIGHT	SE-PRED	STANDARDIZED RESIDUAL
1.000	.3622	.3210	.4121E-01	3.122	.3072E-01	1.731
2.000	1.140	1.121	.1891E-01	.8923	.2568E-01	.2780
3.000	1.700	1.687	.1349E-01	.5930	.2904E-01	.1600
4.000	2.100	2.068	.3210E-01	.4836	.3209E-01	.3437
5.000	2.350	2.306	.4375E-01	.4336	.3308E-01	.4424
6.000	2.400	2.435	-.3459E-01	.4108	.3236E-01	-.3386
7.000	2.500	2.479	.2055E-01	.4033	.3063E-01	.1982
8.000	2.400	2.462	-.6202E-01	.4062	.2849E-01	-.5968
10.00	2.300	2.304	-.4161E-02	.4340	.2474E-01	-.0410
12.00	2.100	2.057	.4276E-01	.4861	.2298E-01	.4465
14.00	1.700	1.779	-.7950E-01	.5619	.2283E-01	-.8961
24.00	.6500	.6689	-.1892E-01	1.495	.1949E-01	-.3592
216.0	.7400	.7041	.3594E-01	1.420	.2217E-01	.6757
216.5	.7000	.6656	.3445E-01	1.502	.2158E-01	.6662
217.0	.7400	.9499	-.2099	1.053	.3639E-01	-3.738
218.0	1.800	1.682	.1181	.5947	.2383E-01	1.376
219.0	2.100	2.186	-.8593E-01	.4575	.2322E-01	-.8695
220.0	2.600	2.512	.8799E-01	.3981	.2599E-01	.8326
221.0	2.600	2.701	-.1007	.3703	.2803E-01	-.9209
222.0	2.800	2.784	.1555E-01	.3591	.2885E-01	.1402
223.0	2.700	2.789	-.8944E-01	.3585	.2885E-01	-.8058
224.0	2.800	2.736	.6359E-01	.3654	.2851E-01	.5783
226.0	2.400	2.519	-.1185	.3970	.2796E-01	-1.125
228.0	2.300	2.224	.7589E-01	.4496	.2810E-01	.7706

```
230.0       1.900       1.909      -.9002E-02  .5238      .2846E-01  -.0994
240.0       .8000       .7041       .9594E-01  1.420      .2217E-01   1.804
```

CORRECTED SUM OF SQUARED OBSERVATIONS = 14.9088
WEIGHTED CORRECTED SUM OF SQUARED OBSERVATIONS = 13.5058
SUM OF SQUARED RESIDUALS =   .146273
SUM OF WEIGHTED SQUARED RESIDUALS = .108498
S = .686827E-01 WITH   23 DEGREES OF FREEDOM
CORRELATION (OBSERVED,PREDICTED) = .9952

AIC criteria =   -51.74656
SC criteria =   -47.97227

AUC (0 to last time) computed by trapezoidal rule = 216.532

SUMMARY OF ESTIMATED SECONDARY PARAMETERS

PARAMETER	ESTIMATE	STANDARD ERROR	CV%
AUC	43.829899	.530442	1.21
K-HL	4.505947	.060280	1.34
CL/F	8.037892	.097374	1.21
Tmax	7.175337	.083331	1.16
Cmax	2.480364	.030260	1.22

PK5 - One-compartment plasma and urine data I – intravenous dosing

Objectives

◆ **To fit plasma and urine data simultaneously**

◆ **To parameterize in terms of *Cl*, *V* and *f$_e$* or *Cl$_R$***

◆ **To apply a user-specified integrated model**

◆ **To apply a user-specified differential equation model**

Problem specification

In this exercise, we demonstrate how to simultaneously fit plasma and urine data in order to get a proper estimate of clearance, volume of distribution and fraction of dose excreted in urine. We will demonstrate how to graphically obtain initial estimates *Cl*, *V* and *f$_e$* or *Cl$_R$*. We will fit the model by means of a constant *CV* weighting scheme i.e., weight is equal to one over the predicted concentration (or amount) squared. We refer the reader to section 2.4 for more information about different weighting schemes. For further reading on urinary excretion, see sections 3.2.12 and 3.2.13.

A volunteer was given an intravenous bolus dose of 250 mg. Plasma concentration-time data and the cumulative amount in urine can be found in Figure 5.1 and in the program output.

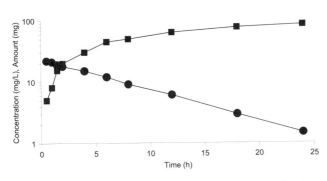

Figure 5.1 Semi-logarithmic plot of concentration (circles) and cumulative amount excreted in urine (squares) versus time. Observed data are connected by straight lines.

We will not estimate the fraction of the dose that is metabolized *f$_m$* here, but it can easily be obtained from *1 − f$_e$*.

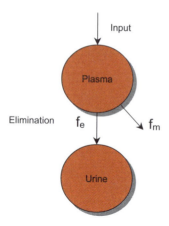

Figure 5.2 *Schematic illustration of the one-compartment model with partial urinary excretion. The model may be parameterized by means of either Cl, V and f_e or Cl, V and Cl_R. The f_m parameter may be estimated as 1-f_e.*

The pharmacokinetics of drug in plasma and urine are represented by means of a system of integrated equations. Plasma equation

$$C = \frac{D_{iv}}{V} \cdot e^{-\frac{Cl}{V} \cdot t} \tag{5:1}$$

The corresponding function for cumulative amount in urine is

$$X_u = f_e \cdot D_{iv} \left[1 - e^{-\frac{Cl}{V} \cdot t} \right] \tag{5:2}$$

The differential equation that describes the plasma compartment is defined as

$$\frac{dC}{dt} = -K \cdot C = -\frac{Cl}{V} \cdot C \tag{5:3}$$

The differential equation that describes the urine compartment is defined as

$$\frac{dA_u}{dt} = f_e \cdot Cl \cdot C = Cl_R \cdot C \tag{5:4}$$

Initial parameter estimates

Parameterize in terms of Cl and V, and fraction of dose excreted in the urine f_e or Cl_R according to Equations 5:1 and 5:2, or Equations 5:3 and 5:4. We calculate the initial estimates of the basal pharmacokinetic parameters from the plot of the concentration-time data on a semi-logarithmic scale. Figure 5.3 illustrates how the initial estimates of K, V and f_e are derived. Clearance is estimated as $K \cdot V$.

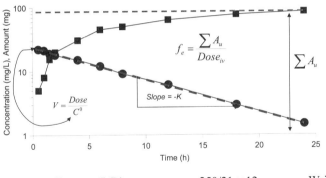

Figure 5.3 *Schematic illustration of how to derive the initial estimates of K, V and fraction of dose excreted into urine f_e.*

V	$= 7$ (L)	$250/21 \approx 12$	Written as Vd in code.
Cl	$= 2$ (L/h)	$250/150 \approx 2$	
f_e	$= 0.3$ (or 30%)	$86/250 \approx 0.3$	

Interpretation of results and conclusions

Some understanding of the mechanisms by which a drug is cleared from the plasma into urine can be acquired by calculation of the renal clearance. We have therefore fit a one-compartment model to plasma and urine data obtained after administration of an intravenous bolus dose of 250 mg. The plasma data in Figure 5.4 clearly demonstrates the mono-exponential decline.

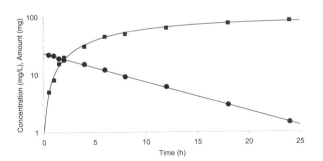

Figure 5.4 *Semi-logarithmic plot of observed (filled symbols) and model-predicted (solid lines) plasma concentration (circles) and cumulative amount in urine (squares) versus time. A typical mono-exponential decline can be seen in the plasma concentrations.*

A model for cumulative amount of drug excreted in urine was selected in combination with the one-compartment plasma model. Equations 5:1 and 5:2 include three parameters, of which Cl and V occur in both the plasma and urine equation. The coefficients of variation for the *plasma* and *urine functions* are proportional expressions when data are weighted by means of a proportional error model (constant CV).

$$CV\% \propto 100 \cdot \sqrt{\frac{WRSS}{df}} \qquad (5{:}5)$$

The estimated coefficients of variation for plasma and urine were 2.84% and 8.96%, respectively. This dataset was originally generated by adding noise with a *CV* of 5 and

10% for plasma and urine data, respectively. The coefficient of variation for all *parameters* was below 5%. Plasma clearance was 1.2 L/h and the fraction of dose excreted via the urine was approximately 35% corresponding to a renal clearance of 0.42 L/h. Virtually no correlation was seen between the parameters, nor were there obvious trends in the residuals.

This exercise demonstrates the benefits of simultaneously utilizing two different sources of data (such as plasma and urine, drug and metabolite, or intravenous and extravascular data). Each separate dataset contributes something that the other dataset does not necessarily include. For example, urine contains information about the fraction of the dose that is excreted. By combining urine and plasma data one therefore obtains e.g., renal clearance. Other situations when combinations of data are useful, include simultaneous fitting of: 1) oral and intravenous data (see PK6), 2) parent compound and metabolite(s), 3) intravenous bolus and constant-rate infusion data, 4) data from several species, 5) two or more dose levels, 6) two or more response levels, and so on.

The solutions to this problem also contain output from fitting the differential equations. Note that the final parameter estimates and their precisions from the two models (i.e., the analytical solution by means of Equations 5:1 and 5:2 vis-a-vis the differential equation system of Equations 5:3 and 5:4) are almost identical.

The next step in the characterization of kinetics would be to give multiple doses via the intended route of administration.

Solution I - Constant CV-model – Integrated model

```
TITLE 1
Simultaneous fit of plasma and urine data
Model
 COMM
NPARM 3
NCON  1
NFUN  2
PNAMES 'Vd', 'Cl','fe'
END
    1: TEMP
    2: DOSE=CON(1)
    3: T=X
    4: END
    5: FUNC1
    6: F = (DOSE/VD)*DEXP(-(CL/VD)*T)
    7: END
    8: FUNC2
    9: F = FE*DOSE*(1. - DEXP(-(CL/VD)*T))
   10: END
   11: EOM
NVARIABLES 3
FNUMBER 3
NPOINTS 100
XNUMBER 1
YNUMBER 2
CONSTANTS 250
METHOD 2  'Gauss-Newton (Levenberg and Hartley)
REWEIGHT -2
ITERATIONS 50
INITIAL 12,2,.3
LOWER BOUNDS 0,0,0
UPPER BOUNDS 20,10,1
```

```
DATA 'WINNLIN.DAT'
BEGIN
PARAMETER  ESTIMATE          STANDARD      CV%            UNIVARIATE C.I.
                             ERROR
  VD       10.664005         .284026      2.66       10.064766      11.263243

  CL        1.227235         .023575      1.92        1.177498       1.276973

  FE         .352917         .008399      2.38         .335197        .370637

*** CORRELATION MATRIX OF THE ESTIMATES ***
PARAMETER  VD        CL          FE
VD      1.00000
CL       .624653 1.00000
FE       .416516 -.07254    1.00000

Condition_number=          43.08

  X         OBSERVED   PREDICTED   RESIDUAL     WEIGHT     SE-PRED  STANDARDIZ
             Y          Y                                           RESIDUAL
  .5000      22.10      22.13     -.3247E-01   .2041E-02   .5712     -.0268
 1.000       21.30      20.89      .4051       .2290E-02   .5226      .3528
 1.500       19.00      19.73     -.7265       .2570E-02   .4783     -.6660
 2.000       18.00      18.62     -.6235       .2883E-02   .4381     -.6020
 4.000       15.10      14.79      .3055       .4569E-02   .3117      .3649
 6.000       12.01      11.75      .2572       .7240E-02   .2294      .3829
 8.000        9.100      9.336    -.2365       .1147E-01   .1783     -.4422
12.00         6.100      5.892     .2080       .2881E-01   .1279      .6265
18.00         2.950      2.954    -.3853E-02   .1146       .9249E-01 -.0252
24.00         1.480      1.481    -.8528E-03   .4560       .6467E-01 -.0138

CORRECTED SUM OF SQUARED OBSERVATIONS =  513.805
WEIGHTED CORRECTED SUM OF SQUARED OBSERVATIONS =  5.73069
SUM OF SQUARED RESIDUALS =            1.34034
SUM OF WEIGHTED SQUARED RESIDUALS =  .564962E-02
S =  .284093E-01 WITH      7 DEGREES OF FREEDOM
CORRELATION (OBSERVED,PREDICTED) =  .9988

  FUNCTION   2

  X         OBSERVED   PREDICTED   RESIDUAL     WEIGHT     SE-PRED  STANDARDIZE
             Y          Y                                           RESIDUAL
  .5000       5.020      4.934      .8650E-01   .4100E-01   .9913E-01  .3076
 1.000        8.110      9.591    -1.481       .1085E-01   .1911    -2.706
 1.500       15.50      13.99      1.512       .5100E-02   .2765     1.892
 2.000       20.00      18.14      1.860       .3033E-02   .3561     1.794
 4.000       30.60      32.55     -1.950       .9420E-03   .6268    -1.046
 6.000       45.20      44.00      1.203       .5156E-03   .8411      .4767
 8.000       50.00      53.09     -3.091       .3541E-03  1.017     -1.016
12.00        63.90      66.05     -2.155       .2287E-03  1.296      -.5706
18.00        77.00      77.11      -.1124      .1678E-03  1.595      -.0257
24.00        85.50      82.66      2.844       .1461E-03  1.793      .6099

CORRECTED SUM OF SQUARED OBSERVATIONS =  7466.61
WEIGHTED CORRECTED SUM OF SQUARED OBSERVATIONS =  5.26622
SUM OF SQUARED RESIDUALS =             35.4942
SUM OF WEIGHTED SQUARED RESIDUALS =  .562148E-01
S =  .896141E-01 WITH      7 DEGREES OF FREEDOM
CORRELATION (OBSERVED,PREDICTED) =  .9976

TOTALS FOR ALL CURVES COMBINED
```

```
SUM OF SQUARED RESIDUALS =              36.8345
SUM OF WEIGHTED SQUARED RESIDUALS = .618644E-01
S = .603248E-01 WITH    17 DEGREES OF FREEDOM

AIC criteria =           -49.65620
SC  criteria =           -46.66900
```

Solution II - Constant CV-model – Differential equation model

```
TITLE 1
Simultaneous fit of plasma and urine data
Model
 COMM
  NPAR 3
  NDER 2
  NCON 1
  NFUN 2
  PNAMES 'Vd', 'Cl','fe'
END
   1: TEMP
   2: DOSE=CON(1)
   3: T=X
   4: END
   5: START
   6: Z(1) = DOSE/VD
   7: Z(2) = 0.0
   8: END
   9: DIFF
  10: DZ(1) = -CL*Z(1)/VD
  11: DZ(2) = FE*CL*Z(1)
  12: END
  13: FUNC1
  14: F = Z(1)
  15: END
  16: FUNC2
  17: F = Z(2)
  18: END
  19: EOM
NVARIABLES 3
FNUMBER 3
NPOINTS 100
XNUMBER 1
YNUMBER 2
CONSTANTS 250
METHOD 2  'Gauss-Newton (Levenberg and Hartley)
REWEIGHT -2
ITERATIONS 50
INITIAL 7,2,.5
LOWER BOUNDS 0,0,0
UPPER BOUNDS 20,10,1
DATA 'WINNLIN.DAT'
BEGIN
```

PARAMETER	ESTIMATE	STANDARD ERROR	CV%	UNIVARIATE C.I.	
VD	10.709546	.285012	2.66	10.108227	11.310864
CL	1.229536	.023666	1.92	1.179606	1.279466
FE	.354099	.008418	2.38	.336338	.371860

```
*** CORRELATION MATRIX OF THE ESTIMATES ***
```

```
PARAMETER  VD          CL          FE
VD         1.00000
CL          .623168   1.00000
FE          .415536   -.07598    1.00000

Condition_number=          43.17
```

FUNCTION   1

X	OBSERVED Y	PREDICTED Y	RESIDUAL	WEIGHT	SE-PRED	STANDARDIZED RESIDUAL
.5000	22.10	22.04	.5862E-01	.2058E-02	.5679	.0487
1.000	21.30	20.81	.4882	.2309E-02	.5197	.4272
1.500	19.00	19.65	-.6507	.2590E-02	.4759	-.5993
2.000	18.00	18.55	-.5545	.2905E-02	.4360	-.5378
4.000	15.10	14.75	.3522	.4598E-02	.3106	.4224
6.000	12.01	11.72	.2878	.7278E-02	.2289	.4302
8.000	9.100	9.317	-.2172	.1152E-01	.1780	-.4074
12.00	6.100	5.886	.2137	.2886E-01	.1278	.6449
18.00	2.950	2.956	-.5859E-02	.1145	.9247E-01	-.0384
24.00	1.480	1.484	-.4296E-02	.4539	.6475E-01	-.0694

```
CORRECTED SUM OF SQUARED OBSERVATIONS =  513.805
WEIGHTED CORRECTED SUM OF SQUARED OBSERVATIONS =  5.76186
SUM OF SQUARED RESIDUALS =              1.27247
SUM OF WEIGHTED SQUARED RESIDUALS =  .559363E-02
S =  .282682E-01 WITH      7 DEGREES OF FREEDOM
CORRELATION (OBSERVED,PREDICTED) =  .9988
```

FUNCTION   2

X	OBSERVED Y	PREDICTED Y	RESIDUAL	WEIGHT	SE-PRED	STANDARDIZED RESIDUAL
.5000	5.020	4.939	.8145E-01	.4100E-01	.9904E-01	.2899
1.000	8.110	9.602	-1.492	.1085E-01	.1909	-2.727
1.500	15.50	14.00	1.496	.5099E-02	.2763	1.873
2.000	20.00	18.16	1.838	.3032E-02	.3558	1.774
4.000	30.60	32.60	-1.997	.9411E-03	.6265	-1.072
6.000	45.20	44.07	1.128	.5149E-03	.8409	.4474
8.000	50.00	53.19	-3.192	.3534E-03	1.017	-1.049
12.00	63.90	66.20	-2.302	.2282E-03	1.297	-.6095
18.00	77.00	77.32	-.3154	.1673E-03	1.596	-.0719
24.00	85.50	82.90	2.604	.1455E-03	1.796	.5579

```
CORRECTED SUM OF SQUARED OBSERVATIONS =  7466.61
WEIGHTED CORRECTED SUM OF SQUARED OBSERVATIONS =  5.25599
SUM OF SQUARED RESIDUALS =              35.4780
SUM OF WEIGHTED SQUARED RESIDUALS =  .562766E-01
S =  .896633E-01 WITH      7 DEGREES OF FREEDOM
CORRELATION (OBSERVED,PREDICTED) =  .9977

TOTALS FOR ALL CURVES COMBINED
SUM OF SQUARED RESIDUALS =              36.7505
SUM OF WEIGHTED SQUARED RESIDUALS =  .618702E-01
S =  .603276E-01 WITH     17 DEGREES OF FREEDOM

AIC criteria =        -49.65433
SC  criteria =        -46.66713
```

PK6 – One-compartment plasma and urine data II – intravenous and oral dosing

Objectives

◆ **To simultaneously fit plasma and urine data of intravenous and oral dosing**

◆ **To parameterize in terms of** Cl**,** V**,** F**,** K_a**,** t_{lag} **and** f_e **or** Cl_R

◆ **To apply an integrated model**

Problem specification

This exercise demonstrates the benefits from *fitting a model simultaneously to several sources of data.* In the example, we use plasma concentrations and cumulative amount in urine after both intravenous and oral administration of drug X (see also Sections 3.2.11 and 3.2.12).

Test solutions for oral (25 mg/10 mL) and intravenous (12.5 mg/5 mL) dosing were given on two separate occasions. The intravenous dose was given with a constant infusion rate of 1 mL/min over 5 min. We will model this as an intravenous bolus due to the extremely long half-life (> 25 h) of the drug relative to the infusion time (5 min). Figure 6.1 displays data obtained from plasma and urine samples that were collected from a subject at the times specified below in Tables 6.1 and 6.2.

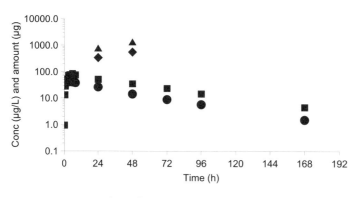

Figure 6.1 *Semi-logarithmic plot of observed concentration versus time data (plasma levels after ● intravenous and ■ oral dosing) and amount versus time data (cumulative urinary levels after ◆ intravenous and ▲ oral dosing), following administration of drug X. The oral dose was 25 mg and the intravenous dose 12.5 mg.*

From a plot of the concentration-time data on log-paper, calculate the basal pharmacokinetic parameters. We parameterize in terms of plasma clearance Cl and volume of distribution V, absorption rate constant K_a, bioavailability F and fraction of (absorbed) dose excreted in urine f_e or renal clearance Cl_R, according to Equations 6:1 and 6:2. Figure 6.2 depicts the one-compartment model.

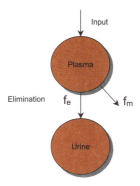

Input

Plasma

Elimination f_e

f_m

Urine

Figure 6.2 Schematic illustration of the one-compartment model with urinary excretion and metabolic elimination.

Assume a constant *CV* error model. The plasma equation describing an intravenous bolus dose into a one-compartment system is as follows

$$C = \frac{D_{iv}}{V} \cdot e^{-\frac{Cl}{V} \cdot t}$$
(6:1)

The corresponding function for the cumulative amount in urine is

$$X_u = f_e \cdot D_{iv}\left[1 - e^{-\frac{Cl}{V} \cdot t}\right]$$
(6:2)

The plasma equation following extravascular administration becomes

$$C = \frac{K_a FD_{po}}{V\left(K_a - \frac{Cl}{V}\right)}\left[e^{-\frac{Cl}{V} \cdot (t - t_{lag})} - e^{-K_a \cdot (t - t_{lag})}\right]$$
(6:3)

The cumulative amount in urine is

$$X_u = f_e \cdot K_a \cdot FD_{po}\left[\frac{1}{K_a} + \frac{e^{-\frac{Cl}{V} \cdot (t - t_{lag})}}{\frac{Cl}{V} - K_a} - \frac{\frac{Cl}{V} \cdot e^{-K_a \cdot (t - t_{lag})}}{K_a \cdot \left(\frac{Cl}{V} - K_a\right)}\right]$$
(6:4)

We refer the reader to Gibaldi and Perrier [1982] for a complete derivation of Equation 6:4. In this exercise, we fit plasma and urine data simultaneously in order to get a proper estimate of *Cl*, *V*, and f_e or Cl_R. The intravenously administered dose was 12.5 mg and the oral dose was 25 mg. Functions 1 to 4 (*FUNC 1, 2, 3 and 4* in program code) are the plasma concentrations after intravenous administration, plasma concentrations after oral administration, cumulative amount in urine after intravenous administration and cumulative amount in urine after oral administration, respectively.

Table 6.1   Plasma data after intravenous and oral dosing

Time iv (h)	C_{iv} (µg/L)	Time po (h)	C_{po} (µg/L)
0.33	47.5	0.33	0.94
0.67	46.2	0.67	13.1
1.00	46.5	1.00	27.9
2.00	42.9	2.00	38.0
3.00	45.9	3.00	71.9
4.00	44.8	4.00	76.1
6.00	40.5	6.00	83.9
8.00	38.0	8.00	75.0
24.32	26.3	23.92	51.6
47.72	14.2	47.97	34.9
71.85	8.8	72.07	23.4
95.67	5.7	96.10	14.5
168.05	1.5	168.13	4.5

Table 6.2   Urine data after intravenous and oral dosing

Route	Time (h)	Urine concentration (µg/L)	Urine volume (mL)
oral	0 – 24	586	1204
	24 – 48	1073	1128
iv	0 – 24	394	863
	24 – 48	346	1591

Initial parameter estimates

V	= 300 (L)	Volume of distribution $Dose_{iv}/C_{max}$
Cl	= 5.0 (L/h)	Clearance $Dose_{iv}/C_{max}/K$
F	= 0.9	Bioavailability, known to be > 0.8
K_a	= 1.0 (h^{-1})	Obtained by deconvolution
t_{lag}	= 0.4 (h)	Lag-time
f_e	= 0.2	Fraction of absorbed dose excreted in urine

The mean residence time *MRT*, time to 90% of steady state t_{90}, half-life $t_{1/2}$, renal clearance Cl_R, time to maximum plasma concentration t_{max}, maximum concentration C_{max}, and half-life for the absorption $t_{1/2Ka}$, will be estimated as secondary parameters. *V, Cl* and *F* are denoted *VD, CLE,* and *BIO*, respectively, in the program code. See PK5 for initial estimates.

Interpretation of results and conclusions

Some understanding of the mechanisms by which a drug is cleared from the plasma into urine can be acquired by calculation of the renal clearance. We have therefore simultaneously fit plasma and urine data after oral and intravenous dosing. This enables us to get a good estimate of bioavailability in parallel with estimating renal clearance. As we will show, this approach resulted in generally high parameter precision and low correlation.

The absorption was rapid and complete (Figure 6.3).

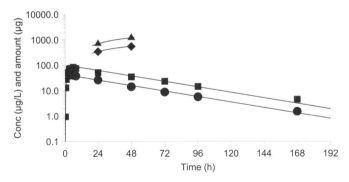

Figure 6.3 *Observed (symbols) and predicted (solid lines) plasma concentration versus time and cumulative urinary amount versus time data following oral and intravenous dosing with drug X. Plasma concentration after ● intravenous and ■ oral dosing. Urinary amount after ◆ intravenous and ▲ oral dosing.*

We were able to estimate the fraction excreted in urine even though we collected urine for only 48/33 half-lives, where 48 is the urine collection interval in h and 33 the estimated half-life in h. Renal clearance was estimated to be 0.42 L/h (6.8 mL/min). Total plasma clearance was 6.0 L/h and volume of 290 L. The parameter precision was very high except for the lag-time which had a CV of 43%. There were trends in the residuals, but they were very marginal. The fit of a two-compartment model (not presented here) improved the fit only marginally, but the model parameters had a low precision. With a half-life of about 33 h, it would take about 120 h to reach 90% of steady state.

You have
☐ analyzed a typical one-compartment (mono-exponential) system
☐ simultaneously fit several sources (plasma and urine) of data
☐ learned to derive the initial estimates
☐ characterized the pharmacokinetics in plasma and urine after single dose administration to one subjects

The next step in the characterization of kinetics would be to give multiple doses (at several dose levels) via the intended route of administration.

Solution
```
TITLE 1
Simultaneous fit of iv- and po data.
Model
 COMM
NPARM 6
NSEC  7
NCON  2
NFUN  4
PNAMES 'Vd', 'Cle','Bio', 'Ka', 'Tlag', 'Fe'
SNAMES 'MRT', 'T90%', 'T1/2', 'CLr', 'Tmax', 'Cmax', 'T1/2ka'
END
    1: TEMP
```

```
 2: DIV=CON(1)
 3: DPO=CON(2)
 4: T=X
 5: KE = CLE/VD
 6: TL = TLAG
 7: END
 8: FUNC1
 9: F = (DIV/VD)*DEXP(-(CLE/VD)*T)
10: END
11: FUNC2
12: F = ((BIO*DPO*KA)/(VD*(KA-KE)))*(DEXP(-KE*(T-TL))-DEXP(-KA*(T-TL)))
13: F = MAX(0,F)
14: DUM = F
15: END
16: FUNC3
17: F = FE*DIV*(1. - DEXP(-(CLE/VD)*T))
18: END
19: FUNC4
20: REST = (CLE/VD)*DEXP(-KA*(T-TL))/(KA*(CLE/VD - KA))
21: F=FE*KA*BIO*DPO*(1/KA+DEXP((-CLE/VD)*(T-TL))/(CLE/VD-KA)-REST)
22: END
23: SECO
24: S(1)=VD/CLE
25: S(2)=3.3*0.693/KE
26: S(3)=0.693/KE
27: S(4)=FE*CLE
28: S(5)=(2.303/(KA - KE))*DLOG(KA/KE) + TLAG
29: S(6)=BIO*(DPO/VD)*DEXP(-KE*S(5))
30: S(7)=0.693/KA
31: END
32: EOM
NOPAGE BREAKS
NVARIABLES 3
FNUMBER 3
NPOINTS 100
XNUMBER 1
YNUMBER 2
CONSTANTS 12500,25000
METHOD 2   'Gauss-Newton (Levenberg and Hartley)
ITERATIONS 50
INITIAL 300,5,.9,1,.4,.2
LOWER BOUNDS 0,0,0,0,0,0
UPPER BOUNDS 500,90,2,5,2,1
MISSING 'Missing'
DATA 'WINNLIN.DAT'
BEGIN
```

PARAMETER	ESTIMATE	STANDARD ERROR	CV%	UNIVARIATE C.I.	
VD	290.344552	9.096435	3.13	271.570583	309.118520
CLE	6.025690	.223322	3.71	5.564781	6.486600
BIO	1.135910	.011069	.97	1.113064	1.158756
KA	.420769	.050657	12.04	.316218	.525320
TLAG	.312919	.089436	28.58	.128333	.497504
FE	.069834	.002056	2.94	.065591	.074077

```
*** CORRELATION MATRIX OF THE ESTIMATES ***
PARAMETER  VD         CLE        BIO        KA         TLAG       FE
VD         1.00000
CLE        .173666    1.00000
BIO        -.204823   .214841    1.00000
KA         .622260    -.207766   -.53301    1.00000
TLAG       .166823    -.052418   -.14713    .53002     1.00000
FE         .509852    -.723123   -.50891    .60586     .15981     1.00000
```

Condition_number=   .4170E+05

FUNCTION   1

X	OBSERVED Y	PREDICTED Y	RESIDUAL	WEIGHT	SE-PRED	STANDARDI RESID
.3330	47.50	42.76	4.744	1.000	1.331	1.023
.6667	46.20	42.46	3.739	1.000	1.315	.8055
1.000	46.50	42.17	4.332	1.000	1.298	.9323
2.000	42.90	41.30	1.598	1.000	1.252	.3430
3.000	45.90	40.45	5.446	1.000	1.207	1.166
4.000	44.80	39.62	5.177	1.000	1.165	1.106
6.000	40.50	38.01	2.488	1.000	1.086	.5293
8.000	38.00	36.47	1.534	1.000	1.015	.3252
24.00	26.30	26.16	.1374	1.000	.6865	.0287
48.00	14.20	15.90	-1.699	1.000	.5871	-.3548
72.00	8.800	9.662	-.8616	1.000	.5296	-.1797
96.00	5.700	5.871	-.1713	1.000	.4398	-.0356
168.0	1.500	1.318	.1824	1.000	.1827	.0378

```
CORRECTED SUM OF SQUARED OBSERVATIONS =  3741.77
WEIGHTED CORRECTED SUM OF SQUARED OBSERVATIONS =  3741.77
SUM OF SQUARED RESIDUALS =              126.532
SUM OF WEIGHTED SQUARED RESIDUALS =  126.532
S =  4.25159     WITH      7 DEGREES OF FREEDOM
CORRELATION (OBSERVED,PREDICTED) =  .9966
```

FUNCTION   2

X	OBSERVED Y	PREDICTED Y	RESIDUAL	WEIGHT	SE-PRED	STANDARDI RESID
.3330	.9400	.8228	.1172	1.000	3.602	.3653E-01
.6667	13.10	13.48	-.3770	1.000	2.632	-.9324E-01
1.000	27.90	24.37	3.527	1.000	2.245	.8259
2.000	38.00	48.75	-10.75	1.000	2.485	-2.601
3.000	71.90	64.09	7.812	1.000	2.540	1.904
4.000	76.10	73.50	2.603	1.000	2.282	.6124
6.000	83.90	82.03	1.871	1.000	1.916	.4227
8.000	75.00	83.66	-8.659	1.000	2.100	-1.994
24.00	51.60	62.92	-11.32	1.000	1.931	-2.561
48.00	34.90	38.24	-3.341	1.000	1.331	-.7204
72.00	23.40	23.24	.1614	1.000	1.149	.3445E-01
96.00	14.50	14.12	.3781	1.000	.9622	.7998E-01
168.0	4.500	3.169	1.331	1.000	.4135	.2769

```
CORRECTED SUM OF SQUARED OBSERVATIONS =  10218.0
WEIGHTED CORRECTED SUM OF SQUARED OBSERVATIONS =  10218.0
SUM OF SQUARED RESIDUALS =              415.830
SUM OF WEIGHTED SQUARED RESIDUALS =  415.830
S =  7.70741     WITH      7 DEGREES OF FREEDOM
```

CORRELATION (OBSERVED,PREDICTED) = .9815

FUNCTION   3

X	OBSERVED Y	PREDICTED Y	RESIDUAL	WEIGHT	SE-PRED	STANDARD RESID
24.00	340.0	342.5	-2.455	1.000	3.162	-.6737
48.00	550.0	550.6	-.5617	1.000	4.328	-.2636

CORRECTED SUM OF SQUARED OBSERVATIONS = 22050.0
WEIGHTED CORRECTED SUM OF SQUARED OBSERVATIONS = 22050.0
SUM OF SQUARED RESIDUALS =   6.34024
SUM OF WEIGHTED SQUARED RESIDUALS = 6.34024
S = .000000   WITH   -4 DEGREES OF FREEDOM
CORRELATION (OBSERVED,PREDICTED) = 1.000

FUNCTION   4

X	OBSERVED Y	PREDICTED Y	RESIDUAL	WEIGHT	SE-PRED	STANDARD RESID
24.00	705.0	707.2	-2.216	1.000	4.208	-.9391
48.00	1210.	1208.	2.238	1.000	4.618	1.604

CORRECTED SUM OF SQUARED OBSERVATIONS = 127513.
WEIGHTED CORRECTED SUM OF SQUARED OBSERVATIONS = 127513.
SUM OF SQUARED RESIDUALS =   9.91975
SUM OF WEIGHTED SQUARED RESIDUALS = 9.91975
S = .000000   WITH   -4 DEGREES OF FREEDOM
CORRELATION (OBSERVED,PREDICTED) = 1.000

TOTALS FOR ALL CURVES COMBINED
SUM OF SQUARED RESIDUALS =   558.622
SUM OF WEIGHTED SQUARED RESIDUALS = 558.622
S = 4.82451   WITH   24 DEGREES OF FREEDOM

AIC criteria =   201.76417
SC criteria =   210.17136

SUMMARY OF ESTIMATED SECONDARY PARAMETERS

PARAMETER	ESTIMATE	STANDARD ERROR	CV%
MRT	48.184447	2.127484	4.42
T90%	110.193011	4.865342	4.42
T1/2	33.391822	1.474346	4.42
CLR	.420798	.010829	2.57
TMAX	17.638664	1.330858	7.55
CMAX	67.825014	1.783319	2.63
T1/2KA	1.646984	.198086	12.03

PK7 – Two-compartment intravenous bolus dosing

Objectives
♦ **To fit mono-, bi- and three-compartment models to intravenous bolus data**
♦ **To discuss various approaches for estimation of initial estimates**
♦ **To apply the model library**
♦ **To apply non-compartmental analysis, curve stripping and nonlinear regression**

Problem specification

This exercise is aimed to show how one can characterize the disposition of a drug given as a bolus dose. Since the drug was given as an intravenous bolus, only distribution and elimination processes are involved. This is called disposition.

The user should fit a mono-, bi-, and three-exponential (compartment) model to the concentration-time data provided. The observed and predicted concentration-time profiles are displayed in Figure 7.1 and in the program output.

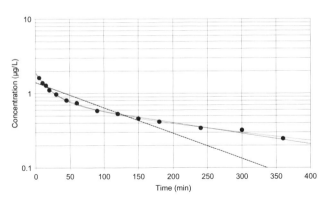

Figure 7.1 *Concentration-time data following intravenous bolus administration of a multi-compartment drug. Mono-, bi- and three-exponential decline of the predicted data are plotted. The bi- and three-exponential models gave almost identical predicted curves.*

How do we analyze concentration-time data that exhibit bi-phasic disposition after an intravenous bolus dose? We will apply three different approaches:

☐ Non-compartmental analysis (*NCA*) by means of the trapezoidal rule
☐ Curve stripping
☐ Nonlinear regression

The first two methods are slightly different. Curve stripping assumes that the concentration-time curve declines in a log-linear fashion, i.e., if data are plotted on log paper, a straight line can be fit to the last few points. On the other hand, NCA uses the trapezoidal method for the incremental areas, and calculates the extrapolated area from the last concentration divided by the terminal slope. The latter approach gives clearance, volume and the terminal half-life.

Noncompartmental analysis

There are several ways of determining the characteristics of a drug that do not require detailed compartmental analysis. Benet and Galeazzi [1979] have described a method which requires the measurement of the area under the plasma concentration curve (AUC) from zero to infinity and also the area under the first moment of the plasma curve ($AUMC$) from zero to infinity following an intravenous bolus dose (non-compartmental analysis, NCA). This approach can also be used for other routes of administration, provided that the input function is known. Thus, if we apply the linear trapezoidal rule (Equation 7.1),

$$AUC_0^\infty = AUC_{0-\infty} = AUC_0^{t_{last}} + \frac{C_{last}}{\lambda_z} = \sum_{i=1}^{n} \frac{C_i + C_{i+1}}{2} \cdot \Delta t \qquad (7:1)$$

$AUC_{0-\infty}$ and λ_z are 250 and 0.003, respectively. Δt is the distance between two observation times and C^* is the last observed concentration, respectively. Since the dose D_{iv} was administered as an intravenous bolus (100 mg), clearance Cl becomes

$$Cl = \frac{D_{iv}}{AUC_0^\infty} = \frac{100}{250} = 0.4 \, L/min \qquad (7:2)$$

Mean residence time MRT is estimated as

$$MRT = \frac{AUMC_0^\infty}{AUC_0^\infty} = 280 \, min \qquad (7:3)$$

Volume of distribution at steady state V_{ss} is

$$V_{ss} = MRT \cdot Cl = \frac{AUMC_0^\infty}{AUC_0^\infty} \cdot \frac{D_{iv}}{AUC_0^\infty} = \frac{D_{iv} \cdot AUMC_0^\infty}{\left[AUC_0^\infty\right]^2} = 110 \, L \qquad (7:4)$$

Curve stripping

The method of residuals is shown schematically in Figure 7.2. If we then apply curve stripping, A, α, B and β are calculated to be 1, 0.05, 0.8, and 0.003, respectively.

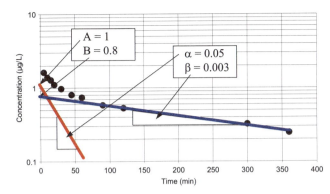

Figure 7.2 *Schematic illustration of the bi-exponential (two-compartment) behavior of a drug administered as an intravenous bolus dose. The red line is obtained by means of the residual method, which is the difference between observed data in the initial phase and the concentration values on the back-extrapolated blue line at corresponding time points.*

The function that obeys bi-exponential decline takes the form

$$C = A \cdot e^{-\alpha \cdot t} + B \cdot e^{-\beta \cdot t} \tag{7:5}$$

λ_z is obtained by log-linear regression of the terminal (seemingly linear) portion of the curve and corresponds to λ_z in the previous approach. B is the intercept of the abscissa by back-extrapolation of the terminal phase. A is the intercept of the residual curve (observed concentrations during the initial phase minus the back-extrapolated concentrations of the terminal phase) and $-\lambda_z$ is its corresponding slope.

Nonlinear regression

In the third approach, we apply nonlinear regression and fit Equations 7:6 to 7:8 to the data. A, α, B, β, G, and γ are the model parameters to be estimated. We can also estimate Cl, MRT, V_{ss} and V_c etc., as secondary model parameters. A to γ are obtained via an iterative process.

If the model fits the data well and the fitted function is the correct model, we will obtain parameters with little bias and high precision, i.e., low standard deviations. The mono-, bi- and three-exponential models are

$$C_p = A \cdot e^{-K \cdot t} \tag{7:6}$$

$$C_p = A \cdot e^{-\alpha \cdot t} + B \cdot e^{-\beta \cdot t} \tag{7:7}$$

$$C_p = A \cdot e^{-\alpha \cdot t} + B \cdot e^{-\beta \cdot t} + C \cdot e^{-\gamma \cdot t} \tag{7:8}$$

Interpretation of results and conclusions

For this particular dataset, the condition numbers were 4778, 446, and 4104 for the mono-, bi- and three-exponential models, respectively. The parameter precision decreases from the simplest model to the most complex. There are 3, 7, and 12 runs in the residuals (from the mono- to the three-exponential model, respectively). Figure 7.3 shows the observed and predicted plasma concentrations *versus* time for all models, and Figure 7.4 the absolute residuals *versus* time for the bi- and three-exponential models.

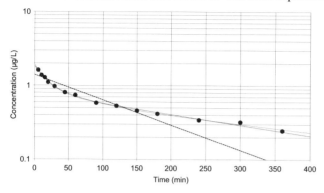

Figure 7.3 Observed (symbols) and predicted (solid lines) concentrations. The bi- and three- exponential models superimpose.

One of the more useful means of examining residuals is via scatter plots. Typically, one would examine a plot of predicted concentration *versus* residual, and time *versus* residual. One should examine such plots to determine if the residuals tend to be randomly scattered within a lower and upper horizontal band. If the residuals do not appear to be randomly scattered, this would suggest that either the model or weighting scheme is incorrect. In Figure 7.4 we have chosen to join the individual absolute residuals with a line in order to discriminate more easily between the two models. Still, the correlation coefficient is high. There is also an obvious trend with time (Figure 7.3). As the concentration declines, the residual decreases, and then continuously increases with time, becoming more and more positive.

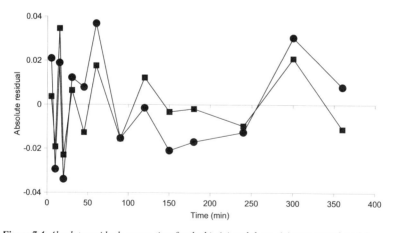

Figure 7.4 *Absolute residuals versus time for the bi- (●) and three- (■) exponential models.*

The bi- and three-exponential models show a more random scatter about the predicted value, with the residuals for the latter model being the more random (Figure 7.4). The frequency of changing +/– signs is larger for the three-exponential model. The absolute magnitude of the residuals, the amplitude, is also smaller for the latter model. We would therefore suggest that the three-exponential model mimics the data better than the bi-exponential model with respect to the relative residual. The three-exponential model also has a slightly lower *AIC* value than the bi-exponential model. In order to help you interpret the results, a summary table is included with the pharmacokinetic parameters and goodness-of-fit criteria (Table 7.1)

Table 7.1 Parameter estimates (mean ± CV%) and goodness-of-fit criteria

Parameter	Mono-exponential model	Bi-exponential model	Three-exponential model
A	1.4 ± 5.9	1.06 ± 4.3	0.66 ± 50
α	0.008 ± 15	0.048 ± 9.8	0.088 ± 51
B	–	0.79 ± 5.4	0.63 ± 43
β	–	0.0031 ± 9.2	0.022 ± 67
C	–	–	0.64 ± 22
γ	–	–	0.0025 ± 30
Condition #	4778	446	4104
WRSS	0.26	0.0065	0.0037
+/–	3	7	12
S	0.15	0.025	0.021
AIC	–14.9	–62.5	–66.5

In spite of the above findings, we would still recommend the use of the bi-exponential model because of its lower condition number, uniformally higher parameter precision relative to the three-exponential model, and *AIC* value that was nearly the smallest.

All three approaches (*NCA*, curve stripping and nonlinear regression) gave consistent results. *NCA* requires very few assumptions and is easy to automate. Nonlinear regression is more assumption rich and the results may depend on the weighting scheme (error model) used. On the other hand, if we fit a regression model to data, we can use the model more readily for simulations with other doses and/or other dosing regimens.

The next step in the characterization of kinetics would be to give multiple doses via the intended route of administration. See also PK4 for a discussion on model discrimination.

Solution I - Mono-exponential model

```
TITLE 1
Mono-exponential model
MODEL 1
NVARIABLES 2
NPOINTS 100
XNUMBER 1
YNUMBER 2
NCONSTANTS 3
CONSTANTS 1,100,0
METHOD 2   'Gauss-Newton (Levenberg and Hartley)
ITERATIONS 50
INITIAL 50,.005
LOWER BOUNDS 0,0
UPPER BOUNDS 200,1
MISSING 'Missing'
```

```
NOBSERVATIONS 14
DATA 'WINNLIN.DAT'
BEGIN
```

PARAMETER	ESTIMATE	STANDARD ERROR	CV%	UNIVARIATE C.I.	
Volume	71.286960	4.209847	5.91	62.114477	80.459442
K10	.007892	.001149	14.56	.005388	.010396

```
*** CORRELATION MATRIX OF THE ESTIMATES ***
PARAMETER      Volume        K10
Volume        1.00000
K10           -.642055     1.00000

  Condition_number=      4778.
```

X	OBSERVED Y	PREDICTED Y	RESIDUAL	WEIGHT	SE-PRED	STANDARDIZED RESIDUAL
5.000	1.625	1.349	.2765	1.000	.7482E-01	2.182
10.00	1.384	1.296	.8767E-01	1.000	.6788E-01	.6716
15.00	1.280	1.246	.3383E-01	1.000	.6196E-01	.2535
20.00	1.105	1.198	-.9295E-01	1.000	.5705E-01	-.6853
30.00	.9730	1.107	-.1340	1.000	.5021E-01	-.9692
45.00	.8060	.9834	-.1774	1.000	.4651E-01	-1.271
60.00	.7400	.8736	-.1336	1.000	.4793E-01	-.9608
90.00	.5820	.6895	-.1075	1.000	.5489E-01	-.7872
120.0	.5300	.5441	-.1410E-01	1.000	.5970E-01	-.1049
150.0	.4580	.4294	.2861E-01	1.000	.6089E-01	.2136
180.0	.4160	.3389	.7713E-01	1.000	.5923E-01	.5727
240.0	.3420	.2110	.1310	1.000	.5106E-01	.9490
300.0	.3210	.1314	.1896	1.000	.4072E-01	1.341
360.0	.2460	.8186E-01	.1641	1.000	.3094E-01	1.141

```
CORRECTED SUM OF SQUARED OBSERVATIONS =  2.49866
  WEIGHTED CORRECTED SUM OF SQUARED OBSERVATIONS =  2.49866
  SUM OF SQUARED RESIDUALS =            .259776
  SUM OF WEIGHTED SQUARED RESIDUALS =  .259776
  S =  .147132      WITH     12 DEGREES OF FREEDOM
  CORRELATION (OBSERVED,PREDICTED) =  .9529

AIC criteria =          -14.87112
SC  criteria =          -13.59301
```

PARAMETER	ESTIMATE	STANDARD ERROR	CV%
AUC	177.741064	20.744254	11.67
K10-HL	87.826046	12.775938	14.55
Cmax	1.402781	.082758	5.90
Cl	.562616	.065729	11.68
AUMC	22520.895007	5786.200831	25.69
MRT	126.706201	18.431782	14.55
Vss	71.286960	4.209847	5.91

Solution II - Bi-exponential model

```
TITLE 1
Bi-exponential model
MODEL 8
NVARIABLES 2
```

```
NPOINTS 100
XNUMBER 1
YNUMBER 2
NCONSTANTS 4
CONSTANTS 100,1,100,0
METHOD 2   'Gauss-Newton (Levenberg and Hartley)
ITERATIONS 50
INITIAL 1,.8,.05,.0033
LOWER BOUNDS 0,0,0,0
UPPER BOUNDS 10,8,.5,.1
MISSING 'Missing'
NOBSERVATIONS 14
DATA 'WINNLIN.DAT'
BEGIN
```

PARAMETER	ESTIMATE	STANDARD ERROR	CV%	UNIVARIATE C.I.	
A	1.056587	.045018	4.26	.956281	1.156893
B	.786837	.042408	5.39	.692346	.881328
Alpha	.048235	.004731	9.81	.037694	.058776
Beta	.003321	.000305	9.19	.002641	.004001

```
*** CORRELATION MATRIX OF THE ESTIMATES ***
PARAMETER    A                 B            Alpha          Beta
A        1.00000
B        -.549699         1.00000
Alpha    -.463495E-01      .813144       1.00000
Beta     -.546605          .909257        .680230        1.00000

 Condition_number=        446.1

FUNCTION   1
```

X	OBSERVED Y	PREDICTED Y	RESIDUAL	WEIGHT	SE-PRED	STANDARDIZED RESIDUAL
5.000	1.625	1.604	.2096E-01	1.000	.2182E-01	1.593
10.00	1.384	1.413	-.2940E-01	1.000	.1323E-01	-1.350
15.00	1.280	1.261	.1892E-01	1.000	.1255E-01	.8532
20.00	1.105	1.139	-.3393E-01	1.000	.1366E-01	-1.578
30.00	.9730	.9608	.1220E-01	1.000	.1382E-01	.5702
45.00	.8060	.7982	.7821E-02	1.000	.1209E-01	.3488
60.00	.7400	.7032	.3684E-01	1.000	.1233E-01	1.653
90.00	.5820	.5973	-.1531E-01	1.000	.1352E-01	-.7089
120.0	.5300	.5314	-.1447E-01	1.000	.1219E-01	-.0647
150.0	.4580	.4789	-.2088E-01	1.000	.1054E-01	-.9003
180.0	.4160	.4330	-.1696E-01	1.000	.9997E-02	-.7238
240.0	.3420	.3546	-.1260E-01	1.000	.1170E-01	-.5568
300.0	.3210	.2905	.3047E-01	1.000	.1395E-01	1.430
360.0	.2460	.2380	.7959E-02	1.000	.1542E-01	.3924

```
CORRECTED SUM OF SQUARED OBSERVATIONS =  2.49866
WEIGHTED CORRECTED SUM OF SQUARED OBSERVATIONS =  2.49866
SUM OF SQUARED RESIDUALS =            .649037E-02
SUM OF WEIGHTED SQUARED RESIDUALS =   .649037E-02
S = .254762E-01 WITH    10 DEGREES OF FREEDOM
CORRELATION (OBSERVED,PREDICTED) =   .9987

AIC criteria =        -62.52410
```

```
SC  criteria =          -59.96787

  PARAMETER       ESTIMATE        STANDARD         CV%
                                   ERROR
AUC             258.828848       12.827922        4.96
K10-HL           97.322453        5.756334        5.91
Alpha-HL         14.370155        1.407995        9.80
Beta-HL         208.713182       19.153285        9.18
K10                .007122         .000422        5.92
K12                .021942         .002158        9.83
K21                .022492         .002788       12.39
Volume           54.246898        1.221989        2.25
Cmax              1.843423         .041547        2.25
Cl                 .386356         .019189        4.97
AUMC          71794.173564     9786.620206       13.63
MRT             277.380880       24.442255        8.81
Vss             107.167683        4.588236        4.28
```

Solution III - Three-exponential model

```
TITLE 1
Three-exponential model
MODEL 18
NVARIABLES 2
NPOINTS 100
XNUMBER 1
YNUMBER 2
NCONSTANTS 4
CONSTANTS 100,1,100,0
METHOD 2  'Gauss-Newton (Levenberg and Hartley)
ITERATIONS 50
INITIAL .6,.6,.7,.08,.025,.003
LOWER BOUNDS 0,0,0,0,0,0
UPPER BOUNDS 6,6,7,.8,.5,.1
MISSING 'Missing'
NOBSERVATIONS 14
DATA 'WINNLIN.DAT'
BEGIN
```

```
PARAMETER  ESTIMATE       STANDARD         CV%          UNIVARIATE C.I.
                           ERROR

  A          .658408       .328798        49.94       -.099810       1.416625

  B          .631106       .271901        43.08        .004095       1.258118

  C          .641149       .142564        22.24        .312392        .969905

  Alpha      .088516       .045261        51.13       -.015857        .192889

  Beta       .022000       .014731        66.96       -.011970        .055970

  Gamma      .002541       .000750        29.53        .000810        .004271
```

```
*** CORRELATION MATRIX OF THE ESTIMATES ***
PARAMETER  A          B          C          Alpha      Beta      Gamma
A          1.00000
B          -.919544   1.00000
C          -.825888    .587083   1.00000
Alpha      -.899161    .955907    .691016   1.00000
Beta       -.965323    .833799    .929179    .868399  1.00000
Gamma      -.762110    .508003    .982783    .630763   .874848   1.00000
```

```
Condition_number=        4104.
```

FUNCTION   1

X	OBSERVED Y	PREDICTED Y	RESIDUAL	WEIGHT	SE-PRED	STANDARDIZED RESIDUAL
5.000	1.625	1.621	.3633E-02	1.000	.2071E-01	.6516
10.00	1.384	1.403	-.1923E-01	1.000	.1393E-01	-1.180
15.00	1.280	1.245	.3458E-01	1.000	.1355E-01	2.081
20.00	1.105	1.128	-.2295E-01	1.000	.1231E-01	-1.307
30.00	.9730	.9665	.6453E-02	1.000	.1410E-01	.3994
45.00	.8060	.8186	-.1265E-01	1.000	.1371E-01	-.7672
60.00	.7400	.7223	.1766E-01	1.000	.1309E-01	1.040
90.00	.5820	.5975	-.1546E-01	1.000	.1444E-01	-.9750
120.0	.5300	.5177	.1228E-01	1.000	.1178E-01	.6855
150.0	.4580	.4613	-.3254E-02	1.000	.1171E-01	-.1811
180.0	.4160	.4179	-.1865E-02	1.000	.1288E-01	-.1088
240.0	.3420	.3517	-.9668E-02	1.000	.1179E-01	-.5397
300.0	.3210	.3000	.2095E-01	1.000	.1218E-01	1.187
360.0	.2460	.2571	-.1111E-01	1.000	.1748E-01	-.8941

```
CORRECTED SUM OF SQUARED OBSERVATIONS =  2.49866
WEIGHTED CORRECTED SUM OF SQUARED OBSERVATIONS =  2.49866
SUM OF SQUARED RESIDUALS =          .367930E-02
SUM OF WEIGHTED SQUARED RESIDUALS =  .367930E-02
S =  .214456E-01 WITH     8 DEGREES OF FREEDOM
CORRELATION (OBSERVED,PREDICTED) =  .9993

AIC criteria =        -66.47046
SC  criteria =        -62.63612
```

SUMMARY OF ESTIMATED SECONDARY PARAMETERS

PARAMETER	ESTIMATE	STANDARD ERROR	CV%
Cmax	1.930662	.077687	4.02
Volume	51.795692	2.083551	4.02
K21	.063124	.045496	72.07
K31	.011711	.006994	59.72
K10	.006692	.000947	14.16
K12	.019493	.006267	32.15
K13	.012036	.007234	60.10
K10-HL	103.570756	14.635102	14.13
Alpha-HL	7.830736	4.000087	51.08
Beta-HL	31.506673	21.075535	66.89
Gamma-HL	272.823372	80.486626	29.50
AUC	288.481552	35.590003	12.34
Cl	.346643	.042900	12.38
AUMC	100715.780255	38370.934840	38.10
MRT	349.123816	90.273993	25.86
Vss	121.021193	16.575203	13.70

PK8 - Two-compartment distribution models

Objectives

◆ **To analyze a dataset exhibiting multiphase decay**

◆ **To apply different parameterizations of this behavior**

◆ **To specify the model as an integrated solution and a set of differential equations**

Problem specification

The data used in this exercise are taken from the literature (Colburn [1983]), and show how to discriminate between five different models that incorporate a bi-phasic decline.

A 100 µg intravenous bolus dose of compound X was given and plasma concentrations were measured at the time points shown by the concentration-time data presented in Figure 8.1 and in the program output. The distribution model is shown in Figure 8.2.

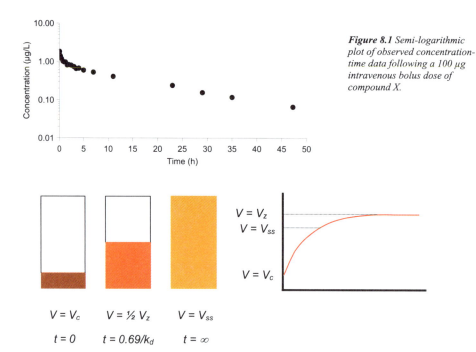

Figure 8.1 Semi-logarithmic plot of observed concentration-time data following a 100 µg intravenous bolus dose of compound X.

Figure 8.2 Volume of distribution as a function of time. Assuming that the drug is administered as an intravenous bolus, it distributes initially in a volume corresponding to V_c and then gradually reaches pseudo-equilibrium at V_z. The boxes to the left represent how the measured apparent volume of distribution increases from time zero to infinity.

Here, we demonstrate and use the simple relationship between distribution volume and time that was originally proposed by Takada and Asada [1981] and Colburn [1983].

Note that the volume increases as a function of time, starting at the central volume, and increasing to V_z. V_z is estimated as $D_{iv}/(AUC \cdot \lambda_z)$ and corresponds to V_β in a two-compartment system, which is calculated as $D_{iv}/(AUC \cdot \beta)$. The true volume of distribution at steady state V_{ss}, is somewhat lower than V_z, since the terminal slope also influences V_z.

One may consider the volume of distribution V as a function of time. Immediately after a bolus dose drug is assumed to distribute momentarily into the central volume. After one distribution half-life (i.e., $t_{1/2d} = 0.693/k_d$) drug has equilibrated into 50% of the distribution volume. At infinite time ($t = \infty$) drug has distributed into the whole body space.

The five models we wish to discriminate between are as follows. The first approach is to specify the model as the traditional bi-exponential model

$$C = A \cdot e^{-\alpha \cdot t} + B \cdot e^{-\beta \cdot t} \tag{8:1}$$

The second method is to specify the volume of distribution as a function of time as in Takada's distribution model

$$C = \frac{D_{iv}}{V_c + V_t} \cdot e^{-\beta \cdot t} \tag{8:2}$$

V_c corresponds to the volume of the central compartment and V_t is defined as

$$V_t = \frac{V_{max} \cdot t}{K_d + t} \tag{8:3}$$

A third approach is to utilize the Colburn distribution model

$$C = \frac{D_{iv}}{V_c + V_t} \cdot e^{-\beta \cdot t} \tag{8:4}$$

where V_t is defined as

$$V_t = V_{max} \cdot \left[1 - e^{-K_v \cdot t}\right] \tag{8:5}$$

Fourthly, one may reparameterize the bi-exponential model with D_{iv} and Cl, and fit this model to the data. The model in Equation 8:6 derives from Equation 8:1

$$Cl = \frac{D_{iv}}{\frac{A}{\alpha} + \frac{B}{\beta}} \tag{8:6}$$

which when rearranged yields

$$A = \alpha \cdot \left[\frac{D_{iv}}{Cl} - \frac{B}{\beta}\right] \tag{8:7}$$

and re-inserted into Equation 8:1 yields

$$C = \alpha \cdot \left[\frac{D_{iv}}{Cl} - \frac{B}{\beta} \right] \cdot e^{-\alpha \cdot t} + B \cdot e^{-\beta \cdot t} \qquad (8{:}8)$$

Finally, the model may be specified in terms of differential equations. A general diagram showing the two-compartment model is drawn in Figure 8.3.

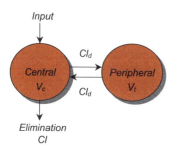

Figure 8.3 Schematic diagram of the two-compartment (bi-exponential) model. The four parameters are V_c, Cl, Cl_d and V_t.

The physiological model parameters are then Cl, V_c, Cl_d, and V_t, which correspond to plasma clearance, the volume of the central compartment, inter-compartmental diffusion, and the volume of the peripheral compartment, respectively. *In* is the input function, which corresponds to the bolus dose in this example. The equation for the central compartment is

$$V_c \frac{dC}{dt} = In - Cl \cdot C - Cl_d \cdot C + Cl_d \cdot C_t \qquad (8{:}9)$$

The equation for the peripheral compartment is

$$V_t \frac{dC_t}{dt} = Cl_d \cdot C - Cl_d \cdot C_t \qquad (8{:}10)$$

Initial parameter estimates

Bi-exponential model

A	$= 2.0\ (\mu g/L)$
α	$= 2.0\ (h^{-1})$
B	$= 1.0\ (\mu g/L)$
β	$= 0.1\ (h^{-1})$

Colburn's model

V_l	$= 100\ (L)$
β	$= 0.1\ (h^{-1})$
V_t	$= 140\ (L)$
k_v	$= 1.0\ (h^{-1})$

Takada's model

V_l	$= 100\ (L)$
β	$= 0.1\ (h^{-1})$
V_{max}	$= 140\ (L)$
k_d	$= 1.0\ (h^{-1})$

Reparameterized Cl-model

Cl	$= 6.0\ (L/h)$
α	$= 2.0\ (h^{-1})$
B	$= 1.0\ (\mu g/L)$
β	$= 0.1\ (h^{-1})$

Differential equation model

V_c	$= 50\ (L)$
Cl	$= 7\ (L \cdot h^{-1})$
Cl_d	$= 50\ (L/h)$
V_t	$= 60\ (L)$

Interpretation of results and conclusions

Observed and predicted (Equation 8:1) data are shown in Figure 8.4.

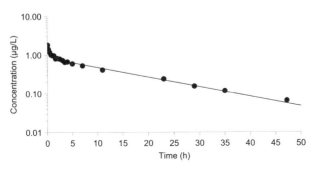

Figure 8.4 _Semi-logarithmic plot of observed (symbols) and predicted (soli line, bi-exponential model) concentration-time data following an intravenous bolus dose._

Five different structural models have been fit to the data. The Poisson error model could either be specified as $WT = 1/F$ in the model file or with the _REWEIGHT –1_ command. Both weighting schemes are equivalent to an iterative reweighted least squares _IRLS_ approach.

In this exercise there is a marginal difference between the quality of fits, and all models generated a low correlation between parameters, and high parameter precision. The difference was in the precision of some of the parameters. Observe that Takada's model gave the lowest _WRSS_, while the differential equation model showed the lowest condition number.

Table 8.1 Final parameter estimates

Model	WRSS	Cond. #
Bi-exponential model	0.0437	125.2
Takada's model	0.0169	3186
Colburn's model	0.0294	2243
Reparameterized model	0.0435	2306
Differential eqn. model	0.0436	29.69

Thus the Takada model would appear to provide the best fit amongst the five models. However, this should be confirmed by an analysis of the residuals. This indicates that the Takada and Colburn models show fewer runs in the residuals than the other models.

In conclusion, in this example we have elaborated a clear bi-exponential system that included enough information (number of observations) during each phase in order to accurately estimate all parameters. One often comes across datasets where the initial phase is hardly discernible, yet it cannot be excluded from analysis. In such a case we would propose the non-compartmental approach (*NCA*), as one does not need to make any assumptions about the number of compartments.

Data were generated with a Poisson error model. This implies that one should use a weighting function according to the formula below, where the exponent λ is equal to unity (1).

$$W_i = \frac{1}{\hat{C}_i}$$

Ideally, to compare the *WRSS*, a weighting scheme with constant weights should have been applied. When this is done, the Takada model is still the model of choice for this dataset.

You have

- analyzed a typical two-compartment (bi-exponential) system
- applied different parameterizations of the two-compartment system
- learned to derive the initial parameter estimates
- characterized the pharmacokinetics after a single bolus dose administration to one subject

The next step in the characterization of kinetics would be to give multiple doses via the intended route of administration.

Solution I - Bi-exponential model

```
TITLE 1
Ordinary bi-exponential model
MODEL
 COMM
  NPARM 4
  NSEC 6
  NCON 1
  PNAMES 'A', 'ALPHA', 'B','BETA'
  SNAMES 'CL','VC','AUC','AUMC','MRT','Vdss'
END
```

```
 1: TEMP
 2: DIV=CON(1)
 3: T=X
 4: END
 5: FUNC1
 6: F=A*DEXP(-ALPHA*T) + B*DEXP(-BETA*T)
 7: WT=1/(F)
 8: END
 9: SECO
10: AUC=A/ALPHA + B/BETA
11: S(1)=DIV/(AUC)
12: S(2)=DIV/(A + B)
13: S(3)=AUC
14: S(4)=A/(ALPHA**2) + B/(BETA**2)
15: S(5)=S(4)/S(3)
16: S(6)=S(1)*S(5)
17: END
18: EOM
```

```
NVARIABLES 2
NPOINTS 100
XNUMBER 1
YNUMBER 2
CONSTANTS 100
METHOD 2  'Gauss-Newton (Levenberg and Hartley)
REWEIGHT -1
ITERATIONS 50
INITIAL 2,2,1,.1
LOWER BOUNDS .1,.1,.1,.01
UPPER BOUNDS 10,5,2,1
NOBSERVATIONS 19
DATA 'WINNLIN.DAT'
BEGIN
```

PARAMETER	ESTIMATE	STANDARD ERROR	CV%	UNIVARIATE C.I. PLANAR C.I.	
A	1.035902	.081171	7.84	.862890	1.208914
ALPHA	1.891659	.260222	13.76	1.337012	2.446307
B	.840451	.026039	3.10	.784950	.895952
BETA	.057496	.002808	4.88	.051510	.063481

```
*** CORRELATION MATRIX OF THE ESTIMATES ***
```

PARAMETER	A	ALPHA	B	BETA
A	1.00000			
ALPHA	.525278	1.00000		
B	-.728035E-01	.600041	1.00000	
BETA	-.774795E-01	.339257	.652658	1.00000

```
Condition_number=      125.2
```

X	OBSERVED Y	PREDICTED Y	RESIDUAL	WEIGHT	SE-PRED	STANDARDIZ RESIDU
.8000E-01	1.810	1.727	.8298E-01	.5790	.6067E-01	2.262
.2500	1.400	1.474	-.7401E-01	.6784	.3561E-01	-1.346
.5000	1.170	1.219	-.4893E-01	.8204	.3072E-01	-.9589
.7500	1.010	1.056	-.4569E-01	.9473	.2940E-01	-.9721
1.000	.9700	.9497	.2027E-01	1.053	.2528E-01	.4397
1.330	.9580	.8623	.9573E-01	1.160	.1946E-01	2.074

1.670	.8000	.8075	-.7495E-02	1.239	.1622E-01	-.1641
2.000	.8190	.7727	.4628E-01	1.294	.1572E-01	1.034
2.500	.7900	.7371	.5292E-01	1.357	.1653E-01	1.223
3.070	.7250	.7076	.1743E-01	1.414	.1703E-01	.4144
3.500	.6470	.6886	-.4163E-01	1.452	.1691E-01	-1.004
4.030	.6630	.6671	-.4133E-02	1.499	.1642E-01	-.1011
5.000	.5910	.6305	-.3955E-01	1.586	.1523E-01	-.9878
7.000	.5240	.5620	-.3798E-01	1.780	.1318E-01	-.9936
11.00	.4060	.4465	-.4052E-01	2.241	.1151E-01	-1.186
23.00	.2370	.2240	.1302E-01	4.471	.1124E-01	.5685
29.00	.1540	.1586	-.4628E-02	6.315	.1039E-01	-.2463
35.00	.1160	.1123	.3654E-02	8.919	.9150E-02	.2346
47.25	.6500E-01	.5555E-01	.9451E-02	18.05	.6374E-02	.8605

```
CORRECTED SUM OF SQUARED OBSERVATIONS =  3.58548
WEIGHTED CORRECTED SUM OF SQUARED OBSERVATIONS =  7.44937
SUM OF SQUARED RESIDUALS =              .384171E-01
SUM OF WEIGHTED SQUARED RESIDUALS =  .436762E-01
S =  .539606E-01 WITH    15 DEGREES OF FREEDOM
CORRELATION (OBSERVED,PREDICTED) =  .9946

AIC criteria =        -51.48808
SC  criteria =        -47.71032

SUMMARY OF ESTIMATED SECONDARY PARAMETERS
```

PARAMETER	ESTIMATE	STANDARD ERROR	CV%
CL	6.594038	.235918	3.58
VC	53.294880	2.368152	4.44
AUC	15.165215	.541957	3.57
AUMC	254.527044	20.558534	8.08
MRT	16.783609	.807726	4.81
VDSS	110.671754	2.829302	2.56

Solution II – Takada's model

```
TITLE 1
Takada model
MODEL
 COMM
  NCON 1
  NPARM 4
  PNAMES 'Vc', 'LAMBD', 'Vmax', 'Kd'
END
   1: TEMP
   2: DOSE=CON(1)
   3: T=X
   4: END
   5: FUNC1
   6: VT = VMAX*T/(KD + T)
   7: F =(DOSE/(VC + VT))*DEXP(-LAMBD*T)
   8: END
   9: EOM
NVARIABLES 2
NPOINTS 100
XNUMBER 1
YNUMBER 2
CONSTANTS 100
METHOD 2  'Gauss-Newton (Levenberg and Hartley)
```

```
REWEIGHT -1
ITERATIONS 50
INITIAL 100,.1,140,1
LOWER BOUNDS 10,.01,10,.01
UPPER BOUNDS 1000,1,1000,5
MISSING 'Missing'
NOBSERVATIONS 19
DATA 'WINNLIN.DAT'
BEGIN
```

PARAMETER	ESTIMATE	STANDARD ERROR	CV%	UNIVARIATE C.I.	
VC	46.916133	2.204572	4.70	42.217217	51.615050
LAMBD	.052328	.002001	3.82	.048063	.056593
VMAX	91.531536	4.138037	4.52	82.711552	100.351519
KD	.772021	.131203	16.99	.492369	1.051672

```
*** CORRELATION MATRIX OF THE ESTIMATES ***
```

PARAMETER	VC	LAMBD	VMAX	KD
VC	1.00000			
LAMBD	-.308644	1.00000		
VMAX	.580365E-01	-.698750	1.00000	
KD	.766153	-.583968	.611471	1.00000

Condition_number=   3186.

X	OBSERVED Y	PREDICTED Y	RESIDUAL	WEIGHT	SE-PRED	STANDARDIZED RESIDUAL
.8000E-01	1.810	1.794	.1606E-01	.5574	.4286E-01	1.185
.2500	1.400	1.424	-.2412E-01	.7020	.2211E-01	-.7223
.5000	1.170	1.175	-.5193E-02	.8507	.1922E-01	-.1681
.7500	1.010	1.045	-.3490E-01	.9568	.1561E-01	-1.142
1.000	.9700	.9628	.7214E-02	1.038	.1275E-01	.2376
1.330	.9580	.8898	.6821E-01	1.124	.1047E-01	2.283
1.670	.8000	.8367	-.3674E-01	1.195	.9446E-02	-1.258
2.000	.8190	.7973	.2167E-01	1.254	.9173E-02	.7595
2.500	.7900	.7508	.3915E-01	1.332	.9289E-02	1.421
3.070	.7250	.7093	.1567E-01	1.410	.9568E-02	.5890
3.500	.6470	.6830	-.3602E-01	1.464	.9725E-02	-1.387
4.030	.6630	.6545	.8466E-02	1.528	.9823E-02	.3345
5.000	.5910	.6099	-.1895E-01	1.639	.9775E-02	-.7792
7.000	.5240	.5360	-.1196E-01	1.866	.9188E-02	-.5250
11.00	.4060	.4246	-.1860E-01	2.355	.7854E-02	-.9114
23.00	.2370	.2215	.1546E-01	4.514	.6777E-02	1.084
29.00	.1540	.1611	-.7129E-02	6.206	.6462E-02	-.6032
35.00	.1160	.1174	-.1368E-02	8.520	.5934E-02	-.1389
47.25	.6500E-01	.6160E-01	.3403E-02	16.23	.4511E-02	.4861

```
CORRECTED SUM OF SQUARED OBSERVATIONS =  3.58548
  WEIGHTED CORRECTED SUM OF SQUARED OBSERVATIONS = 7.25794
  SUM OF SQUARED RESIDUALS =            .129077E-01
  SUM OF WEIGHTED SQUARED RESIDUALS =  .168890E-01
  S = .335550E-01 WITH    15 DEGREES OF FREEDOM
  CORRELATION (OBSERVED,PREDICTED) =  .9982

  AIC criteria =         -69.54071
  SC  criteria =         -65.76296
```

Solution III - Colburn's model

```
TITLE 1
Colburn's model
MODEL
 COMM
  NCON 1
  NPARM 4
  PNAMES 'Vc', 'LAMBD', 'Vmax', 'KV'
END
    1: TEMP
    2: DOSE=CON(1)
    3: T=X
    4: END
    5: FUNC1
    6: VT = VMAX*(1. - DEXP(-KV*T))
    7: F =(DOSE/(VC + VT))*DEXP(-LAMBD*T)
    8: END
    9: EOM
NVARIABLES 2
NPOINTS 100
XNUMBER 1
YNUMBER 2
CONSTANTS 100
METHOD 2  'Gauss-Newton (Levenberg and Hartley)
REWEIGHT -1
ITERATIONS 50
INITIAL 100,.1,140,1
LOWER BOUNDS 10,.01,10,.01
UPPER BOUNDS 1000,1,1000,5
MISSING 'Missing'
NOBSERVATIONS 19
DATA 'WINNLIN.DAT'
BEGIN
```

PARAMETER	ESTIMATE	STANDARD ERROR	CV%	UNIVARIATE C.I.	
VC	50.600938	2.251207	4.45	45.802622	55.399253
LAMBD	.055981	.002411	4.31	.050843	.061120
VMAX	71.915634	3.772498	5.25	63.874774	79.956493
KV	1.087057	.152505	14.03	.762002	1.412113

```
*** CORRELATION MATRIX OF THE ESTIMATES ***
```

PARAMETER	VC	LAMBD	VMAX	KV
VC	1.00000			
LAMBD	-.158571	1.00000		
VMAX	-.310539	-.592481	1.00000	
KV	-.623884	.476112	-.396949	1.00000

```
Condition_number=      2243.
```

X	OBSERVED Y	PREDICTED Y	RESIDUAL	WEIGHT	SE-PRED	STANDARDIZ RESIDUAL
.8000E-01	1.810	1.759	.5083E-01	.5685	.5301E-01	2.012
.2500	1.400	1.456	-.5627E-01	.6868	.2750E-01	-1.229
.5000	1.170	1.204	-.3413E-01	.8307	.2406E-01	-.8088
.7500	1.010	1.057	-.4728E-01	.9461	.2200E-01	-1.187
1.000	.9700	.9622	.7759E-02	1.040	.1905E-01	.1988

1.330	.9580	.8792	.7878E-01	1.138	.1548E-01	2.046
1.670	.8000	.8219	-.2190E-01	1.217	.1314E-01	-.5775
2.000	.8190	.7820	.3705E-01	1.279	.1223E-01	.9963
2.500	.7900	.7382	.5177E-01	1.355	.1244E-01	1.440
3.070	.7250	.7020	.2303E-01	1.425	.1331E-01	.6651
3.500	.6470	.6799	-.3287E-01	1.471	.1380E-01	-.9727
4.030	.6630	.6562	.6811E-02	1.524	.1408E-01	.2065
5.000	.5910	.6185	-.2752E-01	1.617	.1385E-01	-.8616
7.000	.5240	.5518	-.2775E-01	1.812	.1228E-01	-.9097
11.00	.4060	.4409	-.3493E-01	2.268	.1001E-01	-1.264
23.00	.2370	.2252	.1177E-01	4.440	.9184E-02	.6229
29.00	.1540	.1610	-.6971E-02	6.212	.8615E-02	-.4488
35.00	.1160	.1150	.9531E-03	8.692	.7704E-02	.0739
47.25	.6500E-01	.5795E-01	.7051E-02	17.26	.5525E-02	.7736

```
CORRECTED SUM OF SQUARED OBSERVATIONS =  3.58548
WEIGHTED CORRECTED SUM OF SQUARED OBSERVATIONS =  7.35446
SUM OF SQUARED RESIDUALS =            .245902E-01
SUM OF WEIGHTED SQUARED RESIDUALS =  .294043E-01
S = .442751E-01 WITH     15 DEGREES OF FREEDOM
CORRELATION (OBSERVED,PREDICTED) =  .9966

AIC criteria =          -59.00566
SC  criteria =          -55.22790
```

Solution IV– Reparameterized model

```
TITLE 1
Ordinary bi-exponential model (Reparameterized)
MODEL
 COMM
  NPARM 4
  NCON 1
  PNAMES 'CL', 'ALPHA', 'B','BETA'
END
    1: TEMP
    2: DOSE=CON(1)
    3: T=X
    4: END
    5: FUNC1
    6: A=ALPHA*(DOSE/CL - B/BETA)
    7: F=A*DEXP(-ALPHA*T) + B*DEXP(-BETA*T)
    8: F=MAX(0,F)
    9: END
   10: EOM
NVARIABLES 2
NPOINTS 100
XNUMBER 1
YNUMBER 2
CONSTANTS 100
METHOD 2  'Gauss-Newton (Levenberg and Hartley)
REWEIGHT -1
ITERATIONS 50
INITIAL 6,2,1,.1
LOWER BOUNDS .1,0,.1,.01
UPPER BOUNDS 20,5,2,1
MISSING 'Missing'
NOBSERVATIONS 19
DATA 'WINNLIN.DAT'
BEGIN
```

PARAMETER	ESTIMATE	STANDARD ERROR	CV%	UNIVARIATE C.I.

CL	6.595271	.232716	3.53	6.099251	7.091291
ALPHA	1.908416	.270109	14.15	1.332695	2.484137
B	.844412	.026123	3.09	.788732	.900092
BETA	.057752	.002817	4.88	.051748	.063755

```
*** CORRELATION MATRIX OF THE ESTIMATES ***
PARAMETER        CL           ALPHA            B          BETA
CL           1.00000
ALPHA         .827758E-01    1.00000
B             .109473        .623566      1.00000
BETA          .820137        .387165       .655456     1.00000

     Condition_number=        2306.
```

X	OBSERVED Y	PREDICTED Y	RESIDUAL	WEIGHT	SE-PRED	STANDARDIZED RESIDUAL
.8000E-01	1.810	1.727	.8331E-01	.5712	.6115E-01	2.266
.2500	1.400	1.473	-.7295E-01	.6710	.3579E-01	-1.320
.5000	1.170	1.218	-.4795E-01	.8139	.3093E-01	-.9374
.7500	1.010	1.055	-.4534E-01	.9416	.2952E-01	-.9630
1.000	.9700	.9501	.1986E-01	1.048	.2528E-01	.4297
1.330	.9580	.8635	.9445E-01	1.155	.1938E-01	2.041
1.670	.8000	.8094	-.9405E-02	1.234	.1615E-01	-.2054
2.000	.8190	.7750	.4399E-01	1.289	.1570E-01	.9815
2.500	.7900	.7396	.5037E-01	1.352	.1656E-01	1.162
3.070	.7250	.7102	.1483E-01	1.408	.1706E-01	.3522
3.500	.6470	.6912	-.4417E-01	1.447	.1693E-01	-1.064
4.030	.6630	.6696	-.6550E-02	1.494	.1642E-01	-.1599
5.000	.5910	.6327	-.4170E-01	1.581	.1522E-01	-1.040
7.000	.5240	.5636	-.3962E-01	1.775	.1316E-01	-1.035
11.00	.4060	.4474	-.4136E-01	2.237	.1149E-01	-1.210
23.00	.2370	.2237	.1329E-01	4.476	.1124E-01	.5809
29.00	.1540	.1582	-.4197E-02	6.332	.1039E-01	-.2239
35.00	.1160	.1119	.4130E-02	8.957	.9133E-02	.2659
47.25	.6500E-01	.5514E-01	.9860E-02	18.19	.6344E-02	.9013

```
CORRECTED SUM OF SQUARED OBSERVATIONS =   3.58548
WEIGHTED CORRECTED SUM OF SQUARED OBSERVATIONS =  7.41976
SUM OF SQUARED RESIDUALS =            .380368E-01
SUM OF WEIGHTED SQUARED RESIDUALS =  .434788E-01
S =  .538385E-01 WITH      15 DEGREES OF FREEDOM
CORRELATION (OBSERVED,PREDICTED) =   .9947

AIC criteria =        -51.57415
SC  criteria =        -47.79640
```

Solution V - Differential equation model

```
TITLE 1
Differential eqn. model
Model
 COMM
  NDER 2
  NPARM 4
  NCON 1
  CONSTANTS 100.
```

```
    PNAMES 'Vc','CL','Cld','Vt'
END
    1: TEMP
    2: DOSE = CON(1)
    3: END
    4: START
    5: Z(1)=DOSE/VC
    6: Z(2)=0.0
    7: END
    8: DIFF
    9: DZ(1) = (CLD*Z(2) - (CL + CLD)*Z(1))/VC
   10: DZ(2) = (CLD*Z(1) - CLD*Z(2))/VT
   11: END
   12: FUNC 1
   13: F = Z(1)
   14: END
   15: EOM
NVARIABLES 2
NPOINTS 100
XNUMBER 1
YNUMBER 2
METHOD 2  'Gauss-Newton (Levenberg and Hartley)
REWEIGHT -1
ITERATIONS 50
INITIAL 50,7,50,60
LOWER BOUNDS 10,.01,.01,10
UPPER BOUNDS 500,50,250,500
MISSING 'Missing'
NOBSERVATIONS 19
DATA 'WINNLIN.DAT'
BEGIN
```

PARAMETER	ESTIMATE	STANDARD ERROR	CV%	UNIVARIATE C.I.	
VC	53.140381	2.357476	4.44	48.115560	58.165202
CL	6.592733	.237277	3.60	6.086992	7.098474
CLD	51.269621	6.518625	12.71	37.375554	65.163687
VT	57.374961	3.270757	5.70	50.403534	64.346387

```
*** CORRELATION MATRIX OF THE ESTIMATES ***
PARAMETER  VC         CL          CLD        VT
VC        1.00000
CL         .923796E-01  1.00000
CLD       -.674138     -.121239     1.00000
VT        -.559443     -.220229      .270736    1.00000

 Condition_number=       29.69
```

X	OBSERVED Y	PREDICTED Y	RESIDUAL	WEIGHT	SE-PRED	STANDARDIZED RESIDUAL
.8000E-01	1.810	1.730	.8039E-01	.5782	.6060E-01	2.182
.2500	1.400	1.473	-.7279E-01	.6790	.3559E-01	-1.325
.5000	1.170	1.216	-.4556E-01	.8227	.3059E-01	-.8935
.7500	1.010	1.052	-.4223E-01	.9504	.2909E-01	-.8974
1.000	.9700	.9470	.2302E-01	1.056	.2495E-01	.4986
1.330	.9580	.8607	.9729E-01	1.162	.1929E-01	2.108
1.670	.8000	.8070	-.6992E-02	1.239	.1623E-01	-.1532
2.000	.8190	.7730	.4604E-01	1.294	.1576E-01	1.030

```
2.500    .7900     .7380     .5204E-01   1.355    .1648E-01   1.202
3.070    .7250     .7088     .1622E-01   1.411    .1690E-01    .3848
3.500    .6470     .6899    -.4293E-01   1.449    .1675E-01  -1.033
4.030    .6630     .6684    -.5435E-02   1.496    .1628E-01   -.1326
5.000    .5910     .6318    -.4075E-01   1.583    .1519E-01  -1.017
7.000    .5240     .5630    -.3897E-01   1.776    .1332E-01  -1.020
11.00    .4060     .4471    -.4113E-01   2.236    .1181E-01  -1.207
23.00    .2370     .2240     .1299E-01   4.464    .1132E-01    .5676
29.00    .1540     .1586    -.4597E-02   6.305    .1058E-01   -.2459
35.00    .1160     .1122     .3767E-02   8.910    .9315E-02    .2433
47.25    .6500E-01 .5543E-01 .9574E-02   18.04    .6324E-02    .8696
```

```
CORRECTED SUM OF SQUARED OBSERVATIONS =  3.58548
WEIGHTED CORRECTED SUM OF SQUARED OBSERVATIONS =  7.44981
SUM OF SQUARED RESIDUALS =            .377940E-01
SUM OF WEIGHTED SQUARED RESIDUALS =  .436283E-01
S =  .539310E-01 WITH    15 DEGREES OF FREEDOM
CORRELATION (OBSERVED,PREDICTED) =  .9947

AIC criteria =        -51.50894
SC  criteria =        -47.73118
```

PK9 - Two-compartment model discrimination

Objectives
◆ **To analyze a dataset following intravenous administration**
◆ **To use the goodness-of-fit criteria for model discrimination**
◆ **To analyze results with respect to accuracy, precision, and correlation**
◆ **To use the F-test**

Problem specification

The aim of this exercise is to characterize the pharmacokinetics of compound B in a human volunteer. The following plasma concentration-time data are taken from a male volunteer during the 6 h period following an intravenous bolus dose of 100 mg of substance B. Observed data are shown in Figure 9.1 and the program output.

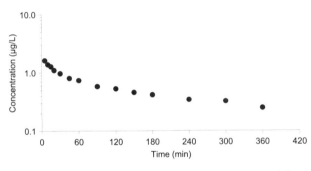

Figure 9.1 *Semi-logarithmic plot of observed concentration-time data following an intravenous bolus dose of 100 mg to a male volunteer.*

We will fit a bi-exponential and three-exponential model to this dataset and analyze the results of the two fits. First, we will therefore implement and fit Equations 9:1 and 9:2.

$$C_p = A \cdot e^{-\alpha \cdot t} + B \cdot e^{-\beta \cdot t} \qquad (9:1)$$

$$C_p = A \cdot e^{-\alpha \cdot t} + B \cdot e^{-\beta \cdot t} + C \cdot e^{-\lambda \cdot t} \qquad (9:2)$$

We will observe the weighted residual sum of squares (*WRSS*) and the correlation coefficient and search for trends in the residual plots. Then we will apply the *Akaike* [1978] and *F*-tests for the bi- and three-exponential models. The *Akaike Information Criterion* (*AIC*) attempts to represent the information content of a given set of parameter estimates by relating the *WRSS* to the number of parameters that were required to obtain the fit. When comparing two models with different numbers of parameters, this criterion places the burden on the model with more parameters to not only have a lower *WRSS*, but also to quantify how much better it must be for the model to be deemed the more appropriate.

$$AIC = N_{obs} \cdot \ln(WRSS) + 2 \cdot N_{par} \qquad (9:3)$$

The *AIC*, as defined above, is dependent on the magnitude of residuals as well as on the number of observations. The most appropriate model is the one with the smallest value of *AIC*. The *Schwarz criterion (SC)* is used in very much the same manner as the *AIC*.

$$SC = N_{obs} \cdot \ln(WRSS) + N_{par} \cdot \ln(N_{obs}) \qquad (9:4)$$

Note: One must be careful when interpreting the result of a fit from only *AIC* and *SC*. Never judge the goodness-of-fit without a battery of statistical tools. We show you the application of these tests in more detail elsewhere in the book (see e.g., PD3).

Initial parameter estimates

$A = 1.1$ (μg/L)
$\alpha = 1.0$ (min^{-1})
$B = 0.2$ (μg/L)
$\beta = 0.01$ (min^{-1})

$A = 1.0$ (μg/L)
$\alpha = 1.0$ (min^{-1})
$B = 0.7$ (μg/L)
$\beta = 0.1$ (min^{-1})
$C = 0.2$ (μg/L)
$\gamma = 0.001$ (min^{-1})

Interpretation of results and conclusions

Let us first compare the fit of the bi- and three-exponential models. Figure 9.2 shows observed and predicted data from the two models. The difference is barely visible within the present concentration and time range. A larger dose and time range would be needed to separate these two models.

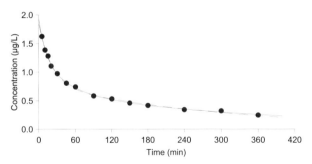

Figure 9.2 *Semi-logarithmic plot of observed (symbols) and predicted (solid line is three-exponential model and dashed line the bi-exponential model) concentration-time data following an intravenous bolus dose of 100 mg to a male volunteer*

For simplicity, we will only use the two runs with constant weight (weight equal to 1 or constant weight). In this case we may apply the F-test. $F(\alpha, \Delta df, df)$, where α, df, and Δdf are the level of significance, degrees of freedom ($N_{obs} - N_{par}$) for the three-exponential model and difference in degrees of freedom between the models, respectively. Δdf corresponds to the column numbers in Table 9.1 and df_{3-exp} is the row number. F_{table} is indicated by the arrows.

Table 9.1 *F-table*

F distribution υ_1, υ_2 degrees of freedom P = 0.05				
υ_1 $\Delta df \rightarrow$	1	2	3	4
υ_2 1	161.45	199.50	215.71	224.58
2	18.513	19.000	19.164	19.247
3	10.128	9.5521	9.2766	9.1172
4	7.7086	6.9443	6.5914	6.3882
5	6.6079	5.7861	5.4095	5.1922
6	5.9874	5.1433	4.7571	4.5337
7	5.5914	4.7374	4.3868	4.1203
8	5.3177	**4.4590**	4.0662	3.8379
9	5.1174	4.2565	3.8626	3.6331
10	4.9646	4.1028	3.7083	3.4780

One approach to testing for the significance of the parameter in any model relates to the following question. *Does the model that includes the new parameter tell us significantly more about the outcome (or response) variable than a model that lacks that parameter?* If we apply an F-test as a measure of whether the three-exponential model is superior to the bi-exponential model we obtain

$$F^* = \frac{\dfrac{|WRSS_1 - WRSS_2|}{df_1 - df_2}}{\dfrac{WRSS_2}{df_2}} = 3.05 < F_{table} \qquad (9:5)$$

Note that F^* is smaller than F_{table} (= 4.459 from column number equal to $\Delta df = 2$ and row equal to $df_2 = 8$, P = 0.05). This suggests that the three-exponential model does not provide a better fit to the data than the bi-exponential model. It does not, however, ensure that the two-compartment model provides an adequate fit.

It is usually a good rule of thumb to choose the simplest model that adequately describes the data. The bi-exponential fit shows a trend in the residuals. The three-exponential fit does not show the same trend in the residuals. However, a high correlation is found between several parameters and the standard errors are generally high for the three-exponential model.

Note: The *F*-test is not appropriate when using non-constant weights such as when weight (*WT*) is equal to *1/F* . Therefore, only the two unweighted runs were used in the example of the *F*-test above.

You have
☐ practised model discrimination
☐ learned to use the F-test
☐ learned to derive the initial estimates

The next step would be to practice this kind of analysis on your own data. See also PK4 for a discussion on model discrimination.

Solution I - Bi-exponential model

```
TITLE 1
Bi-exponential model
MODEL
 COMM
  NPARM 4
  PNAMES 'A', 'Alpha', 'B','Beta'
END
     1: TEMP
     2: T=X
     3: END
     4: FUNC1
     5: F= A*DEXP(-ALPHA*T) + B*DEXP(-BETA*T)
     6: END
     7: EOM
NVARIABLES 2
NPOINTS 100
XNUMBER 1
YNUMBER 2
METHOD 2  'Gauss-Newton (Levenberg and Hartley)
ITERATIONS 50
INITIAL 1.1,1,.2,.01
LOWER BOUNDS 0,0,0,0
UPPER BOUNDS 10,5,1.5,1
NOBSERVATIONS 14
DATA 'WINNLIN.DAT'
BEGIN
```

PARAMETER	ESTIMATE	STANDARD ERROR	CV%	UNIVARIATE C.I.	
A	1.056916	.044977	4.26	.956701	1.157131
ALPHA	.047973	.004700	9.80	.037500	.058447
B	.784751	.042500	5.42	.690054	.879448
BETA	.003308	.000306	9.24	.002626	.003989

```
*** CORRELATION MATRIX OF THE ESTIMATES ***
PARAMETER              A            ALPHA            B            BETA
A             1.00000
ALPHA        -.520257E-01     1.00000
B            -.553873         .813638      1.00000
BETA         -.550489         .681080      .909699      1.00000

  Condition_number=        447.2

   FUNCTION   1

   X          OBSERVED    PREDICTED    RESIDUAL    WEIGHT    SE-PRED    STANDARDI
              Y           Y                                             RESIDUAL
   5.000        1.625       1.603      .2161E-01   1.000    .2179E-01   1.638
  10.00         1.384       1.413     -.2940E-01   1.000    .1323E-01  -1.351
  15.00         1.280       1.261      .1857E-01   1.000    .1252E-01   .8374
  20.00         1.105       1.139     -.3442E-01   1.000    .1364E-01  -1.600
  30.00          .9730       .9612     .1177E-01   1.000    .1382E-01   .5500
  45.00          .8060       .7983     .7739E-02   1.000    .1209E-01   .3453
  60.00          .7400       .7029     .3708E-01   1.000    .1231E-01   1.663
  90.00          .5820       .5968    -.1480E-01   1.000    .1352E-01  -.6856
 120.0           .5300       .5310    -.1007E-02   1.000    .1222E-01  -.0450
 150.0           .4580       .4786    -.2061E-01   1.000    .1056E-01  -.8894
 180.0           .4160       .4329    -.1687E-01   1.000    .9999E-02  -.7203
 240.0           .3420       .3548    -.1281E-01   1.000    .1169E-01  -.5662
 300.0           .3210       .2909     .3006E-01   1.000    .1396E-01   1.411
 360.0           .2460       .2386     .7432E-02   1.000    .1545E-01   .3670

 CORRECTED SUM OF SQUARED OBSERVATIONS =   2.49866
 WEIGHTED CORRECTED SUM OF SQUARED OBSERVATIONS =   2.49866
 SUM OF SQUARED RESIDUALS =            .648679E-02
 SUM OF WEIGHTED SQUARED RESIDUALS =  .648679E-02
 S =  .254692E-01 WITH     10 DEGREES OF FREEDOM
 CORRELATION (OBSERVED,PREDICTED) =   .9987

AIC criteria =        -62.53182
SC  criteria =        -59.97559
```

Solution II - Three-exponential model

```
TITLE 1
Tri-exponential model
MODEL
 COMM
  NPARM 6
  PNAMES 'A', 'Alpha', 'B','Beta','C','Gam'
END
    1: TEMP
    2: T=X
    3: END
    4: FUNC1
    5: F = A*DEXP(-ALPHA*T) + B*DEXP(-BETA*T) + C*DEXP(-GAM*T)
    6: END
    7: EOM
NVARIABLES 2
NPOINTS 100
XNUMBER 1
YNUMBER 2
METHOD 2  'Gauss-Newton (Levenberg and Hartley)
INITIAL 1,1,.7,.1,.2,.001
LOWER BOUNDS 0,0,0,0,0,0
```

```
UPPER BOUNDS 10,5,1,1,1,1
NOBSERVATIONS 14
DATA 'WINNLIN.DAT'
BEGIN
```

PARAMETER	ESTIMATE	STANDARD ERROR	CV%	UNIVARIATE C.I.	
A	.661558	.330992	50.03	-.101720	1.424835
ALPHA	.087959	.044759	50.89	-.015257	.191174
B	.628035	.272446	43.38	-.000232	1.256303
BETA	.021901	.014833	67.73	-.012303	.056105
C	.640430	.144114	22.50	.308099	.972760
GAM	.002537	.000757	29.85	.000791	.004283

```
*** CORRELATION MATRIX OF THE ESTIMATES ***
```

PARAMETER	A	ALPHA	B	BETA	C	GAM
A	1.00000					
ALPHA	-.900329	1.00000				
B	-.919465	.955804	1.00000			
BETA	-.965555	.868745	.833012	1.00000		
C	-.827153	.692605	.587525	.929966	1.00000	
GAM	-.763881	.632763	.508850	.876185	.983019	1.00000

```
Condition_number=        4125.
```

X	OBSERVED Y	PREDICTED Y	RESIDUAL	WEIGHT	SE-PRED	STANDARDIZED RESIDUAL
5.000	1.625	1.621	.3592E-02	1.000	.2070E-01	.6400
10.00	1.384	1.403	-.1942E-01	1.000	.1389E-01	-1.189
15.00	1.280	1.246	.3446E-01	1.000	.1355E-01	2.073
20.00	1.105	1.128	-.2294E-01	1.000	.1231E-01	-1.306
30.00	.9730	.9663	.6663E-02	1.000	.1407E-01	.4116
45.00	.8060	.8184	-.1238E-01	1.000	.1373E-01	-.7518
60.00	.7400	.7222	.1784E-01	1.000	.1308E-01	1.050
90.00	.5820	.5974	-.1544E-01	1.000	.1445E-01	-.9740
120.0	.5300	.5177	.1227E-01	1.000	.1180E-01	.6854
150.0	.4580	.4613	-.3256E-02	1.000	.1170E-01	-.1812
180.0	.4160	.4179	-.1855E-02	1.000	.1289E-01	-.1082
240.0	.3420	.3517	-.9668E-02	1.000	.1183E-01	-.5404
300.0	.3210	.3001	.2092E-01	1.000	.1218E-01	1.185
360.0	.2460	.2572	-.1120E-01	1.000	.1750E-01	-.9034

```
CORRECTED SUM OF SQUARED OBSERVATIONS = 2.49866
WEIGHTED CORRECTED SUM OF SQUARED OBSERVATIONS = 2.49866
SUM OF SQUARED RESIDUALS =          .367925E-02
SUM OF WEIGHTED SQUARED RESIDUALS = .367925E-02
S = .214454E-01 WITH      8 DEGREES OF FREEDOM
CORRELATION (OBSERVED,PREDICTED) =  .9993

AIC criteria =        -66.47065
SC  criteria =        -62.63630
```

PK10 - Simultaneous fitting of intravenous and oral data

Objectives

◆ **To utilize several sources of data**

◆ **To fit intravenous data separately**

◆ **To fit intravenous and oral data simultaneously**

◆ **To refine the combined intravenous/oral model to also include a lag-time**

◆ **To estimate rate and extent of absorption in parallel with the disposition**

Problem specification

A 100 mg intravenous bolus dose and 500 mg oral dose of drug C were given on separate occasions and plasma concentrations were measured (Figure 10.1 and program output).

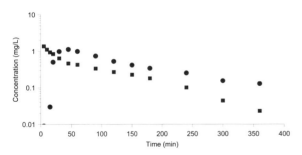

Figure 10.1 *Observed concentration-time data following a 100 mg intravenous bolus (■) and a 500 mg oral (●) dose.*

The goal of this exercise is to separately analyze the intravenous and extravascular (oral) data and then to refine the combined model so as to include a lag-time for the absorption process. By utilizing a combination of several data sources, parameter precision is increased.

In step I we fit only the intravenous data. In step II we fit the intravenous and oral data simultaneously. In step III we fit the intravenous and oral data simultaneously, using a model that includes a lag-time.

Step I: To analyze the intravenous data separately, apply the following equation (Gibaldi and Perrier [1982])

$$C = \frac{D_{iv}}{V_c} \cdot \left\{ \frac{k_{21} - \alpha}{\beta - \alpha} \cdot e^{-\alpha \cdot t} + \frac{k_{21} - \beta}{\alpha - \beta} \cdot e^{-\beta \cdot t} \right\} \tag{10:1}$$

Step II: To fit intravenous and oral data simultaneously, apply Equations 10:1 and 10:2

$$C = \frac{K_a FD_{po}}{V_c}\left\{\frac{(k_{21}-\alpha)\cdot e^{-\alpha\cdot t}}{(K_a-\alpha)(\beta-\alpha)} + \frac{(k_{21}-\beta)\cdot e^{-\beta\cdot t}}{(K_a-\beta)(\alpha-\beta)} + \frac{(k_{21}-K_a)\cdot e^{-K_a\cdot t}}{(\alpha-K_a)(\beta-K_a)}\right\} \qquad (10:2)$$

Step III: The solution to the problem can be found in the behavior of the oral model. We can improve the fit by means of a lag-time as shown in Equation 10:3 and then fit Equations 10:1 and 10:3 simultaneously to the two sets of data.

$$C = \frac{K_a FD_{po}}{V_c}\left\{\frac{(k_{21}-\alpha)\cdot e^{-\alpha\cdot(t-t_{lag})}}{(K_a-\alpha)(\beta-\alpha)} + \frac{(k_{21}-\beta)\cdot e^{-\beta\cdot(t-t_{lag})}}{(K_a-\beta)(\alpha-\beta)} + \frac{(k_{21}-K_a)\cdot e^{-K_a\cdot(t-t_{lag})}}{(\alpha-K_a)(\beta-K_a)}\right\} \quad (10:3)$$

Determine the estimates and coefficient of variation for V_c, k_{21}, α and β for steps I-III above.

Initial parameter estimates for Steps I-III

Step I
$V_c = 60$ (L)
$k_{21} = 0.06$ (min^{-1})
$\alpha = 0.06$ (min^{-1})
$\beta = 0.007$ (min^{-1})

Step II
$V_c = 60$ (L)
$K_a = 0.04$ (min^{-1})
$k_{21} = 0.026$ (min^{-1})
$\alpha = 0.06$ (min^{-1})
$\beta = 0.007$ (min^{-1})
$F = 0.4$

Step III
$V_c = 60$ (L)
$K_a = 0.04$ (min^{-1})
$k_{21} = 0.026$ (min^{-1})
$\alpha = 0.06$ (min^{-1})
$\beta = 0.007$ (min^{-1})
$F = 0.4$
$t_{lag} = 15$ (min)

Interpretation of results and conclusions

In this exercise we modeled intravenous data on its own, followed by intravenous and oral data separately and simultaneously. In Step I, we generated estimates of V_c, k_{21}, α and β that may be somewhat biased since only intravenous data were used. Step II involved simultaneous fit of intravenous and oral data, but C_{max} was poorly fit. The weighted sum of squares, *WRSS* and the estimated average standard deviation of the fit S, were 0.63 and 0.17, respectively. The precision of α and k_{21} were also relatively poor. By including a lag-time into the oral model in Step III, we improved the precision of most parameters and particularly of α and k_{21}. The new function also captured C_{max} appropriately (Figure 10.3).

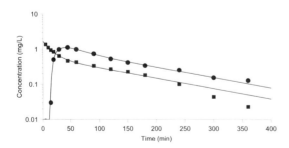

Figure 10.3 Semi-logarithmic plot of observed (symbols) concentration-time data following a 100 mg intravenous bolus (■) and a 500 mg oral (●) dose superimposed on model-predicted (solid lines) data. Note the lag-time indicated by means of the horizontal line at the lower left corner of the plot.

Furthermore, the condition number is very much the same as in Step I, and *WRSS* and *S* were reduced to 0.008 and 0.017, indicating a better fit. The conclusion to be drawn from this exercise is that one should use all available data whenever possible. In this example, this was achieved by simultaneously fitting both models. Using all data simultaneously often improves the parameter precision and avoids bias. Oral data not only contain information about rate K_a and extent F, but also about disposition (V_c, k_{21}, *etc.*). The underlying assumption is that clearance remains constant between the two study occasions. To avoid uncertainties about clearance, a short time interval of only a couple of hours could be used to separate the intravenous and oral dose administration. This approach is discussed in PK12.

Other examples where fitting a combination of data sources is appropriate include multiple dose levels, plasma and urine data, and drug and metabolite data, to mention just a few. See also section 5.3.3 on Combining several sources of data for modeling.

The next step in the characterization of kinetics would be to give multiple doses (at several dose levels) via the intended route of administration.

Solution - Step I: Intravenous data only

```
TITLE 1
IV fit: 2-COMPARTMENT
Model
 COMM
NPARM 4
NCON 1
PNAMES 'V', 'K21', 'Alpha', 'Beta'
END
     1: TEMP
     2: T=X
     3: D1=CON(1)
     4: D=(D1/V)*(ALPHA-K21)/(ALPHA-BETA)
     5: E=(D1/V)*(K21-BETA)/(ALPHA-BETA)
     6: END
     7: FUNC1
     8: F=D*DEXP(-ALPHA*T) + E*DEXP(-BETA*T)
     9: END
    10: EOM
NVARIABLES 2
NPOINTS 100
XNUMBER 1
YNUMBER 2
CONSTANTS 100
METHOD 2  'Gauss-Newton (Levenberg and Hartley)
REWEIGHT -2
ITERATIONS 50
INITIAL 60,.06,.1,.007
LOWER BOUNDS 0,0,0,0
UPPER BOUNDS 80,1,1,.1
MISSING 'Missing'
NOBSERVATIONS 14
DATA 'WINNLIN.DAT'
BEGIN
```

PARAMETER	ESTIMATE	STANDARD ERROR	CV%	UNIVARIATE C.I.	
V	53.161985	13.098688	24.64	23.976107	82.347864
K21	.052279	.016654	31.86	.015171	.089387

| ALPHA | .107071 | .049213 | 45.96 | -.002584 | .216726 |
| BETA | .009348 | .000428 | 4.58 | .008394 | .010302 |

*** CORRELATION MATRIX OF THE ESTIMATES ***

PARAMETER	V	K21	ALPHA	BETA
V	1.00000			
K21	-.547251	1.00000		
ALPHA	-.853644	.893125	1.00000	
BETA	-.251168	.664782	.472280	1.00000

Condition_number=   .6579E+05

FUNCTION   1

X	OBSERVED Y	PREDICTED Y	RESIDUAL	WEIGHT	SE-PRED	STANDARDI RESIDUAL
5.000	1.375	1.406	-.3111E-01	.5058	.1637	-.4243
10.00	1.120	1.114	.5868E-02	.8058	.7782E-01	.0493
15.00	.9440	.9299	.1409E-01	1.157	.6947E-01	.1465
20.00	.8480	.8094	.3863E-01	1.527	.6105E-01	.4639
30.00	.6460	.6668	-.2076E-01	2.251	.3983E-01	-.2763
45.00	.4680	.5511	-.8314E-01	3.296	.3113E-01	-1.319
60.00	.4310	.4733	-.4233E-01	4.469	.2878E-01	-.7980
90.00	.3390	.3564	-.1736E-01	7.890	.2025E-01	-.4269
120.0	.2720	.2692	.2835E-02	13.84	.1341E-01	.0897
150.0	.2290	.2033	.2566E-01	24.26	.9134E-02	1.058
180.0	.1820	.1536	.2839E-01	42.53	.6663E-02	1.543
240.0	.1010	.8767E-01	.1333E-01	130.7	.4468E-02	1.303
300.0	.4400E-01	.5003E-01	-.6034E-02	401.8	.3405E-02	-1.122
360.0	.2300E-01	.2856E-01	-.5555E-02	1235.	.2551E-02	-2.150

CORRECTED SUM OF SQUARED OBSERVATIONS = 2.34247
WEIGHTED CORRECTED SUM OF SQUARED OBSERVATIONS = 9.82941
SUM OF SQUARED RESIDUALS =   .138465E-01
SUM OF WEIGHTED SQUARED RESIDUALS = .163488
S = .127862   WITH   10 DEGREES OF FREEDOM
CORRELATION (OBSERVED,PREDICTED) = .9972

AIC criteria =   -17.35423
SC criteria =   -14.79800

Solution - Step II: Intravenous plus oral data

```
TITLE 1
PO+IV Simultaneous fit without lag-time: 2-COMPARTMENT
Model
 COMM
NPARM 6
NCON 2
NFUN 2
PNAMES 'V', 'Ka', 'K21', 'Alpha', 'Beta','Bio'
END
    1: TEMP
    2: T=X
    3: D1=CON(1)
    4: D2=CON(2)
    5: A=(BIO*D2/V)*KA*(K21-ALPHA)/(ALPHA-BETA)/(ALPHA-KA)
    6: B=-1*(BIO*D2/V)*KA*(K21-BETA)/(ALPHA-BETA)/(BETA-KA)
    7: C=(BIO*D2/V)*KA*(K21-KA)/(BETA-KA)/(ALPHA-KA)
    8: D=(D1/V)*(ALPHA-K21)/(ALPHA-BETA)
```

```
    9: E=(D1/V)*(K21-BETA)/(ALPHA-BETA)
   10: END
   11: FUNC1
   12: F=A*DEXP(-ALPHA*T)+B*DEXP(-BETA*T)+C*DEXP(-KA*T)
   13: END
   14: FUNC2
   15: F=D*DEXP(-ALPHA*T) + E*DEXP(-BETA*T)
   16: END
   17: EOM
NVARIABLES 3
FNUMBER 3
NPOINTS 100
XNUMBER 1
YNUMBER 2
CONSTANTS 100,500
METHOD 2  'Gauss-Newton (Levenberg and Hartley)
ITERATIONS 50
INITIAL 60,.04,.026,.06,.007,.4
LOWER BOUNDS 0,0,0,0,0,.032
UPPER BOUNDS 80,.1,.1,.1,.01,1
MISSING 'Missing'
DATA 'WINNLIN.DAT'
BEGIN
```

PARAMETER	ESTIMATE	STANDARD ERROR	CV%	UNIVARIATE C.I.	
V	72.584651	9.466191	13.04	52.953127	92.216174
KA	.016437	.005321	32.38	.005401	.027473
K21	.009715	.025765	265.21	-.043717	.063147
ALPHA	.025463	.021082	82.79	-.018258	.069185
BETA	.005728	.012751	222.61	-.020715	.032171
BIO	.371910	.079133	21.28	.207799	.536020

```
*** CORRELATION MATRIX OF THE ESTIMATES ***
PARAMETER  V          KA         K21        ALPHA      BETA       BIO
V          1.00000
KA          .191954   1.00000
K21        -.473947    .218040   1.00000
ALPHA      -.661555    .163211    .936984   1.00000
BETA       -.427304    .140748    .983870    .877587   1.00000
BIO        -.054479   -.840777   -.213348   -.135293   -.14677   1.00000

  Condition_number=         .1457E+05


FUNCTION   1
```

X	OBSERVED Y	PREDICTED Y	RESIDUAL	WEIGHT	SE-PRED	STANDARDI RESIDUAL
5.000	.9000E-02	.1916	-.1826	1.000	.3360E-01	-1.103
10.00	.2000E-02	.3488	-.3468	1.000	.5506E-01	-2.172
15.00	.3000E-01	.4766	-.4466	1.000	.6790E-01	-2.888
20.00	.5080	.5792	-.7119E-01	1.000	.7489E-01	-.4702
30.00	.9870	.7232	.2638	1.000	.7929E-01	1.769
45.00	1.124	.8279	.2961	1.000	.7906E-01	1.984
60.00	.9940	.8477	.1463	1.000	.7996E-01	.9834
90.00	.7440	.7651	-.2111E-01	1.000	.8230E-01	-.1431
120.0	.5310	.6316	-.1006	1.000	.8382E-01	-.6859

```
     150.0        .4190        .5036      -.8456E-01  1.000      .8566E-01   -.5809
     180.0        .3420        .3972      -.5522E-01  1.000      .8409E-01   -.3770
     240.0        .2540        .2493       .4664E-02  1.000      .7469E-01    .0307
     300.0        .1540        .1618      -.7831E-02  1.000      .8287E-01   -.0532
     360.0        .1280        .1085       .1946E-01  1.000      .1019        .1444
```

```
CORRECTED SUM OF SQUARED OBSERVATIONS =  1.94848
WEIGHTED CORRECTED SUM OF SQUARED OBSERVATIONS =  1.94848
SUM OF SQUARED RESIDUALS =             .558032
SUM OF WEIGHTED SQUARED RESIDUALS =  .558032
S =  .264110      WITH       8 DEGREES OF FREEDOM
CORRELATION (OBSERVED,PREDICTED) =  .8828
```

```
     FUNCTION   2
```

X	OBSERVED Y	PREDICTED Y	RESIDUAL	WEIGHT	SE-PRED	STANDARDIZ RESIDUAL
5.000	1.375	1.238	.1366	1.000	.1224	1.173
10.00	1.120	1.115	.4930E-02	1.000	.8774E-01	.0341
15.00	.9440	1.006	-.6177E-01	1.000	.7333E-01	-.4060
20.00	.8480	.9089	-.6086E-01	1.000	.7232E-01	-.3987
30.00	.6460	.7465	-.1005	1.000	.8007E-01	-.6760
45.00	.4680	.5646	-.9664E-01	1.000	.8296E-01	-.6568
60.00	.4310	.4360	-.4959E-02	1.000	.7936E-01	-.0332
90.00	.3390	.2774	.6164E-01	1.000	.7975E-01	.4140
120.0	.2720	.1917	.8025E-01	1.000	.8184E-01	.5431
150.0	.2290	.1420	.8701E-01	1.000	.7543E-01	.5757
180.0	.1820	.1105	.7150E-01	1.000	.6487E-01	.4585
240.0	.1010	.7283E-01	.2817E-01	1.000	.5484E-01	.1763
300.0	.4400E-01	.5045E-01	-.6451E-02	1.000	.6093E-01	-.0409
360.0	.2300E-01	.3552E-01	-.1252E-01	1.000	.6541E-01	-.0803

```
CORRECTED SUM OF SQUARED OBSERVATIONS =  2.34247
WEIGHTED CORRECTED SUM OF SQUARED OBSERVATIONS =  2.34247
SUM OF SQUARED RESIDUALS =             .695801E-01
SUM OF WEIGHTED SQUARED RESIDUALS =  .695801E-01
S =  .932605E-01 WITH       8 DEGREES OF FREEDOM
CORRELATION (OBSERVED,PREDICTED) =  .9859
```

```
TOTALS FOR ALL CURVES COMBINED
SUM OF SQUARED RESIDUALS =             .627612
SUM OF WEIGHTED SQUARED RESIDUALS =  .627612
S =  .168902      WITH      22 DEGREES OF FREEDOM
```

```
AIC criteria =          -1.04332
SC  criteria =           6.94991
```

Solution - Step III: The lag-time model

```
TITLE 1
PO+IV Simultaneous fit: 2-COMPARTMENT
Model
 COMM
NPARM 7
NCON 2
NFUN 2
PNAMES 'V', 'Ka', 'K21', 'Alpha', 'Beta','Bio','Tl'
END
    1: TEMP
    2: T=X
    3: D1=CON(1)
```

```
 4: D2=CON(2)
 5: A=(BIO*D2/V)*KA*(K21-ALPHA)/(ALPHA-BETA)/(ALPHA-KA)
 6: B=-1*(BIO*D2/V)*KA*(K21-BETA)/(ALPHA-BETA)/(BETA-KA)
 7: C=(BIO*D2/V)*KA*(K21-KA)/(BETA-KA)/(ALPHA-KA)
 8: D=(D1/V)*(ALPHA-K21)/(ALPHA-BETA)
 9: E=(D1/V)*(K21-BETA)/(ALPHA-BETA)
10: END
11: FUNC1
12: IF T <= TL THEN
13: F=0.0
14: ELSE
15: F=A*DEXP(-ALPHA*(T-TL))+B*DEXP(-BETA*(T-TL))+C*DEXP(-KA*(T-TL))
16: ENDIF
17: END
18: FUNC2
19: F=D*DEXP(-ALPHA*T) + E*DEXP(-BETA*T)
20: END
21: EOM
NVARIABLES 3
FNUMBER 3
NPOINTS 100
XNUMBER 1
YNUMBER 2
CONSTANTS 100,500
METHOD 2  'Gauss-Newton (Levenberg and Hartley)
ITERATIONS 50
INITIAL 60,.04,.026,.06,.007,.4,15
LOWER BOUNDS 0,0,0,0,0,.032,0
UPPER BOUNDS 80,.1,.1,.1,.01,1,25
MISSING 'Missing'
DATA 'WINNLIN.DAT'
BEGIN
```

PARAMETER	ESTIMATE	STANDARD ERROR	CV%	UNIVARIATE C.I.	
V	59.910749	1.259090	2.10	57.292349	62.529150
KA	.047161	.001862	3.95	.043289	.051033
K21	.025720	.002348	9.13	.020837	.030603
ALPHA	.061045	.004125	6.76	.052466	.069624
BETA	.006806	.000382	5.62	.006011	.007602
BIO	.318727	.005648	1.77	.306982	.330472
TL	14.819667	.147757	1.00	14.512391	15.126943

```
*** CORRELATION MATRIX OF THE ESTIMATES ***
```

PARAMETER	V	KA	K21	ALPHA	BETA	BIO	TL
V	1.00000						
KA	.236870	1.00000					
K21	-.487678	.246096	1.00000				
ALPHA	-.753679	.092298	.912874	1.00000			
BETA	-.327654	.056861	.862702	.687091	1.00000		
BIO	-.014122	-.700261	-.267045	-.126133	-.08951	1.00000	
TL	-.126821	.353510	.153950	.180740	.06091	-.153526	1.00000

```
 Condition_number=       .1313E+05

    FUNCTION   1
```

X	OBSERVED Y	PREDICTED Y	RESIDUAL	WEIGHT	SE-PRED	STANDARDI RESIDUAL
5.000	.9000E-02	.0000	.9000E-02	1.000		
10.00	.2000E-02	.0000	.2000E-02	1.000		
15.00	.3000E-01	.2244E-01	.7559E-02	1.000	.1795E-01	.9980
20.00	.5080	.5173	-.9282E-02	1.000	.1268E-01	-.6272
30.00	.9870	.9944	-.7368E-02	1.000	.1442E-01	-.5624
45.00	1.124	1.107	.1713E-01	1.000	.1184E-01	1.107
60.00	.9940	.9939	.5337E-04	1.000	.1163E-01	.0034
90.00	.7440	.7286	.1542E-01	1.000	.1133E-01	.9730
120.0	.5310	.5539	-.2291E-01	1.000	.1006E-01	-1.373
150.0	.4190	.4386	-.1962E-01	1.000	.8275E-02	-1.112
180.0	.3420	.3539	-.1194E-01	1.000	.7578E-02	-.6650
240.0	.2540	.2342	.1977E-01	1.000	.8488E-02	1.127
300.0	.1540	.1556	-.1631E-02	1.000	.8809E-02	-.0938
360.0	.1280	.1034	.2455E-01	1.000	.8109E-02	1.386

CORRECTED SUM OF SQUARED OBSERVATIONS = 1.94848
WEIGHTED CORRECTED SUM OF SQUARED OBSERVATIONS = 1.94848
SUM OF SQUARED RESIDUALS =   .286239E-02
SUM OF WEIGHTED SQUARED RESIDUALS = .286239E-02
S = .202216E-01 WITH   7 DEGREES OF FREEDOM
CORRELATION (OBSERVED,PREDICTED) = .9993

FUNCTION   2

X	OBSERVED Y	PREDICTED Y	RESIDUAL	WEIGHT	SE-PRED	STANDARDI RESIDUAL
5.000	1.375	1.364	.1128E-01	1.000	.1697E-01	1.179
10.00	1.120	1.134	-.1416E-01	1.000	.1012E-01	-.8507
15.00	.9440	.9607	-.1667E-01	1.000	.9624E-02	-.9839
20.00	.8480	.8286	.1937E-01	1.000	.1002E-01	1.159
30.00	.6460	.6487	-.2704E-02	1.000	.9585E-02	-.1594
45.00	.4680	.4982	-.3020E-01	1.000	.9158E-02	-1.756
60.00	.4310	.4148	.1619E-01	1.000	.9487E-02	.9514
90.00	.3390	.3199	.1908E-01	1.000	.8503E-02	1.088
120.0	.2720	.2579	.1409E-01	1.000	.6799E-02	.7718
150.0	.2290	.2098	.1919E-01	1.000	.5927E-02	1.034
180.0	.1820	.1710	.1102E-01	1.000	.5757E-02	.5919
240.0	.1010	.1136	-.1265E-01	1.000	.5777E-02	-.6795
300.0	.4400E-01	.7554E-01	-.3154E-01	1.000	.5386E-02	-1.684
360.0	.2300E-01	.5022E-01	-.2722E-01	1.000	.4665E-02	-1.439

CORRECTED SUM OF SQUARED OBSERVATIONS = 2.34247
WEIGHTED CORRECTED SUM OF SQUARED OBSERVATIONS = 2.34247
SUM OF SQUARED RESIDUALS =   .511032E-02
SUM OF WEIGHTED SQUARED RESIDUALS = .511032E-02
S = .270194E-01 WITH   7 DEGREES OF FREEDOM
CORRELATION (OBSERVED,PREDICTED) = .9989

TOTALS FOR ALL CURVES COMBINED
SUM OF SQUARED RESIDUALS =   .797271E-02
SUM OF WEIGHTED SQUARED RESIDUALS = .797271E-02
S = .194847E-01 WITH   21 DEGREES OF FREEDOM

AIC criteria =   -121.28845
SC criteria =   -111.96302

PK11 – Two-compartment repeated oral dosing

Objectives

◆ **To fit multiple dose kinetic and dynamic data sequentially**

◆ **To fix the kinetics and then fit a dynamic model to the effect-time data**

◆ **To select an appropriate weighting scheme**

Problem specification

The aim of this exercise is to show that there may be more than one way to satisfactorily model a single dataset. A study was undertaken to determine the pharmacokinetics and the pharmacodynamics in human volunteers after repeated administration of a new compound. On Day 1, each subject received a single 400 mg oral dose. On Day 5 (at 96 h), the subjects were started on 400 mg *tid.* (three times daily) for five days (days 5 - 9). On Day 9, they received only the morning dose. Blood samples were taken at various times until 72 h after the last dose. The drug effects were measured prior to each blood sampling. Data are presented in Figure 11.1 and the program output.

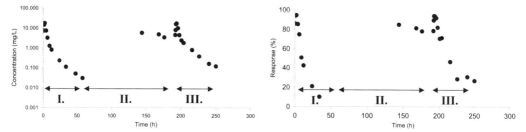

Figure 11.1 *Observed concentration-time data (left) and effect-time data (right) during the first dose (I), repetitive dosing (II), and last dose (III).*

We will fit a two-compartment library model with first-order input plus a lag-time to the kinetic concentration-time data (Gibaldi and Perrier [1982]). The equation for a two-compartment model with first-order input, K_a is

$$C = \frac{K_a F D_{po}}{V_c} \left\{ \frac{(k_{21} - \alpha) \cdot e^{-\alpha \cdot t}}{(K_a - \alpha)(\beta - \alpha)} + \frac{(k_{21} - \beta) \cdot e^{-\beta \cdot t}}{(K_a - \beta)(\alpha - \beta)} + \frac{(k_{21} - K_a) \cdot e^{-K_a \cdot t}}{(\alpha - K_a)(\beta - K_a)} \right\} \quad (11:1)$$

An alternative to Equation 11:1 which includes a lag-time takes the form

$$C = \frac{K_a F D_{po}}{V_c} \left\{ \frac{(k_{21} - \alpha) \cdot e^{-\alpha \cdot (t - t_{lag})}}{(K_a - \alpha)(\beta - \alpha)} + \frac{(k_{21} - \beta) \cdot e^{-\beta \cdot (t - t_{lag})}}{(K_a - \beta)(\alpha - \beta)} + \frac{(k_{21} - K_a) \cdot e^{-K_a \cdot (t - t_{lag})}}{(\alpha - K_a)(\beta - K_a)} \right\} \quad (11:2)$$

After having obtained the kinetic estimates, we will fix them in order to estimate the pertinent dynamic parameters from the effect-time data. Figure 11.2 displays the observed concentration-effect relationship.

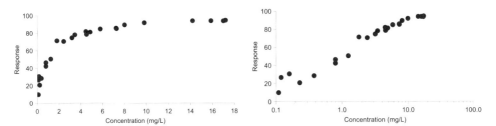

Figure 11.2 *Observed concentration-effect data from the complete collection period on a linear scale (left) and semi-logarithmic scale (right). The approximate EC_{50} value is 1-2 mg/L.*

The effect appears linear in the 20 - 80% interval of maximal effect when plotted against the logarithm of the concentration. Levy [1995] derived the slope at a concentration (C) equal to EC_{50} according to

$$E = \frac{E_{max}C^n}{EC_{50}^n + C^n} = \frac{E_{max}e^{n\ln[C]}}{EC_{50}^n + e^{n\cdot\ln[C]}}$$

(11:3)

And the derivative of E with respect to $\ln[C]$

$$\frac{dE}{d\ln[C]} = \frac{n\cdot E_{max}EC_{50}^n[C]^n}{\{EC_{50}^n + [C]^n\}^2}$$

(11:4)

The slope at $C = EC_{50}$ is then

$$m = \frac{dE}{d\ln[C]} = \frac{n\cdot E_{max}EC_{50}^{2n}}{[2\cdot EC_{50}^n]^2}$$

(11:5)

and rewritten

$$m = \frac{nE_{max}}{4}$$

(11:6)

The m parameter is the slope of the tangent to the curve at EC_{50}.

$$m = \frac{E_2 - E_1}{\ln[C_2] - \ln[C_1]} = \frac{n\cdot E_{max}}{4}$$

(11:7)

which rearranged gives the exponent n

$$n = \frac{4m}{E_{max}} \tag{11:8}$$

The simplest form of the relationship shown in Figure 11.2 and Equation 11:3 is represented by the hyperbolic function ($n = 1$) that we call the ordinary E_{max} model

$$E = \frac{E_{max}C}{EC_{50} + C} \tag{11:9}$$

E_{max} is the maximum effect and EC_{50} is the concentration at 50% of maximal effect. E_{max} is often called the intrinsic activity and EC_{50} the potency of a drug. Throughout this book, E_{max} corresponds to the maximum drug-induced effect.

We propose the following approach to modeling this dataset:
- Write the kinetic command file and utilize a library model
- Treat the kinetic parameters from that run as fixed values in the dynamic model
- Fit the dynamic model to the effect-time data

Initial kinetic and dynamic parameter estimates

A	= 70 (mg/L)
B	= 1 (mg/L)
K_a	= 0.7 (h^{-1})
α	= 0.6 (h^{-1})
β	= 0.07 (h^{-1})
t_{lag}	= 0.4 (h)
E_{max}	= 100 (units)
EC_{50}	= 2 (mg/L)

Interpretation of results and conclusions

This exercise includes the analysis of multiple-compartment behavior, single and multiple dose kinetic and dynamic data, and instantaneous equilibration between an observed effect and plasma concentration. We needed to weight the kinetic data since the concentration range covers three orders of magnitude.

The kinetic data demonstrate that orally administered drug was rapidly absorbed from the gastrointestinal tract. When the drug was repeatedly administered, the plasma concentrations after each dose showed good conformity with the concentration curve estimated from the single dose data. The kinetic model had well-defined high precision parameters, with generally low correlations. Data indicated rapid absorption with a 20 min lag-time (0.3 h). The model-predicted concentrations were consistent with the observed data, as shown in Figure 11.3 (left).

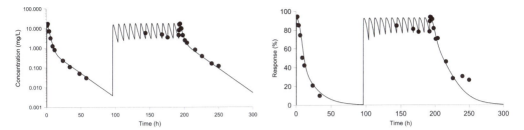

Figure 11.3 *Observed (circles) and model-predicted (curves) concentration-time data (left) and effect-time data (right).*

The kinetic parameters were then fixed in the fit of the E_{max} model to multiple dose-effect data. The original kinetic model (Model #14 in the compiled library) was modified (*Dynamic analysis* in the output) to include the E_{max} model. The maximum (E_{max}) predicted effect was 97% and EC_{50} was 0.8 mg/L. Both parameters had high precision. Figure 11.3 (right) shows the consistency between the observed and model-predicted effect-time data.

The proposed multiple dose regimen of 400 mg every 8 h demonstrates an acceptable regimen for establishment of a good therapeutic effect. Figure 11.4 clearly demonstrates the relationship between effect and concentration. Aiming at a therapeutic concentration of about 5 mg/L will establish 80% of the maximal effect.

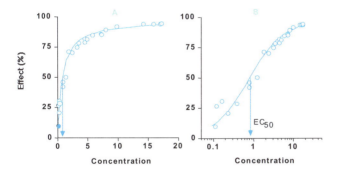

Figure 11.4 *Observed and predicted concentration-effect data from the complete collection period on a linear scale (A) and semi-logarithmic scale (B).*

In conclusion, there are basically three ways of analyzing this dataset. The first is to perform the analysis as we have shown here, by first fitting the kinetics, then fixing the kinetic parameters, and finally fitting the dynamics. A second approach would have been to fit the kinetics and dynamics simultaneously. The third and most simple approach, which is shown Figure 11.4, would have been to fit concentration-effect data directly by means of the E_{max} model.

In this exercise, we have practised
☐ regression modeling of multiple dose kinetic data
☐ multi-compartment modeling
☐ fitting kinetic and dynamic data sequentially
☐ fitting of an ordinary E_{max} model to multiple dose dynamic data
☐ weighting

The next step in the characterization of kinetics would be to give multiple doses at the intended dose level.

Solution I - Kinetic analysis

```
TITLE 1
PK/PD data: kinetics
MODEL 14
NVARIABLES 2
NPOINTS 1000
XNUMBER 1
YNUMBER 2
WEIGHT -2
NCONSTANTS 30
CONSTANTS 400,14,400,0,400,96, &
          400,104,400,112,400,120,400, &
          128,400,136,400,144,400,152, &
          400,160,400,168,400,176,400, &
          184,400,192
METHOD 2  'Gauss-Newton (Levenberg and Hartley)
ITERATIONS 50
INITIAL 50,1,.7,.6,.07,.4
LOWER BOUNDS .1,0,0,0,0,0
UPPER BOUNDS 100,20,5,5,1,1
MISSING 'Missing'
NOBSERVATIONS 26
DATA 'WINNLIN.DAT'
BEGIN
```

PARAMETER	ESTIMATE	STANDARD ERROR	CV%	UNIVARIATE C.I.	
A	27.468673	8.159942	29.71	10.447460	44.489886
B	.893847	.222691	24.91	.429325	1.358369
K01	1.934126	.980975	50.72	-.112136	3.980388
Alpha	.380654	.049524	13.01	.277350	.483958
Beta	.056894	.005193	9.13	.046061	.067727
Tlag	.327536	.087110	26.60	.145829	.509242

```
*** CORRELATION MATRIX OF THE ESTIMATES ***
PARAMETER  A         B         K01       Alpha     Beta      Tlag
A          1.00000
B           .30680   1.00000
K01        -.87145   -.24243   1.00000
Alpha       .90010    .55966   -.71697   1.00000
```

```
Beta      .26501     .95748    -.20324    .50083    1.00000
Tlag    -.58662    -.13773     .84752   -.42913    -.106527    1.00000
 Condition_number=        5578.
```

```
   FUNCTION   1
```

X	OBSERVED Y	PREDICTED Y	RESIDUAL	WEIGHT	SE-PRED	STANDARDIZ RESIDUAL
.5000	7.240	6.291	.9495	.2863E-03	1.121	.9418
1.000	14.21	14.40	-.1906	.7431E-04	2.113	-.0920
2.000	17.20	14.23	2.971	.5072E-04	1.583	.9245
4.000	7.270	7.490	-.2196	.2839E-03	.8580	-.1760
6.000	3.180	3.817	-.6370	.1484E-02	.3168	-1.095
9.000	1.250	1.558	-.3076	.9603E-02	.1523	-1.457
12.00	.8000	.7831	.1691E-01	.2345E-01	.8131E-01	.1163
24.00	.2300	.2358	-.5813E-02	.2836	.2970E-01	-.1547
34.00	.1100	.1317	-.2168E-01	1.240	.1250E-01	-1.129
48.00	.5000E-01	.5934E-01	-.9340E-02	6.002	.4274E-02	-.9835
58.00	.3000E-01	.3359E-01	-.3594E-02	16.67	.3155E-02	-.6664
144.0	5.790	3.032	2.758	.4476E-03	.1974	2.319
168.0	4.860	3.108	1.752	.6353E-03	.1973	1.764
176.0	3.450	3.118	.3321	.1261E-02	.1973	.4807
192.0	4.530	3.128	1.402	.7312E-03	.1973	1.520
192.5	8.010	9.105	-1.095	.2339E-03	1.122	-.8868
193.0	15.87	16.95	-1.079	.5958E-04	2.123	-.4261
194.0	17.01	16.36	.6509	.5186E-04	1.543	.2041
196.0	9.840	9.082	.7582	.1550E-03	.8340	.4050
198.0	4.470	5.094	-.6236	.7510E-03	.3164	-.7123
201.0	2.420	2.551	-.1310	.2562E-02	.1700	-.2760
204.0	1.810	1.594	.2159	.4580E-02	.1190	.6037
216.0	.8000	.6376	.1624	.2345E-01	.5176E-01	1.026
226.0	.3800	.3590	.2097E-01	.1039	.2574E-01	.2802
240.0	.1600	.1619	-.1850E-02	.5861	.1731E-01	-.0649
250.0	.1200	.9163E-01	.2837E-01	1.042	.1370E-01	1.357

```
CORRECTED SUM OF SQUARED OBSERVATIONS = 776.160
WEIGHTED CORRECTED SUM OF SQUARED OBSERVATIONS = .319310
SUM OF SQUARED RESIDUALS =              26.9091
SUM OF WEIGHTED SQUARED RESIDUALS = .130174E-01
S = .255122E-01 WITH    20 DEGREES OF FREEDOM
CORRELATION (OBSERVED,PREDICTED) = .9841
```

```
AIC criteria =       -100.87812
SC  criteria =        -93.32954
```

```
AUC (0 to last time) computed by trapezoidal rule =        662.183
```

PARAMETER	ESTIMATE	STANDARD ERROR	CV%
K10	.313218	.037556	11.99
K12	.055187	.013596	24.64
K21	.069143	.008352	12.08
AUC	73.208285	4.818364	6.58
K10-HL	2.212988	.265166	11.98
K01-HL	.358377	.181585	50.67
Alpha-HL	1.820936	.236670	13.00
Beta-HL	12.183196	1.110967	9.12
Volume/F	17.444296	2.893982	16.59
CL/F	5.463862	.358676	6.56
Tmax	1.390126	.234927	16.90
Cmax	15.539546	1.782462	11.47

Solution II - Dynamic analysis

```
TITLE 1
Milan dynamics - multiple doses
Model
 COMM
NPARM 2
PNAMES 'Emax', 'EC50'
END
    1: TEMP
    2: DOSE1=CON(1)
    3: A=27.47/DOSE1
    4: B=0.894/DOSE1
    5: C=-1*(A+B)
    6: K01=1.934
    7: ALPHA=0.381
    8: BETA=0.0569
    9: TLAG=0.328
   10: END
   11: FUNC1
   12: I=0
   13: J=2
   14: L=CON(2)
   15: GREEN:
   16: I=I+1
   17: J=J+2
   18: IF X <= CON(J) THEN GOTO RED
   19: ELSE IF I < L THEN GOTO GREEN
   20: ENDIF
   21: ENDIF
   22: I=I+1
   23: RED:
   24: L=I-1
   25: SUM=0
   26: I=0
   27: J=2
   28: BLUE:
   29: I=I+1
   30: J=J+2
   31: T=X - CON(J) - TLAG
   32: D=CON(J-1)
   33: AMT=A*DEXP(-ALPHA*T) + B*DEXP(-BETA*T) + C*DEXP(-K01*T)
   34: SUM=SUM + MAX(0,D*AMT)
   35: IF I < L THEN GOTO BLUE
   36: ENDIF
   37: CP = SUM
   38: F= EMAX*CP/(EC50 + CP)
   39: END
   40: EOM
NVARIABLES 2
NPOINTS 1000
XNUMBER 1
YNUMBER 2
NCONSTANTS 30
CONSTANTS 400,14,400,0,400,96, &
          400,104,400,112,400,120,400, &
          128,400,136,400,144,400,152, &
          400,160,400,168,400,176,400, &
          184,400,192
METHOD 2   'Gauss-Newton (Levenberg and Hartley)
ITERATIONS 50
INITIAL 100,2
```

```
LOWER BOUNDS 50,.1
UPPER BOUNDS 110,20
NOBSERVATIONS 24
DATA 'WINNLIN.DAT'
BEGIN
```

PARAMETER	ESTIMATE	STANDARD ERROR	CV%	UNIVARIATE C.I.	
EMAX	97.205001	2.769776	2.85	91.460884	102.949119
EC50	.783805	.108788	13.88	.558195	1.009415

```
*** CORRELATION MATRIX OF THE ESTIMATES ***
```

PARAMETER	EMAX	EC50
EMAX	1.00000	
EC50	.760515	1.00000

```
 Condition_number=        39.25

FUNCTION   1
```

X	OBSERVED Y	PREDICTED Y	RESIDUAL	WEIGHT	SE-PRED	STANDARDIZED RESIDUAL
.5000	85.69	86.41	-.7218	1.000	1.688	-.1150
1.000	93.90	92.18	1.715	1.000	2.167	.2800
2.000	94.43	92.13	2.302	1.000	2.161	.3757
4.000	85.20	87.99	-2.788	1.000	1.792	-.4465
6.000	74.62	80.63	-6.006	1.000	1.500	-.9500
9.000	50.48	64.63	-14.15	1.000	2.001	-2.289
12.00	42.31	48.54	-6.234	1.000	2.487	-1.039
24.00	20.85	22.48	-1.630	1.000	1.954	-.2631
34.00	9.950	13.98	-4.031	1.000	1.381	-.6350
144.0	84.83	77.22	7.613	1.000	1.524	1.205
168.0	81.21	77.61	3.600	1.000	1.517	.5698
176.0	78.14	77.66	.4818	1.000	1.516	.0762
192.0	78.51	77.71	.8021	1.000	1.515	.1270
192.5	89.43	89.49	-.5557E-01	1.000	1.910	-.0089
193.0	93.92	92.91	1.014	1.000	2.244	.1663
194.0	93.71	92.76	.9516	1.000	2.228	.1559
196.0	91.78	89.48	2.304	1.000	1.909	.3711
198.0	81.94	84.23	-2.288	1.000	1.581	-.3631
201.0	70.46	74.34	-3.877	1.000	1.599	-.6157
204.0	71.24	65.15	6.094	1.000	1.979	.9847
216.0	46.33	43.60	2.730	1.000	2.523	.4560
226.0	28.51	30.54	-2.025	1.000	2.313	-.3336
240.0	30.68	16.63	14.05	1.000	1.582	2.229
250.0	26.69	10.17	16.52	1.000	1.059	2.577

```
 CORRECTED SUM OF SQUARED OBSERVATIONS =  16918.8
 WEIGHTED CORRECTED SUM OF SQUARED OBSERVATIONS =  16918.8
 SUM OF SQUARED RESIDUALS =          928.659
 SUM OF WEIGHTED SQUARED RESIDUALS =  928.659
 S =  6.49706     WITH    22 DEGREES OF FREEDOM
 CORRELATION (OBSERVED,PREDICTED) =  .9745

 AIC criteria =       168.00980
 SC  criteria =       170.36591
```

PK12 - Intravenous and oral dosing

Objectives

◆ **To analyze data after semi-simultaneous oral and intravenous administration**

◆ **To write a multi-compartment model in terms of differential equations**

◆ **To obtain initial parameter estimates**

◆ **To discuss various aspects of experimental design**

Problem specification

This exercise demonstrates the semi-simultaneous method of determining bioavailability. When we want to estimate the relative and/or absolute bioavailability, the drug must be given by different routes. The *test* regimen is usually given extravascularly (e.g., orally) and the *reference* dose intravenously. One administers the two regimens at different times separated by a so-called washout period. The washout period is meant to be of sufficient duration for the compound to be washed out (eliminated) from the organism. This time can vary from a day to as long as several months. However, it is important that clearance is the same on the two study occasions.

In order to eliminate the risk of a non-stationary clearance, the *test* and *reference* regimens can be given separated with only a small time window (e.g., 30 min to 1 h). With such a short interval, it is less likely that clearance or any other parameter will have changed (Figure 12.1).

In this exercise, the *test* dose of drug was given extravascularly, 60 min before the intravenous *reference* dose. The *reference* dose was given as a 15 min constant-rate intravenous infusion.

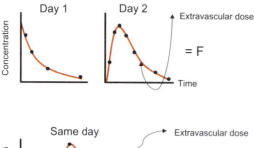

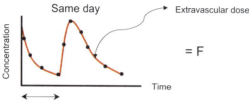

Figure 12.1 Schematic representation of how bioavailability F can be estimated. Top figure shows the traditional approach of separating the reference dose (usually intravenous) and the extravascular dose (usually oral) by means of one or several days or weeks. The bottom figure demonstrates how the reference and extravascular doses may be separated by less than a few hours. The order of the reference and test doses has to be decided on a case-by-case basis. Clearance is assumed to vary less between the two dosing occasions in the latter situation.

Concentration-time data from this experiment are shown in Figure 12.2 and in the program output.

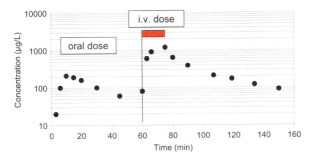

Figure 12.2 *Observed concentration-time data of drug X after an oral dose of 2.5 mg/kg followed by a 15 min constant rate intravenous infusion of 0.5 mg/kg starting at 60 min. The horizontal bar corresponds to the length of the intravenous infusion. Note that the order of dosing of the reference and test doses are different from Figure 12.1*

The concentration in the central compartment is written as

$$V_c \cdot \frac{dC}{dt} = Input - Cl \cdot C - Cl_d \cdot C + Cl_d \cdot C_t \qquad (12:1)$$

Input is the oral and intravenous dosing regimen. The concentration in the peripheral compartment is written as

$$V_t \cdot \frac{dC_t}{dt} = Cl_d \cdot C - Cl_d \cdot C_t \qquad (12:2)$$

Initial parameter estimates

The terminal slope (β) is obtained from

$$slope_\beta = -\beta = \frac{\ln(2)}{t_{1/2\alpha}} = \frac{\ln(640/90)}{80 - 150} \approx -0.03 \, min^{-1} \qquad (12:3)$$

The initial slope (α) is approximately

$$slope_\alpha = -\alpha = \frac{\ln(2)}{t_{1/2\alpha}} \approx \frac{\ln(1200/640)}{75 - 80} \approx -0.13 \, min^{-1} \qquad (12:4)$$

The k_{10} parameter is estimated from

$$k_{10} = \frac{Cl}{V_c} \approx \frac{Cl}{Dose/C_{peak}} \approx 0.18 \, min^{-1} \qquad (12:5)$$

The k_{21} parameter is estimated from α, β and k_{10} according to

$$k_{21} = \frac{\alpha \cdot \beta}{k_{10}} = \frac{0.13 \cdot 0.03}{0.18} \approx 0.02 \ min^{-1} \tag{12:6}$$

An approximate (overestimated) value of V_c is estimated as

$$V_c \approx \frac{Dose_{iv}}{C_{peak}} = \frac{500}{1200} \approx 0.4 \ L/kg \tag{12:7}$$

Bioavailability F is estimated as

$$F = \frac{AUC_{po}}{AUC_{iv}} \cdot \frac{Dose_{iv}}{Dose_{po}} = \frac{8000}{36000} \cdot \frac{0.5}{2.5} \approx 0.04 \tag{12:8}$$

As shown in Figure 12.3, lag-time t_{lag} can be determined from the semi-logarithmic plot of concentration *versus* time. It is approximately 5 min.

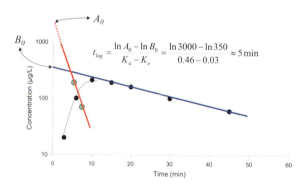

$$t_{lag} = \frac{\ln A_0 - \ln B_0}{K_a - K_e} = \frac{\ln 3000 - \ln 350}{0.46 - 0.03} \approx 5 \min$$

Figure 12.3 *Observed concentration-time data of drug X after the oral dose of 2.5 mg/kg. The blue line corresponds to the terminal slope, the red line is the residual based on the difference between the back-extrapolated blue line and the dashed black curve. The two green symbols are the differences between the back-extrapolated blue line and the dashed upswing on the initial phase. The intersection of the blue and red curves is the initial estimate of the lag-time.*

Implement a two-compartment disposition model parameterized with Cl, V_c, V_t and Cl_d.

Cl	=	0.01-0.02 L/(kg·min)	Obtained from $Dose/AUC$
V_c	=	0.2-0.4 L/kg	Obtained from $Dose/$ initial C_{peak}
V_t	=	0.6 L/kg	Obtained from $3 \cdot V_c$
Cl_d	=	0.01 L/kg	Obtained from $k_{21} \cdot V_t = 0.02 \cdot 0.6$
F	=	0.05	
K_a	=	0.1 min^{-1}	Estimated by means of the residual method.
t_{lag}	=	5 min	Estimated by means of the residual method.

The k_{21} parameter is derived from α, β and k_{10}, where k_{10} is equal to Cl/V_c. One can also assume that Cl_d is generally greater than Cl, since a rapid decline occurs in the plasma concentration-time course after the rapid infusion. Obtain initial parameter estimates and then implement a two-compartment disposition model (written as two differential equations).

Interpretation of results and conclusions

Observed and model-predicted plasma concentrations are shown in Figure 12.4 and the final parameter estimates are given in Table 12.1. The parameter estimates had a high precision and low correlation.

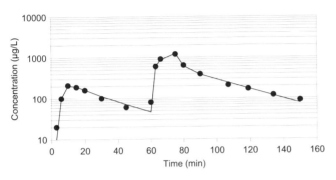

Figure 12.4 _Observed and predicted concentration-time data of drug X after an oral dose of 2.5 mg/kg followed by a 15 min constant rate intravenous infusion of 0.5 mg/kg starting at 60 min._

Table 12.1 Final estimates (mean) and their precision (CV%)

	Mean	CV%		Mean	CV%
K_a	0.103	25	Cl_d	0.021	13
Cl	0.015	4	F	0.046	12
V_c	0.121	8	t_{lag}	4.678	12
V_t	0.276	11			

You have
- [ ] analyzed a typical two-compartment (bi-exponential) system
- [ ] fit data obtained after semi-simultaneous administration
- [ ] learned to derive the initial estimates
- [ ] characterized the pharmacokinetics, including the absorption kinetics, after semi-simultaneous administration in one subject

The next step in the characterization of kinetics would be to give multiple doses via the intended route of administration. We encourage this type of design for both pre-clinical and clinical studies, particulary in situations where changes in e.g., clearance may occur. Also try to reduce the dataset to 10 observation and refit the model to the new reduced sample in order to see how precision and accuracy may change.

Solution

```
Model
 COMM
   nder 2
   nparm 7
   ncon   4
   pnam 'Ka', 'Cl', 'Vc', 'Vt', 'Cld', 'Bio', 'Tlag'
end
     1: TEMP
     2: T=X
```

```
 3: DPO = CON(1)
 4: TD = CON(2)
 5: TINF= CON(3)
 6: DINF= CON(4)
 7: END
 8: START
 9: Z(1)=0
10: Z(2)=0
11: END
12: DIFF
13: IF T LE TD OR T GE TD+TINF THEN
14: INP1 = 0
15: ELSE
16: INP1 = DINF/TINF
17: ENDIF
18: IF T LE TLAG THEN
19: INP2 = 0
20: DZ(1) = 0.0
21: DZ(2) = 0.0
22: ELSE
23: INP2 = BIO*KA*DPO*DEXP(-KA*(T-TLAG))
24: ENDIF
25: INPUT = INP1 + INP2
26: DZ(1)=(INPUT - CL*Z(1) - CLD*Z(1) + CLD*Z(2))/VC
27: DZ(2)=(CLD*Z(1) - CLD*Z(2))/VT
28: END
29: FUNC 1
30: F = Z(1)
31: END
32: EOM
NVARIABLES 2
NPOINTS 1000
XNUMBER 1
YNUMBER 2
CONSTANTS 2500,60,15,500
METHOD 2   'Gauss-Newton (Levenberg and Hartley)
ITERATIONS 50
INITIAL .1,.02,.2,.6,.05,.05,5
LOWER BOUNDS 0,0,.0001,.0001,0,0,0
UPPER BOUNDS 1,.5,1,1,.5,1,10
MISSING 'Missing'
NOBSERVATIONS 17
DATA 'WINNLIN.DAT'
BEGIN
```

PARAMETER	ESTIMATE	STANDARD ERROR	CV%	UNIVARIATE C.I.	
KA	.103040	.025606	24.85	.045986	.160094
CL	.014567	.000543	3.72	.013359	.015776
VC	.120866	.010181	8.42	.098181	.143552
VT	.275932	.028981	10.50	.211358	.340507
CLD	.020868	.002709	12.98	.014831	.026905
BIO	.046498	.005432	11.68	.034394	.058602
TLAG	4.677694	.574165	12.27	3.398368	5.957021

	Ka	CL	VC	VT	CLD	BIO	TLAG
KA	1						
CL	-0.211	1					
VC	-0.262	-0.16140	1				
VT	-0.115	-0.67200	0.2790	1			
CLD	0.2241	8.97E-02	-0.8235	-0.262521	1		
BIO	-0.7611	0.249352	0.2506	0.138407	-0.1448	1	
TLAG	0.5949	-5.59E-02	-0.3049	-7.71E-02	0.2653	-0.3492	1

Condition_number=   1678.

FUNCTION   1

X	OBSERVED Y	PREDICTED Y	RESIDUAL	WEIGHT	SE-PRED	STANDARDI RESIDUAL
3.000	20.00	.0000	20.00	1.000		
6.000	100.0	101.5	-1.495	1.000	25.80	-.5030
10.00	210.0	204.0	5.967	1.000	21.49	.4092
15.00	190.0	194.1	-4.063	1.000	15.43	-.1944
20.00	160.0	162.7	-2.708	1.000	14.22	-.1246
30.00	100.0	113.5	-13.46	1.000	15.75	-.6518
45.00	60.00	71.12	-11.12	1.000	12.08	-.4838
60.00	80.00	46.63	33.37	1.000	8.387	1.357
63.00	600.0	603.3	-3.345	1.000	19.42	-.1940
66.00	910.0	870.5	39.51	1.000	16.46	1.966
75.00	1200.	1237.	-37.36	1.000	20.37	-2.317
80.00	640.0	612.9	27.10	1.000	19.89	1.622
90.00	390.0	379.8	10.18	1.000	21.99	.7367
107.0	210.0	239.0	-28.95	1.000	13.26	-1.296
119.0	170.0	173.6	-3.563	1.000	13.78	-.1618
134.0	120.0	116.4	3.609	1.000	14.34	.1667
150.0	90.00	76.00	14.00	1.000	13.30	.6274

CORRECTED SUM OF SQUARED OBSERVATIONS = .180018E+07
WEIGHTED CORRECTED SUM OF SQUARED OBSERVATIONS = .180018E+07
SUM OF SQUARED RESIDUALS =   6746.80
SUM OF WEIGHTED SQUARED RESIDUALS = 6746.80
S = 25.9746   WITH   10 DEGREES OF FREEDOM
CORRELATION (OBSERVED,PREDICTED) = .9982

AIC criteria =   163.88600
SC criteria =   169.71849

PK13 - Bolus + constant rate infusion

Objectives

◆ To model bolus + infusion regimen

◆ To apply the model library

◆ To implement a two-compartment differential equation model

◆ To apply differential equations parameterized with Cl, V_c, Cl_d and V_t

◆ To fit weighted data according to constant CV error model

Problem specification

The aim of this exercise is to fit a bolus followed by constant rate infusion regimen and to show that different models may provide a suitable fit for the one dataset. We will also practise building a differential equation model that fits the bolus followed by constant rate infusion regimen. The following dataset is taken from a study of a new peptide in human volunteers. The compound was given as a bolus dose followed by an infusion for 26 min. It was known that concentrations above 50 µg/L and below 300 µg/L were needed to establish adequate therapeutic effects. The intravenous bolus dose was therefore 400 µg/kg and the total intravenous infusion dose was 800 µg/kg. Figure 13.1 and the program output show the concentration-time profile following the bolus dose and constant rate infusion.

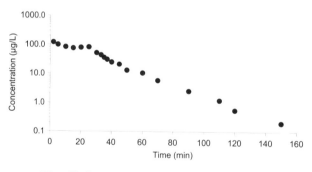

Figure 13.1 Semi-logarithmic plot of observed (•) concentration-time data following a bolus dose of 400 µg/kg and subsequent constant rate intravenous infusion of 800 µg/kg for 26 min.

We will fit a two-compartment model to the data. In Part I, we fit a library model (Model #17) with weighting of data. We assume a constant relative error, which means that we should weight by means of $1/predicted^2$.

In Part II, we express the model as a system of differential equations. The differential equation for the central compartment with linear elimination is

$$V_c \cdot \frac{dC}{dt} = In - Cl \cdot C - Cl_d \cdot C + Cl_d \cdot C_t \qquad (13:1)$$

The corresponding differential equation for the concentration in the peripheral compartment is

$$V_t \cdot \frac{dC_t}{dt} = Cl_d \cdot C - Cl_d \cdot C_t \qquad (13:2)$$

The distribution parameter Cl_d can also be estimated by means of $k_{12}V_c$ or $k_{21}V_t$.

$$Cl_d = k_{12} \cdot V_c = k_{21} \cdot V_t \qquad (13:3)$$

See also section 3.5 on multi-compartment models for a more thorough description of the parameter Cl_d. F_{inf} is the input function modeled as a bolus plus a continuous infusion. The initial condition for the plasma compartment is $C = Bolus\ Dose/V_c$ and for the peripheral compartment $C_t = 0.0$. We will use the simplex algorithm (METHOD 1) when fitting the model to the data. As the data cover three orders of magnitude, some sort of weighting scheme is needed to balance the relative fit of high concentrations against low concentrations.

Initial parameter estimates

Part I – library model
   $A = 120\ (\mu g/L)$
   $B = 30\ (\mu g/L)$
   $\alpha = 0.2\ (h^{-1})$
   $\beta = 0.05\ (h^{-1})$

Part II - differential equation model
   $V_c = 2.0\ (L/kg)$ Obtained from $Dose_{iv}/C^0 = 400/200$
   $Cl = 1.0\ (L/h/kg)$ _Total dose/AUC_
   $Cl_d = 1.0\ (L/h/kg)$ Assumed to be $\leq Cl$
   $V_t = 5.0\ (L/kg)$ Assumed to be $> V_c$

A robust method (Collier [1983]) to estimate the volume of distribution at steady state ($V_{ss} = V_c + V_t$) when the infusion is many (>10) half-lives, is to is to infuse the drug until steady state is achieved and then to measure the area under the fall-off curve ($AUC_{t*-\infty}$) and divide that produc by the steady state concentration squared according to

$$V_{ss} = \frac{k_0 \cdot AUC_{t*-\infty}}{C_{ss}^2} \qquad (13:4)$$

where k_0 is the infusion rate.

Interpretation of results and conclusions

Both models fit the data adequately. The condition number was better for the differential equation model than the library model. However, in the differential equation model, Cl_d had lower precision than the other parameters. Figure 13.2 shows the observed and model-predicted concentration-time data for the differential equation model.

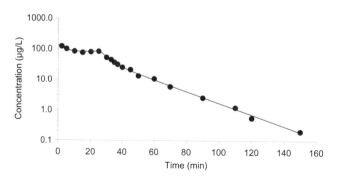

Figure 13.2 *Observed (●) and predicted (solid line) concentration-time data following a bolus dose of 400 μg/kg and a subsequent constant rate intravenous infusion of 800 μg/kg for 26 min.*

Uniform weights (not shown here) rendered a systematic deviation between observed and predicted data. By means of the constant *CV* error model we obtained a more random error distribution about the predicted curve. The results of such an analysis can then be used to provide the basis for a refined study protocol in patients.

In this exercise, we have practised

☐ regression modeling of bolus plus constant rate infusion data

☐ multi-compartment modeling by means of differential equations

☐ weighting

The next step in the characterization of kinetics would be to give multiple doses at the intended dose level.

Solution I - Model library
```
TITLE 1
FITTING A SYSTEM OF Bolus + infusion + wash-out
MODEL 17
NVARIABLES 2
NPOINTS 1000
XNUMBER 1
YNUMBER 2
NCONSTANTS 4
CONSTANTS 400,400,800,26
METHOD 2    'Gauss-Newton (Levenberg and Hartley)
REWEIGHT -2
ITERATIONS 50
INITIAL 120,30,.2,.05
LOWER BOUNDS 1,1,.1,.01
UPPER BOUNDS 500,100,10,10
MISSING 'Missing'
NOBSERVATIONS 19
DATA 'WINNLIN.DAT'
BEGIN
```

PARAMETER   ESTIMATE   STANDARD   CV%   UNIVARIATE C.I.

		ERROR			
A	104.753800	12.897161	12.31	77.264257	132.243344
B	27.594522	2.412388	8.74	22.452658	32.736385
Alpha	.199283	.030654	15.38	.133946	.264621
Beta	.043540	.001004	2.31	.041400	.045680

```
*** CORRELATION MATRIX OF THE ESTIMATES ***
PARAMETER  A          B        Alpha     Beta
A          1.00000
B           .34808   1.00000
Alpha       .79578    .74914   1.00000
Beta        .27522    .92231    .64059   1.00000

Condition_number=       .3410E+05

    FUNCTION    1
```

X	OBSERVED Y	PREDICTED Y	RESIDUAL	WEIGHT	SE-PRED	STANDARDIZED RESIDUAL
2.000	120.0	113.0	7.030	.7836E-04	7.596	1.040
5.000	99.00	95.91	3.085	.1087E-03	4.008	.4034
10.00	83.00	84.26	-1.265	.1409E-03	3.109	-.1828
15.00	75.00	81.41	-6.413	.1509E-03	2.699	-.9413
20.00	79.00	81.52	-2.525	.1505E-03	2.471	-.3655
25.00	82.00	82.50	-.5038	.1470E-03	2.393	-.0716
30.00	52.00	53.61	-1.613	.3483E-03	1.978	-.3665
33.00	44.00	41.03	2.974	.5949E-03	1.596	.8937
35.00	36.00	35.12	.8757	.8119E-03	1.307	.3044
37.00	31.00	30.53	.4700	.1075E-02	1.065	.1857
40.00	25.00	25.30	-.2992	.1566E-02	.8345	-.1414
45.00	21.00	19.26	1.741	.2703E-02	.6902	1.096
50.00	13.00	15.09	-2.089	.4406E-02	.6078	-1.723
60.00	10.50	9.588	.9118	.1092E-01	.4081	1.202
70.00	5.800	6.180	-.3799	.2632E-01	.2431	-.7615
90.00	2.500	2.584	-.8446E-01	.1507	.8701E-01	-.3930
110.0	1.200	1.082	.1182	.8617	.4117E-01	1.346
120.0	.5500	.7000	-.1500	2.060	.3056E-01	-2.738
150.0	.2000	.1896	.1042E-01	28.16	.1270E-01	.9258

```
CORRECTED SUM OF SQUARED OBSERVATIONS =  25881.3
WEIGHTED CORRECTED SUM OF SQUARED OBSERVATIONS =  16.5984
SUM OF SQUARED RESIDUALS =          129.225
SUM OF WEIGHTED SQUARED RESIDUALS = .122294
S = .902937E-01 WITH     15 DEGREES OF FREEDOM
CORRELATION (OBSERVED,PREDICTED) =  .9975

AIC criteria =          -31.92519
SC  criteria =          -28.14743

SUMMARY OF ESTIMATED SECONDARY PARAMETERS
```

PARAMETER	ESTIMATE	STANDARD ERROR	CV%
K10	.114150	.010843	9.50
K12	.052661	.015136	28.74
K21	.076012	.007025	9.24
K10-HL	6.072258	.576503	9.49

Alpha-HL	3.478200	.534490	15.37
Beta-HL	15.919772	.366797	2.30
Volume	3.022328	.317685	10.51

Solution II - Differential equations

```
TITLE 1
Bolus + infusion + wash-out: Differential equations
MODEL
 COMM
  NDER 2
  NPARM 4
  NCON 3
  PNAMES 'Vc','CL','Cld','Vt'
END
    1: TEMP
    2: T = X
    3: TINF = CON(1)
    4: DOSE1 = CON(2)
    5: DOSE2 = CON(3)
    6: IF T <= TINF THEN
    7: FINF = DOSE2/TINF
    8: ELSE
    9: FINF = 0.0
   10: ENDIF
   11: END
   12: START
   13: Z(1)=DOSE1/VC
   14: Z(2)=0.0
   15: END
   16: DIFF
   17: DZ(1)=(FINF - CL*Z(1) + CLD*Z(2) - CLD*Z(1))/VC
   18: DZ(2) = (CLD*Z(1) - CLD*Z(2))/VT
   19: END
   20: FUNC 1
   21: F= Z(1)
   22: END
   23: EOM
NVARIABLES 2
NPOINTS 100
XNUMBER 1
YNUMBER 2
CONSTANTS 26,400,800
METHOD 2  'Gauss-Newton (Levenberg and Hartley)
REWEIGHT -2
ITERATIONS 50
INITIAL 2,1,1,5
LOWER BOUNDS 0,0,0,0
UPPER BOUNDS 10,10,10,10
MISSING 'Missing'
NOBSERVATIONS 19
DATA 'WINNLIN.DAT'
BEGIN
```

PARAMETER	ESTIMATE	STANDARD ERROR	CV%	UNIVARIATE C.I.	
VC	2.933298	.316712	10.80	2.258246	3.608350
CL	.344814	.009348	2.71	.324889	.364738
CLD	.168239	.034745	20.65	.094183	.242296

```
VT       2.164111        .263612     12.18       1.602238        2.725984
```

*** CORRELATION MATRIX OF THE ESTIMATES ***

PARAMETER	VC	CL	CLD	VT
VC	1.00000			
CL	.496224	1.00000		
CLD	-.649384	.171441	1.00000	
VT	-.675354	.225446	.970852	1.00000

```
 Condition_number=        120.4
```

FUNCTION   1

X	OBSERVED Y	PREDICTED Y	RESIDUAL	WEIGHT	SE-PRED	STANDARDI RESIDUAL
2.000	120.0	114.8	5.217	.7590E-04	7.722	.7593
5.000	99.00	96.20	2.797	.1080E-03	4.007	.3642
10.00	83.00	83.96	-.9601	.1419E-03	3.150	-.1397
15.00	75.00	81.17	-6.170	.1518E-03	2.699	-.9083
20.00	79.00	81.41	-2.408	.1509E-03	2.473	-.3489
25.00	82.00	82.47	-.4720	.1470E-03	2.407	-.0672
30.00	52.00	53.12	-1.121	.3544E-03	2.053	-.2595
33.00	44.00	40.62	3.381	.6061E-03	1.611	1.030
35.00	36.00	34.82	1.183	.8249E-03	1.296	.4143
37.00	31.00	30.32	.6791	.1088E-02	1.048	.2693
40.00	25.00	25.21	-.2087	.1574E-02	.8291	-.0987
45.00	21.00	19.28	1.718	.2690E-02	.7020	1.082
50.00	13.00	15.15	-2.151	.4356E-02	.6155	-1.767
60.00	10.50	9.643	.8566	.1075E-01	.4040	1.114
70.00	5.800	6.209	-.4093	.2594E-01	.2379	-.8088
90.00	2.500	2.587	-.8699E-01	.1494	.8587E-01	-.4017
110.0	1.200	1.078	.1215	.8597	.4120E-01	1.382
120.0	.5500	.6964	-.1464	2.062	.3056E-01	-2.673
150.0	.2000	.1874	.1256E-01	28.46	.1254E-01	1.111

```
 CORRECTED SUM OF SQUARED OBSERVATIONS =  25881.3
 WEIGHTED CORRECTED SUM OF SQUARED OBSERVATIONS =  16.5925
 SUM OF SQUARED RESIDUALS =             103.172
 SUM OF WEIGHTED SQUARED RESIDUALS =   .121644
 S =  .900533E-01 WITH    15 DEGREES OF FREEDOM
 CORRELATION (OBSERVED,PREDICTED) =   .9980

 AIC criteria =        -32.02648
 SC  criteria =        -28.24873
```

PK14 - Multi-compartment model oral dosing

Objectives

◆ **To implement a two-compartment first-order absorption model**

◆ **To apply the model library**

◆ **To fit unweighted and weighted data**

◆ **To analyze the impact of leverage in data**

◆ **To fit a model with and without lag-time**

◆ **To analyze residuals with respect to correct structural and variance models**

Problem specification

The aim of this exercise is to study the impact of weighting on parameter precision and to demonstrate the importance of using the correct model and modeling trail. The concentration-time dataset (see program output) is from a single volunteer who received 23158 µg orally of a new cardiovascular drug. A semi-logarithmic plot of concentration-time data is shown in Figure 14.1. Note the three distinct phases of the concentration-time curve, including the rapid absorption phase, the rapid initial downswing and the slow termial phase.

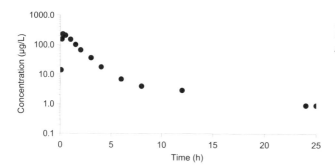

Figure 14.1 Observed concentration time data following an oral dose (23 µg) of a multi-compartment drug.

We will fit a two-compartment model with first-order input K_a and apply both constant and proportional weights. Figure 14.2 shows the two-compartment *micro-constant* model that will be fit to the data. In practice, it is preferable to use the library models since they are available to minimize the chance of user error.

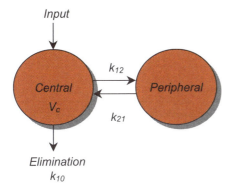

Figure 14.2 *Schematic illustration of the two-compartment micro-constant model.*

Equation 14:1 shows the parameterization of the model that will initially be fit to the data.

$$C = \frac{K_a FD_{po}}{V_c}\left\{ \frac{(k_{21}-\alpha)\cdot e^{-\alpha\cdot t}}{(K_a-\alpha)(\beta-\alpha)} + \frac{(k_{21}-\beta)\cdot e^{-\beta\cdot t}}{(K_a-\beta)(\alpha-\beta)} + \frac{(k_{21}-K_a)\cdot e^{-K_a\cdot t}}{(\alpha-K_a)(\beta-K_a)} \right\} \tag{14:1}$$

Weighting data

By now you know how to build a simple model, how to reparameterize the model, and what to look for in the program output file. In this exercise, you will learn how to weight your data and to see see how this affects your output. You will also learn how to interpret the results. Figure 14.3 shows the various steps in this particular analysis.

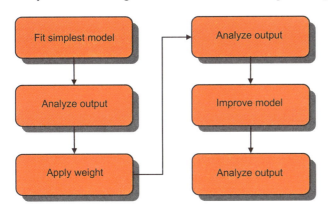

Figure 14.3 *The modeling trail used for this particular exercise. We refer the reader to Figure 14.4 for the preferred modeling approach.*

To weight data correctly, you must have some prior knowledge (assume some behavior) about the error structure of your data. One way to get a feel for the variability is to pool all observations from all individuals into one plot. You may also apply a log-linear graphical presentation, or fit the unweighted data. From the residual plots, you can draw conclusions about the error structure in the dataset.

This example illustrates some of the basics of weighting data. It shows how to:
- select a structural model
- select an error (weighting) model
- analyze the diagnostic (residual) plots
- analyze output with respect to impact of weighting

Since the first observation in both runs (with and without weights) is poorly fit, a good next step would be to try a model including a lag-time and then to do the appropriate tests for goodness-of-fit. In fact, by using the lag-time model in Equation 14:2, the less successful fit of Equation 14:1 is substantially improved.

$$C = \frac{K_a FD_{po}}{V_c}\left\{\frac{(k_{21}-\alpha)\cdot e^{-\alpha\cdot(t-t_{lag})}}{(K_a-\alpha)(\beta-\alpha)} + \frac{(k_{21}-\beta)\cdot e^{-\beta\cdot(t-t_{lag})}}{(K_a-\beta)(\alpha-\beta)} + \frac{(k_{21}-K_a)\cdot e^{-K_a\cdot(t-t_{lag})}}{(\alpha-K_a)(\beta-K_a)}\right\} \tag{14:2}$$

Initial Parameter estimates

No lag-time model:

V_c/F	= 350 (L)	Obtained from *Dose/Cmax*
K_a	= 11 (h^{-1})	Obtained by curve stripping
k_{10}	= 5 (h^{-1})	See section 3.3.2
k_{12}	= 0.9 (h^{-1})	See section 3.3.2
k_{21}	= 0.1 (h^{-1})	See section 3.3.2

Lag-time model:

V_c/F	= 150 (L)
K_a	= 5 (h^{-1})
k_{10}	= 1 (h^{-1})
k_{12}	= 0.25 (h^{-1})
k_{21}	= 0.15 (h^{-1})
t_{lag}	= 0.05 (h)

It can be seen that the above lag-time model initial estimates were slightly changed to incorporate information from the previous fit.

Interpretation of results and conclusions

The aim of this exercise was to study the impact of weighting on the parameter precision. As *a rule of thumb*, one can calculate the ratio of the highest-to-the-lowest concentration values and use that ratio as an indication of the need for weighting. If this ratio is greater than 20-30:1, weights are generally needed. In this example 226/0.9 is greater than 200:1, suggesting the need for weighting.

The concentration-time profile exhibits a bi-exponential decline post-peak. Together with the absorption phase (upswing of the initial portion of the curve), this suggests a three-exponential model. Initially, we do not know anything about the need of a potential lag-time. As a result, the first model to be fit to the data is a two-compartment disposition model (equivalent to a bi-exponential model following a bolus dose) with first-order input, which is the same as a three-exponential model.

Figure 14.3 shows the steps that were followed in our analysis. Even though the structural model was incorrect, we selected this execution order to demonstrate the impact of weighting on parameter precision. As a rule of thumb, we recommend the analyst to work along the lines proposed in Figure 14.4 (rather than those in Figure 14.3). The important difference is that weights are only applied when the correct structural model is known.

First, fit unweighted data with a simple model (two-compartment with first-order absorption and no lag-time). Then, analyze the residual plots. These reveal that there is a systematic trend between fitted and observed data over time, suggesting that weighting may be required. We would not suggest an additional (fourth) exponential term since there is nothing in the data to indicate such a term. In addition, one should always report the simplest model that provides an adequate fit.

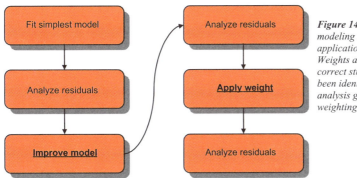

Figure 14.4 The preferred modeling steps for the application of weights. Weights are applied after the correct structural model has been identified. Let residual analysis guide you through the weighting jungle.

After having fit the model using weights, the overall fit looks much better. However, the fit is still poor on the upswing (absorption phase) and for the peak concentration. As a result, a lag-time is included in the model and the model re-run with weighted data. Remember that the overall fit from the previous run was acceptable but showed some minor flaws close to the peak. Consequently, there is nothing to suggest that the weighting scheme should be altered for the lag-time model.

A schematic illustration of the behavior of the *lag-time* and *no lag-time* models is shown in Figure 14.5.

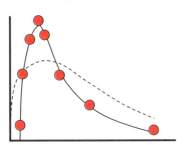

Figure 14.5 Schematic illustration of the principal differences between a lag-time model (solid line) and a model without lag-time (dashed line). The model without a lag-time (dashed line) does not display the rapid upswing. It also probably misses the peak-time and cuts through the higher concentrations with systematic deviations between observed (symbols) and predicted concentrations. The lag-time model captures the observed concentrations well.

From our solutions, the following observations may be made. The unweighted fit (Run #1) shows an obvious trend in the residuals that one may interpret as a need for an additional phase. Figure 14.6 displays the observed and predicted concentrations. Even though the correlation coefficient between the observed and predicted concentrations is reasonably high (0.96) in Run #1, the over-all fit is poor, as the model systematically under-predicts terminal concentrations. Thus, this example illustrates that a high correlation between observed and predicted values may not be a useful tool for assessing the adequacy of a model.

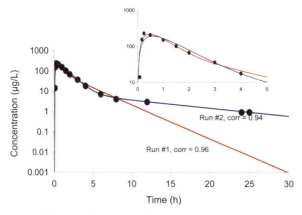

Figure 14.6 Semi-logarithmic plot of predicted (lines) and observed (symbols) concentration-time data following oral dosing. Run #1 (red line) represents the OLS fit and Run #2 (blue line) the IRLS fit using the model without a lag-time. The inset gives the corresponding data over the first 5 h after dosing.

The standard errors of several parameters and the correlation between these are high. This is almost eliminated in the constant *CV* error model (Run #2). The lag-time model improves the fit and parameter precision. The model captures all principal parts of the data - upswing, peak, and wash-out including the terminal observations. The model with a lag-time is such a clear improvement that there is no need to run an *F*-test or to use the *AIC* or *SC* criteria for model discrimination. Parameter precision is presented in Table 14.1 for all parameters from the different runs.

Table 14.1   Final kinetic parameter estimates from the three different runs

	Run # 1, wrong model, no weights		Run # 2, wrong model, weights		Run # 3, correct model, weights	
	Mean	CV%	Mean	CV%	Mean	CV%
V/F	43.2	8778	73.4	31.6	83.0	4.28
K_a	2.32	8752	2.98	46.5	10.0	11.6
k_{10}	1.34	8763	0.74	25.2	0.66	3.04
k_{12}	0.73	12062	0.16	43.7	0.13	6.45
k_{21}	0.68	374	0.11	31.1	0.10	5.91
t_{lag}	-	-	-	-	0.078	0.74

By including a lag-time into the model (Run #3), we were able to improve the fit and remove the trend in the residuals at early time points. We were also able to improve parameter precision (Table 14.1 and Figure 14.7). Thus, the best fit is obtained with the lag-time model using *IRLS*.

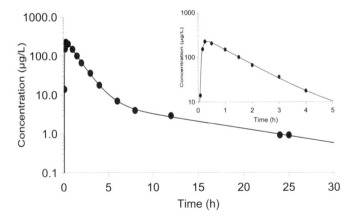

Figure 14.7 *Semi-logarithmic plot of the fit of the lag-time model superimposed on observed data. The inset gives the corresponding data over the first 5 h after dosing.*

In this exercise, we have practised

☐ regression modeling of single extravascular dosing data

☐ multi-compartment modeling

☐ weighting

The next step in the characterization of kinetics would be to give multiple doses via the intended route of administration.

Solution I - Constant absolute variance-model – No lag-time, no weights

```
TITLE 1
2-CMPT ORAL including no Lag-time
MODEL 11
NVARIABLES 2
NPOINTS 500
XNUMBER 1
YNUMBER 2
NCONSTANTS 3
CONSTANTS 1,23158,0
METHOD 2   'Gauss-Newton (Levenberg and Hartley)
ITERATIONS 50
INITIAL 350,11,5,.9,.1
LOWER BOUNDS 0,0,0,0,0
UPPER BOUNDS 500,30,10,5,1
NOBSERVATIONS 14
DATA 'WINNLIN.DAT'
BEGIN
```

PARAMETER	ESTIMATE	STANDARD ERROR	CV%	UNIVARIATE C.I.	
Volume/F	43.205229	3792.41233	8777.67	-8535.900496	8622.310955
K01	2.317172	202.78977	8751.61	-456.429064	461.063409
K10	1.338723	117.31823	8763.44	-264.055809	266.733256

```
K12           .725451      87.50273    12061.83    -197.221162        198.672065

K21           .677709       2.53136      373.52      -5.048689          6.404107

*** CORRELATION MATRIX OF THE ESTIMATES ***
PARAMETER  Volume/F       K01          K10          K12          K21
Volume/F    1.00000
K01          .99999     1.00000
K10         -.99999     -.99998      1.00000
K12         -.99997     -.99996       .99995      1.00000
K21          .85090      .85143      -.84945      -.84982      1.00000

 Condition_number=          .2259E+05

    FUNCTION    1

    X          OBSERVED     PREDICTED    RESIDUAL      WEIGHT      SE-PRED      STANDARD
               Y            Y                                                  RESIDUAL
    .8300E-01  13.90         86.00        -72.10       1.000       15.18       -2.930
    .1670      152.0        144.2           7.771      1.000       18.09         .3446
    .2500      226.0        180.6          45.39       1.000       17.51        1.973
    .5000      204.0        213.1          -9.132      1.000       23.19        -.5289
    1.000      149.0        157.7          -8.719      1.000       21.40        -.4485
    1.500      100.0         98.26          1.736      1.000       20.56         .0854
    2.000       66.00        63.01          2.987      1.000       19.20         .1382
    3.000       36.00        32.84          3.156      1.000       21.24         .1609
    4.000       17.70        20.82         -3.124      1.000       18.21        -.1391
    6.000        6.900        9.482        -2.582      1.000       19.68        -.1219
    8.000        3.960        4.388         -.4284     1.000       17.36        -.0185
    12.00        2.890         .9410        1.949      1.000        7.524        .0698
    24.00         .9000        .9279E-02     .8907     1.000        .1920        .0308
    25.00         .9000        .6315E-02     .8937     1.000        .1375        .0309
```

```
CORRECTED SUM OF SQUARED OBSERVATIONS = 85607.8
WEIGHTED CORRECTED SUM OF SQUARED OBSERVATIONS =  85607.8
SUM OF SQUARED RESIDUALS =          7521.88
SUM OF WEIGHTED SQUARED RESIDUALS = 7521.88
S =  28.9096      WITH      9 DEGREES OF FREEDOM
CORRELATION (OBSERVED,PREDICTED) =  .9560

AIC criteria =          134.95800
SC  criteria =          138.15329

AUC (0 to last time) computed by trapezoidal rule =  419.632
SUMMARY OF ESTIMATED SECONDARY PARAMETERS
```

```
    PARAMETER       ESTIMATE        STANDARD          CV%
                                    ERROR
AUC             400.381378      147.053954         36.73
K10-HL             .517767       45.328898       8754.69
K01-HL             .299135       26.152960       8742.86
Alpha            2.356952      203.374327       8628.70
Beta              .384931         .976712        253.74
Alpha-HL          .294086       25.365009       8625.03
Beta-HL          1.800704        4.616101        256.35
A            -26586.275464  **************     1038647.18
B               95.430635      415.475621        435.37
CL/F            57.839853       21.264937         36.77
Tmax              .480270         .087166         18.15
Cmax           213.294776       23.440485         10.99
```

Solution II - Constant CV model – No lag-time, with weights

```
TITLE 1
2-CMPT ORAL including no Lag-time
MODEL 11
NVARIABLES 2
NPOINTS 500
XNUMBER 1
YNUMBER 2
NCONSTANTS 3
CONSTANTS 1,23158,0
METHOD 2  'Gauss-Newton (Levenberg and Hartley)
REWEIGHT -2
ITERATIONS 50
INITIAL 350,11,5,.9,.1
LOWER BOUNDS 0,0,0,0,0
UPPER BOUNDS 500,30,10,5,1
MISSING 'Missing'
NOBSERVATIONS 14
DATA 'WINNLIN.DAT'
BEGIN
```

PARAMETER	ESTIMATE	STANDARD ERROR	CV%	UNIVARIATE C.I.	
Volume/F	73.404213	23.212264	31.62	20.893977	125.914448
K01	2.979618	1.384397	46.46	-.152132	6.111367
K10	.738127	.185974	25.20	.317421	1.158833
K12	.155199	.067865	43.73	.001677	.308721
K21	.112074	.034831	31.08	.033280	.190868

```
*** CORRELATION MATRIX OF THE ESTIMATES ***
```

PARAMETER	Volume/F	K01	K10	K12	K21
Volume/F	1.00000				
K01	.851160	1.00000			
K10	-.942356	-.86029	1.00000		
K12	-.657237	-.64218	.78973	1.00000	
K21	-.334183	-.30800	.41748	.64628	1.00000

```
Condition_number=       1047.

FUNCTION   1
```

X	OBSERVED Y	PREDICTED Y	RESIDUAL	WEIGHT	SE-PRED	STANDARD RESIDUAL
.8300E-01	13.90	66.52	-52.62	.2260E-03	14.61	-2.905
.1670	152.0	114.2	37.81	.7669E-04	21.65	1.126
.2500	226.0	146.5	79.50	.4659E-04	24.54	1.767
.5000	204.0	186.9	17.12	.2863E-04	27.41	.2884
1.000	149.0	162.4	-13.41	.3791E-04	30.27	-.2790
1.500	100.0	114.7	-14.66	.7606E-04	22.67	-.4431
2.000	66.00	77.09	-11.09	.1683E-03	14.13	-.4826
3.000	36.00	34.94	1.057	.8188E-03	6.815	.1042
4.000	17.70	17.36	.3405	.3317E-02	4.274	.0789
6.000	6.900	6.799	.1009	.2161E-01	1.439	.0532
8.000	3.960	4.441	-.4806	.5067E-01	1.046	-.4179
12.00	2.890	2.896	-.6076E-02	.1192	.6583	-.0079
24.00	.9000	.9761	-.7610E-01	1.051	.2309	-.3030
25.00	.9000	.8917	.8275E-02	1.260	.2260	.0385

```
CORRECTED SUM OF SQUARED OBSERVATIONS =  85607.8
WEIGHTED CORRECTED SUM OF SQUARED OBSERVATIONS =  11.1176
SUM OF SQUARED RESIDUALS =          11331.1
SUM OF WEIGHTED SQUARED RESIDUALS =  1.10150
S =  .349841     WITH      9 DEGREES OF FREEDOM
CORRELATION (OBSERVED,PREDICTED) =  .9393

AIC criteria =          11.35341
SC  criteria =          14.54870

AUC (0 to last time) computed by trapezoidal rule =          419.632
```

SUMMARY OF ESTIMATED SECONDARY PARAMETERS

PARAMETER	ESTIMATE	STANDARD ERROR	CV%
AUC	427.414193	49.272585	11.53
K10-HL	.939062	.236364	25.17
K01-HL	.232630	.107977	46.42
Alpha	.914989	.248832	27.20
Beta	.090411	.024875	27.51
Alpha-HL	.757547	.205889	27.18
Beta-HL	7.666628	2.107034	27.48
A	443.339466	273.342796	61.66
B	8.547820	4.099240	47.96
CL/F	54.181636	6.252340	11.54
Tmax	.579604	.114231	19.71
Cmax	188.623335	28.358793	15.03

Solution III - Constant CV error model – Lag-time, with weights

```
TITLE 1
 Two-compartment ORAL including Lag-time
MODEL 12
NVARIABLES 2
NPOINTS 500
XNUMBER 1
YNUMBER 2
NCONSTANTS 3
CONSTANTS 1,23158,0
METHOD 2  'Gauss-Newton (Levenberg and Hartley)
REWEIGHT -2
ITERATIONS 50
INITIAL 150,5,1,.25,.15,.05
LOWER BOUNDS 0,0,0,0,0,0
UPPER BOUNDS 500,30,5,2,1,.1
NOBSERVATIONS 14
DATA 'WINNLIN.DAT'
BEGIN
```

PARAMETER	ESTIMATE	STANDARD ERROR	CV%	UNIVARIATE C.I.	
Volume/F	82.956782	3.551107	4.28	74.767835	91.145730
K01	10.024833	1.166281	11.63	7.335359	12.714307
K10	.660481	.020110	3.04	.614107	.706854
K12	.126882	.008181	6.45	.108017	.145747
K21	.104775	.006194	5.91	.090491	.119058

| Tlag | .077916 | .000576 | .74 | .076588 | .079244 |

*** CORRELATION MATRIX OF THE ESTIMATES ***

PARAMETER	Volume/F	K01	K10	K12	K21	Tlag
Volume/F	1.00000					
K01	.64331	1.00000				
K10	-.89463	-.640454	1.00000			
K12	-.45178	-.375259	.66560	1.00000		
K21	-.26714	-.202747	.39387	.64610	1.00000	
Tlag	.27164	.782082	-.30874	-.20842	-.10387	1.00000

Condition_number=   .1158E+05

FUNCTION   1

X	OBSERVED Y	PREDICTED Y	RESIDUAL	WEIGHT	SE-PRED	STANDARDI RESIDUAL
.8300E-01	13.90	13.84	.5645E-01	.5377E-02	.8318	7.004
.1670	152.0	158.4	-6.404	.3989E-04	8.227	-1.266
.2500	226.0	210.6	15.39	.2255E-04	7.234	1.450
.5000	204.0	213.1	-9.061	.2203E-04	7.649	-.8623
1.000	149.0	147.4	1.648	.4605E-04	4.981	.2203
1.500	100.0	100.5	-.4638	.9907E-04	2.810	-.0852
2.000	66.00	69.03	-3.030	.2098E-03	1.814	-.7974
3.000	36.00	33.78	2.219	.8762E-03	1.096	1.271
4.000	17.70	17.80	-.9923E-01	.3156E-02	.6877	-.1181
6.000	6.900	6.949	-.4886E-01	.2071E-01	.2449	-.1412
8.000	3.960	4.249	-.2894	.5538E-01	.1712	-1.487
12.00	2.890	2.682	.2084	.1391	.1119	1.746
24.00	.9000	.9504	-.5040E-01	1.107	.3926E-01	-1.182
25.00	.9000	.8722	.2776E-01	1.314	.3865E-01	.7592

CORRECTED SUM OF SQUARED OBSERVATIONS = 85607.8
WEIGHTED CORRECTED SUM OF SQUARED OBSERVATIONS = 10.3040
SUM OF SQUARED RESIDUALS =   377.100
SUM OF WEIGHTED SQUARED RESIDUALS = .297689E-01
S = .610009E-01 WITH   8 DEGREES OF FREEDOM
CORRELATION (OBSERVED,PREDICTED) = .9978

AIC criteria =   -37.20009
SC criteria =   -33.36575

SUMMARY OF ESTIMATED SECONDARY PARAMETERS

PARAMETER	ESTIMATE	STANDARD ERROR	CV%
AUC	422.657896	8.729232	2.07
K10-HL	1.049459	.031921	3.04
K01-HL	.069143	.008036	11.62
Alpha	.806312	.027233	3.38
Beta	.085825	.004559	5.31
Alpha-HL	.859651	.029017	3.38
Beta-HL	8.076307	.428564	5.31
A	295.589935	15.570754	5.27
B	7.405611	.661726	8.94
CL/F	54.791358	1.132748	2.07
Tmax	.353648	.020168	5.70
Cmax	224.800371	6.934733	3.08

PK15 - Toxicokinetics I

Objectives

◆ **To model plasma disposition data following intravenous administration**

◆ **To solve the kinetics by means of differential equations**

◆ **To estimate clearance and volume**

◆ **To predict suitable doses for an exposure study**

◆ **To relate the exposure to the dose**

◆ **To relate the exposure level to the observed toxic effect**

Problem specification

This problem is designed to help you learn and apply toxicokinetic principles by analyzing preclinical kinetic and toxicity data. The study was divided into two parts. In the first part of the study, 12 female Sprague-Dawley rats (225-250 g, 3-4 plasma samples per rat) received an intravenous injection of monomethyl hydrazine (*MMH*) (8.75 mg/kg body weight) as a short (approximately 3 min) constant infusion. *MMH*, a metabolite of gyromitrin (commonly found in certain mushrooms), was measured in plasma at selected time points. *MMH* is used as a biomarker of gyromitrin. Data were then pooled from the 12 rats. The results showed a delayed peak of *MMH* in serum at approximately 60 min after the injection (Figure 15.1 and program output).

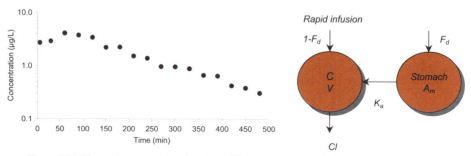

Figure 15.1 *Observed concentration-time data following intravenous bolus administration (left hand figure). The right hand figure shows the kinetic model.*

The plasma concentration-time data will be analyzed by two different approaches: the traditional non-compartmental analysis *NCA* and also by means of a compartmental regression model (Figure 15.1). Due to the limited number of plasma samples from each individual, animal data were pooled for the pharmacokinetic analysis. Each data point is the mean of 3 - 4 animals.

The model parameters (fraction of the intravenous dose that is initially sequestered into the stomach compartment ($1–F_d$ %), the absorption rate constant (K_a min^{-1}), plasma clearance (Cl mL/min), volume of distribution (V mL) and biological half-life ($t_{1/2}$ min)) and their precision were estimated by fitting Equations 15:1 to the data. This information was needed in order to be able to undertake the second part of the study. The equation describing the turnover of *MMH* in the organism is

$$V\frac{dC_p}{dt} = (1 - F_d) \cdot In - Cl \cdot C_p + K_a \cdot A_m \tag{15:1}$$

In is the rate of infusion and A_m is the amount of drug in the stomach compartment. *In* is defined as

$$In = \frac{D_{inf}}{T_{inf}} \tag{15:2}$$

Autoradiographic studies have indicated a pronounced uptake (probably due to ion-trapping) of *MMH* into the stomach. This phenomenon is also well known for bases with a low molecular weight, such as nicotine. We will propose a model using reasonable assumptions that can explain the unusual concentration-time profile of *MMH* in plasma following intravenous administration.

In the second part of the study, the survival frequency (effect) was obtained at five different steady state concentrations of *MMH* in pregnant rats. Six groups of pregnant rats (day 6 of pregnancy through day 13) were infused at one of six rates aiming at 0 (control group), 0.1, 0.2, 0.3, 0.5, and 0.6 µg of *MMH* /mL plasma. The control animals received saline. Figure 15.2 displays the results. The pump was set to deliver (infuse) *MMH* intravenously at a constant rate over a 7 day period. The pump was inserted subcutaneously on the back of the dam on day 6 of gestation and remained *in situ* through to day 13. The animals were then examined on day 21 of gestation.

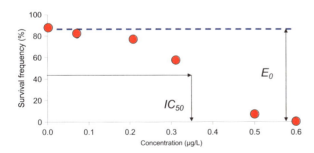

Figure 15.2 Survival frequency (%) versus steady state concentration of MMH (µg/L). Data were obtained at zero and five different steady state plasma concentrations of MMH. Note that the survival frequency is zero at a plasma MMH concentration of 0.6 µg/L.

We will fit the following model to the effect-concentration data

$$Survival\ frequency = E_0 \cdot \left[1 - \frac{C_{ss}^n}{IC_{50}^n + C_{ss}^n}\right] \tag{15:3}$$

C_{ss} is the steady state concentration, IC_{50} is the concentration at 50% of maximal effect, *n* the sigmoidicity factor, and E_0 the baseline survival frequency in control rats. We will estimate E_0, IC_{50} and *n*.

In the program, infusion rate *FINF* is the total dose *DOSE1* divided by the infusion time *TINF*. As long as time *t* is less than or equal to *TINF*, *FINF* = *DOSE1/TINF*. When *t* exceeds *TINF*, *FINF* = 0.

Initial parameter estimates

V = 1200 (mL)
Cl = 10 (mL/min)
K_a = 0.2 (min^{-1})
F_d = 0.5
E_0 = 80 (%)   Obtained from Figure 15.2
IC_{50} = 0.5 (µg/L)   (0.35 µg/L) in Figure 15.2
n = 2.5   Called g1 in program code

Interpretation of results and conclusions

The aim of the first part of the study was to estimate the clearance of the compound. With clearance and the targeted steady state plasma concentrations, appropriate infusion rates for the minipump (*Alzet®*) could be predicted. A correct clearance value was necessary to obtain suitable plasma levels throughout the second part of the study.

Plasma concentrations of *MMH* showed an interesting disposition pattern following short intravenous infusion, with a peak of *MMH* at approximately 60 min (Figure 15.3).

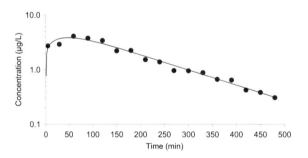

Figure 15.3 Semi-logarithmic plot of observed (symbols) and predicted (line) concentration-time data following a rapid intravenous infusion of 8.75 mg/kg of MMH.

The parameter estimates from the non-compartmental and compartmental analysis were consistent and parameter precision was generally good. The compartment model that was used took into account the potential sequestration of *MMH* into the stomach. The hypothetical fraction of the dose that was initially distributed into the stomach was estimated to be 59%. The absorption rate constant of *MMH* from the gastrointestinal tract into the blood stream was 0.029 min^{-1} ± 37%, with an absorption rate half-life of 24 min. The value of *Cl* and *V* indicated low extraction and high tissue distribution characteristics of 9.8 mL/min ± 4% and 1530 mL ± 7%, respectively. Non-compartmental analysis revealed a clearance, volume, and terminal half-life of 10 mL/min, 1780 mL and 111 min, respectively. Weights were computed as the inverse of the measured mean plasma concentrations squared. The 120-500 min time interval was used for estimation of the terminal slope for the *NCA*.

We could have used the non-compartmental approach straight away for clearance estimation, but we also proposed a differential equation model that could be fit to the data. This model has its limitations as it is derived after single dose experiments with only one dose level. Nevertheless, it can be applied for the design of new experiments and for multiple dose predictions. In addition, we obtained basically the same values of clearance and volume with the two approaches.

The second part of the study dealt with analysis of the relationship between fetal death (survival frequency) and the measured exposure levels in plasma. The pump was set to deliver a constant rate of infusion of *MMH*. Figure 15.4 shows the results of survival frequency (effect) plotted against steady state plasma concentration of *MMH*.

The dynamic model (Equation 15:3) was then fit to the effect-concentration data, and estimated the baseline effect of surviving fetuses (control group) to be approximately 80% in this experimental setting. The IC_{50} value or potency of *MMH* was 0.35 ng/mL (Figure 15.4). The concentration-response curve was very steep, which was manifested in a sigmoidicity factor of about 7 (6.7).

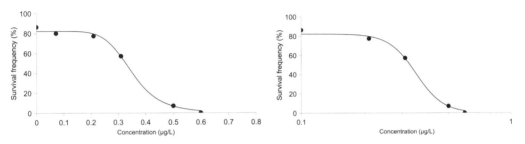

Figure 15.4 *Observed (symbols) and predicted (line) survival frequency (%) versus steady state concentration of MMH (µg/L). Left is a linear and right is a semi-logarithmic concentration scale.*

This model may also be used for *"What-if?"* predictions. We believe that the above example illustrates how one can integrate kinetics and dynamics to obtain a more optimal design of exposure ('toxicokinetics').

In this exercise, we have practised

☐ regression modeling of multiple toxicokinetic data

☐ multi-compartment absorption modeling

☐ fitting kinetic and dynamic data sequentially

The next step in the characterization of the toxicokinetics would be to give multiple doses via the intended route of administration.

Solution I - Kinetic data - regression analysis

```
TITLE 1
MMH in rats after an iv bolus dose
Model
 COMM
NDER 2
```

```
NPARM 4
PNAMES 'Vc','Cle','Ka','Fd'
END
    1: TEMP
    2: DOSE1 = 8750.
    3: TINF = 3.1
    4: T=X
    5: FINF = 0.0
    6: IF T LE TINF THEN
    7: FINF = DOSE1/TINF
    8: ENDIF
    9: END
   10: START
   11: Z(1)= 0.0
   12: Z(2)= 0.0
   13: END
   14: DIFF
   15: DZ(1)=((1. - FD)*FINF - CLE*Z(1) + KA*Z(2))/VC
   16: DZ(2)= FD*FINF - KA*Z(2)
   17: END
   18: FUNC 1
   19: F = Z(1)
   20: END
   21: EOM
NVARIABLES 2
NPOINTS 100
XNUMBER 1
YNUMBER 2
METHOD 1   'Nelder-Mead
REWEIGHT -2
ITERATIONS 500
INITIAL 1200,10,.2,.5
LOWER BOUNDS 0,0,0,0
UPPER BOUNDS 5000,25,1,1
MISSING 'Missing'
NOBSERVATIONS 32
DATA 'WINNLIN.DAT'
BEGIN
```

PARAMETER	ESTIMATE	STANDARD ERROR	CV%	UNIVARIATE C.I.	
VC	1526.826705	105.409055	6.90	1299.104422	1754.548988
CLE	9.806323	.338996	3.46	9.073967	10.538679
KA	.028781	.010602	36.84	.005877	.051686
FD	.590017	.070725	11.99	.437225	.742809

```
*** CORRELATION MATRIX OF THE ESTIMATES ***
PARAMETER  VC        CLE       KA        FD
VC          1.00000
CLE          .52492   1.00000
KA           .46649   -.32509   1.00000
FD          -.23990   -.26576    .31084   1.00000


    FUNCTION   1
```

X	OBSERVED Y	PREDICTED Y	RESIDUAL	WEIGHT	SE-PRED	STANDARD RESIDUAL
1.000	.	.7710	.	.0000	-.6400+152	.
2.500	.	1.975	.	.0000	-.6400+152	.
5.000	2.710	2.613	.9725E-01	.1465	.3150	2.120
7.000	.	2.750	.	.0000	-.6400+152	.
10.00	.	2.934	.	.0000	-.6400+152	.
15.00	.	3.191	.	.0000	-.6400+152	.
20.00	.	3.393	.	.0000	-.6400+152	.
25.00	.	3.548	.	.0000	-.6400+152	.
30.00	2.900	3.663	-.7628	.7454E-01	.3220	-2.470
35.00	.	3.744	.	.0000	-.6400+152	.
40.00	.	3.796	.	.0000	-.6400+152	.
45.00	.	3.823	.	.0000	-.6400+152	.
55.00	.	3.820	.	.0000	-.6400+152	.
60.00	4.080	3.795	.2853	.6945E-01	.2680	.7576
65.00	.	3.758	.	.0000	-.6400+152	.
70.00	.	3.711	.	.0000	-.6400+152	.
75.00	.	3.656	.	.0000	-.6400+152	.
85.00	.	3.527	.	.0000	-.6400+152	.
90.00	3.750	3.456	.2941	.8373E-01	.1805	.7732
120.0	3.400	2.988	.4121	.1120	.1420	1.230
150.0	2.210	2.522	-.3122	.1572	.1234	-1.110
180.0	2.240	2.105	.1353	.2258	.1039	.5773
210.0	1.520	1.746	-.2261	.3280	.8277E-01	-1.154
240.0	1.380	1.444	-.6448E-01	.4793	.6326E-01	-.3927
270.0	.9600	1.193	-.2332	.7024	.4751E-01	-1.697
300.0	.9500	.9848	-.3483E-01	1.031	.3629E-01	-.3046
330.0	.8800	.8126	.6744E-01	1.515	.2936E-01	.7134
360.0	.6600	.6703	-.1029E-01	2.226	.2570E-01	-.1328
390.0	.6400	.5529	.8712E-01	3.271	.2392E-01	1.384
420.0	.4200	.4560	-.3601E-01	4.809	.2290E-01	-.7115
450.0	.3800	.3761	.3899E-02	7.070	.2202E-01	.0971
480.0	.3050	.3102	-.5192E-02	10.39	.2101E-01	-.1653

```
CORRECTED SUM OF SQUARED OBSERVATIONS =  49.0133
WEIGHTED CORRECTED SUM OF SQUARED OBSERVATIONS =  6.85513
SUM OF SQUARED RESIDUALS =          1.16942
SUM OF WEIGHTED SQUARED RESIDUALS =  .192924
S =  .121821     WITH     13 DEGREES OF FREEDOM
CORRELATION (OBSERVED,PREDICTED) =  .7003

AIC criteria =         -19.97281
SC  criteria =         -16.63995
```

Solution II - Exposure data - regression analysis

```
TITLE 1
Modeling tox. data of MMH
MODEL
 COMM
  NPARM 3
  PNAMES 'IC50', 'g1', 'E0'
END
    1: TEMP
    2: C = X
    3: END
    4: FUNC 1
    5: ZZ2 = C**G1
    6: DRUG = ZZ2/(IC50**G1 + ZZ2)
    7: F = E0*(1 - DRUG)
```

```
     8: END
     9: EOM
NVARIABLES 2
NPOINTS 2000
XNUMBER 1
YNUMBER 2
METHOD 2  'Gauss-Newton (Levenberg and Hartley)
ITERATIONS 50
INITIAL .5,2.5,80
LOWER BOUNDS 0,0,1
UPPER BOUNDS 1,10,100
MISSING 'Missing'
NOBSERVATIONS 6
DATA 'WINNLIN.DAT'
BEGIN
```

PARAMETER	ESTIMATE	STANDARD ERROR	CV%	UNIVARIATE C.I.	
IC50	.348883	.010273	2.94	.316179	.381587
G1	6.690441	1.039533	15.54	3.381108	9.999773
E0	82.252064	2.047094	2.49	75.735182	88.768946

```
*** CORRELATION MATRIX OF THE ESTIMATES ***
PARAMETER       IC50           G1             E0
IC50           1.00000
G1             -.229283        1.00000
E0             -.342839       -.271912        1.00000
```

```
     FUNCTION   1
```

X	OBSERVED Y	PREDICTED Y	RESIDUAL	WEIGHT	SE-PRED	STANDARDI RESIDUAL
.0000	86.40	82.25	4.148	1.000	2.047	1.616
.7200E-01	80.00	82.25	-2.250	1.000	2.046	-.8762
.2100	77.30	79.59	-2.286	1.000	1.887	-.8508
.3100	57.10	56.59	.5143	1.000	3.206	.7257
.5000	7.000	6.793	.2070	1.000	2.890	.1329
.6000	.0000	2.130	-2.130	1.000	1.331	-.7096

```
CORRECTED SUM OF SQUARED OBSERVATIONS =  7359.52
WEIGHTED CORRECTED SUM OF SQUARED OBSERVATIONS =  7359.52
SUM OF SQUARED RESIDUALS =            32.3373
SUM OF WEIGHTED SQUARED RESIDUALS =  32.3373
S =  3.28315     WITH     3 DEGREES OF FREEDOM
CORRELATION (OBSERVED,PREDICTED) =  .9979
```

```
AIC criteria =        26.85732
SC  criteria =        26.23260
```

PK16 – Two-compartment plasma and urine data intravenous dosing

Objectives

◆ **To analyze plasma and urine data following multiple intravenous infusions**

◆ **To obtain initial parameter estimates from a multiple-infusion dataset**

◆ **To apply a user specified differential equation model**

◆ **To discuss various aspects of experimental design and weighting**

Problem specification

This exercise highlights some complexities in modeling multiple intravenous infusion plasma kinetics simultaneously with urinary data. Specific emphasis is put on obtaining *initial parameter estimates*, application of *weights*, and simultaneous *fitting* of plasma and urinary data.

A male dog received two consecutive intravenous infusions of a new antibiotic drug X. The doses were 538 (0 to 0.983 h) and 3390 (0.983 to 23.95 h) µmol per kg body weight. Assume a body weight of 15 kg. Plasma and urinary data are plotted in Figure 16.1.

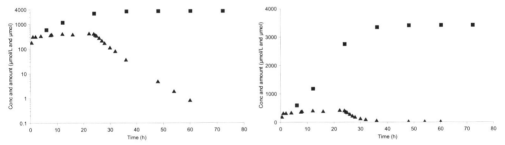

Figure 16.1 *Observed plasma concentration (µmol/L, ▲) and cumulative amount of drug excreted into urine (µmol, ■) versus time for a new antibiotic drug that was given using two consecutive constant rate intravenous infusions. The total infused dose was 3928 µmol/kg. The data are from a male dog, and are shown on a semi-logarithmic scale in the left hand figure and linear scale in the right hand figure.*

Concentration-time and urinary data are given in Table 16.1. We will fit the following system of differential equations to the data. For the concentration in the central compartment, we will use

$$V_c \cdot \frac{dC}{dt} = In - Cl_m \cdot C - Cl_R \cdot C - Cl_d \cdot C + Cl_d \cdot C_t \qquad (16:1)$$

The concentration in the peripheral compartment will be determined from

$$V_t \cdot \frac{dC_t}{dt} = Cl_d \cdot C - Cl_d \cdot C_t \tag{16:2}$$

The cumulative amount in urine is obtained from

$$\frac{dA_u}{dt} = Cl_R \cdot C \tag{16:3}$$

An approximate estimate of V_c is obtained from $Dose/C_{max}$. V_t is equal to V_{ss} - V_c. Cl_d is approximately $\alpha \cdot V_c$-Cl, where total Cl is obtained from the *Dose* divided by the total *AUC*. Renal clearance Cl_R is obtained from the *total amount that was accumulated in urine* divided by the *Dose* and then multiplied by Cl.

Table 16.1   Plasma concentration (µmol/L) and cumulative amount (µmol) in urine of drug X

Time (h)	Conc. (µmol/L)	Time (h)	Conc. (µmol/L)	Time (h)	Conc. (µmol/L)	Time (h)	Amount (µmol)
0	Missing	15.95	369.33	29.97	108.06	6.1	594
0.5	173.64	22.13	403.2	31.94	77.41	12.15	1179
1	301.59	23.89	391.37	35.96	34.78	24.05	2749
2	302.13	24.46	348.36	48	4.59	36.2	3338
4	319.87	24.94	324.17	54	1.8	48.2	3402
7.6	347.83	25.94	264.5	60	0.78	60.2	3411
8.02	367.72	26.96	205.36	72	-	72.2	3413
12.05	390.84	27.95	163.97	-	-	-	-

Discuss the disposition of this compound with your colleague(s), obtain initial parameter estimates, and apply a multi-compartment disposition model in the *WinNonlin* library. The initial parameter estimates for Cl, Cl_R and AUC should be derived graphically.

Initial parameter estimates

We applied non-compartmental analysis to obtain estimates of the terminal slope λ_z (≈ 0.15 h^{-1}, total plasma clearance Cl (0.37 L/h) and volume of distribution V_z (2.3 L). From those data and information about the fraction of the dose excreted in urine (3413/3928 $\approx$ 0.9), we estimated renal clearance Cl_R (0.32 L/h). Figure 16.2 shows how some of the parameters were estimated graphically from the plasma concentration and the cumulative amount in urine over time.

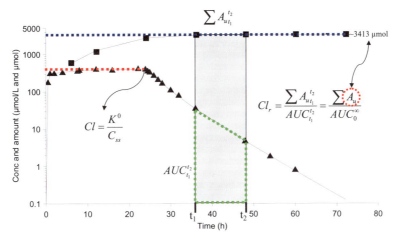

Figure 16.2 *Semi-logarithmic plot of observed plasma concentration and cumulative amount excreted in the urine versus time. The upper curve is urine (μmol) and lower curve plasma (μmol/L). The dashed blue horizontal line corresponds to the total amount excreted via the kidneys. The dotted/shaded area from time t_1 to t_2 corresponds to the cumulative amount excreted within that time frame. The green trapezoidal area is the corresponding AUC for plasma. The dashed red horizontal line is the approximate steady state plasma concentration C_{ss}, used together with the infusion rate K^0 to estimate total plasma clearance Cl.*

Interpretation of results and conclusions

This exercise deals with obtaining *initial parameter estimates*, applying *weights*, and inspecting for the *goodness-of-fit*. It is of considerable value to manually obtain good initial parameter estimates, because it speeds up the fitting and avoids local minima in the parameter space. Remember that the analyst should always be in charge of the analysis, not the program. Figure 16.3 contains the observed and predicted concentrations and cumulative urinary amounts of the compound over time after application of proportional weights.

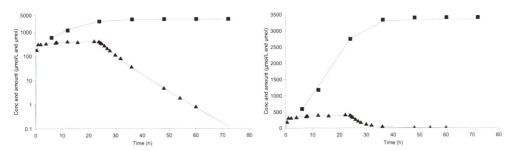

Figure 16.3 *Observed and predicted concentration-time data versus time, plotted on a semi-logarithmic scale (left) and on a linear scale (right). The upper curve is urine (cumulative amount, μmol) and the lower curve plasma concentration (μmol/L). Proportional weights were applied.*

Table 16.2 contains the final parameter estimates and the secondary parameter estimates (V_{ss}, Cl and $t_{1/2}$) from the model, and their precision for three different weighting schemes (unweighted, Poisson, and proportional). The greatest impact of weighting was on Cl_d and V_t, while Cl, Cl_R and V_{ss} were hardly affected at all. Simultaneous fitting of the present data with data obtained from plasma after an intravenous bolus dose might improve the accuracy and precision of Cl_d and V_t. Cl_R had very high precision because we simultaneously fitted both plasma and urinary data (Cl_R is placed between the observed variables (i.e. plasma and urine).

Table 16.2 Comparison of final estimates and precision from the three weighting schemes

Parameter	Unweighted		Poisson		Proportional	
	Estimate	CV%	Estimate	CV%	Estimate	CV%
V_c	1.4	5	1.5	6	1.6	2
Cl_m	0.05	9	0.05	4	0.05	9
CL_d	0.25	64	0.16	65	0.030	26
V_t	0.53	19	0.35	27	0.16	16
Cl_R	0.32	2	0.32	1	0.31	2
V_{ss}	2.0	3	1.87	2	1.8	2
Cl	0.37	3	0.37	1	0.37	1
$t_{1/2}$	3.8	3	3.5	2	3.4	1

Looking at Table 16.2 one might conclude that proportional weighting results provided the best fit and the highest parameter precision. One should also inspect the residuals, however, before assuming this to be the case.

You have
☐ analyzed a typical two-compartment (bi-exponential) system
☐ simultaneously fit plasma and urinary data
☐ learned to derive the initial estimates
☐ characterized the pharmacokinetics in plasma and urine after a multiple infusions regimen to a male dog

The next step in the characterization of kinetics would be to give multiple doses via the intended route of administration.

Solution – proportional weights

```
TITLE 1
Two consecutive infusions + wash-out: Differential equations
MODEL
 COMM
  NFUN 2
  NDER 3
  NPAR 5
  NCON 4
  NSEC 3
  PNAM 'Vc','CLm','Cld','Vt','Clr'
  SNAM 'Vss','CL','T12'
END
    1: TEMP
    2: T = X
    3: T1 = CON(1)
    4: T2 = CON(2)
    5: DOSE1 = CON(3)
    6: DOSE2 = CON(4)
    7: FINF = 0.0
```

```
 8: IF T <= T1 THEN
 9: FINF = DOSE1/T1
10: ENDIF
11: IF T > T1 AND T <= T2 THEN
12: FINF = DOSE2/(T2-T1)
13: ENDIF
14: IF T > T2 THEN
15: FINF = 0.0
16: ENDIF
17: END
18: START
19: Z(1) = 0.0
20: Z(2) = 0.0
21: Z(3) = 0.0
22: END
23: DIFF
24: DZ(1) = (FINF - CLM*Z(1) - CLR*Z(1) + CLD*Z(2) - CLD*Z(1))/VC
25: DZ(2) = (CLD*Z(1) - CLD*Z(2))/VT
26: DZ(3) = CLR*Z(1)
27: END
28: FUNC 1
29: F= Z(1)
30: END
31: FUNC 2
32: F= Z(3)
33: END
34: SECO
35: VSS = VC + VT
36: CL = CLM + CLR
37: T12 = 0.693*VSS/CL
38: END
39: EOM
NOPAGE BREAKS
NVARIABLES 3
FNUMBER 3
NPOINTS 100
XNUMBER 1
YNUMBER 2
CONSTANTS .983,23.95,538,3390
METHOD 2  'Gauss-Newton (Levenberg and Hartley)
REWEIGHT -2
ITERATIONS 50
INITIAL 1.5,.05,.1,.5,.32
LOWER BOUNDS .1,0,0,0,0
UPPER BOUNDS 5,.5,1,5,1
MISSING 'Missing'
DATA 'WINNLIN.DAT'
BEGIN
```

PARAMETER	ESTIMATE	STANDARD ERROR	CV%	UNIVARIATE C.I.	
VC	1.618941	.037772	2.33	1.540804	1.697078
CLM	.054035	.004831	8.94	.044040	.064030
CLD	.030407	.007815	25.70	.014241	.046573
VT	.164742	.026971	16.37	.108949	.220536
CLR	.314900	.005785	1.84	.302932	.326867

```
*** CORRELATION MATRIX OF THE ESTIMATES ***
PARAMETER   VC        CLM        CLD        VT        CLR
VC        1.00000
CLM        -.05964   1.00000
CLD        -.64931    .06511    1.00000
VT         -.68999    .08018     .98890   1.00000
CLR         .31553   -.78158     .06941    .07076   1.00000

  Condition_number=          68.10

    FUNCTION    1
```

X	OBSERVED Y	PREDICTED Y	RESIDUAL	WEIGHT	SE-PRED	STANDARDIZ RESIDUAL
.0000	.	.0000	.	.0000	-.6400+152	.
.5000	173.6	159.0	14.59	.3953E-04	3.410	2.707
1.000	301.6	295.6	6.017	.1145E-04	5.795	.5818
2.000	302.1	312.6	-10.49	.1023E-04	4.798	-.9057
4.000	319.9	338.3	-18.44	.8737E-05	4.131	-1.426
7.600	347.8	366.3	-18.42	.7455E-05	4.079	-1.305
8.020	367.7	368.5	-.7742	.7364E-05	4.078	-.0545
12.05	390.8	383.5	7.329	.6799E-05	4.027	.4937
15.95	369.3	391.1	-21.73	.6539E-05	3.973	-1.432
22.13	403.2	396.6	6.634	.6359E-05	3.934	.4304
23.89	391.4	397.4	-6.017	.6332E-05	3.930	-.3895
24.46	348.4	353.9	-5.542	.7984E-05	3.594	-.4036
24.94	324.2	317.6	6.524	.9911E-05	3.408	.5315
25.94	264.5	254.4	10.08	.1545E-04	3.122	1.037
26.96	205.4	203.8	1.556	.2408E-04	2.799	.2026
27.95	164.0	165.1	-1.081	.3671E-04	2.444	-.1757
29.97	108.1	108.8	-.7113	.8452E-04	1.762	-.1782
31.94	77.41	73.64	3.768	.1844E-03	1.291	1.418
35.96	34.78	34.79	-.1145E-01	.8261E-03	.8047	-.0100
48.00	4.590	4.721	-.1310	.4487E-01	.1346	-.9839
54.00	1.800	1.870	-.7014E-01	.2859	.4331E-01	-1.145
60.00	.7800	.7536	.2642E-01	1.761	.2679E-01	1.887
72.00	.	.1248	.	.0000	-.6400+152	.

```
CORRECTED SUM OF SQUARED OBSERVATIONS =  515709.
WEIGHTED CORRECTED SUM OF SQUARED OBSERVATIONS =  18.7106
SUM OF SQUARED RESIDUALS =            1838.21
SUM OF WEIGHTED SQUARED RESIDUALS =  .278288E-01
S =  .417049E-01 WITH    16 DEGREES OF FREEDOM
CORRELATION (OBSERVED,PREDICTED) =  .8966

    FUNCTION    2
```

X	OBSERVED Y	PREDICTED Y	RESIDUAL	WEIGHT	SE-PRED	STANDARDIZED RESIDUAL
6.100	594.0	580.6	13.35	.2966E-05	9.447	.6272
12.15	1179.	1290.	-111.4	.6005E-06	19.77	-2.329
24.05	2749.	2762.	-12.57	.1311E-06	41.90	-.1226
36.20	3338.	3290.	48.28	.9240E-07	50.25	.3957
48.20	3402.	3343.	58.73	.8947E-07	51.12	.4737
60.20	3411.	3351.	59.85	.8905E-07	51.24	.4817
72.20	3413.	3352.	60.57	.8898E-07	51.26	.4873

```
CORRECTED SUM OF SQUARED OBSERVATIONS =  .857016E+07
```

```
WEIGHTED CORRECTED SUM OF SQUARED OBSERVATIONS = 2.96622
SUM OF SQUARED RESIDUALS =              25778.4
SUM OF WEIGHTED SQUARED RESIDUALS = .917246E-02
S = .677217E-01 WITH      2 DEGREES OF FREEDOM
CORRELATION (OBSERVED,PREDICTED) = .9992


TOTALS FOR ALL CURVES COMBINED
SUM OF SQUARED RESIDUALS =              27616.6
SUM OF WEIGHTED SQUARED RESIDUALS = .370013E-01
S = .401092E-01 WITH     23 DEGREES OF FREEDOM


AIC criteria =        -88.90410
SC  criteria =        -81.89812

SUMMARY OF ESTIMATED SECONDARY PARAMETERS
```

PARAMETER	ESTIMATE	STANDARD ERROR	CV%
VSS	1.783683	.027355	1.53
CL	.368935	.003622	.98
T12	3.350438	.031760	.95

PK17 - Nonlinear kinetics - Capacity I

Objectives

◆ **To fit and discriminate between a linear and a Michaelis-Menten model**

◆ **To estimate terminal half-life and clearance**

◆ **To write the models in terms of differential equations**

◆ **To emphasize the importance of adequate data for model discrimination**

Problem specification

The aim of the present exercise is to demonstrate the importance of having adequate data when attempting to discriminate between models. A new agent X was administered as a rapid injection followed by a slow constant intravenous infusion to a single subject. The rapid infusion dose was 1800 µg and rapid infusion time 0.5 min. The slow infusion total dose was 5484.8 µg and the corresponding infusion time 39.63 min. Data are given in the program output, and a plot of plasma concentration *versus* time is shown in Figure 17.1. In this example, analyze the data with a one-compartment model, using both first-order and Michaelis-Menten elimination.

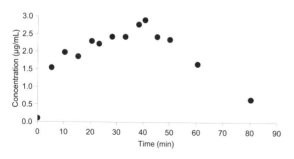

Figure 17.1 *Observed and predicted concentration-time data following a regimen of two consecutive intravenous infusions.*

We will parameterize the model in terms of clearance *Cl* and volume of distribution *V*. The model with linear elimination is defined as

$$V \frac{dC}{dt} = In - Cl \cdot C \qquad (17:1)$$

In is the input function that contains a rapid and slow rate and *Cl* is the clearance parameter.

The model with Michaelis-Menten kinetics uses *Cl* defined as

$$Cl = \frac{V_{max}}{K_m + C} \qquad (17:2)$$

The *ITRIP* variable (switch, either 0 or 1) in the program code, is 0 during the infusion and 1 after cessation of the infusion. Since *ITRIP = 0* during infusion, *D/TI/V* becomes the concentration flow into the system. When time (*T*) is larger than infusion time *(TI) ITRIP = 1*, which turns off the input.

Initial parameter estimates

Linear model:
$V = 1500$ (mL)
$Cl = 50$ (mL/min)

Michaelis-Menten model:
V	= 1500 (mL)	
V_{max}	= 120 (µg/min)	*VM* in model code
K_m	= 1 (µg/mL)	Obtained from biochemical data

There are some residual concentrations (0.105 µg/mL) of the compound at the start of the infusion. Therefore we need to set the *initial condition* of the state variable (concentration) in the program code to:

$$Z(1) = 0.105$$

Interpretation of results and conclusions

Observed and fitted data are shown in Figure 17.2. Although the correlation between some parameters of the Michaelis-Menten model was high (K_m and V = -0.96), the parameter precision is still good.

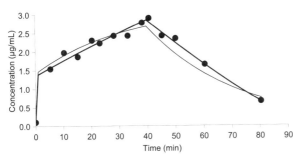

Figure 17.2 *Observed and predicted plasma concentration-time data following a regimen of two-consecutive intravenous infusions. The difference between the linear model (hairthin line) and the Michaelis-Menten model (thick solid line) is clearly demonstrated in that the Michaelis-Menten model does not deviate systematically with time.*

The residuals, the weighted residual sum of squares *WRSS* and the *AIC* value favor the use of the Michaelis-Menten model. The *WRSS* of the Michaelis-Menten model should be equal to or less than the *WRSS* of the linear model (which it is), since the former model has one additional parameter.

Figure 17.3 displays the clearance and half-life *versus* steady state plasma concentration for the Michaelis-Menten model. The lowest value for the half-life is approximately 5 min and the highest clearance value is 190 mL/min. These numbers are based on the estimated V_{max} and K_m parameters.

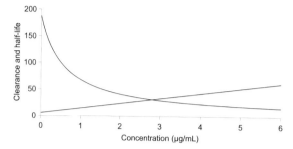

Figure 17.3 *Clearance (curvilinear relationship, mL/min) and half-life (straight line) versus steady state plasma concentration. Plot obtained by means of Equations 17:2 and 17:3.*

Clearance is inversely proportional to $K_m + C$ and half-life increases proportionately with increasing plasma concentrations, as evident from Equations 17:2 and Figure 17.3.

$$t_{1/2} = \ln(2) \cdot \frac{V_{ss}}{V_{max}} [K_m + C] \qquad (17:3)$$

The linear model predicts a half-life of about 20 min whereas the nonlinear model estimates the half-life to be about 5 min (at 0 μg/mL). The latter approaches 20 min at a plasma concentration of about 1.5 ng/mL. The plasma concentrations are above 1.5 ng/mL during most of the observation period.

Using information obtained for V_{max} and K_m, and the infusion rate, we conclude that one would never achieve steady state with the present rate and duration of infusion since the system is saturated. Therefore, this experiment could only be used as a pilot study, and data for V_{max} and K_m should not be applied for extrapolations outside the sampled time and concentration interval. The present design does not lend itself for model discrimination purposes. In order to discriminate between the two models, a higher dose should also be investigated. By doubling or even tripling the highest dose, the potential nonlinear behavior should be discernible.

Solution I - Linear model

```
TITLE 1
Michaelis-Menten vs Linear elimination of hya.
MODEL
 COMM
  NDER 1
  NPARM 2
  NSEC  1
  NCON 4
  PNAMES 'VOLUME', 'CL'
  SNAMES 'T1/2'
END
    1: TEMP
    2: D=CON(1)
    3: TI=CON(2)
    4: IVBD=CON(3)
    5: TBOL=CON(4)
    6: T=X
    7: V=P(1)
    8: CL=P(2)
    9: DENOM=MAX(1.D-20,ABS(T-TI))
   10: ITRIP=MAX(0,(T-TI)/DENOM)
```

```
11: END
12: START
13: Z(1)=0.105
14: END
15: DIFF
16: IF T <= TBOL THEN
17: FINF = IVBD/TBOL
18: ELSE
19: FINF = 0.0
20: ENDIF
21: DZ(1)=(FINF/V + (D/TI/V)*(1.0-ITRIP)) - (CL*Z(1)/V)
22: END
23: FUNC 1
24: F = Z(1)
25: END
26: SECO
27: S(1)=0.693*V/CL
28: END
29: EOM
NVARIABLES 2
NPOINTS 100
XNUMBER 1
YNUMBER 2
WEIGHT -2
CONSTANTS 5484.8,39.63,1800,.5
METHOD 2   'Gauss-Newton (Levenberg and Hartley)
ITERATIONS 50
INITIAL 1500,50
LOWER BOUNDS 1,0
UPPER BOUNDS 3000,300
NOBSERVATIONS 14
DATA 'WINNLIN.DAT'
BEGIN
```

PARAMETER	ESTIMATE	STANDARD ERROR	CV%	UNIVARIATE C.I.	
VOLUME	1377.665874	81.273541	5.90	1200.585779	1554.745970
CL	43.283556	1.535150	3.55	39.938746	46.628366

```
   Condition_number=       54.00
```

FUNCTION   1

X	OBSERVED Y	PREDICTED Y	RESIDUAL	WEIGHT	SE-PRED	STANDARDIZED RESIDUAL
.0000	.1050	.1050	.0000	13.27		
5.380	1.540	1.696	-.1559	.6167E-01	.5832E-01	-1.056
10.33	1.975	1.912	.6279E-01	.3749E-01	.6505E-01	.3253
15.30	1.860	2.098	-.2380	.4227E-01	.7025E-01	-1.334
20.35	2.300	2.259	.4066E-01	.2765E-01	.7439E-01	.1806
23.13	2.225	2.338	-.1128	.2954E-01	.7631E-01	-.5214
28.15	2.425	2.463	-.3825E-01	.2487E-01	.7933E-01	-.1613
33.18	2.425	2.571	-.1456	.2487E-01	.8195E-01	-.6162
38.23	2.775	2.663	.1124	.1899E-01	.8430E-01	.4112
40.62	2.900	2.605	.2949	.1739E-01	.8015E-01	1.023
45.25	2.425	2.252	.1725	.2487E-01	.6881E-01	.7177
50.08	2.350	1.935	.4147	.2648E-01	.6406E-01	1.774
60.42	1.650	1.399	.2515	.5372E-01	.6328E-01	1.592
80.35	.6515	.7477	-.9620E-01	.3446	.5892E-01	-2.981

CORRECTED SUM OF SQUARED OBSERVATIONS = 7.96837

```
WEIGHTED CORRECTED SUM OF SQUARED OBSERVATIONS = 1.62601
SUM OF SQUARED RESIDUALS =          .495764
SUM OF WEIGHTED SQUARED RESIDUALS = .186605E-01
S =  .394341E-01 WITH    12 DEGREES OF FREEDOM
CORRELATION (OBSERVED,PREDICTED) = .9701


AIC criteria =         -51.73882
SC  criteria =         -50.46071
```

PARAMETER	ESTIMATE	STANDARD ERROR	CV%
T1/2	22.057394	1.379748	6.26

Solution II - Michaelis-Menten model

```
TITLE 1
Michaelis-Menten vs Linear elimination of hya.
MODEL
 COMM
  NDER 1
  NPARM 3
  NSEC  2
  NCON 4
  PNAMES 'VOLUME', 'VM', 'KM'
  SNAMES 'T12', 'CL'
END
    1: TEMP
    2: D=CON(1)
    3: TI=CON(2)
    4: IVBD=CON(3)
    5: TBOL=CON(4)
    6: T=X
    7: V=P(1)
    8: VM=P(2)
    9: KM=P(3)
   10: DENOM=MAX(1.D-20,ABS(T-TI))
   11: ITRIP=MAX(0,(T-TI)/DENOM)
   12: END
   13: START
   14: Z(1)=0.105
   15: END
   16: DIFF
   17: IF T <= TBOL THEN
   18: FINF = IVBD/TBOL
   19: ELSE
   20: FINF = 0.0
   21: ENDIF
   22: DZ(1)=(FINF/V + (D/TI/V)*(1.0-ITRIP)) - (VM*Z(1)/V)/(KM+Z(1))
   23: END
   24: FUNC 1
   25: F = Z(1)
   26: END
   27: SECO
   28: S(1)=0.693*V*KM/VM
   29: S(2)=VM/KM
   30: END
   31: EOM
NVARIABLES 2
NPOINTS 100
XNUMBER 1
YNUMBER 2
```

```
WEIGHT -2
CONSTANTS 5484.8,39.63,1800,.5
METHOD 2   'Gauss-Newton (Levenberg and Hartley)
ITERATIONS 50
INITIAL 1500,120,1
LOWER BOUNDS 1,0,0
UPPER BOUNDS 3000,900,10
MISSING 'Missing'
NOBSERVATIONS 14
DATA 'WINNLIN.DAT'
BEGIN
```

PARAMETER	ESTIMATE	STANDARD ERROR	CV%	UNIVARIATE C.I.	
VOLUME	1454.454469	72.458528	4.98	1294.973714	1613.935223
VM	107.353277	1.665983	1.55	103.686460	111.020094
KM	.566206	.026594	4.70	.507673	.624740

```
*** CORRELATION MATRIX OF THE ESTIMATES ***
```

PARAMETER	VOLUME	VM	KM
VOLUME	1.00000		
VM	.485860	1.00000	
KM	-.955597	-.645043	1.00000

```
Condition_number=       .1298E+05

    FUNCTION    1
```

X	OBSERVED Y	PREDICTED Y	RESIDUAL	WEIGHT	SE-PRED	STANDARDIZED RESIDUAL
.0000	.1050	.1050	.0000	13.27		
5.380	1.540	1.576	-.3616E-01	.6167E-01	.6191E-01	-.6676
10.33	1.975	1.774	.2009	.3749E-01	.4873E-01	2.147
15.30	1.860	1.966	-.1055	.4227E-01	.3958E-01	-1.158
20.35	2.300	2.154	.1463	.2765E-01	.3583E-01	1.245
23.13	2.225	2.255	-.3000E-01	.2954E-01	.3644E-01	-.2652
28.15	2.425	2.434	-.9240E-02	.2487E-01	.4151E-01	-.0753
33.18	2.425	2.610	-.1847	.2487E-01	.4987E-01	-1.545
38.23	2.775	2.782	-.7195E-02	.1899E-01	.5992E-01	-.0531
40.62	2.900	2.769	.1307	.1739E-01	.5943E-01	.9134
45.25	2.425	2.488	-.6322E-01	.2487E-01	.5028E-01	-.5296
50.08	2.350	2.201	.1489	.2648E-01	.4216E-01	1.259
60.42	1.650	1.614	.3630E-01	.5372E-01	.3200E-01	.4420
80.35	.6515	.6539	-.2404E-02	.3446	.3347E-01	-.2530

```
CORRECTED SUM OF SQUARED OBSERVATIONS = 7.96837
WEIGHTED CORRECTED SUM OF SQUARED OBSERVATIONS = 1.62601
SUM OF SQUARED RESIDUALS =          .153884
SUM OF WEIGHTED SQUARED RESIDUALS = .458959E-02
S = .204263E-01 WITH    11 DEGREES OF FREEDOM
CORRELATION (OBSERVED,PREDICTED) =  .9905

AIC criteria =         -69.37552
SC  criteria =         -67.45834
```

PARAMETER	ESTIMATE	STANDARD ERROR	CV%
T12	5.316096	.135029	2.54
CL	189.600964	11.026036	5.82

PK18 - Nonlinear kinetics of ethanol – Capacity II

Objectives

◆ To analyze an intravenous infusion dataset with nonlinear behavior

◆ To write a multi-compartment model in terms of differential equations

◆ To obtain initial parameter estimates of V_{max}, K_m, V_c, V_t and Cl_d.

◆ To discuss various aspects of correlation, accuracy and precision

◆ To fit and discuss literature data

Problem specification

The pharmacokinetics of ethanol was characterized following a constant intravenous infusion. Ethanol displayed capacity-limited kinetics and a volume of distribution equal to total body water. This problem highlights some of the complexities in modeling nonlinear (capacity-limited) kinetics. Specific emphasis is placed on obtaining *initial parameter estimates*, implementation of *differential equations*, and assessing the *goodness-of-fit*. At the end of the exercise (Appendix), we have added a dataset from Wilkinson *et al* [1976] and analyzed the data with our model.

A number of volunteers were infused intravenously with a dose of 0.4 g ethanol per kg body weight over 30 min. Assume a body weight of 70 kg. Plasma samples (Figure 18.1) were obtained in parallel with expired air every fifth min for 6 h. The high resolution of data was needed for research purposes, with data being kindly supplied by Dr. Norberg, Huddinge Hospital, Stockholm (Norberg *et al*, [2000]).

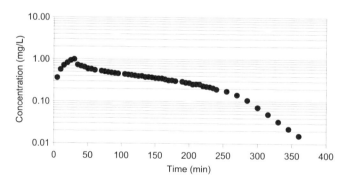

Figure 18.1. Semi-logarithmic plot of observed ethanol concentration versus time following a 30 min constant rate intravenous infusion of ethanol. Data obtained from Norberg et al [2000].

Obtain initial parameter estimates and implement a two-compartment disposition model with nonlinear elimination into *WinNonlin*. Then fit a system of differential equations to the data. For the concentration in the central compartment

$$V_c \cdot \frac{dC}{dt} = In - Cl \cdot C - Cl_d \cdot C + Cl_d \cdot C_t$$

(18:1)

The concentration in the peripheral compartment is

$$V_t \cdot \frac{dC_t}{dt} = Cl_d \cdot C - Cl_d \cdot C_t \qquad (18:2)$$

where Cl is the capacity-limited clearance in Equation 18:3

$$Cl = \frac{V_{max}}{K_m + C} \qquad (18:3)$$

Initial parameter estimates

We assume that the washout curve of ethanol can be approximated by the disposition following an intravenous bolus dose. If so, we can adopt the approach of Gibaldi and Perrier [1982] to obtain initial parameter estimates of V_{max} and K_m. Then, the slope in Figure 18.2A is equal to

$$slope_A = -K' = \frac{\ln(y_2) - \ln(y_1)}{t_2 - t_1} \qquad (18:4)$$

The slope of the terminal phase (Figure 18.2B) is equal to

$$slope_B = -\left[K' + \frac{V'_{max}}{K_m} \right] \qquad (18:5)$$

This equation rearranged gives

$$\frac{V'_{max}}{K_m} = \left| slope_B \right| - \left| K' \right| \qquad (18:6)$$

which together with K' inserted into Equation 18:5 gives

$$K_m = \frac{C_0 K' / (K' + V'_{max} / K_m)}{(C_0^* / C_0)^{K' K_m / V'_{max}} - 1} \qquad (18:7)$$

C_0 and C_0^* correspond to the intercept on the concentration axis in Figures 18.2A and B, respectively. Since V'_{max} contains a volume unit (concentration per time), the corrected value for V_{max} (amount per time) becomes

$$V_{max} = V'_{max} \cdot V_c \qquad (18:8)$$

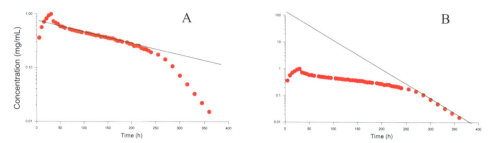

Figure 18.2 *Observed plasma concentration-time data of ethanol following a 30 min constant rate intravenous infusion. The solid line in figure A corresponds to Equation 18:4 and the line in figure B to Equation 18:5.*

When a compound is given as an intravenous bolus dose, an estimate of V_c is obtained from $Dose/C_{max}$ because all of the drug is stored in the central compartment at time zero (Figure 18.3). However, although ethanol was given as a constant rate intravenous infusion in this case, the estimation of V_c by means of $Dose/C_{max}$ would be highly overestimated (c.f. Figure 18.1). In fact, for the present dose of 28 g and C_{max} of 1 g/L, V_c would approximate 28 L.

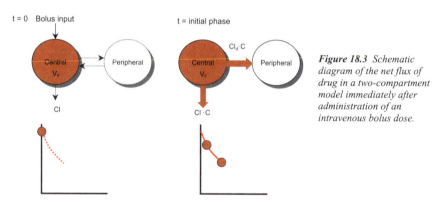

Figure 18.3 *Schematic diagram of the net flux of drug in a two-compartment model immediately after administration of an intravenous bolus dose.*

Using Equations 18:1 and 18:2, V_c for ethanol is approximately 4-5 L rather than 28 L. V_t is estimated from $V_{ss}-V_c$, where V_{ss} can be derived by means of non-compartmental analysis. The rate of elimination from the central compartment is initially equal to

$$\frac{dX}{dt} = Cl_d \cdot C + Cl \cdot C \qquad (18:9)$$

The total clearance

$$Cl(tot) = \frac{1}{C} \cdot \frac{dX}{dt} = Cl_d + Cl \qquad (18:10)$$

The initial rate constant can be approximated with α, which then becomes

$$\alpha \approx \frac{Cl_d}{V_c} + \frac{Cl}{V_c} \qquad (18:11)$$

which rearranged gives

$$\alpha \cdot V_c = Cl_d + Cl \qquad (18:12)$$

Cl_d is then $\alpha \cdot V_c$-Cl, where Cl is the capacity-limited clearance in Equation 18:3. For the purpose of estimating Cl_d, Cl is estimated by means of *Dose/AUC*.

Interpretation of results and conclusions

This exercise deals with obtaining *initial parameter estimates*, implementing *differential equations*, and the *goodness-of-fit* of a nonlinear kinetic system. It is of considerable value to manually obtain the initial parameter estimates for a nonlinear system. Otherwise the program may end up in a local minimum in the parameter space. The selected strategy was taken from Gibaldi and Perrier [1982]. The observed and predicted data are shown in Figure 18.4.

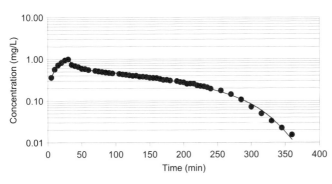

Figure 18.4. Observed (symbols) and predicted (line) plasma concentration-time data of ethanol following a 30 min constant rate intravenous infusion. Data obtained from Norberg et al [2000].

Particularly when fitting data from a single dose level of drug, some parameters are often highly correlated, such as V_{max} and K_m, so that the estimate of one almost governs the estimate of the other. Estimates of these parameters are also often skewed to the right, with many of the estimates being many times larger than the true parameter value. They may also have high variances, in spite of the fact that the variances computed from individual datasets are underestimated. Also, there is a high correlation between the estimates of V_{max} and K_m with their respective standard deviations. Thus, the estimated precision of V_{max} and K_m given by regression programs does not give a true picture of the uncertainty in these estimates (see Metzler and Tong [1981] for a thorough review). To avoid some of these problems we advocate the modeler to fit data obtained from two or more dose levels simultaneously.

The partial derivatives of the plasma concentration model with respect to some of the model parameters are shown in Figure 18.5. Maximum information about V_{max} and K_m is obtained at about 280 and 330 min, respectively.

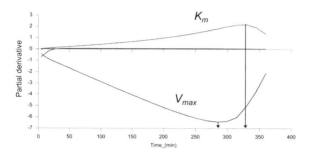

Figure 18.5 Simulated partial derivatives of the plasma concentration with respect to the model parameters. Note that maximum information about K_m can be found at about 330 min and V_{max} at about 280 min.

Ethanol is one of the most self-prescribed drugs. In spite of its long tradition on the market (>3000 years), a thorough picture of its disposition has not been obtained and people generally do not understand its optimal use. What we can say is that it displays nonlinear (capacity- or dose-dependent) kinetics. Because the bioavailability is nonlinear due to saturable hepatic extraction, spirits and beer are a very inefficient and expensive way of utilizing the pharmacological properties of ethanol. This is because low ethanol concentrations in such drinks result in a high and efficient first-pass extraction, even though the volume of the vehicle may be substantial (e.g., the number of beers consumed during an evening). A shot, on the other hand, will be absorbed efficiently and then saturate the hepatic enzymes, leaving a larger fraction of the dose to produce a pharmacological effect. Finally, it must be remembered that ethanol changes renal blood flow and diuresis. A third dimension is the development of acute (pharmacological) tolerance to the stimulating effects of ethanol, which is beyond the scope of this exercise.

We recommend, as a general rule, that you should supply the program with initial parameter estimates. However, in this particular case, we obtained good final parameter estimates with high precision when we let the program estimate the initial parameters but supplied the parameter boundaries. Similar results were obtained when we supplied the initial parameter estimates but let the program select the parameter boundaries.

Appendix - Wilkinson *et al* [1976] data

Data on ethanol kinetics were taken from Wilkinson *et al* [1976] and carried through the same analysis as the Norberg *et al* [2000] data. Table 18.1 contains the 'population' average of each parameter from six subjects and the inter-subject variability (*CV%*).

Table 18.1 'Population' estimates (n = 6)

Parameter	N_{subj}	Mean	CV%
Cl_d	6	1.6	50
K_m	6	0.028	30
V_c	6	18	30
V_{max}	6	0.14	10
V_t	6	27	20

Observed and predicted data of one subject and predicted data of all subjects are shown below (Figure 18.6 left and right, respectively). See also solution section.

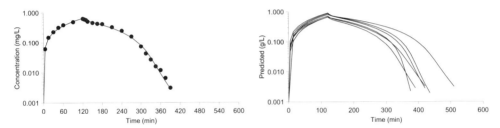

Figure 18.6 *Observed (circles) and predicted (line) plasma concentration-time data of subject # 2 in the Wilkinson et al [1976] study (left). Spaghetti plot of predicted concentration-time courses of all six subjects in the same study (right).*

The results from Norberg *et al* [2000] and Wilkinson *et al* [1976] are surprisingly consistent, in spite of the fact of different study designs and populations were used. The volume of distribution at steady state was approximately 40 and 45 L for the Norberg (individual) and Wilkinson (population mean) studies, respectively. The ratio of V_{max}-to-K_m was about 5 in both studies.

Solution – Norberg *et al* [2000] data

```
MODEL
 COMM
   NDER 2
   NCON 5
   NPAR 5
   NCON 2
   NSEC 3
   PNAM 'Vc', 'CLD','Vt', 'Vmax', 'Km'
   SNAM 'Vss', 'MRT', 'Tz'
END
     1: TEMP
     2: T=X
     3: DOSE1 = CON(1)
     4: TINF1 = CON(2)
     5: VC = P(1)
     6: CLD = P(2)
     7: VT = P(3)
     8: VMAX= P(4)
     9: KM = P(5)
    10: END
    11: START
    12: Z(1)=0
    13: Z(2)=0
    14: END
    15: DIFF
    16: FINF1 = 0.0
    17: IF T LE TINF1 THEN
    18: FINF1 = DOSE1/TINF1
    19: ENDIF
    20: CL = VMAX/(KM + Z(1))
    21: DZ(1)=(FINF1 - CL*Z(1) - CLD*Z(1) + CLD*Z(2))/VC
    22: DZ(2)=(CLD*Z(1) - CLD*Z(2))/VT
    23: END
    24: SECO
    25: S(1) = VC + VT
    26: VSS = S(1)
```

```
   27: CLI = VMAX/KM
   28: S(2) = S(1)/CLI
   29: S(3) = 0.693*VSS/CL
   30: END
   31: FUNC 1
   32: F = Z(1)
   33: END
   34: EOM
NOPAGE BREAKS
NVARIABLES 2
NPOINTS 1500
XNUMBER 1
YNUMBER 2
CONSTANTS 28,30
METHOD 2   'Gauss-Newton (Levenberg and Hartley)
REWEIGHT -1
ITERATIONS 50
INITIAL 5,1,35,.1,.05
LOWER BOUNDS 0,0,0,0,0
UPPER BOUNDS 80,10,90,.9,.1
NOBSERVATIONS 53
BEGIN
```

PARAMETER	ESTIMATE	STANDARD ERROR	CV%	UNIVARIATE C.I.	
VC	8.716385	.426238	4.89	7.859379	9.573391
CLD	1.300156	.052449	4.03	1.194701	1.405611
VT	31.175077	.487033	1.56	30.195834	32.154320
VMAX	.082212	.001349	1.64	.079499	.084925
KM	.014877	.003198	21.50	.008447	.021307

```
*** CORRELATION MATRIX OF THE ESTIMATES ***
PARAMETER   VC          CLD         VT        VMAX       KM
VC          1.00000
CLD         -.77988    1.00000
VT          -.65637     .42317    1.00000
VMAX        -.19256     .33478    -.45002    1.00000
KM          -.15435     .25437    -.40343     .95246   1.00000

 Condition_number=        2030.
```

X	OBSERVED Y	PREDICTED Y	RESIDUAL	WEIGHT	SE-PRED	STANDARDIZ RESIDUAL
5.000	.3620	.3559	.6088E-02	2.810	.9108E-02	.5603
10.00	.5660	.5570	.9039E-02	1.795	.8442E-02	.5795
15.00	.7150	.6999	.1509E-01	1.429	.8432E-02	.8383
20.00	.8230	.8206	.2421E-02	1.219	.9082E-02	.1240
25.00	.9360	.9327	.3322E-02	1.072	.9620E-02	.1594
30.00	.9940	1.041	-.4749E-01	.9602	.9977E-02	-2.148
35.00	.7320	.7648	-.3279E-01	1.308	.1270E-01	-1.993
40.00	.6870	.6505	.3650E-01	1.537	.8904E-02	2.150
45.00	.6520	.6004	.5163E-01	1.666	.5693E-02	2.948
50.00	.5920	.5749	.1709E-01	1.739	.4589E-02	.9805
55.00	.5810	.5590	.2203E-01	1.789	.4471E-02	1.281
60.00	.5520	.5467	.5324E-02	1.829	.4481E-02	.3133
70.00	.5270	.5255	.1547E-02	1.903	.4321E-02	.0927
75.00	.5080	.5153	-.7325E-02	1.941	.4185E-02	-.4429
80.00	.4940	.5053	-.1128E-01	1.979	.4036E-02	-.6878

85.00	.4810	.4953	-.1428E-01	2.019	.3885E-02	-.8777
90.00	.4720	.4853	-.1329E-01	2.061	.3734E-02	-.8241
95.00	.4590	.4753	-.1632E-01	2.104	.3586E-02	-1.021
105.0	.4460	.4554	-.9392E-02	2.196	.3303E-02	-.5984
110.0	.4330	.4454	-.1244E-01	2.245	.3169E-02	-.8006
115.0	.4240	.4355	-.1150E-01	2.296	.3042E-02	-.7475
120.0	.4120	.4256	-.1357E-01	2.350	.2920E-02	-.8911
125.0	.4000	.4156	-.1564E-01	2.406	.2805E-02	-1.039
130.0	.4040	.4057	-.1730E-02	2.465	.2698E-02	-.1161
135.0	.3800	.3958	-.1583E-01	2.526	.2599E-02	-1.075
140.0	.3810	.3859	-.4932E-02	2.591	.2508E-02	-.3390
145.0	.3730	.3760	-.3049E-02	2.659	.2427E-02	-.2122
150.0	.3610	.3662	-.5178E-02	2.731	.2355E-02	-.3650
155.0	.3560	.3563	-.3195E-03	2.806	.2293E-02	-.0228
160.0	.3540	.3465	.7526E-02	2.886	.2242E-02	.5451
165.0	.3350	.3366	-.1641E-02	2.971	.2201E-02	-.1206
170.0	.3180	.3268	-.8824E-02	3.060	.2171E-02	-.6579
175.0	.3190	.3170	.1979E-02	3.154	.2151E-02	.1498
180.0	.3070	.3072	-.2352E-03	3.255	.2140E-02	-.0181
190.0	.2970	.2877	.9284E-02	3.476	.2147E-02	.7388
195.0	.2820	.2780	.4014E-02	3.597	.2163E-02	.3252
200.0	.2770	.2683	.8723E-02	3.727	.2185E-02	.7200
205.0	.2550	.2586	-.3591E-02	3.867	.2212E-02	-.3022
210.0	.2590	.2489	.1007E-01	4.017	.2245E-02	.8650
215.0	.2570	.2393	.1771E-01	4.179	.2280E-02	1.553
220.0	.2340	.2297	.4314E-02	4.354	.2318E-02	.3868
225.0	.2260	.2201	.5890E-02	4.543	.2357E-02	.5405
230.0	.2210	.2106	.1043E-01	4.749	.2396E-02	.9806
235.0	.2100	.2011	.8939E-02	4.974	.2434E-02	.8616
240.0	.1950	.1916	.3405E-02	5.219	.2469E-02	.3370
255.0	.1760	.1635	.1253E-01	6.117	.2554E-02	1.352
270.0	.1430	.1359	.7106E-02	7.359	.2580E-02	.8487
285.0	.1070	.1091	-.2074E-02	9.168	.2525E-02	-.2790
300.0	.7170E-01	.8338E-01	-.1168E-01	11.99	.2371E-02	-1.814
315.0	.4940E-01	.5945E-01	-.1005E-01	16.82	.2152E-02	-1.868
330.0	.3290E-01	.3845E-01	-.5554E-02	26.01	.2007E-02	-1.320
345.0	.2230E-01	.2210E-01	.1989E-03	45.25	.2027E-02	.0687
360.0	.1530E-01	.1169E-01	.3606E-02	85.51	.1789E-02	1.955

```
CORRECTED SUM OF SQUARED OBSERVATIONS =  2.50219
WEIGHTED CORRECTED SUM OF SQUARED OBSERVATIONS =  11.9044
SUM OF SQUARED RESIDUALS =            .119391E-01
SUM OF WEIGHTED SQUARED RESIDUALS =  .271114E-01
S =  .237660E-01 WITH    48 DEGREES OF FREEDOM
CORRELATION (OBSERVED,PREDICTED) =  .9977

AIC criteria =        -181.21337
SC  criteria =        -171.36192
```

PARAMETER	ESTIMATE	STANDARD ERROR	CV%
VSS	39.891462	.382575	.96
MRT	7.218619	1.393725	19.31
TZ	8.946641	.085802	.96

Solution – Wilkinson *et al* [1976]

```
subj=2
MODEL
 COMM
  NDER 2
  NCON 5
```

```
      NPAR 5
      NCON 2
      NSEC 3
      PNAM 'Vc', 'CLD','Vt', 'Vmax', 'Km'
      SNAM 'Vss', 'MRT', 'Tz'
END
    1: TEMP
    2: T=X
    3: DOSE1 = CON(1)
    4: TINF1 = CON(2)
    5: VC = P(1)
    6: CLD = P(2)
    7: VT = P(3)
    8: VMAX= P(4)
    9: KM = P(5)
   10: END
   11: START
   12: Z(1)=0
   13: Z(2)=0
   14: END
   15: DIFF
   16: FINF1 = 0.0
   17: IF T LE TINF1 THEN
   18: FINF1 = DOSE1/TINF1
   19: ENDIF
   20: CL = VMAX/(KM + Z(1))
   21: DZ(1)=(FINF1 - CL*Z(1) - CLD*Z(1) + CLD*Z(2))/VC
   22: DZ(2)=(CLD*Z(1) - CLD*Z(2))/VT
   23: END
   24: SECO
   25: S(1) = VC + VT
   26: VSS = S(1)
   27: CLI = VMAX/KM
   28: S(2) = S(1)/CLI
   29: S(3) = 0.693*VSS/CL
   30: END
   31: FUNC 1
   32: F = Z(1)
   33: END
   34: EOM
NOPAGE BREAKS
NVARIABLES 4
NPOINTS 1500
XNUMBER 2
YNUMBER 3
CONSTANTS 45.77076,120
METHOD 2  'Gauss-Newton (Levenberg and Hartley)
REWEIGHT -1
ITERATIONS 50
INITIAL 10,1,20,.1,.02
LOWER BOUNDS 0,0,0,0,0
UPPER BOUNDS 25,5,75,1,1
NOBSERVATIONS 24
BEGIN
```

PARAMETER	ESTIMATE	STANDARD ERROR	CV%	UNIVARIATE C.I.	
VC	19.037097	3.367686	17.69	11.961898	26.112296
CLD	1.433493	.418105	29.17	.555093	2.311893
VT	30.465163	3.094126	10.16	23.964688	36.965637

VMAX	.151047	.007481	4.95	.135331	.166764
KM	.030385	.010135	33.36	.009091	.051678

```
*** CORRELATION MATRIX OF THE ESTIMATES ***
PARAMETER  VC        CLD        VT        VMAX       KM
VC         1.00000
CLD        -.85640   1.00000
VT         -.88888    .61100   1.00000
VMAX       -.15885    .51153   -.24427   1.00000
KM         -.15498    .49042   -.23413    .98901   1.00000

Condition_number=        5851.
```

X	OBSERVED Y	PREDICTED Y	RESIDUAL	WEIGHT	SE-PRED	STANDARD RESIDUAL
.0000	.	.0000	.	.0000	-.6400+152	.
5.000	.6200E-01	.6723E-01	-.5231E-02	14.87	.6580E-02	-.9795
15.00	.1500	.1439	.6142E-02	6.951	.5908E-02	.5636
30.00	.2300	.2257	.4295E-02	4.431	.6960E-02	.3094
45.00	.3200	.3000	.1999E-01	3.333	.7377E-02	1.225
60.00	.3900	.3729	.1707E-01	2.681	.7482E-02	.9228
90.00	.4900	.5175	-.2748E-01	1.932	.8345E-02	-1.250
120.0	.6300	.6611	-.3107E-01	1.513	.9926E-02	-1.260
125.0	.6000	.6061	-.6098E-02	1.650	.1167E-01	-.2697
130.0	.5700	.5654	.4587E-02	1.769	.1134E-01	.2104
135.0	.5100	.5368	-.2681E-01	1.863	.1023E-01	-1.238
150.0	.4600	.4797	-.1971E-01	2.085	.1067E-01	-.9870
165.0	.4300	.4348	-.4766E-02	2.300	.1040E-01	-.2525
180.0	.4200	.3920	.2799E-01	2.551	.9037E-02	1.525
210.0	.3300	.3084	.2161E-01	3.243	.6622E-02	1.279
240.0	.2500	.2272	.2280E-01	4.401	.6008E-02	1.586
270.0	.1600	.1503	.9729E-02	6.655	.6172E-02	.8793
300.0	.7400E-01	.8164E-01	-.7635E-02	12.25	.5058E-02	-.9726
315.0	.4300E-01	.5325E-01	-.1025E-01	18.78	.3755E-02	-1.566
330.0	.2700E-01	.3129E-01	-.4291E-02	31.96	.2632E-02	-.8337
345.0	.1600E-01	.1703E-01	-.1025E-02	58.74	.2154E-02	-.2786
360.0	.1200E-01	.9150E-02	.2850E-02	109.3	.1593E-02	1.059
375.0	.6600E-02	.5054E-02	.1546E-02	197.9	.1048E-02	.7455
390.0	.3100E-02	.2863E-02	.2368E-03	349.3	.6815E-03	.1470

```
CORRECTED SUM OF SQUARED OBSERVATIONS =  1.10090
WEIGHTED CORRECTED SUM OF SQUARED OBSERVATIONS =  5.54497
SUM OF SQUARED RESIDUALS =          .574055E-02
SUM OF WEIGHTED SQUARED RESIDUALS =  .192278E-01
S =  .326835E-01 WITH    18 DEGREES OF FREEDOM
CORRELATION (OBSERVED,PREDICTED) =   .9656

AIC criteria =         -80.88214
SC  criteria =         -75.20467
```

SUMMARY OF ESTIMATED SECONDARY PARAMETERS

PARAMETER	ESTIMATE	STANDARD ERROR	CV%
VSS	49.502260	1.546144	3.12
MRT	9.957851	2.593375	26.04
TZ	7.559838	.236123	3.12

PK19 - Metabolite kinetics - Capacity III

Objectives

◆ To simultaneously model plasma data of a drug and metabolite

◆ To model capacity-limited elimination

◆ To model formation-rate limited elimination

◆ To set up a system of differential equations

◆ To compare the predictive strenth of different models

Problem specification

The aim of this exercise is to demonstrate the benefit of having experimental data on both sides (drug and metabolite data) of the modeled route (metabolic clearance). The concentration of drug A and metabolite M were measured in plasma at different times after intravenous bolus doses of 10, 50, and 300 µmol/kg (see program output). Assume all elimination occurs from the central compartment by a Michaelis-Menten metabolic pathway.

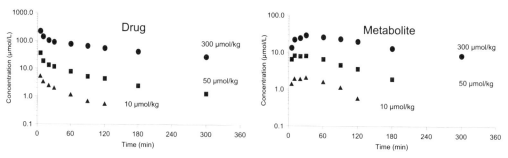

Figure 19.1 *Observed drug (left) and metabolite (right) plasma concentration-time data. Note the change in half-life with increasing concentrations. The intravenous doses of drug were 10, 50 and 300 µmol/kg.*

Figure 19.1 depicts the experimental concentration data for drug (left) and metabolite (right). Notice that the peak concentration of drug increases approximately linearly with dose, whereas the metabolite peak concentration increases in a less than proportional manner and occurs later in time with increasing doses than does the drug peak. The slope of the initial/intermediate portion of the plasma concentration-time profiles increases with dose. The terminal portion obeys first-order kinetics and should therefore be independent of dose (concentration).

We will demonstrate how to set up a system of equations for the parent compound (*drug*) and metabolite (*M*). The model includes a two-compartment drug model with Michaelis-Menten elimination, that provides the formation of the metabolite, and a one-compartment metabolite model (Figure 19.2). In this model, an intravenously administered drug is converted to metabolite through its metabolic clearance and then excreted. Although this model may be rather simplistic, it may easily be expanded to encompass more complex situations encountered in metabolite kinetics.

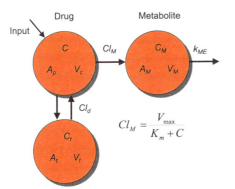

Figure 19.2 *Schematic model of the drug and metabolite disposition. The drug kinetics are described by a two-compartment model with Michaelis-Menten elimination into the one-compartment metabolite model. Metabolite is eliminated by means of a first-order process.*

The equation for the central or plasma compartment is

$$V_c \cdot \frac{dC}{dt} = In - Cl \cdot C - Cl_d \cdot C + Cl_d \cdot C_t \qquad (19:1)$$

The concentration in the peripheral compartment

$$V_t \cdot \frac{dC_t}{dt} = Cl_d \cdot C - Cl_d \cdot C_t \qquad (19:2)$$

The formation of *M* is capacity-limited and can be approximated by the Michaelis-Menten Equation 19:3.

$$V \frac{dC}{dt} = \frac{V_{max} \cdot C}{K_m + C} \qquad (19:3)$$

Metabolite concentration in plasma is

$$V_M \cdot \frac{dC_M}{dt} = \frac{V_{max} \cdot C}{K_m + C} - Cl_{ME} \cdot C_M \qquad (19:4)$$

Cl_{ME} and V_M are the elimination clearance and volume of distribution for the metabolite, respectively. The equation for the metabolite model is then rewritten as

$$\frac{dC_M}{dt} = \frac{V_{max} \cdot C}{K_m + C} \cdot \frac{1}{V_M} - \frac{Cl_{ME} \cdot C_M}{V_M} = \frac{V_{max} \cdot C}{K_m + C} \cdot \frac{1}{V_M} - k_{ME} \cdot C_M \qquad (19:5)$$

The k_{ME} parameter is the first-order elimination rate constant of the metabolite. Note that Cl_m is the metabolic clearance of the drug, which is the same as the formation clearance of the metabolite.

Initial parameter estimates

V_c	= 1 (L/kg)	Obtained from *Dose*/C(0)
V_t	= 1 (L/kg)	Obtained from previous studies
Cl_d	= 0.1 (L/min)	Obtained from previous studies
V_{max}	= 2.0 (µmol/min/kg)	Obtained from *in vitro* data
K_m	= 55 (µmol/L)	Obtained from *in vitro* data
K_{ME}	= 0.15 (min^{-1})	(VIPK10 in program code)
V_M	= 0.3 (L/kg)	(VVC in program code)

We refer the reader to the chapter on nonlinear pharmacokinetic models in Gibaldi and Perrier [1982], in which obtaining initial parameter estimates for a nonlinear system is discussed, and to section 3.6 in this book. There, we explain how one can discriminate graphically between a linear and a nonlinear system. We need nine differential equations for the drug and metabolite model. This includes two equations for the drug (central and tissue compartment) and one equation for the metabolite at each dose level. There are three dose levels.

Interpretation of results and conclusions

The aim of this exercise was to show how one can build a model including drug and metabolite and then fit that model to data. In addition, we have demonstrated that a system with Michaelis-Menten elimination and production can be well-defined. Since the fit of the data was good (Figure 19.3), the parameter precision was high (Table 19.1) and the condition number low. A two-compartment model with capacity-limited elimination could explain the kinetics. The initial dip in plasma concentrations after the rapid infusion is mainly due to distribution into the extravascular space.

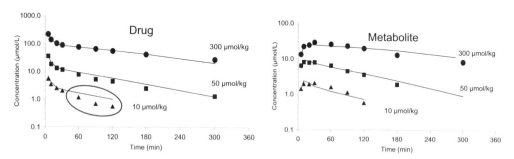

Figure 19.3 *Observed (filled symbols) and predicted (solid lines) plasma concentration-time data of drug and metabolite. Note the slight systematic deviation between the predicted and observed data for the drug at the 10 µmol/kg dose level (encircled area in left hand figure). The model used a two-compartment system with Michaelis-Menten elimination.*

For the parent drug, a slight systematic deviation was observed between the experimental and calculated concentrations for the 10 µmol/kg dose, although the overall fit was good. A parallel linear elimination route model was also tested but resulted in an infinitesimally small value, not different from zero as judged by its confidence intervals. K_m for the Michaelis-Menten elimination was estimated to be 57 µmol/L and V_{max} 1.7 µmol/min/kg. The total volume of distribution at steady state was 3 L/kg, which was consistent with previous studies.

Table 19.1 Parameter estimates and their precision using full model (drug and metabolite).

Parameter	Mean	SD	CV%
V_c (L/kg)	1.07	0.061	5.8
V_t (L/kg)	1.99	0.100	5.0
Cl_d (L/min/kg)	0.128	0.014	10
V_{max} (µmol/min/kg)	1.69	0.16	9.5
K_m (µmol/L)	57.0	8.87	16
k_{me} (min^{-1})	0.14	0.020	14
V_{me} (L/kg)	0.29	0.039	13

Correlation between V_{max} and K_m is 0.92.

The theoretically shortest half-life, which is reached when plasma concentration is zero, was calculated to be 69 min, by means of

$$t_{1/2} = \frac{\ln(2) \cdot V_{ss} \cdot (K_m + C)}{V_{max}}$$

(19:6)

When the concentration was increased to 100 and 200 µmol/L (Equation 19:6) the half-lives became 194 and 317 min, respectively. A typical flip-flop situation was seen at all doses for metabolite, with the elimination rate constant being calculated to be about 0.04 min^{-1} although the formation rate constants were 0.008, 0.003, and 0.002 min^{-1} at 10, 100 and 200 µmol/L, respectively.

One reason why the Michaelis-Menten parameters were so well-defined with a high level of precision is that we had data not only for elimination of drug but also for build-up and elimination of metabolite. Having data *on both sides* of the modeled route is an ideal situation. The predictive strength of this model is also superior to linear elimination models as the model does a better job not only in interpolating data but also in extrapolating outside the dosing range.

In order to evaluate the drug kinetics alone (i.e., without the metabolite data), we also fit a reduced model to the data taking only the drug concentration-time profiles into account (Figure 19.4). Such an approach would be applicable if the concern was adequate systemic exposure of parent compound.

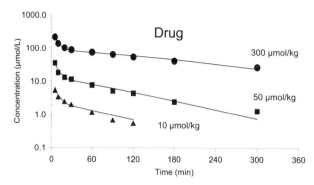

Figure 19.4 *Observed and predicted plasma concentrations of drug following the high (300 µmol/kg), intermediate (50 µmol/kg) and low dose (10 µmol/kg). Note the somewhat better fit of the terminal concentration-time points of the lower dose for this reduced model (drug only) compared with the full model (drug and metabolite, Figure 19.3).*

Except for V_{max} and K_m, which were approximately halved, the values obtained for all parameters were consistent with results from the full model (Table 19.2). The approximately 50% reduction in V_{max} and K_m values may affect predictions at low concentrations (below K_m) but has marginal effects at plasma concentrations of 100 µmol/L and above. In order to elucidate hepatic elimination, further analyses by means of *in vitro* metabolism would be strongly recommended for this compound.

Table 19.2   Parameter estimates and their precision

Parameter	Mean	SD	CV%
V_c (L/kg)	1.10	0.056	5.1
V_t (L/kg)	2.16	0.107	5.0
Cl_d (L/min/kg)	0.12	0.011	9.4
V_{max} (µmol/min/kg)	1.11	0.11	9.9
K_m (µmol/L)	27.5	5.30	19

Correlation between V_{max} and K_m is 0.91.

The reduced model predicts a clearance of 40, 14, 5 mL/min at a drug plasma concentration of 0, 50, and 200 µmol/L, respectively, which is reasonably close to the full model (30, 16, and 7 mL/min, respectively).

Solution I - Drug and metabolite data

```
TITLE 1
simultaneous fit of 10, 50 and 300 umol/kg
MODEL
 COMM
  NPARM 7
  NSEC 14
  NCON 4
  nfun 6
  nder 9
  PNAMES 'CLD','VC','VMAX','KM','VT','VipK10', 'VVC'
  SNAMES 'VSS', 'Cl1', 'Cl10', 'Cl50', 'Cl100', 'Cl200', 'Cl500' &
         't1', 't0', 't50', 't100', 't200', 't500','tm'
END
      1: TEMP
      2: DOS1=CON(1)
      3: DOS2=CON(2)
```

```
 4: DOS3=CON(3)
 5: TINF=CON(4)
 6: CLD=P(1)
 7: VC=P(2)
 8: VMAX=P(3)
 9: KM=P(4)
10: VT=P(5)
11: VIPK10=P(6)
12: VVC=P(7)
13: T=X
14: END
15: START
16: Z(1)=0
17: Z(2)=0
18: Z(3)=0
19: Z(4)=0
20: Z(5)=0
21: Z(6)=0
22: Z(7)=0
23: Z(8)=0
24: Z(9)=0
25: END
26: DIFF
27: IF T LT TINF
28: THEN
29: RINF1=DOS1/TINF
30: RINF2=DOS2/TINF
31: RINF3=DOS3/TINF
32: ELSE
33: RINF1=0
34: RINF2=0
35: RINF3=0
36: ENDIF
37: DZ(1)=RINF1/VC-CLD*Z(1)/VC+CLD*Z(7)/VC-VMAX*Z(1)/(KM+Z(1))/VC
38: DZ(2)=RINF2/VC-CLD*Z(2)/VC+CLD*Z(8)/VC-VMAX*Z(2)/(KM+Z(2))/VC
39: DZ(3)=RINF3/VC-CLD*Z(3)/VC+CLD*Z(9)/VC-VMAX*Z(3)/(KM+Z(3))/VC
40: DZ(4)=(VMAX*Z(1)/(KM+Z(1))/VVC - VIPK10*Z(4))
41: DZ(5)=(VMAX*Z(2)/(KM+Z(2))/VVC - VIPK10*Z(5))
42: DZ(6)=(VMAX*Z(3)/(KM+Z(3))/VVC - VIPK10*Z(6))
43: DZ(7)=CLD*Z(1)/VT - CLD*Z(7)/VT
44: DZ(8)=CLD*Z(2)/VT - CLD*Z(8)/VT
45: DZ(9)=CLD*Z(3)/VT - CLD*Z(9)/VT
46: END
47: FUNC 1
48: F = Z(1)
49: END
50: FUNC 2
51: F = Z(2)
52: END
53: FUNC 3
54: F = Z(3)
55: END
56: FUNC 4
57: F = Z(4)
58: END
59: FUNC 5
60: F = Z(5)
61: END
62: FUNC 6
63: F = Z(6)
64: END
```

```
65: SECO
66: K12 = CLD/VC
67: K21 = CLD/VT
68: VSS=VC*(1+ K12/K21)
69: CL1=VMAX/(KM + 1)
70: CL10=VMAX/(KM + 10)
71: CL50=VMAX/(KM + 50)
72: CL100=VMAX/(KM + 100)
73: CL200=VMAX/(KM + 200)
74: CL500=VMAX/(KM + 500)
75: T1=0.7*VSS/CL1
76: T10=0.7*VSS/CL10
77: T50=0.7*VSS/CL50
78: T100=0.7*VSS/CL100
79: T200=0.7*VSS/CL200
80: T500=0.7*VSS/CL500
81: TM=0.7/VIPK10
82: END
83: EOM
NOPLOTS
NVARIABLES 3
FNUMBER 3
NPOINTS 100
XNUMBER 1
YNUMBER 2
CONSTANTS 300,50,10,5
METHOD 2   'Gauss-Newton (Levenberg and Hartley)
REWEIGHT -1
ITERATIONS 50
INITIAL .1,1,2,55,1,.1,.3
LOWER BOUNDS 0,0,0,0,0,0,0
UPPER BOUNDS 5,5,5,100,10,2,5
MISSING 'Missing'
DATA 'WINNLIN.DAT'
BEGIN
```

PARAMETER	ESTIMATE	STANDARD ERROR	CV%	UNIVARIATE C.I.	
CLD	.128030	.013550	10.58	.100703	.155357
VC	1.067259	.061927	5.80	.942371	1.192147
VMAX	1.694918	.160711	9.48	1.370814	2.019022
KM	56.950321	8.876926	15.59	39.048361	74.852282
VT	1.989753	.099641	5.01	1.788809	2.190697
VIPK10	.144840	.019838	13.70	.104833	.184847
VVC	.294010	.038808	13.20	.215748	.372273

```
*** CORRELATION MATRIX OF THE ESTIMATES ***
PARAMETER  CLD        VC         VMAX       KM         VT         VIPK10     VVC
CLD        1.00000
VC         -.712414   1.00000
VMAX       .110535    -.109695   1.00000
KM         .048212    -.073995   .923234    1.00000
VT         .120699    -.376258   -.325394   -.203531   1.00000
VIPK10     -.029732   .139731    -.235508   -.206280   -.064864   1.00000
VVC        .075862    -.17191    .370584    .261352    -.074649   -.93667
1.000
```

Condition_number=   1911.

  FUNCTION   1

X	OBSERVED Y	PREDICTED Y	RESIDUAL	WEIGHT	SE-PRED	STANDARDIZED RESIDUAL
5.000	216.3	213.8	2.470	.4678E-02	6.660	.8033
10.00	137.0	138.6	-1.556	.7217E-02	4.770	-.4470
20.00	101.1	94.66	6.437	.1056E-01	2.536	1.543
30.00	88.30	85.10	3.197	.1175E-01	2.266	.7921
60.00	75.12	74.45	.6650	.1343E-01	2.073	.1750
90.00	64.51	65.46	-.9536	.1528E-01	1.723	-.2594
120.0	54.38	57.04	-2.661	.1753E-01	1.521	-.7668
180.0	41.12	42.05	-.9300	.2378E-01	1.468	-.3203
300.0	27.10	20.15	6.953	.4964E-01	1.536	4.222

CORRECTED SUM OF SQUARED OBSERVATIONS = 26756.9
WEIGHTED CORRECTED SUM OF SQUARED OBSERVATIONS = 250.250
SUM OF SQUARED RESIDUALS =   117.828
SUM OF WEIGHTED SQUARED RESIDUALS = 3.16803
S = 1.25858   WITH   2 DEGREES OF FREEDOM
CORRELATION (OBSERVED,PREDICTED) = .9982

  FUNCTION   2

X	OBSERVED Y	PREDICTED Y	RESIDUAL	WEIGHT	SE-PRED	STANDARDIZED RESIDUAL
5.000	35.60	34.73	.8666	.2879E-01	1.050	.3135
10.00	18.22	21.33	-3.110	.4688E-01	.7640	-1.422
20.00	13.10	13.60	-.5028	.7351E-01	.3852	-.2778
30.00	11.54	11.74	-.2029	.8516E-01	.3742	-.1209
60.00	7.748	9.203	-1.455	.1087	.3526	-.9825
90.00	5.184	7.261	-2.077	.1377	.3407	-1.587
120.0	4.402	5.697	-1.295	.1755	.3430	-1.129
180.0	2.432	3.463	-1.031	.2888	.3261	-1.179
300.0	1.292	1.239	.5252E-01	.8068	.2145	.1018

CORRECTED SUM OF SQUARED OBSERVATIONS = 917.551
WEIGHTED CORRECTED SUM OF SQUARED OBSERVATIONS = 57.0794
SUM OF SQUARED RESIDUALS =   19.8912
SUM OF WEIGHTED SQUARED RESIDUALS = 1.92488
S = .981040   WITH   2 DEGREES OF FREEDOM
CORRELATION (OBSERVED,PREDICTED) = .9940

FUNCTION   3

X	OBSERVED Y	PREDICTED Y	RESIDUAL	WEIGHT	SE-PRED	STANDARDIZED RESIDUAL
5.000	5.030	6.853	-1.823	.1459	.2036	-1.405
10.00	3.180	4.086	-.9060	.2447	.1535	-.9037
20.00	2.320	2.546	-.2255	.3928	.8714E-01	-.2834
30.00	1.890	2.176	-.2860	.4596	.8878E-01	-.3892
60.00	1.096	1.660	-.5644	.6022	.8657E-01	-.8810
90.00	.6440	1.281	-.6371	.7806	.8371E-01	-1.134
120.0	.5160	.9872	-.4712	1.013	.8063E-01	-.9578

CORRECTED SUM OF SQUARED OBSERVATIONS = 15.4807
WEIGHTED CORRECTED SUM OF SQUARED OBSERVATIONS = 4.43082
SUM OF SQUARED RESIDUALS =   5.22203
SUM OF WEIGHTED SQUARED RESIDUALS = 1.47681

```
S = .000000      WITH      0 DEGREES OF FREEDOM
CORRELATION (OBSERVED,PREDICTED) = .9862


FUNCTION    4

X      OBSERVED    PREDICTED    RESIDUAL      WEIGHT     SE-PRED   STANDARDI
          Y           Y                                           RESIDUAL
5.000     13.15       13.20      -.5251E-01   .7574E-01   1.126    -.0366
10.00     21.90       21.63       .2699       .4623E-01   1.317     .1401
20.00     24.12       24.84      -.7240       .4025E-01   1.058    -.3195
30.00     28.70       24.31      4.388        .4113E-01   1.057    1.962
60.00     25.60       22.83      2.766        .4379E-01   .9707    1.262
90.00     22.96       21.58      1.377        .4633E-01   .8669     .6366
120.0     19.72       20.24      -.5176       .4941E-01   .7847    -.2446
180.0     12.92       17.27      -4.346       .5792E-01   .7089    -2.217
300.0     8.230       10.77      -2.536       .9288E-01   .7609    -1.737


CORRECTED SUM OF SQUARED OBSERVATIONS = 371.246
WEIGHTED CORRECTED SUM OF SQUARED OBSERVATIONS = 22.8716
SUM OF SQUARED RESIDUALS =            54.9872
SUM OF WEIGHTED SQUARED RESIDUALS = 2.94414
S = 1.21329      WITH      2 DEGREES OF FREEDOM
CORRELATION (OBSERVED,PREDICTED) = .9513


FUNCTION    5

X      OBSERVED    PREDICTED    RESIDUAL      WEIGHT     SE-PRED   STANDARD
          Y           Y                                           RESIDUAL
5.000     6.372       5.288      1.084        .1891       .5310    1.058
10.00     8.018       9.004      -.9859       .1111       .6101    -.7163
20.00     7.750       8.648      -.8981       .1156       .4389    -.6376
30.00     7.934       7.445       .4894       .1343       .3953     .3734
60.00     6.458       5.809       .6489       .1721       .2998     .5539
90.00     4.508       4.729      -.2213       .2114       .2296    -.2075
120.0     3.552       3.813      -.2609       .2623       .1848    -.2712
180.0     1.892       2.413      -.5208       .4145       .1495    -.6809
300.0     1.414       .9004       .5136       1.111       .1133    1.111


CORRECTED SUM OF SQUARED OBSERVATIONS = 53.2120
WEIGHTED CORRECTED SUM OF SQUARED OBSERVATIONS = 16.1774
SUM OF SQUARED RESIDUALS =             4.26548
SUM OF WEIGHTED SQUARED RESIDUALS = .961496
S = .693360      WITH      2 DEGREES OF FREEDOM
CORRELATION (OBSERVED,PREDICTED) = .9641


 FUNCTION    6

X      OBSERVED    PREDICTED    RESIDUAL      WEIGHT     SE-PRED   STANDARDI
          Y           Y                                           RESIDUAL
5.000     1.330       1.384      -.5393E-01   .7226       .1669    -.0952
10.00     1.830       2.351      -.5207       .4254       .2100    -.7037
20.00     1.810       2.047      -.2366       .4886       .1350    -.3356
30.00     1.970       1.658       .3119       .6031       .1030     .4891
60.00     1.514       1.199       .3154       .8343       .6339E-01  .5781
90.00     1.074       .9299       .1441       1.075       .4424E-01  .2991
120.0     .5500       .7206      -.1706       1.388       .3588E-01 -.4021


CORRECTED SUM OF SQUARED OBSERVATIONS = 1.51353
WEIGHTED CORRECTED SUM OF SQUARED OBSERVATIONS = 1.38408
SUM OF SQUARED RESIDUALS =             .576680
SUM OF WEIGHTED SQUARED RESIDUALS = .349208
```

```
S = .000000      WITH      0 DEGREES OF FREEDOM
CORRELATION (OBSERVED,PREDICTED) =  .8518


TOTALS FOR ALL CURVES COMBINED
SUM OF SQUARED RESIDUALS =            202.771
SUM OF WEIGHTED SQUARED RESIDUALS =  10.8246
S = .501731      WITH     43 DEGREES OF FREEDOM

AIC criteria =        133.09090
SC  criteria =        146.47506
```

Solution II - Drug data alone

```
TITLE 1
simultaneous fit of 10, 50 and 300 umol/kg
MODEL
 COMM
  NPARM 5
  NSEC 13
  NCON 4
  nfun 3
  nder 6
  PNAMES 'CLD','VC','VMAX','KM','VT'
  SNAMES 'VSS','Cl1','Cl10','Cl50','Cl100','Cl200','Cl500' &
         't1','t10','t50','t100','t200','t500'
END
     1: TEMP
     2: DOS1=CON(1)
     3: DOS2=CON(2)
     4: DOS3=CON(3)
     5: TINF=CON(4)
     6: CLD=P(1)
     7: VC=P(2)
     8: VMAX=P(3)
     9: KM=P(4)
    10: VT=P(5)
    11: T=X
    12: END
    13: START
    14: Z(1)=0
    15: Z(2)=0
    16: Z(3)=0
    17: Z(4)=0
    18: Z(5)=0
    19: Z(6)=0
    20: END
    21: DIFF
    22: IF T LT TINF
    23: THEN
    24: RINF1=DOS1/TINF
    25: RINF2=DOS2/TINF
    26: RINF3=DOS3/TINF
    27: ELSE
    28: RINF1=0
    29: RINF2=0
    30: RINF3=0
    31: ENDIF
    32: DZ(1)=RINF1/VC-CLD*Z(1)/VC+CLD*Z(4)/VC-VMAX*Z(1)/(KM+Z(1))/VC
    33: DZ(2)=RINF2/VC-CLD*Z(2)/VC+CLD*Z(5)/VC-VMAX*Z(2)/(KM+Z(2))/VC
    34: DZ(3)=RINF3/VC-CLD*Z(3)/VC+CLD*Z(6)/VC-VMAX*Z(3)/(KM+Z(3))/VC
```

```
35: DZ(4)=CLD*Z(1)/VT - CLD*Z(4)/VT
36: DZ(5)=CLD*Z(2)/VT - CLD*Z(5)/VT
37: DZ(6)=CLD*Z(3)/VT - CLD*Z(6)/VT
38: END
39: FUNC 1
40: F = Z(1)
41: END
42: FUNC 2
43: F = Z(2)
44: END
45: FUNC 3
46: F = Z(3)
47: END
48: SECO
49: K12 = CLD/VC
50: K21 = CLD/VT
51: VSS=VC*(1+ K12/K21)
52: CL1=VMAX/(KM + 1)
53: CL10=VMAX/(KM + 10)
54: CL50=VMAX/(KM + 50)
55: CL100=VMAX/(KM + 100)
56: CL200=VMAX/(KM + 200)
57: CL500=VMAX/(KM + 500)
58: T1=0.7*VSS/CL1
59: T10=0.7*VSS/CL10
60: T50=0.7*VSS/CL50
61: T100=0.7*VSS/CL100
62: T200=0.7*VSS/CL200
63: T500=0.7*VSS/CL500
64: END
65: EOM
NOPLOTS
NVARIABLES 3
FNUMBER 3
NPOINTS 100
XNUMBER 1
YNUMBER 2
CONSTANTS 300,50,10,5
METHOD 2   'Gauss-Newton (Levenberg and Hartley)
REWEIGHT -1
ITERATIONS 50
INITIAL .1,1,2,55,1
LOWER BOUNDS 0,0,0,0,0
UPPER BOUNDS 5,5,5,100,10
MISSING 'Missing'
DATA 'WINNLIN.DAT'
BEGIN
```

PARAMETER	ESTIMATE	STANDARD ERROR	CV%	UNIVARIATE C.I.	
CLD	.118546	.011154	9.41	.095279	.141814
VC	1.103301	.055820	5.06	.986863	1.219739
VMAX	1.109446	.110251	9.94	.879468	1.339424
KM	27.455744	5.303064	19.31	16.393830	38.517657
VT	2.158556	.106942	4.95	1.935480	2.381632

*** CORRELATION MATRIX OF THE ESTIMATES ***

```
PARAMETER  CLD        VC         VMAX       KM         VT
CLD        1.00000
VC         -.697366   1.00000
VMAX       .189020    -.136200   1.00000
KM         .113862    -.092134   .911226    1.00000
VT         -.013019   -.251388   -.522467   -.36161    1.00000

Condition_number=        698.2
```

FUNCTION   1

X	OBSERVED Y	PREDICTED Y	RESIDUAL	WEIGHT	SE-PRED	STANDARD RESIDUAL
5.000	216.3	212.2	4.074	.4713E-02	6.130	1.389
10.00	137.0	141.2	-4.229	.7081E-02	4.283	-1.201
20.00	101.1	94.16	6.938	.1062E-01	2.588	1.868
30.00	88.30	82.95	5.353	.1206E-01	2.079	1.445
60.00	75.12	73.42	1.703	.1362E-01	2.030	.4944
90.00	64.51	66.17	-1.656	.1511E-01	1.701	-.4881
120.0	54.38	59.17	-4.787	.1690E-01	1.522	-1.473
180.0	41.12	46.01	-4.891	.2173E-01	1.585	-1.786
300.0	27.10	24.11	2.988	.4147E-01	1.955	2.503

```
CORRECTED SUM OF SQUARED OBSERVATIONS =  26756.9
WEIGHTED CORRECTED SUM OF SQUARED OBSERVATIONS =  239.736
SUM OF SQUARED RESIDUALS =        172.684
SUM OF WEIGHTED SQUARED RESIDUALS =  2.42010
S =  .777833     WITH     4 DEGREES OF FREEDOM
CORRELATION (OBSERVED,PREDICTED) =  .9970
```

FUNCTION   2

X	OBSERVED Y	PREDICTED Y	RESIDUAL	WEIGHT	SE-PRED	STANDARD RESIDUAL
5.000	35.60	34.31	1.294	.2915E-01	.9642	.5062
10.00	18.22	21.45	-3.235	.4661E-01	.6871	-1.579
20.00	13.10	12.93	.1663	.7732E-01	.3913	.1019
30.00	11.54	10.69	.8473	.9352E-01	.3658	.5721
60.00	7.748	8.054	-.3058	.1242	.3890	-.2416
90.00	5.184	6.149	-.9655	.1626	.3919	-.8870
120.0	4.402	4.644	-.2425	.2153	.3951	-.2622
180.0	2.432	2.582	-.1498	.3873	.3553	-.2270
300.0	1.292	.7509	.5411	1.332	.1937	1.525

```
CORRECTED SUM OF SQUARED OBSERVATIONS =  917.551
WEIGHTED CORRECTED SUM OF SQUARED OBSERVATIONS =  62.8850
SUM OF SQUARED RESIDUALS =        14.2845
SUM OF WEIGHTED SQUARED RESIDUALS =  1.18025
S =  .543197     WITH     4 DEGREES OF FREEDOM
CORRELATION (OBSERVED,PREDICTED) =  .9924
```

FUNCTION   3

X	OBSERVED Y	PREDICTED Y	RESIDUAL	WEIGHT	SE-PRED	STANDARD RESIDUAL
5.000	5.030	6.688	-1.658	.1495	.1847	-1.391
10.00	3.180	3.952	-.7722	.2530	.1451	-.8430
20.00	2.320	2.254	.6563E-01	.4436	.1084	.0948
30.00	1.890	1.826	.6386E-01	.5476	.1112	.1029
60.00	1.096	1.309	-.2131	.7639	.1110	-.4082
90.00	.6440	.9586	-.3146	1.043	.1042	-.7074

```
   120.0       .5160       .7003      -.1843      1.428      .9482E-01   -.4867
```

CORRECTED SUM OF SQUARED OBSERVATIONS = 15.4807
WEIGHTED CORRECTED SUM OF SQUARED OBSERVATIONS = 4.98011
SUM OF SQUARED RESIDUALS =   3.53237
SUM OF WEIGHTED SQUARED RESIDUALS = .752563
S = .613418   WITH   2 DEGREES OF FREEDOM
CORRELATION (OBSERVED,PREDICTED) = .9854

TOTALS FOR ALL CURVES COMBINED
SUM OF SQUARED RESIDUALS =   190.501
SUM OF WEIGHTED SQUARED RESIDUALS = 4.35291
S = .466525   WITH   20 DEGREES OF FREEDOM

AIC criteria =   46.77113
SC criteria =   52.86551

SUMMARY OF ESTIMATED SECONDARY PARAMETERS

PARAMETER	ESTIMATE	STANDARD ERROR	CV%
VSS	3.261857	.107477	3.29
CL1	.038988	.004056	10.40
CL10	.029620	.001935	6.53
CL50	.014324	.000666	4.65
CL100	.008705	.000555	6.38
CL200	.004878	.000384	7.87
CL500	.002103	.000190	9.03
T1	58.563458	6.089428	10.40
T10	77.085944	5.681067	7.37
T50	159.408105	11.568090	7.26
T100	262.310807	23.301121	8.88
T200	468.116211	47.977229	10.25
T500	1085.532421	122.896157	11.32

PK20 - Nonlinear kinetics - Capacity IV

Objectives
◆ **To analyze nonlinear concentration-time data from two individuals**
◆ **To fit a Michaelis-Menten model simultaneously to two datasets**
◆ **To use the goodness-of-fit criteria for model discrimination**

Problem specification

The purpose of this exercise is to demonstrate how to simultaneously utilize two sets of plasma data for fitting a nonlinear model. Data were obtained from two healthy young individuals who had received an intravenous bolus of a new experimental compound X (Figure 20.1 and program output). Their body size (weight) and age were very similar. The measured plasma concentrations include free and bound drug of this highly plasma protein bound compound. We assume that the major source of variability is in the Michaelis-Menten constant K_m rather than the maximum metabolic rate V_{max} or the volume of distribution V. We also assume that distribution appears momentarily upon administration and that V is directly proportional to body weight. The proposed system is a one-compartment model with Michaelis-Menten kinetics (Figure 20.1).

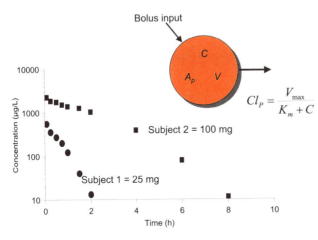

Figure 20.1 *Observed and predicted concentration-time data from two individuals who received 25 mg (Subject 1, circles) or 100 mg (Subject 2, squares) of a new antibiotic. The drug exhibits one-compartment kinetics with a capacity-limited clearance, as shown by the model.*

Figure 20.1 shows the observed concentration-time data and a schematic diagram of the turnover of agent *X*. The differential equation that describes the system is defined as

$$\frac{dC}{dt} = -K \cdot C = -\frac{Cl}{V} \cdot C \qquad (20:1)$$

For subject 1 who received 25 mg, *Cl* in Equation 20:1 is equal to

$$Cl(subject_1) = \frac{V_{max}}{K_{m1} + C} \tag{20:2}$$

and for subject 2 (100 mg)

$$Cl(subject_2) = \frac{V_{max}}{K_{m2} + C} \tag{20:3}$$

V_{max}, K_{m1}, K_{m2} and V are the maximum metabolic rate, Michaelis-Menten parameters for subject 1 and subject 2, and volume of distribution, respectively. The initial condition for the state variable defining subject 1 is

$$C^0 = \frac{D_{iv\,25}}{V} \tag{20:4}$$

and for subject 2

$$C^0 = \frac{D_{iv100}}{V} \tag{20:5}$$

We assume V_{max} and V to be the same for both subjects. However, K_m differs between the two because we have measured total drug concentration in plasma.

Initial parameter estimates

Figure 20.2 shows how to obtain the initial parameter estimates graphically (Gibaldi and Perrier, [1983]). Remember that K_m can be approximately estimated by means of

$$K_m = \frac{C(0)}{\ln\left(\dfrac{C(0)^*}{C(0)}\right)} \tag{20:6}$$

$C(0)^*$ and $C(0)$ are the back-extrapolated initial concentration based on the terminal phase and the initial concentration based on the observed initial concentrations, respectively (Figure 20.2). V_{max} is determined from

$$V_{max} = K_2 \cdot V \cdot K_{m2} = 0.9 \cdot 45 \cdot 1200 \cong 49000 \tag{20:7}$$

or

$$V_{max} = K_1 \cdot V \cdot K_{m1} = 2.3 \cdot 45 \cdot 350 \cong 36000 \tag{20:8}$$

K is the terminal linear slope of the respective terminal portions of the plasma concentration-time curves in Figure 20.2.

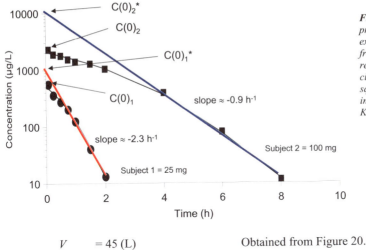

Figure 20.2 Observed, predicted and back-extrapolated concentrations from two individuals who received 25 mg (Subject 1, circles) or 100 mg (Subject 2, squares) of a new antibiotic. K_2 in Equation 20:7 is 0.9 h^{-1} and K_1 in Equation 20:8 is 2.3 h^{-1}.

V	= 45 (L)	Obtained from Figure 20.2
V_{max}	= 45000 (µg/h)	Obtained from Equations 20:7 and 20:8
K_{m1}	= 350 (µg/L)	Obtained from Figure 20.2
K_{m2}	= 1000 (or 1200) (µg/L)	Obtained from Figure 20.2

All units were changed from mg to µg to match the concentration units (µg/L).

Interpretation of results and conclusions

This exercise illustrates how to obtain the initial parameter estimates graphically and then fit a one-compartment model with Michaelis-Menten elimination kinetics. Dose-normalized observed and predicted data are shown in Figure 20.3. There was good consistency between the observed and model-predicted values. The correlation between V_{max}, K_{m1} and K_{m2} was high, being about 0.98 to 0.99 in spite of the high parameter precision.

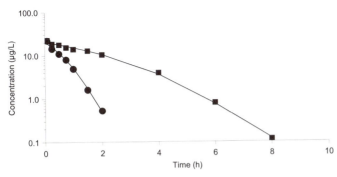

Figure 20.3 Semi-logarithmic plot of dose-normalized observed (circles Subject 1 and squares Subject 2) and predicted (lines) plasma concentrations versus time.

Note that there are some limitations with this particular analysis even though the concentration range covers two orders of magnitude. One is the potential mis-specification of the model in that we assumed a one-compartment model when a two-compartment model may appear more appropriate with another design. Secondly, the numerical estimation of V_{max} and K_m is inherently sensitive to the covered concentration-range in that they are often highly correlated[1]. Thirdly, we assume that K_m contains the largest source of variability since we measured total plasma concentrations. This compound is highly bound to albumin. Prior to this fit, we tried to model a system with two different volume terms, one for each subject. However, their estimates were almost identical. This may be appreciated from the dose-normalized (to 25 mg) $C(0)$ values, which are more or less identical. We also fitted a model with two different V_{max} values without any substantial improvement.

    K_{m1} is estimated to be 282 µg/L and K_{m2} 812 µg/L. If one corrects for protein binding (i.e., only considers the free fraction), which is 0.03 for subject 1 and 0.01 for subject 2, one ends up with two K_m values based on the unbound concentration which are very close (8.5 and 8.1 µg/L, respectively). Using protein binding data to correct for inter-individual differences in the K_m values, we obtained an unbound K_m of 8.75 µg/L, V_{max} of 39000 µg/min, and a V of 50 L.

    Running this problem without parameter boundaries resulted in an increased variability in the parameters (less precision). Data are not included from this run.

Solution I - Uncorrected K_m values

```
TITLE 1
 25- and 100 mg/kg i.v. bolus
MODEL
 COMM
 NDER 2
 NFUN 2
 NCON 2
 NPARM 4
 PNAMES 'VD','Vmax','Km1','Km2'
END
     1: TEMP
     2: DOSE1=CON(1)
     3: DOSE2=CON(2)
     4: END
     5: START
     6: Z(1)=DOSE1/VD
     7: Z(2)=DOSE2/VD
     8: END
     9: DIFF
    10: DZ(1)=(-VMAX*Z(1)/(KM1 + Z(1)))/VD
    11: DZ(2)=(-VMAX*Z(2)/(KM2 + Z(2)))/VD
    12: END
    13: FUNC 1
    14: F = Z(1)
    15: END
    16: FUNC 2
```

---

[1] Some parameters are often highly correlated, such as V_{max} and K_m, so that the estimate of one almost governs the estimate of the other. Estimates of these parameters are also often skewed to the right, with many of the estimates being many times larger than the true parameter value. They may also have high variances, in spite of the fact that the variances computed from individual data sets are underestimated. Also, there is a high correlation between the estimates of V_{max} and K_m with their respective standard deviations. Thus, the estimated precision of V_{max} and K_m, given by regression programs, may not give a true picture of the uncertainty in these estimates (see Metzler and Tong [1981] for a thorough review).

```
    17: F = Z(2)
    18: END
    19: EOM
NVARIABLES 3
FNUMBER 3
NPOINTS 1000
CONSTANTS 25000,100000
METHOD 2   'Gauss-Newton (Levenberg and Hartley)
REWEIGHT -2
ITERATIONS 50
INITIAL 45,45000,350,1000
LOWER BOUNDS 10,10000,100,100
UPPER BOUNDS 100,100000,1000,3000
MISSING 'Missing'
DATA 'WINNLIN.DAT'
BEGIN
```

PARAMETER	ESTIMATE	STANDARD ERROR	CV%	UNIVARIATE C.I.	
VD	48.562294	1.634284	3.37	45.031640	52.092949
VMAX	37775.142162	3167.096493	8.38	30933.050263	44617.234061
KM1	282.521130	41.190477	14.58	193.534568	371.507692
KM2	812.522475	121.675591	14.98	549.658498	1075.386451

```
*** CORRELATION MATRIX OF THE ESTIMATES ***
```

PARAMETER	VD	VMAX	KM1	KM2
VD	1.00000			
VMAX	-.632815	1.00000		
KM1	-.704782	.984833	1.00000	
KM2	-.722405	.987374	.985673	1.00000

```
 Condition_number=      4566.

    FUNCTION   1
```

X	OBSERVED Y	PREDICTED Y	RESIDUAL	WEIGHT	SE-PRED	STANDARDIZED RESIDUAL
.8000E-01	554.4	475.2	79.21	.4429E-05	14.93	2.061
.2500	358.1	395.1	-37.00	.6406E-05	10.73	-1.136
.5000	272.9	289.0	-16.09	.1197E-04	6.851	-.6671
.7500	200.0	199.4	.6153	.2515E-04	5.190	.03728
1.000	122.4	128.7	-6.266	.6040E-04	4.167	-.6050
1.500	40.00	43.85	-3.850	.5201E-03	2.030	-1.196
2.000	13.10	12.37	.7269	.6532E-02	.9266	1.340

```
CORRECTED SUM OF SQUARED OBSERVATIONS =  218764.
  WEIGHTED CORRECTED SUM OF SQUARED OBSERVATIONS =  4.69268
  SUM OF SQUARED RESIDUALS =            7956.86
  SUM OF WEIGHTED SQUARED RESIDUALS =  .531955E-01
  S = .133161     WITH      3 DEGREES OF FREEDOM
  CORRELATION (OBSERVED,PREDICTED) =  .9841

    FUNCTION   2
```

X	OBSERVED Y	PREDICTED Y	RESIDUAL	WEIGHT	SE-PRED	STANDARDI RESIDUAL
.8000E-01	2290.	2015.	275.3	.2464E-06	66.34	1.702

.2500	1876.	1921.	-45.14	.2709E-06	60.49	-.2906
.5000	1776.	1786.	-9.955	.3135E-06	52.61	-.0682
.7500	1531.	1654.	-122.9	.3656E-06	45.72	-.9035
1.000	1395.	1525.	-130.2	.4299E-06	39.92	-1.032
1.500	1279.	1279.	-.2234	.6111E-06	31.90	-.0021
2.000	1038.	1050.	-12.41	.9063E-06	28.27	-.1433
4.000	391.5	361.5	30.04	.7654E-05	19.60	1.227
6.000	79.50	75.73	3.775	.1744E-03	4.264	.7551
8.000	11.40	12.07	-.6703	.6864E-02	.9733	-1.733

```
CORRECTED SUM OF SQUARED OBSERVATIONS =  .546782E+07
WEIGHTED CORRECTED SUM OF SQUARED OBSERVATIONS =  8.66342
SUM OF SQUARED RESIDUALS =            111050.
SUM OF WEIGHTED SQUARED RESIDUALS = .446793E-01
S = .862934E-01 WITH     6 DEGREES OF FREEDOM
CORRELATION (OBSERVED, PREDICTED) =  .9899


TOTALS FOR ALL CURVES COMBINED
SUM OF SQUARED RESIDUALS =            119007.
SUM OF WEIGHTED SQUARED RESIDUALS = .978749E-01
S = .867689E-01 WITH    13 DEGREES OF FREEDOM
```

Solution II - Protein binding corrected K_m values

```
TITLE 1
 25- and 100 mg/kg i.v. bolus
Model
 COMM
 NDER 2
 NFUN 2
 NCON 2
 NPARM 3
 PNAMES 'VD','Vmax', 'Km'
END
    1: TEMP
    2: DOSE1=CON(1)
    3: DOSE2=CON(2)
    4: END
    5: START
    6: Z(1)=DOSE1/VD
    7: Z(2)=DOSE2/VD
    8: END
    9: DIFF
   10: KM1 = KM/0.03
   11: KM2 = KM/0.01
   12: DZ(1)=(-VMAX*Z(1)/(KM1 + Z(1)))/VD
   13: DZ(2)=(-VMAX*Z(2)/(KM2 + Z(2)))/VD
   14: END
   15: FUNC 1
   16: F = Z(1)
   17: END
   18: FUNC 2
   19: F = Z(2)
   20: END
   21: EOM
NVARIABLES 3
FNUMBER 3
NPOINTS 100
XNUMBER 1
YNUMBER 2
CONSTANTS 25000,100000
```

```
METHOD 2   'Gauss-Newton (Levenberg and Hartley)
REWEIGHT -2
ITERATIONS 50
INITIAL 45,45000,10
LOWER BOUNDS 10,10000,1
UPPER BOUNDS 100,100000,100
MISSING 'Missing'
DATA 'WINNLIN.DAT'
BEGIN
```

PARAMETER	ESTIMATE	STANDARD ERROR	CV%	UNIVARIATE C.I.	
VD	47.991053	1.701056	3.54	44.342661	51.639446
VMAX	38961.785354	3582.334730	9.19	31278.459603	46645.111105
KM	8.752904	1.393570	15.92	5.764001	11.741807

```
*** CORRELATION MATRIX OF THE ESTIMATES ***
```

PARAMETER	VD	VMAX	KM
VD	1.00000		
VMAX	-.631818	1.00000	
KM	-.716050	.990134	1.00000

```
   Condition_number=      .3341E+05

FUNCTION   1
```

X	OBSERVED Y	PREDICTED Y	RESIDUAL	WEIGHT	SE-PRED	STANDARDI RESIDUAL
.8000E-01	554.4	479.9	74.49	.4342E-05	15.87	1.810
.2500	358.1	397.1	-39.03	.6341E-05	11.36	-1.125
.5000	272.9	288.0	-15.06	.1206E-04	7.236	-.5915
.7500	200.0	196.5	3.502	.2590E-04	5.308	.2028
1.000	122.4	125.2	-2.753	.6384E-04	3.886	-.2542
1.500	40.00	41.47	-1.473	.5814E-03	1.303	-.4113
2.000	13.10	11.44	1.665	.7647E-02	.6173	1.957

```
CORRECTED SUM OF SQUARED OBSERVATIONS =  218764.
WEIGHTED CORRECTED SUM OF SQUARED OBSERVATIONS =  4.78101
SUM OF SQUARED RESIDUALS =              7323.83
SUM OF WEIGHTED SQUARED RESIDUALS =  .597409E-01
S =  .122210     WITH     4 DEGREES OF FREEDOM
CORRELATION (OBSERVED,PREDICTED) =  .9849


    FUNCTION   2
```

X	OBSERVED Y	PREDICTED Y	RESIDUAL	WEIGHT	SE-PRED	STANDARDI RESIDUAL
.8000E-01	2290.	2038.	251.9	.2407E-06	70.62	1.451
.2500	1876.	1942.	-66.28	.2651E-06	64.20	-.3979
.5000	1776.	1804.	-27.98	.3073E-06	55.61	-.1791
.7500	1531.	1669.	-138.1	.3590E-06	48.14	-.9477
1.000	1395.	1538.	-142.8	.4229E-06	41.91	-1.058
1.500	1279.	1287.	-8.411	.6033E-06	33.49	-.0741
2.000	1038.	1055.	-17.41	.8978E-06	29.86	-.1886
4.000	391.5	363.8	27.66	.7554E-05	20.50	1.046
6.000	79.50	78.83	.6700	.1609E-03	4.267	.1144

```
   8.000        11.40        13.29       -1.891       .5661E-02  .9221       -2.360
```

CORRECTED SUM OF SQUARED OBSERVATIONS = .546782E+07
WEIGHTED CORRECTED SUM OF SQUARED OBSERVATIONS = 8.46303
SUM OF SQUARED RESIDUALS =   109203.
SUM OF WEIGHTED SQUARED RESIDUALS = .585535E-01
S = .914592E-01 WITH   7 DEGREES OF FREEDOM
CORRELATION (OBSERVED,PREDICTED) = .9901

TOTALS FOR ALL CURVES COMBINED
SUM OF SQUARED RESIDUALS =   116527.
SUM OF WEIGHTED SQUARED RESIDUALS = .118294
S = .919217E-01 WITH   14 DEGREES OF FREEDOM

AIC criteria =   -30.28785
SC criteria =   -27.78821

PK21 - Nonlinear kinetics – Induction I

Objectives
◆ **To model the metabolic interaction between a drug and an inducer**

◆ **To model enzyme induction using the enzyme elimination rate constant**

◆ **To estimate oral clearance before and during treatment**

◆ **To estimate *MRT* before and during treatment**

Problem specification

This exercise is designed to demonstrate how to build an induction model and to estimate the apparent fractional turnover rate for the enzyme. A study was conducted to see if the drug metabolizing enzymes of nortriptyline *NT* are inducible by pentobarbital *PB*. *NT* was therefore administered orally as 10 mg every 8 h during a period of 29 days (696 h). After 9 days (216 h), treatment with *PB* (inducer) was initiated and lasted for 12.5 days (300 h), i.e., until 21.5 days (516 h) after the start of *NT* administration.

The observed and predicted plasma concentration time course of *NT* before, during, and after treatment with inducer is depicted in Figure 21.1 and in the program output.

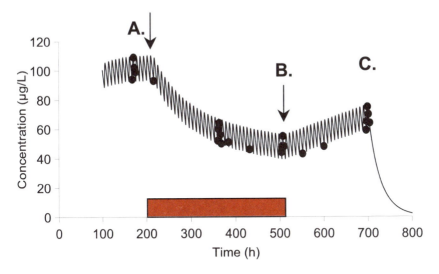

Figure 21.1 *Observed (filled circles) and predicted (solid line) plasma concentrations of nortriptyline (10 mg tid) before (A), during (B), and after (C) pentobarbital treatment. The red horizontal bar represents the induction period.*

Let us approximate the behavior of *NT* after oral administration to a one-compartment model. We also assume that *PB* affects the rate of synthesis of enzyme (i.e., the turnover rate), and that the clearance of *NT* is directly proportional to the size of the enzyme pool. A time-dependent change, due to enzyme induction, occurs in intrinsic clearance since intrinsic clearance is assumed to be proportional to the amount of enzyme.

What needs to be done? and How do we accomplish it with available software?

- Plot the data
- Decide what the data show
- Obtain initial estimates
- Derive the necessary pharmacokinetic relationships
- Fit the data
- Interpret the output
- Characterize the drug's disposition

We will model the time-dependent change of the elimination rate constant of drug, as well as the rate of change in the responsible metabolic enzyme. Absorption rate and lag-time will also be estimated. We will estimate oral clearance of *NT* pre- and post-induction and the mean residence time pre- and post-induction. For further reading about time-dependency in kinetic variables and turnover concepts, see Lassen and Perl [1979], Levy [1979], Gibaldi and Perrier [1982], and Rowland and Tozer [1995].

Induction may be an effect of either increased rate of synthesis (increased turnover rate) of the responsible enzyme or decreased loss of enzyme (decreased fractional turnover rate). Both processes increase the amount of enzyme, also called the enzyme pool. Figure 21.2 depicts the situation where *PB* affects the pool size of the responsible enzyme(s) for the nortriptyline metabolism. *[Ass]* corresponds to the pool size of the enzyme.

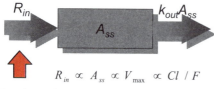

$$R_{in} \propto A_{ss} \propto V_{max} \propto Cl\ /\ F$$

Pb treatment

Figure 21.2 Schematic representation of the turnover model. Inducer (pentobarbital) stimulates the production R_{in} of enzyme A_{ss}. The half-life of A is not affected. We assume that an increase in R_{in}, A_{ss}, and V_{max} increase Cl/F (or Cl_o) proportionately.

Initial parameter estimates

The absorption rate constant and lag-time of drug absorption were estimated simultaneously with oral clearance Cl_0 for *NT* pre- and post-induction together with mean residence time pre- and post-induction. Cl_0 (pre-induction) is estimated as $Dose/AUC_{ss}$ (pre-induction), and Cl_0 (peri-induction) is calculated as $Dose/AUC_{ss}$ (peri-induction). The k_{out} parameter is derived visually from the graph. It takes approximately 150 h to lower the drug level 50%. V is estimated as $Dose/C_{max}$ from the dosing interval pre-induction. We assume that an increase in R_{in}, A_{ss}, and V_{max} leads to a proportional increase in Cl_0. For a complete derivation of the relationship between Cl_0 for *NT* and time, see Gibaldi and Perrier [1982].

The final model that we propose is a one-compartment oral model taken from the *ASCII*-library and modified according to the following specification:

```
               Time = X
               RKpre = CLpre/V
               RKss = CLss/V

Pre-induced state
    If TIME LT TBP THEN
           K10=RKpre
    ENDIF

Peri-induced state
    IF TIME GE TBP AND TIME LT TBP2 THEN
           K10=RKss - (RKss - RKpre)•DEXP(-RKout•(X - TBP))
    ENDIF

Post-induced state
    IF TIME GE TBP2, THEN
           a=RKss - (RKss - RKpre)•DEXP(-RKout•(TBP2 - TBP))
           K10=RKpre - (RKpre - a)•DEXP(-RKout•(X - TBP2))
    ENDIF
```

The k_{10}, Cl_{pre}, Cl_{ss}, k_{out}, TBP, and TBP_2 parameters are the time-dependent variables for the elimination rate constant (drug), oral or intrinsic clearance pre-treatment, intrinsic clearance post-treatment, the elimination rate constant of the induced enzyme (fractional turnover rate), start of inducer treatment and cessation of inducer treatment, respectively, which were used in the model.

K_a	= 3 (h^{-1})	Absorption rate constant
Cl_{ss}	= 140 (L/h)	Induced Cl_0 ($Dose/AUC_{ss}$)
t_{lag}	= 0.7 (h)	Lag-time
Cl_{pre}	= 40 (L/h)	Pre-induced Cl_0 ($Dose/AUC_{pre}$)
k_{out}	= 0.005 (h^{-1})	Fractional turnover rate, *RKD* in program
V	= 1260 (L)	Volume of distribution, actually *V/F*

In the *WinNonlin* code, the number of constants is 177. This is obtained from the number of doses times 2 plus 1 ($2•N_{dose} + 1$). Each dosing occasion contains two records, namely information about the dose size and the dosing time.

Interpretation of results and conclusions

Although enzyme induction is perhaps among the most common causes of time-dependent kinetics, there are many reasons for such kinetics. These include diurnal variations in renal function, urine pH, plasma protein concentration and hormone levels, to mention just a few (Rowland and Tozer [1995]).

The model does not explicitly include the pharmacokinetics of the inducing substance. We assume that the fractional turnover rate (k_{out}[2]) of the induced enzyme $_t$ has a longer half-life than the drug or the inducer. In addition, we assume that the time course from one level of enzyme activity to another will be governed by the fractional turnover rate of the enzyme. *PB* is assumed not to change the fractional turnover rate or the rate of loss of enzyme. We do not include a lag-time between initiation of *PB* and the appearance and altered plasma concentration of *NT*.

The pre-induction oral clearance was estimated to be 46 L/h and the induced oral clearance was about 118 L/h, which corresponds to a 150% increase. This is equivalent to a corresponding increase in the enzyme level, which will have consequences for bioavailability and hepatic extraction. We are aware that *F* appears as a constant in the *V/F*-ratio. The volume of distribution *V/F* is ≈1700 (1681) L. Figure 21.3 shows the concentration-time course of *NT* within three dosing intervals before, during and after induction.

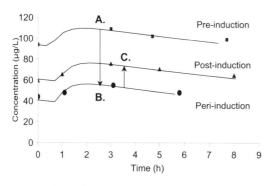

Figure 21.3 Observed and predicted plasma concentrations of NT during a dosing interval pre-, peri-, and post-induction. The arrows indicate the time order of the curves. A, B and C corresponds to the same time-points in Figure 21.1.

The volume of distribution and clearance (Equation (21:1) determine the half-life, and half-life governs the time to steady state. Since the half-life is shortened during induction (i.e., *Cl* increases), the time to establish steady state is shortened compared to the situation if clearance were decreased.

$$t_{1/2} = \frac{\ln(2) \cdot V}{Cl} \qquad (21:1)$$

When the inducer is removed (i.e., *Cl* decreases), the half-life returns (increases) to its pre-induction value. This is longer compared to the half-life during the induction phase. As a result, the return of the induced state towards normal levels is significantly prolonged (Figure 21.4).

---

[2]Note that the estimate of k_{out} is the apparent change in enzyme content, including transcription, translation and loss of enzyme.

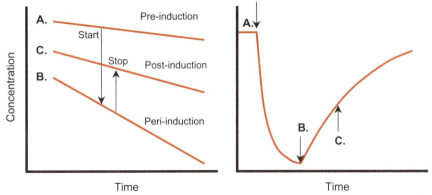

Figure 21.4 *Schematic illustration of the log-concentration-time profile pre- (A), peri- (B), and post-induction (C) (left) . The simulated concentration-time profile following a constant rate input is depicted in the right hand graph. The time to steady state is shortened during induction (half-life is shortened) compared to post-induction (half-life increases and returns to its pre-induction value).*

Mean residence times *MRT* pre- and post-induction were 36 h and 14 h, respectively. The absorption (K_a 2.1 h^{-1}) is preceded by a lag-time of 50 min. The fractional turnover rate constant for the induced enzyme k_{out} and its corresponding half-life could be estimated. The half-life was about 158 h, which is what one would expect from the data. The average plasma concentration is halved within this time frame which is consistent with a 100% increase in the induced enzyme. However, one important limitation of this model is that the kinetics of the inducer or of some intermediate biochemical message was not taken into account. These may in part be rate-limiting. This difficulty can be overcome by measuring the kinetics of the inducing agent itself (Abramson [1986]). We can probably rule out that *PB*, with a half-life of 15-48 h, was responsible for the rate-limiting step. This experiment does not predict the maximum induction possible, since only one dose level of inducer was given. Ideally, two or more dosing rates would have had to be used to obtain that prediction.

In spite of the crude sampling schedule, the model parameters had good precision. Non-compartmental analysis (*NCA*) gave only a rough estimate of Cl_{pre} and Cl_{ss}, because of the sparse number of data points. In this exercise, the regression approach is superior to *NCA*, since the regression model uses all data to estimate all parameters.

Solution

```
TITLE 1
Drug and inducer interaction
Model
 COMM
NPARM 6
NSEC 4
PNAMES  'Ka', 'CLss', 'TLAG', 'CLpre', 'RKd', 'V'
SNAMES 'MRT1', 'MRT2', 'MRTenz', 'T12enz'
END
    1: TEMP
    2: END
    3: FUNC1
    4: I=0
    5: J=1
```

```
 6: L=CON(1)
 7: GREEN:
 8: I=I+1
 9: J=J+2
10: IF X <= CON(J) THEN GOTO RED
11: ELSE IF I < L THEN GOTO GREEN
12: ENDIF
13: ENDIF
14: I=I+1
15: RED:
16: L=I-1
17: SUM=0
18: I=0
19: J=1
20: BLUE:
21: I=I+1
22: J=J+2
23: T=X - CON(J) - TLAG
24: D=CON(J-1)
25: DKOR= 3796.94
26: TBP=216.
27: TBP2=512.
28: RKPRE = CLPRE/V
29: RKSS = CLSS/V
30: IF X LT TBP THEN
31: K10=RKPRE
32: ENDIF
33: IF X GE TBP AND X LT TBP2 THEN
34: K10=RKSS - (RKSS - RKPRE)*DEXP(-RKD*(X - TBP))
35: ENDIF
36: IF X GE TBP2 THEN
37: ASS=RKSS - (RKSS - RKPRE)*DEXP(-RKD*(TBP2 - TBP))
38: K10=RKPRE - (RKPRE - ASS)*DEXP(-RKD*(X - TBP2))
39: ENDIF
40: COEF=D*DKOR*KA/((KA-K10)*V)
41: AMT=MAX(0,COEF*(DEXP(-K10*T)-DEXP(-KA*T)))
42: SUM=SUM + AMT
43: IF I < L THEN GOTO BLUE
44: ENDIF
45: F=SUM
46: END
47: SECO
48: S(1)=1/KA + 1/RKPRE - TLAG
49: S(2)=1/KA + 1/RKSS - TLAG
50: S(3)=1/RKD
51: S(4)=0.693/RKD
52: END
53: EOM
NVARIABLES 2
NPOINTS 2000
XNUMBER 1
YNUMBER 2
NCONSTANTS 177
CONSTANTS 88,10,0,10,8,10, &
          16,10,24,10,32,10,40, &
          10,48,10,56,10,64,10, &
          72,10,80,10,88,10,96, &
          10,104,10,112,10,120,10, &
          128,10,136,10,144,10,152, &
          10,160,10,168,10,176,10, &
          184,10,192,10,200,10,208, &
```

```
              10,216,10,224,10,232,10, &
              240,10,248,10,256,10,264, &
              10,272,10,280,10,288,10, &
              296,10,304,10,312,10,320, &
              10,328,10,336,10,344,10, &
              352,10,360,10,368,10,376, &
              10,384,10,392,10,400,10, &
              408,10,416,10,424,10,432, &
              10,440,10,448,10,456,10, &
              464,10,472,10,480,10,488, &
              10,496,10,504,10,512,10, &
              520,10,528,10,536,10,544, &
              10,552,10,560,10,568,10, &
              576,10,584,10,592,10,600, &
              10,608,10,616,10,624,10, &
              632,10,640,10,648,10,656, &
              10,664,10,672,10,680,10, &
              688,10,696
METHOD 2   'Gauss-Newton (Levenberg and Hartley)
ITERATIONS 50
INITIAL 3,140,.7,40,.005,1260
LOWER BOUNDS .1,20,0,4,0,100
UPPER BOUNDS 5,500,3,200,1,3000
MISSING 'Missing'
NOBSERVATIONS 23
DATA 'WINNLIN.DAT'
BEGIN
```

PARAMETER	ESTIMATE	STANDARD ERROR	CV%	UNIVARIATE C.I.	
KA	2.114475	1.358796	64.26	-.752318	4.981267
CLSS	118.148120	2.153121	1.82	113.605458	122.690783
TLAG	.837049	.147303	17.60	.526269	1.147829
CLPRE	46.170463	.435150	.94	45.252381	47.088545
RKD	.004395	.000213	4.85	.003945	.004845
V	1681.495891	146.853659	8.73	1371.663581	1991.328202

```
*** CORRELATION MATRIX OF THE ESTIMATES ***
PARAMETER  KA         CLSS       TLAG       CLPRE      RKD        V
KA         1.00000
CLSS       -.339297   1.00000
TLAG       .957517    -.396507   1.00000
CLPRE      -.371791   -.185765   -.319261   1.00000
RKD        .348423    -.662647   .362778    .010970    1.00000
V          .469602    .159670    .412396    -.252997   .124261    1.00000

 Condition_number=        .9739E+06

   FUNCTION   1
```

X	OBSERVED Y	PREDICTED Y	RESIDUAL	WEIGHT	SE-PRED	STANDARDI RESIDUAL
168.0	94.00	94.35	-.3484	1.000	.8904	-.2055
171.0	109.0	108.2	.7894	1.000	1.122	.5085
172.7	102.0	103.6	-1.579	1.000	.9726	-.9571

175.7	99.00	95.40	3.605	1.000	.8949	2.129
216.0	93.00	95.04	-2.041	1.000	1.028	-1.263
360.0	52.00	51.96	.4448E-01	1.000	.7273	.2510E-01
361.2	60.00	60.31	-.3132	1.000	1.669	-.3334
363.2	64.00	64.81	-.8063	1.000	.9716	-.4885
365.4	56.00	58.07	-2.068	1.000	.7804	-1.182
368.0	50.00	50.99	-.9933	1.000	.7190	-.5595
384.0	51.00	49.25	1.749	1.000	.7028	.9814
432.0	46.00	45.16	.8392	1.000	.6591	.4666
504.0	44.00	41.12	2.880	1.000	.6141	1.587
505.1	48.00	47.57	.4295	1.000	1.131	.2778
507.1	55.00	54.41	.5862	1.000	.9176	.3487
509.8	48.00	46.62	1.380	1.000	.6374	.7641
552.0	43.00	45.27	-2.270	1.000	.5938	-1.246
600.0	48.00	50.63	-2.634	1.000	.6673	-1.467
696.0	59.00	60.84	-1.836	1.000	.9517	-1.105
697.0	65.00	65.03	-.3105E-01	1.000	1.766	-.4185E-01
699.0	75.00	74.87	.1332	1.000	1.165	.8762E-01
701.0	70.00	69.37	.6316	1.000	.9000	.3736
704.0	64.00	61.64	2.364	1.000	.9754	1.434

```
CORRECTED SUM OF SQUARED OBSERVATIONS =  9198.00
WEIGHTED CORRECTED SUM OF SQUARED OBSERVATIONS =  9198.00
SUM OF SQUARED RESIDUALS =          62.3679
SUM OF WEIGHTED SQUARED RESIDUALS =  62.3679
S =  1.91538      WITH     17 DEGREES OF FREEDOM
CORRELATION (OBSERVED,PREDICTED) =  .9966

AIC criteria =        107.06016
SC  criteria =        113.87313
```

PARAMETER	ESTIMATE	STANDARD ERROR	CV%
MRT1	36.091600	.446679	1.24
MRT2	13.882214	.446679	3.22
MRTENZ	227.522479	11.025726	4.85
T12ENZ	157.673078	7.640828	4.85

PK22 - Nonlinear kinetics - Induction II

Objectives

◆ **To predict the steady-state concentration of drug at equilibrium**

◆ **To estimate basal and induced clearance**

◆ **To estimate the half-life of the overall induction process**

◆ **To model enzyme induction and estimate the fractional turnover rate**

◆ **To apply the feedback model for autoinduction**

Problem specification

This exercise is designed to demonstrate how to build a feedback model for induction and to estimate the apparent fractional turnover rate for the enzyme. A constant rate intravenous infusion of 120 mg of drug X was followed by 9 consecutive constant-rate intravenous infusions of 40 mg each. The duration of each infusion was 30 min except for the first, which lasted 1 h. The infusions were commenced at 8 hourly intervals. Plasma data including peak and trough values were collected (see Figure 22.1 and the program output). Note that the maximum concentrations of the second and third regimens were equal to or larger than for the first but then gradually decreased over time.

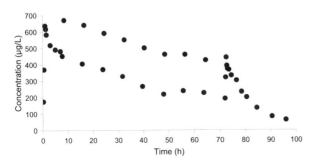

Figure 22.1 Observed plasma concentrations versus time data of drug X following repeated intravenous infusions at 8 hourly intervals.

In order to be able to design an appropriate dosage regimen, we also need to estimate the induced clearance and the time to induction equilibrium. Unfortunately, we only have one dose level which may not suffice for a complete assessment of the induction phenomenon. For further reading about time-dependent processes with respect to enzyme kinetics, we refer the reader to Abramson [1986]. The proposed model that we elaborate in this example is shown in Figure 22.2. The model involves feedback, with drug concentration C and clearance Cl providing the link between the two-compartment drug model and the one-compartment enzyme model.

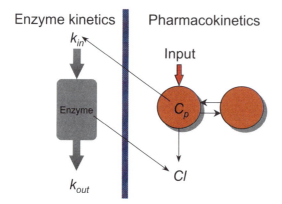

Figure 22.2 *Proposed feedback model, showing the relationship between the two-compartment drug model and the one-compartment enzyme model. Enzyme kinetics and pharmacokinetics are linked by drug concentration (C) and the clearance term (Cl). The left hand diagram shows the assumed linear relationship at steady state between the enzyme level and plasma concentration.*

We will analyze this dataset (Table 22.1) with a two-compartment model which includes a time- dependent change in clearance.

Table 22.1   Concentration-time data of compound X.

Time (h)	Conc (μg/L)	Time (h)	Conc (μg/L)	Time (h)	Conc (μg/L)	Time (h)	Conc (μg/L)
0.25	174	8.5	670	47.99	217	72.75	392
0.5	370	15.75	405	48.5	461	73	373
1	636	16.5	640	55.75	238	73.5	368
1.25	616	23.99	370	56.5	460	74.5	333
1.5	582	24.5	590	63.75	226	76.5	302
3	518	31.75	327	64.5	426	78.5	232
5	491	32.5	550	72	191	80.5	199
7	481	39.75	265	72.25	320	84.5	134
7.75	452	40.5	500	72.5	442	90.5	82
						96	61.4

Some people usually parameterize their compartment models in terms of micro-constants, such as V_c, $Cl(E)$ and Cl_d. Cl_d is the inter-compartmental distribution term. The differential equations for a two-compartment model then become

$$V_c \frac{dC_p}{dt} = In - Cl \cdot C_p - Cl_d \cdot C_p + Cl_d \cdot C_t \qquad (22:1)$$

$$V_t \frac{dC_t}{dt} = Cl_d \cdot C_p - Cl_d \cdot C_t \qquad (22:2)$$

The change in plasma clearance is described by Equation 22:3.

$$Cl = Cl_0 - a \cdot C_p \qquad (22:3)$$

The autoinduction may be due to increased transcription or translation, or to decreased loss of enzyme. For simplicity, we assume that the rate constants for input and output from the *[E]* compartment are the same

$$\frac{dE}{dt} = k_{out} \cdot [E_0 + C_p] - k_{out} \cdot E \qquad (22:4)$$

At equilibrium

$$\frac{dE}{dt} = k_{out} \cdot E_0 - k_{out} \cdot E = 0 \qquad (22:5)$$

which after rearrangement yields $E = E_0$. With no induction, the baseline value of *Cl* is expressed as

$$Cl(E_0) = a \cdot [E_0 + 0] \qquad (22:6)$$

The relationship between the enzyme level and drug plasma concentration is schematically shown in Figure 22.3. When no drug is present, the enzyme level corresponds to E_0. E_0 is fed into Equation 22:6 to give the baseline value of *Cl*.

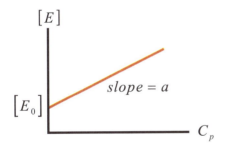

Figure 22.3 Schematic representation of the enzyme concentration as a function of the plasma concentration of drug. We assume for simplicity that the enzyme concentration increases linearly with C_p with slope a. With no drug in plasma the enzyme concentration is equal to its baseline value E_0.

Using the plasma concentrations, we will model this process and obtain an overall measure of the induction process. However, had data been available, the more appropriate way would have been to measure the individual processes shown in Figure 22.4.

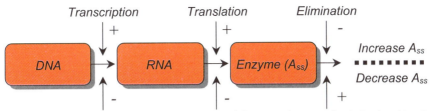

Figure 22.4 Schematic illustration of possible causes of changes in the enzyme level, A_{ss}. A positive sign indicates increased synthesis (production of enzyme) or an increased loss of enzyme, and a negative sign indicates the opposite. Enhanced input and reduced loss lead to build-up of the enzyme level. Reduced input (e.g., inhibition) and enhanced loss reduce the enzyme level.

Interpretation of results and conclusions

This exercise has shown that the feedback model also suits itself for mimicking autoinduction data. The nice feature of the feedback model is that concentration C_p is counterbalanced by the enzymatic activity E. This enables one to use the backflow model after multiple dose regimens, which is a prerequisite for mechanistically oriented models. In addition, two or more optimally spaced dose levels probably help to define the appropriate behavior of the induction process. Figure 22.5 shows observed and model- predicted concentrations of drug X.

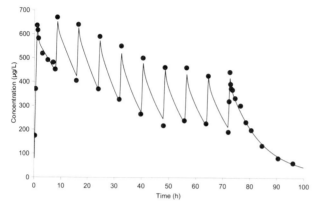

Figure 22.5 Observed and predicted plasma concentrations versus time using the feedback model.

Since no information is available about gene transcription or translation, or about the fractional turnover rate of *[E]*, k_{out} is a measure of the rate-limiting step.

Table 22.2 contains information about the final parameter estimates of V_c, a, Cl_d, V_t, E_0 and k_{out}. The parameters were estimated with reasonable precision and low correlation. No trends were seen in the residual plots.

Table 22.2 Final parameter estimates

Parameter	Model I mean ± CV%	Model II mean ± CV%
V_c (L)	145 ± 5	144 ± 5
a (L/h)	0.041 ± 6	0.041 ± 5
Cl_d (L/h)	–	123 ± 30
V_t (L)	–	61.7 ± 12
k_{12} (h^{-1})	0.82 ± 33	–
k_{21} (h^{-1})	1.94 ± 25	–
k_{out} (h^{-1})	0.024 ± 15	0.023 ± 12
E_0	135 ± 21	135 ± 16

k_{out} coded k_{tol} in program. Model II shown in output.

The induction half-life was estimated to be 30 h, and the enzyme baseline 135 units. Volume of distribution was 144 L for the central compartment and volume of distribution at steady state was 206 L. A slope value a of 0.041 means that clearance increased by approximately 4% for every µg/L increase in the concentration. Uninduced clearance was 5.5 L/h (0.041(1+135)), and the clearance values at 10, 50, and 100 µg/L were 5.95, 7.59 and 9.64 L/h, respectively.

The proposed model is suitable for multiple dosing and for different modes of administration. The enzyme level adjusts itself with respect to drug concentration. Ideally, different modes of administration and two or more dose levels, as well as metabolite data, would be needed to fully describe the underlying time-dependent process. Simultaneous measurement of the metabolite(s) and responsible enzyme would also have strengthened the model.

Dr. D. McCarthy, AstraZeneca, kindly provided concentration-time data. These data were then slightly adapted. The model was proposed by JG.

Solution - Distribution parameters

```
MODEL
 COMM
  NDER   3
  NPARM 6
  PNAMES 'Vc','a', 'Cld','Vt', 'ktol', 'E0'
END
    1: TEMP
    2: T=X
    3: VC = P(1)
    4: a = P(2)
    5: CLD = P(3)
    6: VT = P(4)
    7: KTOL = P(5)
    8: E0 = P(6)
    9: END
   10: START
   11: Z(1)=0.0
   12: Z(2)=0.0
   13: Z(3)=E0
   14: END
   15: DIFF
   16: J = 0
   17: NDOSE = CON(1)
   18: DO I = 1 TO NDOSE
   19: J = J+3
   20: IF X <= CON(J) THEN GOTO RED
   21: ENDIF
   22: NEXT
   23: RED:
   24: NDOSE = I-1
   25: SUM=0
   26: J=0
   27: IF X > 0 THEN
   28: DO I = 1 TO NDOSE
   29: J = J + 3
   30: T = X - CON(J)
   31: DOSE = CON(J-1)*1000
   32: TI = CON(J+1) - CON(J)
   33: DEL = T - TI
   34: DENOM = MAX(1.D-20,ABS(DEL))
   35: TSTAR = MAX(0,DEL/DENOM)
   36: IF X>0 AND T = 0 THEN
   37: Z(1) = Z(1) + DOSE / VC
   38: ENDIF
   39: NEXT
   40: ELSE
   41: TI = CON(4) - CON(3)
```

```
   42: Z(1) = 0
   43: DOSE = CON(2)
   44: TSTAR = 0
   45: ENDIF
   46: DZ(1)= (DOSE/TI/VC)*(1-TSTAR)-(CLD*Z(1)-CLD*Z(2)+a*Z(1)*(1+Z(3)))/VC
   47: DZ(2)= CLD*Z(1)/VT - CLD*Z(2)/VT
   48: DZ(3)= KTOL*(E0 + Z(1)) - KTOL*Z(3)
   49: END
   50: FUNC 1
   51: F = Z(1)
   52: END
   53: EOM
NVARIABLES 2
NPOINTS 200
NCONSTANTS 31
CONSTANTS 10,120,0,1,40,8, 8.5,40,16,16.5,40,24,24.5,40,32,32.5,40,40,40.5,40, &
          48,48.5,40,56,56.5,40,64,64.5,40,72,72.5
METHOD 2   'Gauss-Newton (Levenberg and Hartley)
ITERATIONS 50
INITIAL 148,.06,130,200,.01,100
LOWER BOUNDS 0,0,0,0,0,0
UPPER BOUNDS 500,1,600,2000,1,600
MISSING 'Missing'
NOBSERVATIONS 37
DATA 'WINNLIN.DAT'
BEGIN
```

PARAMETER	ESTIMATE	STANDARD ERROR	CV%	UNIVARIATE C.I.	
VC	144.253531	7.528603	5.22	128.898951	159.608111
a	.040823	.002138	5.24	.036462	.045184
CLD	122.845252	37.323538	30.38	46.723925	198.966579
VT	61.656922	7.341104	11.91	46.684747	76.629097
KTOL	.023317	.002887	12.38	.017428	.029205
E0	135.033984	21.763652	16.12	90.647033	179.420935

```
CORRECTED SUM OF SQUARED OBSERVATIONS =  949961.
WEIGHTED CORRECTED SUM OF SQUARED OBSERVATIONS =  949961.
SUM OF SQUARED RESIDUALS =            5378.40
SUM OF WEIGHTED SQUARED RESIDUALS =  5378.40
S =  13.1718      WITH     31 DEGREES OF FREEDOM
CORRELATION (OBSERVED,PREDICTED) =  .9972

AIC criteria =       329.83540
SC  criteria =       328.66816
```

PK23 - Nonlinear kinetics - Flow I

Objectives

◆ **To analyze the well-stirred, parallel-tube, distributed and dispersion models**

◆ **To analyze a dataset with Q_H and C_{In} as independent variables**

◆ **To estimate intrinsic clearance Cl_{int} for each model**

Problem specification

The *well-stirred, parallel-tube, distributed,* and *dispersion models* have been used to describe hepatic elimination of drugs. Table 23.1 gives data from a study of nicotine disposition after constant rate infusion (at two rates) in human volunteers. Arterial blood C_{in}, venous hepatic blood C_{out}, and hepatic blood flow Q_H were measured independently.

Table 23.1 Experimental data

Arterial concentration (µg/L)	Concentration difference (µg/L)	Hepatic blood flow (L/min)
10.7	7.50	1.79
10.2	6.60	1.72
14.4	7.30	1.73
15.3	10.9	1.42
12.4	8.10	1.69
15.3	11.1	1.02
27.4	15.7	1.73
33.6	25.0	1.87
34.6	24.8	1.74
37.9	25.1	1.51
34.5	25.1	1.53
41.6	27.2	1.55

Columns I-III are denoted as C_A, C_{AV} and Q_H, respectively.

Applying the theory for the *well-stirred model*, the relationship between the arterio-venous concentration difference C_{AV}, the incoming arterial and venous splanchnic blood C_A, and blood flow Q_H becomes

$$C_{AV} = C_A \cdot E_H = C_A \cdot [1 - F_H] = C_A \cdot \frac{Cl_{int}}{Q_H + Cl_{int}} \qquad (23{:}1)$$

where E_H and F_H are the hepatic extraction ratio and availability, respectively. Cl_{int}, is the model parameter to be estimated. C_{AV}, is implemented in the model file as

```
FUNC 1
        F=CA• (CLINT/(CLINT + QH))
END
```

where CLINT is the model parameter. $Cl_{int}/(Cl_{int} + Q_H)$ in Equation 23:1 is the extraction ration, C_A is the concentration in incoming arterial and venous splanchnic blood, and Q_H is the second independent variable. All the C_A, C_{AV} and Q_H data can be found in the first (CA=DTA(1)), second (CAV = YNUM 2) and third (QH=DTA(3)) column, respectively of the dataset file (Table 1). They are also defined in the TEMP-block in the program output.

Assuming the *parallel-tube model*, we derived the following relationship between the dependent variable C_{AV}, the two independent variables (C_A and Q_H), and intrinsic clearance Cl_{int}

$$C_{AV} = C_A \cdot E_H = C_A \cdot (1 - F_H) = C_A \cdot \left[1 - e^{-\frac{Cl_{int}}{Q_H}} \right] \qquad (23:2)$$

where E_H and F_H are the hepatic extraction ratio and availability, respectively. C_{AV} is implemented in the model file as

```
FUNC 1
        F=CA• (1 – DEXP(–CLINT/QH))
END
```

where the parameters are defined as above.
The *distributed model* is expressed in the following formulas

$$C_{AV} = C_A \cdot E_H = C_A \cdot (1 - F_H) = C_A \cdot \left[1 - e^{-\left\{ \frac{Cl_{int}}{Q_H} + \frac{1}{2}\varepsilon^2 \left[\frac{Cl_{int}}{Q_H} \right]^2 \right\}} \right] \qquad (23:3)$$

$$Cl_H = Q_H \cdot \left[1 - e^{-\left\{ \frac{Cl_{int}}{Q_H} + \frac{1}{2}\varepsilon^2 \left(\frac{Cl_{int}}{Q_H} \right)^2 \right\}} \right] \qquad (23:4)$$

ε is the variance for each sinusoid in the whole liver. ε is a parameter like Cl_{int}, which can be estimated during the fitting procedure. The efficiency number R_N, is equal to Cl_{int}/Q_H. When blood flow is large or ε is close to zero, the distributed model collapses into the parallel-tube model. These two models are therefore nested models and the F-test can be applied for model discrimination.

The *dispersion model* is expressed in the following formulas

$$C_{AV} = C_A \cdot E_H = C_A \cdot \left[1 - \frac{4a}{(1+a)^2 e^{\left[\frac{a-1}{2D_N}\right]} - (1-a)^2 e^{-\left[\frac{a+1}{2D_N}\right]}} \right] \qquad (23:5)$$

$$Cl_H = Q_H \cdot \left[1 - \frac{4a}{(1+a)^2 e^{\left[\frac{a-1}{2D_N}\right]} - (1-a)^2 e^{-\left[\frac{a+1}{2D_N}\right]}} \right] \qquad (23:6)$$

where $a = (1 + 4 \cdot R_N \cdot D_N)^{1/2}$. Cl_{int} and D_N are parameters to be estimated during the fitting procedure. R_N is Cl_{int}/Q_H.

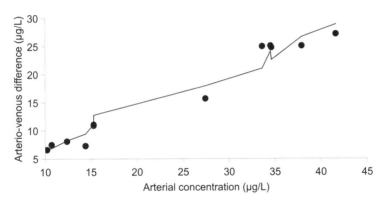

Figure 23.1 *Plot of observed (filled circles) and model-predicted (line) arterio-venous concentration difference versus arterial concentration. The reason for the discontinuity is that different blood flows are applied.*

Initial parameter estimates

The initial estimate of intrinsic clearance is obtained as follows. We select a value of the arterio-venous concentration difference, arterial concentration and hepatic blood flow from Table 23.1. The extraction ratio E_H is estimated as

$$E_H = \frac{C_A - C_V}{C_A} = \frac{C_{AV}}{C_A} = \frac{7.5}{10.7} = 0.7 \qquad (23:7)$$

Extraction ratio E_H can also be expressed in terms of intrinsic clearance and hepatic blood flow

$$E_H = \frac{f_u \cdot Cl_{int}}{Q_H + f_u \cdot Cl_{int}} \approx \frac{Cl_{int}}{Q_H + Cl_{int}} = \frac{Cl_{int}}{1.79 + Cl_{int}} = 0.7 \qquad (23:8)$$

which when rearranged gives intrinsic clearance

$$Cl_{int} = \frac{1.79 \cdot 0.7}{0.3} = 4.2 \ L/\min \qquad (23:9)$$

We have selected to use 5.0 L·min^{-1} as our initial estimate for intrinsic clearance.

Cl_{int} = 5.0 (L/min)
ε = 0.3
D_N = 100

Interpretation of results and conclusions

The *well-stirred* model (*WRSS* 31.4) fits the data better than the *parallel-tube* model (*WRSS* 39.8) with respect to goodness-of-fit criteria, although parameter precision is lower for Cl_{int} in the *well-stirred* model. According to the objective function *WRSS*, the *distributed* model was not better than the *parallel-tube* model, and the parameter precision of the *distributed* model was low. Epsilon ε was predicted to be about 3%, indicating that this model collapses into the *parallel-tube* model.

    The *dispersion* model could not be fit to the data with acceptable precision of the model parameters. The high dispersion number D_N indicates that the *dispersion* model collapses into the *well-stirred* model.

    We have practiced building and implementing four (*well-stirred, parallel-tube, distributed* and *dispersion*) models in this exercise. We have also derived initial estimates of Cl_{int}. Section 3.4.5 covers these models more thoroughly.

Solution I - Well-stirred model

```
MODEL
 COMM
  nparm 1
  PNAMES 'CLint'
END
    1: TEMP
    2: CA=DTA(1)
    3: QH=DTA(3)
    4: CLINT=P(1)
    5: END
    6: FUNC1
    7: EH = CLINT/(QH + CLINT)
    8: CAV = CA*EH
    9: F = CAV
   10: END
   11: EOM
DATE
TIME
NVARIABLES 3
NPOINTS 10
```

```
XNUMBER 1
YNUMBER 2
METHOD 2  'Gauss-Newton (Levenberg and Hartley)
ITERATIONS 50
INITIAL 5
LOWER BOUNDS .001
UPPER BOUNDS 20
MISSING 'Missing'
NOBSERVATIONS 12
DATA 'WINNLIN.DAT'
BEGIN
```

PARAMETER	ESTIMATE	STANDARD ERROR	CV%	UNIVARIATE C.I.	
CLINT	3.443000	.291268	8.46	2.801920	4.084081

```
 CORRECTED SUM OF SQUARED OBSERVATIONS =  797.840
 WEIGHTED CORRECTED SUM OF SQUARED OBSERVATIONS =  797.840
 SUM OF SQUARED RESIDUALS =          31.3894
 SUM OF WEIGHTED SQUARED RESIDUALS =  31.3894
 S =  1.68926     WITH     11 DEGREES OF FREEDOM
 CORRECTION (OBSERVED,PREDICTED) =  .9807
```

```
 AIC criteria =         43.35765
 SC  criteria =         42.60010
```

Solution II - Parallel-tube model

```
MODEL
 COMM
  nparm 1
  PNAMES 'CLint'
END
    1: TEMP
    2: CA=DTA(1)
    3: QH=DTA(3)
    4: CLINT=P(1)
    5: END
    6: FUNC1
    7: EH = 1. - DEXP(-CLINT/QH)
    8: CAV = CA*EH
    9: F = CAV
   10: END
   11: EOM
DATE
TIME
NVARIABLES 3
NPOINTS 10
METHOD 2  'Gauss-Newton (Levenberg and Hartley)
INITIAL 5
LOWER BOUNDS .001
UPPER BOUNDS 20
MISSING 'Missing'
NOBSERVATIONS 12
DATA 'WINNLIN.DAT'
BEGIN
```

PARAMETER	ESTIMATE	STANDARD ERROR	CV%	UNIVARIATE C.I.	
CLINT	1.840971	.105107	5.71	1.609631	2.072311

```
       CORRECTED SUM OF SQUARED OBSERVATIONS =  797.840
       WEIGHTED CORRECTED SUM OF SQUARED OBSERVATIONS =  797.840
       SUM OF SQUARED RESIDUALS =            39.8374
       SUM OF WEIGHTED SQUARED RESIDUALS =  39.8374
       S =  1.90305      WITH     11 DEGREES OF FREEDOM
       CORRELATION (OBSERVED,PREDICTED) =  .9752

  AIC criteria =          46.21768
  SC  criteria =          45.46014
```

Solution III - Distributed model

```
TITLE 1
Distributed model
MODEL
 COMM
  nparm 2
  PNAMES 'CLint', 'eps'
END
    1: TEMP
    2: CA=DTA(1)
    3: QH=DTA(3)
    4: CLINT=P(1)
    5: EPS =P(2)
    6: END
    7: FUNC1
    8: A = CLINT/QH
    9: B = 0.5*(EPS**2)*(A**2)
   10: EH = 1. - DEXP(- (A + B))
   11: CAV = CA*EH
   12: F = CAV
   13: END
   14: EOM
DATE
TIME
NOPAGE BREAKS
NVARIABLES 3
NPOINTS 2000
XNUMBER 1
YNUMBER 2
METHOD 2   'Gauss-Newton (Levenberg and Hartley)
ITERATIONS 50
INITIAL 1.839471,.032897
LOWER BOUNDS .001,.0001
UPPER BOUNDS 10,.2
MISSING 'Missing'
NOBSERVATIONS 12
DATA 'WINNLIN.DAT'
BEGIN
```

PARAMETER	ESTIMATE	STANDARD ERROR	CV%	UNIVARIATE C.I.	
CLINT	1.842654	.998811	54.21	-.382850	4.068157
EPS	.027007	17.434591	64554.64	-38.819927	38.873942

```
 *** CORRELATION MATRIX OF THE ESTIMATES ***
 PARAMETER  CLINT     EPS
 CLINT         1.00000
 EPS          -.993880        1.00000
```

```
Condition_number=        158.5
```

```
    FUNCTION    1

    X          OBSERVED    PREDICTED    RESIDUAL    WEIGHT    SE-PRED    STANDARD
               Y           Y                                            RESIDUAL
   10.20        6.600       6.707       -.1074      1.000     .2580     -.5423E-01
   10.70        7.500       6.879        .6207      1.000     .3169      .3149
   12.40        8.100       8.234       -.1341      1.000     .2952     -.6792E-01
   14.40        7.300       9.439      -2.139       1.000     .3723     -1.090
   15.30       10.90       11.12        -.2230      1.000     .5094     -.1155
   15.30       11.10       12.79       -1.690       1.000    1.441      -1.223
   27.40       15.70       17.96       -2.259       1.000     .7083     -1.211
   33.60       25.00       21.06        3.938       1.000    1.182       2.449
   34.50       25.10       24.16        .9406       1.000     .8260      .5176
   34.60       24.80       22.61        2.195       1.000     .9146      1.237
   37.90       25.10       26.72       -1.620       1.000     .9533     -.9238
   41.60       27.20       28.94       -1.736       1.000     .9564     -.9906
```

```
    CORRECTED SUM OF SQUARED OBSERVATIONS =  797.840
    WEIGHTED CORRECTED SUM OF SQUARED OBSERVATIONS =  797.840
    SUM OF SQUARED RESIDUALS =           39.8503
    SUM OF WEIGHTED SQUARED RESIDUALS =  39.8503
    S =  1.99625     WITH      10 DEGREES OF FREEDOM
    CORRELATION (OBSERVED,PREDICTED) =  .9752
```

```
 AIC criteria =          48.22157
 SC  criteria =          49.19138
```

Solution IV - Dispersion model

```
TITLE 1
Dispersion model
MODEL
 COMM
  nparm 2
  PNAMES 'CLint', 'Dn'
END
    1: TEMP
    2: CA=DTA(1)
    3: QH=DTA(3)
    4: CLINT=P(1)
    5: DN =P(2)
    6: END
    7: FUNC1
    8: RN = CLINT/QH
    9: A = (1 + 4*RN*DN)**0.5
   10: B = (1 + A)**2
   11: C = (A - 1)/(2*DN)
   12: D = (1 - A)**2
   13: E = -(A + 1)/(2*DN)
   14: BIO = 4*A/(B*DEXP(C) - D*DEXP(E))
   15: EH = 1. - BIO
   16: CAV = CA*EH
   17: F = CAV
   18: END
   19: EOM
DATE
TIME
```

```
NVARIABLES 3
METHOD 2  'Gauss-Newton (Levenberg and Hartley)
INITIAL 5,100
LOWER BOUNDS .001,0
UPPER BOUNDS 10,3000
NOBSERVATIONS 12
DATA 'WINNLIN.DAT'
BEGIN
```

PARAMETER	ESTIMATE	STANDARD ERROR	CV%	UNIVARIATE C.I.	
CLINT	3.444448	2.624369	76.19	-2.331770	9.220667
DN	1355.481886	3920215.957250	289211.98	*************	8629725.580760

```
 CORRECTED SUM OF SQUARED OBSERVATIONS =  797.840
 WEIGHTED CORRECTED SUM OF SQUARED OBSERVATIONS =  797.840
 SUM OF SQUARED RESIDUALS =           31.3921
 SUM OF WEIGHTED SQUARED RESIDUALS =  31.3921
 S =  1.68933    WITH    11 DEGREES OF FREEDOM
 CORRELATION (OBSERVED,PREDICTED) =  .9807

 AIC criteria =        45.35867
 SC  criteria =        43.84358
```

PK24 - Nonlinear kinetics - Flow II

Objectives

◆ **To analyze intravenous infusion data with potential flow-dependent clearance**

◆ **To write a multi-compartment model in terms of differential equations**

◆ **To obtain initial parameter estimates**

◆ **To discriminate between rival models**

Problem specification

The aim of this exercise is to demonstrate how to model nonlinear kinetics with respect to *flow*. When a high (e.g., hepatic) extraction drug changes the perfusion of the eliminating organ, depending on the drug concentration, we have a nonlinear system with respect to clearance.

A dose of 10 mg/kg of a new drug was infused for 2 h to a human volunteer. Blood samples of 250 µl were drawn frequently during and up to 5 h after the infusion (Figure 24.1 and Table 24.1).

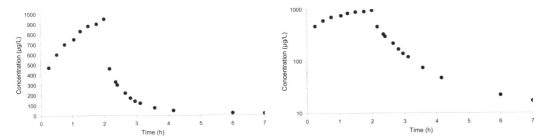

Figure 24.1 *The left hand figure displays the observed plasma concentration-time data on a linear scale and the right hand figure displays the same data on a semi-logarithmic scale. The infused dose was 10 mg/kg.*

Table 24.1 Concentration-time data

Time (h)	Conc (µg/L)	Time (h)	Conc (µg/L)
0.25	470	2.4	300
0.5	600	2.65	220
0.75	700	2.81	170
1.05	750	2.95	140
1.25	830	3.11	120
1.49	880	3.56	74
1.75	900	4.15	47
1.99	950	6	22
2.16	460	7	17
2.35	330	-	-

The compound displayed multi-compartment disposition kinetics. The elimination of the compound displayed a perfusion-limited behavior. The drug is known to reduce cardiac output and hepatic blood flow when the plasma concentration increases. This should be taken into account in the model.

Figure 24.2 shows schematically the three-compartment model that is proposed to mimic the system. The shallow compartment represents well-perfused (rapidly equilibrated) tissues such as liver and kidneys, and the deep compartment represents poorly perfused (slowly equilibrated) tissues such as adipose tissue.

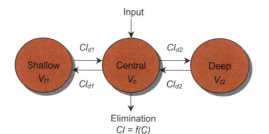

Figure 24.2 Schematic representation of a static or flow-dependent three-compartment model. In a static model, clearance Cl is constant. In a flow-dependent model it is concentration-dependent.

The turnover of drug in each compartment is determined as

$$V_c \frac{dC_p}{dt} = In - Cl \cdot C_p - Cl_{d1} \cdot C_p + Cl_{d1} \cdot C_{t1} - Cl_{d2} \cdot C_p + Cl_{d2} \cdot C_{t2} \qquad (24{:}1)$$

$$V_{t1} \frac{dC_{t1}}{dt} = Cl_{d1} \cdot C_p - Cl_{d1} \cdot C_{t1} \qquad (24{:}2)$$

$$V_{t2} \frac{dC_{t2}}{dt} = Cl_{d2} \cdot C_p - Cl_{d2} \cdot C_{t2} \qquad (24{:}3)$$

The change in plasma clearance is described by Equation 24:4

$$Cl = Cl_0 - a \cdot C_p \qquad (24{:}4)$$

Initial parameter estimates

Figure 24.3 shows how to obtain initial estimates from the wash-out curve after the cessation of infusion.

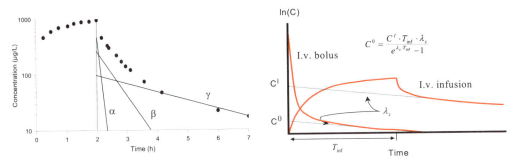

Figure 24.3 *The left figure displays observed concentration-time data on a log-linear scale and includes the alpha, beta and gamma slopes. The right hand figure displays schematically the kinetics of a multi-compartment drug after a bolus dose and constant rate infusion. This figure also relates the intercept of the back-extrapolated terminal slope (C^I) after an infusion to the intercept of the same compound after a bolus dose ($C^{\lambda z}$ or C^0). In the case of the alpha, beta and gamma phase, $C^{\lambda i}$ correspond to A, B and C, respectively where 'i' denotes the respective slope.*

How can one derive initial parameter estimates for this system? Can it be done by a graphical method? How can one discriminate between a linear and a nonlinear model? What assumptions have to be made about the relationship between flow and plasma concentration? What is the appropriate weighting method?

Model the disposition of the drug with a static multi-compartment model, and with a model that includes concentration-dependent clearance, and obtain the initial parameter estimates. Implement the multi-compartment model in terms of differential equations into *WinNonlin*. The slope of the terminal phase is

$$slope = -\lambda = \frac{\ln(180) - \ln(100)}{0 - 2} = -0.29 \, h^{-1} \qquad (24:5)$$

The slope of the intermediate phase is

$$-\beta = \frac{\ln(250) - \ln(10)}{0 - 1.75} = -1.83 \, h^{-1} \qquad (24:6)$$

The slope of the initial phase is approximately

$$-\alpha = \frac{\ln(750) - \ln(10)}{0 - 0.25} \approx -13 \, h^{-1} \qquad (24:7)$$

The α value is connected with a high degree of uncertainty, because of the small number of observations during that phase. Consequently, the value $C(\alpha)$ or A will also be poorly estimated.

$$C^0 = \frac{C^I T_{inf} \lambda_z}{e^{\lambda_z \cdot T_{inf}} - 1} \qquad (24:8)$$

C^0, C^I, T_{inf} and λ_z are the estimated bolus intercept, the back-extrapolated intercept at time zero (0) after the constant rate infusion, the length of the infusion and the slope of the actual phase (e.g., α, β or γ), respectively

$$C(\gamma) = C = \frac{180 \cdot 2 \cdot 0.29}{e^{0.29 \cdot 2} - 1} \approx 130 \ \mu g/L \tag{24:9}$$

$$C(\beta) = B = \frac{6500 \cdot 2 \cdot 1.83}{e^{1.83 \cdot 2} - 1} \approx 630 \ \mu g/L \tag{24:10}$$

$$C(\alpha) = A = \frac{\infty}{e^{13 \cdot 2} - 1} \approx 20000 \ \mu g/L \tag{24:11}$$

As already pointed out, remember that $C(\alpha)$ will be poorly estimated. We had to assume a reasonable value of 20000 µg/L.
Total clearance is estimated from

$$Cl = \frac{D_{iv}}{AUC} = \frac{D_{iv}}{\dfrac{A}{\alpha} + \dfrac{B}{\beta} + \dfrac{C}{\gamma}} = \frac{10000}{\dfrac{20000}{13} + \dfrac{630}{1.83} + \dfrac{130}{0.29}} = 4.3 \ L/h/kg \tag{24:12}$$

The volume of the central compartment can be estimated by means of

$$V_c = \frac{D_{iv}}{A + B + C} = \frac{10000}{20000 + 630 + 130} \approx 0.5 \ L/kg \tag{24:13}$$

and Cl_d is approximately

$$Cl_d = \alpha \cdot V_c - Cl = 13 \cdot 0.5 - 4.3 \approx 2.2 \ \ L/h/kg \tag{24:14}$$

However, we have two Cl_d parameters, one of which is more likely to be greater than plasma clearance (4.3 L/h/kg). Therefore, we set one Cl_d value to 2.2 L/h/kg (slow) and the other to 2.2 + 4.3 L/h/kg (rapid). This gives Cl_d (rapid) of 6.5 L/h/kg. The volume of distribution at steady state V_{ss} is predicted by means of the mean residence time *MRT* and *Cl* according to

$$V_{ss} = \frac{AUMC}{AUC} \cdot \frac{D_{iv}}{AUC} = MRT \cdot Cl$$

$$= \frac{\dfrac{A}{\alpha^2} + \dfrac{B}{\beta^2} + \dfrac{C}{\gamma^2}}{\dfrac{A}{\alpha} + \dfrac{B}{\beta} + \dfrac{C}{\gamma}} \cdot \frac{D_{iv}}{\dfrac{A}{\alpha} + \dfrac{B}{\beta} + \dfrac{C}{\gamma}} = 0.79 \cdot 4.3 = 3.4 \ L/kg \tag{24:15}$$

The parameter a that relates the decrease in plasma clearance (and, indirectly, the decrease in cardiac output and hepatic perfusion) to an increase in plasma concentration is difficult to estimate graphically. We therefore assume that total plasma clearance is reduced by a factor of 10 to 15% at 1000 µg/L. This means that a will be approximately

$$a = \frac{Cl_0 - Cl}{C_p} = \frac{4.3 - 3.8}{1000} \approx 0.0005 \; L^2/(h \cdot \mu g \cdot kg) \qquad (24:16)$$

Cl_0 is approximated by D_{iv}/AUC (4.3 L/h/kg) and Cl (3.8 L/h/kg) is a value 10-15% below that value for a plasma concentration (C_p) of approximately 1000 µg/L.

It should be noted that you may derive initial estimates for the parameters of this model that do not necessarily correspond to our estimates. This is acceptable. The important point is that you understand our reasoning.

Interpretation of results and conclusions

The observed and model-predicted concentration-time data for both the static and flow-dependent models are shown in Figure 24.4, and the final parameter estimates in Table 24.2.

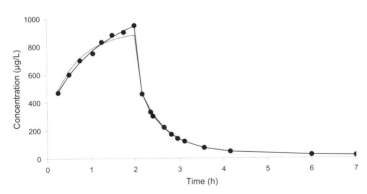

Figure 24.4 Observed and predicted concentration-time data following a 2 h constant rate intravenous infusion. The solid line represents the concentration-dependent clearance model. The line which deviates systematically from the experimental data is the static model.

Table 24.2 Parameter estimates of the static and flow-dependent models

Parameter	Static model Estimate	CV%	Flow-dependent model Estimate	CV%
V_c, L/kg	0.80	41	0.68	14
Cl_2, L/kg/h	5.10	1.2	6.61	2.9
Cl_{d1}, L/h/kg	6.33	36	5.94	9.2
Cl_{d2}, L/h/kg	0.72	6.8	0.93	4.1
V_{t1}, L/kg	1.50	18	1.77	4.7
V_{t2}, L/kg	2.40	13	3.18	7.4
a, $L^2 \cdot h/\mu g/kg$	-	-	0.0025	12

The weighted residuals *versus* time of the static clearance model are plotted in Figure 24.5 and for the flow-dependent model in Figure 24.6.

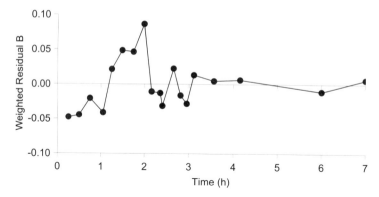

Figure 24.5 *Plot of the weighted residuals versus time for the static (time-invariant) clearance model. Note the lack of randomness (systematic trends) over time.*

The plot of the weighted residuals *versus* time for the flow-dependent model (Figure 24.6), indicates that a proportional error model might be more appropriate as a weighting scheme than constant absolute weights.

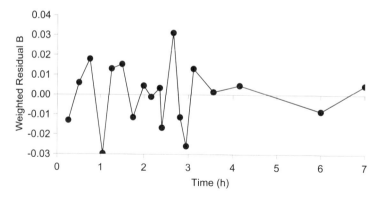

Figure 24.6 *Plot of weighted residual plotted versus time for the flow-dependent (time-variant) clearance model, where clearance is a function of plasma concentrations. Note the randomness (no systematic trends more than a dampening of the amplitude).*

In light of the more random residuals and the better parameter precision we select the flow-dependent model as the model of choice.

You have learned to:
☐ analyze the behavior of a compound displaying nonlinear (flow-dependent) kinetics
☐ write and implement a system of differential equations
☐ obtain initial parameter estimates
☐ discriminate between rival models

To fully characterize the kinetics of this compound, one would need to study its behavior during extended exposure (e.g., multiple dosing).

Solution I – Static *Cl* model

```
Model
 COMM
  NDER 3
  NCON 2
  NPARM 6
  PNAMES 'Vc', 'Cl2', 'cld1', 'cld2', 'Vt1', 'vt2'
END
    1: TEMP
    2: T=X
    3: DOSE1 = CON(1)
    4: TINF1 = CON(2)
    5: END
    6: START
    7: Z(1)=0.0
    8: Z(2)=0.0
    9: Z(3)=0.0
   10: END
   11: DIFF
   12: IF T LE TINF1 THEN
   13: FINF = DOSE1/TINF1
   14: ELSE
   15: FINF = 0.0
   16: ENDIF
   17: CL1 = CL2
   18: DZ(1) = (FINF-CL1*Z(1)-CLD1*Z(1)+CLD1*Z(2)-CLD2*Z(1)+CLD2*Z(3))/VC
   19: DZ(2) = (CLD1*Z(1) - CLD1*Z(2))/VT1
   20: DZ(3) = (CLD2*Z(1) - CLD2*Z(3))/VT2
   21: END
   22: FUNC 1
   23: F=Z(1)
   24: END
   25: EOM
NVARIABLES 2
NPOINTS 1000
XNUMBER 1
YNUMBER 2
CONSTANTS 10000,2
METHOD 2  'Gauss-Newton (Levenberg and Hartley)
REWEIGHT -2
ITERATIONS 50
INITIAL .5,4.3,6.5,2.2,1.7,1.7
LOWER BOUNDS .1,.5,.5,.1,.2,.3
UPPER BOUNDS 10,10,15,10,20,30
MISSING 'Missing'
NOBSERVATIONS 19
DATA 'WINNLIN.DAT'
BEGIN
```

PARAMETER	ESTIMATE	STANDARD ERROR	CV%	UNIVARIATE C.I.	
VC	.801475	.329262	41.08	.090149	1.512801
CL2	5.098760	.062466	1.23	4.963811	5.233709
CLD1	6.328982	2.271758	35.89	1.421149	11.236814
CLD2	.722152	.048987	6.78	.616322	.827982

VT1	1.495898	.274443	18.35	.903000	2.088796
VT2	2.397024	.315956	13.18	1.714443	3.079605

```
*** CORRELATION MATRIX OF THE ESTIMATES ***
PARAMETER  VC          CL2         CLD1        CLD2        VT1         VT2
VC         1.00000
CL2        .011499     1.00000
CLD1       -.956678    .135489     1.00000
CLD2       -.344349    .515063     .474714     1.00000
VT1        -.957814    .051123     .919689     .196626     1.00000
VT2        .268439     -.270609    -.333286    -.621824    -.06244     1.00000
```

```
    FUNCTION   1
```

X	OBSERVED Y	PREDICTED Y	RESIDUAL	WEIGHT	SE-PRED	STANDARDIZED RESIDUAL
.2500	470.0	493.5	-23.51	.4106E-05	12.14	-1.444
.5000	600.0	627.9	-27.94	.2536E-05	10.77	-1.189
.7500	700.0	714.7	-14.70	.1958E-05	10.65	-.5362
1.050	750.0	782.1	-32.06	.1635E-05	10.96	-1.059
1.250	830.0	812.6	17.41	.1514E-05	11.01	.5512
1.490	880.0	839.3	40.66	.1419E-05	11.00	1.242
1.750	900.0	860.1	39.91	.1352E-05	10.96	1.186
1.990	950.0	874.2	75.78	.1308E-05	10.93	2.211
2.160	460.0	465.0	-4.983	.4625E-05	18.73	-1.260
2.350	330.0	334.2	-4.238	.8951E-05	8.269	-.3854
2.400	300.0	309.7	-9.660	.1043E-04	7.404	-.9312
2.650	220.0	215.1	4.859	.2160E-04	3.809	.6078
2.810	170.0	172.8	-2.794	.3349E-04	3.099	-.4365
2.950	140.0	144.0	-4.042	.4820E-04	2.822	-.7751
3.110	120.0	118.4	1.572	.7130E-04	2.529	.3773
3.560	74.00	73.66	.3417	.1843E-03	1.715	.1367
4.150	47.00	46.71	.2920	.4584E-03	1.478	.2374
6.000	22.00	22.24	-.2378	.2022E-02	.6734	-.3835
7.000	17.00	16.90	.1048	.3503E-02	.6395	.3836

```
CORRECTED SUM OF SQUARED OBSERVATIONS =  .198786E+07
WEIGHTED CORRECTED SUM OF SQUARED OBSERVATIONS =  13.6914
SUM OF SQUARED RESIDUALS =            12057.0
SUM OF WEIGHTED SQUARED RESIDUALS =  .220250E-01
S =  .411610E-01 WITH     13 DEGREES OF FREEDOM
CORRELATION (OBSERVED,PREDICTED) =  .9974

AIC criteria =        -60.49598
SC  criteria =        -54.82935
```

Solution II – Flow-dependent model (i.e., *Cl* as a function of *C*)

```
Model
 COMM
  NDER 3
  NCON 2
  NPARM 7
  PNAMES 'Vc', 'Cl2', 'cld1', 'cld2', 'Vt1', 'vt2', 'a'
END
    1: TEMP
```

```
 2: T=X
 3: DOSE1 = CON(1)
 4: TINF1 = CON(2)
 5: END
 6: START
 7: Z(1)=0.0
 8: Z(2)=0.0
 9: Z(3)=0.0
10: END
11: DIFF
12: IF T LE TINF1 THEN
13: FINF = DOSE1/TINF1
14: ELSE
15: FINF = 0.0
16: ENDIF
17: CL1 = CL2 - A*Z(1)
18: DZ(1) = (FINF-CL1*Z(1)-CLD1*Z(1)+CLD1*Z(2)-CLD2*Z(1)+CLD2*Z(3))/VC
19: DZ(2) = (CLD1*Z(1) - CLD1*Z(2))/VT1
20: DZ(3) = (CLD2*Z(1) - CLD2*Z(3))/VT2
21: END
22: FUNC 1
23: F=Z(1)
24: END
25: EOM
NVARIABLES 2
NPOINTS 1000
XNUMBER 1
YNUMBER 2
CONSTANTS 10000,2
METHOD 2  'Gauss-Newton (Levenberg and Hartley)
REWEIGHT -2
ITERATIONS 50
INITIAL .5,4.3,6.5,2.2,1.7,1.7,.0005
LOWER BOUNDS .1,.5,.5,.1,.2,.3,.000001
UPPER BOUNDS 10,10,15,10,20,30,.1
MISSING 'Missing'
NOBSERVATIONS 19
DATA 'WINNLIN.DAT'
BEGIN
```

PARAMETER	ESTIMATE	STANDARD ERROR	CV%	UNIVARIATE C.I.	
VC	.681976	.094048	13.79	.477064	.886889
CL2	6.608924	.194511	2.94	6.185121	7.032727
CLD1	5.938948	.548427	9.23	4.744026	7.133870
CLD2	.929928	.038078	4.09	.846962	1.012894
VT1	1.770018	.083787	4.73	1.587462	1.952574
VT2	3.184762	.234509	7.36	2.673809	3.695714
A	.002480	.000310	12.48	.001805	.003155

```
*** CORRELATION MATRIX OF THE ESTIMATES ***
PARAMETER  VC        CL2       CLD1      CLD2      VT1       VT2       A
VC         1.00000
```

```
CL2    -.076466  1.00000
CLD1   -.900704  -.116370  1.00000
CLD2   -.304550  .697280   .274169  1.00000
VT1    -.788988  .324898   .641291  .149241  1.00000
VT2    .206273   .398336   -.361760 -.157254 .334654  1.00000
A      -.070556  .992086   -.148549 .640652  .346169  .476958 1.00000
```

FUNCTION   1

X	OBSERVED Y	PREDICTED Y	RESIDUAL	WEIGHT	SE-PRED	STANDARDI RESIDUAL
.2500	470.0	476.2	-6.192	.4410E-05	5.363	-.8510
.5000	600.0	596.4	3.596	.2811E-05	6.140	.3780
.7500	700.0	687.6	12.36	.2115E-05	6.131	1.073
1.050	750.0	772.9	-22.92	.1674E-05	5.433	-1.682
1.250	830.0	819.2	10.76	.1490E-05	5.296	.7361
1.490	880.0	866.7	13.34	.1331E-05	6.458	.8817
1.750	900.0	910.5	-10.46	.1206E-05	9.358	-.7197
1.990	950.0	945.7	4.341	.1118E-05	13.01	.3509
2.160	460.0	460.5	-.5161	.4715E-05	8.693	-.5573
2.350	330.0	328.9	1.132	.9246E-05	3.828	.2296
2.400	300.0	305.1	-5.077	.1074E-04	3.415	-1.085
2.650	220.0	213.3	6.659	.2197E-04	1.770	1.828
2.810	170.0	172.0	-1.956	.3382E-04	1.422	-.6658
2.950	140.0	143.7	-3.716	.4842E-04	1.288	-1.545
3.110	120.0	118.4	1.561	.7129E-04	1.159	.8102
3.560	74.00	73.89	.1075	.1831E-03	.8008	.0933
4.150	47.00	46.78	.2203	.4570E-03	.6787	.3846
6.000	22.00	22.19	-.1865	.2032E-02	.3157	-.6690
7.000	17.00	16.92	.8010E-01	.3493E-02	.2969	.6540

```
CORRECTED SUM OF SQUARED OBSERVATIONS =  .198786E+07
WEIGHTED CORRECTED SUM OF SQUARED OBSERVATIONS = 13.6654
SUM OF SQUARED RESIDUALS =           1243.57
SUM OF WEIGHTED SQUARED RESIDUALS = .432443E-02
S = .189834E-01 WITH     12 DEGREES OF FREEDOM
CORRELATION (OBSERVED,PREDICTED) =  .9997

AIC criteria =       -89.42604
SC  criteria =       -82.81497
```

PK25 - Inhalation kinetics

Objectives
◆ **To simultaneously fit first-order pulmonary input and intravenous infusion data**

Problem specification

This problem is designed to help you learn and apply the concepts of pulmonary absorption. A study was designed where intravenous infusion and an extravascular dose (pulmonary deposition with assumed first-order absorption) of drug A were given. The intravenous dose of 700 µg was given as a short infusion over 2 min. The extravascular dose was 1000 µg. We will estimate the absorption rate constant K_a and the bioavailability.

Data are shown in Figure 25.1 and in the program output. We will simultaneously fit both datasets to determine the basal kinetics, including the absorption parameters. The baseline concentration at the beginning of the intravenous experiment was 2.45 µg/L and prior to the extravascular administration it was 2.8 µg/L.

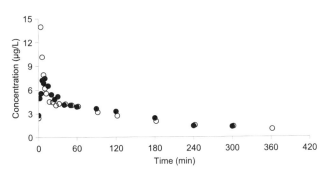

Figure 25.1 Observed plasma concentrations of drug A following a constant rate intravenous infusion (open circles) and pulmonary deposition (filled circles).

The model is specified in terms of a system of differential equations. The model parameters are Cl, V_c, Cl_d and V_t which correspond to plasma clearance, the volume of the central compartment, inter-compartmental diffusion and the volume of the peripheral compartment, respectively. *In* is the input function, which in this case is either the rapid intravenous infusion or pulmonary input. The equation for the central compartment

$$V_c \cdot \frac{dC}{dt} = In - Cl \cdot C - Cl_d \cdot C + Cl_d \cdot C_t \qquad (25:1)$$

The equation for the peripheral compartment is

$$V_t \cdot \frac{dC_t}{dt} = Cl_d \cdot C - Cl_d \cdot C_t \qquad (25:2)$$

The extravascular input function was successfully modeled with a classical first-order absorption model (Gibaldi and Perrier [1982]) as

$$In = AMG = K_a \cdot F \cdot D_1 \cdot e^{-K_a \cdot t} \tag{25:3}$$

AMG is the input function in the program code and D_1 is the extravascular dose. *In* enters the central compartment of the second (pulmonary) function according to Equation 25:4

$$V_c \cdot \frac{dC}{dt} = In - Cl \cdot C - Cl_d \cdot C + Cl_d \cdot C_t \tag{25:4}$$

The expression for the peripheral compartment in the extravascular model is written as Equation 25:2.

Initial parameter estimates

Initial parameter estimates were obtained according to Gorrod and Wahren [1993]. The bioavailability *F* is called BIO in the program code. *'F'* is the default name for the predicted function(s) values in *WinNonlin*.

Interpretation of results and conclusions

This exercise has shown how to simultaneously fit plasma data following intravenous and extravascular administration by means of a set of differential equations. Since the disposition parameters V_c, V_t, Cl_d and Cl are the same for both the intravenous and extravascular data, they could be estimated precisely, along with absorption parameters such as K_a and F.

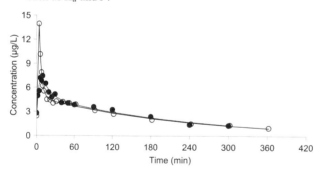

Figure 25.2 Observed (symbols) and predicted (lines) plasma concentrations of drug A. Filled circles represent data following the intravenous infusion and open circles the extravascular (first-order) input.

Analysis of the plasma concentration data for drug A after intravenous and pulmonary administration suggest that the correlations between the parameters were low and that parameter precision was high. The extravascularly administered dose was rapidly absorbed ($t_{1/2Ka}$ 12 min), though absorption was incomplete ($F = 65\%$).

You have learned to:
- ☐ simultaneously analyze the behavior of a compound after intravenous and pulmonary delivery
- ☐ write and implement a system of differential equations
- ☐ characterize the absorption kinetics after pulmonary administration

To fully characterize the kinetics of this compound one would need to study its behavior during extended exposure (e.g., multiple dosing).

Solution

```
TITLE 1
Rapid intra- and extravascular administration
MODEL
 COMM
  NFUN 2
  NDER 4
  NCON 3
  nparm 6
  PNAMES 'Vc','Cl','Cld','Vt','Ka', 'Bio'
END
    1: TEMP
    2: DOSE1=CON(1)
    3: DOSE2=CON(2)
    4: TINF=CON(3)
    5: T=X
    6: FINF = 0.0
    7: IF T LE TINF THEN
    8: FINF = DOSE2/TINF
    9: ELSE
   10: FINF = 0.0
   11: ENDIF
   12: END
   13: START
   14: Z(1)=2.45
   15: Z(2)=2.8
   16: Z(3)=2.45
   17: Z(4)=2.8
   18: END
   19: DIFF
   20: AMG = BIO*DOSE1*KA*DEXP(-KA*T)
   21: DZ(1)=(FINF - CL*Z(1) - CLD*Z(1) + CLD*Z(3))/VC
   22: DZ(2)=(AMG - CL*Z(2) - CLD*Z(2) + CLD*Z(4))/VC
   23: DZ(3)=(CLD*Z(1) - CLD*Z(3))/VT
   24: DZ(4)=(CLD*Z(2) - CLD*Z(4))/VT
   25: END
   26: FUNC 1
   27: F = Z(1)
   28: END
   29: FUNC 2
   30: F = Z(2)
   31: END
   32: EOM
DATE
TIME
NVARIABLES 3
FNUMBER 3
NPOINTS 100
XNUMBER 1
YNUMBER 2
```

```
CONSTANTS 1000,700,2
METHOD 2  'Gauss-Newton (Levenberg and Hartley)
REWEIGHT -1
ITERATIONS 50
INITIAL 82,.7,1.5,43,.5,.6
LOWER BOUNDS 0,0,0,0,0,0
UPPER BOUNDS 300,19,80,500,15,1
MISSING 'Missing'
DATA 'WINNLIN.DAT'
BEGIN
```

PARAMETER	ESTIMATE	STANDARD ERROR	CV%	UNIVARIATE C.I.	
VC	31.062014	2.628335	8.46	25.701526	36.422502
CL	1.060350	.033651	3.17	.991718	1.128981
CLD	6.235548	.240989	3.86	5.744051	6.727045
VT	179.654178	8.199722	4.56	162.930850	196.377505
KA	.082290	.007395	8.99	.067207	.097372
BIO	.646639	.027493	4.25	.590568	.702710

```
*** CORRELATION MATRIX OF THE ESTIMATES ***
PARAMETER  VC          CL          CLD         VT          KA          BIO
VC         1.00000
CL          .272140    1.00000
CLD         .119843    -.042943     1.00000
VT          .141374     .168575    -.165755    1.00000
KA          .489430    -.156956     .309871    -.150827    1.00000
BIO         .148527     .547714    -.004140     .343000    -.517534    1.00000

Condition_number=        1919.

   FUNCTION   1
```

X	OBSERVED Y	PREDICTED Y	RESIDUAL	WEIGHT	SE-PRED	STANDARDIZ RESIDUAL
.0000	2.450	2.450	.0000	.4082		
4.000	13.95	13.91	.3970E-01	.7189E-01	.4447	.1975
6.000	10.15	10.08	.6673E-01	.9917E-01	.2184	.1888
8.000	7.900	7.814	.8568E-01	.1280	.1944	.2766
10.00	6.200	6.462	-.2624	.1547	.1727	-.9230
12.00	5.550	5.650	-.1003	.1770	.1422	-.3625
17.00	4.500	4.740	-.2398	.2110	.9836E-01	-.8971
22.00	4.450	4.426	.2358E-01	.2259	.9746E-01	.0915
27.00	4.050	4.273	-.2232	.2340	.9913E-01	-.8868
32.00	4.250	4.164	.8597E-01	.2402	.9742E-01	.3458
42.00	4.200	3.977	.2228	.2514	.9050E-01	.9105
52.00	4.050	3.803	.2470	.2630	.8353E-01	1.024
62.00	3.900	3.637	.2632	.2750	.7750E-01	1.110
92.00	3.150	3.180	-.3045E-01	.3144	.6522E-01	-.1359
122.0	2.700	2.781	-.8139E-01	.3595	.5991E-01	-.3879
182.0	2.000	2.127	-.1272	.4701	.5951E-01	-.7016
242.0	1.500	1.627	-.1269	.6147	.6109E-01	-.8170
302.0	1.350	1.244	.1058	.8037	.6027E-01	.7956
362.0	1.000	.9516	.4840E-01	1.051	.5703E-01	.4239

```
CORRECTED SUM OF SQUARED OBSERVATIONS =  184.214
WEIGHTED CORRECTED SUM OF SQUARED OBSERVATIONS =  30.3628
SUM OF SQUARED RESIDUALS =            .440867
SUM OF WEIGHTED SQUARED RESIDUALS =  .118828
S =  .956066E-01 WITH    13 DEGREES OF FREEDOM
CORRELATION (OBSERVED,PREDICTED) =  .9988
```

FUNCTION   2

X	OBSERVED Y	PREDICTED Y	RESIDUAL	WEIGHT	SE-PRED	STANDARDIZ RESIDUAL
.0000	2.800	2.800	.0000	.3571		
2.000	4.950	5.163	-.2133	.1937	.1348	-.8051
4.000	5.550	6.310	-.7602	.1585	.1676	-2.689
6.000	7.200	6.776	.4244	.1476	.1639	1.421
8.000	6.850	6.869	-.1927E-01	.1456	.1503	-.0625
10.00	7.450	6.769	.6810	.1477	.1364	2.184
15.00	6.500	6.227	.2734	.1606	.1149	.8945
20.00	5.400	5.670	-.2698	.1764	.1108	-.9267
25.00	4.800	5.225	-.4245	.1914	.1123	-1.532
30.00	5.150	4.886	.2642	.2047	.1122	.9911
40.00	4.100	4.424	-.3238	.2260	.1048	-1.273
50.00	4.050	4.120	-.7043E-01	.2427	.9441E-01	-.2837
60.00	3.850	3.892	-.4197E-01	.2569	.8516E-01	-.1723
90.00	3.600	3.371	.2288	.2966	.6827E-01	.9934
120.0	3.250	2.945	.3046	.3395	.6192E-01	1.411
180.0	2.400	2.252	.1475	.4440	.6164E-01	.7913
240.0	1.400	1.723	-.3227	.5805	.6367E-01	-2.023
300.0	1.300	1.318	-.1752E-01	.7590	.6307E-01	-.1285

```
CORRECTED SUM OF SQUARED OBSERVATIONS =  58.4411
WEIGHTED CORRECTED SUM OF SQUARED OBSERVATIONS =  17.1331
SUM OF SQUARED RESIDUALS =            2.04815
SUM OF WEIGHTED SQUARED RESIDUALS =  .411903
S =  .185271     WITH    12 DEGREES OF FREEDOM
CORRELATION (OBSERVED,PREDICTED) =  .9824
```

```
TOTALS FOR ALL CURVES COMBINED
SUM OF SQUARED RESIDUALS =            2.48902
SUM OF WEIGHTED SQUARED RESIDUALS =  .530731
S =  .130845     WITH    31 DEGREES OF FREEDOM
```

```
AIC criteria =        -11.43947
SC  criteria =         -1.77396
```

PK26 - Toxicokinetics II

Objectives

◆ **To simultaneously fit data at three dose levels**

◆ **To modify a library model of differential equations to include several dose levels**

Problem specification

Synthetic plasma concentration-time data on *Quitpain*® were obtained from a multiple dose study in the rat for a compound A which was known from *in vitro* experiments to display Michaelis-Menten behavior. The concentration-time data for the three dose levels are plotted in Figure 26.1 and given in Table 26.1.

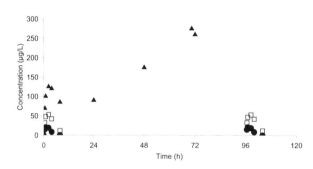

Figure 26.1 Observed plasma concentrations of drug A after the first dose and at steady-state for the two lowest doses and after the third administration of the highest dose. Lowest dose (●), intermediate dose (□), high dose (▲) Note how the highest dose drifts off between 24 and 72 h. The high dose animal was heavily sedated and, prior to the end of the study, two plasma sample were drawn at 70.4 and 72 h.

Table 26.1   Concentration data (µg/L) at different times after administration of 200, 400 and 800 mg per day of drug A

	Dose (mg)		
Time (h)	200	400	800
0	0	0	0
0.6	15	32	69
1	20	48	100
2.5	19	53	125
4	8.6	42	120
8	0.03	12	85
24	-	-	90
32	-	-	-
48	-	-	175
56	-	-	-
70.4	-	-	275
72	-	-	260
80	-	-	-
96	-	-	-
96.6	15	33	-
97	21	47	-
98.5	19	53	-
100	8.6	42	-
104	0.03	12	-

Notes for the reader

Initial parameter estimates were taken from previous studies in the rat. This exercise is aimed to demonstrate how the user can modify a *WinNonlin* library file to include multiple dose data at several (3) dose levels. Data also contain *missing values* shown as '-' in the program output. This exercise contains a limited analysis which shows the reader how to implement a model of this kind into *WinNonlin*. Observe that we simultaneously fit three different datasets that were originally generated by means of their respective models and error was then added to the data. This was done in order to simulate a typical toxicokinetic dataset. Toxicokinetic data analysis after multiple dose administration, capacity-limited elimination, and pooled data from several animals and dose levels, are all commonly encountered situations in the pharmaceutical industry.

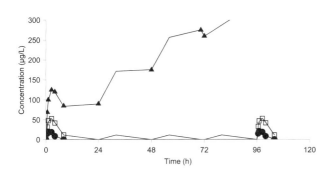

Figure 26.2 Observed (symbols) and predicted (lines) plasma concentrations of drug A after the first dose and at steady-state for the two lowest doses ana after the third administration of the highest dose. Lowest dose (●), intermediate dose (□), high dose (▲) Note how the highest dose drifts off between 24 and 72 h. The animal was heavily sedated and, prior to the end of the experiment, two plasma sample were drawn at 70.4 and 72 h.

We have not commented on the fits because data are synthetic and no comparisons are made between models. We refer readers to the section 3.6 on nonlinear models and section 5.4 on initial estimates for further information. Other references are Gibaldi and Perrier [1982] and Levy [1983].

Solution

```
model
 COMM
nder   3
NFUN   3
nparm 4
nsec  1
pnames 'volume', 'k01', 'vm', 'km'
snames 'k01-hl'
end
    1: START
    2: Z(1) = 0
    3: Z(2) = 0
    4: Z(3) = 0
    5: END
    6: DIFF
    7: J = 1
    8: NDOSE = CON(1)
    9: DO I = 1 TO NDOSE
   10: J = J + 2
   11: IF X <= CON(J) THEN GOTO RED
   12: ENDIF
```

```
13: NEXT
14: RED:
15: NDOSE = I-1
16: J=1
17: IF X GT 0 THEN
18: DO I = 1 TO NDOSE
19: J = J + 2
20: T = X - CON(J)
21: DOSE1 = CON(J-1)
22: DOSE2 = 2*CON(J-1)
23: DOSE3 = 4*CON(J-1)
24: IF X>0 AND T=0 THEN
25: Z(1) = Z(1) + DOSE1 / VOLUME
26: Z(2) = Z(2) + DOSE2 / VOLUME
27: Z(3) = Z(3) + DOSE3 / VOLUME
28: ENDIF
29: NEXT
30: ELSE
31: DOSE1 = CON(2)
32: Z(1) = 0
33: DOSE2 = 2*CON(2)
34: Z(2) = 0
35: DOSE3 = 4*CON(2)
36: Z(3) = 0
37: T = 0
38: ENDIF
39: DZ(1) = K01 * (DOSE1/VOLUME) * EXP(-K01*T) - VM * Z(1) / (KM+Z(1))
40: DZ(2) = K01 * (DOSE2/VOLUME) * EXP(-K01*T) - VM * Z(2) / (KM+Z(2))
41: DZ(3) = K01 * (DOSE3/VOLUME) * EXP(-K01*T) - VM * Z(3) / (KM+Z(3))
42: END
43: FUNC 1
44: F = Z(1)
45: END
46: FUNC 2
47: F = Z(2)
48: END
49: FUNC 3
50: F = Z(3)
51: END
52: SECO
53: S(1) = -DLOG(.5) / K01
54: END
55: EOM
NVARIABLES 3
FNUMBER 3
NPOINTS 100
XNUMBER 1
YNUMBER 2
NCONSTANTS 21
CONSTANTS 10,200,0,200,8,200, &
          24,200,32,200,48,200,56, &
          200,72,200,80,200,96,200,104
METHOD 2   'Gauss-Newton (Levenberg and Hartley)
ITERATIONS 50
INITIAL 5.2,1.1,45,5.3
LOWER BOUNDS 0,0,0,0
UPPER BOUNDS 10,10,100,100
BEGIN
```

PARAMETER	ESTIMATE	STANDARD ERROR	CV%	UNIVARIATE C.I.	
VOLUME	5.032365	.044857	.89	4.940480	5.124250

```
K01        1.104263         .021670        1.96        1.059874        1.148651

VM         9.883385         .123473        1.25        9.630464       10.136306

KM         4.736815         .448709        9.47        3.817683        5.655947

*** CORRELATION MATRIX OF THE ESTIMATES ***
PARAMETER   VOLUME         K01            VM            KM
VOLUME     1.00000
K01         .777930       1.00000
VM         -.974949       -.78682        1.00000
KM          .064053       -.09359         .13897        1.00000

 Condition_number=         114.0

  FUNCTION    1

  X          OBSERVED     PREDICTED    RESIDUAL      WEIGHT      SE-PRED STANDARDI
             Y            Y                                              RESIDUAL
   .0000       .0000        .0000         .0000       1.000
   .6000      15.00        15.66         -.6553       1.000       .1818      -.5284
  1.000       20.00        19.84          .1575       1.000       .2046       .1274
  2.500       19.00        18.46          .5422       1.000       .3013       .4457
  4.000        8.600        9.475        -.8751       1.000       .4689      -.7530
  8.000        .3000E-01    .2583E-01     .4172E-02   1.000       .1533E-01   .0033
 24.00        .            .9386E-06     .            .0000      -.6400+152   .
 32.00        .            .2582E-01     .            .0000      -.6400+152   .
 48.00        .            .9433E-06     .            .0000      -.6400+152   .
 56.00        .            .2582E-01     .            .0000      -.6400+152   .
 70.40        .            .5577E-05     .            .0000      -.6400+152   .
 72.00        .            .9474E-06     .            .0000      -.6400+152   .
 80.00        .            .2583E-01     .            .0000      -.6400+152   .
 96.00        .            .9433E-06     .            .0000      -.6400+152   .
 96.60       15.00        15.65         -.6537       1.000       .1817      -.5272
 97.00       21.00        19.84         1.159        1.000       .2046       .9374
 98.50       19.00        18.46          .5435       1.000       .3013       .4468
100.0         8.600        9.474        -.8740       1.000       .4689      -.7520
104.0         .3000E-01    .2582E-01     .4179E-02   1.000       .1533E-01   .0333

CORRECTED SUM OF SQUARED OBSERVATIONS =   1321.89
WEIGHTED CORRECTED SUM OF SQUARED OBSERVATIONS =   711.687
SUM OF SQUARED RESIDUALS =               4.34397
SUM OF WEIGHTED SQUARED RESIDUALS =    4.34397
S =  .787761      WITH      7 DEGREES OF FREEDOM
CORRELATION (OBSERVED,PREDICTED) =  .7317

  FUNCTION    2

  X          OBSERVED     PREDICTED    RESIDUAL      WEIGHT      SE-PRED STANDARDI
             Y            Y                                              RESIDUAL
   .0000       .0000        .0000         .0000       1.000
   .6000      32.00        34.10         -2.096       1.000       .3395      -1.737
  1.000       48.00        45.19         2.807        1.000       .3434       2.329
  2.500       53.00        52.93          .7174E-01   1.000       .2710       .0586
  4.000       42.00        43.49         -1.488       1.000       .3075      -1.225
  8.000       12.00        11.38          .6185       1.000       .5861       .5583
 24.00        .            .6963E-04     .            .0000      -.6400+152   .
 32.00        .           11.38          .            .0000      -.6400+152   .
 48.00        .            .7257E-04     .            .0000      -.6400+152   .
 56.00        .           11.38          .            .0000      -.6400+152   .
```

```
70.40         .         .2140E-02      .       .0000    -.6400+152      .
72.00         .         .7525E-04      .       .0000    -.6400+152      .
80.00         .        11.38           .       .0000    -.6400+152      .
96.00         .         .7258E-04      .       .0000    -.6400+152      .
96.60       33.00      34.09        -1.092     1.000     .3395       -.9053
97.00       47.00      45.19         1.811     1.000     .3434       1.502
98.50       53.00      52.92          .7507E-01 1.000    .2711        .0613
100.0       42.00      43.49        -1.485     1.000     .3076      -1.222
104.0       12.00      11.38          .6210    1.000     .5862        .5606
```

```
CORRECTED SUM OF SQUARED OBSERVATIONS =  8698.11
WEIGHTED CORRECTED SUM OF SQUARED OBSERVATIONS =  3344.00
SUM OF SQUARED RESIDUALS =            21.9452
SUM OF WEIGHTED SQUARED RESIDUALS =   21.9452
S =  1.77060    WITH    7 DEGREES OF FREEDOM
CORRELATION (OBSERVED,PREDICTED) =  .6180
```

```
    FUNCTION   3
```

X	OBSERVED Y	PREDICTED Y	RESIDUAL	WEIGHT	SE-PRED	STANDARDI RESIDUAL
.6000	69.00	72.02	-3.023	1.000	.6837	-2.879
1.000	100.0	97.54	2.459	1.000	.6988	2.364
2.500	125.0	125.9	-.9304	1.000	.6959	-.8927
4.000	120.0	119.8	.2125	1.000	.8118	.2226
8.000	85.00	83.92	1.084	1.000	.4966	.9425
24.00	90.00	89.66	.3426	1.000	.3348	.2836
32.00	.	171.5	.	.0000	-.6400+152	.
48.00	175.0	175.4	-.3856	1.000	.5331	-.3400
56.00	.	256.6	.	.0000	-.6400+152	.
70.40	275.0	275.2	-.2403	1.000	.7872	-.2464
72.00	260.0	259.7	.2979	1.000	.8122	.3121
96.00	.	343.2	.	.0000	-.6400+152	.
96.60	.	414.4	.	.0000	-.6400+152	.
97.00	.	439.7	.	.0000	-.6400+152	.
98.50	.	467.7	.	.0000	-.6400+152	.
100.0	.	461.2	.	.0000	-.6400+152	.
104.0	.	423.9	.	.0000	-.6400+152	.

```
CORRECTED SUM OF SQUARED OBSERVATIONS =  140217.
WEIGHTED CORRECTED SUM OF SQUARED OBSERVATIONS =  65220.9
SUM OF SQUARED RESIDUALS =            17.6876
SUM OF WEIGHTED SQUARED RESIDUALS =   17.6876
S =  1.71696    WITH    6 DEGREES OF FREEDOM
CORRELATION (OBSERVED,PREDICTED) =  .6819
```

```
TOTALS FOR ALL CURVES COMBINED
SUM OF SQUARED RESIDUALS =            43.9767
SUM OF WEIGHTED SQUARED RESIDUALS =   43.9767
S =  1.25324    WITH   28 DEGREES OF FREEDOM
```

```
AIC criteria =        189.61573
SC  criteria =        197.10053
```

PARAMETER	ESTIMATE	STANDARD ERROR	CV%
K01-HL	.627701	.012306	1.96

PK27 - Toxicokinetics III

Objectives
◆ **To analyze a 12 month, repeated dose safety study in dogs**
◆ **To analyze plasma data by means of the non-compartmental approach**
◆ **To relate exposure to dose, gender and time**
◆ **To discuss aspects of exploratory data analysis and experimental design**

Problem specification and exploratory data analysis

The aim of this exercise is to organize and explore data from a toxicokinetic study so that an appropriate study report can be prepared. The analysis we do is of an exploratory nature and is aimed to lay out the concentration-time data by means of various graphs. Studying a new anti-inflammatory drug *Unflame®*, you and your colleagues obtained the following plasma concentrations from a 12 month, once daily, oral, repeated dose study in beagle dogs. Three dose groups of 2, 4 and 6 mg/kg including both sexes were used. Data were collected on Day 0 (first day of dosing and denoted Period = 0d in Figures 27.1 and 27.2) and 14 (Period = 14d), and after 6 (Period = 6m) and 12 (Period = 12m) months of dosing. Blood samples were drawn at 0 and 30 min, 1, 2, 4, 8 and 24 h post- dosing.

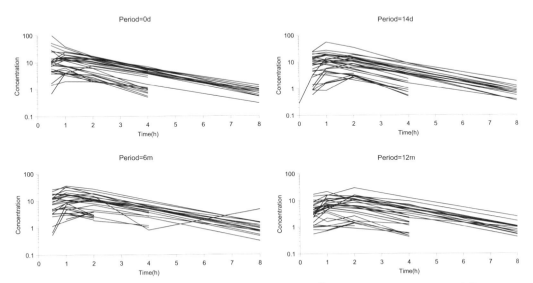

Figure 27.1 *Dose-normalized concentration-time data of Unflame® after the first dose Day 0 (upper left), on Day 14 (upper right), and at 6 months (bottom left) and 1 year (bottom right) in male and female dogs.*

The data file, *TK12dog.WDO* stored on the enclosed diskette, includes data for time, concentration, subject, gender, period and dose.

Figure 27.1 is obtained by means of the plotting function in *WinNonlin*. Click on the GRAPH-option, set TIME and CONCENTRATION as the X and Y variables, respectively. Go to the sorting option on the same page (Define Graphing Columns) and move PERIOD (from variable collection) to sort variables and SUBJECT ID to group variables. Then click ok. You will then have data presented as four spaghetti plots sorted by means of PERIOD as Day 0, Day 14, 6-months and 1 year. The mean for each dose group is plotted in Figure 27.2.

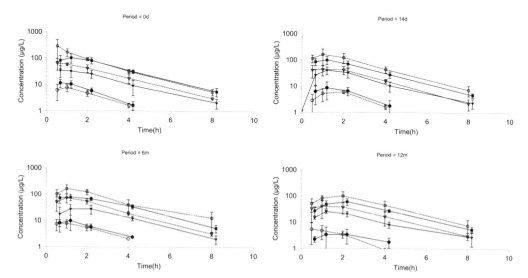

Figure 27.2 *Mean ± SD plasma concentration-time data of each dose group and sex (male dogs solid lines and filled symbols, and females open symbols) after the first dose Day 0 (upper left), on Day 14 (upper right), and at 6 months (bottom left) and 1 year(bottom right).*

You should graphically analyze the kinetics and exposure to *Unflame*® prior to non-compartmental analysis. Summarize the data with appropriate spaghetti plots, mean plots (including SD), dose-normalized variables etc., and perform the descriptive statistics you find necessary for the report.

Given the exposure data, relate your findings on half-life, C_{max}, AUC etc., to dose, gender and time. Try to fit a reduced dataset that includes repeated dosing at several dose levels.

Simulate the appropriate dosing regimen (e.g., *bid*) by means of *WinNonlin*. Based on the analysis, design an appropriate oral dosing regimen for a cancer study in the dog.

Initial parameter estimates

No initial estimates were derived for this exercise, since it was a non-compartmental analysis.

Interpretation of results and conclusions

The exposure to *Unflame*® was assessed by means of C_{max} and *AUC*, which were based on total plasma concentrations.

Dose: Exposure to the compound increased disproportionately with dose. Dose-normalized C_{max} and *AUC* increased disproportionately with dose, although the terminal half-life remained the same throughout the studied dose range. This might be explained by the fact that the bioavailability increased with dose. Plots of *AUC/dose* and C_{max} *versus* dose are shown in Figures 27.3 and 27.4, respectively, for periods 3 (6 months) and 4 (12 months).

Gender: The mean exposure to the compound was 50 to 100% higher in female animals than in males at the 4 and 6 mg/kg dose levels. No differences were seen at the 2 mg/kg dose level. Data for *AUC/dose* and C_{max} for the males and females are shown in Figures 27.3 and 27.4, respectively. This difference in exposure could be partly explained by the lower unbound clearance in female dogs compared to males.

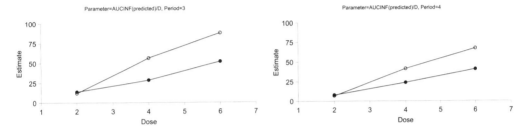

Figure 27.3 *Mean dose-normalized AUC from periods 3 (6 months, left) and 4 (1 year right) in male (●) and female (○) dogs.*

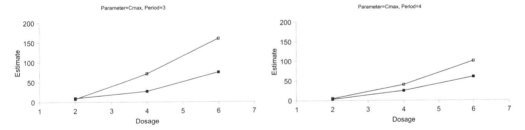

Figure 27.4. *Mean C_{max} from period 3 (6 months, left) and 4 (1 year, right) in male (■) and female (□) dogs.*

Time: Plots of *AUC/dose* and C_{max} *versus* period (time), and of mean residence time *MRT* and terminal half-life $t_{1/2z}$ *versus* period are shown in Figures 27.5 and 27.6, respectively. Exposure to the compound decreased over time. The ratio of $AUC_{12month}$ to AUC_{day14} (i.e., period 4 to period 2) was about 60%. Comparisons with the Day 1 *AUC* should be done cautiously, because the Day 1 area is most probably underestimated. The fact that both C_{max} and *AUC* decreased over time while terminal half-life remained the same indicates that a decrease in bioavailability occurred with time rather than an increase in metabolic capacity. *MRT* (*AUMC/AUC*) increased slightly over time suggesting that absorption was slightly delayed.

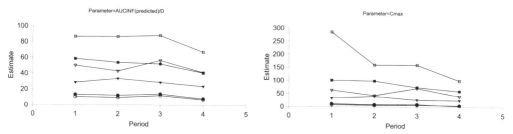

Figure 27.5. *Mean dose-normalized AUC (left) and C_{max} (right), for all dose groups and both sexes. Males are shown by closed symbols, females by open symbols. Periods 1-4 represent Day 0, Day 14, 6 months and 12 months, respectively.*

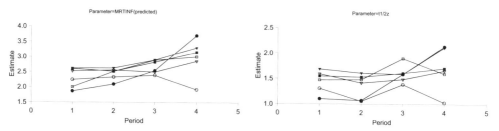

Figure 27.6 *Mean MRT- and $t_{1/2z}$-time data of each dose group and sex (males closed symbols and females open symbols). Periods 1-4 represent Day 0, Day 14, 6 months and 12 months, respectively.*

All plots, descriptive statistics and data manipulations should be clearly documented with output listings, the data history trail and intermediate and final tables in toxicokinetic reporting. Printouts should be dated and signed by the analyst. This will then be appropriate as the *GLP (Good Laboratory Practice)* documentation for such an analysis. Store the documentation from the analysis in the report appendices.

Design: Based on the present results, it is concluded that all animals were exposed to the parent compound at all dose levels. Since the dog has a metabolic profile that resembles that in humans it was decided to use dogs in a two year cancer study rather than rats or mice which deviate substantially from humans and primates in their metabolic patterns and oral bioavailability (less than 2%).

The same dose range (2 to 6 mg/kg) and route of administration (gavage) were selected for toxicological and practical reasons. It was decided to split the daily dose into two equal doses separated by an 8 h dosing interval and a 16 h washout period to establish 24 h exposure above 0.1 to 1 µg/L in the intermediate and high dose groups. It was decided to not dose adjust for the change (decrease) in bioavailability since the exposure to parent compound at the intermediate and high doses was adequate throughout the 12 month study period.

PK28 – Allometric scaling - Elementary Dedrick plot

Objectives

◆ To simultaneously fit an allometric model to data from mouse, rat, and man

◆ To apply the allometric constants *a, b, c,* and *d* as parameters

◆ To model *Cl* and body weight by means of *a* and *b*

◆ To model *V* and body weight by means of *c* and *d*

◆ To model *dose, AUC,* and body weight

◆ To construct the elementary Dedrick plot

Problem specification

The aim of this exercise is to analyze data from three different species simultaneously in order to obtain the allometric constants and exponents. Plasma concentration-time data were obtained for methadone from mouse, rat, and man after an intravenous bolus dose of 25, 500, or 100,000 µg, respectively (Figure 28.1). We assume instantaneous distribution upon administration. The body weights *BW* were 23 g (mouse), 250 g (rat), and 70 kg (man), covering a 3000-fold range.

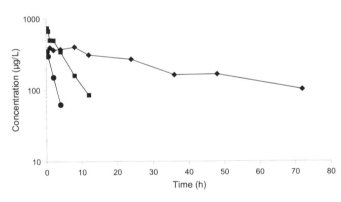

Figure 28.1 Observed plasma concentrations in mouse (●), rat (■), and man (◆) following an intravenous bolus dose of methadone.

Table 28.1   Concentration-time data for mouse, rat and man

Time (h)	Conc (µg/L)	Species	Time (h)	Conc (µg/L)	Species	Time (h)	Conc (µg/L)	Species
0.167	347.	mouse	1	499.	rat	4	371.	man
0.5	298.		2	496.		8	399.	
2	151.		4	342.		12	309.	
4	62.0		8	159.		24	268.	
0.167	742.	rat	12	85.2		36	162.	
0.33	679.		1	393.	man	48	166.	
0.5	665.		2	363.		72	102.	

We will model the disposition of methadone by means of the following allometric expressions for clearance *Cl* and volume of distribution *V*

$$Cl_i = a \cdot BW_i^b \tag{28:1}$$

$$V_i = c \cdot BW_i^d \tag{28:2}$$

where i represents mouse, rat, or man, as appropriate. As can be seen, Cl and V are not estimated *per se* but rather the allometric parameters a, b, c, and d are determined. The generic disposition model for the different species is specified as

$$C_i = \frac{Dose_i}{V_i} e^{-\frac{Cl_i}{V_i} \cdot t} \tag{28:3}$$

where Cl_i and V_i are specified as above. Cl_i and AUC_i are estimated as secondary parameters for each species, together with the corresponding half-lives $t_{1/2}$.

Table 28.2   Comparison of parameters

Species	Cl	V	AUC	$t_{1/2}$*
Mouse	$a\,23^b$	$c23^d$	$\dfrac{Dose_i}{a \cdot 23^b}$	$\dfrac{c \cdot 23^d}{a \cdot 23^b}$
Rat	$a\,250^b$	$c\,250^d$	$\dfrac{Dose_i}{a \cdot 250^b}$	$\dfrac{c \cdot 250^d}{a \cdot 250^b}$
Man	$a\,70000^b$	$c70000^d$	$\dfrac{Dose_i}{a \cdot 70000^b}$	$\dfrac{c \cdot 70000^d}{a \cdot 70000^b}$

*All ratios of $t_{1/2}$ should be multiplied by $\ln(2)$ to correct for the exponential term.

Equations 28:1 and 28:2 are inserted into Equation 28:3 to give Equation 28:4

$$C_i = \frac{Dose_i}{V_i} e^{-\frac{Cl_i}{V_i} \cdot t} = \frac{Dose_i}{c \cdot BW_i^d} \cdot e^{-\frac{a \cdot BW_i^b}{c \cdot BW_i^d} \cdot t} \tag{28:4}$$

This is the generic form of the model that will be fit to the three data sets. In the output below, we have based the models on kilogram body weight rather than gram body weight. If one applies gramunits, the a and c parameters are 1/1000 of the values used for the corresponding a and c parameters when kilogram body weights are used. Initial estimates of Cl and V are derived by means of non-compartmental analysis for each species. One obtains initial estimates of a and c by means of plotting the apparent Cl and V *versus* body weight. When body weight is set to 1 (unity) the regressed Cl value corresponds to a. A similar approach is used for obtaining c in the case of V.

    In the program output, the data for mouse, rat and man correspond to function 1, 2 and 3, respectively.

Interpretation of results and conclusions

This exercise has shown how to simultaneously fit an allometric model to concentration-time data obtained from three different species (mouse, rat, and man). The terminal half-lives were found to be 1.5, 3.9, and 35 h for mouse, rat, and man, respectively. Based on the plasma-to-blood concentration ratio (unity) and the hepatic blood flow for each animal, this drug is classified as a low-extraction drug with a hepatic extraction of less than 10%. Figure 28.2 contains observed and predicted concentration-time data from each species.

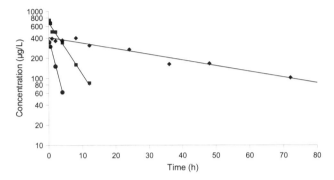

Figure 28.2 *Observed and predicted concentration-time data following rapid intravenous injection to the mouse (●), rat (■), and man (◆).*

The plot of dose- and body weight normalized plasma concentration $C/(Dose/BW)$ *versus* t/BW^{1-b} is a straight line and allows the superposition of the curves (the elementary Dedrick plot, Figure 28.3). The abscissa represents the *kallynochrons*, which is defined as a unit of pharmacokinetic time during which the different species clear the same volume of plasma per kilogram of body weight. The area under the curve is equal to $1/a,$ where a is the allometric coefficient of clearance. Since $a = 0.319,$ *AUC* becomes 3.13, which is consistent with 3.04 calculated from the intercept of the ordinate ($C_0 = 0.35$) divided by the slope of the curve ($K \sim 0.12$).

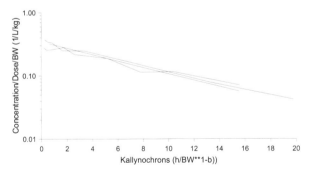

Figure 28.3 *The elementary Dedrick plot of dose and body weight normalized plasma concentrations versus body weight normalized time (kallynochrons).*

The absolute values of clearance ranged from 30 mL/h/20 g (mouse) to 4.78 L/h/70 kg (man). This ratio is about 160 in comparison to the weight ratio which is close to 3000. The relative values of clearance ranged from 1.5 L/h/kg (mouse) to 0.068 L/h/kg (man). This is a 22-fold range. Even though man weighs about 3000 times more than the rat, man's clearance increases only 160 times. Based on a per kg basis, the mouse clears the drug 22 times more efficiently than man. Some limitations are

- Potential model misspecification, e.g., we assume a one-compartment model when a two-compartment may be visible with another design. Methadone is known in other studies to display multiple compartment kinetics

- Only one dose level was applied for each species

- Only single dose administration was applied, whereas multiple dosing may reveal induction or inhibition patterns

Nevertheless, the data still serve as a good example of allometric analysis and of how to write and implement such a system into *WinNonlin*. Data were taken from Gabrielsson and Weiner [1997].

Solution

```
MODEL
 COMM
NFUN   3
NPARM 4
NSEC   9
NCON   6
PNAMES 'A', 'B', 'C', 'D'
SNAMES 'AUCM','AUCR','AUCH','T12M','T12R','T12H','CLM','CLR','CLH'
END
    1: TEMP
    2: BWM=CON(1)
    3: BWR=CON(2)
    4: BWH=CON(3)
    5: DM=CON(4)
    6: DR=CON(5)
    7: DH=CON(6)
    8: T=X
    9: A=P(1)
   10: B=P(2)
   11: C=P(3)
   12: D=P(4)
   13: CLM = A*(BWM**B)
   14: CLR = A*(BWR**B)
   15: CLH = A*(BWH**B)
   16: VM = C*(BWM**D)
   17: VR = C*(BWR**D)
   18: VH = C*(BWH**D)
   19: END
   20: FUNC 1
   21: F = (DM/VM)*DEXP(- (CLM/VM)*T)
   22: END
   23: FUNC 2
   24: F = (DR/VR)*DEXP(- (CLR/VR)*T)
   25: END
   26: FUNC 3
   27: F = (DH/VH)*DEXP(- (CLH/VH)*T)
   28: END
   29: SECO
   30: S(1)=(DM/CLM)
   31: S(2)=(DR/CLR)
```

```
    32: S(3)=(DH/CLH)
    33: S(4)=0.693*VM/CLM
    34: S(5)=0.693*VR/CLR
    35: S(6)=0.693*VH/CLH
    36: S(7)= CLM
    37: S(8)= CLR
    38: S(9)= CLH
    39: END
    40: EOM
NVARIABLES 3
FNUMBER 3
NPOINTS 100
XNUMBER 1
YNUMBER 2
CONSTANTS .023,.25,70,25,500,100000
METHOD 2  'Gauss-Newton (Levenberg and Hartley)
REWEIGHT -2
ITERATIONS 50
INITIAL .5,.8,3.5,1.5
LOWER BOUNDS .0001,.01,.1,.01
UPPER BOUNDS 1,3,10,3
BEGIN
```

PARAMETER	ESTIMATE	STANDARD ERROR	CV%	UNIVARIATE C.I.	
A	.318955	.007665	2.40	.302783	.335128
B	.636769	.007740	1.22	.620438	.653100
C	3.067420	.082130	2.68	2.894143	3.240698
D	1.030688	.008200	.80	1.013387	1.047988

```
*** CORRELATION MATRIX OF THE ESTIMATES ***
```

PARAMETER	A	B	C	D
A	1.00000			
B	.159935	1.00000		
C	.224206	-.199352	1.00000	
D	-.233292	.968911E-01	-.177460	1.00000

```
    FUNCTION   1
```

X	OBSERVED Y	PREDICTED Y	RESIDUAL	WEIGHT	SE-PRED	STANDARDIZED RESIDUAL
.1670	347.0	368.5	-21.46	.7313E-05	15.51	-.7567
.5000	298.0	316.2	-18.18	.9944E-05	11.98	-.7272
2.000	151.0	158.7	-7.702	.3968E-04	5.292	-.6004
4.000	62.00	63.31	-1.307	.2512E-03	3.848	-.3307

```
CORRECTED SUM OF SQUARED OBSERVATIONS = 51817.0
WEIGHTED CORRECTED SUM OF SQUARED OBSERVATIONS = 1.25635
SUM OF SQUARED RESIDUALS =          851.797
SUM OF WEIGHTED SQUARED RESIDUALS = .943456E-02
S = .000000     WITH      0 DEGREES OF FREEDOM
CORRELATION (OBSERVED,PREDICTED) = 1.000
```

```
    FUNCTION   2
```

X	OBSERVED Y	PREDICTED Y	RESIDUAL	WEIGHT	SE-PRED	STANDARDIZED RESIDUAL
.1670	742.0	660.3	81.74	.2288E-05	19.99	1.508
.3300	679.0	641.2	37.78	.2426E-05	19.02	.7156

.5000	665.0	621.9	43.06	.2579E-05	18.05	.8385
1.000	499.0	568.5	-69.55	.3086E-05	15.52	-1.471
2.000	496.0	475.1	20.88	.4422E-05	11.63	.5233
4.000	342.0	331.8	10.21	.9076E-05	7.444	.3639
8.000	159.0	161.8	-2.811	.3824E-04	5.349	-.2148
12.00	85.20	78.91	6.288	.1611E-03	4.114	1.139

CORRECTED SUM OF SQUARED OBSERVATIONS = 417307.
 WEIGHTED CORRECTED SUM OF SQUARED OBSERVATIONS = 4.06969
 SUM OF SQUARED RESIDUALS = 15388.4
 SUM OF WEIGHTED SQUARED RESIDUALS = .480035E-01
 S = .109548   WITH   4 DEGREES OF FREEDOM
 CORRELATION (OBSERVED,PREDICTED) = .9856

    FUNCTION   3

X	OBSERVED Y	PREDICTED Y	RESIDUAL	WEIGHT	SE-PRED	STANDARDIZED RESIDUAL
1.000	393.0	400.9	-7.900	.6276E-05	15.66	-.2534
2.000	363.0	393.2	-30.16	.6524E-05	15.04	-.9811
4.000	371.0	378.1	-7.115	.7051E-05	13.88	-.2384
8.000	399.0	349.7	49.26	.8236E-05	11.86	1.756
12.00	309.0	323.5	-14.49	.9621E-05	10.22	-.5517
24.00	268.0	256.0	12.02	.1533E-04	7.298	.5697
36.00	162.0	202.6	-40.56	.2444E-04	6.524	-2.468
48.00	166.0	160.3	5.709	.3894E-04	6.557	.4612
72.00	102.0	100.4	1.628	.9891E-04	6.498	.2752

CORRECTED SUM OF SQUARED OBSERVATIONS = 101670.
 WEIGHTED CORRECTED SUM OF SQUARED OBSERVATIONS = 2.05076
 SUM OF SQUARED RESIDUALS = 5484.26
 SUM OF WEIGHTED SQUARED RESIDUALS = .726400E-01
 S = .120532   WITH   5 DEGREES OF FREEDOM
 CORRELATION (OBSERVED,PREDICTED) = .9732

 TOTALS FOR ALL CURVES COMBINED
 SUM OF SQUARED RESIDUALS = 21724.4
 SUM OF WEIGHTED SQUARED RESIDUALS = .130078
 S = .874737E-01 WITH   17 DEGREES OF FREEDOM

AIC criteria =   -34.83204
SC criteria =   -30.65395

PARAMETER	ESTIMATE	STANDARD ERROR	CV%
AUCM	865.786849	30.073005	3.47
AUCR	3789.773144	93.539810	2.47
AUCH	20958.919544	915.282916	4.37
T12M	1.508102	.068264	4.53
T12R	3.860237	.122642	3.18
T12H	35.529983	2.173583	6.12
CLM	.028875	.001002	3.47
CLR	.131934	.003259	2.47
CLH	4.771238	.208795	4.38

PK29 – Allometric scaling - Complex Dedrick plot

Objectives

◆ To simultaneously fit data from mouse, rat, monkey, dog, and man

◆ To apply multi-compartment allometric scaling

◆ To apply the allometric constants *a, b, c, d, e,* and *g* as parameters

◆ To model *Cl* and body weight by means of *a* and *b*

◆ To model V_c and body weight by means of *c* and *d*

◆ To model V_t and body weight by means of *e* and *d*

◆ To model *dose, AUC* and *body weight*

◆ To construct the complex Dedrick plot

Problem specification

Plasma concentration-time data were simulated for the mouse, rat, monkey, dog and man after an intravenous bolus dose of 10, 125, 200, 6000, or 12000 μg, respectively, of a new compound (Figure 29.1). The body weights were 20 g (mouse), 250 g (rat), 3.5 kg (monkey), 14 kg (dog), and 70 kg (man), covering a 3000-fold weight range.

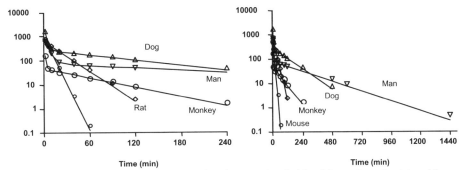

Figure 29.1 *Data for each individual species plotted against time. Left hand figure shows time interval 0-240 min and right hand figure 0-1440 min. Note the large interspecies differences in the plasma kinetics.*

When the volume of distribution *V* is not proportional to body weight *BW* (d > 1 see Equations 29:2 and 29:4), the curves can no longer be superimposed in an elementary Dedrick plot (see PK28). Rather, plasma concentration *C* is divided by the ratio *Dose/BW^d* and time *t* is divided by BW^{d-b}. The curve generated of $C/Dose/BW^d$ (ordinate) *versus* t/BW^{d-b} (abscissa) is called the complex Dedrick plot. The *X* axis represents the *apolysichron* (equal to t/BW^{d-b}), and is defined as the unit of pharmacokinetic time during which animal species eliminate the same fraction of drug and clear the same volume of plasma per BW^d of body weight. In chronological time, the apolysichron is equivalent to BW^{d-b} units. The area under the curve in the complex

Dedrick plot is still proportional to $1/a$, where a is the allometric coefficient of clearance. We will model the disposition of the compound using the following allometric expressions

$$Cl_i = a \cdot BW_i^b \tag{29:1}$$

$$V_i = c \cdot BW_i^d \tag{29:2}$$

$$Cl_{di} = g \cdot BW_i^b \tag{29:3}$$

$$V_{ti} = e \cdot BW_i^d \tag{29:4}$$

The i represents mouse, rat, monkey, dog, or man, as appropriate. Note the use of an intercompartmental distribution parameter Cl_d, scaled with parameters g and b. The extravascular tissue volume V_t is scaled with parameters e and d. The reason is that the intercompartmental distribution parameter is similar to clearance and blood flow with respect to the exponent b, whereas the coefficient g is assumed to be different for different parameters. The same reasoning is assumed to be valid for the extravascular tissue volume which scales like a volume term with exponent d similar to the central volume, although e is a new coefficient.

This model can either be expressed as a system of differential equations or as sums of exponentials. Drug concentration in the central compartment is given by

$$V_c \frac{dC}{dt} = In - Cl \cdot C - Cl_d \cdot C + Cl_d \cdot C_t \tag{29:5}$$

where In is the bolus input function. The kinetics of drug in the peripheral compartment

$$V_t \frac{dC_t}{dt} = Cl_d \cdot C - Cl_d \cdot C_t \tag{29:6}$$

The inter-compartmental diffusion parameter, which is equal in both directions, and the volume of the central compartment can be used for calculations of the micro-constants k_{12} and k_{21}

$$k_{12} = \frac{Cl_d}{V_c} \tag{29:7}$$

$$k_{21} = \frac{Cl_d}{V_t} \tag{29:8}$$

k_{12} and k_{21} are the (fractional) inter-compartmental rate constants which determine the rate of drug movement from the central to the peripheral compartment and from the peripheral to the central compartment, respectively. The k_{10} is

$$k_{10} = \frac{Cl}{V_c} \tag{29:9}$$

and the initial slope factor α

$$\alpha = \frac{k_{21} \cdot k_{10}}{\beta} \tag{29:10}$$

The terminal slope factor β is a function of the micro-constants

$$\beta = \frac{1}{2}\left[k_{12} + k_{21} + k_{10} - \sqrt{\left(k_{12} + k_{21} + k_{10}\right)^2 - 4k_{21} \cdot k_{10}}\,\right] \tag{29:11}$$

In this exercise we will fit the plasma concentrations by means of an intravenous bolus model, defined as a sum of exponentials

$$C = \frac{D_{iv}}{V_c} \cdot \left\{ \frac{\left(k_{21} - \alpha\right) \cdot e^{-\alpha t}}{\beta - \alpha} + \frac{\left(k_{21} - \beta\right) \cdot e^{-\beta t}}{\alpha - \beta} \right\} \tag{29:12}$$

where each parameter is defined as above for each species.

Interpretation of results and conclusions

By means of information about total clearance and dose, one is able to estimate the appropriate exposure level of this compound in experimental animals, in order to mimic the human situation. The body weight rule offers the investigator a very powerful tool with which to obtain general guidelines for making reasonable inter-species predictions.

The final parameter estimates of the allometric parameters were obtained by fitting a bi-exponential expression to the concentration-time data of each species. The macro- and micro-constants in Equations 29:7 to 29:12 were defined in terms of Cl, V_c, Cl_d and V_t. The latter four parameters were then expressed as allometric functions for each species, taking body weight into account. The resulting parameter values of a-g (f not included as a parameter name) were estimated to be 0.021, 0.74, 0.076, 1.18, 0.56, and 0.075.

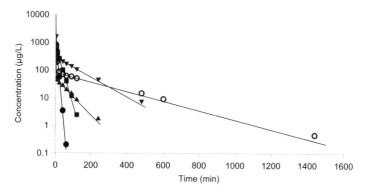

Figure 29.2 *Observed and predicted concentration-time data for the mouse (●), rat (■), monkey (▲), dog (▼) and man (○) after an intravenous bolus dose of 10, 125, 200, 6000 and 12000 μg, respectively.*

A complex Dedrick plot that includes the observed data of all species is shown in Figure 29.3. One can observe a distinct bi-exponential behavior, with perfectly super-imposed data.

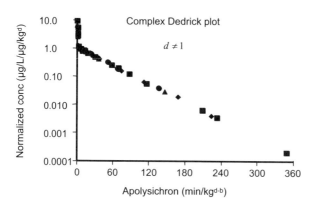

Figure 29.3 *Complex Dedrick plot of dose and body weight normalized concentration-time data plotted for each individual species. Data superimpose nicely, since they were simulated with a very low added error. The symbols represent the same species as in Figure 29.2*

This exercise illustrates how to simultaneously fit a system of equations to plasma concentration-time data from five different species. There was a high consistency between observed and predicted concentrations, and well-defined parameters with high precision (the relative standard deviations were less than 10% CV). Data are semi-synthetic.

Solution

```
MODEL
 COMM
  nfun   5
  NPARM  6
  NCON   10
  PNAMES 'a','b', 'c','d', 'e', 'g'
END
```

```
 1: TEMP
 2: T = X
 3: DOSE1 = CON(1)
 4: DOSE2 = CON(2)
 5: DOSE3 = CON(3)
 6: DOSE4 = CON(4)
 7: DOSE5 = CON(5)
 8: BWMSE = CON(6)
 9: BWRAT = CON(7)
10: BWMKY = CON(8)
11: BWDOG = CON(9)
12: BWMAN = CON(10)
13: A = P(1)
14: B = P(2)
15: C = P(3)
16: D = P(4)
17: E = P(5)
18: G = P(6)
19: CL1 = A*BWMSE**B
20: CL2 = A*BWRAT**B
21: CL3 = A*BWMKY**B
22: CL4 = A*BWDOG**B
23: CL5 = A*BWMAN**B
24: CLD1 = G*BWMSE**B
25: CLD2 = G*BWRAT**B
26: CLD3 = G*BWMKY**B
27: CLD4 = G*BWDOG**B
28: CLD5 = G*BWMAN**B
29: VC1 = C*BWMSE**D
30: VC2 = C*BWRAT**D
31: VC3 = C*BWMKY**D
32: VC4 = C*BWDOG**D
33: VC5 = C*BWMAN**D
34: VT1 = E*BWMSE**D
35: VT2 = E*BWRAT**D
36: VT3 = E*BWMKY**D
37: VT4 = E*BWDOG**D
38: VT5 = E*BWMAN**D
39: K10M = CL1/VC1
40: K12M = CLD1/VC1
41: K21M = CLD1/VT1
42: R1M=DSQRT((K12M + K21M + K10M)**2 - (4*K21M*K10M))
43: ALPHM=((K12M+K21M+K10M) + R1M)/2
44: BETAM=((K12M+K21M+K10M) - R1M)/2
45: K10R = CL2/VC2
46: K12R = CLD2/VC2
47: K21R = CLD2/VT2
48: R1R=DSQRT((K12R + K21R + K10R)**2 - (4*K21R*K10R))
49: ALPHR=((K12R+K21R+K10R) + R1R)/2
50: BETAR=((K12R+K21R+K10R) - R1R)/2
51: K10Y = CL3/VC3
52: K12Y = CLD3/VC3
53: K21Y = CLD3/VT3
54: R1Y=DSQRT((K12Y + K21Y + K10Y)**2 - (4*K21Y*K10Y))
55: ALPHY=((K12Y+K21Y+K10Y) + R1Y)/2
56: BETAY=((K12Y+K21Y+K10Y) - R1Y)/2
57: K10D = CL4/VC4
58: K12D = CLD4/VC4
59: K21D = CLD4/VT4
60: R1D=DSQRT((K12D + K21D + K10D)**2 - (4*K21D*K10D))
61: ALPHD=((K12D+K21D+K10D) + R1D)/2
```

```
 62: BETAD=((K12D+K21D+K10D) - R1D)/2
 63: K10N = CL5/VC5
 64: K12N = CLD5/VC5
 65: K21N = CLD5/VT5
 66: R1N=DSQRT((K12N + K21N + K10N)**2 - (4*K21N*K10N))
 67: ALPHN=((K12N+K21N+K10N) + R1N)/2
 68: BETAN=((K12N+K21N+K10N) - R1N)/2
 69: END
 70: FUNC1
 71: AM=(DOSE1/VC1)*(ALPHM-K21M)/(ALPHM-BETAM)
 72: BM=-1*(DOSE1/VC1)*(BETAM-K21M)/(ALPHM-BETAM)
 73: AMTM=(AM*DEXP(-ALPHM*T)) + (BM*DEXP(-BETAM*T))
 74: F=AMTM
 75: END
 76: FUNC2
 77: AR=(DOSE2/VC2)*(ALPHR-K21R)/(ALPHR-BETAR)
 78: BR=-1*(DOSE2/VC2)*(BETAR-K21R)/(ALPHR-BETAR)
 79: AMTR=(AR*DEXP(-ALPHR*T)) + (BR*DEXP(-BETAR*T))
 80: F=AMTR
 81: END
 82: FUNC3
 83: AY=(DOSE3/VC3)*(ALPHY-K21Y)/(ALPHY-BETAY)
 84: BY=-1*(DOSE3/VC3)*(BETAY-K21Y)/(ALPHY-BETAY)
 85: AMTY=(AY*DEXP(-ALPHY*T)) + (BY*DEXP(-BETAY*T))
 86: F=AMTY
 87: END
 88: FUNC4
 89: AD=(DOSE4/VC4)*(ALPHD-K21D)/(ALPHD-BETAD)
 90: BD=-1*(DOSE4/VC4)*(BETAD-K21D)/(ALPHD-BETAD)
 91: AMTD=(AD*DEXP(-ALPHD*T)) + (BD*DEXP(-BETAD*T))
 92: F=AMTD
 93: END
 94: FUNC5
 95: AN=(DOSE5/VC5)*(ALPHN-K21N)/(ALPHN-BETAN)
 96: BN=-1*(DOSE5/VC5)*(BETAN-K21N)/(ALPHN-BETAN)
 97: AMTN=(AN*DEXP(-ALPHN*T)) + (BN*DEXP(-BETAN*T))
 98: F=AMTN
 99: END
100: EOM
NVARIABLES 3
FNUMBER 3
NPOINTS 100
XNUMBER 1
YNUMBER 2
CONSTANTS 10,125,200,6000,12000,.02,.25,3.5,14,73
METHOD 2   'Gauss-Newton (Levenberg and Hartley)
ITERATIONS 50
INITIAL .02,.7,.1,1.2,.8,.1
LOWER BOUNDS 0,0,0,0,0,0
UPPER BOUNDS .2,2,2,2,10,1
MISSING 'Missing'
DATA 'WINNLIN.DAT'
BEGIN
```

PARAMETER	ESTIMATE	STANDARD ERROR	CV%	UNIVARIATE C.I.	
A	.020630	.000246	1.19	.020131	.021128
B	.740007	.004451	.60	.730996	.749017
C	.075592	.001066	1.41	.073435	.077750

D	1.184543	.003678	.31	1.177099	1.191988
E	.564886	.008808	1.56	.547055	.582717
G	.074842	.001388	1.85	.072033	.077651

Condition_number=   66.80

FUNCTION   1

X	OBSERVED Y	PREDICTED Y	RESIDUAL	WEIGHT	SE-PRED	STANDARDI RESIDUAL
2.000	810.0	796.4	13.61	1.000	6.711	3.105
5.000	485.0	507.3	-22.31	1.000	4.532	-3.375
10.00	247.0	239.2	7.753	1.000	5.375	1.304
20.00	55.00	53.21	1.790	1.000	2.861	.2391
40.00	3.400	2.632	.7680	1.000	.3091	.0959
60.00	.2000	.1302	.6981E-01	1.000	.2359E-01	.0087

CORRECTED SUM OF SQUARED OBSERVATIONS = 528384.
WEIGHTED CORRECTED SUM OF SQUARED OBSERVATIONS = 528384.
SUM OF SQUARED RESIDUALS =   746.671
SUM OF WEIGHTED SQUARED RESIDUALS = 746.671
S = .000000   WITH   0 DEGREES OF FREEDOM
CORRELATION (OBSERVED,PREDICTED) = .9993

FUNCTION   2

X	OBSERVED Y	PREDICTED Y	RESIDUAL	WEIGHT	SE-PRED	STANDARDI RESIDUAL
2.000	673.0	661.4	11.57	1.000	6.051	2.202
5.000	500.0	528.5	-28.54	1.000	3.664	-4.004
10.00	420.0	413.9	6.145	1.000	2.805	.8185
20.00	255.0	253.8	1.237	1.000	3.102	.1675
40.00	99.00	95.41	3.592	1.000	2.678	.4755
60.00	39.00	35.87	3.129	1.000	1.605	.3985
90.00	11.50	8.270	3.230	1.000	.5791	.4041
120.0	2.400	1.906	.4936	1.000	.1819	.0616

CORRECTED SUM OF SQUARED OBSERVATIONS = 455864.
WEIGHTED CORRECTED SUM OF SQUARED OBSERVATIONS = 455864.
SUM OF SQUARED RESIDUALS =   1021.18
SUM OF WEIGHTED SQUARED RESIDUALS = 1021.18
S = 22.5963   WITH   2 DEGREES OF FREEDOM
CORRELATION (OBSERVED,PREDICTED) = .9989

FUNCTION   3

X	OBSERVED Y	PREDICTED Y	RESIDUAL	WEIGHT	SE-PRED	STANDARD RESIDUAL
2.000	165.0	161.1	3.944	1.000	1.574	.5018
5.000	50.00	54.88	-4.881	1.000	.6030	-.6107
10.00	42.00	40.96	1.043	1.000	.4630	.1303
20.00	33.00	35.02	-2.020	1.000	.3710	-.2523
40.00	27.00	25.87	1.126	1.000	.3316	.1406
60.00	19.00	19.12	-.1173	1.000	.3536	-.0146
90.00	14.00	12.14	1.859	1.000	.3528	.2321
120.0	8.300	7.711	.5891	1.000	.3113	.0735
240.0	1.700	1.254	.4455	1.000	.1094	.0556

```
CORRECTED SUM OF SQUARED OBSERVATIONS =   19535.8
WEIGHTED CORRECTED SUM OF SQUARED OBSERVATIONS =   19535.8
SUM OF SQUARED RESIDUALS =            49.8202
SUM OF WEIGHTED SQUARED RESIDUALS =  49.8202
S =   4.07514    WITH      3 DEGREES OF FREEDOM
CORRELATION (OBSERVED,PREDICTED) =   .9989
```

```
    FUNCTION    4
```

X	OBSERVED Y	PREDICTED Y	RESIDUAL	WEIGHT	SE-PRED	STANDARDI RESIDUAL
2.000	1650.	1646.	3.976	1.000	7.114	1.078
5.000	640.0	650.2	-10.15	1.000	6.439	-2.128
10.00	310.0	300.1	9.935	1.000	3.415	1.370
20.00	230.0	234.4	-4.412	1.000	3.429	-.6090
40.00	205.0	198.5	6.500	1.000	2.704	.8616
60.00	170.0	168.6	1.427	1.000	2.446	.1870
90.00	140.0	131.9	8.075	1.000	2.501	1.061
120.0	108.0	103.2	4.755	1.000	2.616	.6277
240.0	46.20	38.73	7.472	1.000	2.159	.9681
480.0	7.500	5.449	2.051	1.000	.6551	.2567

```
CORRECTED SUM OF SQUARED OBSERVATIONS =  .215579E+07
WEIGHTED CORRECTED SUM OF SQUARED OBSERVATIONS =   .215579E+07
SUM OF SQUARED RESIDUALS =            429.229
SUM OF WEIGHTED SQUARED RESIDUALS =  429.229
S =   10.3589    WITH      4 DEGREES OF FREEDOM
CORRELATION (OBSERVED,PREDICTED) =   .9999
```

```
    FUNCTION    5
```

X	OBSERVED Y	PREDICTED Y	RESIDUAL	WEIGHT	SE-PRED	STANDARD RESIDUAL
2.000	680.0	681.5	-1.457	1.000	5.572	-.2530
5.000	395.0	404.7	-9.680	1.000	2.882	-1.294
10.00	202.0	193.7	8.330	1.000	2.812	1.110
20.00	85.00	87.53	-2.527	1.000	1.212	-.3190
40.00	69.00	66.80	2.196	1.000	1.268	.2775
60.00	60.00	61.52	-1.522	1.000	1.119	-.1918
90.00	57.00	54.69	2.310	1.000	.9571	.2903
120.0	49.00	48.62	.3801	1.000	.8850	.0477
480.0	14.50	11.85	2.651	1.000	.7994	.3324
600.0	9.000	7.401	1.599	1.000	.6438	.2001
1440.	.4600	.2746	.1854	1.000	.6214E-01	.0231

```
CORRECTED SUM OF SQUARED OBSERVATIONS =   441892.
WEIGHTED CORRECTED SUM OF SQUARED OBSERVATIONS =   441892.
SUM OF SQUARED RESIDUALS =            193.840
SUM OF WEIGHTED SQUARED RESIDUALS =  193.840
S =   6.22640    WITH      5 DEGREES OF FREEDOM
CORRELATION (OBSERVED,PREDICTED) =   .9998
```

```
TOTALS FOR ALL CURVES COMBINED
SUM OF SQUARED RESIDUALS =            2440.75
SUM OF WEIGHTED SQUARED RESIDUALS =  2440.75
S =   8.01437    WITH     38 DEGREES OF FREEDOM
```

PK30 – Turnover I – Subcutaneous dosing of an endogenous compound

Objectives

◆ **To model the turnover of an endogenous compound given subcutaneously**

◆ **To estimate turnover rate and clearance simultaneously**

◆ **To estimate absorption half-life, input time, and turnover time**

◆ **To set up a system of differential equations**

Problem specification

This exercise demonstrates the turnover concept including turnover rate and turnover time, which was introduced in Section 3.5 and is further developed in exercises PK31 and PK32.

A healthy volunteer received a 40 µg/kg dose D of *IgX* subcutaneously (*sc*). The plasma concentrations of *IgX*, which were measured before and after the dose, are shown in Figure 30.1 together with the model-predicted concentrations.

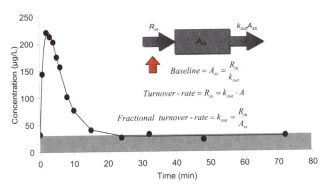

Figure 30.1 Observed concentration-time data of IgX after a 40 µg/kg subcutaneous dose. A schematic representation of the turnover model is also shown.

The pre-dose average concentration was 32 µg/L. We will estimate the basic kinetic parameters such as clearance Cl/F, volume of distribution V/F, and half-life. In addition, mean residence time MRT, which corresponds to turnover time, mean input time MIT, and the absorption rate constant K_a will be determined. The underlying differential equation that takes endogenous turnover rate (synthesis which is expressed as $SYNT$ in the program code) and elimination Cl, together with the subcutaneous application, into account is written as

$$V \frac{dC_{IgX}}{dt} = R_{in} + Input_{sc} - Cl_{IgX} \cdot C_{IgX} \tag{30:1}$$

Input$_{sc}$ is defined as

$$Input_{sc} = K_a \cdot FD_{sc} \cdot e^{-Ka \cdot t} \tag{30:2}$$

where D_{sc} is the subcutaneous dose. *Turnover time* is defined as

$$t = \frac{1}{K} = \frac{V}{Cl} \tag{30:3}$$

Equation 30:1 assumes that 100% of the dose is absorbed, that no feedback occurs with this dose, regimen and time frame, and that there is parameter stationarity, and linear kinetics.

The initial condition for the system is C^0 = *turnover rate/Cl* = $C_{baseline}$. The secondary parameters are the baseline concentration, half-life, time to steady state, mean residence time and mean input time. Concentration-time data were kindly supplied by Dr. Calleberg, Pharmacia Upjohn AB, and were then slightly adapted. The model was proposed by JG.

Initial parameter estimates

We assume that the baseline concentration (initial condition) can be written as R_{in}/Cl (*synthesis* or *turnover rate* divided by *Cl*).

V/F	=	0.2 (L/kg)	*Dose/C_{max}* = 40/200
Cl	=	0.03 (L/h/kg)	Previous estimate
K_a	=	0.5 (h⁻¹)	
Turnover =		1 (µg/h/kg)	Expressed as SYNT in the code

Interpretation of results and conclusions

The turnover of *IgX* was modeled in a healthy volunteer by means of a one-compartment model with first-order absorption/elimination and zero-order production of *IgX*. Without additional information obtained from a subcutaneous dose, we would not have been able to discriminate between clearance and rate of synthesis, since they form a ratio in the equation (see initial condition or C^0). However, clearance, rate of synthesis and volume of distribution also include the bioavailability. We simplify the reasoning and assume that bioavailability is 100% and that the kinetics is linear for the concentration and time frame used. One also has to be cautious about the potential for feedback, as such mechanisms govern the majority of endogenous compounds, such as hormones. Figure 30.2 shows the observed and model-predicted concentration-time data for this exercise.

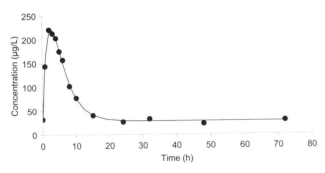

Figure 30.2 *Observed and predicted concentration-time data of IgX after a 40 µg/kg subcutaneous dose.*

Clearance and volume of distribution were 0.03L/h/kg and 0.10 L/kg body weight, respectively. Rate of synthesis or turnover rate was 0.78 µg/h/kg body weight. Turnover time (*1/K*) and mean input time were estimated as 3.6 h and 1.8 h, respectively. The terminal half-life was 2.5 h, with a corresponding fractional turnover rate of 0.27 h^{-1}. The pool size of *IgX* at steady state was 3.3 µg/kg^{-1}, calculated from the product of turnover rate and fractional turnover rate.

The conclusions to be drawn from this exercise are that while parameter precision was generally high, the correlation between the absorption rate constant and volume was also high (0.98). This can be partly explained by the fact that we had few observations on the upswing of the concentration-time curve, which governs K_a. The volume also determines C_{max} (approximately *Dose/V*), and therefore K_a and *V/F* are highly correlated. We applied a homoscedastic (constant absolute error) weighting scheme which, according to the residuals, was acceptable, since no obvious trend was seen over time.

Solution

```
MODEL
 COMM
  NDER   1
  NPARM  4
  NSEC   5
  NCON   2
  PNAMES 'VF','Ka','Synt','Cl'
  SNAMES 'Css', 'T12', 'Tss', 'MRT1', 'MRT2'
END
     1: TEMP
     2: DOSE1= CON(1)
     3: TLAG = CON(2)
     4: T=X
     5: VF=P(1)
     6: KA=P(2)
     7: SYNT=P(3)
     8: CL=P(4)
     9: IF T LE TLAG THEN
    10: FINF = 0.0
    11: ELSE
    12: FINF = DOSE1*KA*DEXP(-KA*(T-TLAG))
    13: ENDIF
    14: END
    15: START
    16: Z(1)=SYNT/CL
```

```
17: END
18: DIFF
19: DZ(1)=(FINF + SYNT - CL*Z(1))/VF
20: END
21: FUNC 1
22: F = Z(1)
23: END
24: SECO
25: S(1)=SYNT/CL
26: S(2)=0.693*VF/CL
27: S(3)=3.5*S(2)
28: S(4)=VF/CL
29: S(5)=1/(CL/VF) + 1/KA
30: END
31: EOM
NVARIABLES 2
NPOINTS 200
XNUMBER 1
YNUMBER 2
CONSTANTS 40,0
METHOD 2   'Gauss-Newton (Levenberg and Hartley)
ITERATIONS 50
INITIAL .2,.5,1,.03
LOWER BOUNDS 0,0,0,0
UPPER BOUNDS 50,5,10,1
MISSING 'Missing'
NOBSERVATIONS 14
DATA 'WINNLIN.DAT'
BEGIN
```

PARAMETER	ESTIMATE	STANDARD ERROR	CV%	UNIVARIATE C.I.	
VF	.101941	.011139	10.93	.077122	.126759
KA	.534893	.079275	14.82	.358256	.711529
SYNT	.781530	.082003	10.49	.598815	.964245
CL	.028010	.001057	3.77	.025654	.030366

```
*** CORRELATION MATRIX OF THE ESTIMATES ***
PARAMETER   VF        KA        SYNT       CL
VF        1.00000
KA         .97846   1.00000
SYNT      -.53941   -.58268   1.00000
CL        -.75705   -.76948    .88164    1.00000
```

FUNCTION   1

X	OBSERVED Y	PREDICTED Y	RESIDUAL	WEIGHT	SE-PRED	STANDARDI RESIDUAL
.0000	32.00	27.90	4.098	1.000	2.061	1.015
.7500	144.0	144.3	-.2748	1.000	3.336	-.0895
2.000	221.0	216.8	4.187	1.000	2.906	1.203
3.000	213.0	219.6	-6.598	1.000	2.419	-1.721
4.000	203.0	201.8	1.242	1.000	2.451	.3256
5.000	175.0	176.5	-1.516	1.000	2.374	-.3925
6.000	157.0	150.5	6.509	1.000	2.255	1.655
8.000	102.0	106.3	-4.291	1.000	2.526	-1.140
10.00	77.00	75.77	1.233	1.000	2.897	.3536
15.00	40.00	40.72	-.7245	1.000	2.275	-.1848

24.00	26.00	29.00	-3.003	1.000	1.822	-.7236
32.00	32.00	28.02	3.976	1.000	2.010	.9785
48.00	22.00	27.90	-5.903	1.000	2.060	-1.462
72.00	29.00	27.90	1.098	1.000	2.061	.2720

```
CORRECTED SUM OF SQUARED OBSERVATIONS =  78430.4
WEIGHTED CORRECTED SUM OF SQUARED OBSERVATIONS =  78430.4
SUM OF SQUARED RESIDUALS =           205.482
SUM OF WEIGHTED SQUARED RESIDUALS =  205.482
S =  4.53301     WITH    10 DEGREES OF FREEDOM
CORRELATION (OBSERVED,PREDICTED) =  .9987

AIC criteria =        82.55503
SC  criteria =        85.11126
```

PK31 – Turnover II – Intravenous dosing of an endogenous hormone

Objectives

◆ To analyze an intravenous infusion dataset with parallel endogenous production

◆ To write a multi-compartment model in terms of differential equations

◆ To obtain initial parameter estimates of R_{in}, Cl and V_{ss}

◆ To discuss various aspects of experimental design

Problem specification

This problem highlights some complexities in modeling *turnover data*. Specific emphasis is placed on obtaining *initial parameter estimates*, implementing *turnover equations*, and applying *weights*. See Section 3.5 and exercises PK30 and PK32 for further reading on turnover concepts.

In recent years, controlled input devices have become increasingly popular for administration of sexual hormones, nitroglycerin, nicotine, etc. These provide convenient delivery systems that result in relatively uniform serum levels. By bypassing the liver and maintaining a continuous availability of drug to the target tissues, such devices produce more stable exposure than does the oral route. This also raises the possibility, that unwanted metabolites that appear in too high a concentration during oral (e.g., estrogen) therapy, will occur at lower concentration when a controlled input (e.g., intravaginal) device is used.

In order to optimize the input rate for a new controlled release device containing estradiol, a study on the estradiol disposition in post-menopausal women was carried out. Data in Figure 31.1 and Table 31.1 were obtained from one previously untreated subject who received estradiol as a rapid intravenous infusion (total dose 36630 pmol given over 1 min, mw 272.37) (Gabrielsson *et al* [1995]).

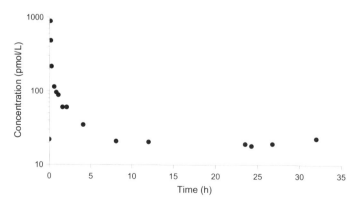

Figure 31.1 Observed estradiol concentration-time data after a rapid intravenous infusion of estradiol. The observed concentration of 3683 pM (pmol/L) at 1 min is not included in this plot.

Table 31.1 Concentration-time data of estradiol

Time (h)	Concentration (pmol/L)	Time (h)	Concentration (pmol/L)
0.0167	3683	2.083	60.1728
0.1167	884.729	4.083	34.8914
0.167	481.056	8.083	20.9927
0.25	215.643	12	20.5364
0.583	113.983	23.5	19.2779
0.833	95.8037	24.25	18.1825
1.083	87.8946	26.75	19.3928
1.583	60.1948	32	22.7245

The values represent the untreated readout from the analytical instrument.
Ideally, estradiol concentrations of 3683 or 22.7245, for example, would
be presented as 3700 and 23, respectively.

Some marketed drugs, including estradiol, are endogenous to the body. With these, the basal concentration that exists in plasma must be taken into account when attempting to define the pharmacokinetics of the administered compound. We did so in this example by regressing the model simultaneously to baseline and exogenous intravenous dose data. The two-compartment model (Figure 31.2) was parameterized in terms of clearance Cl, volume of the central compartment V_c, volume of the peripheral compartment V_t, inter-compartmental distribution Cl_d, and endogenous turnover rate of estradiol R_{in}.

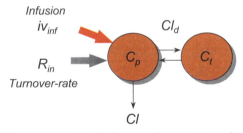

Figure 31.2 Schematic illustration of the two-compartment turnover model taking simultaneous endogenous and exogenous inputs into account.

The concentration in the central compartment is

$$V_c \cdot \frac{dC}{dt} = In + R_{in} - Cl \cdot C - Cl_d \cdot C + Cl_d \cdot C_t \qquad (31{:}1)$$

where *In* represents the rapid constant intravenous infusion, and R_{in} the turnover rate or endogenous synthesis of estradiol (called *SYNT* in the program code). The concentration in the peripheral compartment is

$$V_t \cdot \frac{dC_t}{dt} = Cl_d \cdot C - Cl_d \cdot C_t \qquad (31{:}2)$$

Throughout this analysis, the actual time after the start of dosing was used. We will model the rapid intravenous input in parallel with the endogenous production.

Initial parameter estimates

The impact that different initial estimates had on the final parameter estimates was thoroughly investigated in order to avoid local minima. An approximate estimate of V_c is obtained from

$$V_c = \frac{Dose_{iv}}{C_{max}} \tag{31:3}$$

V_t is equal to

$$V_c = V_{ss} - V_c \cdot Cl \tag{31:4}$$

V_{ss} is calculated by means of non-compartmental analysis of the curve resulting from the intravenous dose corrected for baseline values. Cl_d is approximately

$$Cl_d = \alpha \cdot V_c - Cl \tag{31:5}$$

Cl is obtained from

$$Cl = \frac{Dose_{iv}}{AUC_{above\ baseline}} \tag{31:6}$$

$AUC_{above\ baseline}$ is the area contribution from the intravenous dose ($Dose_{iv}$), and baseline concentration is about 20 pmol/L. The number *10* corresponds to the time elapsed after the intravenous dose until the concentration returns to baseline. We used literature data as starting values for most parameters. The following values were used in this exercise.

R_{in} or *Synthesis* = 1600(pmol/h)		Expressed as *SYNT* in program code
Cl	= 90 (L/h)	Plasma clearance (literature data)
Cl_d	= 110 (L/h)	Equation 31:5
V_c	= 13 (L)	Volume of central compartment
V_t	= 65 (L)	Volume of peripheral compartment

Interpretation of results and conclusions

This example of endogenous drugs covers the estimation of turnover rate, fractional turnover rate and turnover time, along with clearance and volume of distribution, of estradiol in post-menopausal women. In this population, the endogenous level of estradiol is suppressed and the biological rhythm of estradiol in plasma is negligible. The assessment we have made of turnover rate is successful because basal turnover rate is unaffected by the additional input of estradiol. In the case of some endogenous compounds, this approach may be inappropriate because feedback control adjusts the endogenous turnover rate to maintain equilibrium. In order to estimate the turnover and disposition parameters of estradiol in this exercise, a two-compartment model was regressed using data following baseline measurements and administration of exogenous estradiol.

The time course of estradiol in the subjects displayed a more rapid washout than was expected from compiled literature data, although the predicted turnover rate of estradiol was consistent with previously reported data (Figure 31.3).

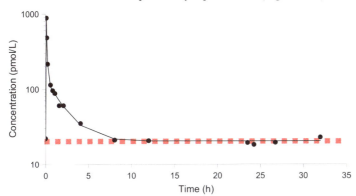

Figure 31.3 *Observed and superimposed model-predicted concentration-time data of estradiol. The dashed horizontal line corresponds to the baseline.*

Clearance was similar to that found in fertile women. The turnover rate varied between less than 1 and 36 μg per 24 h, with a mean of 19 μg per 24 h. The fractional turnover rate was on average 3% per min, ranging between 1 and 15% per min. This resulted in a pool size of 0.4 μg estradiol/70 kg body weight (Lassen and Perl [1979], Gabrielsson *et al* [1995]).

Using data obtained from intravenous administration, the natural disposition of the hormone could be established in terms of clearance, volume, and endogenous production rate. We propose a flexible routine that may be used for estimation of the endogenous production rate of e.g., a hormone. This routine also includes a bolus regimen and may be adapted to mimic data obtained from controlled input in order to estimate the bioavailability. For this particular dataset, data were successfully fit by means of the proposed model, in that the parameters were well-defined and the corresponding condition number was low. Table 31.2 contains a summary of the data from 15 subjects.

Table 31.2   Mean kinetic data for estradiol from 15 post-menopausal women

	Turnover (μg/24 h)	C_{ss} (pmol/L)	Cl (L/min)	V_{ss} (L)	$t_{1/2z}$ (min)	MRT (min)
Mean	19	–	1.6	50	26	18
SD	8.1	–	0.58	30	29	20
CV%	43	–	36	60	110	110
Max	36	41	2.9	140	120	85
Min	<1	<20	0.76	20	9.0	6.3

Table adapted from Gabrielsson *et al* [1995].

By means of fitting a regression model to individual data, we were able to characterize the estradiol kinetics for each subject. By means of clearance and turnover rate, we were also able to explain why some women with high or low estradiol levels may still have comparable clearance or turnover rates. A high baseline value of estradiol may be due to a high turnover rate or a low clearance value. In the studied population, we obtained a 40-fold range in turnover rates, a 3.5-fold range in clearance values and a more than 2-fold variability in baseline estradiol.

Figure 31.4 shows a plot of the weighted residual *versus* time. The predicted mean standard deviation of the function is about 7% (S = 0.074). This seems to be a reasonable value since the majority of data points lie within the ± 10% boundaries for the weighted residuals in Figure 31.4.

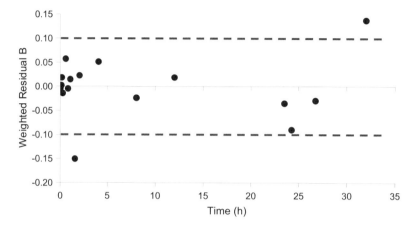

Figure 31.4 *Plot of weighted residual versus time. The horizontal dashed lines correspond to the ± 10% boundaries. All but two of the residuals fall within that interval.*

The model may easily be adapted for estimation of the bioavailability following exogenous input via drug delivery devices such as the nicotine patch or the intravaginal estrogen ring.

Solution - rapid infusion model

```
TITLE 1
 This is a model for E2-disposition incl. Synt
MODEL
 COMM
    NDER  2
    NPARM 5
    NCON  2
    NSEC  3
    PNAMES 'Synt','Cl', 'ClD', 'Vc', 'Vt'
    SNAMES 'MRT','Ass', 'T1/2'
 END
    1: TEMP
    2: T=X
    3: TINS = CON(1)
    4: DOSE1 = CON(2)
    5:
    6: IF T LE TINS THEN
    7: RTE1 = DOSE1/TINS
    8: ELSE
    9: RTE1 = 0.0
   10: ENDIF
```

```
11: SYNT=P(1)
12: CL=P(2)
13: CLD=P(3)
14: VC=P(4)
15: VT=P(5)
16: END
17: START
18: Z(1) = SYNT/CL
19: CSS = SYNT/CL
20: Z(2) = CSS
21: END
22: DIFF
23: DZ(1)= (RTE1 + SYNT - CL*Z(1) - CLD*Z(1) + CLD*Z(2))/VC
24: DZ(2)= (CLD*Z(1) - CLD*Z(2))/VT
25: END
26: FUNC 1
27: F = Z(1)
28: END
29: SECO
30: S(1) = (VC + VT)/CL
31: S(2) = (SYNT/CL)*(VC + VT)
32: S(3) = 0.693*(VC + VT)/CL
33: END
34: EOM
NVARIABLES 9
NPOINTS 1000
XNUMBER 1
YNUMBER 2
CONSTANTS .01667,36630
METHOD 2  'Gauss-Newton (Levenberg and Hartley)
REWEIGHT -2
ITERATIONS 50
INITIAL 1600,90,110,13,65
LOWER BOUNDS 100,10,10,1,5
UPPER BOUNDS 3000,300,300,50,99
MISSING 'Missing'
NOBSERVATIONS 16
DATA 'WINNLIN.DAT'
BEGIN
```

PARAMETER	ESTIMATE	STANDARD ERROR	CV%
SYNT	1523.530034	85.116492	5.59
CL	76.259378	2.872065	3.
CLD	56.960355	4.646024	8.16
VC	8.804455	.648475	7.37
VT	58.766983	6.502097	11.06

```
*** CORRELATION MATRIX OF THE ESTIMATES ***
```

PARAMETER	SYNT	CL	CLD	VC	VT
SYNT	1.00000				
CL	.83949	1.00000			
CLD	.18214	.39403	1.00000		
VC	.40934	.61733	.72838	1.00000	
VT	-.25879	-.13454	.62491	.44084	1.00000

```
 Condition_number=       243.1


FUNCTION   1
```

X	OBSERVED	PREDICTED	RESIDUAL	WEIGHT	SE-PRED	STANDARDI

Y		Y				RESIDUAL
.1670E-01	3683.	3697.	-13.62	.7318E-07	255.2	-.1413
.1167	884.7	882.9	1.841	.1283E-05	37.57	.0345
.1670	481.1	472.4	8.645	.4481E-05	22.42	.3238
.2500	215.6	218.8	-3.133	.2089E-04	10.49	-.2552
.5830	114.0	107.8	6.177	.8604E-04	4.778	.9710
.8330	95.80	96.27	-.4682	.1079E-03	3.688	-.0770
1.083	87.89	86.64	1.259	.1332E-03	2.919	.2212
1.583	60.19	70.88	-10.68	.1991E-03	2.274	-2.268
2.083	60.17	58.84	1.331	.2888E-03	2.227	.3570
4.083	34.89	33.19	1.702	.9078E-03	1.866	1.073
8.083	20.99	21.50	-.5121	.2162E-02	.6731	-.3563
12.00	20.54	20.16	.3737	.2460E-02	.6070	.2751
23.50	19.28	19.98	-.7007	.2505E-02	.6340	-.5264
24.25	18.18	19.98	-1.796	.2505E-02	.6344	-1.350
26.75	19.39	19.98	-.5855	.2505E-02	.6341	-.4399
32.00	22.72	19.98	2.746	.2505E-02	.6343	2.063

```
CORRECTED SUM OF SQUARED OBSERVATIONS =  .125355E+08
WEIGHTED CORRECTED SUM OF SQUARED OBSERVATIONS =  6.42335
SUM OF SQUARED RESIDUALS =           444.219
SUM OF WEIGHTED SQUARED RESIDUALS =  .599053E-01
S =  .737966E-01 WITH    11 DEGREES OF FREEDOM
CORRELATION (OBSERVED,PREDICTED) =  1.000

AIC criteria =         -35.03984
SC  criteria =         -31.17690
```

SUMMARY OF ESTIMATED SECONDARY PARAMETERS

PARAMETER	ESTIMATE	STANDARD ERROR	CV%
MRT	.886074	.097507	11.00
ASS	1349.960061	130.591630	9.67
T1/2	.614049	.067572	11.00

PK32 - Turnover III – Nonlinear disposition

Objectives

◆ To characterize turnover of an endogenous compound

◆ To analyze a multiple intravenous infusion dataset with non-linear disposition

◆ To write a multiple intravenous input model in terms of differential equations

◆ To obtain initial estimates of turnover rate, clearance and volume

◆ To obtain estimates of half-life and turnover time

◆ To discuss various aspects of correlation, accuracy, and precision

◆ To discuss various aspects of experimental design

Problem specification

The aim of this exercise is to analyze the turnover of an endogenous compound hyaluronan (*HA*). The following introductory section about *HA* is adapted from the Ph.D. thesis by Lebel [1989]. *HA* is synthesized by cells in the interstitium, is transported by lymph to the blood and is finally catabolized in the liver by the endothelial cells. Highly elevated plasma concentrations of *HA* have been found during different disease states in man. *HA* has therefore been proposed as an early marker for different pathological conditions (Lebel [1989]). This is a recommended text about turnover concepts.

The major route of transport of *HA* from the tissues to the circulation is via the lymphatics. It has been calculated that 10-100 mg of *HA* enters the blood stream every day in the adult human. However, a considerable amount of *HA* is catabolized by the lymph nodes, indicating that the normal total daily turnover of *HA* in the body may be very high. The turnover rates of *HA* in the pig, sheep and human are 40-50, 37, and 19-49 µg/min, respectively. The half-lives of *HA* in the circulation of the rat, rabbit, sheep and human are 1.4-1.9, 2.4-4.5, 2-19, and 3-9 min, respectively. Figure 32.1 contains a schematic diagram of the turnover model.

Figure 32.1 Schematic representation of the turnover of HA with the endogenous input rate (k_{in}), exogenous infusion (Inf) and Michaelis-Menten type of catabolism (V_{max} and K_m).

The elimination of *HA* that reaches the circulation mainly takes place in the liver, via receptor-mediated endocytosis in the endothelial cells. *HA* is subsequently degraded in the lysosomes and the final end-products are lactate and acetate (Lebel [1989]).

A volunteer was given multiple intravenous infusions of *HA* according to the dosing scheme shown in Table 32.1.

Table 32.1　Dosing information about hyaluronan for the multiple dose regimen

	Rapid iv 1	Infusion 1	Rapid iv 2	Infusion 2	Rapid iv 3	Infusion 3
Dose, µg	1669	1131.8	1701	1884.4	1733	6300
Interval, min	Bolus at t=0	0-30.1	125-126	125.2-154.3	260-261	260.1-290.1

Plasma samples were obtained over 7.5 h (Figure 32.2).

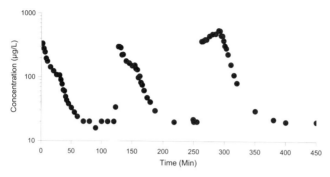

Figure 32.2　*Observed plasma concentration-time data of HA following a multiple bolus + constant rate intravenous infusion regimen.*

In this exercise, you need to determine how to derive initial parameter estimates, whether such derivation can be done by a graphical method and how to discriminate between a linear and nonlinear model

Initial parameter estimates

The estimates which need to be obtained are

$$
\begin{array}{lll}
V_{max} & = & (\text{µg/min}) \qquad \text{Maximum metabolic capacity} \\
K_m & = & (\text{µg/L}) \qquad \text{Michaelis-Menten constant} \\
V & = & (\text{L}) \qquad \text{Volume of distribution} \\
k_{in} & = & (\text{µg/min}) \qquad \text{Rate of synthesis, turnover rate} \\
Cl & = & (\text{L/min}) \qquad \text{Plasma clearance (linear model)}
\end{array}
$$

The nonlinear model contains the model parameters V_{max}, K_m, V and k_{in} and the linear model Cl, V and k_{in}. You now need to decide

1. what the relationship between the baseline concentration, k_{in} and Cl is
2. what determines the peak concentration of *HA* after a rapid infusion if you assume a one-compartment turnover model
3. if it is possible to discriminate between clearance and turnover rate

Cl is obtained from *Dose/AUC* of one of the rapid plus slow infusion regimens (e.g., where the *AUC* is calculated between 260-360 min and the *Dose* is 1733 + 6300 µg). The turnover equation of *HA* is then written as

$$V_c \cdot \frac{dC}{dt} = In + k_{in} - Cl \cdot C \qquad (32:1)$$

In and k_{in} are the intravenous infusions and turnover rate, respectively. *Cl* is the capacity limited clearance

$$Cl = \frac{V_{max}}{K_m + C} \qquad (32:2)$$

In order to estimate V_{max}, set *C* equal to *AUC/tau* (where *tau* is 360-250 min = 110 min), and give K_m a value similar to *C*. Rearrange the equation to get V_{max}. The turnover rate (k_{in}) is then the baseline value (e.g., 20.2 µg/L) multiplied by the previously derived estimate of *Cl*.

Conclusions and interpretation of output

In this study, it was shown that the elimination of *HA* occurs by means of a saturable Michaelis-Menten type of process. The disposition of *HA* was studied by means of a multiple intravenous infusion regimen. Increasing infusion doses covering a 20 to 600 µg/L concentration interval were applied in order to elucidate the saturable catabolism in parallel with the turnover rate (Lebel [1989]). The fit of the one-compartment turnover model superimposed on the experimental data can be seen in Figure 32.3.

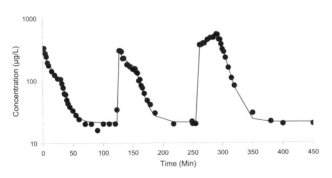

Figure 32.3 *Observed and predicted concentration-time data of HA following a multiple bolus + constant rate intravenous infusion regimen.*

Below are summarized the final parameter estimates and their precision (Table 32.2), the correlation matrix (Table 32.3a nonlinear model and Table 32.3b, linear model) and the estimates and precision of the secondary parameters (Table 32.4).

Table 32.2 Final parameter estimates and their precision

	Nonlinear model			Linear model		
	Mean	S.D.	CV%	Mean	S.D.	CV%
V_c	6.15	0.19	3.1	5.70	0.32	5.7
V_{Max}	356	26.4	7.4	-	-	-
K_m	490	61.6	13	-	-	-
Cl	-	-	-	0.436	0.014	3.3
k_{in}	14.9	1.00	6.7	8.52	0.63	7.4

The model parameters were obtained with a CV% generally less than 13%. The 490 µg/L value of K_m is about 10 times higher than the normal plasma concentration of *HA*.

This reinforces that the plasma kinetics of *HA* is linear under normal physiological conditions. During pathological states in man, plasma concentrations of *HA* exceeding K_m are commonly seen in patients, which complicates the sparse data analysis (Lebel [1989]). The square root of mean predicted variance σ^2 was less than 0.5% of the predicted concentrations.

Table 32.3a   Correlation matrix of the nonlinear model

	V_c	V_{max}	K_m	k_{in}
V_c	1			
V_{max}	-0.321405	1		
K_m	-0.394622	0.982094	1	
k_{in}	0.343636	-0.759301	-0.845127	1

Table 32.3b   Correlation matrix of the linear model

	V_c	Cl	k_{in}
V_c	1		
Cl	0.59072	1	
k_{in}	0.13602	0.641637	1

The correlation between two or more parameters was low, except for the Michaelis-Menten parameters, V_{max} and K_m, which had a correlation of 0.98. This correlation is probably due to the fact that the plasma concentrations were less than 50% higher (600 µg/L) than the final estimate of K_m (490 µg/L). In spite of the high correlation, the precision was good for these parameters.

Parameter estimates and their precision were obtained for the half-life, turnover time (*MRT*) and the corresponding first-order rate constant k_{out} by means of the secondary model parameter option in *WinNonlin*. Observe that these estimates were based on a plasma concentration of *HA* set to zero. At higher plasma concentrations, the half-life and turnover time will be longer (Table 32.4).

Table 32.4 Nonlinear and linear secondary model parameters

	Nonlinear model			Linear model		
	Mean	S.D.	CV%	Mean	S.D.	CV%
$T_{1/2}$	5.9	0.29	4.9	9.3	0.42	4.6
k_{out}	0.11	0.006	4.8	0.07	0.0034	4.6
MRT	8.5	0.41	4.9	13.	0.61	4.6

The disposition of *HA* is characterized by means of a one-compartment system with a zero-order turnover rate of 15µg/min and a saturable catabolism with a V_{max} of 356 µg/min and K_m of about 490 µg/L. The daily turnover rate for this particular individual is 22 mg based on an estimated synthesis (turnover) rate of 15 µg/min. The theoretically shortest half-life in this individual is 5.9 min, which corresponds well with literature data of 3-9 min (Lebel [1989]).

Solution I – Nonlinear turnover model
```
TITLE 1
```

```
IV-BOLUS + INFUSION
MODEL
 COMM
 NDER 1
 NPAR 4
 NCON 14
 PNAM 'Vc', 'VM', 'KM','Kin'
 NSEC 3
 SNAM 'T12', 'Kout', 'MRT'
END
     1: TEMP
     2: T3=CON(1)
     3: TI3=CON(2)
     4: DI3=CON(3)
     5: T2=CON(4)
     6: TI2=CON(5)
     7: DI2=CON(6)
     8: TI1=CON(7)
     9: DI1=CON(8)
    10: BD3=CON(9)
    11: BD2=CON(10)
    12: ETT=CON(11)
    13: TB3=CON(12)
    14: TB2=CON(13)
    15: BD1=CON(14)
    16: T=X
    20: DOSI = 0.0
    21: DOSB = 0.0
    22: IF T GT T3 AND T LT T3+TI3 THEN
    23: DOSI = DI3/TI3
    24: ELSE
    25: IF T GT T2 AND T LT T2+TI2 THEN
    26: DOSI = DI2/TI2
    27: ELSE
    28: IF T LT TI1 THEN
    29: DOSI = DI1/TI1
    30: ENDIF
    31: ENDIF
    32: ENDIF
    33: IF T GT TB3 AND T LT TB3 + ETT THEN
    34: DOSB = BD3/ETT
    35: ELSE
    36: IF T GT TB2 AND T LT TB2 + ETT THEN
    37: DOSB = BD2/ETT
    38: ENDIF
    39: ENDIF
    40: END
    41: START
    42: Z(1)=BD1/VC + Kin/(VM/(KM + 20.2))
    43: END
    44: DIFF
    45: DZ(1)= DOSI/VC + DOSB/VC - (VM*Z(1))/(KM+Z(1))/VC + Kin/VC
    46: END
    47: FUNC 1
    48: F = Z(1)
    49: END
    50: SECO
    51: T12 = 0.693*VC/(VM/(KM + 0))
    52: KOUT = (VM/(KM + 0))/VC
    53: MRT = 1/KOUT
    54: END
```

```
    55: EOM
NVARIABLES 2
NPOINTS 100
XNUMBER 1
YNUMBER 2
CONSTANTS 260.1,30,6300,125.2,29.1,1884.4, &
        30.1,1131.8,1733,1701,1,260,125,1669
METHOD 2   'Gauss-Newton (Levenberg and Hartley)
REWEIGHT -2
ITERATIONS 50
INITIAL 5,300,450,15
LOWER BOUNDS 2,50,50,5
UPPER BOUNDS 12,1000,550,50
MISSING 'Missing'
NOBSERVATIONS 72
DATA 'WINNLIN.DAT'
BEGIN
```

PARAMETER	ESTIMATE	STANDARD ERROR	CV%	UNIVARIATE C.I.	
VC	6.150351	.191778	3.12	5.767663	6.533039
VM	356.414361	26.405738	7.41	303.722438	409.106285
KM	490.831335	61.643483	12.56	367.823462	613.839208
Kin	14.920927	.998976	6.70	12.927498	16.914357

```
    *** CORRELATION MATRIX OF THE ESTIMATES ***
PARAMETER  VC            VM            KM            Kin
VC         1.00000
VM         -.321405      1.00000
KM         -.394622      .982094       1.00000
Kin        .343636       -.759301      -.845127      1.00000

Condition_number=        450.4
```

X	OBSERVED Y	PREDICTED Y	RESIDUAL	WEIGHT	SE-PRED	STANDARDI RESIDUAL
2.230	334.0	265.0	68.98	.1424E-04	7.049	2.284
4.200	278.0	242.9	35.06	.1694E-04	5.972	1.261
6.050	245.0	224.2	20.81	.1990E-04	5.120	.8088
8.030	195.0	206.1	-11.15	.2353E-04	4.370	-.4699
10.00	176.0	190.2	-14.16	.2765E-04	3.777	-.6457
15.00	143.0	157.4	-14.42	.4035E-04	2.852	-.7923
20.00	124.0	134.0	-10.05	.5565E-04	2.490	-.6488
25.00	108.0	117.8	-9.792	.7207E-04	2.370	-.7211
30.00	106.0	106.7	-.7080	.8782E-04	2.316	-.0576
32.00	90.00	92.59	-2.590	.1166E-03	2.164	-.2440
34.10	77.00	79.57	-2.567	.1580E-03	2.031	-.2824
36.10	62.00	69.20	-7.196	.2088E-03	1.910	-.9145
38.10	60.00	60.55	-.5452	.2728E-03	1.783	-.0795
40.10	49.00	53.37	-4.369	.3511E-03	1.647	-.7252
42.00	43.00	47.72	-4.716	.4392E-03	1.513	-.8774
45.10	38.00	40.49	-2.486	.6101E-03	1.295	-.5455
50.00	33.00	32.81	.1902	.9290E-03	.9853	.0512
55.00	28.00	28.12	-.1163	.1265E-02	.7576	-.0363
60.00	24.00	25.35	-1.347	.1557E-02	.6328	-.4647
70.00	20.20	22.77	-2.572	.1928E-02	.5878	-.9895
80.00	20.20	21.90	-1.695	.2086E-02	.6164	-.6816

90.20	16.00	21.59	-5.595	.2144E-02	.6378	-2.288
100.0	20.20	21.50	-1.297	.2164E-02	.6475	-.5337
110.0	20.20	21.46	-1.263	.2171E-02	.6517	-.5208
120.0	20.20	21.45	-1.252	.2173E-02	.6534	-.5164
122.8	34.00	21.45	12.55	.2173E-02	.6536	5.178
127.0	302.0	286.8	15.25	.1216E-04	7.815	.4673
129.0	301.0	270.7	30.30	.1365E-04	6.913	.9801
131.0	292.0	256.1	35.86	.1524E-04	6.133	1.222
133.0	225.0	243.0	-18.00	.1694E-04	5.467	-.6450
135.0	228.0	231.2	-3.174	.1871E-04	4.904	-.1193
140.0	180.0	206.7	-26.75	.2340E-04	3.889	-1.120
145.1	167.0	188.2	-21.18	.2824E-04	3.315	-.9730
150.0	152.0	175.1	-23.08	.3262E-04	3.058	-1.139
154.0	155.0	167.0	-12.04	.3584E-04	2.976	-.6231
156.0	137.0	146.8	-9.839	.4638E-04	2.463	-.5785
158.0	132.0	126.5	5.508	.6250E-04	2.218	.3763
160.0	99.00	109.0	-9.964	.8422E-04	2.068	-.7918
162.0	104.0	94.00	10.00	.1132E-03	1.964	.9242
164.0	83.00	81.32	1.679	.1512E-03	1.874	.1800
166.0	76.00	70.67	5.335	.2003E-03	1.779	.6606
169.0	62.00	57.90	4.100	.2983E-03	1.615	.6231
174.0	48.00	43.26	4.737	.5343E-03	1.300	.9681
179.0	41.00	34.36	6.642	.8471E-03	.9983	1.705
186.8	30.00	27.07	2.931	.1365E-02	.6835	.9474
218.0	20.00	21.64	-1.639	.2136E-02	.6341	-.6686
249.0	22.00	21.45	.5473	.2173E-02	.6532	.2258
250.0	20.20	21.45	-1.252	.2173E-02	.6533	-.5165
255.0	20.20	21.45	-1.249	.2174E-02	.6537	-.5155
262.2	370.0	338.0	31.97	.8752E-05	10.12	.8359
264.2	371.0	362.9	8.117	.7594E-05	10.15	.1968
265.9	382.0	382.5	-.5286	.6834E-05	10.17	-.0121
268.0	395.0	405.1	-10.14	.6092E-05	10.20	-.2191
270.0	394.0	425.2	-31.17	.5532E-05	10.24	-.6401
275.1	448.0	470.6	-22.59	.4516E-05	10.45	-.4178
280.0	480.0	508.0	-27.99	.3875E-05	10.92	-.4789
285.0	485.0	541.1	-56.11	.3415E-05	11.72	-.9016
290.0	548.0	570.1	-22.09	.3077E-05	12.85	-.3374
292.0	540.0	517.4	22.61	.3736E-05	12.20	.3812
294.1	455.0	461.7	-6.749	.4690E-05	11.72	-.1279
296.2	450.0	409.7	40.35	.5959E-05	11.37	.8663
298.1	380.0	365.7	14.29	.7477E-05	11.07	.3457
300.0	315.0	324.9	-9.878	.9475E-05	10.71	-.2708
302.4	288.0	277.9	10.14	.1295E-04	10.14	.3281
305.2	232.0	229.5	2.510	.1899E-04	9.279	.0995
310.1	160.0	161.2	-1.208	.3848E-04	7.439	-.0697
315.2	109.0	110.3	-1.266	.8225E-04	5.460	-.1083
320.0	82.00	77.70	4.298	.1656E-03	3.875	.5224
350.1	30.00	23.84	6.161	.1760E-02	.5974	2.261
380.0	22.00	21.54	.4600	.2155E-02	.6433	.1887
400.0	20.20	21.46	-1.257	.2172E-02	.6527	-.5183
450.0	20.20	21.45	-1.246	.2174E-02	.6544	-.5142

```
CORRECTED SUM OF SQUARED OBSERVATIONS =  .171323E+07
WEIGHTED CORRECTED SUM OF SQUARED OBSERVATIONS =  36.0332
SUM OF SQUARED RESIDUALS = 22288.4
SUM OF WEIGHTED SQUARED RESIDUALS =  .931327
S =  .117030  WITH  68 DEGREES OF FREEDOM
CORRELATION (OBSERVED,PREDICTED) =  .9935

AIC criteria =  2.87756
SC  criteria = 11.98422
```

SUMMARY OF ESTIMATED SECONDARY PARAMETERS

PARAMETER	ESTIMATE	STANDARD ERROR	CV%
T12	5.869623	.285030	4.86
KOUT	.118065	.005713	4.84
MRT	8.469875	.411299	4.86

Solution II – Linear turnover model

```
TITLE 1
IV-BOLUS + INFUSION
MODEL
 COMM
 NDER 1
 NPAR 3
 NCON 14
 PNAM 'Vc', 'Clp','Kin'
 NSEC 3
 SNAM 'T12', 'Kout', 'MRT'
END
    1: TEMP
    2: T3=CON(1)
    3: TI3=CON(2)
    4: DI3=CON(3)
    5: T2=CON(4)
    6: TI2=CON(5)
    7: DI2=CON(6)
    8: TI1=CON(7)
    9: DI1=CON(8)
   10: BD3=CON(9)
   11: BD2=CON(10)
   12: ETT=CON(11)
   13: TB3=CON(12)
   14: TB2=CON(13)
   15: BD1=CON(14)
   16: T=X
   17: FI1 = 0.0
   18: FI2 = 0.0
   19: FI3 = 0.0
   20: DOSI = 0.0
   21: DOSB = 0.0
   22: IF T GT T3 AND T LT T3+TI3 THEN
   23: DOSI = DI3/TI3
   24: ELSE
   25: IF T GT T2 AND T LT T2+TI2 THEN
   26: DOSI = DI2/TI2
   27: ELSE
   28: IF T LT TI1 THEN
   29: DOSI = DI1/TI1
   30: ENDIF
   31: ENDIF
   32: ENDIF
   33: IF T GT TB3 AND T LT TB3 + ETT THEN
   34: DOSB = BD3/ETT
   35: ELSE
   36: IF T GT TB2 AND T LT TB2 + ETT THEN
   37: DOSB = BD2/ETT
   38: ENDIF
   39: ENDIF
   40: END
```

```
41: START
42: Z(1)=BD1/VC + Kin/CLP
43: END
44: DIFF
45: DZ(1)= DOSI/VC + DOSB/VC - (CLP*Z(1))/VC + Kin/VC
46: END
47: FUNC 1
48: F = Z(1)
49: END
50: SECO
51: T12 = 0.693*VC/CLP
52: KOUT = CLP/VC
53: MRT = 1/KOUT
54: END
55: EOM
NVARIABLES 2
NPOINTS 100
XNUMBER 1
YNUMBER 2
CONSTANTS 260.1,30,6300,125.2,29.1,1884.4, &
          30.1,1131.8,1733,1701,1,260,125,1669
METHOD 2   'Gauss-Newton (Levenberg and Hartley)
REWEIGHT -2
ITERATIONS 50
INITIAL 5,1,15
LOWER BOUNDS 2,.1,0
UPPER BOUNDS 12,10,50
NOBSERVATIONS 72
BEGIN
```

PARAMETER	ESTIMATE	STANDARD ERROR	CV%	UNIVARIATE C.I.	
VC	5.700589	.323130	5.67	5.055961	6.345216
CLP	.426071	.013876	3.26	.398389	.453753
Kin	8.522350	.628335	7.37	7.268854	9.775846

```
*** CORRELATION MATRIX OF THE ESTIMATES ***
```

PARAMETER	VC	CLP	Kin
VC	1.00000		
CLP	.590725	1.00000	
Kin	.136028	.641637	1.00000

```
 Condition_number=        79.11

 CORRECTED SUM OF SQUARED OBSERVATIONS =  .171323E+07
 WEIGHTED CORRECTED SUM OF SQUARED OBSERVATIONS =  37.4048
 SUM OF SQUARED RESIDUALS = 90966.2
 SUM OF WEIGHTED SQUARED RESIDUALS =  3.01599
 S =  .209069  WITH 69 DEGREES OF FREEDOM
 CORRELATION (OBSERVED,PREDICTED) =  .9756

AIC criteria =        85.48272
SC  criteria =        92.31272
```

PK33 - Transdermal input and kinetics

Objectives
◆ To analyze plasma data following transdermal input
◆ To fit a one-compartment model with zero-order input to transdermal data

Problem specification

The present exercise is designed to help you learn and apply the concepts of modeling zero-order input functions in order to mimic a transdermal delivery system. In recent years, transdermal therapeutic systems have become increasingly popular for administration of sex hormones, nitroglycerine, nicotine (*Nicorette*®), and so on, as such dosage forms provide a reliable and convenient delivery system. For some drugs, the patch is changed twice weekly, and for others daily. By bypassing the liver and maintaining a continuous supply of drug to the target tissues, patches produce a more stable exposure of drug than, e.g., the oral route. After *release* of the active substance from the transdermal therapeutic system (patch), the absorption process involves penetration into the horny layer (stratum corneum), permeation (diffusion) through the epidermis and then uptake by the blood capillaries. An initial burst of active substance from the patch is seen after application of the device onto the skin. The release of the active substance is governed by the release from the reservoir across the control membrane.

In a study of a new transdermal therapeutic system in humans, data were obtained for nicotine after the first application of a patch. (Figure 33.1).

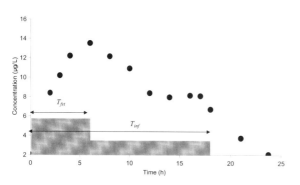

Figure 33.1 Observed plasma concentration-time data after the application of a nicotine patch. The shaded area corresponds to the initial burst (T_{fst}) of approximately 6 h overlapped by the maintenance release (T_{inf}) over 17-18 h. The length of the rapid and prolonged input time will be estimated in parallel with the relative amount released during the two release periods.

A nicotine patch was applied to volunteers for 16 h for 5 consecutive days. Plasma concentration-time data of nicotine were obtained on day 1 and day 5. Data from day 5 will be analyzed in this exercise. The estimated total dose released from each patch was on average 15890 μg. We assume that a multiple zero order input could mimic the data and that a one-compartment model could describe the disposition data of nicotine in this experimental setting.

Figure 33.2 demonstrates conceptually the instantaneous rapid release or burst from the patch upon application to the skin, and the parallel controlled maintenance infusion across the skin. The transport processes across the skin are, of course, more complex than we have demonstrated in this example. However, high resolution data about the release rate from the patch, uptake into various layers of the skin, and appearance in and disappearance from the plasma would be needed to elaborate a more mechanistic model. We will model burst and maintenance processes as two constant infusions running in parallel, one of which is short and rapid, while the other occurs at a slower rate. Even though the patch is removed at 16 h, the plasma concentration is maintained for some time due to the *depot* of drug in the skin. We will estimate the relative contribution (amount and length of time) of each infusion and the total input time over which drug is being absorbed into the circulation.

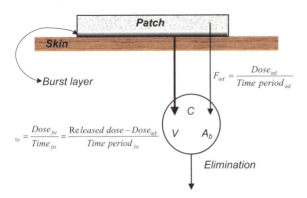

Figure 33.2 Schematic diagram of the proposed model for the patch-skin-plasma system. The horizontal black band and the thick arrow represent the initial burst process.

Concentration data will be modeled as

$$V \cdot \frac{dC}{dt} = F_{inf} + F_{fst} - Cl \cdot C \qquad (33:1)$$

F_{fst} (burst infusion) and F_{inf} (maintenance infusion) are defined as

$$F_{fst} = \frac{Dose_{fst}}{Time_{fst}} = \frac{Released\ dose - Dose_{inf}}{Time\ period_{fst}} \qquad (33:2)$$

$$F_{inf} = \frac{Dose_{inf}}{Time\ period_{inf}} \qquad (33:3)$$

The maintenance dose $Dose_{inf}$ is a model parameter together with burst period T_{fst} and the maintenance infusion time T_{inf}. We apply the secondary parameter option in order to estimate the fraction of the total dose that is released during both the burst FRC_1 and the maintenance period FRC_2. Note also that the baseline value of nicotine was 2 µg/L. Therefore, the initial condition (START) in the program has to be set to $Z(1) = 2.0$. We will fit the model using a variance proportional to the predicted mean value. Dr. Molander, Pharmacia Upjohn AB, Sweden, kindly supplied the data.

Initial parameter estimates

V/F = 140 (L) V in the model
Cl/F = 78 (L/h) Cl in the model
$Dose_{inf}$ = 10000 (µg)   Dose of slow infusion
T_{fst} = 6 (h)   Length of rapid infusion
T_{inf} = 17 (h)   Length of slow infusion (ideally 18 h)

Values on Cl and V were derived from Gorrod and Wahren [1993].

Interpretation of results and conclusions

We succeeded in fitting a simple one-compartment model with multiple controlled input rates to the data. The relative contribution of the burst and maintenance infusions was also estimated. Analysis of the plasma data suggests that the fraction of the total released dose that was released during the burst was approximately 37%, with a 63% release occurring during the maintenance infusion. The burst lasted for about 6.5 h and the maintenance infusion for 18 h (Figure 33.3). Even though the patch was removed after 16 h, nicotine continued to be released from the skin *depot* for two h. This meant that the skin area functions as a depot for nicotine release during the remaining period. Cl and V were estimated to be 80 L/h and 337 L, respectively.

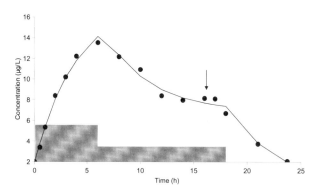

Figure 33.3 *Observed and predicted plasma concentration-time data during the 24 h period following application of the first patch. The arrow indicates the time (16 h) at which the patch was removed.*

This model produced a good fit with low correlation between parameters and high precision. This exercise demonstrated how to model systemic (plasma) data after application of the transdermal patch. By means of the proposed model, we have estimated the relative contribution of the dose from the burst and maintenance infusions.

Solution

```
TITLE 1
Transdermal input model
MODEL
 COMM
  NDER 1
  NCON 1
  nparm 5
  nsec  2
```

```
    PNAMES 'V','Cl','DINF', 'TFST', 'Tinf'
    SNAMES 'FRC1', 'FRC2'
END
     1: TEMP
     2: DOSE1=CON(1)
     3: T=X
     4: FINF = 0.0
     5: IF T LE TINF THEN
     6: FINF = DINF/TINF
     7: ELSE
     8: FINF = 0.0
     9: ENDIF
    10: BINF = 0.0
    11: IF T LE TFST THEN
    12: BINF = (DOSE1 - DINF)/TFST
    13: ELSE
    14: BINF = 0.0
    15: ENDIF
    16: END
    17: START
    18: Z(1)=2.0
    19: END
    20: DIFF
    21: DZ(1)=(FINF + BINF - CL*Z(1))/V
    22: END
    23: FUNC 1
    24: F = Z(1)
    25: END
    26: SECO
    27: FRC1=(DOSE1 - DINF)/DOSE1
    28: FRC2=DINF/DOSE1
    29: S(1)=FRC1
    30: S(2)=FRC2
    31: END
    32: EOM
DATE
TIME
NOPAGE BREAKS
NVARIABLES 2
NPOINTS 100
XNUMBER 1
YNUMBER 2
CONSTANTS 15890
METHOD 2  'Gauss-Newton (Levenberg and Hartley)
REWEIGHT -1
ITERATIONS 50
INITIAL 140,78,10000,6,17
LOWER BOUNDS 0,0,500,0,0
UPPER BOUNDS 500,190,15000,8,30
NOBSERVATIONS 16
BEGIN
```

PARAMETER	ESTIMATE	STANDARD ERROR	CV%	UNIVARIATE C.I.	
V	336.918037	49.940253	14.82	226.999862	446.836211
CL	80.163090	1.521618	1.90	76.814019	83.512160
DINF	10086.135967	721.475598	7.15	8498.172817	11674.099117
TFST	6.543292	.515666	7.88	5.408315	7.678270

```
TINF        18.252699        .543921        2.98        17.055533        19.449865

*** CORRELATION MATRIX OF THE ESTIMATES ***
PARAMETER   V            CL          DINF        TFST        TINF
V           1.00000
CL           .208457    1.00000
DINF        -.822572     .05229     1.00000
TFST        -.670040    -.36293      .254729   1.00000
TINF        -.850566    -.26233      .644824    .592046    1.00000

   Condition_number=          2777.

    FUNCTION    1

    X           OBSERVED    PREDICTED   RESIDUAL    WEIGHT      SE-PRED  STANDARDI
                Y           Y                                            RESIDUAL
    .0000       2.000       2.000        .0000       .5000
    .5000       3.400       3.790       -.3899       .2639      .9920E-01  -1.216
   1.000        5.360       5.379       -.1901E-01   .1859      .1649      -.0522
   2.000        8.400       8.043        .3575       .1243      .2298       .8286
   3.000       10.20       10.14         .5791E-01   .9860E-01  .2570       .1194
   4.000       12.22       11.80         .4229       .8477E-01  .2976       .8265
   6.000       13.53       14.13        -.6000       .7077E-01  .4743      -1.360
   8.000       12.17       12.34        -.1717       .8103E-01  .5107      -.5281
  10.00        10.93       10.28         .6514       .9729E-01  .2891      1.383
  12.00         8.390       8.997       -.6068       .1112      .2177      -1.294
  14.00         7.970       8.200       -.2303       .1219      .2063      -.5136
  16.00         8.150       7.705        .4447       .1298      .2255      1.054
  17.00         8.100       7.533        .5666       .1327      .2405      1.391
  18.00         6.700       7.398       -.6978       .1352      .2568      -1.779
  21.00         3.740       3.830       -.9042E-01   .2611      .3003      -.5885
  23.73         2.070       2.001        .6945E-01   .4999      .1892       .4519

CORRECTED SUM OF SQUARED OBSERVATIONS =   197.318
WEIGHTED CORRECTED SUM OF SQUARED OBSERVATIONS =   35.7459
SUM OF SQUARED RESIDUALS =                2.71598
SUM OF WEIGHTED SQUARED RESIDUALS =  .326732
S =  .172345      WITH      11 DEGREES OF FREEDOM
CORRELATION (OBSERVED,PREDICTED) =   .9931

AIC criteria =           -7.89784
SC  criteria =           -4.03490

SUMMARY OF ESTIMATED SECONDARY PARAMETERS

   PARAMETER       ESTIMATE        STANDARD        CV%
                                   ERROR
   FRC1             .365253         .045404       12.43
   FRC2             .634747         .045404        7.15
```

PK34 - Reversible metabolism

Objectives
◆ **To implement a set of differential equations mimicking reversible metabolism**
◆ **To estimate the kinetic parameters governing the disposition and interconversion between the two compounds**

Problem specification

The aim of this exercise is to demonstrate how to implement a system of differential equations describing reversible metabolism. Data were taken from the literature on cisplatin kinetics (Andersson [1995]) and used to generate synthetic concentration-time values for cisplatin (p) and its monohydrate (m). This information was then used to obtain initial parameter estimates for a model of reversible metabolism. This exercise demonstrates how to implement the equations for this model into the *WinNonlin* program. For further reading on reversible metabolism, see Ebling and Jusko [1986].

In an infusion solution of cisplatin, equilibrium is established between cisplatin and its monohydrate complex. Therefore, the input rate can be split into In_p (*cisplatin infusion rate*) and In_m (monohydrate infusion rate). We have also chosen to describe the equilibrium process *in vivo* between cisplatin and its monohydrate complex by means of two clearance rates Cl_{pd} and Cl_{md}. The model for reversible metabolism that takes account of such a split input rate and clearance is depicted in Figure 34.1.

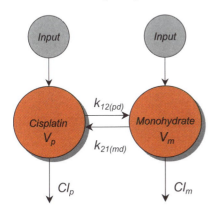

Figure 34.1 *Two-compartment model mimicking reversible metabolism.*

The equation for cisplatin (p) is

$$V_p \cdot \frac{dC_p}{dt} = Input_p - Cl_p \cdot C_p - Cl_{pd} \cdot C_p + Cl_{md} \cdot C_m \tag{34:1}$$

The conversion of cisplatin to the monohydrate occurs with a rate constant derived according to Equation 34:2

$$k_{12(pd)} = Cl_{pd} \cdot V_p \tag{34:2}$$

The reversal rate for conversion of the monohydrate to cisplatin is

$$k_{21(md)} = Cl_{md} \cdot V_m \tag{34:3}$$

The equation for the monohydrate (m) is

$$V_m \cdot \frac{dC_m}{dt} = Input_m - Cl_m \cdot C_m - Cl_{md} \cdot C_m + Cl_{pd} \cdot C_p \tag{34:4}$$

Initial parameter estimates

Procedures involved in obtaining initial parameter estimates, *AUC, MRT,* and so on for a system of reversible processes are covered in depth in Ebling and Jusko [1986]. The initial parameter estimates were selected that differed a little from the values used to generate the data. The model contains four differential equations, two defining the disposition at each of two dose levels. Data are available for both parent compound (cisplatin) and its monohydrate in the program output.

Interpretation of results and conclusions

Procedures involved in obtaining initial parameter estimates, *AUC, MRT,* and so on for a system of reversible processes are covered in depth in Ebling and Jusko [1986]. The observed and model predicted plasma concentrations, and the final parameter estimates are shown in Figure 34.2 and Table 34.1, respectively. *WRSS* in Table 34.1 denotes the weighted residual sum of squares.

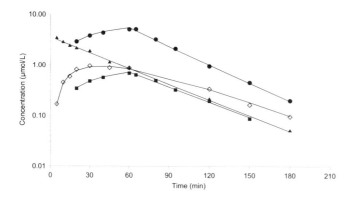

Figure 34.2 Observed and predicted data of cisplatin and its monohydrate. The symbols correspond to cisplatin (●) and the monohydrate (■) after infusion of cisplatin which also contains the monohydrate, and cisplatin (◊) and monohydrate (▲) after monohydrate infusion.

Table 34.1 Parameter estimates of the micro-constant and interconversion clearance models

Microconstant model			Interconversion clearance model		
Parameter	Estimate	CV%	Parameter	Estimate	CV%
V_c	14.1	3	V_c	14.1	3
Cl_m	0.0085	22	Cl_m	0.0084	22
V_m	2.96	2	V_m	2.97	2
Cl_p	0.446	2	Cl_p	0.445	2
k_{12}	0.00021	34	Cl_{d1}	0.0031	33
k_{21}	0.021	3	Cl_{d2}	0.063	4
WRSS	0.0089	-	*WRSS*	0.0089	-

Solution I – Microconstant model

```
Model
 COMM
 NDER   4
 NFUN   4
 NPARM  6
 NCON   6
 PNAMES 'Vc','CLm','Vm','CLp', 'k12', 'k21'
END
    1: TEMP
    2: DOSE=CON(1)
    3: TINF=CON(2)
    4: TINF2=CON(3)
    5: DOSE2=CON(4)
    6: TINF3=CON(5)
    7: DOSE3=CON(6)
    8: T=X
    9: VC=P(1)
   10: CLM=P(2)
   11: VM=P(3)
   12: CLP=P(4)
   13: K12 = P(5)
   14: K21 = P(6)
   15: FINF1=0.0
   16: IF T GE 0. AND T LE TINF THEN
   17: FINF1 = DOSE/TINF
   18: ENDIF
   19: FINF2=0.0
   20: IF T GE 0. AND T LE TINF2 THEN
   21: FINF2 = DOSE2/TINF2
   22: ENDIF
   23: FINF3=0.0
   24: IF T GE 0. AND T LE TINF3 THEN
   25: FINF3 = DOSE3/TINF3
   26: ENDIF
   27: END
   28: START
   29: Z(1)=0.0
   30: Z(2)=0.0
   31: Z(3)=0.0
   32: Z(4)=0.0
   33: END
   34: DIFF
   35: DZ(1) = FINF1/VC - Z(1)*K12 - Z(1)*CLP/VC + Z(2)*K21
   36: DZ(2) = FINF2/VM + Z(1)*K12 - Z(2)*CLM/VM - Z(2)*K21
   37: DZ(3) = - Z(3)*K12 - Z(3)*CLP/VC + Z(4)*K21
   38: DZ(4) = FINF3/VM + Z(3)*K12 - Z(4)*CLM/VM - Z(4)*K21
```

```
39: END
40: FUNC 1
41: F = Z(1)
42: END
43: FUNC 2
44: F = Z(2)
45: END
46: FUNC 3
47: F = Z(3)
48: END
49: FUNC 4
50: F = Z(4)
51: END
52: EOM
NOPAGE BREAKS
NVARIABLES 3
FNUMBER 3
NPOINTS 100
XNUMBER 1
YNUMBER 2
CONSTANTS 164.1,60,60,3.9,5,10
METHOD 2  'Gauss-Newton (Levenberg and Hartley)
REWEIGHT -2
ITERATIONS 50
INITIAL 15,.01,3,.5,.0002,.02
LOWER BOUNDS 0,0,0,0,0,0
UPPER BOUNDS 30,.5,30,5,.001,.1
MISSING 'Missing'
DATA 'WINNLIN.DAT'
BEGIN
```

PARAMETER	ESTIMATE	STANDARD ERROR	CV%	UNIVARIATE C.I.	
VC	14.130960	.365617	2.59	13.386229	14.875692
CLM	.008522	.001842	21.61	.004770	.012274
VM	2.964480	.054095	1.82	2.854293	3.074668
CLP	.445677	.008758	1.97	.427838	.463517
K12	.000209	.000071	34.00	.000064	.000354
K21	.021217	.000574	2.71	.020047	.022387

```
*** CORRELATION MATRIX OF THE ESTIMATES ***
PARAMETER  VC         CLM        VM         CLP        K12        K21
VC         1.00000
CLM         .348897   1.00000
VM         -.142129   -.67857    1.00000
CLP         .858306    .03934    -.027305   1.00000
K12         .076771   -.07875     .299136    .06591   1.00000
K21        -.278892   -.90406     .616634   -.15502    .34623   1.00000

 Condition_number=        7692.

   FUNCTION   1
```

X	OBSERVED Y	PREDICTED Y	RESIDUAL	WEIGHT	SE-PRED	STANDARDI RESIDUAL

20.00	2.898	2.931	-.3302E-01	.1164	.6766E-01	-.2375
30.00	3.796	3.869	-.7298E-01	.6681E-01	.8529E-01	-.3937
40.00	4.349	4.572	-.2227	.4784E-01	.9691E-01	-1.008
60.00	5.083	5.497	-.4137	.3309E-01	.1097	-1.542
65.00	5.111	4.761	.3496	.4412E-01	.9033E-01	1.492
80.00	3.222	3.094	.1276	.1045	.5326E-01	.8269
95.00	2.113	2.019	.9417E-01	.2453	.3486E-01	.9358
120.0	.9499	1.001	-.5101E-01	.9982	.2038E-01	-1.047
150.0	.4529	.4380	.1491E-01	5.213	.1142E-01	.7424
180.0	.1996	.1949	.4700E-02	26.33	.6220E-02	.5742

```
CORRECTED SUM OF SQUARED OBSERVATIONS =  30.2923
WEIGHTED CORRECTED SUM OF SQUARED OBSERVATIONS =  6.65884
SUM OF SQUARED RESIDUALS =              .377379
SUM OF WEIGHTED SQUARED RESIDUALS =  .221244E-01
S =  .743714E-01 WITH     4 DEGREES OF FREEDOM
CORRELATION (OBSERVED,PREDICTED) =  .9942
```

FUNCTION   2

X	OBSERVED Y	PREDICTED Y	RESIDUAL	WEIGHT	SE-PRED	STANDARDI RESIDUAL
20.00	.3432	.3537	-.1050E-01	7.993	.5540E-02	-.5895
30.00	.4787	.4792	-.5207E-03	4.354	.7165E-02	-.0214
40.00	.5655	.5794	-.1387E-01	2.979	.8414E-02	-.4722
60.00	.6874	.7228	-.3540E-01	1.914	.1028E-01	-.9644
65.00	.6351	.6463	-.1116E-01	2.394	.9327E-02	-.3405
80.00	.4933	.4603	.3299E-01	4.720	.7684E-02	1.432
95.00	.3230	.3272	-.4247E-02	9.338	.6624E-02	-.2665
120.0	.1947	.1847	.1002E-01	29.32	.4957E-02	1.195
150.0	.8730E-01	.9254E-01	-.5238E-02	116.8	.3199E-02	-1.421

```
CORRECTED SUM OF SQUARED OBSERVATIONS =  .324428
WEIGHTED CORRECTED SUM OF SQUARED OBSERVATIONS =  3.67952
SUM OF SQUARED RESIDUALS =              .291534E-02
SUM OF WEIGHTED SQUARED RESIDUALS =  .156059E-01
S =  .721246E-01 WITH     3 DEGREES OF FREEDOM
CORRELATION (OBSERVED,PREDICTED) =  .9965
```

FUNCTION   3

X	OBSERVED Y	PREDICTED Y	RESIDUAL	WEIGHT	SE-PRED	STANDARDI RESIDUAL
5.000	.1678	.1631	.4682E-02	37.58	.3411E-02	.5928
10.00	.4507	.4328	.1791E-01	5.339	.8799E-02	.8502
15.00	.5905	.6296	-.3910E-01	2.523	.1242E-01	-1.270
20.00	.8120	.7680	.4396E-01	1.695	.1472E-01	1.165
30.00	.9562	.9154	.4082E-01	1.193	.1665E-01	.9007
45.00	.8809	.9345	-.5357E-01	1.145	.1600E-01	-1.149
60.00	.8715	.8360	.3553E-01	1.431	.1386E-01	.8488
120.0	.3371	.3308	.6258E-02	9.136	.6349E-02	.3850
150.0	.1646	.1842	-.1965E-01	29.46	.4124E-02	-2.233
180.0	.9650E-01	.9888E-01	-.2381E-02	102.3	.2588E-02	-.5258

```
CORRECTED SUM OF SQUARED OBSERVATIONS =  1.00063
WEIGHTED CORRECTED SUM OF SQUARED OBSERVATIONS =  4.39453
SUM OF SQUARED RESIDUALS =              .100333E-01
SUM OF WEIGHTED SQUARED RESIDUALS =  .290595E-01
S =  .852342E-01 WITH     4 DEGREES OF FREEDOM
CORRELATION (OBSERVED,PREDICTED) =  .9951
```

```
     FUNCTION    4

     X              OBSERVED    PREDICTED    RESIDUAL      WEIGHT      SE-PRED   STANDARDI
                    Y           Y                                                RESIDUAL
     5.000          3.391       3.178        .2130         .9901E-01   .5703E-01  1.351
     10.00          2.810       2.821        -.1149E-01    .1256       .4910E-01  -.0818
     15.00          2.381       2.502        -.1204        .1598       .4245E-01  -.9641
     20.00          2.133       2.218        -.8534E-01    .2032       .3687E-01  -.7685
     30.00          1.871       1.745        .1258         .3284       .2825E-01  1.436
     45.00          1.122       1.218        -.9591E-01    .6737       .1978E-01  -1.569
     60.00          .8564       .8512        .5237E-02     1.380       .1451E-01  .1233
     120.0          .2119       .2039        .7968E-02     24.05       .5155E-02  .8440
     180.0          .5070E-01   .4917E-01    .1530E-02     413.6       .1813E-02  .8249
```

```
     CORRECTED SUM OF SQUARED OBSERVATIONS = 10.7274
     WEIGHTED CORRECTED SUM OF SQUARED OBSERVATIONS =  7.05531
     SUM OF SQUARED RESIDUALS =            .924222E-01
     SUM OF WEIGHTED SQUARED RESIDUALS =   .222357E-01
     S = .860924E-01 WITH      3 DEGREES OF FREEDOM
     CORRELATION (OBSERVED,PREDICTED) =  .9959

     TOTALS FOR ALL CURVES COMBINED
     SUM OF SQUARED RESIDUALS =             .482749
     SUM OF WEIGHTED SQUARED RESIDUALS =   .890255E-01
     S = .527451E-01 WITH     32 DEGREES OF FREEDOM

     AIC criteria =          -79.91564
     SC  criteria =          -70.09012
```

Solution II – Interconversion clearance (*I*)

```
Model
 COMM
 NDER  4
 NFUN  4
 NPARM 6
 NCON  6
 PNAMES 'Vc','CLm','Vm','CLp', 'Cld1', 'Cld2'
END
    1: TEMP
    2: DOSE=CON(1)
    3: TINF=CON(2)
    4: TINF2=CON(3)
    5: DOSE2=CON(4)
    6: TINF3=CON(5)
    7: DOSE3=CON(6)
    8: T=X
    9: K12 = CLD1/VC
   10: K21 = CLD2/VM
   11: FINF1=0.0
   12: IF T GE 0. AND T LE TINF THEN
   13: FINF1 = DOSE/TINF
   14: ENDIF
   15: FINF2=0.0
   16: IF T GE 0. AND T LE TINF2 THEN
   17: FINF2 = DOSE2/TINF2
   18: ENDIF
   19: FINF3=0.0
   20: IF T GE 0. AND T LE TINF3 THEN
```

```
 21: FINF3 = DOSE3/TINF3
 22: ENDIF
 23: END
 24: START
 25: Z(1)=0.0
 26: Z(2)=0.0
 27: Z(3)=0.0
 28: Z(4)=0.0
 29: END
 30: DIFF
 31: DZ(1) = FINF1/VC - Z(1)*K12 - Z(1)*CLP/VC + Z(2)*K21
 32: DZ(2) = FINF2/VM + Z(1)*K12 - Z(2)*CLM/VM - Z(2)*K21
 33: DZ(3) = - Z(3)*K12 - Z(3)*CLP/VC + Z(4)*K21
 34: DZ(4) = FINF3/VM + Z(3)*K12 - Z(4)*CLM/VM - Z(4)*K21
 35: END
 36: FUNC 1
 37: F = Z(1)
 38: END
 39: FUNC 2
 40: F = Z(2)
 41: END
 42: FUNC 3
 43: F = Z(3)
 44: END
 45: FUNC 4
 46: F = Z(4)
 47: END
 48: EOM
NOPAGE BREAKS
NVARIABLES 3
FNUMBER 3
NPOINTS 100
XNUMBER 1
YNUMBER 2
CONSTANTS 164.1,60,60,3.9,5,10
METHOD 2   'Gauss-Newton (Levenberg and Hartley)
REWEIGHT -2
ITERATIONS 50
INITIAL 15,.01,3,.5,.003,.3
LOWER BOUNDS 0,0,0,0,0,0
UPPER BOUNDS 30,.5,30,5,.5,1
MISSING 'Missing'
DATA 'WINNLIN.DAT'
BEGIN
```

PARAMETER	ESTIMATE	STANDARD ERROR	CV%	UNIVARIATE C.I.	
VC	14.130675	.365461	2.59	13.386260	14.875090
CLM	.008444	.001847	21.88	.004681	.012207
VM	2.967338	.054105	1.82	2.857130	3.077547
CLP	.445804	.008755	1.96	.427970	.463638
CLD1	.003133	.001018	32.50	.001059	.005208
CLD2	.063130	.002583	4.09	.057869	.068391

```
*** CORRELATION MATRIX OF THE ESTIMATES ***
```

```
PARAMETER  VC         CLM        VM         CLP        CLD1       CLD2
VC         1.00000
CLM         .34929    1.00000
VM         -.14218    -.678483   1.00000
CLP         .85798     .038953   -.02702    1.00000
CLD1        .15836    -.049863    .28389     .13696    1.00000
CLD2       -.24781    -.900671    .85357    -.02174     .33852    1.00000


     Condition_number=          1064.

      FUNCTION    1

     X          OBSERVED   PREDICTED   RESIDUAL    WEIGHT    SE-PRED  STANDARDI
                Y          Y                                          RESIDUAL
     20.00       2.898      2.931      -.3264E-01   .1164     .6762E-01  -.2350
     30.00       3.796      3.868      -.7229E-01   .6684E-01 .8523E-01  -.3903
     40.00       4.349      4.571      -.2217       .4787E-01 .9684E-01 -1.005
     60.00       5.083      5.495      -.4123       .3311E-01 .1096     -1.537
     65.00       5.111      4.759       .3511       .4415E-01 .9025E-01  1.500
     80.00       3.222      3.093       .1290       .1046     .5321E-01   .8368
     95.00       2.113      2.018       .9521E-01   .2455     .3483E-01   .9472
     120.0        .9499     1.000      -.5053E-01   .9991     .2036E-01 -1.039
     150.0        .4529      .4379      .1502E-01  5.215      .1141E-01   .7484
     180.0        .1996      .1949      .4666E-02  26.32      .6214E-02   .5701


CORRECTED SUM OF SQUARED OBSERVATIONS = 30.2923
WEIGHTED CORRECTED SUM OF SQUARED OBSERVATIONS =  6.66264
SUM OF SQUARED RESIDUALS =             .377183
SUM OF WEIGHTED SQUARED RESIDUALS =  .221606E-01
S =  .744322E-01 WITH      4 DEGREES OF FREEDOM
CORRELATION (OBSERVED,PREDICTED) =  .9942

      FUNCTION    2

     X          OBSERVED   PREDICTED   RESIDUAL    WEIGHT    SE-PRED  STANDARDI
                Y          Y                                          RESIDUAL
     20.00       .3432      .3536      -.1042E-01  7.997     .5532E-02  -.5853
     30.00       .4787      .4792      -.5452E-03  4.354     .7158E-02  -.0225
     40.00       .5655      .5795      -.1404E-01  2.977     .8410E-02  -.4781
     60.00       .6874      .7233      -.3589E-01  1.911     .1029E-01  -.9777
     65.00       .6351      .6469      -.1180E-01  2.390     .9339E-02  -.3599
     80.00       .4933      .4612       .3214E-01  4.702     .7705E-02   1.394
     95.00       .3230      .3281      -.5093E-02  9.290     .6644E-02  -.3190
     120.0        .1947      .1853      .9358E-02 29.11      .4969E-02   1.112
     150.0        .8730E-01 .9295E-01 -.5653E-02 115.7      .3205E-02  -1.525


CORRECTED SUM OF SQUARED OBSERVATIONS =  .324428
WEIGHTED CORRECTED SUM OF SQUARED OBSERVATIONS =  3.66737
SUM OF SQUARED RESIDUALS =            .291204E-02
SUM OF WEIGHTED SQUARED RESIDUALS =  .155971E-01
S =  .721042E-01 WITH      3 DEGREES OF FREEDOM
CORRELATION (OBSERVED,PREDICTED) =  .9966

      FUNCTION    3

     X          OBSERVED   PREDICTED   RESIDUAL    WEIGHT    SE-PRED  STANDARDI
                Y          Y                                          RESIDUAL
     5.000       .1678      .1634       .4405E-02  37.46     .3417E-02   .5571
```

10.00	.4507	.4335	.1722E-01	5.322	.8811E-02	.8168
15.00	.5905	.6305	-.4002E-01	2.515	.1244E-01	-1.299
20.00	.8120	.7691	.4293E-01	1.691	.1473E-01	1.137
30.00	.9562	.9164	.3981E-01	1.191	.1666E-01	.8779
45.00	.8809	.9352	-.5428E-01	1.143	.1600E-01	-1.164
60.00	.8715	.8363	.3516E-01	1.430	.1386E-01	.8402
120.0	.3371	.3306	.6482E-02	9.148	.6339E-02	.3993
150.0	.1646	.1840	-.1944E-01	29.52	.4115E-02	-2.213
180.0	.9650E-01	.9873E-01	-.2231E-02	102.6	.2582E-02	-.4937

```
CORRECTED SUM OF SQUARED OBSERVATIONS =  1.00063
WEIGHTED CORRECTED SUM OF SQUARED OBSERVATIONS =  4.38898
SUM OF SQUARED RESIDUALS =         .995345E-02
SUM OF WEIGHTED SQUARED RESIDUALS = .285266E-01
S =  .844491E-01 WITH     4 DEGREES OF FREEDOM
CORRELATION (OBSERVED,PREDICTED) =  .9952
```

```
   FUNCTION   4
```

X	OBSERVED Y	PREDICTED Y	RESIDUAL	WEIGHT	SE-PRED	STANDARDI RESIDUAL
5.000	3.391	3.175	.2163	.9921E-01	.5694E-01	1.374
10.00	2.810	2.818	-.8181E-02	.1259	.4903E-01	-.0583
15.00	2.381	2.498	-.1172	.1602	.4240E-01	-.9398
20.00	2.133	2.215	-.8216E-01	.2038	.3683E-01	-.7415
30.00	1.871	1.742	.1287	.3295	.2823E-01	1.473
45.00	1.122	1.216	-.9350E-01	.6764	.1977E-01	-1.534
60.00	.8564	.8493	.7148E-02	1.387	.1449E-01	.1688
120.0	.2119	.2033	.8570E-02	24.19	.5138E-02	.9111
180.0	.5070E-01	.4901E-01	.1691E-02	416.3	.1806E-02	.9151

```
CORRECTED SUM OF SQUARED OBSERVATIONS =  10.7274
WEIGHTED CORRECTED SUM OF SQUARED OBSERVATIONS =  7.07871
SUM OF SQUARED RESIDUALS =         .927773E-01
SUM OF WEIGHTED SQUARED RESIDUALS = .226360E-01
S =  .868639E-01 WITH     3 DEGREES OF FREEDOM
CORRELATION (OBSERVED,PREDICTED) =  .9959
```

```
TOTALS FOR ALL CURVES COMBINED
SUM OF SQUARED RESIDUALS =         .482826
SUM OF WEIGHTED SQUARED RESIDUALS = .889203E-01
S =  .527139E-01 WITH    32 DEGREES OF FREEDOM
```

```
AIC criteria =        -79.96057
SC  criteria =        -70.13505
```

PK35 - Bayesian model - Digoxin

Objectives
◆ To set up a Bayesian model
◆ To implement the model into *WinNonlin*

Problem specification

The aim of this exercise is to demonstrate how to implement a Bayesian model into nonlinearregression software. Minimization of the Bayesian objective function (Equation 35:1) results in estimates of pharmacokinetic parameters that are unique to a patient. These take into account measured and predicted drug concentration along with information about measurement error and the typical variability of values of pharmacokinetic parameters in the population (Peck and Rodman [1986]). In Equation 35:1, P_{pop} is defined as the population average of parameter P, *P-hat* is the individual expected average of parameter P, and var(*P-hat*) is the variance of the estimated parameter P. C_{pop} is the observed individual concentration value, *C-hat* is the corresponding predicted concentration value and, finally, var(*C-hat*) is the variance of the predicted concentration.

$$Bayes(LS) = \sum \frac{\left[P_{pop} - \hat{P}\right]^2}{\mathrm{var}(\hat{P})} + \sum \frac{\left[C_{obs} - \hat{C}\right]^2}{\mathrm{var}(\hat{C})} \qquad (35{:}1)$$

The relationship between concentration and P_{pop} and *P-hat* where P is clearance Cl, is shown in Figure 35.1. The blue observation below the dashed population mean curve in the concentration-time plot is partly due to the predicted high individual clearance value (*Cl-hat*). *Cl-hat* is greater than the population mean (Cl_{pop}), which results in a lower concentration than predicted from the population mean parameters.

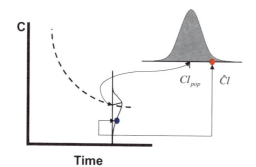

Figure 35.1 *Schematic representation of the relationship between Cl_{pop}, Cl-hat (red filled circle), predicted mean concentration (dashed line) and the observed individual concentration (blue filled circle).*

The Bayesian objective function encompasses a wide variety of the usual methods for estimating an individual's pharmacokinetic parameters (assuming normally distributed population parameters). When no drug concentrations are available in an individual (prior to starting therapy), the usual basis for assigning parameter values is to assume population average values. Since we have no such concentrations, Bayes' objective function for our example reduces to only the first summation term of which the minimum is the set of population pharmacokinetic parameters. When drug concentrations are available, but no population-based prior expectations are admitted, the Bayesian objective function reduces to the maximum-likelihood estimate. When prior population parameters and drug levels are available, the complete Bayesian method is expressed as shown in Equation 35:1 (Peck and Rodman [1986]).

A 55 year old, 60 kg man with congestive heart failure (CHF) stated that he took his Lanoxicap (0.2 mg) each day at 9:00 a.m. Since he was doing well, he was given a refill for the Lanoxicap. However, he was told that he needed to have some blood tests done. The digoxin level was 2.5 µg/L at 11:05 a.m., the time of the blood sample on the day of the hospital visit. Forty six h (and 2 doses) later, the level was 0.9 µg/L. Assume the patient had been taking medication long enough that his plasma levels are at steady-state.

The two plasma concentrations of 2.5 µg/L and 0.9 µg/L were obtained at 458 and 479 h, respectively. The clearance and volume of distribution were approximated to 1.8 L/h and 500 L, with a corresponding variance proportional to 3.24 and 250000, respectively.

The underlying model, for which we assume that absorption is much faster than elimination, and could be approximated with a bolus input, is shown in Figure 35.2 and the fit Equation is 35:2.

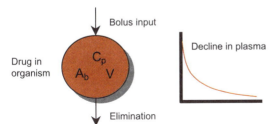

Figure 35.2 Schematic illustration of the one-compartment model.

$$C = C^0 \cdot e^{-K \cdot t} = \frac{D}{V} \cdot e^{-\frac{Cl}{V} \cdot t} \qquad (35:2)$$

Equations 35:3 and 35:4 give the objective function of the difference between population values for clearance Cl_{pop} and volume of distribution $V_{pop,}$ and the model-predicted values.

$$Objective\ function\ (Cl) = Cl_{pop} - \hat{Cl} \qquad (35:3)$$

$$Objective\ function\ (V) = V_{pop} - \hat{V} \qquad (35:4)$$

The complete objective function for plasma concentration, clearance and volume becomes

$$Bayes(LS) = \sum \frac{\left[Cl_{pop} - \hat{Cl}\right]^2}{var(\hat{Cl})} + \sum \frac{\left[V_{pop} - \hat{V}\right]^2}{var(\hat{V})} + \sum \frac{\left[C_{obs} - \hat{C}\right]^2}{var(\hat{C})} \qquad (35:5)$$

Initial parameter estimates

$V = 350$ (L)

Cl (oral) $= 2$ (L/h)

Interpretation of results and conclusions

We have fit plasma concentration data as a function of clearance Cl and volume of distribution V. Cl and V for this subject were estimated to 5.7 L/h and 119.6 L, respectively. The resulting half-life was 14.5 h. The predicted volume of distribution deviates substantially from the population average. This may be explained by the fact that the subject had a low body weight (63 kg). Volume of distribution of digoxin is highly correlated with body weight.

Solution

```
TITLE 1
TDM of digoxin in a CHF patient: From case report #2
MODEL
 COMM
NPARM 2
NSEC 3
NFUN 3
PNAMES 'V', 'CL'
SNAMES 'AUC', 'T1/2', 'CMAX'
END
      1: TEMP
      2: V=P(1)
      3: CL=P(2)
      4: K=P(2)/P(1)
      5: END
      6: FUNC1
      7: I=0
      8: J=1
      9: L=CON(1)
     10: GREEN:
     11: I=I+1
     12: J=J+2
     13: IF X <= CON(J) THEN GOTO RED
     14: ELSE IF I < L THEN GOTO GREEN
     15: ENDIF
     16: ENDIF
     17: I=I+1
     18: RED:
     19: L=I-1
     20: SUM=0
     21: I=0
     22: J=1
```

```
23: BLUE:
24: I=I+1
25: J=J+2
26: T=X - CON(J)
27: D=CON(J-1)
28: AMT= (D/V)*DEXP(-K*T)
29: SUM=SUM + AMT
30: IF I < L THEN GOTO BLUE
31: ENDIF
32: F=SUM
33: WT=1
34: END
35: FUNC2
36: F = V
37: WT = 1/250000.
38: END
39: FUNC3
40: F = CL
41: WT = 1/3.24
42: END
43: SECO
44: D=CON(2)
45: S(1)=D/V/K
46: S(2)=-DLOG(.5)/K
47: S(3) = D/V
48: END
49: EOM
NOPLOTS
NOPAGE BREAKS
NVARIABLES 3
FNUMBER 3
NPOINTS 100
XNUMBER 1
YNUMBER 2
NCONSTANTS 41
CONSTANTS 20,200,0,200,24,200,48,200,72,200,96,200,120, &
          200,144,200,168,200,192,200,216,200,240,200,264,200,288,&
          200,312,200,336,200,360,200,384,200,408,200,432,200,456
METHOD 3   'Gauss-Newton
ITERATIONS 50
INITIAL 350,2
LOWER BOUNDS 0,0
UPPER BOUNDS 1000,10
BEGIN
```

PARAMETER	ESTIMATE	STANDARD ERROR	CV%	UNIVARIATE C.I.	
V	119.599269	70.433002	58.89	-185.205818	424.404356
CL	5.719890	.907550	15.87	1.792385	9.647394

```
*** CORRELATION MATRIX OF THE ESTIMATES ***
PARAMETER  V          CL
V      1.00000
CL     -.125364   1.00000

 Condition_number=        78.23

   FUNCTION   1
```

X	OBSERVED	PREDICTED	RESIDUAL	WEIGHT	SE-PRED	STANDARDI

```
              Y           Y                                             RESIDUAL
     458.0    2.500      2.226     .2739      1.000      .5080      .9420
     479.0    .9000      .8154     .8459E-01  1.000      .3912      .1943

CORRECTED SUM OF SQUARED OBSERVATIONS = 1.28000
WEIGHTED CORRECTED SUM OF SQUARED OBSERVATIONS = 1.28000
SUM OF SQUARED RESIDUALS =              .821614E-01
SUM OF WEIGHTED SQUARED RESIDUALS = .821614E-01
S = .000000    WITH     0 DEGREES OF FREEDOM
CORRELATION (OBSERVED,PREDICTED) = 1.000


   FUNCTION   2

 X          OBSERVED   PREDICTED   RESIDUAL    WEIGHT    SE-PRED STANDARDI
              Y           Y                                             RESIDUAL
    .0000     500.0      119.6      380.4      .4000E-05 70.43      1.339

CORRECTED SUM OF SQUARED OBSERVATIONS = .000000
WEIGHTED CORRECTED SUM OF SQUARED OBSERVATIONS = -.868446E-16
SUM OF SQUARED RESIDUALS =              144705.
SUM OF WEIGHTED SQUARED RESIDUALS = .578819
S = .000000    WITH    -1 DEGREES OF FREEDOM
CORRELATION (OBSERVED,PREDICTED) = .0000


   FUNCTION   3

 X          OBSERVED   PREDICTED   RESIDUAL    WEIGHT    SE-PRED  STANDARDI
              Y           Y                                             RESIDUAL
    .0000     6.000      5.720      .2801      .3086     .9076      .5234

CORRECTED SUM OF SQUARED OBSERVATIONS = .000000
WEIGHTED CORRECTED SUM OF SQUARED OBSERVATIONS = .155431E-14
SUM OF SQUARED RESIDUALS =              .784617E-01
SUM OF WEIGHTED SQUARED RESIDUALS = .242166E-01
S = .000000    WITH    -1 DEGREES OF FREEDOM
CORRELATION (OBSERVED,PREDICTED) = .0000


TOTALS FOR ALL CURVES COMBINED
SUM OF SQUARED RESIDUALS =              144705.
SUM OF WEIGHTED SQUARED RESIDUALS = .685197
S = .585319    WITH     2 DEGREES OF FREEDOM

AIC criteria =         2.48780
SC  criteria =         1.26039

SUMMARY OF ESTIMATED SECONDARY PARAMETERS

PARAMETER      ESTIMATE        STANDARD       CV%
                                ERROR
AUC            34.965708       5.542316       15.85
T1/2           14.493268       9.112820       62.88
CMAX           1.672251        .983819        58.83
```

PK36 - Time-controlled drug delivery

Objectives
◆ **To practise using a model of time-controlled delivery**
◆ **To implement a pulsatile input system into a one-compartment model**
◆ **To use a cosine-function that includes peak shift, amplitude and frequency**

Problem specification

This exercise is aimed to show how to implement a system of pulsatile input and to demonstrate how to model that input. Oscillating functions have been applied for a variety of phenomena such as changes in body temperature, immunological responses and turnover of nicotine (Urquhart [1969], Jusko [1992], Haefner [1996]). The following exercise demonstrates the versatility of the cosine function to mimic time-controlled drug delivery.

A new subcutaneous device for pulsatile delivery was being tested in rats. The microdevice could be pre-programmed with respect to rate (flow), frequency, and amplitude. This type of device has potential for e.g., delivery for hormones and/or compounds (such as nicotine or cocaine) where one aims to obtain narrow plasma concentration peaks with a desired pre-programmed frequency. Figure 36.1 shows experimental data over a 6 h observational period in the rat.

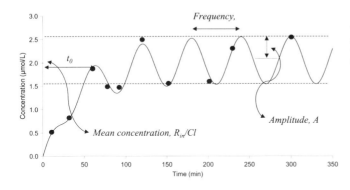

Figure 36.1 Observed and predicted plasma concentration-time data obtained from an oscillating input-device. The time shift t_0 is indicated by the horizontal arrow to the left of the figure and the amplitude (A) by the vertical arrow to the right. Mean input is equivalent to C_{ss}·Cl.

A total volume of less than 2 mL blood was drawn from each animal throughout the study. Data are based on venous blood. The following expression was used to mimic the input

$$Input = Mean + A \cdot Mean \cdot \cos[\varpi \cdot (t - t_0)]$$ (36:1)

Mean is the average input rate, *A* the amplitude and t_0 the shifted peak (phase shift, 60 min). ω scales the frequency of oscillations of the function to the physical frequency $(2\pi/60)$, i.e., the frequency is 60 min and π is 3.14. The implemented *Input function* is written as

$$Input = R_{in} + A \cdot R_{in} \cdot \cos\left[\varpi \cdot (t - t_0)\right] \tag{36:2}$$

The amplitude A is a fraction of the mean rate of input R_{in}. The differential equation model for the pharmacokinetic system employs volume of distribution V and clearance Cl as model parameters as follows

$$V \cdot \frac{dC}{dt} = Input - Cl \cdot C \tag{36:3}$$

Initial parameter estimates

V	=	0.1 (L)	
$Cl\ (sc)$	=	2.5 (L/h)	
A	=	0.4	range 0 – 1, written as *AMPLT* in program code.

Interpretation of results and conclusions

Results from fitting Equation 36:2 to the data are shown in Figure 36.1. Note in the program output how well the parameters were estimated. The amplitude A was about 25% of R_{in}, as shown in Figure 36.1. Note also the 60 min time shift (t_0).

Solution

```
MODEL
 COMM
  NPARM 3
  NCON  3
  PNAMES  'Cl', 'V', 'Amplt'
END
     1: TEMP
     2: DOSE1=CON(1)
     3: TINF =CON(2)
     4: CYCLE=CON(3)
     5: RIN = DOSE1/TINF
     6: CL =P(1)
     7: V =P(2)
     8: AMPLT=P(3)
     9: T=X
    10: END
    11: FUNC 1
    12: RINT = RIN + AMPLT*RIN*COS((2*3.14/CYCLE)*(T - 60))
    13: CP = (RINT/CL)*(1 - DEXP(-(CL/V)*T))
    14: F = CP
    15: END
    16: EOM
NOPAGE BREAKS
NVARIABLES 2
NPOINTS 500
XNUMBER 1
YNUMBER 2
CONSTANTS 100,1000,60
```

```
METHOD 2  'Gauss-Newton (Levenberg and Hartley)
ITERATIONS 50
INITIAL .1,2.5,.4
LOWER BOUNDS 0,0,0
UPPER BOUNDS 5,3,1
MISSING 'Missing'
NOBSERVATIONS 10
DATA 'WINNLIN.DAT'
BEGIN
```

PARAMETER	ESTIMATE	STANDARD ERROR	CV%	UNIVARIATE C.I.	
CL	.048638	.001051	2.16	.046152	.051123
V	2.119573	.149770	7.07	1.765421	2.473725
AMPLT	.248013	.019322	7.79	.202324	.293703

```
*** CORRELATION MATRIX OF THE ESTIMATES ***
PARAMETER  CL              V               AMPLT
CL      1.00000
V       -.512662        1.00000
AMPLT   .254137         -.747809E-01     1.00000

 Condition_number=        171.5

    FUNCTION    1
```

X	OBSERVED Y	PREDICTED Y	RESIDUAL	WEIGHT	SE-PRED	STANDARDI RESIDUAL
10.61	.5173	.4928	.2449E-01	1.000	.3059E-01	.3057
31.82	.8214	.8060	.1543E-01	1.000	.4064E-01	.2044
60.10	1.869	1.920	-.5063E-01	1.000	.6204E-01	-.8550
77.78	1.487	1.590	-.1026	1.000	.3666E-01	-1.324
91.92	1.473	1.367	.1054	1.000	.4491E-01	1.443
120.2	2.496	2.403	.9272E-01	1.000	.4666E-01	1.289
152.0	1.558	1.509	.4811E-01	1.000	.5041E-01	.6934
201.5	1.599	1.720	-.1216	1.000	.4595E-01	-1.680
229.8	2.307	2.286	.2103E-01	1.000	.4488E-01	.2878
300.5	2.546	2.563	-.1704E-01	1.000	.5850E-01	-.2717

```
CORRECTED SUM OF SQUARED OBSERVATIONS =  4.03337
WEIGHTED CORRECTED SUM OF SQUARED OBSERVATIONS =  4.03337
SUM OF SQUARED RESIDUALS =         .514819E-01
SUM OF WEIGHTED SQUARED RESIDUALS =  .514819E-01
S =  .857587E-01 WITH     7 DEGREES OF FREEDOM
CORRELATION (OBSERVED,PREDICTED) =  .9936

AIC criteria =        -23.66524
SC  criteria =        -22.75749
```

PK37 - *In vitro/in vivo* extrapolation I – Transformation plots

Objectives
◆ **To fit a rate equation to *in vitro* data**
◆ **To predict intrinsic and organ clearance, extraction ratio and bioavailability**
◆ **To study different weighting schemes and their impact on the parameters**
◆ **To propose an alternative design using *VIF* and partial derivatives**

Problem specification

This exercise highlights some complexities in modeling *in vitro concentrations versus metabolic rate data*. Specific emphasis is placed on obtaining *initial parameter estimates*, implementing *metabolic rate equations*, and practicing *simulation* when the model parameters have been estimated.

A new compound was being developed for the treatment of hypertension. In order to predict the bioavailability in man, an *in vitro* experiment was performed using human microsomes to obtain V_{max} and K_m. Data are given in Figure 37.1 and program output. The compound is eliminated via hepatic metabolism and no secretion of the parent compound occurs into urine.

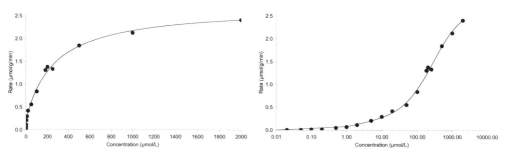

Figure 37.1 *Observed and predicted metabolic rate versus substrate concentration. Linear scale in the left hand figure and semi-logarithmic scale in the right hand figure.*

We propose the following relationship between rate (*Rate*) and the substrate concentration C,

$$Rate = \frac{V_{max1} \cdot C}{K_{m1} + C} + \frac{V_{max2} \cdot C}{K_{m2} + C} \tag{37:1}$$

since a single-enzyme system (Equation 37:2) resulted in systematic devations between observed and model predicted data

$$Rate = \frac{V_{max} \cdot C}{K_m + C} \tag{37:2}$$

Scale the *in vitro* data to a 1500 g liver. Assuming 77 mg of microsomal protein per gram liver, the product 77•1500 converts μmol/min/g microsomal liver protein to μmol•min/1.5 kg liver. Therefore we can obtain the human hepatic intrinsic clearance (L/min).

$$Cl_{int} = \frac{V_{max1} \cdot 77 \cdot 1.5}{K_{m1} + C} + \frac{V_{max2} \cdot 77 \cdot 1.5}{K_{m2} + C} \tag{37:3}$$

Organ clearance is

$$Cl_H = Q_H \cdot \frac{Cl_{int}}{Q_H + Cl_{int}} \tag{37:4}$$

and extraction ratio

$$E_H = \frac{Cl_{int}}{Q_H + Cl_{int}} \tag{37:5}$$

Bioavailability across the liver is

$$F_H = \frac{Q_H}{Q_H + Cl_{int}} \tag{37:6}$$

or rearranged

$$F_H = 1 - E_H \tag{37:7}$$

We assume that no compensation for tissue and/or plasma protein binding is needed for this loosely bound compound (f_u ~0.5). We assume that the hepatic blood flow can be approximated to 1.4 L/min. Figure 37.2 shows plots of Cl_{int} and Cl_H *versus* C for an appropriate concentration interval. These plots were generated by means of the final parameter estimates from fitting Equation 37:1 to the experimental data.

We will fit the data by means of three different weighting schemes: proportional error (*weight* = $1/C^2$), Poisson error (*weight* = $1/C$) and constant absolute error (*weight* = 1). AstraZeneca kindly supplied the original V_{max} and K_m data. New simulated data were then generated by means of adapted parameter values with a 3% noise level added.

Initial parameter estimates

V_{max1}	= 3	(μmol/g/min)	Maximum metabolic capacity
K_{m1}	= 250	(μmol/L)	Michaelis-Menten constant
V_{max2}	= 0.5	(μmol/g/min)	Maximum metabolic capacity
K_{m2}	= 2	(μmol/L)	Michaelis-Menten constant

How do you derive initial parameter estimates for this system? Can you do it by a graphical method? The *Lineweaver-Burke plot* is applicable here. Transformation plots have their limitations because the error structure is also transformed, albeit differently from the primary variables. We therefore only recommend this transformation to obtain the initial parameter estimates.

Interpretation of results and conclusions

Observed and predicted concentration-rate data are shown in Figure 37.2 (A). Because of the high quality of the experimental data, that included well-spaced substrate concentrations and low variability in the measured rate, precise parameter estimates were obtained for V_{max1}, K_{m1}, V_{max2} and K_{m2} (see Table 37.1). The highest parameter correlations were obtained for V_{max2} and K_{m2}.

Clearly, the three weighting schemes had very little or no impact on the parameter estimates, though they did affect precision.

Table 37.1 Final parameter estimates (mean) ± precision (CV%)

Parameter	Proportional error (mean ± CV%)	Poisson error (mean ± CV%)	Constant error (mean ± CV%)
V_{max1}	2.5 ± 10	2.4 ± 3.5	2.4 ± 2.9
K_{m1}	270 ± 23	270 ± 11	260 ± 11
V_{max2}	0.24 ± 15	0.24 ± 16	0.23 ± 29
K_{m2}	2.4 ± 17	2.7 ± 29	2.3 ± 84

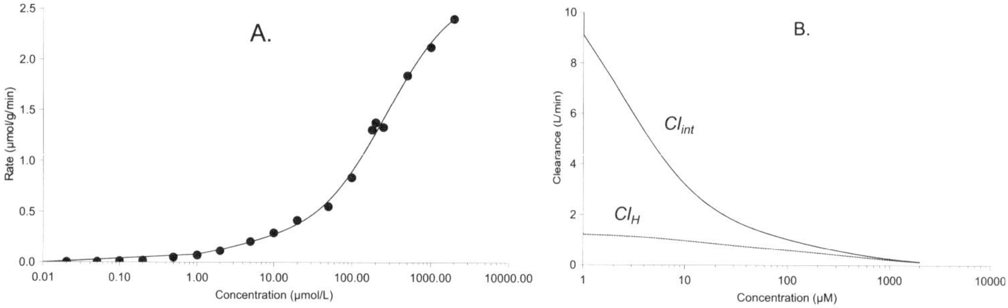

Figure 37.2 *Observed and predicted metabolic rate versus substrate concentration (A). Simulated Cl_{int} and Cl_H versus concentration (B).*

It is interesting to note how the clearance of this drug changes from high extraction characteristics at plasma concentrations less than 10 μmol/L, where clearance is highly dependent on the hepatic blood flow, to low extraction characteristics when plasma concentration exceeds 100 μmol/L (Figure 37.2 (B)). The variability in clearance (and plasma concentrations) will depend on variability in hepatic blood flow at low concentrations. Plasma concentrations exceeding 100 μmol/L will be highly variable because of the nonlinear clearance.

The last solution in the program code includes instructions about how to implement Cl_{int}, Cl_H, E_H, and F_H for simulation purposes.

Apply *VIF* (Variance Inflation Factor, in the simulation mode) and the partial derivatives of the dependent variable (rate) with respect to each parameter to improve the experimental design.

Solution I - Proportional error

```
MODEL
 COMM
  NPARM 4
  PNAME 'Vmax1', 'Km1', 'Vmax2', 'Km2'
END
    1: FUNC 1
    2: C = X
    3: F = VMAX1*C/(KM1 + C) + VMAX2*C/(KM2 + C)
    4: END
    5: EOM
NVARIABLES 2
NPOINTS 2000
XNUMBER 1
YNUMBER 2
METHOD 2  'Gauss-Newton (Levenberg and Hartley)
REWEIGHT -2
ITERATIONS 50
INITIAL 3,250,.5,2
LOWER BOUNDS 0,0,0,0
UPPER BOUNDS 20,1000,1,10
NOBSERVATIONS 18
DATA 'WINNLIN.DAT'
BEGIN
```

PARAMETER	ESTIMATE	STANDARD ERROR	CV%	UNIVARIATE C.I.	
VMAX1	2.461589	.237919	9.67	1.951306	2.971873
KM1	270.222916	60.676562	22.45	140.084940	400.360892
VMAX2	.235761	.035591	15.10	.159425	.312097
KM2	2.378050	.403244	16.96	1.513180	3.242919

```
*** CORRELATION MATRIX OF THE ESTIMATES ***
```

PARAMETER	VMAX1	KM1	VMAX2	KM2
VMAX1	1.00000			
KM1	.82530	1.00000		
VMAX2	.25874	.63861	1.00000	
KM2	.19668	.54855	.95335	1.00000

X	OBSERVED Y	PREDICTED Y	RESIDUAL	WEIGHT	SE-PRED	STANDARDI RESIDUAL
.2000E-01	.2400E-02	.2148E-02	.2516E-03	.2170E+06	.1041E-03	1.305
.5000E-01	.5300E-02	.5310E-02	-.1033E-04	.3552E+05	.2509E-03	-.0215
.1000	.1150E-01	.1042E-01	.1075E-02	9218.	.4730E-03	1.130
.2000	.1570E-01	.2011E-01	-.4410E-02	2477.	.8488E-03	-2.363
.5000	.4860E-01	.4550E-01	.3095E-02	483.7	.1666E-02	.7147
1.000	.6820E-01	.7887E-01	-.1067E-01	161.0	.2848E-02	-1.418
2.000	.1131	.1258	-.1269E-01	63.26	.5551E-02	-1.097
5.000	.2067	.2045	.2208E-02	23.93	.1146E-01	.1267
10.00	.2929	.2783	.1459E-01	12.91	.1512E-01	.6069
20.00	.4157	.3803	.3536E-01	6.914	.1774E-01	1.024
50.00	.5516	.6094	-.5781E-01	2.693	.2948E-01	-1.056
100.0	.8368	.8952	-.5838E-01	1.248	.4470E-01	-.7328
180.0	1.307	1.217	.8987E-01	.6754	.5338E-01	.8015
200.0	1.378	1.280	.9792E-01	.6104	.5436E-01	.8245
250.0	1.332	1.416	-.8499E-01	.4984	.5656E-01	-.6389
500.0	1.844	1.833	.1138E-01	.2978	.7737E-01	.0668
1000.	2.123	2.173	-.5022E-01	.2118	.1245	-.2736

```
   2000.      2.401      2.404      -.3569E-02  .1730       .1722       -.0204
```

```
   CORRECTED SUM OF SQUARED OBSERVATIONS =  11.0548
   WEIGHTED CORRECTED SUM OF SQUARED OBSERVATIONS =  14.8055
   SUM OF SQUARED RESIDUALS =          .360751E-01
   SUM OF WEIGHTED SQUARED RESIDUALS =  .145953
   S =  .102104      WITH      14 DEGREES OF FREEDOM
   CORRECTED (OBSERVED,PREDICTED) =  .9984
```

```
 AIC criteria =        -26.64043
 SC  criteria =        -28.85969
```

Solution II - Poisson error

```
MODEL
 COMM
  NPARM 4
  PNAME 'Vmax1', 'Km1', 'Vmax2', 'Km2'
END
    1: FUNC 1
    2: C = X
    3: F = VMAX1*C/(KM1 + C) + VMAX2*C/(KM2 + C)
    4: END
    5: EOM
NVARIABLES 2
NPOINTS 2000
XNUMBER 1
YNUMBER 2
METHOD 2  'Gauss-Newton (Levenberg and Hartley)
REWEIGHT -1
ITERATIONS 50
INITIAL 3,250,.5,2
LOWER BOUNDS 0,0,0,0
UPPER BOUNDS 20,1000,1,10
MISSING 'Missing'
NOBSERVATIONS 18
DATA 'WINNLIN.DAT'
BEGIN
```

PARAMETER	ESTIMATE	STANDARD ERROR	CV%	UNIVARIATE C.I.	
VMAX1	2.437105	.085368	3.50	2.254009	2.620200
KM1	268.010473	30.319352	11.31	202.982084	333.038861
VMAX2	.244040	.039690	16.26	.158913	.329167
KM2	2.685466	.781575	29.10	1.009158	4.361774

```
*** CORRELATION MATRIX OF THE ESTIMATES ***
```

PARAMETER	VMAX1	KM1	VMAX2	KM2
VMAX1	1.00000			
KM1	.57856	1.00000		
VMAX2	-.05051	.70188	1.00000	
KM2	-.09822	.55206	.87155	1.00000

X	OBSERVED Y	PREDICTED Y	RESIDUAL	WEIGHT	SE-PRED	STANDARDI RESIDUAL
.2000E-01	.2400E-02	.1986E-02	.4141E-03	504.0	.3084E-03	.1832
.5000E-01	.5300E-02	.4915E-02	.3847E-03	203.6	.7500E-03	.1096

.1000	.1150E-01	.9670E-02	.1830E-02	103.5	.1434E-02	.3791
.2000	.1570E-01	.1873E-01	-.3032E-02	53.43	.2625E-02	-.4667
.5000	.4860E-01	.4284E-01	.5757E-02	23.36	.5124E-02	.6204
1.000	.6820E-01	.7528E-01	-.7076E-02	13.29	.7154E-02	-.5852
2.000	.1131	.1222	-.9121E-02	8.186	.8397E-02	-.5768
5.000	.2067	.2034	.3299E-02	4.918	.1132E-01	.1638
10.00	.2929	.2800	.1286E-01	3.571	.1557E-01	.5795
20.00	.4157	.3844	.3131E-01	2.602	.1777E-01	1.190
50.00	.5516	.6148	-.6318E-01	1.627	.1781E-01	-1.755
100.0	.8368	.8999	-.6310E-01	1.111	.2217E-01	-1.459
180.0	1.307	1.220	.8708E-01	.8199	.2657E-01	1.743
200.0	1.378	1.282	.9562E-01	.7799	.2698E-01	1.862
250.0	1.332	1.418	-.8613E-01	.7054	.2739E-01	-1.580
500.0	1.844	1.829	.1463E-01	.5466	.2921E-01	.2328
1000.	2.123	2.165	-.4248E-01	.4618	.4249E-01	-.6822
2000.	2.401	2.393	.7675E-02	.4179	.6046E-01	.1498

```
CORRECTED SUM OF SQUARED OBSERVATIONS =  11.0548
WEIGHTED CORRECTED SUM OF SQUARED OBSERVATIONS =  12.6132
SUM OF SQUARED RESIDUALS =            .355300E-01
SUM OF WEIGHTED SQUARED RESIDUALS =  .367424E-01
S =  .512294E-01 WITH    14 DEGREES OF FREEDOM
CORRELATION (OBSERVED,PREDICTED) =  .9984
```

```
AIC criteria =        -51.46884
SC  criteria =        -53.68810
```

Solution III - Constant absolute error

```
MODEL
 COMM
  NPARM 4
  PNAME 'Vmax1', 'Km1', 'Vmax2', 'Km2'
END
     1: FUNC 1
     2: C = X
     3: F = VMAX1*C/(KM1 + C) + VMAX2*C/(KM2 + C)
     4: END
     5: EOM
NVARIABLES 2
NPOINTS 2000
XNUMBER 1
YNUMBER 2
METHOD 2  'Gauss-Newton (Levenberg and Hartley)
ITERATIONS 50
INITIAL 3,250,.5,2
LOWER BOUNDS 0,0,0,0
UPPER BOUNDS 20,1000,1,10
MISSING 'Missing'
NOBSERVATIONS 18
DATA 'WINNLIN.DAT'
BEGIN
```

PARAMETER	ESTIMATE	STANDARD ERROR	CV%	UNIVARIATE C.I.	
VMAX1	2.432628	.069796	2.87	2.282930	2.582325
KM1	256.813373	28.761399	11.20	195.126452	318.500293
VMAX2	.224998	.065560	29.14	.084386	.365610

```
KM2        2.280348      1.925516       84.44       -1.849464        6.410160
```

```
*** CORRELATION MATRIX OF THE ESTIMATES ***
PARAMETER     VMAX1        KM1         VMAX2       KM2
VMAX1        1.00000
KM1          -.00614     1.00000
VMAX2        -.54286      .79858     1.00000
KM2          -.48948      .59958      .81806    1.00000
```

X	OBSERVED Y	PREDICTED Y	RESIDUAL	WEIGHT	SE-PRED	STANDARDI RESIDUAL
.2000E-01	.2400E-02	.2146E-02	.2544E-03	1.000	.1226E-02	.00508
.5000E-01	.5300E-02	.5301E-02	-.1090E-05	1.000	.2976E-02	-.00002
.1000	.1150E-01	.1040E-01	.1101E-02	1.000	.5670E-02	.02213
.2000	.1570E-01	.2004E-01	-.4335E-02	1.000	.1032E-01	-.08849
.5000	.4860E-01	.4519E-01	.3411E-02	1.000	.1982E-01	.0741
1.000	.6820E-01	.7803E-01	-.9825E-02	1.000	.2694E-01	-.2328
2.000	.1131	.1239	-.1083E-01	1.000	.2883E-01	-.2646
5.000	.2067	.2010	.5718E-02	1.000	.2421E-01	.1305
10.00	.2929	.2744	.1851E-01	1.000	.2632E-01	.4346
20.00	.4157	.3777	.3797E-01	1.000	.2963E-01	.9408
50.00	.5516	.6116	-.6002E-01	1.000	.2535E-01	-1.390
100.0	.8368	.9017	-.6495E-01	1.000	.2172E-01	-1.440
180.0	1.307	1.225	.8209E-01	1.000	.2365E-01	1.860
200.0	1.378	1.288	.9040E-01	1.000	.2401E-01	2.058
250.0	1.332	1.423	-.9143E-01	1.000	.2439E-01	-2.091
500.0	1.844	1.831	.1287E-01	1.000	.2296E-01	.2893
1000.	2.123	2.160	-.3714E-01	1.000	.2800E-01	-.8947
2000.	2.401	2.381	.1995E-01	1.000	.4020E-01	.6684

```
CORRECTED SUM OF SQUARED OBSERVATIONS =  11.0548
WEIGHTED CORRECTED SUM OF SQUARED OBSERVATIONS =  11.0548
SUM OF SQUARED RESIDUALS =             .350950E-01
SUM OF WEIGHTED SQUARED RESIDUALS =    .350950E-01
S = .500678E-01 WITH    14 DEGREES OF FREEDOM
CORRELATION (OBSERVED,PREDICTED) =  .9984
```

```
AIC criteria =        -52.29455
SC  criteria =        -54.51381
```

Solution IV - Simulation model for Clint, CIH, FH, EH

```
 MODEL
COMM
  NFUN   2                  <- Varies depending on no. of functions
  NPARM 4
  NCON   1
  PNAME 'Vmax1', 'Km1', 'Vmax2', 'Km2'
END
TEMP
  QH = CON(1)
END

FUNC 1
rema
ema   This function simulates Clint
rema
  C  = x
  F  = (Vmax1/(Km1 + C)  + Vmax2/(Km2 + C))*77*1.5
```

```
end
FUNC 2
rema
ema    This function simulates ClH
rema
  C  = x
  Clint  = (Vmax1/(Km1 + C)   + Vmax2/(Km2 + C))*77*1.5
  F  = QH*Clint/(QH + Clint)
end
rema FUNC 1
rema
ema    This function simulates Bioavailability (across the liver)
rema
rema   C  = x
rema   Clint  = (Vmax1/(Km1 + C)   + Vmax2/(Km2 + C))*77*1.5
rema   F  =  QH/(Clint + QH)
rema end
rema FUNC 2
rema
ema    This function simulates Extration(across the liver)
rema
rema   C  = x
rema   Clint  = (Vmax1/(Km1 + C)   + Vmax2/(Km2 + C))*77*1.5
rema   F  = Clint/(QH + Clint)
rema end

EOM

rema    Vmax1    Km1    Vmax2    Km2
init     2.5    270     .24     2.4
lowe      0      0       0       0
uppe     20    1000      1       10

cons 1.4

NPOINT 1000
NOBS 18, 18
simulate

data
     0.0200
     0.0500
     0.1000
     0.2000
     0.5000
     1.0000
     2.0000
     5.0000
    10.0000
    20.0000
    50.0000
   100.0000
   180.0000
   200.0000
   250.0000
   500.0000
  1000.0000
  2000.0000
     0.0200
     0.0500
     0.1000
```

```
   0.2000
   0.5000
   1.0000
   2.0000
   5.0000
  10.0000
  20.0000
  50.0000
 100.0000
 180.0000
 200.0000
 250.0000
 500.0000
1000.0000
2000.0000
output plot data
graph view plot data
begin
finish
```

PK38 – *In vitro/in vivo* extrapolation II – Differential equations

Objectives

◆ To fit a two-enzyme model to concentration-time data

◆ To predict intrinsic clearance, systemic clearance and extraction ratio *in vivo*

◆ To predict bioavailability

◆ To scrutinize the 'single point' *in vitro* metabolism method

Problem specification

The aim of this exercise is to analyze the metabolic rate (*in vitro*) of a new compound by means of its differential equation rather than transforming the data into a rate *versus* concentration plot. The limitations of the 'single point' approach are also discussed at the end of this section.

A new compound was being developed for treatment of hypertension. In order to predict the bioavailability in man, an *in vitro* experiment was performed using human microsomes to obtain V_{max} and K_m. Data are given in Table 38.1 and shown in Figure 38.1. Simultaneous nonlinear regression analysis was applied to the enzyme kinetic data at five different starting concentrations. The data clearly displayed saturable kinetics. The compound is eliminated via hepatic metabolism by means of two parallel routes (two different P450 isozymes) and no excretion of the parent compound occurs via the urine.

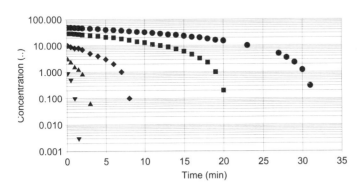

Figure 38.1 Semi-logarithmic plot of observed concentration-time data in vitro following five different starting concentrations of a compound that is metabolized by two parallel enzymes.

We propose the following relationship between rate of decline of the parent compound dC/dt and the actual concentration C for the two enzyme model.

$$V_{medium} \cdot \frac{dC}{dt} = -\sum \frac{V_{max,i} \cdot C}{K_{m,i} + C} = -\frac{V_{max,1} \cdot C}{K_{m,1} + C} - \frac{V_{max,2} \cdot C}{K_{m,2} + C} \tag{38:1}$$

The volume of the incubation medium (V_{medium}) is 1 mL. The concentration-time data of parent compound is given in Table 38.1.

Table 38.1   Concentration-time data of the two-enzyme system. The '#' column denotes the function number of the differential equation

Time (min)	Conc (µmol/L)	#	Time (min)	Conc (µmol/L)	#	Time (min)	Conc (µmol/L)	#
0.1	49	1	28	3.5	1	19	1	2
0.5	49.1	1	29	2.3	1	20	0.2	2
1	48	1	30	1.2	1	0.1	10	3
1.5	47	1	31	0.3	1	0.5	9	3
2	46	1	0.1	29	2	1	8	3
3	44.5	1	0.5	29	2	1.5	8	3
4	43	1	1	28	2	2	7	3
5	40	1	1.5	27	2	3	5	3
6	39	1	2	26	2	4	4	3
7	37	1	3	24	2	5	3	3
8	35	1	4	23	2	6	2	3
9	33	1	5	21	2	7	1	3
10	32	1	6	20	2	8	0.1	3
11	30	1	7	18	2	0.1	3	4
12	28	1	8	16	2	0.5	2.3	4
13	26	1	9	14	2	1	1.5	4
14	25	1	10	13	2	1.5	1.2	4
15	23	1	11	11	2	2	0.8	4
16	21	1	12	10	2	3	0.06	4
17	20	1	13	8.9	2	0.1	0.9	5
18	18	1	14	7.3	2	0.5	0.5	5
19	16	1	15	6	2	1	0.1	5
20	15	1	16	4.6	2	1.5	0.003	5
23	10	1	17	3.3	2			
27	5	1	18	2.3	2			

Initial parameter estimates

V_{max1}	=	(nmol/min)	Maximum metabolic capacity
K_{m1}	=	(µmol/L)	Michaelis-Menten constant
V_{max2}	=	(nmol/min)	Maximum metabolic capacity
K_{m2}	=	(µmol/L)	Michaelis-Menten constant

Interpretation of results and conclusions

This exercise demonstrates the use of *in vitro* concentration-time data as such for estimation of enzyme kinetics. Figure 38.2 shows observed and model-predicted data.

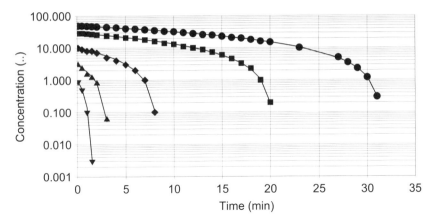

Figure 38.2. *Semi-logarithmic plot of observed and model-predicted concentration-time data following five different starting concentrations of a compound that is metabolized by two different enzymes.*

The final parameter estimates obtained from fitting Equation 38:1 to the original concentration (C, µunits/L) *versus* time data in Figure 38.1 can be found in Table 38.2.

Table 38.2 Parameter estimates

	Mean	CV%
V_{max1}	0.97	1.4
K_{m1}	0.091	3.5
V_{max2}	1.01	0.7
K_{m2}	8.95	5.9

One can scale the *in vitro* data to a 1500 g liver. Assuming 77 mg of microsomal protein per gram liver, the product 77·1500 converts micro units/g microsomal liver protein/min to micro units /1.5 kg liver/min. Hence, one obtains the human hepatic intrinsic clearance (L/min)

$$Cl_H = Q_H \cdot E_H = \frac{Q_H \cdot f_u \cdot Cl_{int}}{Q_H + f_u \cdot Cl_{int}} \tag{38:2}$$

Intrinsic clearance is defined as

$$Cl_{int} = \frac{V_{maxA}}{K_{mA} + C} + \frac{V_{maxB}}{K_{mB} + C} + \ldots\ldots + \frac{V_{maxi}}{K_{mi} + C} \tag{38:3}$$

Figure 38.3 demonstrates the consequences of using the often-hailed *'single point'* method for selection of a compound with appropriate first-pass characteristics (e.g., highest bioavailability).

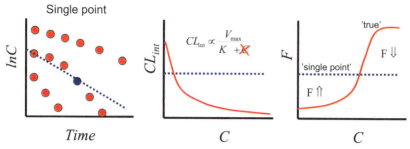

Figure 38.3 *Schematic illustration of the 'single point' method for estimation of intrinsic clearance Cl_{int}. The left hand figure shows a single (blue point) versus multiple (red points) measurements. The middle figure shows the consequences of assuming a concentration-independent clearance. Cl_{int}, which is usually concentration-dependent (red line), results in a constant value that is independent of concentration when the 'single point' method is applied. The right hand figure demonstrates the high risk of miscalculating the bioavailability for the 'single point' (blue dotted line) versus the multiple measurement method (red line). At low concentrations, the 'single point' approach will overestimate bioavailability whereas at higher concentrations the bioavailability, will be underestimated.*

Figures 38.4 and 38.5 illustrate two other limitations with the approach we have used for estimation of V_{max} and K_m. In Figure 38.4, we show how data are not fully utilized. Only a portion of the experimental data is used for back-extrapolation and estimation of v_0 which is then plotted against C_0 according to the rate plot in Figure 38.4 (right). We call this condensation of data. The red dots in the data tails (Figure 38.4 (left)) are not used.

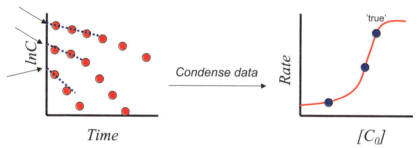

Figure 38.4 *Schematic illustration of the estimation of v_0 in vitro by means of back-extrapolation to time zero. The slope of each line represents the respective v_0 (enzymatic rate) at each C_0 concentration (left). These v_0 values are then plotted against their corresponding C_0 values in the right hand figure. Observe that not all experimental data are used in the left hand figure. In other words, an information-rich dataset (left) is condensed into a less information-rich and transformed dataset (right). The latter is then used to obtain the apparent V_{max} and K_m parameters.*

Figure 38.5 illustrates the problems with transforming data from the rate *versus* time plot (left) into a rate *versus* concentration plot. A rich dataset (left) is condensed into an information-sparse dataset (right).

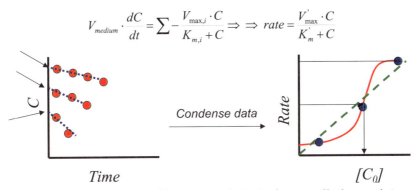

$$V_{medium} \cdot \frac{dC}{dt} = \sum - \frac{V_{max,i} \cdot C}{K_{m,i} + C} \Rightarrow \Rightarrow rate = \frac{V'_{max} \cdot C}{K'_{m} + C}$$

Figure 38.5 *Schematic illustration of the estimation of v_0 in vitro by means of back-extrapolation to time zero. The slope of each line represents the respective v_0 (enzymatic rate) at each C_0 concentration (left). These v_0 values are then plotted against their corresponding C_0 values in the right hand figure. Observe that nonlinearities that are obvious in the left hand figure are condensed, due to data reduction, into an apparently linear relationship between rate (v) and concentration (C_0). In other words, an information rich dataset (left) is condensed into a less information-rich and transformed dataset (right). The latter three-point plot cannot be used to obtain the apparent V_{max} and K_m parameters since it is a straight line. By fitting a nonlinear model to the rate data in the left hand figure, one has a greater chance of obtaining V_{max} and K_m.*

The point we would like to make from Figures 38.4 and 38.5 is that a more robust approach to obtain V_{max} and K_m is to analyze the untransformed concentration-time data, rather than to use data reduction and transformation.

If we had applied the traditional approach (data reduction and transformation into a rate *versus* concentration plot) we would have obtained a rate plot (v_0 *versus* C_0) based on only five data points, since we had five different starting concentrations. There is very little chance that we could obtain the original two sets of V_{max} and K_m values that the original concentration-time data (Table 38.2) were generated with.

In this exercise we have
- covered *in vitro/in vivo* extrapolation of enzyme kinetic data
- fitted a system of differential equations to concentration-time data
- discussed problems and pitfalls of the 'single point' approach

Solution

```
MODEL
 COMM
  NFUN 5
  NDER 5
  NPAR 4
  PNAM 'VM1', 'KM1','Vm2', 'Km2'
END
     1: TEMP
     2: T = X
     3: END
     4: START
     5: Z(1)= 50
     6: Z(2)= 30
```

```
 7: Z(3)= 10
 8: Z(4)= 3
 9: Z(5)= 1
10: END
11: DIFF
12: DZ(1)= -VM1*Z(1)/(KM1 + Z(1)) - VM2*Z(1)/(KM2 + Z(1))
13: DZ(2)= -VM1*Z(2)/(KM1 + Z(2)) - VM2*Z(2)/(KM2 + Z(2))
14: DZ(3)= -VM1*Z(3)/(KM1 + Z(3)) - VM2*Z(3)/(KM2 + Z(3))
15: DZ(4)= -VM1*Z(4)/(KM1 + Z(4)) - VM2*Z(4)/(KM2 + Z(4))
16: DZ(5)= -VM1*Z(5)/(KM1 + Z(5)) - VM2*Z(5)/(KM2 + Z(5))
17: END
18: FUNC 1
19: F = Z(1)
20: END
21: FUNC 2
22: F = Z(2)
23: END
24: FUNC 3
25: F = Z(3)
26: END
27: FUNC 4
28: F = Z(4)
29: END
30: FUNC 5
31: F = Z(5)
32: END
33: EOM
NVARIABLES 3
FNUMBER 3
NPOINTS 100
XNUMBER 1
YNUMBER 2
METHOD 2   'Gauss-Newton (Levenberg and Hartley)
REWEIGHT -2
ITERATIONS 50
INITIAL .8,.1,.8,10
LOWER BOUNDS 0,0,0,0
UPPER BOUNDS 3,1,3,20
MISSING 'Missing'
DATA 'WINNLIN.DAT'
BEGIN
```

PARAMETER	ESTIMATE	STANDARD ERROR	CV%	UNIVARIATE C.I.	
VM1	.967711	.013156	1.36	.941466	.993955
KM	.091461	.003185	3.48	.085108	.097815
VM2	1.016493	.007312	.72	1.001906	1.031080
KM2	8.949636	.530550	5.93	7.891215	10.008057

```
*** CORRELATION MATRIX OF THE ESTIMATES ***
```

PARAMETER	VM1	KM1	VM2	KM2
VM1	1.00000			
KM1	.947838	1.00000		
VM2	-.342134	-.434938	1.00000	
KM2	.948662	.855417	-.342213E-01	1.00000

FUNCTION   1

X	OBSERVED Y	PREDICTED Y	RESIDUAL	WEIGHT	SE-PRED	STANDARDI RESIDUAL
.1000	49.00	49.82	-.8172	.4029E-03	.4988E-03	-.2746
.5000	49.10	49.09	.1344E-01	.4150E-03	.2473E-02	.0045
1.000	48.00	48.17	-.1744	.4309E-03	.4894E-02	-.0605
1.500	47.00	47.26	-.2635	.4477E-03	.7262E-02	-.0933
2.000	46.00	46.35	-.3539	.4654E-03	.9575E-02	-.1278
3.000	44.50	44.54	-.3887E-01	.5041E-03	.1403E-01	-.0146
4.000	43.00	42.73	.2703	.5477E-03	.1825E-01	.1059
5.000	40.00	40.93	-.9268	.5970E-03	.2223E-01	-.3791
6.000	39.00	39.13	-.1306	.6531E-03	.2594E-01	-.0558
7.000	37.00	37.34	-.3415	.7172E-03	.2939E-01	-.1531
8.000	35.00	35.56	-.5601	.7908E-03	.3255E-01	-.2637
9.000	33.00	33.79	-.7870	.8760E-03	.3541E-01	-.3900
10.00	32.00	32.02	-.2291E-01	.9752E-03	.3796E-01	-.0119
11.00	30.00	30.27	-.2685	.1091E-02	.4018E-01	-.1485
12.00	28.00	28.52	-.5246	.1229E-02	.4206E-01	-.3080
13.00	26.00	26.79	-.7921	.1393E-02	.4359E-01	-.4951
14.00	25.00	25.07	-.7216E-01	.1591E-02	.4475E-01	-.0482
15.00	23.00	23.37	-.3659	.1832E-02	.4552E-01	-.2623
16.00	21.00	21.67	-.6748	.2129E-02	.4591E-01	-.5215
17.00	20.00	20.00	-.3433E-03	.2500E-02	.4591E-01	-.0003
18.00	18.00	18.34	-.3443	.2972E-02	.4550E-01	-.3145
19.00	16.00	16.71	-.7089	.3582E-02	.4471E-01	-.7110
20.00	15.00	15.10	-.9644E-01	.4388E-02	.4354E-01	-.1071
23.00	10.00	10.43	-.4280	.9196E-02	.3814E-01	-.6884
27.00	5.000	4.770	.2301	.4395E-01	.2873E-01	.8118
28.00	3.500	3.503	-.2852E-02	.8150E-01	.2619E-01	-.0137
29.00	2.300	2.317	-.1681E-01	.1863	.2349E-01	-.1233
30.00	1.200	1.232	-.3153E-01	.6593	.2066E-01	-.4466
31.00	.3000	.3043	-.4328E-02	10.80	.1574E-01	-.4760

CORRECTED SUM OF SQUARED OBSERVATIONS = 7050.01
WEIGHTED CORRECTED SUM OF SQUARED OBSERVATIONS = 25.5032
SUM OF SQUARED RESIDUALS =   5.33094
SUM OF WEIGHTED SQUARED RESIDUALS = .114416E-01
S = .213931E-01 WITH   25 DEGREES OF FREEDOM
CORRELATION (OBSERVED,PREDICTED) = .9998

FUNCTION   2

X	OBSERVED Y	PREDICTED Y	RESIDUAL	WEIGHT	SE-PRED	STANDARDI RESIDUAL
.1000	29.00	29.83	-.8253	.1124E-02	.2330E-03	-.4632
.5000	29.00	29.13	-.1275	.1179E-02	.1137E-02	-.0732
1.000	28.00	28.26	-.2578	.1252E-02	.2204E-02	-.1527
1.500	27.00	27.39	-.3910	.1333E-02	.3201E-02	-.2390
2.000	26.00	26.53	-.5272	.1421E-02	.4129E-02	-.3327
3.000	24.00	24.81	-.8093	.1625E-02	.5788E-02	-.5461
4.000	23.00	23.11	-.1053	.1873E-02	.7198E-02	-.0762
5.000	21.00	21.42	-.4166	.2180E-02	.8395E-02	-.3257
6.000	20.00	19.74	.2552	.2565E-02	.9426E-02	.2164
7.000	18.00	18.09	-.9184E-01	.3055E-02	.1036E-01	-.0849
8.000	16.00	16.46	-.4598	.3691E-02	.1129E-01	-.4677
9.000	14.00	14.85	-.8511	.4534E-02	.1229E-01	-.9595
10.00	13.00	13.27	-.2687	.5680E-02	.1347E-01	-.3391
11.00	11.00	11.72	-.7160	.7285E-02	.1488E-01	-1.023
12.00	10.00	10.20	-.1972	.9617E-02	.1650E-01	-.3238

13.00	8.900	8.717	.1831	.1316E-01	.1827E-01	.3520
14.00	7.300	7.281	.1917E-01	.1886E-01	.2002E-01	.0441
15.00	6.000	5.896	.1040	.2877E-01	.2149E-01	.2959
16.00	4.600	4.571	.2937E-01	.4787E-01	.2231E-01	.1079
17.00	3.300	3.315	-.1510E-01	.9099E-01	.2201E-01	-.0767
18.00	2.300	2.143	.1570	.2178	.2007E-01	1.242
19.00	1.000	1.076	-.7606E-01	.8636	.1608E-01	-1.222
20.00	.2000	.1921	.7918E-02	27.10	.8192E-02	.9857

```
CORRECTED SUM OF SQUARED OBSERVATIONS = 2086.63
WEIGHTED CORRECTED SUM OF SQUARED OBSERVATIONS = 20.0286
SUM OF SQUARED RESIDUALS =          3.74349
SUM OF WEIGHTED SQUARED RESIDUALS = .245941E-01
S = .359781E-01 WITH    19 DEGREES OF FREEDOM
CORRELATION (OBSERVED,PREDICTED) = .9996
```

FUNCTION   3

X	OBSERVED Y	PREDICTED Y	RESIDUAL	WEIGHT	SE-PRED	STANDARDI RESIDUAL
.1000	10.00	9.851	.1493	.1031E-01	.3139E-03	.2538
.5000	9.000	9.257	-.2573	.1167E-01	.1575E-02	-.4654
1.000	8.000	8.525	-.5252	.1376E-01	.3150E-02	-1.031
1.500	8.000	7.804	.1957	.1642E-01	.4702E-02	.4199
2.000	7.000	7.095	-.9545E-01	.1986E-01	.6200E-02	-.2252
3.000	5.000	5.718	-.7179	.3059E-01	.8892E-02	-2.103
4.000	4.000	4.401	-.4011	.5163E-01	.1089E-01	-1.527
5.000	3.000	3.156	-.1557	.1004	.1178E-01	-.8276
6.000	2.000	1.996	.4146E-02	.2510	.1121E-01	.0349
7.000	1.000	.9457	.5431E-01	1.118	.9299E-02	.9746
8.000	.1000	.1111	-.1113E-01	80.97	.4981E-02	-2.536

```
CORRECTED SUM OF SQUARED OBSERVATIONS = 116.609
WEIGHTED CORRECTED SUM OF SQUARED OBSERVATIONS = 8.85074
SUM OF SQUARED RESIDUALS =         1.11539
SUM OF WEIGHTED SQUARED RESIDUALS = .454450E-01
S = .805739E-01 WITH     7 DEGREES OF FREEDOM
CORRELATION (OBSERVED,PREDICTED) = .9965
```

FUNCTION   4

X	OBSERVED Y	PREDICTED Y	RESIDUAL	WEIGHT	SE-PRED	STANDARDI RESIDUAL
.1000	3.000	2.881	.1190	.1205	.1874E-03	.6914
.5000	2.300	2.414	-.1141	.1716	.9702E-03	-.7915
1.000	1.500	1.853	-.3526	.2914	.2115E-02	-3.187
1.500	1.200	1.319	-.1191	.5747	.3541E-02	-1.513
2.000	.8000	.8201	-.2008E-01	1.487	.5163E-02	-.4123
3.000	.6000E-01	.5217E-01	.7832E-02	367.4	.2737E-02	5.255

```
CORRECTED SUM OF SQUARED OBSERVATIONS = 5.54033
WEIGHTED CORRECTED SUM OF SQUARED OBSERVATIONS = 4.04460
SUM OF SQUARED RESIDUALS =          .166195
SUM OF WEIGHTED SQUARED RESIDUALS = .714635E-01
S = .189028    WITH     2 DEGREES OF FREEDOM
CORRELATION (OBSERVED,PREDICTED) = .9884
```

FUNCTION   5

X	OBSERVED Y	PREDICTED Y	RESIDUAL	WEIGHT	SE-PRED	STANDARDI RESIDUAL
.1000	.9000	.9020	-.1965E-02	1.229	.4318E-03	-.0364
.5000	.5000	.5297	-.2975E-01	3.563	.2065E-02	-.9422
1.000	.1000	.1429	-.4286E-01	49.00	.2468E-02	-5.247
1.500	.3000E-02	.3026E-02	-.2569E-04	.1092E+06	.1803E-03	-2.122

CORRECTED SUM OF SQUARED OBSERVATIONS =   .505257
WEIGHTED CORRECTED SUM OF SQUARED OBSERVATIONS =   2.32967
SUM OF SQUARED RESIDUALS =   .272621E-02
SUM OF WEIGHTED SQUARED RESIDUALS = .932499E-01
S = .000000   WITH   0 DEGREES OF FREEDOM
CORRELATION (OBSERVED,PREDICTED) =   .9987

TOTALS FOR ALL CURVES COMBINED
SUM OF SQUARED RESIDUALS =   10.3587
SUM OF WEIGHTED SQUARED RESIDUALS = .246194
S = .597330E-01 WITH   69 DEGREES OF FREEDOM

AIC criteria =   -94.31935
SC criteria =   -85.15751

PK39 – Two-compartment plasma data – Issues on experimental design and data pollution

Objectives

◆ **To analyze a multiple intravenous infusion data set**

◆ **To fit a model to three sets of observations**

◆ **To write a multi-compartment model in terms of differential equations**

◆ **To obtain initial parameter estimates**

◆ **To discuss various aspects of correlation, accuracy, and precision**

◆ **To discuss various aspects of experimental design**

Problem specification

A multiple constant rate intravenous infusion regimen was designed in order to characterize the kinetics of a new compound in man. The first dose was 26.9 µg/kg given over 15 min. The second dose of 139 µg/kg was given from 15 min to 8 h. The third dose of 138.95 µg/kg was given between 8 and 24 h. The concentration-time course in one subject for two different designs is shown in Figure 39.1 and the actual values are given in Table 39.1.

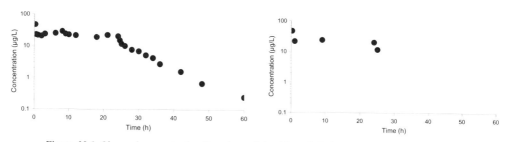

Figure 39.1 *Observed concentration-time data of the 24- and 5-observations designs (left to right, respectively. The 14-observations design is shown in Figure 39.3.*

Table 39.1 Concentration-time data for three different designs

Time (h)	Concentration (µg/L)	Time (h)	Concentration (µg/L)	Time (h)	Concentration (µg/L)
0.25	53.21**	10	25.81*	28	8.74
0.5	25.67*	12	24.48	30	7.89*
1	24.59**	18	21.79	32	5.92
2	23.36*	21	25.20*	34	4.86*
3	26.94	24	23.10**	36	3.15
6	28.55*	24.5	17.22*	42	1.82*
8	32.94	25	13.55**	48	0.771
9	27.38**	26	11.49*	60	0.288

* excluded from the 14-observations run.
** included in the 5-observations run.

To study the impact of various experimental designs, the complete plasma concentration-time dataset (24 observations) was reduced in two steps. A naive data analyst selected what he assumed to be an appropriate number of data points for the regression analysis. This dataset contained 14 observations.

The data were further reduced by selecting the number of data points (5 observations) which appeared to be most appropriate according to the partial derivative plot obtained from fitting Equations 39:1 and 39:2 to the original (24 observations) dataset. One can either fit *WinNonlin* library model number 10 or the two-compartment differential equation model described by Equations 39:1 and 39:2. The concentration in the central compartment is written as

$$V_c \cdot \frac{dC}{dt} = In - Cl \cdot C - Cl_d \cdot C + Cl_d \cdot C_t \tag{39:1}$$

In is the three consecutive infusions regimen. The concentration in the peripheral compartment is written as

$$V_t \cdot \frac{dC_t}{dt} = Cl_d \cdot C - Cl_d \cdot C_t \tag{39:2}$$

Initial parameter estimates

The terminal slope (β) is obtained from

$$slope_\beta = -\beta = \frac{\ln(2)}{t_{1/2\alpha}} = \frac{\ln(7.89/0.288)}{30 - 60} = -0.11\,h^{-1} \tag{39:3}$$

The initial slope (α) is approximately

$$slope_\alpha = -\alpha = \frac{\ln(2)}{t_{1/2\alpha}} \approx \frac{\ln(53.21/25.67)}{0.25 - 0.5} = -2.9\,h^{-1} \tag{39:4}$$

The total area under the curve from zero to infinity (*AUC*) is obtained by means of the non-compartmental routines in *WinNonlin*. *AUC* is approximately 740 µg·h/L, which gives a good estimate of clearance (0.4 L/kg/h) by means of dividing the total dose (304.85 µg·h/kg) by means of *AUC*. The k_{10} parameter is then estimated from

$$k_{10} = \frac{Cl}{V_c} \approx \frac{Cl}{Dose/C_{peak}} = \frac{0.4}{26.9/53.21} \approx 0.8\,h^{-1} \tag{39:5}$$

The k_{21} parameter is estimated from α, β and k_{10} according to

$$k_{21} = \frac{\alpha \cdot \beta}{k_{10}} = \frac{2.9 \cdot 0.11}{0.8} \approx 0.4\,h^{-1} \tag{39:6}$$

The k_{12} parameter is estimated from α, β, k_{21} and k_{10} according to

$$k_{12} = \alpha + \beta - k_{21} - k_{10} = 2.9 + 0.11 - 0.4 - 0.8 \approx 1.8\,h^{-1} \tag{39:7}$$

An approximate value of V_c is estimated as

$$V_c \approx \frac{Dose}{C_{peak}} = \frac{26.9}{53.21} \approx 0.5\,L/kg \tag{39:8}$$

An approximate value of V_t is estimated as

$$V_t = V_c \cdot \frac{k_{12}}{k_{21}} \approx 2\,L/kg \tag{39:9}$$

Implement a two-compartment disposition model parameterized with Cl, V_c, V_t and Cl_d.

Cl	=	0.4 L/kg/h	*Dose/AUC*
V_c	=	0.4 L/kg	Equation 38:8
V_t	=	2.0 L/kg	Equation 38:9
Cl_d	=	0.8 L/kg	Obtained from $k_{21} \cdot V_t$

One can also assume that Cl_d is greater than Cl, since a rapid decline occurs in the plasma concentration-time course after the first rapid infusion.

Interpretation of results and conclusions

The results from fitting *WinNonlin* library model number 10 to the three datasets are shown in Figures 39.2 to 39.4. The final parameter estimates and their precision are shown in Table 39.2.

Table 39.2   Kinetic estimates from fitting *WinNonlin* library model number 10

Parameter	24-observations		14-observations		5-observations	
	Estimate	CV%	Estimate	CV%	Estimate	CV%
$t_{1/2\alpha}$	0.15	12.5	0.22	6.9	0.21	9.0
AUC	65	1.8	65	1.2	66.8	2.5
$AUMC$	390	4.8	370	3.4	380	13
$t_{1/2\beta}$	5.4	4.8	5.7	3.7	5.5	13
Cl	0.41	1.8	0.41	1.2	0.40	2.6
C_{max}	53.4	4.9	53.2	2.4	53.3	2.6
MRT	5.8	4.1	5.6	2.9	5.6	12
V_{ss}	2.4	4.1	2.3	2.9	2.2	10

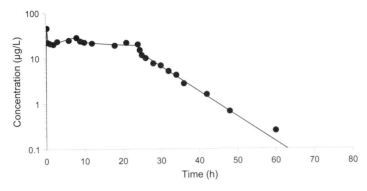

Figure 39.2 Observed and predicted concentration-time data of the 24-observations run.

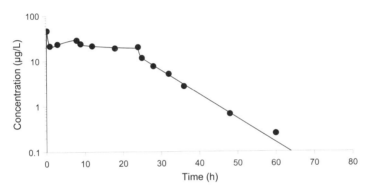

Figure 39.3 Observed and predicted concentration-time data of the 14-observations run.

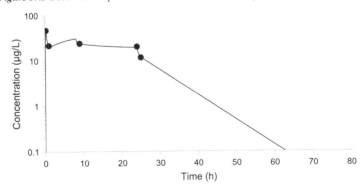

Figure 39.4 Observed and predicted concentration-time data of the 5-observations run.

The 14-observation design gave in general the highest parameter precision. The primary kinetic parameters (Cl and V_{ss}) had a CV% of less than 10%. There were surprisingly few differences between parameter estimates of the three designs. Taking the biological variability into account, the most feasible design is probably closer to the 14-observations dataset than it is to either the 24- or 5-observations dataset. This is because parameter precision is made up from the objective function and the Variance inflation factor. The objective function is written as

$$WRSS = \sum_{i=1}^{n} w_i \cdot (Y_i - \hat{Y}_i)^2 \tag{39:10}$$

The mean predicted variance then becomes

$$\sigma^2 = \frac{WRSS}{N_{obs} - N_{par}} = \frac{WRSS}{df} \tag{39:11}$$

The variance inflation factor denotes the contribution to the variance of a model parameter that is dependent on the sampling times. The variance of, e.g., clearance is proportional to product of the mean predicted variance and the *'design'* hence

$$var(Cl) \propto \sigma^2 \cdot VIF \tag{39:12}$$

In other words, both the experimental design and experimental noise contribute to the variance of an estimated parameter. As shown in this exercise, *'fertilizing'* with too many plasma concentrations does not necessarily improve the results, particularly when the concentrations approaches the limit of quantification (*LOQ*).

Solution – Full dataset with user-specified model

```
MODEL
 COMMANDS
  NFUN  1
  NDER  2
  NPAR  4
  PNAM   'Vc', 'Cl', 'Cld', 'Vt'
  NCON  6
  NSEC  1
  SNAM 'Vss'
END
     1: TEMPORARY
     2: T = X
     3: DOSE1 = CON(1)
     4: TINF1 = CON(2)
     5: DOSE2 = CON(3)
     6: TINF2 = CON(4)
     7: DOSE3 = CON(5)
     8: TINF3 = CON(6)
     9: FINF = 0.0
    10: IF T LE TINF1 THEN
    11: FINF = DOSE1/TINF1
    12: ENDIF
    13: IF T LE TINF2 AND T GT TINF1 THEN
    14: FINF = DOSE2/(TINF2 - TINF1)
    15: ENDIF
```

```
16: IF T LE TINF3 AND T GT TINF2 THEN
17: FINF = DOSE3/(TINF3 - TINF2)
18: ENDIF
19: END
20: START
21: Z(1) = 0.
22: Z(2) = 0.
23: END
24: DIFF
25: DZ(1)= (FINF - CL*Z(1) - CLD*Z(1) + CLD*Z(2))/VC
26: DZ(2)= (CLD*Z(1) - CLD*Z(2))/VT
27: END
28: FUNC 1
29: F = Z(1)
30: END
31: SECO
32: VSS = VT + VC
33: END
34: EOM
NVARIABLES 4
NPOINTS 100
XNUMBER 1
YNUMBER 2
CONSTANTS 26.9,.25,139,8,138.95,24
METHOD 2  'Gauss-Newton (Levenberg and Hartley)
REWEIGHT -1
ITERATIONS 50
INITIAL .5,.4,.8,2
LOWER BOUNDS .01,.01,.01,.01
UPPER BOUNDS 2.5,2,4,10
MISSING 'Missing'
NOBSERVATIONS 24
DATA 'WINNLIN.DAT'
BEGIN
```

PARAMETER	ESTIMATE	STANDARD ERROR	CV%	UNIVARIATE C.I.	
VC	.322762	.031372	9.72	.257322	.388203
CL	.417287	.008259	1.98	.400060	.434515
CLD	.898564	.080483	8.96	.730682	1.066446
VT	2.141773	.113255	5.29	1.905529	2.378017

```
*** CORRELATION MATRIX OF THE ESTIMATES ***
PARAMETER  VC        CL        CLD       VT
VC         1.00000
CL          .024772  1.00000
CLD        -.413662   .043481  1.00000
VT         -.211111   .205128   .08110  1.00000

 Condition_number=        14.16

   FUNCTION   1
```

X	OBSERVED Y	PREDICTED Y	RESIDUAL	WEIGHT	SE-PRED	STANDARDI RESIDUAL
.2500	53.21	52.87	.3381	.1891E-01	2.748	.7338
.5000	25.67	30.16	-4.487	.3316E-01	1.639	-3.401
1.000	24.59	21.13	3.465	.4734E-01	.8971	2.286

2.000	23.36	22.52	.8436	.4441E-01	.5517	.4869
3.000	26.94	24.89	2.051	.4018E-01	.5843	1.126
6.000	28.55	30.51	-1.961	.3277E-01	.7028	-.9823
8.000	32.94	33.25	-.3113	.3007E-01	.7529	-.1499
9.000	23.81	26.49	-2.680	.3775E-01	.7581	-1.472
10.00	25.81	25.76	.4998E-01	.3882E-01	.6936	.0275
12.00	24.48	24.67	-.1920	.4053E-01	.5839	-.1059
18.00	21.79	22.65	-.8554	.4416E-01	.4604	-.4848
21.00	25.20	22.08	3.124	.4530E-01	.4435	1.790
24.00	23.10	21.68	1.417	.4612E-01	.4343	.8189
24.50	17.22	15.64	1.579	.6393E-01	.4545	1.092
25.00	13.35	14.17	-.8179	.7058E-01	.4536	-.5973
26.00	11.49	12.45	-.9620	.8031E-01	.4122	-.7469
28.00	8.740	9.715	-.9752	.1029	.3555	-.8552
30.00	7.890	7.581	.3095	.1319	.3328	.3091
32.00	5.920	5.915	.5101E-02	.1691	.3164	.0581
34.00	4.860	4.615	.2447	.2167	.2971	.3187
36.00	3.150	3.601	-.4511	.2777	.2736	-.6696
42.00	1.820	1.711	.1092	.5845	.1929	.2361
48.00	.7710	.8127	-.4169E-01	1.230	.1228	-.1291
60.00	.2880	.1834	.1046	5.452	.4200E-01	.6593

```
CORRECTED SUM OF SQUARED OBSERVATIONS =  3677.78
WEIGHTED CORRECTED SUM OF SQUARED OBSERVATIONS = 369.933
SUM OF SQUARED RESIDUALS =            66.2599
SUM OF WEIGHTED SQUARED RESIDUALS = 2.93717
S =  .383221     WITH     20 DEGREES OF FREEDOM
CORRELATION (OBSERVED,PREDICTED) =  .9910
```

```
AIC criteria =        33.85871
SC  criteria =        38.57093
```

PARAMETER	ESTIMATE	STANDARD ERROR	CV%
VSS	2.464535	.110954	4.50

PK40 - Enterohepatic recirculation

Objectives
◆ **To model enterohepatic recirculation after intravenous administration**
◆ **To assess the goodness-of-fit**

Problem specification

This exercise demonstrates how to set up an enterohepatic model and how to fit the model to experimental data. A subject was given an intravenous bolus dose of 5617.3 µg of a new compound exhibiting pronounced enterohepatic recirculation (*EHC*). Plasma concentrations were measured over a period of 36 h. The concentration-time course is depicted in Figure 40.1.

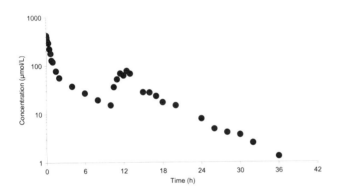

Figure 40.1 Observed concentrations following an intravenous bolus dose of 5617 µg.

The model that we propose for this dataset is based on a two-compartment model parameterized with a volume of the central and peripheral compartments, clearance and an inter-compartmental diffusion parameter. A first-order rate constant for absorption of drug from the gastrointestinal tract, a first-order rate constant for drug excreted into the bile K_{lg}, and a term for gall bladder release are also incorporated into the model (see schematic model in Figure 40.2). The gall bladder is emptied intermittently into the gut. Drug is then absorbed from the gut into the central compartment by a first-order process K_a. We will estimate the time it takes after dosing until bile is released into the gut. This model takes only one release of bile into account, even though cyclic releases are more physiologically realistic.

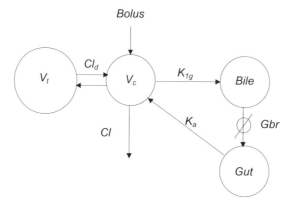

Figure 40.2 *Schematic illustration of the model for enterohepatic recirculation. The model is based on either first or zero-order parameters and the parameters are independent of capacity and time.*

The model parameters are *Cl, V_c, Cl_d* and *V_t*, which correspond to plasma clearance, the volume of the central compartment, inter-compartmental diffusion, and the volume of the peripheral compartment, respectively. *In* is the input function, which in this case is bolus input. The equation for the central or plasma compartment is

$$V_c \cdot \frac{dC}{dt} = In + K_a \cdot A_g - Cl \cdot C - Cl_d \cdot C + Cl_d \cdot C_t - K_{1g} \cdot C \cdot V_c \qquad (40:1)$$

where Ag is the amount in the gastrointestinal compartment.

Equation 40:1 corresponds to *DZ(1)* in the program code. The equation for the peripheral compartment is

$$V_t \cdot \frac{dC_t}{dt} = Cl_d \cdot C - Cl_d \cdot C_t \qquad (40:2)$$

Equation 40:2 corresponds to *DZ(2)* in the program code. The diffusion parameter, which is equal in both directions, can be derived from the expressions

$$Cl_d = k_{12} \cdot V_c = k_{21} \cdot V_t \qquad (40:3)$$

The *k_{12}* and *k_{21}* parameters are the inter-compartmental rate constants which determine the rate of drug movement from the central to the peripheral and the peripheral to the central compartment, respectively. For the bile compartment

$$\frac{dA_{bile}}{dt} = K_{1g} \cdot C \cdot V_c - \frac{A_{bile}}{\tau} \qquad (40:4)$$

Equation 40:4 corresponds to *DZ(4)* in the program code. For the gastrointestinal compartment

$$\frac{dA_g}{dt} = \frac{A_{bile}}{\tau} - K_a \cdot A_g \qquad (40:5)$$

Equation 40:5 corresponds to _DZ(3)_ in the program code. A_{bile}/τ is the zero-order release rate of drug from the bile compartment.

Initial parameter estimates

V_c	= 13 (L)	Central volume
Cl	= 1.2 (L/h)	Plasma clearance
Cl_d	= 10 (L/h)	Diffusion parameter
V_t	= 35 (L)	Peripheral volume
K_a	= 2.3 (h^{-1})	Absorption rate constant
k_{lg}	= 0.7 (h^{-1})	Bile excretion rate constant
t_{tom}	= 2.8 (h)	Bile emptying interval

Interpretation of results and conclusions

We propose a model that mimics _EHC_. The observed and model-predicted data that were obtained after an intravenous bolus experiment are shown in Figure 40.3. Special attention should be made to the frequent sampling schedule. It is very rare to obtain as many as 32 samples from a single subject, and it is possible to obtain similar parameter accuracy and precision with fewer but more optimally selected design points (see also exercise PK39 on experimental design). We therefore recommend that the experimenter invest extra time in study design in situations of complex systems.

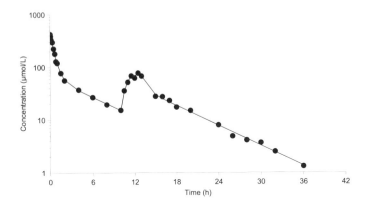

Figure 40.3 _Observed (filled circles) and predicted (solid line) concentrations following an intravenous bolus dose._

The data lend themselves to prediction of the volumes of the central and peripheral compartments, clearance, and an inter-compartmental diffusion parameter. The first-order rate constant for absorption of drug from the gastrointestinal tract, a first-order rate constant for drug excreted into the bile, and a term for gall bladder release are also estimated with good precision and low correlation. There is no obvious trend in the residuals.

We would like to give a warning about the assumption of zero-order release from bile. Release from bile may not necessarily be a zero-order or a very regular process, which may cause problems when one tries to model multiple dose data with _EHC_.

Studies investigating complex phenomena, such as *EHC*, usually involve fitting a
hypothetical model that is poorly validated to plasma concentration-time data.

Solution

```
TITLE 1
Oral solution subject Borje
MODEL
 COMM
  NDER 4
  NCON 1
  nparm 7
  PNAMES 'Vc','Cl','Cld','Vt','Ka','K1g' 'ttom'
END
    1: TEMP
    2: DOSE=CON(1)
    3: T=X
    4: VC=P(1)
    5: CL=P(2)
    6: CLD=P(3)
    7: VT=P(4)
    8: KA=P(5)
    9: K1G=P(6)
   10: TTOM=P(7)
   11: RI = 10.
   12: END
   13: START
   14: Z(1)=DOSE/VC
   15: Z(2)=0.0
   16: Z(3)=0.0
   17: Z(4)=0.0
   18: END
   19: DIFF
   20: DZ(1)=(KA*Z(3) - CL*Z(1) - CLD*Z(1) + CLD*Z(2) - K1G*Z(1)*VC)/VC
   21: DZ(2)=(CLD*Z(1) - CLD*Z(2))/VT
   22: IF T GE RI AND T LE (RI+TTOM) THEN
   23: DZ(3)=(Z(4)/TTOM - KA*Z(3))
   24: DZ(4)=(K1G*Z(1)*VC - Z(4)/TTOM)
   25: ELSE
   26: DZ(3)=(-KA*Z(3))
   27: DZ(4)=K1G*Z(1)*VC
   28: ENDIF
   29: END
   30: FUNC 1
   31: F = Z(1)
   32: END
   33: EOM
DATE
TIME
NOPAGE BREAKS
NVARIABLES 2
NPOINTS 100
XNUMBER 1
YNUMBER 2
CONSTANTS 5617.3
METHOD 2  'Gauss-Newton (Levenberg and Hartley)
REWEIGHT -2
ITERATIONS 50
INITIAL 13,1.2,10,35,2.3,.7,2.8
LOWER BOUNDS 0,0,0,0,0,0,0
UPPER BOUNDS 35,19,20,50,10,10,40
```

```
MISSING 'Missing'
NOBSERVATIONS 32
DATA 'WINNLIN.DAT'
BEGIN
```

PARAMETER	ESTIMATE	STANDARD ERROR	CV%	UNIVARIATE C.I.	
VC	12.786294	.535024	4.18	11.684399	13.888189
CL	.871675	.286416	32.86	.281794	1.461555
CLD	11.356138	.705471	6.21	9.903205	12.809070
VT	29.075335	.981436	3.38	27.054046	31.096624
KA	2.916200	.827313	28.37	1.212330	4.620070
K1G	.605135	.037464	6.19	.527976	.682294
TTOM	2.769508	.221118	7.98	2.314110	3.224906

```
*** CORRELATION MATRIX OF THE ESTIMATES ***
PARAMETER  VC          CL          CLD         VT          KA          K1G         TTOM
VC     1.00000
CL     -.147424    1.00000
CLD    -.191194    -.067837    1.00000
VT     -.152171    -.103208    .550784     1.00000
KA     .165372     .325682     -.067249    -.033382    1.00000
K1G    -.486656    -.746262    .226711     .353316     -.293514    1.00000
TTOM   -.010282    .149604     -.182578    -.136265    .751909     -.03428     1.00000

Condition_number=        157.8

    FUNCTION    1
```

X	OBSERVED Y	PREDICTED Y	RESIDUAL	WEIGHT	SE-PRED	STANDARDI RESIDUAL
.3000E-01	424.6	419.3	5.317	.5688E-05	16.64	.1935
.8300E-01	385.6	386.4	-.7997	.6698E-05	13.99	-.0306
.1500	331.8	349.0	-17.20	.8209E-05	11.30	-.7096
.1700	308.6	338.7	-30.11	.8717E-05	10.62	-1.272
.3300	297.9	268.1	29.76	.1391E-04	6.966	1.540
.5000	223.6	212.0	11.61	.2224E-04	5.473	.7590
.6700	179.6	170.4	9.126	.3443E-04	4.969	.7558
.8300	128.6	141.0	-12.49	.5026E-04	4.637	-1.280
1.000	119.8	117.5	2.300	.7239E-04	4.220	.2891
1.500	77.17	77.02	.1463	.1686E-03	2.831	.0282
2.000	56.09	58.54	-2.449	.2918E-03	1.903	-.6030
4.000	37.34	36.47	.8643	.7517E-03	1.281	.3480
6.000	26.90	27.06	-.1614	.1365E-02	.9227	-.0869
8.000	19.37	20.22	-.8484	.2445E-02	.6636	-.6060
10.00	15.19	15.11	.7622E-01	.4378E-02	.4930	.0727
10.50	35.94	34.53	1.405	.8385E-03	2.231	.9882
11.00	51.91	55.21	-3.301	.3280E-03	2.264	-.9239
11.50	68.49	65.78	2.713	.2311E-03	2.442	.6154
12.00	62.25	70.17	-7.915	.2031E-03	3.013	-1.778
12.50	77.34	71.73	5.615	.1944E-03	3.319	1.282
13.00	67.85	68.45	-.5976	.2134E-03	3.854	-.1680
15.00	28.09	30.50	-2.410	.1075E-02	1.073	-1.161

16.00	27.67	25.32	2.358	.1560E-02	.7646	1.322
17.00	23.31	21.71	1.602	.2122E-02	.6109	1.036
18.00	17.36	18.74	-1.373	.2849E-02	.4932	-1.019
20.00	14.90	14.00	.9051	.5103E-02	.3266	.8859
24.00	7.924	7.820	.1040	.1635E-01	.1706	.1810
26.00	4.854	5.844	-.9900	.2928E-01	.1381	-2.324
28.00	4.085	4.368	-.2828	.5241E-01	.1170	-.9019
30.00	3.664	3.265	.3995	.9382E-01	.1008	1.745
32.00	2.491	2.440	.5073E-01	.1680	.8688E-01	.3065
36.00	1.311	1.363	-.5244E-01	.5382	.6288E-01	-.6289

```
CORRECTED SUM OF SQUARED OBSERVATIONS =  467421.
WEIGHTED CORRECTED SUM OF SQUARED OBSERVATIONS =  24.8765
SUM OF SQUARED RESIDUALS =            2636.26
SUM OF WEIGHTED SQUARED RESIDUALS =  .146763
S =  .766193E-01 WITH     25 DEGREES OF FREEDOM
CORRELATION (OBSERVED,PREDICTED) =  .9972

AIC criteria =        -47.40596
SC  criteria =        -37.14581
```

PK41 - Multiple intravenous infusions - NCA *versus* regression

Objectives

◆ **To analyze plasma data following multiple intravenous infusions**

◆ **To obtain initial parameter estimates from a multiple infusions dataset**

◆ **To apply a user-specified differential equation model**

◆ **To discuss various aspects of experimental design and weighting**

Problem specification

This exercise highlights some complexities in modeling multiple intravenous infusion plasma kinetics simultaneously with urinary data. Specific emphasis is put on obtaining *initial parameter estimates*, application of *weights*, and *simultaneous fitting* of several plasma profiles. We also show that fitting data of several sources makes it possible to utilize a limited number (3 to 5) of samples per individual dataset.

One volunteer received a 5 h infusion regimen of a new potent life saver (Turbojoint®). The doses were 310, 520 and 780 µg/kg body weight infused over 5 h. Assume a body weight of 70 kg. Plasma data from one subject are plotted in Figure 41.1 and are shown in the program output.

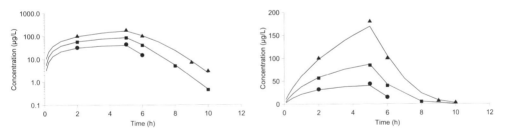

Figure 41.1 *Observed and model-predicted plasma concentrations of Turbojoint plotted on a semi-logarithmic scale (left) and linear scale (right). The infusion doses were 310, 520 and 780 µg/kg body weight infused over 5 h.*

Non-compartmental analysis *NCA* revealed that plasma clearance *Cl* decreased with dose (from 310 to 780 µg/kg), as shown in Figure 41.2. Volume of distribution *V* decreased from 2.3 to 1.8 L/kg and mean residence time *MRT* increased from 1.4 to 1.9 h. Since *Cl* and *V* changed with dose, we will propose a nonlinear elimination model.

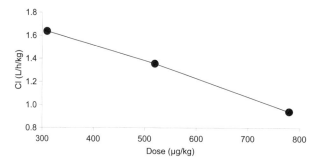

Figure 41.2 *Plasma clearance Cl (L/h/kg) versus infused Dose (μg/kg). Results were obtained from non-compartmental analysis of the 5 h constant rate intravenous infusion data.*

Plasma concentration-time data are given in Table 41.1. We will fit the following system of differential equations to the data. The plasma concentration is

$$V \cdot \frac{dC}{dt} = In - Cl \cdot C \qquad (41:1)$$

In is the constant rate intravenous infusion. The plasma clearance is modeled as

$$Cl = \frac{V_{max}}{K_m + C} \qquad (41:2)$$

Approximate estimates of clearance *Cl* and volume of distribution *V* were obtained from non-compartmental analysis.

Table 41.1 Dose, time and concentration (μmol/L) data of drug X

Dose (μg)	Time (h)	Conc (μmol/L)	Dose (μg)	Time (h)	Conc (μmol/L)	Dose (μg)	Time (h)	Conc (μmol/L)
310	0.1	-	520	0.1	-	780	1	-
	0.25	-		0.25	-		2	100.0
	0.5	-		0.5	-		3.5	-
	1	-		1	-		4	-
	2	32.0		2	57.0		5	180.0
	3.5	-		3.5	-		6	100.0
	5	44.0		5	85.0		7	-
	6	15.0		6	40.0		8	-
	-	-		8	5.0		9	7.0
	-	-		10	0.47		10	3

We will practise non-compartmental analysis for the estimation of initial parameter estimates and then fit Equations 41:1 and 41:2 to the three plasma concentration datasets simultaneously. Derivations of V_{max} and K_m follow below.

Initial parameter estimates

The initial parameter estimate of *V* was derived from non-compartmental analysis. The V_{max} and K_m parameters were obtained by means of inserting clearance values (1.6 and 0.94 L/h/kg from the non-compartmental analysis), and steady state plasma concentrations (44 and 180 μg/L from the lowest and highest infusion regimens) into

Equations 41:3 and 41:4.

$$1.6 = \frac{V_{max}}{K_m + 44}$$

(41:3)

$$0.94 = \frac{V_{max}}{K_m + 180}$$

(41:4)

We solve Equation 41:3 for V_{max} and insert this expression into Equation 41:4. This gives an approximate value of K_m of 150 µg/L. V_{max} can then be calculated to about 170 µg/h/kg from Equation 41:3. We have selected values of V_{max} and K_m of 200 µg/h/kg and 100 µg/L, respectively, as initial parameter estimates in order to challenge the program. Remember that *NCA* will underestimate the *AUC* during the upswing of the plasma concentration-time curve and overestimate *AUC* during the downswing if the time intervals between the observed plasma concentration-time data are large (Figure 41.3).

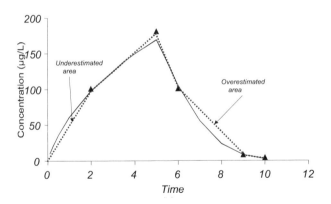

Figure 41.3 Schematic illustration of the prediction of the area under the curve AUC by means of a regression model (solid black line) and NCA (dotted blue line).

We use a weight of one over the predicted value squared, in order to balance the fit to the high and low plasma concentrations.

Interpretation of results and conclusions

This exercise has dealt with obtaining *initial parameter estimates*, applying *weights*, and inspection of the *goodness-of-fit*. It is of considerable value to invest time to obtain good initial parameter estimates. Figure 41.1 contains the observed and predicted plasma concentrations of the compound. The program estimated V_{max} and K_m to be 184 µg/h/kg (c.f. initial estimate 200 µg/h/kg) and 83 µg/L (initial estimate of 100 µg/h/kg) and the volume of distribution is 1.8 L/kg.

We have shown that fitting data of several sources makes it possible to utilize very few samples per individual dataset. We were able to estimate the Michaelis-Menten parameters with high precision, since concentration-time data of several doses were combined.

Solution – Proportional error

```
MODEL
 COMM
  NFUN  3
  NDER  3
  NPAR  3
  ncon  4
  PNAM  'V', 'Vmax', 'Km'

END
    1: TEMP
    2: T = X
    3: DOSE1 = CON(1)
    4: DOSE2 = CON(2)
    5: DOSE3 = CON(3)
    6: TINF = CON(4)
    7: END
    8: START
    9: Z(1)= 0
   10: Z(2)= 0
   11: END
   12: DIFF
   13: FINF1 = 0
   14: FINF2 = 0
   15: FINF3 = 0
   16: IF T LE TINF THEN
   17: FINF1 = DOSE1/TINF
   18: FINF2 = DOSE2/TINF
   19: FINF3 = DOSE3/TINF
   20: ELSE
   21: FINF1 = 0
   22: FINF2 = 0
   23: FINF3 = 0
   24: ENDIF
   25: CL1 = VMAX/(KM + Z(1))
   26: DZ(1) = (FINF1 - CL1*Z(1))/V
   27: CL2 = VMAX/(KM + Z(2))
   28: DZ(2) = (FINF2 - CL2*Z(2))/V
   29: CL3 = VMAX/(KM + Z(3))
   30: DZ(3) = (FINF3 - CL3*Z(3))/V
   31: END
   32: FUNC 1
   33: F = Z(1)
   34: END
   35: FUNC 2
   36: F = Z(2)
   37: END
   38: FUNC 3
   39: F = Z(3)
   40: END
   41: EOM
NOPAGE BREAKS
NVARIABLES 11
FNUMBER 1
NPOINTS 100
XNUMBER 2
YNUMBER 3
CONSTANTS 310,520,780,5
METHOD 2  'Gauss-Newton (Levenberg and Hartley)
REWEIGHT -2
```

```
ITERATIONS 50
INITIAL 2,200,100
LOWER BOUNDS .2,20,10
UPPER BOUNDS 10,1000,500
MISSING 'Missing'
DATA 'WINNLIN.DAT'
BEGIN
```

PARAMETER	ESTIMATE	STANDARD ERROR	CV%	UNIVARIATE C.I.	
V	1.786282	.071675	4.01	1.626579	1.945986
VMAX	183.880766	7.697861	4.19	166.728754	201.032778
KM	82.831770	6.564542	7.93	68.204967	97.458573

```
*** CORRELATION MATRIX OF THE ESTIMATES ***
PARAMETER   V         VMAX       KM
V           1.00000
VMAX        -.25170   1.00000
KM          -.59339    .92036    1.00000


 Condition_number=        486.4

    FUNCTION   1
```

X	OBSERVED Y	PREDICTED Y	RESIDUAL	WEIGHT	SE-PRED	STANDARD RESIDUAL
.1000	.	3.269	.	.0000	-.6400+152	.
.2500	.	7.524	.	.0000	-.6400+152	.
.5000	.	13.31	.	.0000	-.6400+152	.
1.000	.	21.60	.	.0000	-.6400+152	.
2.000	32.00	31.11	.8866	.1033E-02	1.153	.3221
3.500	.	37.52	.	.0000	-.6400+152	.
5.000	44.00	40.14	3.857	.6206E-03	1.369	1.072
6.000	15.00	15.59	-.5850	.4117E-02	.5215	-.4176

```
CORRECTED SUM OF SQUARED OBSERVATIONS =  2149.88
WEIGHTED CORRECTED SUM OF SQUARED OBSERVATIONS =  .601332
SUM OF SQUARED RESIDUALS =            16.0056
SUM OF WEIGHTED SQUARED RESIDUALS =  .114533E-01
S =  .000000     WITH      0 DEGREES OF FREEDOM
CORRELATION (OBSERVED,PREDICTED) =   .4438

    FUNCTION   2
```

X	OBSERVED Y	PREDICTED Y	RESIDUAL	WEIGHT	SE-PRED	STANDARD RESIDUAL
.1000	.	5.489	.	.0000	-.6400+152	.
.2500	.	12.69	.	.0000	-.6400+152	.
.5000	.	22.70	.	.0000	-.6400+152	.
1.000	.	37.93	.	.0000	-.6400+152	.
2.000	57.00	58.10	-1.097	.2963E-03	2.101	-.2125
3.500	.	75.98	.	.0000	-.6400+152	.
5.000	85.00	86.65	-1.651	.1332E-03	2.874	-.2118
6.000	40.00	42.58	-2.579	.5516E-03	1.288	-.6655
8.000	5.000	5.545	-.5451	.3252E-01	.2005	-1.107
10.00	.4700	.4909	-.2085E-01	4.150	.4153E-01	-.9406

CORRECTED SUM OF SQUARED OBSERVATIONS = 8584.72
WEIGHTED CORRECTED SUM OF SQUARED OBSERVATIONS = 3.41842
SUM OF SQUARED RESIDUALS =   10.8795
SUM OF WEIGHTED SQUARED RESIDUALS = .158584E-01
S = .890460E-01 WITH   2 DEGREES OF FREEDOM
CORRELATION (OBSERVED,PREDICTED) = .7683

   FUNCTION   3

X	OBSERVED Y	PREDICTED Y	RESIDUAL	WEIGHT	SE-PRED	STANDARD RESIDUAL
.1000	.	10.48	.	.0000	-.6400+152	.
.2500	.	21.11	.	.0000	-.6400+152	.
.5000	.	36.31	.	.0000	-.6400+152	.
1.000	.	60.91	.	.0000	-.6400+152	.
2.000	100.0	97.78	2.217	.1046E-03	3.005	.2496
3.500	.	138.0	.	.0000	-.6400+152	.
4.000	.	149.0	.	.0000	-.6400+152	.
5.000	180.0	168.7	11.34	.3515E-04	5.750	.7498
6.000	100.0	105.0	-4.997	.9071E-04	4.039	-.5417
7.000	.	55.25	.	.0000	-.6400+152	.
8.000	.	23.41	.	.0000	-.6400+152	.
9.000	7.000	8.126	-1.126	.1514E-01	.5394	-2.002
10.00	3.000	2.510	.4903	.1588	.1608	2.737

CORRECTED SUM OF SQUARED OBSERVATIONS = 40758.0
WEIGHTED CORRECTED SUM OF SQUARED OBSERVATIONS = 3.13891
SUM OF SQUARED RESIDUALS =   159.940
SUM OF WEIGHTED SQUARED RESIDUALS = .646818E-01
S = .179836   WITH   2 DEGREES OF FREEDOM
CORRELATION (OBSERVED,PREDICTED) = .7337

TOTALS FOR ALL CURVES COMBINED
SUM OF SQUARED RESIDUALS =   186.825
SUM OF WEIGHTED SQUARED RESIDUALS = .919935E-01
S = .959132E-01 WITH   10 DEGREES OF FREEDOM

AIC criteria =   -48.87886
SC criteria =   -45.47237

PK42 - Saturable absorption kinetics

Objectives

◆ **To analyze three oral datasets with saturable input and linear disposition**

◆ **To write a multi-compartment model in terms of differential equations**

◆ **To obtain initial parameter estimates**

◆ **To discuss various aspects of experimental design**

Problem specification

This problem highlights some complexities in modeling nonlinear absorption kinetics. Nonlinear absorption is mostly a combination of saturable first-pass and first-order absorption. In this particular exercise, we assume that the nonlinear absorption is due to saturable transport during the absorption step, not to an effect of metabolic processes. Specific emphasis is put on obtaining *initial parameter estimates*, implementation of *differential equations*, and assessing the *goodness-of-fit*.

The compound for which you will characterize the absorption parameters after oral dosing has a low molecular weight, is actively absorbed from the gastrointestinal tract and is eliminated via expired air and the kidneys by means of filtration and active tubular secretion. The plasma protein binding is negligible. The disposition kinetics shows a highly cleared compound, with a volume of distribution close to body water and an efficient half-life of less than 20 min.

One volunteer received a dose of 10, 30 and 90 mg of X on three different occasions. Frequent blood sampling of 300 μL was done. Assume a body weight of 70 kg. Plasma data are plotted in Figures 42.1 and 42.2 and listed in the program output.

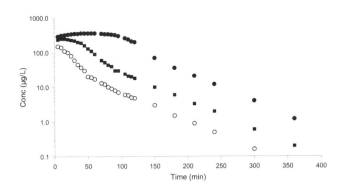

Figure 42.1 Semi-logarithmic plot of observed plasma concentration (μg/L) of compound X versus time following three different oral doses.

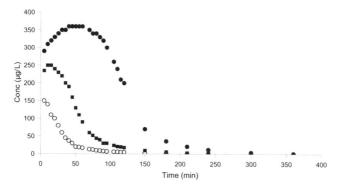

Figure 42.2 Linear plot of observed plasma concentration (µg/L) of compound X versus time following three different oral doses

The compound is rapidly and completely absorbed by means of a saturable transport process. The saturable absorption rate is evident at higher doses. We assume that we can *approximate* the disposition with a linear renal clearance. However, the input rate is saturable and is determined by two parameters V_{max} and K_m. The input rate is mathematically expressed as

$$Input\ rate = \frac{V_{max} \cdot A_g}{K_m + A_g} \tag{42:1}$$

The A_g variable is the amount of drug in the gastrointestinal tract. The input rate to the plasma compartment is equal to the elimination rate from the gastrointestinal tract. We assume that the extent of absorption is 100% in spite of the fact that the drug uptake is capacity limited. There is no first-pass effect. We will fit the following system of differential equations to the data. The concentration in the central compartment is

$$V_c \cdot \frac{dC}{dt} = Input - Cl \cdot C - Cl_d \cdot C + Cl_d \cdot C_t \tag{42:2}$$

and in the peripheral compartment

$$V_t \cdot \frac{dC_t}{dt} = Cl_d \cdot C - Cl_d \cdot C_t \tag{42:3}$$

Initial parameter estimates

Discuss the disposition of the compound with your colleague(s), obtain initial parameter estimates, and implement a two-compartment disposition model with saturable absorption. The initial parameter estimates for V_c, Cl_d, V_t and Cl of 5.2 L, 1.1 L/min, 40 L and 2.2 L/min, respectively, were obtained from a previous study.

At the lowest dose (10 mg), there is very little information about nonlinearity (Figure 42.1). Let us therefore assume that K_m is less than or equal to 10000 µg (10 mg). Observe that we have expressed K_m in terms of amount and not in concentration units in this particular example. This is because we do not know the volume of the gastrointestinal tract. When the amount of drug in the gastrointestinal tract is much less than K_m, absorption occurs as a first-order process

$$Input\ rate = \frac{V_{max} \cdot A_g}{K_m \gg A_g} \approx \frac{V_{max} \cdot A_g}{K_m} = k' \cdot A_g \qquad (42{:}4)$$

When the amount of drug in the gastrointestinal tract is much greater than K_m, absorption occurs as a zero-order (constant-rate) process

$$Input\ rate = \frac{V_{max} \cdot A_g}{K_m \ll A_g} \approx \frac{V_{max} \cdot A_g}{A_g} = V_{max} \qquad (42{:}5)$$

The latter situation can be compared to a constant rate intravenous infusion. There is an approximate steady state (C'_{ss}) at the highest dose between 30 and 100 min. With a clearance of 2.2 L/min and a C'_{ss} of about 350 µg/L, the input rate can be approximated with Equation 42:5 and becomes

$$Input\ rate = V_{max} = C'_{ss} \cdot Cl = 350 \cdot 2.2 = 770 \mu g\,/\min \qquad (42{:}6)$$

When the amount of drug in the gastrointestinal tract is much greater than K_m, the input rate approaches V_{max}.

Interpretation of results and conclusions

This exercise has dealt with obtaining *initial parameter estimates*, implementing *differential equations*, and the *goodness-of-fit* of nonlinear absorption kinetics. Figures 42.3 and 42.4 contain the observed and predicted concentrations of X after each of the three doses. Note the shift in the peak plasma concentration to the right with increasing dose.

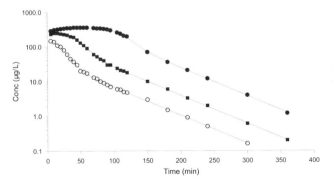

Figure 42.3 Semi-logarithmic plot of observed and model-predicted plasma concentration-time data.

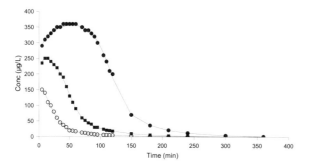

Figure 42.4 Observed and model-predicted plasma concentration-time data plotted on a linear scale.

The disposition parameters such as V_c, Cl, Cl_d and V_t were 4.4 L, 2.0 L/min, 0.99 L/min and 35.0 L, respectively. V_{ss} was 39.4 L and *MRT* about 20 min. Simulate the concentration-time course following an oral dose of 500 mg using the final parameter estimates.

Table 42.1   Parameter estimates of clearance and micro-constant models

	Clearance model		Micro-constant model	
V_c	4.4	16	4.3	17
Cl_d/k_{12}	0.99	2	0.23	18
V_t/k_{21}	35.0	1	0.028	1
V_{max}	980	1	980	1
K_m	9570	3	9580	3
Cl/k_{10}	2.0	1	0.47	17

The absorption parameters V_{max} and K_m were 981 µg/min and 9568 µg, respectively. The area under the plasma concentration-time curve was 4947, 14726 and 45069 µg·h/L for the 10, 30 and 90 mg dose levels, respectively, when obtained by non-compartmental analysis. When dose-normalized, these areas were 0.49, 0.49 and 0.50 h/L for the three respective doses. This shows that for a system with linear disposition and nonlinear absorption, the dose-normalized areas should be the same, provided that absorption is complete. Figure 42.5 shows the cumulative input rate plotted against time for each dose.

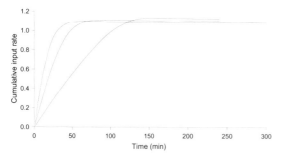

Figure 42.5 The cumulative input (bioavailability) versus time for the three oral doses. Because of the instability of the numerical deconvolution method applied, the total cumulative input is greater than 1 (or 100%). In fact, the estimated bioavailability is 105-110%. The most rapid and slowest upswing are for the 10 and 90 mg doses, respectively.

The data in Figure 42.5 are obtained by numerical deconvolution of the three oral datasets by means of an intravenous dataset (not included in this exercise). Observe that the time to complete absorption increases with an increase in dose. The predicted bioavailability is also somewhat larger than 1.0 (or 100%), which can be explained by the lack of early time points. The first observation originates from a 5 min sample in all three datasets, and the early plasma concentrations were very high. Numerical deconvolution is known for instabilities in such cases, depending on the method used. Analytical deconvolution by means of sums of exponentials, which may be a more numerically stable method, is not applicable for nonlinear systems. We would therefore recommend one or two earlier time points (prior to the 5 min sample) in a future study of this drug and regimen.

Solution I – Clearance model

```
MODEL
 COMM
  NFUN   3
  NDER   9
  NPARM 6
  NCON   3
  PNAMES 'Vc', 'CLD','Vt', 'Vmax', 'Km', 'Cl'
END
     1: TEMP
     2: T=X
     3: DOSE1 = CON(1)
     4: DOSE2 = CON(2)
     5: DOSE3 = CON(3)
     6: END
     7: START
     8: Z(1)=0
     9: Z(2)=0
    10: Z(3)=0
    11: Z(4)=0
    12: Z(5)=0
    13: Z(6)=0
    14: Z(7)=DOSE1
    15: Z(8)=DOSE2
    16: Z(9)=DOSE3
    17: END
    18: DIFF
    19: INPUT1 = VMAX*Z(7)/(KM + Z(7))
    20: INPUT2 = VMAX*Z(8)/(KM + Z(8))
    21: INPUT3 = VMAX*Z(9)/(KM + Z(9))
    22: DZ(1) = (INPUT1 - CL*Z(1) - CLD*Z(1) + CLD*Z(4))/VC
    23: DZ(2) = (INPUT2 - CL*Z(2) - CLD*Z(2) + CLD*Z(5))/VC
    24: DZ(3) = (INPUT3 - CL*Z(3) - CLD*Z(3) + CLD*Z(6))/VC
    25: DZ(4) = (CLD*Z(1) - CLD*Z(4))/VT
    26: DZ(5) = (CLD*Z(2) - CLD*Z(5))/VT
    27: DZ(6) = (CLD*Z(3) - CLD*Z(6))/VT
    28: IF Z(7) LE 0.0001 THEN
    29: Z(7) = 0.0
    30: DZ(7) = 0.0
    31: ELSE
    32: DZ(7) = -VMAX*Z(7)/(KM + Z(7))
    33: ENDIF
    34: IF Z(8) LE 0.0001 THEN
    35: Z(8) = 0.0
    36: DZ(8) = 0.0
    37: ELSE
    38: DZ(8) = -VMAX*Z(8)/(KM + Z(8))
    39: ENDIF
```

```
40: IF Z(9) LE 0.0001 THEN
41: Z(9) = 0.0
42: DZ(9) = 0.0
43: ELSE
44: DZ(9) = -VMAX*Z(9)/(KM + Z(9))
45: ENDIF
46: DZ(8) = -VMAX*Z(8)/(KM + Z(8))
47: DZ(9) = -VMAX*Z(9)/(KM + Z(9))
48: END
49: FUNC 1
50: F=Z(1)
51: END
52: FUNC 2
53: F=Z(2)
54: END
55: FUNC 3
56: F=Z(3)
57: END
58: EOM
NOPAGE BREAKS
NVARIABLES 3
FNUMBER 3
NPOINTS 1500
XNUMBER 1
YNUMBER 2
CONSTANTS 10000,30000,90000
METHOD 2  'Gauss-Newton (Levenberg and Hartley)
REWEIGHT -2
ITERATIONS 50
INITIAL 5.2,1.1,40,900,11000,2.2
LOWER BOUNDS 0,0,0,0,0,0
UPPER BOUNDS 80,10,90,2000,50000,10
BEGIN
```

PARAMETER	ESTIMATE	STANDARD ERROR	CV%	UNIVARIATE C.I.	
VC	4.395538	.708753	16.12	2.984225	5.806851
CLD	.988144	.022835	2.31	.942674	1.033614
VT	35.028032	.492345	1.41	34.047645	36.008420
VMAX	981.325253	8.096713	.83	965.202586	997.447920
KM	9568.428185	252.415662	2.64	9065.802772	10071.053597
CL	2.002183	.010841	.54	1.980595	2.023770

```
*** CORRELATION MATRIX OF THE ESTIMATES ***
PARAMETER  VC       CLD       VT       VMAX      KM        CL
VC        1.0000
CLD       -.1322   1.00000
VT        -.2317    .93662   1.00000
VMAX       .3048    .00261    .02930  1.00000
KM        -.2011   -.42947   -.28642   .57647   1.00000
CL         .4523    .31500    .42975   .25578    .05883   1.00000

 Condition_number=        .5480E+05

   FUNCTION    1
```

X	OBSERVED	PREDICTED	RESIDUAL	WEIGHT	SE-PRED	STANDARDI

	Y		Y				RESIDUAL
5.000	150.0	150.0	.1955E-01	.4446E-04	2.400	.4011E-02	
10.00	140.0	137.9	2.064	.5256E-04	1.489	.4327	
15.00	110.0	118.0	-8.009	.7181E-04	1.350	-1.974	
20.00	100.0	96.88	3.121	.1065E-03	1.048	.9318	
25.00	80.00	76.82	3.182	.1695E-03	.7858	1.192	
30.00	60.00	59.54	.4564	.2821E-03	.6285	.2212	
35.00	45.00	45.85	-.8534	.4756E-03	.5340	-.5425	
40.00	37.00	35.64	1.356	.7871E-03	.4444	1.118	
45.00	30.00	28.30	1.704	.1249E-02	.3496	1.769	
50.00	20.00	23.06	-3.063	.1880E-02	.2623	-3.861	
55.00	19.00	19.30	-.3004	.2685E-02	.1935	-.4471	
60.00	17.00	16.53	.4735	.3661E-02	.1465	.8157	
70.00	13.00	12.74	.2611	.6162E-02	.1020	.5800	
75.00	12.00	11.37	.6296	.7735E-02	.9206E-01	1.568	
80.00	10.00	10.22	-.2175	.9579E-02	.8451E-01	-.6036	
85.00	9.000	9.224	-.2238	.1175E-01	.7765E-01	-.6887	
90.00	8.000	8.353	-.3526	.1433E-01	.7097E-01	-1.199	
95.00	7.000	7.579	-.5793	.1741E-01	.6444E-01	-2.170	
105.0	6.000	6.264	-.2636	.2549E-01	.5229E-01	-1.194	
110.0	5.800	5.700	.1000E+00	.3078E-01	.4685E-01	.4972	
115.0	5.000	5.189	-.1892	.3714E-01	.4189E-01	-1.032	
120.0	4.800	4.725	.7460E-01	.4478E-01	.3744E-01	.4466	
150.0	3.000	2.700	.2999	.1372	.1959E-01	3.129	
180.0	1.500	1.544	-.4412E-01	.4194	.1146E-01	-.8058	
210.0	.9000	.8831	.1691E-01	1.282	.7464E-02	.5435	
240.0	.5000	.5051	-.5051E-02	3.920	.5095E-02	-.2874	
300.0	.1600	.1652	-.5193E-02	36.65	.2358E-02	-.9441	

```
CORRECTED SUM OF SQUARED OBSERVATIONS =   50636.0
  WEIGHTED CORRECTED SUM OF SQUARED OBSERVATIONS =  23.7008
  SUM OF SQUARED RESIDUALS =            104.874
  SUM OF WEIGHTED SQUARED RESIDUALS =  .621781E-01
  S =  .544138E-01 WITH     21 DEGREES OF FREEDOM
  CORRELATION (OBSERVED,PREDICTED) =  .9990
```

FUNCTION   2

X	OBSERVED Y	PREDICTED Y	RESIDUAL	WEIGHT	SE-PRED	STANDARDI RESIDUAL
5.000	235.0	240.0	-4.951	.1737E-04	4.496	-.6654
10.00	250.0	248.5	1.458	.1619E-04	1.856	.1655
15.00	250.0	246.9	3.106	.1641E-04	1.806	.3545
20.00	240.0	241.7	-1.742	.1711E-04	1.762	-.2031
25.00	230.0	233.0	-2.978	.1842E-04	1.719	-.3604
30.00	220.0	220.3	-.3112	.2060E-04	1.654	-.3985E-01
35.00	200.0	203.6	-3.623	.2412E-04	1.541	-.5022
40.00	190.0	183.2	6.809	.2980E-04	1.369	1.049
45.00	160.0	159.9	.5755E-01	.3909E-04	1.172	.1014E-01
50.00	130.0	135.5	-5.531	.5444E-04	1.021	-1.152
55.00	110.0	112.0	-2.036	.7967E-04	.9478	-.5160
60.00	90.00	91.29	-1.290	.1200E-03	.8892	-.4051
70.00	60.00	61.07	-1.068	.2681E-03	.6415	-.5044
75.00	52.00	51.07	.9282	.3834E-03	.4934	.5205
80.00	44.00	43.54	.4637	.5276E-03	.3715	.3025
85.00	40.00	37.77	2.232	.7011E-03	.2879	1.669
90.00	30.00	33.23	-3.233	.9055E-03	.2384	-2.739
95.00	30.00	29.56	.4421	.1145E-02	.2105	.4211
105.0	24.00	23.87	.1262	.1755E-02	.1778	.1491
110.0	21.00	21.59	-.5924	.2145E-02	.1636	-.7744
115.0	20.00	19.58	.4228	.2609E-02	.1495	.6099

```
120.0       18.00       17.78       .2208       .3164E-02   .1357       .3507
150.0       10.00       10.11       -.1105      .9783E-02   .7074E-01   -.3074
180.0       6.000       5.780       .2204       .2994E-01   .3898E-01   1.071
210.0       3.300       3.305       -.5308E-02  .9153E-01   .2458E-01   -.4529E-01
240.0       2.000       1.890       .1097       .2798       .1681E-01   1.652
300.0       .6000       .6183       -.1827E-01  2.616       .8050E-02   -.8742
360.0       .2000       .2022       -.2233E-02  24.45       .3578E-02   -.3492
```

```
CORRECTED SUM OF SQUARED OBSERVATIONS =  237597.
WEIGHTED CORRECTED SUM OF SQUARED OBSERVATIONS =  25.6228
SUM OF SQUARED RESIDUALS =           162.687
SUM OF WEIGHTED SQUARED RESIDUALS =  .259965E-01
S =  .343753E-01 WITH     22 DEGREES OF FREEDOM
CORRELATION (OBSERVED,PREDICTED) =  .9997
```

FUNCTION   3

X	OBSERVED Y	PREDICTED Y	RESIDUAL	WEIGHT	SE-PRED	STANDARDI RESIDUAL
5.000	290.0	291.9	-1.880	.1174E-04	6.062	-.2170
10.00	310.0	311.7	-1.672	.1029E-04	2.879	-.1532
15.00	320.0	321.5	-1.542	.9672E-05	2.754	-.1362
20.00	330.0	329.9	.1107	.9189E-05	2.692	.9511E-02
25.00	340.0	337.0	2.956	.8803E-05	2.641	.2480
30.00	350.0	343.1	6.937	.8497E-05	2.598	.5708
35.00	350.0	348.0	2.015	.8258E-05	2.559	.1633
40.00	360.0	351.8	8.172	.8079E-05	2.520	.6540
45.00	360.0	354.6	5.402	.7953E-05	2.479	.4286
50.00	360.0	356.3	3.720	.7878E-05	2.435	.2935
55.00	360.0	356.8	3.163	.7853E-05	2.385	.2490
60.00	360.0	356.2	3.794	.7881E-05	2.330	.2989
70.00	350.0	351.0	-.9760	.8118E-05	2.202	-.7794E-01
75.00	340.0	346.1	-6.072	.8350E-05	2.136	-.4915
80.00	340.0	339.4	.6421	.8683E-05	2.076	.5299E-01
85.00	330.0	330.5	-.5474	.9152E-05	2.037	-.4639E-01
90.00	320.0	319.3	.7040	.9809E-05	2.037	.6183E-01
95.00	300.0	305.2	-5.216	.1073E-04	2.100	-.4805
105.0	260.0	267.2	-7.172	.1401E-04	2.472	-.7664
110.0	240.0	243.0	-3.000	.1694E-04	2.739	-.3585
115.0	210.0	216.0	-6.030	.2143E-04	2.966	-.8326
120.0	200.0	187.6	12.38	.2841E-04	3.053	2.039
150.0	70.00	70.70	-.6992	.2001E-03	.9371	-.2933
180.0	36.00	36.48	-.4781	.7515E-03	.3383	-.3743
210.0	21.00	20.67	.3318	.2341E-02	.1830	.4570
240.0	12.00	11.81	.1886	.7168E-02	.1051	.4547
300.0	4.000	3.863	.1370	.6701E-01	.4255E-01	1.027
360.0	1.200	1.264	-.6353E-01	.6264	.1870E-01	-1.521

```
CORRECTED SUM OF SQUARED OBSERVATIONS =  458548.
WEIGHTED CORRECTED SUM OF SQUARED OBSERVATIONS =  25.8206
SUM OF SQUARED RESIDUALS =           521.063
SUM OF WEIGHTED SQUARED RESIDUALS =  .128788E-01
S =  .241950E-01 WITH     22 DEGREES OF FREEDOM
CORRELATION (OBSERVED,PREDICTED) =  .9994
```

```
TOTALS FOR ALL CURVES COMBINED
SUM OF SQUARED RESIDUALS =           788.624
SUM OF WEIGHTED SQUARED RESIDUALS =  .101053
S =  .362268E-01 WITH     77 DEGREES OF FREEDOM
```

```
AIC criteria =       -178.24480
```

```
 SC  criteria =        -163.73176
```

Micro constant absorption model code

```
MODEL
REMA Modeling of saturable absorption with micro constants
COMM
  NFUN  3
  NDER  9
  NPARM 6
  NCON  3
  PNAMES 'Vc', 'k12','k21', 'Vmax', 'Km', 'k10'
END
TEMP
  T=X
  Dose1 = CON(1)
  Dose2 = CON(2)
  Dose3 = CON(3)
END
START
  Z(1)=0
  Z(2)=0
  Z(3)=0
  Z(4)=0
  Z(5)=0
  Z(6)=0
  Z(7)=Dose1
  Z(8)=Dose2
  Z(9)=Dose3
END
DIFF
Input1 = Vmax*Z(7)/(Km + Z(7))
  Input2 = Vmax*Z(8)/(Km + Z(8))
  Input3 = Vmax*Z(9)/(Km + Z(9))
  DZ(1) = (Input1 - k10*Z(1) - k12*Z(1) + k21*Z(4))
  DZ(2) = (Input2 - k10*Z(2) - k12*Z(2) + k21*Z(5))
  DZ(3) = (Input3 - k10*Z(3) - k12*Z(3) + k21*Z(6))
  DZ(4) = (k12*Z(1) - k21*Z(4))
  DZ(5) = (k12*Z(2) - k21*Z(5))
  DZ(6) = (k12*Z(3) - k21*Z(6))
  DZ(7) = -Vmax*Z(7)/(Km + Z(7))
  DZ(8) = -Vmax*Z(8)/(Km + Z(8))
  DZ(9) = -Vmax*Z(9)/(Km + Z(9))
END
func 1
  F=Z(1)/Vc
end
func 2
  F=Z(2)/Vc
end
func 3
  F=Z(3)/Vc
end
EOM
```

PK43 - Multiple absorption routes

Objectives
◆ To analyze a dataset obtained from sublingual dosing
◆ To write a compartment model in terms of a differential equation
◆ To discuss various aspects of correlation, accuracy, and precision

Problem specification

This exercise examines some of the features of data obtained after sublingual dosing with a new analgesic compound X in man. The analysis includes aspects of writing a differential equation model for a drug that is partly absorbed from the buccal cavity and partly via the gastrointestinal tract. It is particularly suitable for modeling multiple peaks in the plasma concentration that appear after extravascular dosing. Indeed, we show that it is possible to write and fit a multiple absorption sites model that mimics the kinetics of drug X after sublingual dosing. The parameter estimates had high precision and low correlation.

The concentration-time course after a 2 mg sublingual dose to a male subject is shown in Figure 43.1 and Table 43.1. The first peak appears within half an hour after dosing and the second peak at about 5 h. One of the goals of the exercise is to estimate the fraction of the dose that is absorbed via the sublingual and via the gastrointestinal route, provided that no substantial first-pass elimination occurs via the gastrointestinal route.

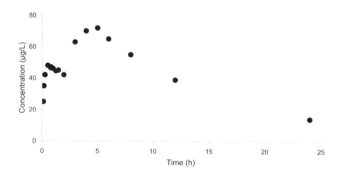

Figure 43.1. Observed concentration-time data following a 2 mg sublingual dose. Note the twin-peak phenomenon.

Table 43.1 Concentration-time data

Time (h)	Concentration (µg/L)	Time (h)	Concentration (µg/L)	Time (h)	Concentration (µg/L)	Time (h)	Concentration (µg/L)
0.0833	-	0.5	48	2	42.1	6	65
0.1	25	0.6	-	2.25	-	8	55
0.125	-	0.75	46.5	2.5	-	10	-
0.15	35	0.8	47	3	63	12	39
0.2	-	0.9	-	3.5	-	16	-
0.25	42	1	46	4	70	24	14
0.3	-	1.25	44.5	5	72	-	-
0.45	-	1.5	45	5.5	-	-	-

This dataset contains 30 time points, and includes 17 observations and 13 missing values. In order to fully discriminate between the fraction of the dose that escapes buccal absorption and the actual bioavailability from the gastrointestinal tract, both intravenous and oral dosing are necessary.

Figure 43.2 is a schematic diagram of the two-site absorption model. A portion of the dose escapes absorption from the buccal cavity and is swallowed. This fraction is then absorbed from the gastrointestinal tract. A first-order absorption/elimination model is assumed to correctly describe the experimental data.

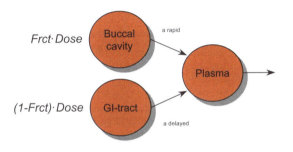

Figure 43.2 Schematic illustration of the two-site absorption model following a 2 mg sublingual dose of compound X.

The first-order input rate from the buccal cavity is given by

$$Rapid\ input = Frct \cdot Dose \cdot K_{a\,rapid} \cdot e^{-K_{a\,rapid}\cdot t} \tag{43:1}$$

The *Frct* parameter is the fraction of the dose that is absorbed via the buccal route. The input rate from the gastrointestinal tract is given by

$$Delayed\ input = (1 - Frct) \cdot Dose \cdot K_{a\,delayed} \cdot e^{-K_{a\,delayed}\cdot(t-t_{lag})} \tag{43:2}$$

The concentration in the plasma compartment is described by

$$\frac{dC}{dt} = \frac{Rapid\ input_1 + Delayed\ input_2}{(V/F)} - K \cdot C \tag{43:3}$$

where K is the elimination rate constant from plasma.

Initial parameter estimates

Implement a one-compartment disposition model parameterized with $K_{a1}, K_{a2}, t_{lag}, V/F, K$ and fraction of dose absorbed sublingually (*Frct*) *versus* orally (*1-Frct*). K_{a1} and K_{a2} are the rapid and delayed absorption rate constants, respectively.

$$K_{a1} = \frac{\ln(35/25)}{0.1-0.15} \approx 7\,\mathrm{h}^{-1} \tag{43:4}$$

$$K_{a2} = \frac{\ln(63/42.1)}{2-3.5} \approx 0.3\,\mathrm{h}^{-1} \tag{43:5}$$

K_{a1}	=	$10\,\mathrm{h}^{-1}$	Obtained from Equation 43:4
K_{a2}	=	$0.5\,\mathrm{h}^{-1}$	Obtained from Equation 43:5
V/F	=	$20\,\mathrm{L}$	Approximated by $Dose/C_{peak}$ (2000/(50+50))
K	=	$0.1\,\mathrm{h}^{-1}$	Obtained from terminal slope ($0.0858 \approx 0.1$)
$Frct$	=	0.5	Obtained from AUC_s/AUC_0 (see below)
t_{lag}	=	$2.0\,\mathrm{h}$	Obtained from Figure 43.1

Frct is obtained from the ratio of the area under the *sublingual portion* (AUC_S) divided by the total area under the curve and AUC_S/AUC_{all} is 0.5 (Figure 43.3).

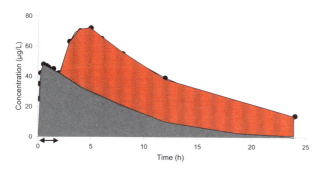

Figure 43.3 *Schematic illustration of the relative area contributions to the total area under the plasma concentration-time curve. The grey area represents the sublingual dose and the shaded red area the gastrointestinal dose. The extrapolated areas (not shown) should also be included.*

AUC_S is estimated by means of the trapezoidal method up to the trough concentration (C_{trough}) prior to the second absorption phase, plus the extrapolated area. The extrapolated area derives from C_{trough} divided by the terminal slope λ_z, where λ_z is estimated from time interval 5 to 24 h.

$$AUC_s = AUC^2_{0\,(trapezoidal)} + \frac{C_2}{\lambda_z} = 85.6 + \frac{42.1}{0.0858} = 576\,\mu g/L \tag{43:6}$$

The total area under the curve is

$$AUC_0^\infty = AUC^{24}_{0\,(trapezoidal)} + \frac{C_{24}}{\lambda_z} = 970 + \frac{14.0}{0.0858} = 1133\,\mu g/L \tag{43:7}$$

Interpretation of output and conclusions

The fit of the two parallel first-order input rates is shown in Figure 43.4. The final parameter estimates and their precision are shown in Table 43.2. All parameters except for the absorption rate constant for delayed input had high precision. The lag-time for the gastrointestinal absorption of the studied compound was substantial (2.3 h). The time between t_{max} of the second peak and its time lag was about 2 h.

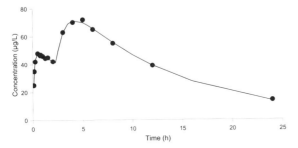

Table 43.2 Kinetic estimates

Parameter	Estimate	CV%
K_{a1}	7.6	4
K_{a2}	1.1	20
V/F	20.6	2
K	0.089	5
$Frct$	0.52	2
t_{lag}	2.3	6

Figure 43.4 *Observed and predicted plasma concentration-time data resulting from the sublingual/gastrointestinal absorption process.*

The fraction of the dose absorbed via the sublingual route was approximately 50% of the totally absorbed dose.

Solution

```
MODEL
  COMMANDS
    NFUN 1
    NDER 1
    NPAR 6
    PNAM  'Ka1', 'Ka2', 'V', 'K', 'frct', 'tlag'
    NCON 1
END
      1: TEMP
      2: T = X
      3: DOSE1 = CON(1)
      4: END
      5: START
      6: Z(1) = 0.
      7: END
      8: DIFF
      9: INPT1 = DOSE1*FRCT*KA1*EXP(-KA1*T)
     10: IF T LE TLAG THEN
     11: INPT2 = 0.0
     12: ELSE
     13: INPT2 = DOSE1*(1-FRCT)*KA2*EXP(-KA2*(T-TLAG))
     14: ENDIF
     15: DZ(1)= (INPT1 + INPT2)/V - K*Z(1)
     16: END
     17: FUNC 1
     18: F = Z(1)
     19: END
     20: EOM
```

```
NVARIABLES 2
NPOINTS 100
XNUMBER 1
YNUMBER 2
CONSTANTS 2000
METHOD 2   'Gauss-Newton (Levenberg and Hartley)
ITERATIONS 50
INITIAL 10,.5,20,.1,.7,2
LOWER BOUNDS 0,0,0,0,0,0
UPPER BOUNDS 15,5,40,2,1,5
MISSING 'Missing'
NOBSERVATIONS 30
DATA 'WINNLIN.DAT'
BEGIN
```

PARAMETER	ESTIMATE	STANDARD ERROR	CV%	UNIVARIATE C.I.	
KA1	7.624903	.321143	4.21	6.918069	8.331738
KA2	1.071771	.217438	20.29	.593191	1.550350
V	20.611842	.416909	2.02	19.694228	21.529455
K	.088823	.004054	4.56	.079900	.097746
FRCT	.514667	.009466	1.84	.493832	.535501
TLAG	2.296174	.127317	5.54	2.015950	2.576398

*** CORRELATION MATRIX OF THE ESTIMATES ***

PARAMETER	KA1	KA2	V	K	FRCT	TLAG
KA1	1.0000					
KA2	.117014	1.0000				
V	.180697	.685444	1.0000			
K	-.179177	-.653064	-.932832	1.0000		
FRCT	-.055611	.613886	.876854	-.811745	1.0000	
TLAG	.056564	.907836	.466054	-.464690	.439526	1.0000

X	OBSERVED Y	PREDICTED Y	RESIDUAL	WEIGHT	SE-PRED	STANDARDI RESIDUAL
.8330E-01	.	23.38	.	.0000	-.6400+152	.
.1000	25.00	26.51	-1.509	1.000	.6600	-1.788
.1250	.	30.49	.	.0000	-.6400+152	.
.1500	35.00	33.76	1.241	1.000	.6664	1.479
.2000	.	38.64	.	.0000	-.6400+152	.
.2500	42.00	41.91	.9253E-01	1.000	.5238	.09899
.3000	.	44.07	.	.0000	-.6400+152	.
.4500	.	46.91	.	.0000	-.6400+152	.
.5000	48.00	47.22	.7837	1.000	.3865	.7842
.6000	.	47.38	.	.0000	-.6400+152	.
.7500	46.50	47.11	-.6053	1.000	.4176	-.6134
.8000	47.00	46.95	.5159E-01	1.000	.4197	.0523
.9000	.	46.59	.	.0000	-.6400+152	.
1.000	46.00	46.21	-.2084	1.000	.4194	-.2114
1.250	44.50	45.21	-.7141	1.000	.4135	-.7224
1.500	45.00	44.22	.7758	1.000	.4100	.7837
2.000	42.10	42.30	-.2036	1.000	.4152	-.2061
2.250	.	41.37	.	.0000	-.6400+152	.
2.500	.	49.62	.	.0000	-.6400+152	.
3.000	63.00	62.79	.2051	1.000	1.068	2.410
3.500	.	69.04	.	.0000	-.6400+152	.
4.000	70.15	71.29	-1.141	1.000	.9436	-2.248
5.000	72.00	69.96	2.038	1.000	.6270	2.345

5.500	.	67.97	.	.0000	-.6400+152	.
6.000	65.00	65.64	-.6374	1.000	.6755	-.7663
8.000	55.00	55.65	-.6522	1.000	.6988	-.8029
10.00	.	46.68	.	.0000	-.6400+152	.
12.00	39.00	39.09	-.8856E-01	1.000	.6977	-.1089
16.00	.	27.40	.	.0000	-.6400+152	.
24.00	14.00	13.46	.5364	1.000	.8315	.7937

```
CORRECTED SUM OF SQUARED OBSERVATIONS =  19930.2
WEIGHTED CORRECTED SUM OF SQUARED OBSERVATIONS =  3647.21
SUM OF SQUARED RESIDUALS =        12.6291
SUM OF WEIGHTED SQUARED RESIDUALS =  12.6291
S =  1.07149     WITH     11 DEGREES OF FREEDOM
CORRELATION (OBSERVED,PREDICTED) =  .4270

AIC criteria =        55.11201
SC  criteria =        60.11129
```

PK44 - Estimation of the inhibitory constant (K_i) by means of simultaneous nonlinear regression

Objectives

◆ To analyze literature enzyme inhibition data by means of nonlinear regression

◆ To obtain initial parameter estimates competitive and non-competitive models

◆ To discuss various aspects of experimental design

Problem specification

The complete *in vitro* characterization of substrate metabolism involves both a *qualitative* determination of responsible isozymes and a *quantitative* analysis of the Michaelis-Menten parameters. The goal of this analysis is to relate the *in vitro* findings to the *in vivo* situation. This is done in order to predict the potential impact of enzyme inhibition or induction and, thus, if an interaction is likely to occur. The interaction potential is assessed by comparison of observed plasma concentrations with the enzyme inhibition constant K_i (Nimmo and Atkins [1976], Kakkar *et al* [1999, 2000]).

Data were scanned from a recent study on simultaneous nonlinear regression analysis of *in vitro* metabolism inhibition data (Figure 44.1 and program output) (Kakkar *et al* [1999]). It is generally well known that linearization of nonlinear relationships often provides biased and imprecise parameter estimates (Leatherbarrow [1990]). Linearization methods also require larger sets of experimental data and are less direct and not so easy to implement.

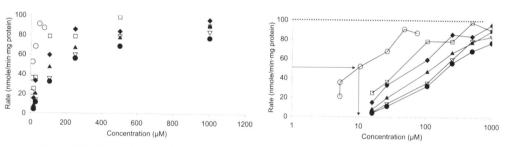

Figure 44.1 *Observed metabolic rate versus substrate concentration, obtained by scanning data presented by Kakkar et al [1999]. Data are shown on a Cartesian scale (left) and semi-logarithmic scale (right). The right hand figure also shows how to obtain V_{max} (horizontal stippled line) and K_m (substrate concentration at 50% of maximal rate in the control experiment). The concentrations of the inhibitor are 0, 10, 25, 50, 75 and 100 μM.*

The metabolite formation rates were analyzed using both a *competitive* and a *noncompetitive* enzyme inhibition model, with the assumption that a single metabolite is produced from the substrate. The *competitive* inhibition model is

$$Rate = v = \frac{V_{max} \cdot [C]}{K_m \cdot \left(1 + \dfrac{[I]}{K_i}\right) + [C]} \tag{44:1}$$

The *noncompetitive* inhibition model

$$Rate = v = \frac{V_{max} \cdot [C]}{K_m + [C]} \cdot \frac{K_i}{K_i + [I]} \tag{44:2}$$

The rate of metabolite formation is denoted v, V_{max} is the maximum metabolic rate, K_m is the Michaelis-Menten constant, K_i is the inhibitor constant, $[C]$ is the drug/substrate concentration and $[I]$ the inhibitor concentration. Figure 44.2 shows two plots illustrating the difference between competitive and noncompetitive inhibition.

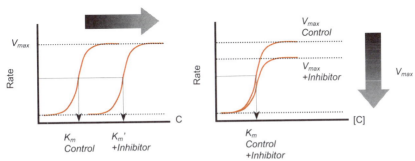

Figure 44.2 *Metabolic rate versus substrate concentration. This picture gives a schematic illustration of competitive (left) and noncompetitive (right) inhibition. In competitive inhibition, the apparent K_m value seems to have changed. In noncompetitive inhibition, the apparent V_{max} seems to have shifted.*

Initial parameter estimates

The initial estimates for V_{max} and K_m are obtained from Figure 44.1 In the control experiment, maximum metabolic rate is extrapolated to about 100 nmol/min/mg protein, which is a good estimate of V_{max}. K_m is the substrate concentration that corresponds to 50% of that rate, which is 10 µmol/L. K_i is set to the same value 10 µmol/L. An approximate value of K_i (to be used as an initial parameter estimate only) might be obtained by means of the Dixon plot. Since the simultaneous nonlinear regression approach is very robust, we simply used a value for K_i which is close to K_m in this particular case.

Interpretations of results and conclusions

The original data in this exercise were scanned from a paper by Kakkar *et al* [1999] that we highly recommend to the reader. The simultaneous nonlinear regression analysis provides accurate and precise parameter estimates. A wide range of enzyme parameter values is less prone to affect the estimation of K_i. It requires less information and experimentation than linearization methods and it is easier and more rapid to implement. Figure 44.3 shows the high consistency between observed and model-predicted data (*competitive* inhibition, Equation 44:1). The model-predicted parameter estimates of V_{max}, K_m and K_i were consistent with the literature data (Kakkar *et al* [1999]).

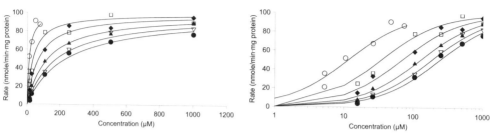

Figure 44.3 *Observed and model-predicted rate versus substrate concentration using the competitive model. Data are shown on a Cartesian scale (left) and semi-logarithmic scale (right).*

Figure 44.4 shows the observed and model-predicted data (*noncompetitive* inhibition, Equation 44:2). This system displays systematic deviations from the observed data, and we therefore conclude that this model is inferior to the model-predicted data shown in Figure 44.3. In addition, program output from the two analyses shows that the parameter precision is lower for the *noncompetitive* model than for the *competitive* model.

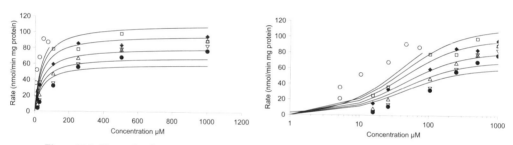

Figure 44.4. *Observed and model predicted rate plotted versus substrate concentration using the noncompetitive model. Data are shown on a Cartesian scale (left) and semi-logarithmic scale (right).*

Figure 44.5 shows the partial derivatives of metabolic rate with respect to K_m, K_i and V_{max}, for an inhibitory concentration of 10 (left plot) and 100 (right plot) µM in the competitive inhibition model. These derivatives are plotted against substrate concentration.

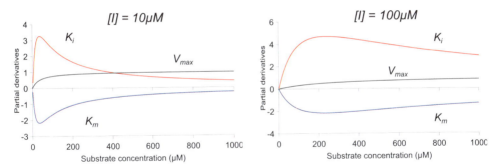

Figure 44.5 *Model-predicted (competitive inhibition) partial derivatives of metabolic rate with respect to each model parameter versus substrate concentration. The left hand plot shows the partial derivatives at an inhibitor concentration of 10 µM and the right hand plot the partial derivatives when inhibitor concentration is 100 µM. The blue, red and black curves correspond to the partial derivative of rate of metabolism with respect to K_m, K_i and V_{max}, respectively.*

The presented analysis is not only applicable to inhibition of enzymatic rate but also to systems of pharmacological response (receptor binding models) as discussed in PD1 and PD2. This exercise demonstrates the simultaneous use of several sources of data.

Solution I – Competitive model

```
MODEL
 COMM
  NFUN  6
  NPAR  3
  PNAM  'Vmax', 'Km', 'Ki'
  NCON  6
END
     1:  TEMP
     2:  C = X
     3:  IN0  = CON(1)
     4:  IN10 = CON(2)
     5:  IN25 = CON(3)
     6:  IN50 = CON(4)
     7:  IN75 = CON(5)
     8:  IN100 = CON(6)
     9:  END
    10:  FUNC 1
    11:  F = (VMAX * C) / (KM*(1 + IN0/KI) + C)
    12:  END
    13:  FUNC 2
    14:  F = (VMAX * C) / (KM*(1 + IN10/KI) + C)
    15:  END
    16:  FUNC 3
    17:  F = (VMAX * C) / (KM*(1 + IN25/KI) + C)
    18:  END
    19:  FUNC 4
    20:  F = (VMAX * C) / (KM*(1 + IN50/KI) + C)
    21:  END
    22:  FUNC 5
    23:  F = (VMAX * C) / (KM*(1 + IN75/KI) + C)
```

```
24: END
25: FUNC 6
26: F = (VMAX * C) / (KM*(1 + IN100/KI) + C)
27: END
28: EOM
NVARIABLES 4
FNUMBER 4
NPOINTS 100
XNUMBER 1
YNUMBER 2
CONSTANTS 0,10,25,50,75,100
METHOD 2  'Gauss-Newton (Levenberg and Hartley)
ITERATIONS 50
INITIAL 100,10,10
LOWER BOUNDS 0,.001,.001
UPPER BOUNDS 500,250,250
MISSING 'Missing'
DATA 'WINNLIN.DAT'
BEGIN
```

PARAMETER	ESTIMATE	STANDARD ERROR	CV%	UNIVARIATE C.I.	
VMAX	99.845708	2.117546	2.12	95.537556	104.153861
KM	11.396562	1.278049	11.21	8.796368	13.996757
KI	5.090384	.632711	12.43	3.803131	6.377636

```
*** CORRELATION MATRIX OF THE ESTIMATES ***
PARAMETER   VMAX        KM          KI
VMAX      1.00000
KM         .518125   1.00000
KI        -.019765    .74223    1.00000

Condition_number=        8.562
```

```
    FUNCTION   1
```

X	OBSERVED Y	PREDICTED Y	RESIDUAL	WEIGHT	SE-PRED	STANDARDI RESIDUAL
5.236	21.10	31.43	-10.33	1.000	2.145	-2.412
5.236	35.60	31.43	4.174	1.000	2.145	.9746
10.47	52.09	47.81	4.277	1.000	2.428	1.036
26.18	67.91	69.56	-1.650	1.000	2.038	-.3807
47.12	90.99	80.40	10.59	1.000	1.699	2.364
73.30	87.03	86.41	.6223	1.000	1.607	.1379

```
CORRECTED SUM OF SQUARED OBSERVATIONS =  3920.09
WEIGHTED CORRECTED SUM OF SQUARED OBSERVATIONS =  3920.09
SUM OF SQUARED RESIDUALS =          257.687
SUM OF WEIGHTED SQUARED RESIDUALS =  257.687
S =  9.26800      WITH      3 DEGREES OF FREEDOM
CORRELATION (OBSERVED,PREDICTED) =  .9735
```

```
    FUNCTION   2
```

X	OBSERVED Y	PREDICTED Y	RESIDUAL	WEIGHT	SE-PRED	STANDARDI RESIDUAL
15.71	25.05	31.69	-6.632	1.000	1.164	-1.427
26.18	36.26	43.59	-7.326	1.000	1.244	-1.584
104.7	78.46	75.49	2.972	1.000	.9780	.6339

251.3	78.46	88.01	-9.552	1.000	1.332	-2.076
502.6	98.24	93.56	4.685	1.000	1.654	1.042
1000.	89.67	96.58	-6.912	1.000	1.865	-1.567

CORRECTED SUM OF SQUARED OBSERVATIONS = 4453.95
WEIGHTED CORRECTED SUM OF SQUARED OBSERVATIONS = 4453.95
SUM OF SQUARED RESIDUALS =   267.457
SUM OF WEIGHTED SQUARED RESIDUALS = 267.457
S = 9.44206   WITH   3 DEGREES OF FREEDOM
CORRELATION (OBSERVED,PREDICTED) = .9814

   FUNCTION   3

X	OBSERVED Y	PREDICTED Y	RESIDUAL	WEIGHT	SE-PRED	STANDARDI RESIDUAL
15.71	15.16	18.88	-3.713	1.000	.9162	-.7897
26.18	32.97	27.94	5.026	1.000	1.168	1.082
104.7	59.34	60.76	-1.416	1.000	1.181	-.3051
251.3	85.71	78.74	6.976	1.000	1.058	1.493
502.6	83.74	88.04	-4.309	1.000	1.343	-.9371
1000.	95.60	93.54	2.060	1.000	1.655	.4584

CORRECTED SUM OF SQUARED OBSERVATIONS = 5207.56
WEIGHTED CORRECTED SUM OF SQUARED OBSERVATIONS = 5207.56
SUM OF SQUARED RESIDUALS =   112.522
SUM OF WEIGHTED SQUARED RESIDUALS = 112.522
S = 6.12433   WITH   3 DEGREES OF FREEDOM
CORRELATION (OBSERVED,PREDICTED) = .9896

   FUNCTION   4

X	OBSERVED Y	PREDICTED Y	RESIDUAL	WEIGHT	SE-PRED	STANDARDI RESIDUAL
15.71	7.912	11.28	-3.367	1.000	.6487	-.7094
26.18	19.78	17.48	2.299	1.000	.9207	.4890
104.7	46.81	45.85	.9679	1.000	1.430	.2117
251.3	66.59	66.98	-.3819	1.000	1.186	-.8229E-01
502.6	77.14	80.17	-3.029	1.000	1.130	-.6508
1000.	89.67	88.88	.7873	1.000	1.398	.1718

CORRECTED SUM OF SQUARED OBSERVATIONS = 5270.16
WEIGHTED CORRECTED SUM OF SQUARED OBSERVATIONS = 5270.16
SUM OF SQUARED RESIDUALS =   27.4976
SUM OF WEIGHTED SQUARED RESIDUALS = 27.4976
S = 3.02752   WITH   3 DEGREES OF FREEDOM
CORRELATION (OBSERVED,PREDICTED) = .9975

   FUNCTION   5

X	OBSERVED Y	PREDICTED Y	RESIDUAL	WEIGHT	SE-PRED	STANDARDI RESIDUAL
15.71	4.615	8.042	-3.426	1.000	.4954	-.7192
26.18	13.85	12.72	1.126	1.000	.7361	.2380
104.7	35.60	36.81	-1.206	1.000	1.435	-.2640
251.3	60.00	58.27	1.730	1.000	1.369	.3769
502.6	79.12	73.59	5.529	1.000	1.145	1.189
1000.	83.74	84.66	-.9283	1.000	1.245	-.2007

CORRECTED SUM OF SQUARED OBSERVATIONS = 5571.50
WEIGHTED CORRECTED SUM OF SQUARED OBSERVATIONS = 5571.50
SUM OF SQUARED RESIDUALS =   48.8893

```
SUM OF WEIGHTED SQUARED RESIDUALS =  48.8893
S =  4.03689     WITH     3 DEGREES OF FREEDOM
CORRELATION (OBSERVED,PREDICTED) =  .9968
```

```
   FUNCTION   6
```

X	OBSERVED Y	PREDICTED Y	RESIDUAL	WEIGHT	SE-PRED	STANDARDI RESIDUAL
15.71	3.956	6.248	-2.292	1.000	.3994	-.4802
26.18	11.21	9.997	1.212	1.000	.6087	.2551
104.7	32.31	30.75	1.557	1.000	1.364	.3391
251.3	56.04	51.57	4.477	1.000	1.474	.9822
502.6	68.57	68.01	.5617	1.000	1.237	.1214
1000.	77.14	80.83	-3.686	1.000	1.173	-.7936

```
CORRECTED SUM OF SQUARED OBSERVATIONS =  4626.40
WEIGHTED CORRECTED SUM OF SQUARED OBSERVATIONS =  4626.40
SUM OF SQUARED RESIDUALS =           43.0862
SUM OF WEIGHTED SQUARED RESIDUALS =  43.0862
S =  3.78973     WITH     3 DEGREES OF FREEDOM
CORRELATION (OBSERVED,PREDICTED) =  .9955
```

```
TOTALS FOR ALL CURVES COMBINED
SUM OF SQUARED RESIDUALS =              757.141
SUM OF WEIGHTED SQUARED RESIDUALS =  757.141
S =  4.78995     WITH    33 DEGREES OF FREEDOM
```

```
AIC criteria =        244.66376
SC  criteria =        249.41432
```

Solution II – Noncompetitive model

```
MODEL
 COMM
  NFUN 6
  NPAR 3
  PNAM 'Vmax', 'Km', 'Ki'
  NCON 6
END
    1: TEMP
    2: C = X
    3: IN0 = CON(1)
    4: IN10 = CON(2)
    5: IN25 = CON(3)
    6: IN50 = CON(4)
    7: IN75 = CON(5)
    8: IN100 = CON(6)
    9: END
   10: FUNC 1
   11: F = ((VMAX * C) / (KM + C))*(KI/(KI + IN0))
   12: END
   13: FUNC 2
   14: F = ((VMAX * C) / (KM + C))*(KI/(KI + IN10))
   15: END
   16: FUNC 3
   17: F = ((VMAX * C) / (KM + C))*(KI/(KI + IN25))
   18: END
   19: FUNC 4
   20: F = ((VMAX * C) / (KM + C))*(KI/(KI + IN50))
   21: END
   22: FUNC 5
   23: F = ((VMAX * C) / (KM + C))*(KI/(KI + IN75))
```

```
    24: END
    25: FUNC 6
    26: F = ((VMAX * C) / (KM + C))*(KI/(KI + IN100))
    27: END
    28: EOM
NVARIABLES 3
FNUMBER 3
NPOINTS 100
XNUMBER 1
YNUMBER 2
CONSTANTS 0,10,25,50,75,100
METHOD 2  'Gauss-Newton (Levenberg and Hartley)
ITERATIONS 50
INITIAL 100,10,10
LOWER BOUNDS 0,.001,.001
UPPER BOUNDS 500,250,250
MISSING 'Missing'
DATA 'WINNLIN.DAT'
BEGIN
```

PARAMETER	ESTIMATE	STANDARD ERROR	CV%	UNIVARIATE C.I.	
VMAX	122.874883	10.325075	8.40	101.868496	143.881270
KM	42.748485	8.957516	20.95	24.524399	60.972571
KI	98.725388	24.324182	24.64	49.237789	148.212988

```
*** CORRELATION MATRIX OF THE ESTIMATES ***
PARAMETER  VMAX       KM          KI
VMAX       1.00000
KM          .63045   1.00000
KI         -.77428   -.24997     1.00000

 Condition_number=       6.753
```

```
    FUNCTION   1
```

X	OBSERVED Y	PREDICTED Y	RESIDUAL	WEIGHT	SE-PRED	STANDARDI RESIDUAL
5.236	21.10	13.41	7.692	1.000	1.992	.5708
5.236	35.60	13.41	22.20	1.000	1.992	1.647
10.47	52.09	24.18	27.91	1.000	3.201	2.108
26.18	67.91	46.67	21.24	1.000	4.706	1.662
47.12	90.99	64.43	26.56	1.000	5.166	2.108
73.30	87.03	77.61	9.422	1.000	5.399	.7534

```
CORRECTED SUM OF SQUARED OBSERVATIONS = 3920.09
WEIGHTED CORRECTED SUM OF SQUARED OBSERVATIONS = 3920.09
SUM OF SQUARED RESIDUALS =           2576.61
SUM OF WEIGHTED SQUARED RESIDUALS = 2576.61
S = 29.3065    WITH    3 DEGREES OF FREEDOM
CORRELATION (OBSERVED,PREDICTED) = .9517
```

```
    FUNCTION   2
```

X	OBSERVED Y	PREDICTED Y	RESIDUAL	WEIGHT	SE-PRED	STANDARDI RESIDUAL
15.71	25.05	29.98	-4.925	1.000	3.494	-.3741

26.18	36.26	42.38	-6.111	1.000	4.066	-.4701
104.7	78.46	79.23	-.7671	1.000	4.015	-.5893E-01
251.3	78.46	95.35	-16.89	1.000	4.930	-1.330
502.6	98.24	102.8	-4.586	1.000	5.945	-.3742
1000.	89.67	107.0	-17.33	1.000	6.672	-1.459

```
CORRECTED SUM OF SQUARED OBSERVATIONS =  4453.95
WEIGHTED CORRECTED SUM OF SQUARED OBSERVATIONS =  4453.95
SUM OF SQUARED RESIDUALS =           668.860
SUM OF WEIGHTED SQUARED RESIDUALS =  668.860
S =  14.9316      WITH      3 DEGREES OF FREEDOM
CORRELATION (OBSERVED,PREDICTED) =  .9790
```

FUNCTION   3

X	OBSERVED Y	PREDICTED Y	RESIDUAL	WEIGHT	SE-PRED	STANDARDI RESIDUAL
15.71	15.16	26.34	-11.18	1.000	3.128	-.8433
26.18	32.97	37.24	-4.271	1.000	3.610	-.3252
104.7	59.34	69.62	-10.28	1.000	2.975	-.7736
251.3	85.71	83.79	1.921	1.000	3.267	.1453
502.6	83.74	90.36	-6.625	1.000	4.042	-.5093
1000.	95.60	94.03	1.577	1.000	4.644	.1232

```
CORRECTED SUM OF SQUARED OBSERVATIONS =  5207.56
WEIGHTED CORRECTED SUM OF SQUARED OBSERVATIONS =  5207.56
SUM OF SQUARED RESIDUALS =           299.035
SUM OF WEIGHTED SQUARED RESIDUALS =  299.035
S =  9.98390      WITH      3 DEGREES OF FREEDOM
CORRELATION (OBSERVED,PREDICTED) =  .9898
```

FUNCTION   4

X	OBSERVED Y	PREDICTED Y	RESIDUAL	WEIGHT	SE-PRED	STANDARDI RESIDUAL
15.71	7.912	21.92	-14.00	1.000	2.832	-1.051
26.18	19.78	30.98	-11.20	1.000	3.340	-.8480
104.7	46.81	57.92	-11.11	1.000	3.124	-.8377
251.3	66.59	69.71	-3.115	1.000	3.196	-.2352
502.6	77.14	75.17	1.971	1.000	3.645	.1502
1000.	89.67	78.22	11.45	1.000	4.035	.8800

```
CORRECTED SUM OF SQUARED OBSERVATIONS =  5270.16
WEIGHTED CORRECTED SUM OF SQUARED OBSERVATIONS =  5270.16
SUM OF SQUARED RESIDUALS =           589.534
SUM OF WEIGHTED SQUARED RESIDUALS =  589.534
S =  14.0183      WITH      3 DEGREES OF FREEDOM
CORRELATION (OBSERVED,PREDICTED) =  .9865
```

FUNCTION   5

X	OBSERVED Y	PREDICTED Y	RESIDUAL	WEIGHT	SE-PRED	STANDARDI RESIDUAL
15.71	4.615	18.76	-14.15	1.000	2.646	-1.059
26.18	13.85	26.52	-12.67	1.000	3.198	-.9572
104.7	35.60	49.58	-13.98	1.000	3.510	-1.062
251.3	60.00	59.68	.3233	1.000	3.702	.2467E-01
502.6	79.12	64.35	14.77	1.000	4.035	1.135

| 1000. | 83.74 | 66.97 | 16.77 | 1.000 | 4.313 | 1.298 |

CORRECTED SUM OF SQUARED OBSERVATIONS = 5571.50
WEIGHTED CORRECTED SUM OF SQUARED OBSERVATIONS = 5571.50
SUM OF SQUARED RESIDUALS =   1055.65
SUM OF WEIGHTED SQUARED RESIDUALS = 1055.65
S = 18.7586   WITH   3 DEGREES OF FREEDOM
CORRELATION (OBSERVED,PREDICTED) = .9706

   FUNCTION   6

X	OBSERVED Y	PREDICTED Y	RESIDUAL	WEIGHT	SE-PRED	STANDARDI RESIDUAL
15.71	3.956	16.40	-12.45	1.000	2.488	-.9294
26.18	11.21	23.18	-11.98	1.000	3.062	-.9023
104.7	32.31	43.35	-11.04	1.000	3.706	-.8422
251.3	56.04	52.17	3.875	1.000	4.000	.2976
502.6	68.57	56.26	12.31	1.000	4.299	.9527
1000.	77.14	58.54	18.60	1.000	4.530	1.448

CORRECTED SUM OF SQUARED OBSERVATIONS = 4626.40
WEIGHTED CORRECTED SUM OF SQUARED OBSERVATIONS = 4626.40
SUM OF SQUARED RESIDUALS =   932.836
SUM OF WEIGHTED SQUARED RESIDUALS = 932.836
S = 17.6336   WITH   3 DEGREES OF FREEDOM
CORRELATION (OBSERVED,PREDICTED) = .9740

TOTALS FOR ALL CURVES COMBINED
SUM OF SQUARED RESIDUALS =   6122.53
SUM OF WEIGHTED SQUARED RESIDUALS = 6122.53
S = 13.6210   WITH   33 DEGREES OF FREEDOM

AIC criteria =   319.91031
SC criteria =   324.66086

PK45 – Simulation and optimization of plasma sampling in a toxicokinetic study

Objectives

◆ To apply a library model

◆ To simulate different sampling designs

◆ To compare the Variance Inflation Factor from different designs

◆ To propose a sampling design

Problem specification

The following exercise is aimed to show the potential of using the *Variance Inflation Factor VIF* in designing a repeated dose toxicokinetic study. Figures 45.1 and 45.2 show two alternative simulated dosing schedules, with a daily dose of 400 µg being given as a single dose each day or being split into two 200 µg doses separated by a 7 h dosing interval. Figures 45.1 and 45.2 show the two study designs plotted on a linear and semi-logarithmic scale, respectively.

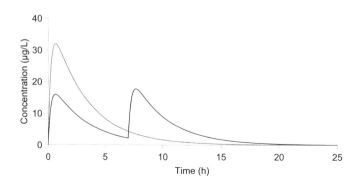

Figure 45.1 Comparison of a single daily dose of 400 µg with two 200 µg doses separated by a 7 h dosing interval.

It is obvious from Figure 45.2 that the duration of exposure is increased when the twice a day 200 µg dosing regimen is applied. This suits the purpose of the study, to mimic human exposure to the compound.

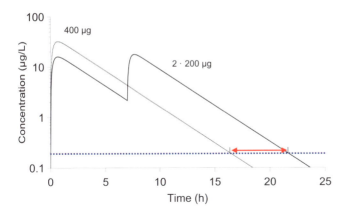

Figure 45.2 Semi-logarithmic plot of the two dosing regimens. The dotted line represents the single daily 400 µg dose and solid line the two daily 200 µg doses separated by a 7 h dosing interval. The red arrow shows the extra 5 h duration above 0.2 µg/L obtained from the twice a day 200 µg regimen. 0.2 µg/L was the minimum effective therapeutic plasma concentration.

Initial parameter estimates

This case study is a simulation exercise and does not involve estimation of initial parameter values.

Interpretation of results and conclusions

We would reemphasize the use of the *Variance Inflation Factor VIF* option in *WinNonlin*. Six to 7 blood samples could be drawn at each study occasion from each individual dog. The study director recommended one blood sample at the time of maximum plasma concentration of the drug (i.e., at 0.5 h after dosing). We found that under these conditions, a sampling protocol according to Design 3 in Table 45.1 was the most appropriate design, which resulted in the lowest *VIF*. The 7.5 h plasma concentration will contain information about the highest level of exposure within a 24 h period.

Table 45.1   Summary table of *VIF* for *V/F*, K_a and *K* obtained from three tentative designs

Parameter	Design 1 (0.5, 1, 2, 4, 8, 24 h) VIF%	Design 2 (0.25, 0.5, 3, 7, 7.5, 24 h) VIF%	Design 3 (0.25, 0.5, 1, 3, 7, 7.5 h) VIF%
V/F	9.93	15.98	8.77
K_a	36.12	35.42	20.87
K	20.77	21.76	16.33

Solution – Design 3

```
MODEL 3
NOPAGE BREAKS
SIMULATE
NVARIABLES 10
NPOINTS 500
XNUMBER 9
NCONSTANTS 3
CONSTANTS 1,400,0
INITIAL 10,4,.33
NOBOUNDS
MISSING 'Missing'
```

```
NOBSERVATIONS 6
DATA 'WINNLIN.DAT'
BEGIN
```

PARAMETER	ESTIMATE	VAR - INF FACTOR	[SQRT(VIF)/P]%
Volume/F	10.000000	.169076	4.11
K01	4.000000	.174157	10.43
K10	.330000	.000599	7.42

```
*** CORRELATION MATRIX OF THE ESTIMATES ***
```

PARAMETER	Volume/F	K01	K10
Volume/F	1.00000		
K01	.81583	1.00000	
K10	-.77154	-.60732	1.00000

```
*** EIGENVALUES OF (Var - Cov) MATRIX ***
```

NUMBER	EIGENVALUE
1	4150.
2	31.60
3	3.206

```
Condition_number=        35.98

    FUNCTION   1
```

X	CALCULATED Y	VAR - INF FACTOR
.2500	24.11	.7623
.5000	31.07	.4119
1.000	30.54	.7399
3.000	16.20	.4898
7.000	4.327	.3213
7.500	3.669	.2748

```
SUMMARY OF ESTIMATED SECONDARY PARAMETERS
```

PARAMETER	ESTIMATE	VAR - INF FACTOR	[SQRT(VIF)/P]%
AUC	121.212121	36.444216	4.98
K01-HL	.17328	.000326	10.42
K10-HL	2.10044	.024219	7.41
CL/F	3.30000	.027067	4.99
Tmax	.67982	.001701	6.07
Cmax	31.96166	.506788	2.23

PD1 - Receptor binding models
Part I – One- and two-site models
Part II - Specific and non-specific binding

Objectives
◆ **To analyze binding data for the one- and two-site models**
◆ **To analyze data containing specific and non-specific binding**
◆ **To discriminate between competing models**
◆ **To study the impact of variability in data on the parameter estimates**

Problem specification Part I – One- and two-site binding model

The aim of Part I of this exercise is to compare the one- and two-site receptor binding models using a single dataset. The one- and two-site models have been used to describe the binding of some experimental compounds to certain structures of the brain. The observed and predicted data are shown in Figure 1.1 in the program output.

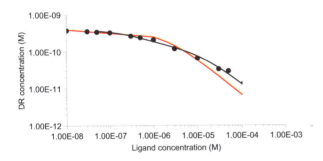

Figure 1.1 Labeled drug-receptor binding complex [D-R] versus cold ligand concentration. As the concentration of cold ligand increases it displaces hot ligand from the receptor. The solid black line represents the two-site binding model and the solid red line the one-site binding model. Dr. Sundqvist, Pharmacia Upjohn AB, kindly supplied the data.

The one- and two-site models are described by Equations 1:1to 1:3

$$A_s = \frac{A}{1 + K_a \cdot D} \tag{1:1}$$

$$[D - R] = R_T \cdot \frac{1 - K_i \cdot A_s}{1 + K_i \cdot A_s} \tag{1:2}$$

$$[D - R] = R_T \cdot \left[1 - \frac{F_1 \cdot K_1 \cdot A_s}{1 + K_1 \cdot A_s} + \frac{(1 - F_1) \cdot K_2 \cdot A_s}{1 + K_2 \cdot A_s} \right] \tag{1:3}$$

A, A_s, K_a, D, R_T, F_1, K_1 and K_2 represents the cold ligand that displaces the hot ligand

(D = *Drug*) from the receptor R, the shifted ligand concentration, and affinity constant, drug concentration, receptor concentration, fraction of binding to site one affinity constant for site one, and site two, respectively. R_T, K_a and D are held constant.

Initial parameter estimate

One-site model   Two-site model

$\qquad K_i = 7.0 \qquad\qquad F_1 = 0.8$

$\qquad\qquad\qquad\qquad\qquad K_1 = 9.5$

$\qquad\qquad\qquad\qquad\qquad K_2 = 1.2$

Interpretation of results and conclusions of Part I

The overall fit of the two-site model is better than the fit of the one-site model, as can be seen in Figure 1.1. This is also evident from the sum of weighted squared residuals for the one-site model ($9.8 \cdot 10^{-11}$) compared with the two-site model ($9.4 \cdot 10^{-12}$), as well as from the more random scatter in the residuals for the two-site model.

You have learned to

☐ analyze single and two-site binding models

☐ implement the equations

☐ characterize the binding function

Problem specification Part II - Specific and non-specific binding

In pharmacological binding studies, the quantity of radioactive ligand that is bound to protein is measured. It is assumed in these types of studies that the ligand binds to a specific population of receptors and to a number of nonreceptor sites referred to collectively as nonspecific binding *NSB* (Kenakin [1993]).

Figure 1.2 shows the fit of Equation 1:4 to experimental data of a compound. Observe the specific and non-specific binding trajectories for the B_{max} model where $n = 1$. For a more in-depth discussion about simultaneous fitting of total and non-specific binding see Kenakin [1993].

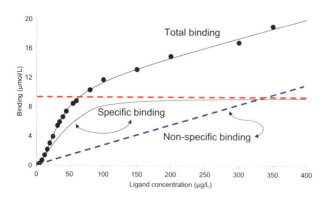

Figure 1.2 Observed (●) and predicted total binding data for a compound exhibiting binding characteristics that can be modeled by means of the ordinary B_{max} (Equation 1:4) model (specific binding). The blue dashed line represents non-specific binding (NSB, Equation 1:5). The dashed red line is the maximum specific binding capacity B_{max}.

The aim is to accurately estimate the relative quantity of ligand bound to the receptors *versus* non-specific binding. The total binding can be estimated by Equation 1:4, which assumes that the specific binding is saturable and non-specific binding is non-saturable. The nonspecific binding is a linear function of ligand concentration (Equation 1:5). Under these circumstances, a curve representing total binding of ligand to membrane preparation can be dissected into the receptor and non-specific site binding portions (Kenakin [1993]). Simultaneous curve fitting of the total binding and non-specific binding to the following equations yields estimates of B_{max}, n, K_d and a.

$$Total\ binding = \frac{B_{max} \cdot C^n}{K_d^n + C^n} + a \cdot C \qquad (1:4)$$

$$Non\text{-}specific\ binding (NSB) = a \cdot C \qquad (1:5)$$

Interpretation of results and conclusions of Part II

This exercise (Part II) contains a limited analysis and demonstrates for the reader how to implement a model of specific and nonspecific binding into *WinNonlin*. Also, we have not commented on the fits, since the data are synthetic. Hence, no comparisons between models were done. We refer the reader to Kenakin [1993] and Matthews [1993] for further information on receptor binding models.

    Christopoulos [1999] concluded that the finding that some parameters conform more closely to normal distribution when converted to logarithms does not necessarily mean that all parameters do so. Parameters such as B_{max} are adequately described by a normal distribution. The logarithm of the n parameter in Equation 1:4 often satisfies the normality assumption, though the non-transformed parameter does not. It is generally well accepted that EC_{50} (and K_d) values are better described by a log-normal distribution (Christopoulos [1999]). See also section PD2 for parameterization of the operational model.

    Rovati [1999] warned about analyzing receptor binding results with many commonly used graphical techniques (e.g., Scatchard, Lineweaver-Burke, Eadie-Hofstee). Each method provides slightly different results of the same data. Most of these methods are based on linearizations of the binding isotherm, which do not correctly take the experimental error into account. All of these methods imply subtraction of non-specific binding (*NSB*), which may affect the accuracy and precision of parameters. In the presence of experimental error, such linearizations become biased or inefficient (Rovati [1999]).

    You have practiced fitting specific and non-specific binding and implemented user-specified models. We have also discussed normal and log-normal distribution of some parameters, and limitations of different linearization techniques.

Solution I Part I - One-site model

```
TITLE 1
One site model
MODEL
 COMM
  NCON 3
```

```
     NPARM 1
     PNAMES 'Ki'
  END
      1: TEMP
      2: KI=P(1)
      3: RT=CON(1)
      4: KA=CON(2)
      5: D=CON(3)
      6: A=X
      7: END
      8: FUNC1
      9: AC=A/(1.0 + KA*D)
     10: F= RT*(1.0 - KI*1.0E+07*AC/(1. + KI*1.0E+07*AC))
     11: END
     12: EOM
  NVARIABLES 2
  CONSTANTS .000000000396,31250000000,.00000000201
  METHOD 2  'Gauss-Newton (Levenberg and Hartley)
  REWEIGHT -1
  ITERATIONS 50
  INITIAL 7
  LOWER BOUNDS 1
  UPPER BOUNDS 100
  NOBSERVATIONS 12
  DATA 'WINNLIN.DAT'
  BEGIN
```

PARAMETER	ESTIMATE	STANDARD ERROR	CV%	UNIVARIATE C.I.	
KI	3.730111	.794173	21.29	1.982141	5.478081

X	OBSERVED Y	PREDICTED Y	RESIDUAL	WEIGHT	SE-PRED	STANDARDI RESIDUAL
.1000E-07	.3740E-09	.3937E-09	-.1970E-10	.2540E+10	.4871E-12	-.3323
.3000E-07	.3530E-09	.3892E-09	-.3618E-10	.2570E+10	.1428E-11	-.6139
.5000E-07	.3390E-09	.3848E-09	-.4575E-10	.2599E+10	.2326E-11	-.7813
.1000E-06	.3240E-09	.3741E-09	-.5013E-10	.2673E+10	.4399E-11	-.8699
.3000E-06	.2620E-09	.3369E-09	-.7492E-10	.2969E+10	.1070E-10	-1.393
.5000E-06	.2370E-09	.3064E-09	-.6944E-10	.3264E+10	.1475E-10	-1.384
.1000E-05	.2060E-09	.2499E-09	-.4391E-10	.4003E+10	.1962E-10	-1.022
.3000E-05	.1180E-09	.1438E-09	-.2581E-10	.6958E+10	.1949E-10	-.8587
.1000E-04	.6500E-10	.5785E-10	.7151E-11	.1730E+11	.1051E-10	.3551
.3000E-04	.3300E-10	.2136E-10	.1164E-10	.4685E+11	.4299E-11	.8871
.5000E-04	.2900E-10	.1310E-10	.1590E-10	.7640E+11	.2694E-11	1.519
.1000E-03	.1500E-10	.6661E-11	.8339E-11	.1503E+12	.1393E-11	1.100

```
  CORRECTED SUM OF SQUARED OBSERVATIONS =  .209766E-18
  WEIGHTED CORRECTED SUM OF SQUARED OBSERVATIONS =  .152039E-08
  SUM OF SQUARED RESIDUALS =          .198409E-19
  SUM OF WEIGHTED SQUARED RESIDUALS =  .982552E-10
  S =  .298869E-05 WITH    11 DEGREES OF FREEDOM
  CORRELATION (OBSERVED,PREDICTED) =  .9910

  AIC criteria =        -274.52144
  SC  criteria =        -275.27898
```

Solution II Part I - Two-site model

```
TITLE 1
Two site model
MODEL
 COMM
```

```
   NCON 3
   NPARM 3
   PNAMES 'F1', 'K1', 'K2'
END
    1: TEMP
    2: F1=P(1)
    3: K1=P(2)
    4: K2=P(3)
    5: RT=CON(1)
    6: KA=CON(2)
    7: D=CON(3)
    8: A=X
    9: END
   10: FUNC1
   11: AC=A/(1.0 + KA*D)
   12: I=F1*K1*1.0E+07*AC/(1+K1*1.0E+07*AC)+(1-
F1)*K2*1.0E+06*AC/(1+K2*1.0E+06*AC)
   13: F= RT*(1.0 - I)
   14: END
   15: EOM
NVARIABLES 2
CONSTANTS .000000000396,31250000000,.00000000201
METHOD 2   'Gauss-Newton (Levenberg and Hartley)
REWEIGHT -1
ITERATIONS 50
INITIAL .8,9.5,1.2
LOWER BOUNDS 0,1,.1
UPPER BOUNDS 1,300,100
NOBSERVATIONS 12
DATA 'WINNLIN.DAT'
BEGIN
```

PARAMETER	ESTIMATE	STANDARD ERROR	CV%	UNIVARIATE C.I.	
F1	.674115	.079773	11.83	.493654	.854576
K1	19.087796	5.549854	29.08	6.533046	31.642545
K2	6.253476	2.222717	35.54	1.225298	11.281654

```
   Condition_number=        273.0

     FUNCTION   1
```

X	OBSERVED Y	PREDICTED Y	RESIDUAL	WEIGHT	SE-PRED	STANDARDI RESIDUAL
.1000E-07	.3740E-09	.3881E-09	-.1412E-10	.2577E+10	.1485E-11	-.7027
.3000E-07	.3530E-09	.3736E-09	-.2064E-10	.2676E+10	.3889E-11	-1.065
.5000E-07	.3390E-09	.3606E-09	-.2164E-10	.2773E+10	.5694E-11	-1.165
.1000E-06	.3240E-09	.3333E-09	-.9282E-11	.3000E+10	.8432E-11	-.5572
.3000E-06	.2620E-09	.2661E-09	-.4059E-11	.3759E+10	.9728E-11	-.2995
.5000E-06	.2370E-09	.2300E-09	.7010E-11	.4348E+10	.8380E-11	.5371
.1000E-05	.2060E-09	.1844E-09	.2158E-10	.5423E+10	.7808E-11	1.879
.3000E-05	.1180E-09	.1265E-09	-.8496E-11	.7907E+10	.8520E-11	-1.099
.1000E-04	.6500E-10	.7381E-10	-.8813E-11	.1355E+11	.4854E-11	-1.204
.3000E-04	.3300E-10	.3570E-10	-.2697E-11	.2803E+11	.3478E-11	-.5370
.5000E-04	.2900E-10	.2365E-10	.5354E-11	.4232E+11	.2766E-11	1.296
.1000E-03	.1500E-10	.1284E-10	.2161E-11	.7795E+11	.1758E-11	.6725

```
   CORRECTED SUM OF SQUARED OBSERVATIONS =  .209766E-18
   WEIGHTED CORRECTED SUM OF SQUARED OBSERVATIONS =  .155890E-08
```

```
SUM OF SQUARED RESIDUALS =              .190173E-20
SUM OF WEIGHTED SQUARED RESIDUALS = .941703E-11
S = .102291E-05 WITH      9 DEGREES OF FREEDOM
CORRELATION (OBSERVED,PREDICTED) = .9972

AIC criteria =        -298.66202
SC  criteria =        -300.93466
```

Solution Part II - Exponent > 1

```
MODEL
 COMM
  NPARM 4
  PNAMES 'BMAX', 'EC50', 'n','A'
END
     1: TEMP
     2: BMAX=P(1)
     3: EC50=P(2)
     4: n=P(3)
     5: A=P(4)
     6: C=X
     7: END
     8: FUNC1
     9: F= BMAX*(C**n)/((EC50**n)+(C**n)) + A*C
    10: END
    11: EOM
NVARIABLES 2
NPOINTS 100
XNUMBER 1
YNUMBER 2
METHOD 2  'Gauss-Newton (Levenberg and Hartley)
REWEIGHT -2
ITERATIONS 50
INITIAL 15,45,3,.035
LOWER BOUNDS 0,0,0,0
UPPER BOUNDS 100,200,5,1
MISSING '-'
NOBSERVATIONS 20
DATA 'WINNLIN.DAT'
BEGIN
```

PARAMETER	ESTIMATE	STANDARD ERROR	CV%	UNIVARIATE C.I.	
BMAX	9.769109	.274766	2.81	9.186635	10.351583
EC50	34.079058	.864438	2.54	32.246540	35.911575
n	2.030501	.032298	1.59	1.962033	2.098969
A	.025698	.001044	4.06	.023485	.027911

X	OBSERVED Y	PREDICTED Y	RESIDUAL	WEIGHT	SE-PRED	STANDARDI RESIDUAL
1.000	.3320E-01	.3325E-01	-.4575E-04	903.4	.4946E-03	-.1076
2.000	.8230E-01	.8216E-01	.1418E-03	148.0	.8694E-03	.1045
4.000	.2254	.2273	-.1859E-02	19.34	.2768E-02	-.5324
8.000	.6782	.6949	-.1666E-01	2.070	.7198E-02	-1.440
12.00	1.398	1.356	.4171E-01	.5437	.1121E-01	1.731
16.00	2.172	2.143	.2938E-01	.2178	.1708E-01	.7657

20.00	3.027	2.987	.4019E-01	.1121	.2464E-01	.7569
25.00	3.929	4.039	-.1108	.6127E-01	.3325E-01	-1.542
32.00	5.427	5.395	.3193E-01	.3434E-01	.3994E-01	.3261
35.00	5.920	5.916	.3605E-02	.2856E-01	.4112E-01	.0332
40.00	6.608	6.700	-.9238E-01	.2227E-01	.4225E-01	-.7428
45.00	7.375	7.384	-.9421E-02	.1833E-01	.4392E-01	-.0683
55.00	8.432	8.501	-.6917E-01	.1383E-01	.5218E-01	-.4371
60.00	8.798	8.959	-.1610	.1245E-01	.5824E-01	-.9717
80.00	10.33	10.36	-.3035E-01	.9318E-02	.8263E-01	-.1637
100.0	11.72	11.35	.3690	.7756E-02	.9850E-01	1.849
150.0	13.12	13.16	-.4754E-01	.5767E-02	.1097	-.2035
200.0	14.91	14.65	.2644	.4658E-02	.1104	.9973
300.0	16.80	17.36	-.5658	.3314E-02	.1432	-1.832
350.0	19.03	18.68	.3493	.2864E-02	.1768	1.089

```
CORRECTED SUM OF SQUARED OBSERVATIONS =  645.375
WEIGHTED CORRECTED SUM OF SQUARED OBSERVATIONS =  17.5557
SUM OF SQUARED RESIDUALS =         .708521
SUM OF WEIGHTED SQUARED RESIDUALS =  .614286E-02
S =  .195941E-01 WITH     16 DEGREES OF FREEDOM
CORRELATION (OBSERVED,PREDICTED) =  .9995

AIC criteria =        -93.84928
SC  criteria =        -95.85782
```

PD2 - Operational binding model

Objectives

◆ To analyze several sources of receptor binding data simultaneously

◆ To characterize affinity, efficacy and maximum response

◆ To apply a user-specified model

◆ To obtain initial parameter estimates

Problem specification

This problem highlights some issues related to the *operational model* that demonstrates an explicit relation between agonist concentration and pharmacological effect. The *operational model* provides a means of directly analyzing agonist concentration effect data, in order to estimate affinity (K_A) and efficacy (τ) values (Black and Leff [1983]). Van der Graaf and Danhof [1997] have highlighted issues related to simultaneous fitting of the *operational model* of antagonism to concentration-effect curves from control and irreversible antagonist-treated tissues. They concluded that the lack of robustness of the parameter estimates showed that, under standard experimental conditions, the outcomes of simultaneous model fitting are highly dependent on between-tissue variations of the upper asymptotes of the control curves and, therefore, may be unreliable. They also concluded that, whenever possible, a multiple curve design should be adopted, with control and treated curves being obtained from one tissue and providing enough information for independent estimation of affinity and efficacy that is free from inter-tissue differences (van der Graaf and Danhof [1997]).

Black and Leff's [1983] operational model of agonism describes agonist concentration-effect curves in terms of four parameters:

$$E = \frac{E_m \cdot \tau^n \cdot [A]^n}{\left(K_A + [A]\right)^n + \tau^n \cdot [A]^n} \tag{2:1}$$

K_A is the agonist dissociation constant, the reciprocal of which defines agonist affinity. E_m is the maximum response achievable in the system, and τ is the model definition of the efficacy of an agonist in a tissue. n is a 'slope' parameter for the hyperbolic function and it is a measure of the sensitivity with which a particular system transduces AR (the ligand-receptor complex) into E. Algebraically, τ is described according to Equation 2:2

$$\tau = \frac{[R_0]}{K_E} \tag{2:2}$$

[R_0] is the total receptor concentration and K_E is the value of [AR] for half-saturation of the transducer relation (Black and Leff [1983]).

The experimental data (Figure 2.1) and model for this exercise were kindly supplied by Dr. Van der Graaf, Pfizer Central Research, Sandwich, UK. Data are also tabulated in the program output (see *Solution*) from *WinNonlin*. Implement Equation 2:1 into the program.

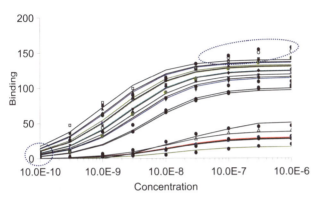

Figure 2.1. Semi-logarithmic plot of observed and predicted concentration versus binding of compound X. Note the systematic deviations between observed and predicted binding for the two highest binding tissues at high concentrations, which is indicated by the ellipsoid. This is also seen at low concentrations. Data kindly supplied by Dr. van der Graaf.

Initial parameter estimates

Figure 2.2 gives an estimate of the maximum achievable response (E_m).

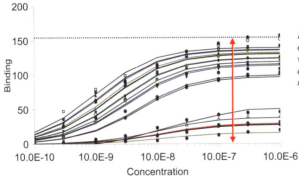

Figure 2.2. Semi-logarithmic plot of observed and predicted concentration versus binding of compound X.The dashed horizontal line indicates the maximum response E_m.

Equation 2:3 shows the parameterization and model that were implemented and run in this excercise.

$$E = \frac{E_m \cdot 10^{\tau \cdot n} \cdot [A]^n}{\left(10^{K_A} + [A]\right)^n + 10^{\tau \cdot n} \cdot [A]^n}$$ (2:3)

This gives you the pK_A and $p\tau$. The *n* parameter and τ for each tissue were supplied by Dr. Van der Graaf. There will be one equation per tissue.

Interpretation of results and conclusions

Figure 2.1 and the program output show generally a good fit, with good consistency between the observed and model-predicted data. The parameter estimates demonstrate in general high precision. However, the model overpredicts the binding at the lowest concentration. Weighting the data did not solve this. The conclusion of Van der Graaf and Danhof [1997] is noteworthy, namely that parameter precision may still be high in spite of a poor fit or a fit with systematic deviations between observed and model-predicted data for some tissues.

Solution - Operational model

```
MODEL
COMM
 NFUN  19
 NPAR  22
 PNAM  'Emax','N','Tau1','Tau2','Tau4','Tau5','Tau6','Tau7', &
       'Tau8','Tau9','Tau10','Tau11','Tau12','Tau13', &
       'Tau14','Tau15','Tau16','Tau17','Tau18','Tau19','Tau20', &
       'Ka'
END
   1: TEMP
   2: C = X
   3: END
   4: FUNC 1
   5: F=(EMAX*((10**(TAU1))**N)*(C**N))/((((10**KA)+C)**N) + &
   6: (((10**(TAU1))**N)*(C**N)))
   7: END
   8: FUNC 2
   9: F=(EMAX*((10**(TAU2))**N)*(C**N))/((((10**KA) + C)**N) + &
  10: (((10**(TAU2))**N)*(C**N)))
  11: END
  12: FUNC 3
  13: F=(EMAX*((10**(TAU4))**N)*(C**N))/((((10**KA) + C)**N) + &
  14: (((10**(TAU4))**N)*(C**N)))
  15: END
  16: FUNC 4
  17: F =(EMAX*((10**(TAU5))**N)*(C**N))/((((10**KA) + C)**N) + &
  18: (((10**(TAU5))**N)*(C**N)))
  19: END
  20: FUNC 5
  21: F =(EMAX*((10**(TAU6))**N)*(C**N))/((((10**KA) + C)**N) + &
  22: (((10**(TAU6))**N)*(C**N)))
  23: END
  24: FUNC 6
  25: F =(EMAX*((10**(TAU7))**N)*(C**N))/((((10**KA) + C)**N) + &
  26: (((10**(TAU7))**N)*(C**N)))
  27: END
  28: FUNC 7
  29: F =(EMAX*((10**(TAU8))**N)*(C**N))/((((10**KA) + C)**N) + &
  30: (((10**(TAU8))**N)*(C**N)))
  31: END
  32: FUNC 8
  33: F =(EMAX*((10**(TAU9))**N)*(C**N))/((((10**KA) + C)**N) + &
```

```
34: (((10**(TAU9))**N)*(C**N)))
35: END
36: FUNC 9
37: F =(EMAX*((10**(TAU10))**N)*(C**N))/((((10**KA) + C)**N) + &
38: (((10**(TAU10))**N)*(C**N)))
39: END
40: FUNC 10
41: F =(EMAX*((10**(TAU11))**N)*(C**N))/((((10**KA) + C)**N) + &
42: (((10**(TAU11))**N)*(C**N)))
43: END
44: FUNC 11
45: F =(EMAX*((10**(TAU12))**N)*(C**N))/((((10**KA) + C)**N) + &
46: (((10**(TAU12))**N)*(C**N)))
47: END
48: FUNC 12
49: F =(EMAX*((10**(TAU13))**N)*(C**N))/((((10**KA) + C)**N) + &
50: (((10**(TAU13))**N)*(C**N)))
51: END
52: FUNC 13
53: F =(EMAX*((10**(TAU14))**N)*(C**N))/((((10**KA) + C)**N) + &
54: (((10**(TAU14))**N)*(C**N)))
55: END
56: FUNC 14
57: F =(EMAX*((10**(TAU15))**N)*(C**N))/((((10**KA) + C)**N) + &
58: (((10**(TAU15))**N)*(C**N)))
59: END
60: FUNC 15
61: F =(EMAX*((10**(TAU16))**N)*(C**N))/((((10**KA) + C)**N) + &
62: (((10**(TAU16))**N)*(C**N)))
63: END
64: FUNC 16
65: F =(EMAX*((10**(TAU17))**N)*(C**N))/((((10**KA) + C)**N) + &
66: (((10**(TAU17))**N)*(C**N)))
67: END
68: FUNC 17
69: F =(EMAX*((10**(TAU18))**N)*(C**N))/((((10**KA) + C)**N) + &
70: (((10**(TAU18))**N)*(C**N)))
71: END
72: FUNC 18
73: F =(EMAX*((10**(TAU19))**N)*(C**N))/((((10**KA) + C)**N) + &
74: (((10**(TAU19))**N)*(C**N)))
75: END
76: FUNC 19
77: F =(EMAX*((10**(TAU20))**N)*(C**N))/((((10**KA) + C)**N) + &
78: (((10**(TAU20))**N)*(C**N)))
79: END
80: EOM
NOPAGE BREAKS
NVARIABLES 4
FNUMBER 4
NPOINTS 100
XNUMBER 1
YNUMBER 2
METHOD 2  'Gauss-Newton (Levenberg and Hartley)
ITERATIONS 50
INITIAL 210,1,.8,.8,.8,.8, &
        .8,.8,.8,.8,.8,.3,.3, &
        .3,.3,.3,.3,.3,.3,.3, &
        .3,-7
LOWER BOUNDS 120,.5,-.8,-.8,-.8,-.8, &
          -.8,-.8,-.8,-.8,-.8,-1.3,-1.3, &
```

```
             -1.3,-1.3,-1.3,-1.3,-1.3,-1.3,-1.3, &
             -1.3,-8
UPPER BOUNDS 400,5,2,2,2,2, &
             2,2,2,2,2,3,3, &
             3,3,3,3,3,3,3, &
             3,10
MISSING 'Missing'
DATA 'WINNLIN.DAT'
BEGIN
```

PARAMETER	ESTIMATE	STANDARD ERROR	CV%	UNIVARIATE C.I.	
EMAX	152.115302	4.898810	3.22	142.432346	161.798257
N	.861202	.052247	6.07	.757930	.964474
TAU1	.972642	.114080	11.73	.747152	1.198132
TAU2	1.223740	.126655	10.35	.973394	1.474086
TAU4	1.110660	.121572	10.95	.870362	1.350958
TAU5	1.079406	.119995	11.12	.842225	1.316587
TAU6	.658458	.093582	14.21	.473485	.843432
TAU7	.758788	.100374	13.23	.560389	.957186
TAU8	.925803	.111253	12.02	.705900	1.145705
TAU9	.908880	.110202	12.13	.691055	1.126705
TAU10	.776581	.101572	13.08	.575813	.977348
TAU11	.591673	.089125	15.06	.415509	.767837
TAU12	.558694	.086974	15.57	.386781	.730606
TAU13	.325725	.073638	22.61	.180173	.471278
TAU14	.290565	.072024	24.79	.148203	.432926
TAU15	-.754396	.106257	14.09	-.964422	-.544369
TAU16	-1.067287	.159394	14.93	-1.382344	-.752230
TAU17	-.734039	.103766	14.14	-.939142	-.528935
TAU18	-.707210	.100630	14.23	-.906115	-.508306
TAU19	-.549556	.085193	15.50	-.717948	-.381164
TAU20	-.411442	.075384	18.32	-.560445	-.262438
KA	-6.751499	.091576	1.36	-6.932507	-6.570490

```
   Condition_number=      281.3

     FUNCTION    1

   X          OBSERVED   PREDICTED   RESIDUAL    WEIGHT    SE-PRED STANDARDI
```

	Y	Y				RESIDUAL
.1000E-08	1.820	11.17	-9.353	1.000	1.631	-1.259
.3000E-08	9.090	25.59	-16.50	1.000	2.639	-2.312
.1000E-07	43.64	54.10	-10.46	1.000	3.742	-1.579
.3000E-07	84.54	86.05	-1.515	1.000	4.019	-.2345
.1000E-06	110.9	112.7	-1.792	1.000	3.149	-.2587
.3000E-06	127.3	125.0	2.253	1.000	2.424	.3125
.1000E-05	142.7	130.3	12.41	1.000	2.189	1.704
.3000E-05	150.0	132.0	18.04	1.000	2.155	2.472
.1000E-04	151.8	132.6	19.26	1.000	2.149	2.640

```
CORRECTED SUM OF SQUARED OBSERVATIONS =  28514.1
WEIGHTED CORRECTED SUM OF SQUARED OBSERVATIONS =  28514.1
SUM OF SQUARED RESIDUALS =          1330.00
SUM OF WEIGHTED SQUARED RESIDUALS =  1330.00
S = .000000    WITH  -13 DEGREES OF FREEDOM
CORRELATION (OBSERVED,PREDICTED) = .9944
```

FUNCTION   2

X	OBSERVED Y	PREDICTED Y	RESIDUAL	WEIGHT	SE-PRED	STANDARDI RESIDUAL
.1000E-08	8.970	17.55	-8.581	1.000	2.345	-1.186
.3000E-08	47.44	37.98	9.463	1.000	3.570	1.409
.1000E-07	79.49	72.40	7.095	1.000	4.411	1.145
.3000E-07	99.36	103.7	-4.362	1.000	3.875	-.6663
.1000E-06	115.4	125.5	-10.07	1.000	2.681	-1.415
.3000E-06	128.9	134.4	-5.558	1.000	2.285	-.7660
.1000E-05	137.8	138.1	-.2587	1.000	2.297	-.3568E-01
.3000E-05	141.0	139.2	1.823	1.000	2.335	.2518
.1000E-04	142.3	139.6	2.713	1.000	2.353	.3751

```
CORRECTED SUM OF SQUARED OBSERVATIONS =  17442.0
WEIGHTED CORRECTED SUM OF SQUARED OBSERVATIONS =  17442.0
SUM OF SQUARED RESIDUALS =          375.590
SUM OF WEIGHTED SQUARED RESIDUALS =  375.590
S = .000000    WITH  -13 DEGREES OF FREEDOM
CORRELATION (OBSERVED,PREDICTED) = .9896
```

FUNCTION   3

X	OBSERVED Y	PREDICTED Y	RESIDUAL	WEIGHT	SE-PRED	STANDARDI RESIDUAL
.1000E-08	2.860	14.36	-11.50	1.000	1.998	-1.566
.3000E-08	25.71	31.95	-6.241	1.000	3.138	-.9006
.1000E-07	61.43	63.97	-2.538	1.000	4.157	-.3985
.3000E-07	90.00	96.04	-6.041	1.000	4.008	-.9344
.1000E-06	114.3	120.2	-5.876	1.000	2.881	-.8346
.3000E-06	134.3	130.6	3.693	1.000	2.280	.5089
.1000E-05	145.7	134.9	10.76	1.000	2.172	1.476
.3000E-05	154.3	136.3	17.99	1.000	2.178	2.469
.1000E-04	157.1	136.8	20.37	1.000	2.186	2.796

```
CORRECTED SUM OF SQUARED OBSERVATIONS =  26199.8
WEIGHTED CORRECTED SUM OF SQUARED OBSERVATIONS =  26199.8
SUM OF SQUARED RESIDUALS =          1116.84
SUM OF WEIGHTED SQUARED RESIDUALS =  1116.84
S = .000000    WITH  -13 DEGREES OF FREEDOM
CORRELATION (OBSERVED,PREDICTED) = .9914
   FUNCTION    4
```

X	OBSERVED Y	PREDICTED Y	RESIDUAL	WEIGHT	SE-PRED	STANDARDI RESIDUAL
.1000E-08	3.120	13.57	-10.45	1.000	1.909	-1.419
.3000E-08	30.66	30.41	.2453	1.000	3.020	.3513E-01
.1000E-07	69.65	61.68	7.967	1.000	4.071	1.240
.3000E-07	93.56	93.83	-.2697	1.000	4.025	-.4178E-01
.1000E-06	111.2	118.6	-7.333	1.000	2.942	-1.045
.3000E-06	125.8	129.4	-3.636	1.000	2.300	-.5014
.1000E-05	131.0	134.0	-3.002	1.000	2.159	-.4115
.3000E-05	139.8	135.4	4.423	1.000	2.155	.6063
.1000E-04	141.9	135.9	5.999	1.000	2.160	.8225

CORRECTED SUM OF SQUARED OBSERVATIONS = 19930.6
WEIGHTED CORRECTED SUM OF SQUARED OBSERVATIONS = 19930.6
SUM OF SQUARED RESIDUALS = 304.417
SUM OF WEIGHTED SQUARED RESIDUALS = 304.417
S = .000000 WITH -13 DEGREES OF FREEDOM
CORRELATION (OBSERVED,PREDICTED) = .9929

FUNCTION 5

X	OBSERVED Y	PREDICTED Y	RESIDUAL	WEIGHT	SE-PRED	STANDARDI RESIDUAL
.1000E-08	7.590	6.204	1.386	1.000	1.016	.1839
.3000E-08	19.31	14.88	4.426	1.000	1.754	.5979
.1000E-07	46.90	34.74	12.16	1.000	2.781	1.717
.3000E-07	70.34	62.56	7.776	1.000	3.599	1.160
.1000E-06	86.62	92.08	-5.456	1.000	3.512	-.8085
.3000E-06	97.24	108.3	-11.09	1.000	2.952	-1.582
.1000E-05	109.0	116.0	-6.994	1.000	2.694	-.9831
.3000E-05	118.6	118.4	.2155	1.000	2.649	.3021E-01
.1000E-04	121.4	119.3	2.084	1.000	2.641	.2921

CORRECTED SUM OF SQUARED OBSERVATIONS = 14293.1
WEIGHTED CORRECTED SUM OF SQUARED OBSERVATIONS = 14293.1
SUM OF SQUARED RESIDUALS = 435.847
SUM OF WEIGHTED SQUARED RESIDUALS = 435.847
S = .000000 WITH -13 DEGREES OF FREEDOM
CORRELATION (OBSERVED,PREDICTED) = .9900

FUNCTION 6

X	OBSERVED Y	PREDICTED Y	RESIDUAL	WEIGHT	SE-PRED	STANDARDI RESIDUAL
.1000E-08	3.600	7.502	-3.902	1.000	1.183	-.5192
.3000E-08	10.09	17.78	-7.688	1.000	1.997	-1.047
.1000E-07	32.42	40.36	-7.943	1.000	3.061	-1.141
.3000E-07	63.04	70.00	-6.959	1.000	3.769	-1.053
.1000E-06	93.66	99.14	-5.475	1.000	3.446	-.8073
.3000E-06	113.1	114.3	-1.156	1.000	2.792	-.1633
.1000E-05	127.5	121.2	6.370	1.000	2.503	.8868
.3000E-05	136.2	123.3	12.83	1.000	2.447	1.782
.1000E-04	142.3	124.1	18.16	1.000	2.434	2.520

CORRECTED SUM OF SQUARED OBSERVATIONS = 23851.6
WEIGHTED CORRECTED SUM OF SQUARED OBSERVATIONS = 23851.6
SUM OF SQUARED RESIDUALS = 752.326
SUM OF WEIGHTED SQUARED RESIDUALS = 752.326
S = .000000 WITH -13 DEGREES OF FREEDOM
CORRELATION (OBSERVED,PREDICTED) = .9923

FUNCTION   7

X	OBSERVED Y	PREDICTED Y	RESIDUAL	WEIGHT	SE-PRED	STANDARDI RESIDUAL
.1000E-08	6.320	10.25	-3.929	1.000	1.520	-.5271
.3000E-08	36.21	23.67	12.54	1.000	2.483	1.744
.1000E-07	75.29	50.91	24.38	1.000	3.589	3.635
.3000E-07	91.09	82.56	8.525	1.000	3.987	1.316
.1000E-06	104.6	109.9	-5.330	1.000	3.231	-.7739
.3000E-06	109.8	122.9	-13.12	1.000	2.498	-1.825
.1000E-05	113.2	128.5	-15.31	1.000	2.235	-2.105
.3000E-05	114.4	130.3	-15.92	1.000	2.190	-2.185
.1000E-04	115.5	130.9	-15.40	1.000	2.181	-2.113

CORRECTED SUM OF SQUARED OBSERVATIONS = 12290.3
WEIGHTED CORRECTED SUM OF SQUARED OBSERVATIONS = 12290.3
SUM OF SQUARED RESIDUALS =   1765.00
SUM OF WEIGHTED SQUARED RESIDUALS = 1765.00
S = .000000   WITH   -13 DEGREES OF FREEDOM
CORRELATION (OBSERVED,PREDICTED) = .9649

FUNCTION   8

X	OBSERVED Y	PREDICTED Y	RESIDUAL	WEIGHT	SE-PRED	STANDARDI RESIDUAL
.1000E-08	9.850	9.932	-.8249E-01	1.000	1.482	-.1106E-01
.3000E-08	27.01	23.01	4.002	1.000	2.429	.5551
.1000E-07	63.87	49.78	14.09	1.000	3.534	2.092
.3000E-07	85.77	81.30	4.474	1.000	3.972	.6895
.1000E-06	102.2	108.9	-6.709	1.000	3.258	-.9760
.3000E-06	113.9	122.1	-8.216	1.000	2.527	-1.145
.1000E-05	121.2	127.8	-6.680	1.000	2.256	-.9195
.3000E-05	124.5	129.7	-5.200	1.000	2.208	-.7144
.1000E-04	126.3	130.3	-4.019	1.000	2.198	-.5519

CORRECTED SUM OF SQUARED OBSERVATIONS = 15145.1
WEIGHTED CORRECTED SUM OF SQUARED OBSERVATIONS = 15145.1
SUM OF SQUARED RESIDUALS =   434.975
SUM OF WEIGHTED SQUARED RESIDUALS = 434.975
S = .000000   WITH   -13 DEGREES OF FREEDOM
CORRELATION (OBSERVED,PREDICTED) = .9923

FUNCTION   9

X	OBSERVED Y	PREDICTED Y	RESIDUAL	WEIGHT	SE-PRED	STANDARDI RESIDUAL
.1000E-08	4.170	7.757	-3.587	1.000	1.215	-.4777
.3000E-08	14.88	18.34	-3.459	1.000	2.044	-.4721
.1000E-07	50.60	41.42	9.182	1.000	3.114	1.323
.3000E-07	82.44	71.33	11.11	1.000	3.797	1.685
.1000E-06	100.0	100.3	-.3468	1.000	3.429	-.5107E-01
.3000E-06	108.8	115.3	-6.510	1.000	2.761	-.9184
.1000E-05	114.3	122.0	-7.731	1.000	2.470	-1.075
.3000E-05	117.9	124.1	-6.230	1.000	2.413	-.8636
.1000E-04	120.8	124.9	-4.094	1.000	2.400	-.5672

CORRECTED SUM OF SQUARED OBSERVATIONS = 16363.6
WEIGHTED CORRECTED SUM OF SQUARED OBSERVATIONS = 16363.6

```
SUM OF SQUARED RESIDUALS =              390.338
SUM OF WEIGHTED SQUARED RESIDUALS =    390.338
S =  .000000     WITH   -13 DEGREES OF FREEDOM
CORRELATION (OBSERVED,PREDICTED) =  .9902
```

FUNCTION 10

X	OBSERVED Y	PREDICTED Y	RESIDUAL	WEIGHT	SE-PRED	STANDARDI RESIDUAL
.1000E-08	.9500	5.462	-4.512	1.000	.9187	-.5975
.3000E-08	4.280	13.20	-8.918	1.000	1.610	-1.200
.1000E-07	31.43	31.32	.1080	1.000	2.613	.1511E-01
.3000E-07	69.52	57.75	11.77	1.000	3.483	1.740
.1000E-06	83.33	87.20	-3.872	1.000	3.533	-.5748
.3000E-06	102.9	104.1	-1.230	1.000	3.043	-.1764
.1000E-05	109.5	112.2	-2.658	1.000	2.816	-.3761
.3000E-05	114.3	114.8	-.5219	1.000	2.785	-.7373E-01
.1000E-04	115.2	115.8	-.5193	1.000	2.783	-.7335E-01

```
CORRECTED SUM OF SQUARED OBSERVATIONS =  17401.4
WEIGHTED CORRECTED SUM OF SQUARED OBSERVATIONS =  17401.4
SUM OF SQUARED RESIDUALS =                262.540
SUM OF WEIGHTED SQUARED RESIDUALS =  262.540
S =  .000000     WITH   -13 DEGREES OF FREEDOM
CORRELATION (OBSERVED,PREDICTED) =  .9930
```

FUNCTION 11

X	OBSERVED Y	PREDICTED Y	RESIDUAL	WEIGHT	SE-PRED	STANDARDI RESIDUAL
.1000E-08	.7700	5.128	-4.358	1.000	.8738	-.5766
.3000E-08	10.77	12.43	-1.661	1.000	1.543	-.2230
.1000E-07	38.08	29.73	8.353	1.000	2.535	1.165
.3000E-07	64.62	55.43	9.194	1.000	3.427	1.354
.1000E-06	85.38	84.76	.6223	1.000	3.538	.9240E-01
.3000E-06	101.5	101.9	-.3753	1.000	3.082	-.5396E-01
.1000E-05	103.1	110.2	-7.142	1.000	2.873	-1.014
.3000E-05	108.5	112.9	-4.470	1.000	2.851	-.6338
.1000E-04	108.9	113.9	-5.069	1.000	2.852	-.7188

```
CORRECTED SUM OF SQUARED OBSERVATIONS =  14655.0
WEIGHTED CORRECTED SUM OF SQUARED OBSERVATIONS =  14655.0
SUM OF SQUARED RESIDUALS =                273.261
SUM OF WEIGHTED SQUARED RESIDUALS =  273.261
S =  .000000     WITH   -13 DEGREES OF FREEDOM
CORRELATION (OBSERVED,PREDICTED) =  .9921
```

FUNCTION 12

X	OBSERVED Y	PREDICTED Y	RESIDUAL	WEIGHT	SE-PRED	STANDARDI RESIDUAL
.1000E-08	.0000	3.271	-3.271	1.000	.6094	-.4314
.3000E-08	.0000	8.076	-8.076	1.000	1.136	-1.074
.1000E-07	6.620	20.19	-13.57	1.000	2.037	-1.851
.3000E-07	47.68	40.36	7.318	1.000	3.030	1.049
.1000E-06	68.21	67.27	.9429	1.000	3.495	.1396
.3000E-06	84.10	85.37	-1.272	1.000	3.269	-.1851
.1000E-05	92.05	94.88	-2.829	1.000	3.198	-.4098
.3000E-05	100.0	98.09	1.908	1.000	3.243	.2773
.1000E-04	103.3	99.28	4.029	1.000	3.273	.5868

CORRECTED SUM OF SQUARED OBSERVATIONS = 15191.7
WEIGHTED CORRECTED SUM OF SQUARED OBSERVATIONS = 15191.7
SUM OF SQUARED RESIDUALS =   343.973
SUM OF WEIGHTED SQUARED RESIDUALS = 343.973
S = .000000   WITH   -13 DEGREES OF FREEDOM
CORRELATION (OBSERVED,PREDICTED) = .9924

   FUNCTION 13

X	OBSERVED Y	PREDICTED Y	RESIDUAL	WEIGHT	SE-PRED	STANDARDI RESIDUAL
.1000E-08	.0000	3.055	-3.055	1.000	.5764	-.4028
.3000E-08	.6900	7.559	-6.869	1.000	1.083	-.9123
.1000E-07	4.170	19.00	-14.83	1.000	1.967	-2.018
.3000E-07	38.19	38.33	-.1383	1.000	2.968	-.1975E-01
.1000E-06	65.28	64.66	.6174	1.000	3.478	.9127E-01
.3000E-06	81.94	82.75	-.8097	1.000	3.285	-.1180
.1000E-05	93.75	92.37	1.381	1.000	3.235	.2006
.3000E-05	98.61	95.64	2.971	1.000	3.289	.4331
.1000E-04	101.4	96.85	4.539	1.000	3.324	.6633

CORRECTED SUM OF SQUARED OBSERVATIONS = 15214.3
WEIGHTED CORRECTED SUM OF SQUARED OBSERVATIONS = 15214.3
SUM OF SQUARED RESIDUALS =   308.801
SUM OF WEIGHTED SQUARED RESIDUALS = 308.801
S = .000000   WITH   -13 DEGREES OF FREEDOM
CORRELATION (OBSERVED,PREDICTED) = .9959

   FUNCTION 14

X	OBSERVED Y	PREDICTED Y	RESIDUAL	WEIGHT	SE-PRED	STANDARDI RESIDUAL
.1000E-08	.2400	.3916	-.1516	1.000	.1068	-.1993E-01
.3000E-08	.7300	.9950	-.2650	1.000	.2388	-.3486E-01
.1000E-07	2.430	2.685	-.2554	1.000	.5717	-.3367E-01
.3000E-07	5.580	6.189	-.6093	1.000	1.198	-.8110E-01
.1000E-06	12.14	12.96	-.8160	1.000	2.219	-.1121
.3000E-06	16.99	19.87	-2.875	1.000	3.053	-.4127
.1000E-05	26.70	24.79	1.913	1.000	3.617	.2859
.3000E-05	27.67	26.73	.9352	1.000	3.861	.1427
.1000E-04	27.67	27.50	.1710	1.000	3.963	.2633E-01

CORRECTED SUM OF SQUARED OBSERVATIONS = 1113.82
WEIGHTED CORRECTED SUM OF SQUARED OBSERVATIONS = 1113.82
SUM OF SQUARED RESIDUALS =   14.0278
SUM OF WEIGHTED SQUARED RESIDUALS = 14.0278
S = .000000   WITH   -13 DEGREES OF FREEDOM
CORRELATION (OBSERVED,PREDICTED) = .9943

   FUNCTION 15

X	OBSERVED Y	PREDICTED Y	RESIDUAL	WEIGHT	SE-PRED	STANDARDI RESIDUAL
.3000E-08	.0000	.5367	-.5367	1.000	.1746	-.7057E-01
.1000E-07	2.790	1.456	1.334	1.000	.4445	.1757
.3000E-07	3.780	3.392	.3883	1.000	.9869	.5147E-01
.1000E-06	4.880	7.252	-2.372	1.000	1.984	-.3230
.3000E-06	7.960	11.37	-3.408	1.000	2.947	-.4859
.1000E-05	12.94	14.41	-1.473	1.000	3.632	-.2204

```
.3000E-05  16.92        15.65       1.273       1.000       3.914      .1952
.1000E-04  19.90        16.13       3.765       1.000       4.027      .5835
```

CORRECTED SUM OF SQUARED OBSERVATIONS = 360.927
WEIGHTED CORRECTED SUM OF SQUARED OBSERVATIONS = 360.927
SUM OF SQUARED RESIDUALS =   37.4296
SUM OF WEIGHTED SQUARED RESIDUALS = 37.4296
S = .000000   WITH   -14 DEGREES OF FREEDOM
CORRELATION (OBSERVED,PREDICTED) = .9481

FUNCTION 16

X	OBSERVED Y	PREDICTED Y	RESIDUAL	WEIGHT	SE-PRED	STANDARDI RESIDUAL
.3000E-08	.0000	1.036	-1.036	1.000	.2448	-.1362
.1000E-07	.6000	2.794	-2.194	1.000	.5835	-.2893
.3000E-07	5.420	6.433	-1.013	1.000	1.217	-.1350
.1000E-06	11.44	13.44	-2.003	1.000	2.239	-.2754
.3000E-06	19.28	20.57	-1.293	1.000	3.061	-.1857
.1000E-05	25.30	25.64	-.3356	1.000	3.615	-.5015E-01
.3000E-05	29.52	27.64	1.884	1.000	3.857	.2873
.1000E-04	29.52	28.42	1.100	1.000	3.958	.1693

CORRECTED SUM OF SQUARED OBSERVATIONS = 1082.73
WEIGHTED CORRECTED SUM OF SQUARED OBSERVATIONS = 1082.73
SUM OF SQUARED RESIDUALS =   17.4668
SUM OF WEIGHTED SQUARED RESIDUALS = 17.4668
S = .000000   WITH   -14 DEGREES OF FREEDOM
CORRELATION (OBSERVED,PREDICTED) = .9970

FUNCTION 17

X	OBSERVED Y	PREDICTED Y	RESIDUAL	WEIGHT	SE-PRED	STANDARDI RESIDUAL
.3000E-08	2.000	1.092	.9081	1.000	.2531	.1194
.1000E-07	2.400	2.944	-.5437	1.000	.5998	-.7170E-01
.3000E-07	6.000	6.769	-.7693	1.000	1.244	-.1025
.1000E-06	12.40	14.11	-1.709	1.000	2.267	-.2353
.3000E-06	19.60	21.54	-1.938	1.000	3.073	-.2785
.1000E-05	27.20	26.79	.4103	1.000	3.613	.6129E-01
.3000E-05	29.60	28.86	.7404	1.000	3.851	.1129
.1000E-04	31.20	29.67	1.530	1.000	3.951	.2353

CORRECTED SUM OF SQUARED OBSERVATIONS = 1047.60
WEIGHTED CORRECTED SUM OF SQUARED OBSERVATIONS = 1047.60
SUM OF SQUARED RESIDUALS =   11.4449
SUM OF WEIGHTED SQUARED RESIDUALS = 11.4449
S = .000000   WITH   -14 DEGREES OF FREEDOM
CORRELATION (OBSERVED,PREDICTED) = .9950

FUNCTION 18

X	OBSERVED Y	PREDICTED Y	RESIDUAL	WEIGHT	SE-PRED	STANDARDI RESIDUAL
.3000E-08	1.330	1.489	-.1588	1.000	.3126	-.0208
.1000E-07	2.560	3.996	-1.436	1.000	.7151	-.1896
.3000E-07	7.780	9.105	-1.325	1.000	1.427	-.1773
.1000E-06	13.56	18.65	-5.092	1.000	2.453	-.7072
.3000E-06	27.22	27.99	-.7682	1.000	3.145	-.1109
.1000E-05	35.56	34.40	1.162	1.000	3.596	.1733
.3000E-05	38.89	36.88	2.007	1.000	3.811	.3048

| .1000E-04 | 38.89 | 37.85 | 1.040 | 1.000 | 3.905 | .1593 |

CORRECTED SUM OF SQUARED OBSERVATIONS = 1847.24
WEIGHTED CORRECTED SUM OF SQUARED OBSERVATIONS = 1847.24
SUM OF SQUARED RESIDUALS =   36.8197
SUM OF WEIGHTED SQUARED RESIDUALS =   36.8197
S = .000000   WITH   -14 DEGREES OF FREEDOM
CORRELATION (OBSERVED,PREDICTED) = .9930

FUNCTION   19

X	OBSERVED Y	PREDICTED Y	RESIDUAL	WEIGHT	SE-PRED	STANDARDI RESIDUAL
.3000E-08	.0000	1.952	-1.952	1.000	.3819	-.2569
.1000E-07	.0000	5.211	-5.211	1.000	.8469	-.6894
.3000E-07	7.040	11.75	-4.712	1.000	1.629	-.6341
.1000E-06	19.72	23.62	-3.896	1.000	2.644	-.5463
.3000E-06	31.00	34.79	-3.789	1.000	3.210	-.5494
.1000E-05	41.55	42.23	-.6774	1.000	3.574	-.1009
.3000E-05	49.30	45.06	4.239	1.000	3.768	.6414
.1000E-04	50.70	46.16	4.544	1.000	3.857	.6930

CORRECTED SUM OF SQUARED OBSERVATIONS = 3161.26
WEIGHTED CORRECTED SUM OF SQUARED OBSERVATIONS = 3161.26
SUM OF SQUARED RESIDUALS =   121.781
SUM OF WEIGHTED SQUARED RESIDUALS = 121.781
S = .000000   WITH   -14 DEGREES OF FREEDOM
CORRELATION (OBSERVED,PREDICTED) = .9923

TOTALS FOR ALL CURVES COMBINED
SUM OF SQUARED RESIDUALS =   8332.88
SUM OF WEIGHTED SQUARED RESIDUALS = 8332.88
S = 7.60705   WITH   144 DEGREES OF FREEDOM

AIC criteria =   1542.64209
SC criteria =   1611.10583

PD3 - Inhibitory I_{max} model

Objectives

◆ To analyze data with an inhibitory I_{max} model

◆ To discriminate between competing models

◆ To discuss and perform a simulation study for experimental design

◆ To study the impact of variability in the data on parameter estimates

◆ To discuss the delta function and variance inflation factor

Problem specification

The goals of this exercise are to demonstrate some simple ways of discriminating between rival inhibitory I_{max} models, and to propose an alternative design in order to select the most appropriate model.

The effect is blood pressure and covers an interval between 170 mm Hg (approximate baseline level) and 140 mm Hg. The data are taken from a study involving measurements at steady state, and are shown in Figure 3.1 and in the program output. We will model the lowering of blood pressure using two inhibitory I_{max} models. We observe a baseline effect E_0 when no drug is present. By increasing the plasma concentration of drug A, the blood pressure is gradually lowered. The maximum drug induced effect corresponds to I_{max}. The plasma concentration at 50% of I_{max} is called the potency (IC_{50}) of the drug. E_0, I_{max} and IC_{50} are shown graphically in Figure 3.2.

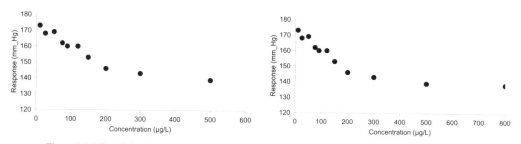

Figure 3.1 *Effect (fall in blood pressure) versus steady state plasma concentration of drug X. The design for Part I (left) includes data from 0-500 μg/L, and that for Part II (right) is expanded to include data at 800 μg/L.*

We plot the effect-concentration data and estimate the basal effect parameters E_0, I_{max}, IC_{50} and n according to Figure 3.2. The n parameter is the sigmoidicity factor that affects the curvature of Equation 3:2.

The models are shown in the program output below. During Part I of this exercise, we will fit the ordinary (Equation 3:1) and sigmoid (Equation 3:2) inhibitory I_{max} models to the data using constant absolute weights. We will then select the most efficient model by using a battery of statistical tools that you will find in the program output. The model uses I_{max}, IC_{50} and E_0 in Equation 3:1 and the additional parameter n, in Equation 3:2.

$$Model_1 = E_0 - \frac{I_{max} \cdot C}{IC_{50} + C} \tag{3:1}$$

$$Model_2 = E_0 - \frac{I_{max} \cdot C^n}{IC_{50}^n + C^n} \tag{3:2}$$

In Part II, the dataset is increased by the addition of a blood pressure measurement of 138 mm Hg at 800 µg/L. We will include that observation in our data file and re-fit both models (Equations 3:1 and 3:2). We then discriminate between them using appropriate statistical methods such as *AIC, F*-test and runs in the residuals.

Initial parameter estimates

The initial parameter estimates for E_0 (baseline), I_{max}, IC_{50} and the sigmoidicity factor n were obtained graphically according to Figure 3.2. I_{max} is estimated to about 35 mm Hg, IC_{50} is about 120 µg/L, E_0 is 175 mm Hg and n is approximated to be 2.5.

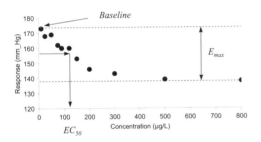

Figure 3.2 Plot of observed effect (fall in blood pressure) versus plasma concentration to demonstrate the estimation of E_0, E_{max}, EC_{50}, and n graphically. These parameters correspond to baseline, I_{max}, IC_{50} and n in Equations 3:1-3:2.

Interpretation of results and conclusions

Comparisons of the observed and model-predicted data of both models and designs are displayed in Figure 3.3.

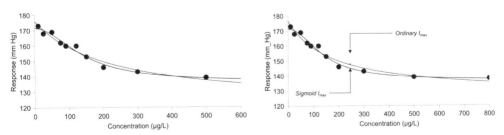

Figure 3.3 Observed and predicted effects obtained using two models and designs. The predicted effects of the sigmoid I_{max} model and ordinary I_{max} model are indicated by arrows in the right hand figure. The left hand figure represents the first run (0-500 µg/L) and right hand the expanded concentration range (0-800 µg/L).

Figure 3.3 (left) does not indicate which function describes the data best. The *Akaike Information Criteria AIC* were 42.1 and 44.2 for the sigmoid I_{max} and ordinary I_{max} models, respectively, which marginally favors the former model, for the 0 to 500 µg/L concentration range.

However, using an *F*-test for comparison of the models for the first data set, the sigmoid I_{max} model is not superior to the ordinary I_{max} model, F^* (= 3.5255) < F_{table} (= 5.9874 in *F*-table $v_1 = 1$ and $v_2 = 6$, $\alpha = 0.05$).

Table 3.1 *F*-table

The *F* distribution v_1, v_2 Degrees of Freedom $P = 0.05$				
v_1 Δdf ->	1	2	3	4
v_2 1	161.45	199.50	215.71	224.58
2	18.513	19.000	19.164	19.247
3	10.128	9.5521	9.2766	9.1172
4	7.7086	6.9443	6.5914	6.3882
5	6.6079	5.7861	5.4095	5.1922
6	*a) 5.9874*	5.1433	4.7571	4.5337
7	*b) 5.5914*	4.7374	4.3868	4.1203
8	5.3177	4.4590	4.0662	3.8379
9	5.1174	4.2565	3.8626	3.6331

a) is the F-test for n in the 0-500 µg/L dataset
b) is the F-test for n in the 0-800 µg/L dataset

Note that it is only appropriate to apply the *F*-test with constant weights such as *OLS* (weight equal to 1 or a constant value) or *WLS* (weight equal to 1 over the observed value raised to a power = $1/Y^n$).

Figure 3.3 (right) demonstrates that the sigmoid I_{max} model exhibits more flexibility and captures the curvature in the data better than the ordinary I_{max} model. The *Akaike Information Criteria AIC* were 45.6 and 50.8 for the sigmoid I_{max} and ordinary I_{max} models, respectively, which favors the former model, for the 0 to 800 µg/L (expanded) concentration range. The following estimates and their precision were obtained for the two models and concentration ranges (Table 3.2)

Table 3.2   Dynamic estimates of the 0-500 and 0-800 µg/L designs (mean +/- CV%)

Parameter	0-500 µg/L		0-800 µg/L	
	Ordinary I_{max}	Sigmoid I_{max}	Ordinary I_{max}	Sigmoid I_{max}
I_{max}	56.1 ± 13.1	35.7 ± 15.2	49.8 ± 7.95	34.7 ± 9.53
IC_{50}	231 ± 37.6	143 ± 16.0	175 ± 30.6	140 ± 10.7
n	-	1.94 ± 27.6	-	2.03 ± 21.3
E_0	176 ± 1.40	172 ± 1.14	177 ± 1.52	171 ± 1.01

If we apply an *F*-test as a measure of whether n is different from unity or not, you will find that the sigmoid I_{max} model (full model, where n is significantly different from 1, in this case 2.03) is superior to the ordinary I_{max} model (reduced model, n set to 1). F^* calculated as

$$F^* = \frac{\dfrac{|WRSS_1 - WRSS_2|}{df_1 - df_2}}{\dfrac{WRSS_2}{df_2}} = 6.4664 > 5.5914 = F_{table} \tag{3:3}$$

$WRSS_1$ and $WRSS_2$ are the objective functions of the ordinary I_{max} (reduced) and sigmoid I_{max} (full) models, and df_1 and df_2 are the degrees of freedom for the reduced and full models, respectively. Degrees of freedom for each model are calculated as the number of observations minus the number of parameters. F^* is larger than F_{table} (= 5.5914, $p = 0.05$ $\upsilon_1 = 1$ and $\upsilon_2 = 7$, $\alpha = 0.05$). In other words, we were able to discriminate between the models with the new design by including an additional data point.

Figure 3.4 shows the absolute residual (filled squares correspond to the sigmoid I_{max} model and open circles to the ordinary I_{max} model) plotted against the steady state concentration. Note that the amplitude of the scatter is larger and the deviating trend is more obvious for the ordinary I_{max} model than for the sigmoid I_{max} model.

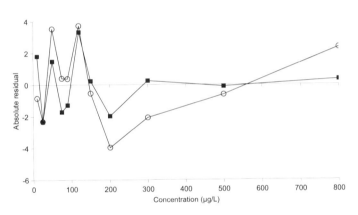

Figure 3.4 *Absolute residual versus steady state concentration. Filled squares are the sigmoid I_{max} model and open circles the ordinary I_{max} model.*

The final parameter estimates for the sigmoid I_{max} model are relatively constant for the two designs. However, the IC_{50} values of the ordinary I_{max} model differs substantially between designs. The previous analysis has shown that residual analysis, F-test and AIC values all favor the sigmoid I_{max} model provided that the 0-800 µg/L dataset is used.

Remember that E_0 includes the disease model, which may not be constant over concentration and/or time.

Delta function

One can use the delta function to determine the plasma concentrations that will optimally discriminate between the two models. The ordinary I_{max} *(reduced)* model will have the following parameterization

$$Model_1 = 176.6 - \frac{49.84 \cdot C}{175.1 + C} \qquad (3:4)$$

and the sigmoid I_{max} (*full*) model

$$Model_2 = 171.36 - \frac{34.66 \cdot C^{2.03}}{139.61^{2.03} + C^{2.03}} \qquad (3:5)$$

We subtract the *full* model from the *reduced* and then square that difference according to

$$\Delta = \left(Model_1 - Model_2\right)^2 \qquad (3:6)$$

Equation 3:6 is then simulated for the 0 to 800 µg/L concentration interval in order to get the time points where the difference between the models is greatest (Godfrey [1983]). This difference is then plotted against the plasma concentration as shown in Figure 3.5.

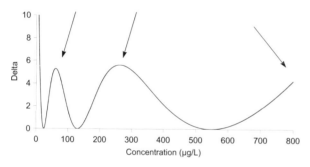

Figure 3.5 The difference function (delta E squared) as a function of plasma concentration. The arrows indicate concentrations where the models deviate the most.

Figure 3.5 indicates the design points that need to be included in the experiment to optimally discriminate between the two models. The maximum difference between the models seems to be at approximately 70 µg/L, 275 µg/L, and concentrations of 800 µg/L or more. Thus, these points should be included as well as other effect-concentration measurements where the models do not differ to such a great extent.

Partial derivatives

The partial derivatives with respect to the parameters were obtained from the fit of the sigmoid I_{max} model to the data of the 0-800 µg/L concentration range. It is interesting to note that maximum information about the sigmoidicity factor is obtained at approximately the 20 and 80% effect level. Observe that the partial derivative for IC_{50} peaks at the actual value of IC_{50}, namely 140 µg/L. As expected, maximum information about I_{max} is at the higher concentration levels (Figure 3.6).

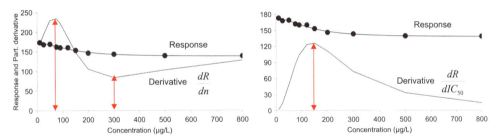

Figure 3.6 *(Left) Predicted effect (solid line with filled circles) versus steady state concentration superimposed upon the partial derivative with respect to the exponent (n) of the sigmoid I_{max} model. (Right) Predicted effect (solid line with filled circles) versus steady state concentration superimposed upon the partial derivative with respect to IC_{50} of the sigmoid I_{max} model.*

Simulation exercise

In order to simulate differences in experimental error, a dataset with 0.5, 2, 3, or 5% error (constant *CV*) was fit by means of the sigmoid I_{max} model from the previous analysis. The parameter values for I_{max}, IC_{50}, *n* and E_0, were 34.66, 139.6, 2.031 and 171.4, respectively (see program output in Solution IV below). The sampled effect *versus* steady state concentration data taken from the previous run were logarithmically equally spaced within the 0-800 µg/L concentration range (Figure 3.7).

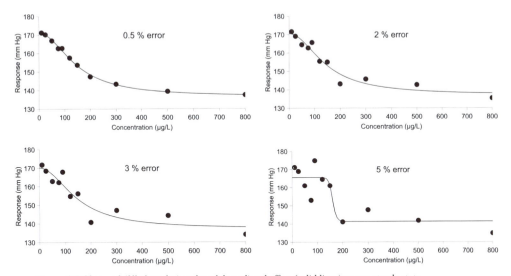

Figure 3.7 *Observed (filled circles) and model-predicted effect (solid lines) versus steady state concentration for four different regressions. Data were generated with 0.5, 2, 3, or 5% relative error.*

It is obvious from Table 3.3 and Figure 3.8 that experimental error in the effect-concentration data transforms directly into the precision of the parameter estimates. An increase from 0.5 to 3% in the coefficient of variation of the experimental data results in a corresponding increase in the standard deviation of the parameter estimate.

Table 3.3   Mean estimates and their precision *(CV%)* at different error levels *(CV%)*

Parameter	0.5% CV	2% CV	3% CV	5% CV
I_{max}	34.7 ± 1.39	33.5 ± 5.71	33.4 ± 8.81	24.5 ± 5.4
IC_{50}	146. ± 6.6	144. ± 27.1	147. ± 43.8	162. ± 40.7
n	1.99 ± 0.18	2.02 ± 0.75	1.96 ± 1.12	17.9 ± 51.7
E_0	171. ± 0.70	171. ± 2.90	170.± 4.37	166. ± 3.09

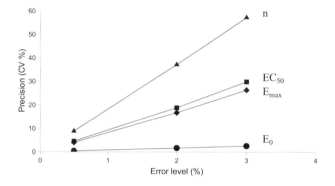

Figure 3.8 Precision versus simulated error level (%). This plot demonstrates how the experimental error inflates the parameter precision provided everything else is held constant.

When the variability in mean experimental data is increased to 5%, not only the parameter precision is affected but also the accuracy of the parameters. I_{max}, IC_{50} and particularly n, differ by a factor of 20, 15 and several thousand per cent, respectively, compared with results from fitting datasets with lower noise (e.g., 0.5 – 3% error). In other words, the estimates are biased. Below we demonstrate how the experimental error inflates the variance of a parameter *var(par)*, via the predicted mean variance σ^2 and the degrees of freedom *df*.

$$Residual \propto Error \tag{3:7}$$

$$WRSS = \sum w_i \cdot (E_i - \hat{E}_i)^2 \tag{3:8}$$

$$\hat{\sigma}^2 = \frac{\sum w_i \cdot (E_i - \hat{E}_i)^2}{df} \tag{3:9}$$

$$\hat{Var}(par) = \hat{\sigma}^2 \cdot VIF \tag{3:10}$$

Clearly, this model and the re-fitting of simulated data with these particular parameter values was very sensitive to the error level. This may, of course, not always be the case with other types of models and datasets.

Variance inflation factor (_VIF_)

A pharmacodynamic study (see problem specification above) is to be conducted on a drug to be administered as a constant intravenous infusion and whose dynamics can be represented by a sigmoid I_{max} model. Samples need to be collected over a 0 - 800 µg/L concentration range. If samples are to be collected above 500 µg/L, then the study will jeopardize the safety of the subjects. That is, subjects would need to be supervised by an anesthetist. This would dramatically increase the cost of the study.

The question is, how much information are we losing by not collecting a sample at e.g., 800 µg/L? To put it another way, would the _value_ of collecting a sample at say, 800 µg/L, exceed the cost required to collect the sample (e.g., the cost of extra staff)? To address this question, we must define _value_. For this example, let us assume that _value_ can be measured by an increase in the precision with which we can estimate the parameters. Thus, the question we are asking is _Would a sample collected at 800 µg/L produce enough of an increase in precision of the estimated parameters to justify the cost of its collection?_ To address this question we will use the simulation option in _WinNonlin_.

The variance of a model parameter can be expressed as the product of the underlying variance of the residuals σ^2 and a multiplier, which we will call the variance inflation factor, or _VIF_ (SCI [1994]).

$$Var(E_{max}) = \sigma^2 \cdot VIF \qquad (3:11)$$

For example, the variance of the mean of a sample is well known to be σ^2/n, where n is the sample size. In this instance, the _VIF_ is $1/n$. Note that _VIF does not depend on the actual data_ which are measured; it is _only dependent on the experimental design_. Although the _VIFs_ which we require are much more complicated to compute than that of a sample mean, the program, when using the _SIMULATION_ option, automatically computes these values for model parameters, secondary parameters and the predicted values (Tables 3.4 and 3:5).

Table 3.4	Comparison of variance inflation factors of the two designs.			
		Variance inflation factor (_VIF_)		
Parameter	Estimate	0-500 µg/L	0-800 µg/L	_VIF_ RATIO
I_{max}	34.660	4.951	2.488	1.99
IC_{50}	139.610	85.78	50.88	1.69
n	2.031	.060664	.042771	1.42
E_0	171.360	.718233	.67891	1.06

Table 3.5   Comparison of variance inflation factors of the two designs

		Variance inflation factor (*VIF*)		
X	Calculated Y	0-500 µg/L	0-800 µg/L	*VIF* RATIO
10.00	171.2	.6025	.5806	1.04
25.00	170.3	.3553	.3552	1.00
50.00	167.5	.2869	.2642	1.10
75.00	163.7	.3228	.3054	1.06
90.00	161.3	.2937	.2895	1.01
120.0	156.7	.2563	.2486	1.03
150.0	152.8	.3143	.2823	1.11
200.0	148.0	.3958	.3710	1.07
300.0	142.8	.3526	.3276	1.08
500.0	139.1	.8199	.3485	2.35
800.0	137.7	–	.6271	–

In examining the output we see that the design which did not include a 800 µg/L sample was associated with *VIFs* which were *inflated* by 6% for E_0, 42% for *n*, 69% for IC_{50} and 99% for I_{max}. Note that the effect of an 800 µg/L collection on *n* and IC_{50} is not intuitively obvious since we had already included five samples beyond 140 µg/L.

The way to interpret the *VIF* results is as follows. If we had run two studies, one using an 800 µg/L sample and one without the 800 µg/L sample, determined the resultant effects and fit the sigmoid I_{max} model to the data, the parameters estimated from the study without the extra sample would have had variances of IC_{50} and I_{max} which were 69 - 99% larger than those estimated from the study which included the 800 µg/L sample. One must now decide if the increase in precision of the three parameters is "worth" the cost of collection of the additional sample.

As a final observation, we note that the inclusion of an 800 µg/L sample would have very little impact on the precision with which we can estimate plasma concentrations less than 500 µg/L.

Overall conclusions

Although there are no absolute criteria regarding the selection of a model, two goals are generally sought when selection is for kinetic/dynamic data. First, it is necessary that the scatter of the residuals of the observed data is randomly distributed around the fitted model, and secondly, that the sums of the weighted squared residuals (*WRSS*) have been sufficiently reduced to justify fitting with additional parameters. This exercise has shown how to fit both an ordinary and a sigmoid inhibitory I_{max} model to concentration-effect data, and how to interpret the output by means of residuals, *AIC/SC* and the *F*-test. In addition, we have elaborated an approach for discriminating between the two models by looking at the *optimal* design points for model discrimination. This was done by simulating the squared difference (Delta-response) between the models. We have also demonstrated how experimental noise inflate parameter precision, and how to use the *VIF* for experimental design.

Solution I - Ordinary E_{max} model – (0 - 500 µg/L data)

```
MODEL
  COMM
   NPARM  3
```

```
     PNAMES 'EMAX', 'EC50','E0'
END
     1: TEMP
     2: C=X
     3: END
     4: FUNC1
     5: F= E0-(EMAX*(C**1)/((EC50**1)+(C**1)))
     6: END
     7: EOM
NOPAGE BREAKS
NVARIABLES 4
NPOINTS 100
XNUMBER 1
YNUMBER 2
METHOD 2  'Gauss-Newton (Levenberg and Hartley)
ITERATIONS 50
INITIAL 35,120,175
LOWER BOUNDS 0,0,0
UPPER BOUNDS 75,500,250
MISSING 'Missing'
NOBSERVATIONS 11
DATA 'WINNLIN.DAT'
BEGIN
```

PARAMETER	ESTIMATE	STANDARD ERROR	CV%	UNIVARIATE C.I.	
EMAX	56.095345	7.367932	13.13	38.672815	73.517875
EC50	230.863117	86.900291	37.64	25.374931	436.351302
E0	175.764401	2.461572	1.40	169.943661	181.585142

```
*** CORRELATION MATRIX OF THE ESTIMATES ***
PARAMETER  EMAX       EC50        E0
EMAX       1.00000
EC50        .82370   1.00000
E0         -.31070    -.76183    1.00000

Condition_number=        114.0
```

X	OBSERVED Y	PREDICTED Y	RESIDUAL	WEIGHT	SE-PRED	STANDARDI RESIDUAL
10.00	173.0	173.4	-.4355	1.000	1.938	-.2624
25.00	168.0	170.3	-2.283	1.000	1.399	-1.070
50.00	169.0	165.8	3.222	1.000	1.033	1.381
75.00	162.0	162.0	-.9389E-02	1.000	1.033	-.0040
90.00	160.0	160.0	-.3002E-01	1.000	1.075	-.0129
120.0	160.0	156.6	3.421	1.000	1.136	1.498
150.0	153.0	153.7	-.6717	1.000	1.149	-.2949
200.0	146.0	149.7	-3.726	1.000	1.131	-1.629
300.0	143.0	144.1	-1.064	1.000	1.279	-.4821
500.0	139.0	137.4	1.612	1.000	2.236	1.312
.	.	.	.	.	.	.

```
CORRECTED SUM OF SQUARED OBSERVATIONS =  23714.0
  WEIGHTED CORRECTED SUM OF SQUARED OBSERVATIONS =  1220.10
  SUM OF SQUARED RESIDUALS =          45.5503
  SUM OF WEIGHTED SQUARED RESIDUALS =  45.5503
  S =  2.55092     WITH      7 DEGREES OF FREEDOM
  CORRELATION (OBSERVED,PREDICTED) =   .2226
```

```
AIC criteria =        44.18817
SC  criteria =        45.09592
```

Solution II - Sigmoid E_{max} model - (0 - 500 µg/L data)

```
MODEL
 COMM
  NPARM 4
  PNAMES 'EMAX', 'EC50','E0', 'n1'
END
    1: TEMP
    2: C=X
    3: END
    4: FUNC1
    5: F= E0-(EMAX*(C**N1)/((EC50**N1)+(C**N1)))
    6: END
    7: EOM
NOPAGE BREAKS
NVARIABLES 6
NPOINTS 100
XNUMBER 1
YNUMBER 2
METHOD 2  'Gauss-Newton (Levenberg and Hartley)
ITERATIONS 50
INITIAL 35,120,175,2.5
LOWER BOUNDS 0,0,0,0
UPPER BOUNDS 75,500,250,10
MISSING 'Missing'
NOBSERVATIONS 11
DATA 'WINNLIN.DAT'
BEGIN
```

PARAMETER	ESTIMATE	STANDARD ERROR	CV%	UNIVARIATE C.I.	
EMAX	35.661611	5.428655	15.22	22.378161	48.945061
EC50	143.081929	22.896067	16.00	87.057232	199.106626
E0	171.509507	1.958718	1.14	166.716693	176.302321
N1	1.946246	.537938	27.64	.629958	3.262533

```
*** CORRELATION MATRIX OF THE ESTIMATES ***
PARAMETER  EMAX         EC50        E0          N1
EMAX        1.00000
EC50         .647939      1.00000
E0           .681314     -.407023E-01  1.00000
N1          -.885538     -.563235     -.639722    1.00000


  Condition_number=       94.37

    FUNCTION   1
```

X	OBSERVED Y	PREDICTED Y	RESIDUAL	WEIGHT	SE-PRED	STANDARDI RESIDUAL
10.00	173.0	171.3	1.690	1.000	1.764	1.215
25.00	168.0	170.4	-2.353	1.000	1.331	-1.300
50.00	169.0	167.4	1.571	1.000	1.207	.8292
75.00	162.0	163.6	-1.612	1.000	1.266	-.8681
90.00	160.0	161.2	-1.218	1.000	1.204	-.6422

120.0	160.0	156.7	3.298	1.000	1.124	1.695
150.0	153.0	152.9	.1400	1.000	1.243	.074
200.0	146.0	148.1	-2.065	1.000	1.410	-1.180
300.0	143.0	142.7	.3264	1.000	1.341	.1811
500.0	139.0	138.7	.2802	1.000	2.051	.3051
.	.	.	.	.	.	.

```
CORRECTED SUM OF SQUARED OBSERVATIONS =  23714.0
  WEIGHTED CORRECTED SUM OF SQUARED OBSERVATIONS =  1220.10
  SUM OF SQUARED RESIDUALS =            30.2892
  SUM OF WEIGHTED SQUARED RESIDUALS =  30.2892
  S =  2.24682      WITH      6 DEGREES OF FREEDOM
  CORRELATION (OBSERVED,PREDICTED) =  .2240

 AIC criteria =          42.10791
 SC  criteria =          43.31825
```

Solution III - Ordinary E_{max} model - (0 - 800 μg/L data)

```
MODEL
 COMM
  NPARM 3
  PNAMES 'EMAX', 'EC50','E0'
END
    1: TEMP
    2: C=X
    3: END
    4: FUNC1
    5: F= E0-(EMAX*C)/(EC50+C))
    6: END
    7: EOM
NOPAGE BREAKS
NVARIABLES 6
NPOINTS 100
XNUMBER 1
YNUMBER 2
METHOD 2  'Gauss-Newton (Levenberg and Hartley)
ITERATIONS 50
INITIAL 35,120,175
LOWER BOUNDS 0,0,0
UPPER BOUNDS 75,500,250
MISSING 'Missing'
NOBSERVATIONS 11
DATA 'WINNLIN.DAT'
BEGIN
```

PARAMETER	ESTIMATE	STANDARD ERROR	CV%	UNIVARIATE C.I.	
EMAX	49.838139	3.961775	7.95	40.702182	58.974095
EC50	175.094239	53.541545	30.58	51.626022	298.562455
E0	176.539041	2.684978	1.52	170.347411	182.730671

```
*** CORRELATION MATRIX OF THE ESTIMATES ***
PARAMETER EMAX        EC50        E0
EMAX          1.00000
EC50          .471355     1.00000
E0            .116444    -.778778      1.00000
```

```
Condition_number=        71.43
```

X	OBSERVED	PREDICTED	RESIDUAL	WEIGHT	SE-PRED
STANDARDIZED					
	Y	Y			RESIDUAL
10.00	173.0	173.8	-.8465	1.000	2.077	-.4844
25.00	168.0	170.3	-2.312	1.000	1.476	-1.015
50.00	169.0	165.5	3.531	1.000	1.082	1.418
75.00	162.0	161.6	.4068	1.000	1.072	.1631
90.00	160.0	159.6	.3811	1.000	1.108	.1538
120.0	160.0	156.3	3.728	1.000	1.163	1.520
150.0	153.0	153.5	-.5435	1.000	1.176	-.2221
200.0	146.0	150.0	-3.965	1.000	1.149	-1.612
300.0	143.0	145.1	-2.069	1.000	1.133	-.8386
500.0	139.0	139.6	-.6271	1.000	1.488	-.2762
800.0	138.0	135.7	2.350	1.000	2.125	1.391

```
CORRECTED SUM OF SQUARED OBSERVATIONS =  1558.73
WEIGHTED CORRECTED SUM OF SQUARED OBSERVATIONS =  1558.73
SUM OF SQUARED RESIDUALS =          58.9534
SUM OF WEIGHTED SQUARED RESIDUALS =  58.9534
S =  2.71462     WITH     8 DEGREES OF FREEDOM
CORRELATION (OBSERVED,PREDICTED) =  .9809

AIC criteria =        50.84423
SC  criteria =        52.03791
```

Solution IV - Sigmoid E_{max} model - (0 - 800 µg/L data)

```
MODEL
 COMM
  NPARM 4
  PNAMES 'EMAX', 'EC50','E0', 'n1'
END
    1: TEMP
    2: C=X
    3: END
    4: FUNC1
    5: F= E0-(EMAX*(C**N1)/((EC50**N1)+(C**N1)))
    6: END
    7: EOM
NOPAGE BREAKS
NVARIABLES 6
NPOINTS 100
XNUMBER 1
YNUMBER 2
METHOD 2  'Gauss-Newton (Levenberg and Hartley)
ITERATIONS 50
INITIAL 35,120,175,2.5
LOWER BOUNDS 0,0,0,0
UPPER BOUNDS 75,500,250,10
MISSING 'Missing'
NOBSERVATIONS 11
DATA 'WINNLIN.DAT'
BEGIN
```

PARAMETER	ESTIMATE	STANDARD ERROR	CV%	UNIVARIATE C.I.	
EMAX	34.659731	3.301427	9.53	26.853034	42.466428
EC50	139.607423	14.929720	10.69	104.303963	174.910884

| EO | 171.358699 | 1.724292 | 1.01 | 167.281362 | 175.436035 |
| N1 | 2.030527 | .432678 | 21.31 | 1.007398 | 3.053656 |

```
*** CORRELATION MATRIX OF THE ESTIMATES ***
PARAMETER  EMAX        EC50       EO         N1
EMAX           1.00000
EC50            .283770    1.00000
EO              .748289    -.317657   1.00000
N1             -.824721    -.273312   -.609477   1.00000
```

Condition_number=   61.80

X	OBSERVED Y	PREDICTED Y	RESIDUAL	WEIGHT	SE-PRED	STANDARDI RESIDUAL
10.00	173.0	171.2	1.805	1.000	1.594	1.332
25.00	168.0	170.3	-2.335	1.000	1.247	-1.390
50.00	169.0	167.5	1.474	1.000	1.076	.8210
75.00	162.0	163.7	-1.710	1.000	1.156	-.9804
90.00	160.0	161.3	-1.279	1.000	1.126	-.7253
120.0	160.0	156.7	3.329	1.000	1.043	1.835
150.0	153.0	152.8	.2322	1.000	1.112	.1310
200.0	146.0	148.0	-1.971	1.000	1.274	-1.187
300.0	143.0	142.8	.2488	1.000	1.198	.1450
500.0	139.0	139.1	-.1166	1.000	1.235	-.0690
800.0	138.0	137.7	.3284	1.000	1.657	.2571

```
CORRECTED SUM OF SQUARED OBSERVATIONS =  1558.73
WEIGHTED CORRECTED SUM OF SQUARED OBSERVATIONS =  1558.73
SUM OF SQUARED RESIDUALS =          30.6446
SUM OF WEIGHTED SQUARED RESIDUALS =  30.6446
S =  2.09232      WITH      7 DEGREES OF FREEDOM
CORRELATION (OBSERVED,PREDICTED) =  .9901
```

```
AIC criteria =        45.64701
SC  criteria =        47.23859
```

Solution V - Delta function

```
TITLE 1
Effect model "Inhibiting SIGMOID E-MAX"
MODEL
 COMM
NPARM 4
PNAMES 'EMAX', 'EC50', 'n','EO'
END
    1: TEMP
    2: C=X
    3: END
    4: FUNC1
    5: E1 = 176.5-(49.84*(C**1)/((175.1**1)+(C**1)))
    6: E2= 171.36-(34.66*(C**2.03)/((139.61**2.03)+(C**2.03)))
    7: F = (E1 - E2)**2
    8: END
    9: EOM
NOPAGE BREAKS
SIMULATE
NVARIABLES 2
NPOINTS 500
XNUMBER 1
INITIAL 35,120,2.5,175
```

```
NOBSERVATIONS 11
BEGIN

The following default parameter boundaries were generated.
Parameter    Lower Bound    Upper Bound
  EMAX           .0000          350.0
  EC50           .0000          1200.
  N              .0000          25.00
  E0             .0000          1750.
```

X	CALCULATED Y	VAR - INF FACTOR
10.00	6.817	.0000
25.00	.3914E-02	.0000
50.00	4.396	.0000
75.00	4.645	.0000
90.00	2.888	.0000
120.0	.1925	.0000
150.0	.5394	.0000
200.0	3.812	.0000
300.0	5.172	.0000
500.0	.2188	.0000
800.0	4.259	.0000

Solution VI - Variance inflation factor

```
MODEL
 COMM
NPARM 4
PNAMES 'EMAX', 'EC50', 'N','E0'
END
    1: TEMP
    2: EMAX=P(1)
    3: EC50=P(2)
    4: N=P(3)
    5: E0=P(4)
    6: C=X
    7: END
    8: FUNC1
    9: F= E0-(EMAX*(C**N)/((EC50**N)+(C**N)))
   10: END
   11: EOM
SIMULATE
NVARIABLES 1
NPOINTS 100
XNUMBER 1
INITIAL 34.66,139.61,2.031,171.36
NOBOUNDS
NOBSERVATIONS 10
BEGIN
```

PARAMETER	ESTIMATE	VAR - INF FACTOR	[SQRT(VIF)/P]%
EMAX	34.660000	4.955975	6.42
EC50	139.610000	85.775751	6.63
N	2.031000	.060664	12.13
E0	171.360000	.718233	.49

X	CALCULATED Y	VAR - INF FACTOR
10.00	171.2	.6025

```
25.00      170.3      .3553
50.00      167.5      .2869
75.00      163.7      .3228
90.00      161.3      .2937
120.0      156.7      .2563
150.0      152.8      .3143
200.0      148.0      .3958
300.0      142.8      .3526
500.0      139.1      .8199
```

THE SAME MODEL BUT WITH AN EXTRA OBSERVATION AT 800 µG/L.

NOBS 11

PARAMETER	ESTIMATE	VAR - INF FACTOR	[SQRT(VIF)/P]%
EMAX	34.660000	2.488349	4.55
EC50	139.610000	50.884576	5.11
N	2.031000	.042771	10.18
E0	171.360000	.678907	.48

   FUNCTION   1

X	CALCULATED Y	VAR - INF FACTOR
10.00	171.2	.5806
25.00	170.3	.3552
50.00	167.5	.2642
75.00	163.7	.3054
90.00	161.3	.2895
120.0	156.7	.2486
150.0	152.8	.2823
200.0	148.0	.3710
300.0	142.8	.3276
500.0	139.1	.3485
800.0	137.7	.6271

PD4 - Indirect response model I – Bolus dosing

Objectives

◆ **To demonstrate an indirect pharmacological response model**

◆ **To model the warfarin - PCA interaction**

◆ **To elaborate on a model with and without a lag-time in the dynamics**

◆ **To compare the two different analyses with lag-time**

Problem specification

In the following example, we will demonstrate how to model the interaction between the plasma concentration of (s)-warfarin (following an intravenous bolus dose) and the prothrombin complex activity (*PCA*). Figure 4.1 shows schematically the action of warfarin on vitamin K reductase. The indirect action of warfarin on the prothrombin complex activity R is also presented in Figure 4.3.

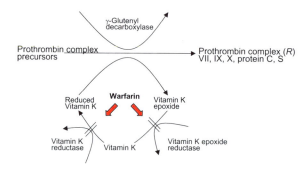

Figure 4.1 *Schematic illustration of the action of warfarin on vitamin K (epoxide) reductase.*

Figure 4.2 shows the time course of observed *PCA* after a bolus dose of warfarin (see also Nagashima *et al* [1969], Pitsui *et al* [1990]).

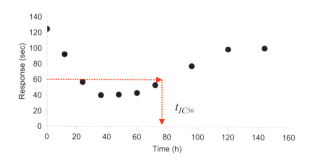

Figure 4.2 *Observed PCA time course following administration of an intravenous bolus dose of (s)-warfarin (C_w). The red dashed line indicates the approximate time at which the plasma concentration is equal to IC_{50}. That time value is inserted into Equation 4:1 to obtain an initial estimate of IC_{50}.*

The plasma kinetics of (s)-warfarin C_w, can be described by the following mono-exponential expression after an intravenous bolus dose

$$C_w = 1.05 \cdot e^{-0.0228 \cdot t} \qquad (4:1)$$

The inhibitory function of warfarin on the synthesis rate of PCA is written as

$$I(C_w) = 1 + \left[\frac{C_w}{IC_{50}} \right]^n \qquad (4:2)$$

The turnover rate of PCA is given by Equation 4:3

$$\frac{dPCA}{dt} = \frac{dP}{dt} = \frac{k_{in}}{I(C_w)} - k_{out} \cdot P \qquad (4:3)$$

where P is the response (i.e., PCA).

The fractional turnover rate k_{out} is equivalent to k_d in the article by Pitsui et al [1993]. At baseline, P_0, Equation 4:3 becomes

$$\frac{dP}{dt} = k_{in} - k_{out} \cdot P = 0 \qquad (4:4)$$

$$P_0 = \frac{k_{in}}{k_{out}} \qquad (4:5)$$

Equation 4:5 rearranged gives

$$k_{in} = P_0 \cdot k_{out} \qquad (4:6)$$

The expression for k_{in} is then inserted into Equation 4:3, which gives

$$\frac{dP}{dt} = \frac{P_0 \cdot k_{out}}{I(C_w)} - k_{out} \cdot P = k_{out} \cdot \left[\frac{P_0}{I(C_w)} - P \right] \qquad (4:7)$$

The k_{out} parameter can be obtained from a plot of the slope of the natural logarithm of P, denoted as $ln(P)$, versus time after administration of a synthesis blocking dose of warfarin (Nagashima et al [1969], Pitsui et al [1993]). P_0 is the baseline value of the prothrombin time, C_w the concentration of (s)-warfarin and IC_{50} the concentration of warfarin at 50% of the maximal blocking effect. We will also estimate the half-life of the apparent first-order degradation rate constant of prothrombin, $t_{1/2}$ of k_{out}. Professor Leon Aarons, University of Manchester, UK kindly supplied the data and model. Data were then slightly adapted.

An alternative model that includes a lag-time to allow for the observed time delay in the onset of the effect after warfarin administration was published by Pitsui *et al* [1993]. In their analysis, the baseline value of clotting factor activity in the absence of warfarin (P_0) was set to a fixed mean of three pre-dose measurements, whereas we let the program estimate that parameter. Figure 4.3 shows a schematic diagram of the indirect response model of the warfarin-*PCA* interaction, and of the action of warfarin.

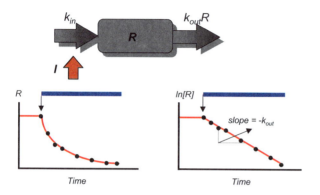

Figure 4.3 Schematic illustration of the indirect response model. R denotes the prothombin complex activity and corresponds to P in Equations 4:3-7. Note that total blockage by means of warfarin of the production (k_{in}) of R (or P) results in a first-order decline determined by means of the fractional turnover rate constant k_{out}. The slope of the line is $-k_{out}$.

The intensity of a pharmacological response is not necessarily due to a direct effect of the drug on the receptor. Rather, it may be the net result of several processes, only one of which is influenced by the drug. The process that is influenced by the drug must be identified and an attempt made to relate plasma drug concentration to changes in this process. A good example is the anticoagulant (hypothrombinemic) effect of the coumarin drugs (warfarin). These inhibit the synthesis of certain vitamin K-dependent clotting factors, and thus indirectly affect blood clotting. They do not exert any direct effect on the clotting cascade. Therefore, as the real effect of these drugs is inhibition of synthesis rate, any correlation with plasma concentration should be based on this effect rather than on the degree of inhibition of clotting time (Nagashima *et al* [1969], Gibaldi and Perrier [1982]). The work of Nagashima *et al* [1969] is covered in more detail in section 4.6.2.

Initial parameter estimates

From the intercept of the response-time curve with the effect-axis (Figure 4.4), the baseline value (120 sec) can be obtained. This value (P_0) is the ratio of k_{in}/k_{out}, if we assume Equation 4:5 to be an appropriate description of the drug-free state. From the intercept and slope $-k_{out}$, k_{in} can then be calculated (3.6 sec/h) according to Equation 4:6. The plasma concentration at the trough of the effect corresponds approximately to the IC_{50}-value.

IC_{50}	$= 0.35$ (mg/L)	
k_{out}	$= 0.3$ (h^{-1})	Deviates from Equation 4:7 estimate
n	$= 3.5$	Gamma in model code
P_0	$= 130$ (sec)	Expressed as *F0* in model code
t_{lag}	$= 1$ (h)	

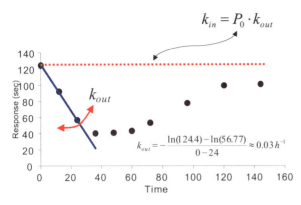

$$k_{in} = P_0 \cdot k_{out}$$

Figure 4.4 *Observed (●) PCA time-course following the administration of an intravenous bolus dose of warfarin. Solid blue line illustrates how the initial estimate of the k_{out} parameter is derived. The curved arrow indicates how k_{out} determines the slope.*

$$k_{out} = -\frac{\ln(124.4) - \ln(56.77)}{0 - 24} \approx 0.03\,h^{-1}$$

The k_{out} parameter is obtained from the slope of the downswing of the curve

$$k_{out} = -\frac{\ln(124.4) - \ln(56.77)}{0 - 24} \approx 0.03\,\text{h}^{-1} \qquad (4:8)$$

Note that although a more acceptable initial parameter value for k_{out} is 0.03 h^{-1}, we deliberately chose 0.3 in order to demonstrate that the program can handle an initial parameter estimate that errs by a factor 10. IC_{50} is either obtained from a plot of plasma concentration *versus* response, or by insertion t_{IC50} from Figure 4.1 into Equation 4:1.

Interpretation of results and conclusions

Figure 4.5 displays observed data together with predicted curves of the lag-time and non-lag-time models and shows schematically how the addition of a lag-time will tilt k_{out} as demonstrated by the blue line. Table 4.1 gives the final parameter estimates of the two models together with the parameter precision. Note that the parameter precision increased particularly for n when we included the lag-time.

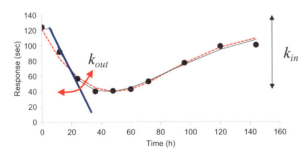

Table 4.1 Final parameter estimates

	No lag-time mean ± CV%	Lag-time mean ± CV%
P_0	120 ± 4	120 ± 4
IC_{50}	0.26 ± 9	0.24 ± 8
k_{out}	0.03 ± 18	0.06 ± 17
n	1.6 ± 44	1.6 ± 17
t_{lag}	-	6.6 ± 18

Figure 4.5 *Schematical illustration of the impact of a change in k_{in} and k_{out}. A change in k_{in} will shift the whole curve vertically. An increase in k_{out} will increase the slope (blue line) of the downswing. The red dashed line represents the model without a lag-time and the solid black line the model with a lag-time included.*

Since we used constant weights (weight = 1) and the models are nested (setting lag-time to zero in the lag-time (full) model reduces it to the simpler model without a lag-time), one could apply the *F*-test. However, this is not necessary, as we can in this example discriminate between the models by inspecting the residuals and parameter precision. The +/- signs of the residuals changed three times in the model without lag-time compared to seven times in the lag-time model, indicating that the latter has a more random error pattern. This, together with its higher parameter precision (i.e., reduced CVs), suggests that the lag-time model provided a better fit of the data than did the model without lag-time.

An alternative model, including a lag-time to allow for the observed time delay in the onset of the effect after warfarin administration, was first published by Pitsui *et al* [1993]. The lag-time may be interpreted as the time for the clotting cascade to develop (Ideally, this lag-time should not affect the kinetics, a fact that could easily have been incorporated into the model. We reasoned that t_{lag} was very small in comparison to the 31h half-life of warfarin in this subject). Their baseline value of clotting factor activity P_0 in the absence of warfarin was set to a fixed mean of three pre-dose measurements, whereas we let the program estimate that parameter to 121 seconds. Even though their parameterization was somewhat different, the results from our analysis are consistent with their analysis.

Solution I - No lag-time on dynamics

```
MODEL
 COMM
 NDER   1
 NPARM 4
 NSEC   1
 PNAMES 'IC50', 'kout', 'n', 'P0'
 SNAMES 'T12kd'
END
     1: TEMP
     2: CW0 = CON(1)
     3: K = CON(2)
     4: T=X
     5: END
     6: START
     7: Z(1)=P0
     8: END
     9: DIFF
    10: CW=CW0*DEXP(-K*T)
    11: DZ(1)=KOUT*((P0/(1. + (CW/IC50)**N)) - Z(1))
    12: END
    13: FUNC 1
    14: F = Z(1)
    15: END
    16: SECO
    17: T12KD = 0.693/KOUT
    18: END
    19: EOM
NVARIABLES 2
NPOINTS 100
XNUMBER 1
YNUMBER 2
NCONSTANTS 2
CONSTANTS 1.0511,.0228
METHOD 2   'Gauss-Newton (Levenberg and Hartley)
ITERATIONS 50
INITIAL .35,.3,3.5,130
```

```
LOWER BOUNDS 0,0,1,100
UPPER BOUNDS .5,1,5,150
MISSING 'Missing'
NOBSERVATIONS 10
DATA 'WINNLIN.DAT'
BEGIN
```

PARAMETER	ESTIMATE	STANDARD ERROR	CV%	UNIVARIATE C.I.	
IC50	.261436	.022207	8.49	.207099	.315774
KOUT	.030793	.005374	17.45	.017645	.043942
N	2.983054	1.316700	44.14	-.238797	6.204905
P0	120.972022	4.532766	3.75	109.880736	132.063307

```
*** CORRELATION MATRIX OF THE ESTIMATES ***
PARAMETER  IC50       KOUT       N          P0
IC50       1.00000
KOUT       -.48913    1.00000
N           .55799    -.91243    1.00000
P0         -.75143     .57125    -.58259    1.00000

 Condition_number=        2141.

    FUNCTION    1
```

X	OBSERVED Y	PREDICTED Y	RESIDUAL	WEIGHT	SE-PRED	STANDARDI RESIDUAL
.0000	124.4	121.0	3.438	1.000	4.533	.9934
12.00	92.00	84.51	7.491	1.000	3.466	1.654
24.00	56.77	60.39	-3.623	1.000	3.753	-.8437
36.00	40.01	45.95	-5.935	1.000	2.874	-1.205
48.00	40.78	40.05	.7299	1.000	3.120	.1529
60.00	42.98	42.26	.7217	1.000	3.689	.1659
72.00	53.00	51.20	1.796	1.000	3.436	.3945
96.00	77.49	76.96	.5316	1.000	4.236	.1392
120.0	99.22	97.46	1.758	1.000	3.301	.3780
144.0	100.8	109.2	-8.438	1.000	3.347	-1.827

```
CORRECTED SUM OF SQUARED OBSERVATIONS =  8174.91
WEIGHTED CORRECTED SUM OF SQUARED OBSERVATIONS =  8174.91
SUM OF SQUARED RESIDUALS =          195.145
SUM OF WEIGHTED SQUARED RESIDUALS =  195.145
S =  5.70299    WITH     6 DEGREES OF FREEDOM
CORRELATION (OBSERVED,PREDICTED) =   .9881

AIC criteria =         60.73741
SC  criteria =         61.94775
```

Solution II - Including lag-time on dynamic

```
MODEL
 COMM
 NDER  1
 NPARM 5
 NSEC  1
 PNAMES 'IC50', 'kout', 'n', 'P0', 'Tlag'
 SNAMES 'T12kd'
```

```
END
   1: TEMP
   2: CW0 = CON(1)
   3: K = CON(2)
   4: T=X
   5: END
   6: START
   7: Z(1)=P0
   8: END
   9: DIFF
  10: IF T LE TLAG THEN
  11: CW = CW0
  12: DZ(1) = 0.0
  13: ELSE
  14: CW=CW0*DEXP(-K*T)
  15: DZ(1)=KOUT*((P0/(1. + (CW/IC50)**N)) - Z(1))
  16: ENDIF
  17: END
  18: FUNC 1
  19: F = Z(1)
  20: END
  21: SECO
  22: T12KD = 0.693/KOUT
  23: END
  24: EOM
NVARIABLES 2
NPOINTS 100
XNUMBER 1
YNUMBER 2
WEIGHT -1
NCONSTANTS 2
CONSTANTS 1.0511,.0228
METHOD 2   'Gauss-Newton (Levenberg and Hartley)
ITERATIONS 50
INITIAL .35,.3,3.5,130,1
LOWER BOUNDS 0,0,1,110,0
UPPER BOUNDS .5,1,5,150,10
MISSING 'Missing'
NOBSERVATIONS 10
DATA 'WINNLIN.DAT'
BEGIN
```

PARAMETER	ESTIMATE	STANDARD ERROR	CV%	UNIVARIATE C.I.	
IC50	.237748	.019291	8.11	.188159	.287337
KOUT	.058276	.009674	16.60	.033409	.083143
N	1.582863	.266528	16.84	.897740	2.267986
P0	121.425114	4.545675	3.74	109.740254	133.109974
TLAG	6.581771	1.154190	17.54	3.614873	9.548669

```
*** CORRELATION MATRIX OF THE ESTIMATES ***
PARAMETER  IC50       KOUT       N          P0         TLAG
IC50       1.00000
KOUT      -.440217    1.00000
N          .705175   -.810646    1.00000
P0        -.842869    .275210   -.567705    1.00000
TLAG       .203306    .472107   -.038038   -.451855    1.00000
```

```
Condition_number=          1297.

  FUNCTION    1

X      OBSERVED   PREDICTED   RESIDUAL    WEIGHT    SE-PRED   STANDARDI
          Y          Y                                          RESIDUAL
 .0000     124.4      121.4      2.985      .4990     4.546      1.675
 12.00     92.00      92.42     -.4241      .6748     4.118     -.5167
 24.00     56.77      55.58      1.188     1.094      2.455      .5393
 36.00     40.01      41.32     -1.311     1.552      1.844     -.6348
 48.00     40.78      39.35      1.426     1.522      1.557      .6140
 60.00     42.98      44.40     -1.423     1.444      1.789     -.6339
 72.00     53.00      53.43     -.4324     1.171      1.888     -.1684
 96.00     77.49      75.63      1.862      .8012     2.340      .6082
 120.0     99.22      95.05      4.166      .6257     2.712      1.220
 144.0     100.8      107.9     -7.075      .6159     2.671     -2.027
```

CORRECTED SUM OF SQUARED OBSERVATIONS = 8174.91
WEIGHTED CORRECTED SUM OF SQUARED OBSERVATIONS = 6620.31
SUM OF SQUARED RESIDUALS =   87.3447
SUM OF WEIGHTED SQUARED RESIDUALS = 59.4836
S = 3.44916   WITH   5 DEGREES OF FREEDOM
CORRELATION (OBSERVED,PREDICTED) = .9946

AIC criteria =   50.85700
SC criteria =   52.36993

PARAMETER	ESTIMATE	STANDARD ERROR	CV%
T12KD	11.891701	1.972042	16.58

PD5 - Indirect response model II – Topical dosing

Objectives

◆ **To set up a model in which drug stimulates production of the response**

◆ **To compare results with a direct (instantaneous) response model**

Problem specification

The aim of this exercise is to practise indirect response modeling using a biomarker for the exposure to compound X in the eye. Data were obtained from a study on the miotic effects in the cat (see also PD22 for more information).

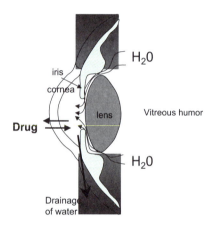

Figure 5.1 *Schematic illustration of the action of drug X in removing water from the different intraocular spaces. The cross sectional diameter of the pupil is a surrogate marker for the presence of drug in the eye and, indirectly, for the action of the drug.*

Data are shown in Figure 5.2 and program output.

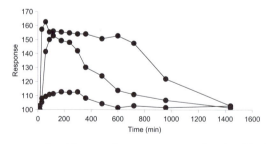

Figure 5.2 *Observed response (●) versus time data after three doses (0.1, 1.0 and 10 μg) of drug X instilled into the cat eye. Observed data points are connected by means of straight lines.*

In this exercise, we assume that instillation of the drug into the eye can be approximated by means of a bolus model (Equation 5:1). We will assume a first-order input-output model in a later exercise (PD22*)*.

$$C = \frac{D}{V} \cdot e^{-K \cdot t} = A = D \cdot e^{-K \cdot t} \tag{5:1}$$

V is ignored because we assume the volume to be equal to unity. Drug is then assumed to stimulate the factors controlling drug response through the stimulation function $H(A)$ in Equation 5:2.

$$H(A) = 1 + \frac{E_{max} \cdot A^n}{ED_{50}^n + A^n} \tag{5:2}$$

E_{max} is the maximum effect attributed to the drug and EC_{50} the dose producing 50% of the maximum stimulation achieved at the effect site. Drug response R, which originates from stimulation of input or through production of the response, is as follows

$$\frac{dR}{dt} = k_{in} \cdot H(A) - k_{out} \cdot R \tag{5:3}$$

Equation 5:3 was then simultaneously fit to effect-time data obtained at three dosage levels.

Drug could also inhibit the factors controlling loss of response through the inhibition function $H(A)$ in Equation 5:4.

$$H(A) = 1 - \frac{I_{max} \cdot A^n}{ID_{50}^n + A^n} \tag{5:4}$$

I_{max} is the maximum fractional inhibition attributed to the drug and ID_{50} the dose producing 50% of the maximum inhibition achieved at the effect site. Drug response then becomes

$$\frac{dR}{dt} = k_{in} - k_{out} \cdot R \cdot H(A) \tag{5:5}$$

The time to pharmacodynamic steady state, assuming a constant drug level (A_{ss}), will then be governed by $I(A_{ss}) \cdot k_{out}$. Equation 5:5 was then simultaneously fit to effect-time data obtained at three dosage levels.

Initial parameter estimates

Model 2: Stimulatory model
$k_{in} = 3$ (units/min)
$k_{out} = 0.1$ (min^{-1})
$E_{max} = 0.8$
$ED_{50} = 0.01$ (µg)
$n = 1.5$
$K = 0.01$ Decay of dose (min^{-1})

Model 3: Inhibitory model
$k_{in} = 3$ (units/min)
$k_{out} = 0.1$ (min^{-1})
$I_{max} = 0.2$
$IC_{50} = 0.01$ (µg)
$n = 1.5$
$K = 0.01$ Decay of dose (min^{-1})

Interpretation of results and conclusions

Figure 5.3 displays the observed and predicted effects using the stimulatory indirect response model (Model 2, Table 5.1) and 'direct' response model (Model 1, Table 5.1). The 'direct' response model will be discussed in more detail in a later chapter (PD22). We obtained parameter estimates with reasonable precision. For this dataset, the 'direct' response model seemed to mimic data somewhat better than the indirect response models (Table 5.1).

An inhibitory indirect response model was also fit to the data (Model 3, Table 5.1). The inhibitory function works on the loss of response including a maximum fractional inhibition parameter I_{max}.

Table 5.1 Final parameter estimates of the three models.

Parameter	'Direct' response model: 1 (mean ± CV%)	Indirect response model: 2 (mean ± CV%)	Indirect response model: 3 (mean ± CV%)
E_{max}/I_{max}	0.56 ± 2.1	0.60 ± 12	0.36 ± 8
ED_{50}/ID_{50}	0.17 ± 12	0.22 ± 28	0.15 ± 30
k_{in}	-	2.5 ± 22	3.8 ± 17
k_{out}	-	0.024 ± 18	0.037 ± 18
K	-	0.0046 ± 11	0.0045 ± 11
n	1.30 ± 7.5	1.1 ± 22	1.27 ± 23.3
$WRSS$	421	2284	2382

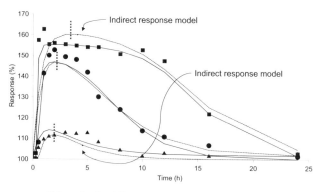

Figure 5.3 Observed (symbols) and model-predicted miotic effect. The 'direct' response model peaks at the same time for each dose. The blue vertical bars indicate the shift in peak values for the indirect response model.

Although the 'direct' response model seemed to mimic these data better than the indirect response model, it may be concluded that *indirect response* (turnover) models may be relevant in modeling where production or loss of an endogenous substance or mediator is immediately or sequentially responsible for the action of drugs. Starting points in such modeling should include an understanding of the mechanism of action of the drug, a search for the appropriate biophase, and consideration of the types of basic indirect response models shown to be useful for diverse drugs.

Solution - Indirect response model 2 (Stimulatory model)

```
MODEL
  COMM
    NDER   3
    NFUN   3
```

```
  NPAR   6
  NCON   4
PNAMES 'Kin','Kout','Emax','EC50', 'N', 'K'
END
    1: TEMP
    2: DOSE1=CON(1)
    3: DOSE2=CON(2)
    4: DOSE3=CON(3)
    5: E0=CON(4)
    6: KIN=P(1)
    7: KOUT=P(2)
    8: EMAX=P(3)
    9: EC50=P(4)
   10: N=P(5)
   11: K=P(6)
   12: TL = X
   13: END
   14: START
   15: Z(1)= E0
   16: Z(2)= E0
   17: Z(3)= E0
   18: END
   19: DIFF
   20: C1 = DOSE1*DEXP(-K*TL)
   21: DRUG1 = 1. + EMAX*(C1**N)/(EC50**N + C1**N)
   22: DZ(1)=KIN*DRUG1 - KOUT*Z(1)
   23: C2 = DOSE2*DEXP(-K*TL)
   24: DRUG2 = 1. + EMAX*(C2**N)/(EC50**N + C2**N)
   25: DZ(2)=KIN*DRUG2 - KOUT*Z(2)
   26: C3 = DOSE3*DEXP(-K*TL)
   27: DRUG3 = 1. + EMAX*(C3**N)/(EC50**N + C3**N)
   28: DZ(3)=KIN*DRUG3 - KOUT*Z(3)
   29: END
   30: FUNC 1
   31: F = Z(1)
   32: END
   33: FUNC 2
   34: F = Z(2)
   35: END
   36: FUNC 3
   37: F = Z(3)
   38: END
   39: EOM
NVARIABLES 4
FNUMBER 3
NPOINTS 100
XNUMBER 1
YNUMBER 4
CONSTANTS .1,1,10,100
METHOD 2  'Gauss-Newton (Levenberg and Hartley)
ITERATIONS 50
INITIAL 3,.1,.8,.01,1.5,.01
LOWER BOUNDS 0,0,0,0,0,0
UPPER BOUNDS 20,3,2,2,4,1
MISSING '-'
DATA 'WINNLIN.DAT'
BEGIN
```

PARAMETER	ESTIMATE	STANDARD ERROR	CV%	UNIVARIATE C.I.	
KIN	2.462731	.452788	18.39	1.548972	3.376490

KOUT	.024187	.004360	18.03	.015388	.032986
EMAX	.595855	.073753	12.38	.447016	.744694
EC50	.217319	.061905	28.49	.092390	.342248
N	1.147041	.255773	22.30	.630872	1.663209
K	.004627	.000525	11.35	.003567	.005687

FUNCTION   1

X	OBSERVED Y	PREDICTED Y	RESIDUAL	WEIGHT	SE-PRED	STANDARDI RESIDUAL
.1000E-02	100.0	100.0	-.4711E-03	1.000	.1091E-03	-.00006
5.000	102.6	102.2	.3706	1.000	.4970	.05036
15.00	102.6	105.8	-3.198	1.000	1.248	-.4399
30.00	105.3	109.5	-4.198	1.000	1.952	-.5903
60.00	109.4	113.1	-3.768	1.000	2.581	-.5454
90.00	110.8	114.0	-3.222	1.000	2.771	-.4714
120.0	111.3	113.6	-2.368	1.000	2.780	-.3466
180.0	112.7	111.6	1.042	1.000	2.563	.1507
240.0	112.2	109.5	2.760	1.000	2.271	.3934
300.0	112.6	107.6	5.000	1.000	2.051	.7059
360.0	108.2	106.2	2.046	1.000	1.957	.2877
480.0	104.3	104.2	.3871E-01	1.000	2.088	.0054
600.0	101.4	103.1	-1.711	1.000	2.378	-.2451
720.0	102.8	102.5	.2797	1.000	2.638	.0406
960.0	101.5	102.0	-.5249	1.000	2.943	-.0776
1440.	102.8	101.8	.9346	1.000	3.103	.1397

CORRECTED SUM OF SQUARED OBSERVATIONS = 318.500
WEIGHTED CORRECTED SUM OF SQUARED OBSERVATIONS = 318.500
SUM OF SQUARED RESIDUALS =   100.208
SUM OF WEIGHTED SQUARED RESIDUALS = 100.208
S = 3.16557   WITH   10 DEGREES OF FREEDOM
CORRELATION (OBSERVED,PREDICTED) = .8502

FUNCTION   2

X	OBSERVED Y	PREDICTED Y	RESIDUAL	WEIGHT	SE-PRED	STANDARDI RESIDUAL
.1000E-02	100.0	100.0	-.1294E-02	1.000	.2087E-03	-.00017
5.000	100.0	106.1	-6.084	1.000	.9204	-.8315
15.00	100.0	116.2	-16.18	1.000	2.150	-2.294
30.00	108.0	127.2	-19.25	1.000	2.972	-2.852
60.00	141.3	139.7	1.611	1.000	2.999	.2391
90.00	150.0	145.0	5.009	1.000	2.690	.7295
120.0	152.7	146.7	6.009	1.000	2.660	.8736
180.0	149.3	145.5	3.781	1.000	2.870	.5566
240.0	148.0	142.0	5.972	1.000	2.873	.8792
300.0	141.9	137.7	4.190	1.000	2.711	.6110
360.0	130.1	133.0	-2.875	1.000	2.548	-.4155
480.0	124.0	123.7	.2945	1.000	2.575	.0426
600.0	113.7	115.8	-2.143	1.000	2.670	-.3117
720.0	110.8	110.2	.6249	1.000	2.419	.0896
960.0	106.7	104.4	2.295	1.000	2.148	.3253
1440.	101.4	102.0	-.6740	1.000	2.929	-.0995

CORRECTED SUM OF SQUARED OBSERVATIONS = 6445.73

```
WEIGHTED CORRECTED SUM OF SQUARED OBSERVATIONS =  6445.73
SUM OF SQUARED RESIDUALS =              819.684
SUM OF WEIGHTED SQUARED RESIDUALS =  819.684
S = 9.05364      WITH     10 DEGREES OF FREEDOM
CORRELATION (OBSERVED,PREDICTED) =  .9439
```

```
  FUNCTION   3
```

X	OBSERVED Y	PREDICTED Y	RESIDUAL	WEIGHT	SE-PRED	STANDARDI RESIDUAL
.1000E-02	100.0	100.0	-.1493E-02	1.000	.2318E-03	-.0002
5.000	100.0	107.0	-7.033	1.000	1.021	-.9629
15.00	102.7	118.8	-16.08	1.000	2.381	-2.304
30.00	157.4	131.8	25.54	1.000	3.278	3.865
60.00	162.7	147.2	15.50	1.000	3.273	2.346
90.00	155.4	154.5	.9033	1.000	2.918	.1334
120.0	156.0	158.0	-1.952	1.000	2.891	-.2877
180.0	155.4	160.1	-4.736	1.000	3.137	-.7096
240.0	154.8	160.2	-5.382	1.000	3.154	-.8074
300.0	154.1	159.6	-5.519	1.000	2.994	-.8188
360.0	154.1	158.6	-4.544	1.000	2.780	-.6652
480.0	150.7	155.6	-4.880	1.000	2.499	-.7034
600.0	152.7	150.6	2.088	1.000	2.793	.3059
720.0	147.4	143.4	3.943	1.000	3.650	.6153
960.0	121.9	125.0	-3.096	1.000	5.260	-.5990
1440.	102.8	104.7	-1.889	1.000	2.508	-.2724

```
CORRECTED SUM OF SQUARED OBSERVATIONS =  8782.03
WEIGHTED CORRECTED SUM OF SQUARED OBSERVATIONS =  8782.03
SUM OF SQUARED RESIDUALS =              1364.36
SUM OF WEIGHTED SQUARED RESIDUALS =  1364.36
S = 11.6806      WITH     10 DEGREES OF FREEDOM
CORRELATION (OBSERVED,PREDICTED) =  .9194
```

```
TOTALS FOR ALL CURVES COMBINED
SUM OF SQUARED RESIDUALS =              2284.25
SUM OF WEIGHTED SQUARED RESIDUALS =  2284.25
S = 7.37475      WITH     42 DEGREES OF FREEDOM
```

Solution - Indirect response model 3 (Inhibitory model)

```
MODEL
 COMM
  NDER   3
  NFUN   3
  NPAR   6
  NCON   4
PNAMES 'Kin','Kout','Imax','IC50', 'N', 'K'
END
    1: TEMP
    2: DOSE1=CON(1)
    3: DOSE2=CON(2)
    4: DOSE3=CON(3)
    5: E0=CON(4)
    6: KIN=P(1)
    7: KOUT=P(2)
    8: IMAX=P(3)
    9: IC50=P(4)
   10: N=P(5)
   11: K=P(6)
```

```
12: TL = X
13: END
14: START
15: Z(1)= E0
16: Z(2)= E0
17: Z(3)= E0
18: END
19: DIFF
20: C1 = DOSE1*DEXP(-K*TL)
21: DRUG1 = 1. - IMAX*(C1**N)/(IC50**N + C1**N)
22: DZ(1)=KIN - DRUG1*KOUT*Z(1)
23: C2 = DOSE2*DEXP(-K*TL)
24: DRUG2 = 1. - IMAX*(C2**N)/(IC50**N + C2**N)
25: DZ(2)=KIN - DRUG2*KOUT*Z(2)
26: C3 = DOSE3*DEXP(-K*TL)
27: DRUG3 = 1. - IMAX*(C3**N)/(IC50**N + C3**N)
28: DZ(3)=KIN - DRUG3*KOUT*Z(3)
29: END
30: FUNC 1
31: F = Z(1)
32: END
33: FUNC 2
34: F = Z(2)
35: END
36: FUNC 3
37: F = Z(3)
38: END
39: EOM
NVARIABLES 4
FNUMBER 3
NPOINTS 100
XNUMBER 1
YNUMBER 4
CONSTANTS .1,1,10,100
METHOD 2  'Gauss-Newton (Levenberg and Hartley)
ITERATIONS 50
INITIAL 3,.1,.2,.01,1.5,.01
LOWER BOUNDS 0,0,0,0,0,0
UPPER BOUNDS 20,3,2,2,4,1
MISSING '-'
DATA 'WINNLIN.DAT'
BEGIN
```

PARAMETER	ESTIMATE	STANDARD ERROR	CV%	UNIVARIATE C.I.	
KIN	3.806746	.664851	17.47	2.465028	5.148463
KOUT	.037065	.006532	17.62	.023881	.050248
IMAX	.361471	.027894	7.72	.305180	.417762
EC50	.150775	.044484	29.50	.061002	.240548
N	1.271819	.295638	23.25	.675200	1.868438
K	.004531	.000506	11.18	.003509	.005553

```
    FUNCTION   1
```

X	OBSERVED Y	PREDICTED Y	RESIDUAL	WEIGHT	SE-PRED	STANDARDI RESIDUAL
.1000E-02	100.0	100.0	-.5991E-03	1.000	.1343E-03	-.00007

5.000	102.6	102.7	-.1760	1.000	.5979	-.0234
15.00	102.6	107.0	-4.382	1.000	1.447	-.5930
30.00	105.3	110.9	-5.608	1.000	2.183	-.7781
60.00	109.4	113.9	-4.559	1.000	2.796	-.6520
90.00	110.8	114.0	-3.215	1.000	2.938	-.4636
120.0	111.3	113.1	-1.812	1.000	2.866	-.2603
180.0	112.7	110.7	1.985	1.000	2.505	.2795
240.0	112.2	108.6	3.614	1.000	2.170	.5011
300.0	112.6	107.0	5.617	1.000	1.987	.7733
360.0	108.2	105.8	2.402	1.000	1.959	.3304
480.0	104.3	104.3	-.5445E-01	1.000	2.171	-.0075
600.0	101.4	103.5	-2.117	1.000	2.448	-.2972
720.0	102.8	103.1	-.3247	1.000	2.658	-.0460
960.0	101.5	102.8	-1.319	1.000	2.875	-.1895
1440.	102.8	102.7	.5764E-01	1.000	2.971	.0083

```
CORRECTED SUM OF SQUARED OBSERVATIONS =   318.500
WEIGHTED CORRECTED SUM OF SQUARED OBSERVATIONS =   318.500
SUM OF SQUARED RESIDUALS =              145.734
SUM OF WEIGHTED SQUARED RESIDUALS =    145.734
S =  3.81752     WITH     10 DEGREES OF FREEDOM
CORRELATION (OBSERVED,PREDICTED) =  .7760
```

FUNCTION   2

X	OBSERVED Y	PREDICTED Y	RESIDUAL	WEIGHT	SE-PRED	STANDARD RESIDUAL
.1000E-02	100.0	100.0	-.1329E-02	1.000	.2165E-03	-.0001
5.000	100.0	106.2	-6.244	1.000	.9512	-.8358
15.00	100.0	116.6	-16.59	1.000	2.204	-2.303
30.00	108.0	127.9	-19.87	1.000	3.002	-2.877
60.00	141.3	140.5	.8381	1.000	2.934	.1208
90.00	150.0	145.7	4.255	1.000	2.616	.6026
120.0	152.7	147.4	5.314	1.000	2.665	.7545
180.0	149.3	146.1	3.204	1.000	3.023	.4645
240.0	148.0	142.5	5.528	1.000	3.097	.8054
300.0	141.9	137.9	3.954	1.000	2.949	.5707
360.0	130.1	133.0	-2.827	1.000	2.763	-.4036
480.0	124.0	123.1	.9269	1.000	2.766	.1323
600.0	113.7	115.0	-1.292	1.000	2.793	-.1848
720.0	110.8	109.5	1.260	1.000	2.417	.1767
960.0	106.7	104.6	2.152	1.000	2.181	.2986
1440.	101.4	102.8	-1.467	1.000	2.860	-.2106

```
CORRECTED SUM OF SQUARED OBSERVATIONS =   6445.73
WEIGHTED CORRECTED SUM OF SQUARED OBSERVATIONS =   6445.73
SUM OF SQUARED RESIDUALS =              831.360
SUM OF WEIGHTED SQUARED RESIDUALS =    831.360
S =  9.11790     WITH     10 DEGREES OF FREEDOM
CORRELATION (OBSERVED,PREDICTED) =  .9420
```

FUNCTION   3

X	OBSERVED Y	PREDICTED Y	RESIDUAL	WEIGHT	SE-PRED	STANDARDI RESIDUAL
.1000E-02	100.0	100.0	-.1434E-02	1.000	.2283E-03	-.0001
5.000	100.0	106.8	-6.759	1.000	1.007	-.9057
15.00	102.7	118.1	-15.39	1.000	2.352	-2.151
30.00	157.4	130.8	26.61	1.000	3.243	3.915
60.00	162.7	145.8	16.86	1.000	3.230	2.479
90.00	155.4	153.1	2.275	1.000	2.864	.3266

120.0	156.0	156.7	-.6554	1.000	2.853	-.0940
180.0	155.4	159.0	-3.631	1.000	3.172	-.5316
240.0	154.8	159.3	-4.473	1.000	3.247	-.6583
300.0	154.1	158.9	-4.825	1.000	3.120	-.7040
360.0	154.1	158.2	-4.102	1.000	2.914	-.5907
480.0	150.7	155.7	-5.045	1.000	2.567	-.7126
600.0	152.7	151.4	1.292	1.000	2.781	.1846
720.0	147.4	144.6	2.799	1.000	3.702	.4268
960.0	121.9	125.3	-3.379	1.000	5.610	-.6727
1440.	102.8	104.9	-2.107	1.000	2.434	-.2956

```
CORRECTED SUM OF SQUARED OBSERVATIONS =  8782.03
WEIGHTED CORRECTED SUM OF SQUARED OBSERVATIONS =  8782.03
SUM OF SQUARED RESIDUALS =            1404.62
SUM OF WEIGHTED SQUARED RESIDUALS =  1404.62
S =  11.8517     WITH     10 DEGREES OF FREEDOM
CORRELATION (OBSERVED,PREDICTED) =  .9168

TOTALS FOR ALL CURVES COMBINED
SUM OF SQUARED RESIDUALS =            2381.72
SUM OF WEIGHTED SQUARED RESIDUALS =  2381.72
S =  7.53044     WITH     42 DEGREES OF FREEDOM
```

PD6 – Indirect response model III - Repeated oral dosing

Objectives

◆ To analyze a multiple dose response-time dataset

◆ To write an indirect response model in terms of differential equations

◆ To obtain initial parameter estimates of k_{in}, k_{out} and IC_{50} graphically

◆ To discuss various aspects of indirect response modeling

◆ To discuss various aspects of experimental design

Problem specification

The aim of this exercise is to elucidate the pharmacodynamics of a new anxiolytic compound *Zooparc®* after repeated dosing. Data were collected during a phase II study in psychiatric patients. The goal was to generate pivotal information that could serve as a basis for a large multi-center clinical trial design that included a competitor's drug. The investigators aimed to establish the turnover characteristics of the biomarker (k_{in}, k_{out}), and drug-specific information (IC_{50}). The observed response-time data are shown in Figure 6.1 and in the program output.

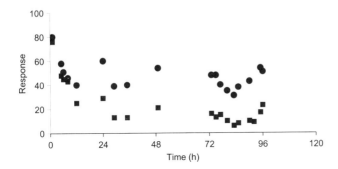

Figure 6.1 Observed response-time data of Zooparc® following the administration of four oral doses of 5 (●) and 25 (■) mg, both given once every 24 h.

The pharmacokinetic model of the drug in plasma takes the following form

$$C = \frac{K_a D_{po}}{(V/F) \cdot (K_a - K)} \left[e^{-K \cdot t} - e^{-K_a \cdot t} \right] \qquad (6:1)$$

The compound is known to fully block the response at high concentrations *in vitro*. Therefore it is reasonable to assume that the inhibitory effect of the drug on the production of the response *I(C)* can take the following form

$$I(C) = 1 - \frac{C_{ss}^n}{IC_{50}^n + C_{ss}^n} \qquad (6:2)$$

The turnover of the response is described by means of a zero-order production k_{in} and first-order loss k_{out} equation according to

$$\frac{dR}{dt} = k_{in} \cdot I(C) - k_{out} \cdot R \qquad (6:3)$$

The drug acts by means of blocking the production of the response. By setting this equation equal to zero

$$\frac{dR}{dt} = k_{in} \cdot I(C_{ss}) - k_{out} \cdot R_{ss} = 0 \qquad (6:4)$$

and then solving for R, one gets the response level at steady state R_{ss} by means of

$$R_{ss} = \frac{k_{in}}{k_{out}} \cdot I(t) = \frac{k_{in}}{k_{out}} \cdot \left[1 - \frac{C_{ss}^n}{IC_{50}^n + C_{ss}^n} \right] \qquad (6:5)$$

Depending on the steady state drug concentration C_{ss}, this or R_{ss} will vary between the baseline value (k_{in}/k_{out}) and zero. The latter value corresponds to high concentrations of the drug.

$$R_{min} = 0 \qquad (6:6)$$

The baseline response is given by Equation 6:7 below

$$R_0 = \frac{k_{in}}{k_{out}} \qquad (6:7)$$

A related goal of this exercise is to select a library model and then to customize that model so as to include an indirect response model that was expressed in terms of differential equations.

Initial parameter estimates

Derive the initial parameter estimates prior to implementing an indirect response model parameterized with k_{in}, k_{out}, IC_{50} and n. The response obeys first-order kinetics when the input of response is totally blocked by the drug ($I(C)$). The dashed lines in Figure 6.2 illustrate this. In this figure, the response-time data are plotted on a semi-logarithmic scale, and slope of the initial downswing corresponds to $-k_{out}$.

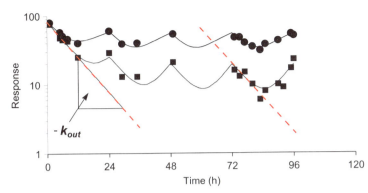

Figure 6.2 Semi-logarithmic plot of observed and predicted response-time data of Zooparc® following the administration of four oral doses of 5 (●) and 25 (■) mg, both given once every 24 h. The dashed lines represent the slope, $-k_{out}$.

Equation 6:8 shows how to obtain an initial estimate of the slope based on the experimental data.

$$k_{out} = \frac{\ln(R_1 / R_2)}{t_1 - t_2} = \frac{\ln(80/48)}{0 - 8} \approx 0.06\,\text{h}^{-1} \qquad (6:8)$$

k_{in} was obtained by rearrangement of Equation 6:7

$$k_{in} = k_{out} \cdot R_0 = 0.06 \cdot 80 = 4.8\,\text{units} \cdot \text{h}^{-1} \qquad (6:9)$$

IC_{50} is the plasma concentration at half-maximal response. The closest one gets to equilibrium between plasma concentration and response is during the upswing of the response-time curve, and not during the downswing of the curve where response lags behind concentration. Since we have steady state response data, we will use that. As can be seen in Figure 6.3, the effect is approximately 50% of maximal response at steady state at about 86 h. The corresponding plasma concentration is obtained by using this time value in Equation 6:1, which gives approximately 0.25 µg/L.

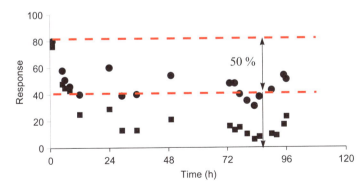

Figure 6.3 Observed response-time data of Zooparc® following the administration of four oral doses of 5 (●) and 25 (■) mg, both given once every 24 hour. The upper horizontal dashed line is the baseline value (80 units). The lower dashed line is 50 % of baseline (40 units). The corresponding time point at the upswing of the 5 mg dose curve is used to predict the IC_{50} value.

Figure 6.4 demonstrates how one translates the terminal response data of Figure 6.3 at 50% (approx. 40 units) of maximal response (approx. 80 units at baseline) into plasma concentrations (IC_{50}, approx. 0.25 µg/L) at the same time point. Figure 6.4 shows both response (observed) and concentration (simulated) data on a semi-logarithmic scale.

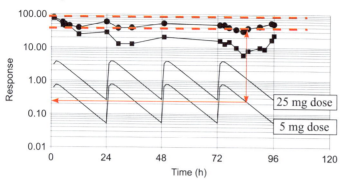

Figure 6.4 Observed response-time data and simulated plasma concentrations (lower solid lines) following administration of four oral doses of 5 (●) and 25 (■) mg, both given once every 24 h. The vertical arrow connects the 50% of maximum response on the upswing of the response-time curve with the corresponding time point on the concentration-time curve. The horizontal arrow indicates the corresponding concentration (Y axis).

The baseline value was set to 80 units. The plasma concentration giving a response at 50% of the baseline value corresponds to the IC_{50} value. The initial parameter estimates that were used in the modeling are shown below.

$$
\begin{aligned}
k_{out} &= 0.06 \text{ h}^{-1} && \text{Obtained from Equation 6:8.} \\
k_{in} &= 4.8 \text{ units} && \text{Obtained from Equation 6:9.} \\
IC_{50} &= 0.25 \text{ µg/L} && \text{From Figure 6.4} \\
n &= 2.0
\end{aligned}
$$

The K_a, K and V/F parameters were estimated by means of fitting Equation 6:1 to plasma concentration-time data, and were found to be 1.1 h^{-1}, 0.128 h^{-1} and 5.0 L/kg, respectively. Since the drug kinetics is unaffected by the response, these should be fixed in the model code.

Interpretation of output and conclusions

The fitted response-time courses are shown in Figure 6.2. It is obvious from these predicted response-time courses that the peak of the response (i.e., the trough value) is shifted to the right with an increase in dose. This can be seen after the first dose as well as at steady state. The fluctuation in response within a dosing interval is less pronounced at the higher dose level.

Table 6.1 contains the final parameter estimates and precision from fitting Equation 6:3 to the observed response-time data.

Table 6.1   Final estimates

Parameter	Mean	CV%
k_{in}	8.8	7.6
k_{out}	0.11	5.6
IC_{50}	0.25	6.4
n	1.40	7.4

The predicted baseline value is about 80 units (k_{in}/k_{out}) and the potency of drug

(IC_{50}) is 0.25 µg/L. The half-life of the response is approximately 7 h, compared to the plasma half-life of the drug, which is 5.4 h. A dose of 25 mg per day would probably suffice in a larger study, since the drug is nontoxic at the present dose levels.

Figures 6.5 and 6.6 show two simulations when drug is acting on the loss of response and production of response, respectively. A number of observations with *'missing'* response were added to the dataset in order to generate smooth predicted curves.

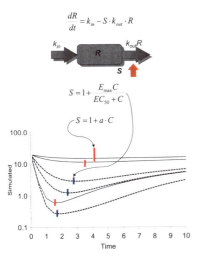

Figure 6.5 Schematic illustration of the response time course at several dose levels when drug is acting on the loss of response. The response is modeled by means of a linear (solid lines) and nonlinear (dashed lines) transduction function. Note the shift in the trough values with increasing doses in both transduction functions when drug is acting on loss (k_{out}) of response.

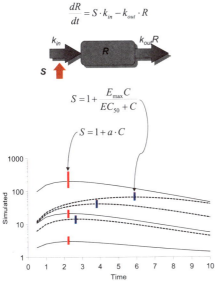

Figure 6.6 Schematic illustration of the response time course at several dose levels when drug is acting on the production of response. The response is modeled by means of a linear (solid lines) and nonlinear (dashed lines) transduction function. Note the shift in the peak values with increasing doses when drug acts via a nonlinear transduction function (Equation 6:10).

The nonlinear transduction function used in the simulations is

$$S = 1 + \frac{E_{max} \cdot C}{EC_{50} + C} \qquad (6{:}10)$$

and the linear transduction function is

$$S = 1 + a \cdot C \qquad (6{:}11)$$

This exercise has shown how to obtain initial parameter estimates and then fit an inhibitory I_{max} model to response-time data after multiple dosing, and how to interpret the output. In addition, we have simulated the response-time course of different doses and transduction-functions (linear *versus* nonlinear), in order to demonstrate how one may anticipate the response-time courses in certain situations.

Solution

```
Model
 COMM
    nfun 2
    nder 2
    ncon 9
    npar 4
    pnam  'Kin', 'kout', 'IC50', 'n1'
end
     1: TEMP
     2: V = 5.
     3: K01 = 1.1
     4: K10 = 0.128
     5: J=1
     6: NDOSE=CON(1)
     7: DO I = 1 TO NDOSE
     8: J=J+2
     9: IF X <= CON(J) THEN GOTO RED
    10: ENDIF
    11: NEXT
    12: RED:
    13: NDOSE = I-1
    14: SUM=0
    15: J=1
    16: DO I = 1 TO NDOSE
    17: J=J+2
    18: T=X - CON(J)
    19: D=CON(J-1)
    20: COEF= D*K01/(V*(K01-K10))
    21: AMT=COEF*(EXP(-K10*T)-EXP(-K01*T))
    22: SUM=SUM + AMT
    23: CP1 = SUM
    24: CP2 = 5*SUM
    25: NEXT
    26: END
    27: START
    28: Z(1) = KIN/KOUT
    29: Z(2) = KIN/KOUT
    30: END
    31: DIFF
    32: DRUG1 = 1 - ((CP1**N1)/((IC50**N1) + (CP1**N1)))
    33: DRUG2 = 1 - ((CP2**N1)/((IC50**N1) + (CP2**N1)))
```

```
   34: DZ(1) = KIN*DRUG1 - KOUT*Z(1)
   35: DZ(2) = KIN*DRUG2 - KOUT*Z(2)
   36: END
   37: FUNC1
   38: F = Z(1)
   39: END
   40: FUNC 2
   41: F = Z(2)
   42: END
   43: EOM
NOPAGE BREAKS
NVARIABLES 3
FNUMBER 3
NPOINTS 100
XNUMBER 1
YNUMBER 2
CONSTANTS 4,5,0,5,24,5, 48,5,72
METHOD 2  'Gauss-Newton (Levenberg and Hartley)
ITERATIONS 50
INITIAL 4.8,.06,.25,2
LOWER BOUNDS 0,0,0,0
UPPER BOUNDS 15,2,1,4
BEGIN
```

PARAMETER	ESTIMATE	STANDARD ERROR	CV%	UNIVARIATE C.I.	
KIN	8.769031	.661682	7.55	7.422837	10.115224
KOUT	.105585	.005898	5.59	.093586	.117584
IC50	.245523	.015603	6.36	.213778	.277267
N1	1.385957	.101981	7.36	1.178476	1.593439

```
*** CORRELATION MATRIX OF THE ESTIMATES ***
PARAMETER  KIN        KOUT       IC50       N1
KIN        1.00000
KOUT        .97118    1.00000
IC50       -.74093     -.61601    1.00000
N1         -.63882     -.62541     .65145    1.00000

 Condition_number=      665.8

   FUNCTION   1
```

X	OBSERVED Y	PREDICTED Y	RESIDUAL	WEIGHT	SE-PRED	STANDARDI RESIDUAL
1.000	80.00	78.14	1.863	1.000	1.691	.8719
2.000	.	71.87	.	.0000	-.6400+152	.
3.000	.	66.14	.	.0000	-.6400+152	.
4.000	.	61.10	.	.0000	-.6400+152	.
5.000	58.00	56.76	1.241	1.000	.8758	.4809
6.000	51.00	53.10	-2.099	1.000	.8411	-.8098
7.000	.	50.08	.	.0000	-.6400+152	.
8.000	46.00	47.67	-1.672	1.000	.8267	-.6440
9.000	.	45.83	.	.0000	-.6400+152	.
11.00	.	43.69	.	.0000	-.6400+152	.
12.00	40.00	43.32	-3.317	1.000	.7911	-1.272
13.00	.	43.35	.	.0000	-.6400+152	.
14.00	.	43.74	.	.0000	-.6400+152	.
15.00	.	44.45	.	.0000	-.6400+152	.
18.00	.	48.03	.	.0000	-.6400+152	.

20.00	.	51.19	.	.0000	-.6400+152	.
23.00	.	56.37	.	.0000	-.6400+152	.
24.00	60.00	58.09	1.907	1.000	.9904	.7514
25.00	.	55.28	.	.0000	-.6400+152	.
26.00	.	51.20	.	.0000	-.6400+152	.
27.00	.	47.46	.	.0000	-.6400+152	.
28.00	.	44.21	.	.0000	-.6400+152	.
29.00	39.00	41.47	-2.474	1.000	.7386	-.9434
30.00	.	39.25	.	.0000	-.6400+152	.
31.00	.	37.52	.	.0000	-.6400+152	.
32.00	.	36.26	.	.0000	-.6400+152	.
33.00	.	35.44	.	.0000	-.6400+152	.
35.00	40.00	35.04	4.959	1.000	.7788	1.899
36.00	.	35.40	.	.0000	-.6400+152	.
37.00	.	36.10	.	.0000	-.6400+152	.
38.00	.	37.09	.	.0000	-.6400+152	.
39.00	.	38.34	.	.0000	-.6400+152	.
42.00	.	43.28	.	.0000	-.6400+152	.
44.00	.	47.17	.	.0000	-.6400+152	.
47.00	.	53.25	.	.0000	-.6400+152	.
48.00	.	55.23	.	.0000	-.6400+152	.
49.00	54.00	52.69	1.312	1.000	.9522	.5137
50.00	.	48.87	.	.0000	-.6400+152	.
51.00	.	45.35	.	.0000	-.6400+152	.
52.00	.	42.31	.	.0000	-.6400+152	.
53.00	.	39.76	.	.0000	-.6400+152	.
54.00	.	37.71	.	.0000	-.6400+152	.
55.00	.	36.12	.	.0000	-.6400+152	.
56.00	.	35.00	.	.0000	-.6400+152	.
57.00	.	34.30	.	.0000	-.6400+152	.
59.00	.	34.11	.	.0000	-.6400+152	.
60.00	.	34.56	.	.0000	-.6400+152	.
61.00	.	35.33	.	.0000	-.6400+152	.
62.00	.	36.39	.	.0000	-.6400+152	.
63.00	.	37.71	.	.0000	-.6400+152	.
66.00	.	42.80	.	.0000	-.6400+152	.
68.00	.	46.78	.	.0000	-.6400+152	.
71.00	.	52.95	.	.0000	-.6400+152	.
72.00	.	54.96	.	.0000	-.6400+152	.
73.00	48.00	52.45	-4.447	1.000	.9624	-1.745
74.00	.	48.65	.	.0000	-.6400+152	.
75.00	48.00	45.16	2.841	1.000	.7336	1.083
76.00	.	42.13	.	.0000	-.6400+152	.
77.00	40.00	39.60	.3962	1.000	.6710	.1500
78.00	.	37.56	.	.0000	-.6400+152	.
79.00	.	36.00	.	.0000	-.6400+152	.
80.00	35.00	34.88	.1197	1.000	.6974	.4546E-01
81.00	.	34.20	.	.0000	-.6400+152	.
83.00	31.00	34.02	-3.023	1.000	.7338	-1.152
84.00	.	34.48	.	.0000	-.6400+152	.
85.00	38.00	35.26	2.740	1.000	.7521	1.046
86.00	.	36.33	.	.0000	-.6400+152	.
87.00	.	37.65	.	.0000	-.6400+152	.
90.00	43.00	42.76	.2402	1.000	.8516	.9281E-01
92.00	.	46.74	.	.0000	-.6400+152	.
95.00	'54.00	52.93	1.072	1.000	1.042	.4259
96.00	51.00	54.94	-3.942	1.000	1.084	-1.577

```
CORRECTED SUM OF SQUARED OBSERVATIONS =   32745.1
WEIGHTED CORRECTED SUM OF SQUARED OBSERVATIONS =  2214.44
SUM OF SQUARED RESIDUALS =            120.704
```

```
SUM OF WEIGHTED SQUARED RESIDUALS =  120.704
S =  2.93628     WITH    14 DEGREES OF FREEDOM
CORRELATION (OBSERVED,PREDICTED) =  .2529
```

FUNCTION   2

X	OBSERVED Y	PREDICTED Y	RESIDUAL	WEIGHT	SE-PRED	STANDARDI RESIDUAL
1.000	76.00	75.64	.3637	1.000	1.641	.1672
2.000	.	68.26	.	.0000	-.6400+152	.
3.000	.	61.61	.	.0000	-.6400+152	.
4.000	.	55.64	.	.0000	-.6400+152	.
5.000	48.00	50.30	-2.301	1.000	.9894	-.9065
6.000	45.00	45.54	-.5397	1.000	1.012	-.2133
7.000	.	41.31	.	.0000	-.6400+152	.
8.000	43.00	37.56	5.444	1.000	1.068	2.172
9.000	.	34.25	.	.0000	-.6400+152	.
11.00	.	28.86	.	.0000	-.6400+152	.
12.00	25.00	26.73	-1.726	1.000	1.038	-.6850
13.00	.	24.93	.	.0000	-.6400+152	.
14.00	.	23.47	.	.0000	-.6400+152	.
15.00	.	22.33	.	.0000	-.6400+152	.
18.00	.	20.72	.	.0000	-.6400+152	.
20.00	.	21.13	.	.0000	-.6400+152	.
23.00	.	23.81	.	.0000	-.6400+152	.
24.00	29.00	25.22	3.781	1.000	.6402	1.428
25.00	.	23.31	.	.0000	-.6400+152	.
26.00	.	21.16	.	.0000	-.6400+152	.
27.00	.	19.21	.	.0000	-.6400+152	.
28.00	.	17.48	.	.0000	-.6400+152	.
29.00	13.00	15.95	-2.951	1.000	.6037	-1.111
30.00	.	14.62	.	.0000	-.6400+152	.
31.00	.	13.46	.	.0000	-.6400+152	.
32.00	.	12.48	.	.0000	-.6400+152	.
33.00	.	11.66	.	.0000	-.6400+152	.
35.00	13.00	10.51	2.491	1.000	.5747	.9354
36.00	.	10.17	.	.0000	-.6400+152	.
37.00	.	9.986	.	.0000	-.6400+152	.
38.00	.	9.965	.	.0000	-.6400+152	.
39.00	.	10.11	.	.0000	-.6400+152	.
42.00	.	11.61	.	.0000	-.6400+152	.
44.00	.	13.55	.	.0000	-.6400+152	.
47.00	.	17.97	.	.0000	-.6400+152	.
48.00	.	19.83	.	.0000	-.6400+152	.
49.00	21.00	18.45	2.553	1.000	.8006	.9801
50.00	.	16.78	.	.0000	-.6400+152	.
51.00	.	15.28	.	.0000	-.6400+152	.
52.00	.	13.94	.	.0000	-.6400+152	.
53.00	.	12.76	.	.0000	-.6400+152	.
54.00	.	11.75	.	.0000	-.6400+152	.
55.00	.	10.88	.	.0000	-.6400+152	.
56.00	.	10.15	.	.0000	-.6400+152	.
57.00	.	9.569	.	.0000	-.6400+152	.
59.00	.	8.812	.	.0000	-.6400+152	.
60.00	.	8.639	.	.0000	-.6400+152	.
61.00	.	8.608	.	.0000	-.6400+152	.
62.00	.	8.723	.	.0000	-.6400+152	.
63.00	.	8.990	.	.0000	-.6400+152	.
66.00	.	10.78	.	.0000	-.6400+152	.
68.00	.	12.87	.	.0000	-.6400+152	.
71.00	.	17.46	.	.0000	-.6400+152	.

```
72.00          .         19.37         .          .0000      -.6400+152        .
73.00        16.00       18.03       -2.030       1.000       .8561         -.7848
74.00          .         16.41         .          .0000      -.6400+152        .
75.00        13.00       14.94       -1.940       1.000       .6856         -.7355
76.00          .         13.63         .          .0000      -.6400+152        .
77.00        15.00       12.49        2.509       1.000       .5884          .9431
78.00          .         11.50         .          .0000      -.6400+152        .
79.00          .         10.66         .          .0000      -.6400+152        .
80.00        10.00        9.955       .4521E-01   1.000       .5437          .1693E-01
81.00          .          9.390        .          .0000      -.6400+152        .
83.00         6.000       8.666      -2.666       1.000       .5788         -1.001
84.00          .          8.508        .          .0000      -.6400+152        .
85.00         8.000       8.490       -.4903      1.000       .6368         -.1851
86.00          .          8.617        .          .0000      -.6400+152        .
87.00          .          8.895        .          .0000      -.6400+152        .
90.00        10.00       10.71        -.7098      1.000       .8380         -.2738
92.00         9.000      12.82        -3.816      1.000       .9019         -1.484
95.00        17.00       17.42        -.4156      1.000       .9407         -.1625
96.00        23.00       19.33        3.673       1.000       .9390          1.436
```

```
CORRECTED SUM OF SQUARED OBSERVATIONS =   13359.1
WEIGHTED CORRECTED SUM OF SQUARED OBSERVATIONS =   5858.53
SUM OF SQUARED RESIDUALS =            124.322
SUM OF WEIGHTED SQUARED RESIDUALS =   124.322
S =   2.87891      WITH     15 DEGREES OF FREEDOM
CORRELATION (OBSERVED,PREDICTED) =   .6552

TOTALS FOR ALL CURVES COMBINED
SUM OF SQUARED RESIDUALS =            245.026
SUM OF WEIGHTED SQUARED RESIDUALS =   245.026
S =   2.72489      WITH     33 DEGREES OF FREEDOM

AIC criteria =          508.62426
SC  criteria =          518.66770
```

PD7 - Indirect response model IV – Repeated intravenous dosing

Objectives

◆ **To analyze response-time data obtained after multiple intravenous infusions**

◆ **To obtain initial estimates**

◆ **To derive and fit two different turnover models to response-time data**

Problem specification

The aim of this exercise is to implement a previously derived kinetic function (with fixed parameter values) that drives the indirect response model. When there are both kinetic and dynamic data, we highly recommend three types of plots: *concentration versus time, response versus time,* and *response versus concentration.* We also recommend transforming the concentration axis so that response *versus* log-concentration is plotted.

In the example presented below, there are both concentration-time (Figure 7.1) and response-time (Figure 7.2 upper panel) data. From the various plots, initial estimates are derived for the model. Thus, from Figure 7.1, one can draw conclusions about the kinetics of the tested compound. Two rapid infusions resulted in a rapid increase and washout of the plasma concentration.

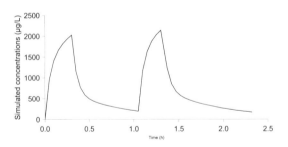

Figure 7.1 Predicted concentration-time course after two consecutive intravenous infusions of 6400 and 5900 µg.

In the response *versus* time graph (Figure 7.2), response is an arbitrary measure. Note the rapid decline in response and the flat portions at maximal response. As the response levels off at a constant fraction of the baseline value, it will not be possible to increase the response by increasing the concentration or dose. Delta (Δ) is obtained from the difference between the baseline value and the drug-induced flat portion of the data. I_{max} (Model 1, Equation 7:1) is derived from the ratio of $\Delta/(k_{in}/k_{out})$. The two lower graphs in Figure 7.2 show response *versus* concentration. This plot enables one to determine the IC_{50} value. From the semi-logarithmic plot of response *versus* time, the slope of the downswing of the data gives $-k_{out}$. The k_{in} parameter is then estimated by means of k_{out} and the baseline value R_0, as shown in Equation 7:3.

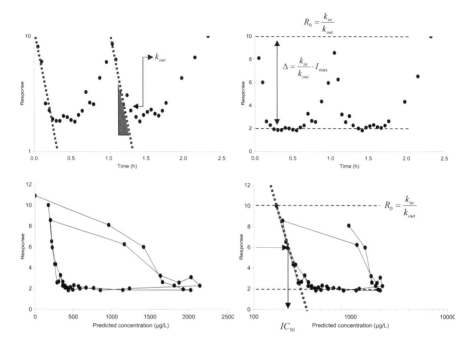

Figure 7.2 *Observed log-response versus time (upper left), response versus time (upper right), response versus predicted concentration (lower left) and response versus log-predicted concentration (lower right) data following two consecutive intravenous infusions of a new experimental compound. The dashed blue lines in the upper left graph represent the slope (-k_{out}) of the response-time curve. The horizontal dashed lines in the upper right graph correspond to the baseline value (upper) and trough value (lower). Delta is the maximum distant between baseline and trough. The vertical arrow in the lower right graph indicates the IC_{50} value (i.e., concentration causing 50% of maximum (10-2=8) response).*

It is obvious from Figure 7.2 that the maximum response occurs at about 500 µg/L. At higher concentrations it will not be possible to further increase this response. Note that the response-concentration loops of both infusions virtually superimpose. This indicates that there is no tolerance development with the present dosages and regimen. We have therefore selected two of the four indirect response models that were discussed in section 4.6.2. Of these models, models 1 and 4 are feasible and capture the decline of response in this example.

Model 1
$$\frac{dR}{dt} = k_{in} \cdot \left(1 - \frac{I_{max} \cdot C^n}{IC_{50}^n + C^n}\right) - k_{out} \cdot R \qquad (7:1)$$

Model 4
$$\frac{dR}{dt} = k_{in} - k_{out} \cdot \left(1 + \frac{E_{max} \cdot C^n}{EC_{50}^n + C^n}\right) \cdot R \qquad (7:2)$$

Initial parameter estimates

The following relationship shows the derivation of response at baseline R_0 and steady state R_{ss} and the approximate estimates of EC_{50} and of the exponent n using model 4. The baseline R_0 is obtained from

$$R_0 = \frac{k_{in}}{k_{out}} \approx 10 \tag{7:3}$$

Equation 7:3 can then be rewritten as

$$k_{in} = R_0 \cdot k_{out} \tag{7:4}$$

Based on model 1 (Equation 7:1), R_{ss} becomes

$$R_{ss} = R_{min} = \frac{k_{in}}{k_{out}} \cdot (1 - I_{max}) \approx 2 \tag{7:5}$$

when $C \rightarrow \infty$. This equation rearranged for I_{max} gives

$$I_{max} = 1 - \frac{R_{min}}{\dfrac{k_{in}}{k_{out}}} = 1 - \frac{2}{10} = 0.8 \tag{7:6}$$

Based on Model 4 (Equation 7:2) R_{ss} becomes

$$R_{ss} = R_{min} = \frac{k_{in}}{k_{out}} \cdot \frac{1}{1 + E_{max}} \approx 2 \tag{7:7}$$

This equation rearranged for E_{max} gives

$$E_{max} = \frac{R_0}{R_{ss}} - 1 = \frac{10}{2} - 1 = 4 \tag{7:8}$$

The k_{out} parameter is determined from the downswing of the response-time curve as

$$k_{out} = \left[\ln\left(\frac{7.5}{2.5} \right) \right] / (0 - 0.25) \approx 4 \tag{7:9}$$

Depending on the time points selected for estimation of k_{out}, the initial estimate may vary between 4 to 10. The k_{in} parameter is obtained as

$$k_{in} = R_0 \cdot k_{out} = 10 \cdot 4 = 40 \tag{7:10}$$

The exponent n is most easily derived from m and E_{max} (Levy [1995]). The m

parameter is the slope of the response *versus* log-concentration curve at EC_{50}. See section 4.4.4 on pharmacodynamic concepts for derivation

$$m = \frac{E_1 - E_2}{\ln C_1 - \ln C_2} = \frac{n \cdot E_{max}}{4} > 4 \tag{7:11}$$

which then rearranged gives

$$n = \frac{m \cdot 4}{E_{max}} > 4 \tag{7:12}$$

EC_{50}/IC_{50} is most readily read from a graph of response *versus* log-concentration. A rough estimate of EC_{50}/IC_{50} is obtained from the upswing of the curve, which is indicated with arrows in Figure 7.2 above (lower right hand graph).

Interpretation of results and conclusions

The final model-predicted response-time data for models 1 and 4 (Equations 7:1 and 7:2, respectively) are superimposed on the observed data in Figure 7.3.

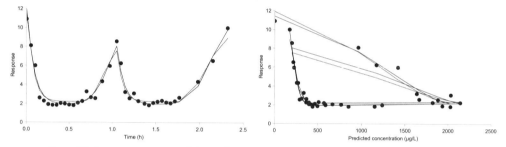

Figure 7.3 *Response versus time of observed and predicted data (left). Response versus predicted plasma concentration of observed and predicted data (right). Model 1 and model 4 predict the data equally well.*

Drug blocks the response only partially, though maximum blockage is reached. Table 7.1 compares the final parameter estimates and their precision.

Table 7.1   Parameter estimates and initial estimates

Parameter	Model 1 mean ± CV%	Model 4 mean ± CV%	Initial estimates
k_{in}	143 ± 10	34.2 ± 13	30-40
k_{out}	12.5 ± 9	2.87 ± 13	6-10
E_{max}	–	4.41 ± 10	4–6
I_{max}	0.83 ± 1.5	–	0.8
EC_{50}	–	312 ± 4.2	250-350
IC_{50}	244 ± 2.6	–	250
n	7.0 ± 13	19 ± 74	>10
WRSS	7.3	12	
AIC	83	102	

The model of choice is the *inhibitory* model (i.e., Model 1) due to its higher parameter precision and lower *WRSS*. Ideally, the mechanism of action should drive which model to use.There seems to be no obvious tolerance development with the present dosage regimen, as shown in Figure 7.3. Thus, the two hysteresis loops more or less superimpose, regardless of model. We would, however, recommend that a feedback model be considered, as tolerance to this compound may develop slowly.

Solution I - Inhibition of build-up of response (Model 1)

```
TITLE 1
Modelling of hysteresis data
MODEL
 COMM
  NDER 4
  NCON 5
  NPARM 5
  PNAMES 'Kin', 'Kout', 'IC50', 'Imax', 'n'
END
    1: TEMP
    2: T=X
    3: D1 = CON(1)
    4: D2 = CON(2)
    5: TI1 = CON(3)
    6: TI2 = CON(4)
    7: T2 = CON(5)
    8: VC = 0.7633
    9: CL = 6.2417
   10: CLD1= 5.4595
   11: CLD2= 0.85806
   12: VT1= 1.72876
   13: VT2= 3.43857
   14: END
   15: START
   16: Z(1)=KIN/KOUT
   17: Z(2)=0.0
   18: Z(3)=0.0
   19: Z(4)=0.0
   20: END
   21: DIFF
   22: FI1 = 0.0
   23: FI2 = 0.0
   24: IF T LE TI1 THEN
   25: FI1 = D1/TI1
   26: FI2 = 0.0
   27: ENDIF
   28: IF T GE T2 AND T LE T2+TI2 THEN
   29: FI1 = 0.0
   30: FI2 = D2/TI2
   31: ENDIF
   32: IF Z(2) LE 0 THEN
   33: ZZ2 = 0.0
   34: ELSE
   35: ZZ2 = Z(2)**N
   36: ENDIF
   37: DRUG = (1. - (IMAX*ZZ2/(IC50**N + ZZ2)))
   38: DZ(1) = KIN*DRUG - KOUT*Z(1)
   39: DZ(2)=(FI1+FI2-CL*Z(2)-CLD1*Z(2)-CLD2*Z(2)+CLD1*Z(3)+CLD2*Z(4))/VC
   40: DZ(3)=(CLD1*Z(2) - CLD1*Z(3))/VT1
   41: DZ(4)=(CLD2*Z(2) - CLD2*Z(4))/VT2
   42: END
   43: FUNC 1
```

```
   44: F = Z(1)
   45: END
   46: EOM
NOPAGE BREAKS
NVARIABLES 2
NPOINTS 100
XNUMBER 1
YNUMBER 2
CONSTANTS 6400,5900,.3,.2667,1.051
METHOD 2  'Gauss-Newton (Levenberg and Hartley)
ITERATIONS 50
INITIAL 140,13,300,.82,6.6
LOWER BOUNDS .1,.1,100,.1,.01
UPPER BOUNDS 200,30,1500,1,12
NOBSERVATIONS 37
BEGIN
```

PARAMETER	ESTIMATE	STANDARD ERROR	CV%	UNIVARIATE C.I.	
KIN	142.977414	14.760967	10.32	112.910510	173.044319
KOUT	12.513445	1.070735	8.56	10.332443	14.694446
IC50	244.281879	6.397336	2.62	231.251019	257.312738
IMAX	.831446	.012582	1.51	.805817	.857075
N	6.962154	.871890	12.52	5.186183	8.738124

```
*** CORRELATION MATRIX OF THE ESTIMATES ***
PARAMETER  KIN        KOUT       IC50       IMAX       N
KIN        1.00000
KOUT        .94581    1.00000
IC50       -.72937    -.583329   1.00000
IMAX       -.96364    -.306368    .022641   1.00000
N          -.74082     .065249    .205426   -.505318   1.00000

 Condition_number=        1959.
```

X	OBSERVED Y	PREDICTED Y	RESIDUAL	WEIGHT	SE-PRED	STANDARD RESIDUAL
.0000	10.90	11.43	-.5259	1.000	.4075	-2.117
.5000E-01	8.090	7.679	.4111	1.000	.2649	1.036
.1000	5.980	5.003	.9768	1.000	.2470	2.392
.1500	2.600	3.572	-.9720	1.000	.1963	-2.235
.2000	2.270	2.806	-.5364	1.000	.1441	-1.179
.2500	1.910	2.397	-.4869	1.000	.1129	-1.050
.3000	1.840	2.178	-.3378	1.000	.1045	-.7255
.3500	1.840	2.061	-.2207	1.000	.1079	-.4747
.4000	2.030	1.999	.3139E-01	1.000	.1136	.0677
.4500	1.980	1.970	.1008E-01	1.000	.1169	.0217
.5000	1.870	1.970	-.9994E-01	1.000	.1159	-.2159
.5500	1.810	2.003	-.1926	1.000	.1118	-.4152
.6000	2.070	2.077	-.6563E-02	1.000	.1096	-.0141
.6500	2.250	2.207	.4278E-01	1.000	.1174	.0924
.7000	3.270	2.421	.8492	1.000	.1407	1.862
.7500	2.650	2.755	-.1050	1.000	.1753	-.2366
.8000	2.560	3.254	-.6941	1.000	.2082	-1.617
.8800	4.330	4.472	-.1420	1.000	.2251	-.3375
.9700	5.940	6.346	-.4063	1.000	.2187	-.9580
1.050	8.560	8.048	.5120	1.000	.2536	1.267
1.100	6.240	5.396	.8439	1.000	.1827	1.914

1.150	3.210	3.782	-.5721	1.000	.1657	-1.278
1.200	2.550	2.919	-.3688	1.000	.1339	-.8051
1.250	3.070	2.457	.6130	1.000	.1110	1.321
1.300	2.260	2.210	.5003E-01	1.000	.1045	.1074
1.370	1.990	2.044	-.5421E-01	1.000	.1101	-.1168
1.420	1.820	1.989	-.1695	1.000	.1156	-.3661
1.470	2.070	1.962	.1079	1.000	.1192	.2334
1.520	2.200	1.953	.2468	1.000	.1201	.5344
1.570	2.300	1.960	.3403	1.000	.1183	.7361
1.620	2.120	1.982	.1378	1.000	.1149	.2975
1.670	2.040	2.025	.1507E-01	1.000	.1114	.0324
1.720	2.250	2.097	.1530	1.000	.1112	.3296
1.770	2.630	2.214	.4159	1.000	.1194	.9003
1.980	4.320	3.848	.4721	1.000	.2280	1.126
2.150	6.520	7.064	-.5436	1.000	.2320	-1.304
2.320	10.00	9.810	.1903	1.000	.2893	.5014

```
CORRECTED SUM OF SQUARED OBSERVATIONS =  218.512
WEIGHTED CORRECTED SUM OF SQUARED OBSERVATIONS =  218.512
SUM OF SQUARED RESIDUALS =             7.28753
SUM OF WEIGHTED SQUARED RESIDUALS =  7.28753
S = .477216      WITH     32 DEGREES OF FREEDOM
CORRELATION (OBSERVED,PREDICTED) =  .9832

AIC criteria =         83.48807
SC  criteria =         91.54266
```

Solution II - Stimulation of loss of response (Model 4)

```
TITLE 1
Modelling of hysteresis data
MODEL
 COMM
  NDER 4
  NCON 5
  NPARM 5
  PNAMES 'Kin', 'Kout', 'EC50', 'Emax', 'n'
END
    1: TEMP
    2: T=X
    3: D1 = CON(1)
    4: D2 = CON(2)
    5: TI1 = CON(3)
    6: TI2 = CON(4)
    7: T2 = CON(5)
    8: VC = 0.7633
    9: CL = 6.2417
   10: CLD1= 5.4595
   11: CLD2= 0.85806
   12: VT1= 1.72876
   13: VT2= 3.43857
   14: END
   15: START
   16: Z(1)=KIN/KOUT
   17: Z(2)=0.0
   18: Z(3)=0.0
   19: Z(4)=0.0
   20: END
   21: DIFF
   22: FI1 = 0.0
```

```
23: FI2 = 0.0
24: IF T LE TI1 THEN
25: FI1 = D1/TI1
26: FI2 = 0.0
27: ENDIF
28: IF T GE T2 AND T LE T2+TI2 THEN
29: FI1 = 0.0
30: FI2 = D2/TI2
31: ENDIF
32: IF Z(2) LE 0 THEN
33: ZZ2 = 0.0
34: ELSE
35: ZZ2 = Z(2)**N
36: ENDIF
37: DRUG = (1. + (EMAX*ZZ2/(EC50**N+ZZ2)))
38: DZ(1) = KIN - DRUG*KOUT*Z(1)
39: DZ(2)=(FI1+FI2-CL*Z(2)-CLD1*Z(2)-CLD2*Z(2)+CLD1*Z(3)+CLD2*Z(4))/VC
40: DZ(3)=(CLD1*Z(2) - CLD1*Z(3))/VT1
41: DZ(4)=(CLD2*Z(2) - CLD2*Z(4))/VT2
42: END
43: FUNC 1
44: F = Z(1)
45: END
46: EOM
NOPAGE BREAKS
NVARIABLES 2
NPOINTS 100
XNUMBER 1
YNUMBER 2
CONSTANTS 6400,5900,.3,.2667,1.051
METHOD 2  'Gauss-Newton (Levenberg and Hartley)
ITERATIONS 50
INITIAL 40,3,350,6,13
LOWER BOUNDS .1,.1,.1,.1,.1
UPPER BOUNDS 80,20,1500,10,20
NOBSERVATIONS 37
BEGIN
```

PARAMETER	ESTIMATE	STANDARD ERROR	CV%	UNIVARIATE C.I.	
KIN	34.181497	4.448336	13.01	25.120593	43.242400
KOUT	2.865828	.378834	13.22	2.094173	3.637483
EC50	312.305061	13.241545	4.24	285.333097	339.277024
EMAX	4.409200	.413423	9.38	3.567091	5.251308
N	18.808816	14.010723	74.49	-9.729900	47.347533

```
*** CORRELATION MATRIX OF THE ESTIMATES ***
PARAMETER  KIN        KOUT       EC50       EMAX       N
KIN        1.00000
KOUT        .944478   1.00000
EC50       -.528771   -.519938   1.00000
EMAX       -.419891   -.636775    .415986   1.00000
N          -.382220   -.285004   -.395753   -.072263   1.00000

Condition_number=        247.2
```

X	OBSERVED Y	PREDICTED Y	RESIDUAL	WEIGHT	SE-PRED	STANDARD RESIDUAL

.0000	10.90	11.93	-1.027	1.000	.5215	-3.221
.5000E-01	8.090	7.637	.4532	1.000	.3550	.9105
.1000	5.980	4.707	1.273	1.000	.3153	2.430
.1500	2.600	3.358	-.7577	1.000	.2309	-1.339
.2000	2.270	2.736	-.4660	1.000	.1619	-.7905
.2500	1.910	2.450	-.5396	1.000	.1282	-.9027
.3000	1.840	2.318	-.4777	1.000	.1195	-.7967
.3500	1.840	2.257	-.4169	1.000	.1199	-.6955
.4000	2.030	2.229	-.1989	1.000	.1220	-.3320
.4500	1.980	2.216	-.2360	1.000	.1236	-.3942
.5000	1.870	2.210	-.3401	1.000	.1245	-.5683
.5500	1.810	2.208	-.3981	1.000	.1253	-.6654
.6000	2.070	2.212	-.1418	1.000	.1280	-.2372
.6500	2.250	2.236	.1397E-01	1.000	.1540	.0236
.7000	3.270	2.340	.9301	1.000	.2702	1.696
.7500	2.650	2.673	-.2300E-01	1.000	.4001	-.0497
.8000	2.560	3.361	-.8010	1.000	.3500	-1.598
.8800	4.330	4.854	-.5235	1.000	.3006	-.9834
.9700	5.940	6.420	-.4797	1.000	.2714	-.8757
1.050	8.560	7.544	1.016	1.000	.2744	1.860
1.100	6.240	4.902	1.338	1.000	.1740	2.283
1.150	3.210	3.448	-.2375	1.000	.1714	-.4048
1.200	2.550	2.777	-.2274	1.000	.1423	-.3825
1.250	3.070	2.469	.6013	1.000	.1238	1.004
1.300	2.260	2.326	-.6646E-01	1.000	.1191	-.1108
1.370	1.990	2.246	-.2560	1.000	.1209	-.4272
1.420	1.820	2.224	-.4039	1.000	.1228	-.6744
1.470	2.070	2.214	-.1437	1.000	.1241	-.2401
1.520	2.200	2.209	-.9013E-02	1.000	.1248	-.0150
1.570	2.300	2.207	.9311E-01	1.000	.1252	.1556
1.620	2.120	2.206	-.8617E-01	1.000	.1255	-.1440
1.670	2.040	2.207	-.1670	1.000	.1261	-.2791
1.720	2.250	2.212	.3773E-01	1.000	.1294	.0631
1.770	2.630	2.234	.3957	1.000	.1537	.6687
1.980	4.320	4.181	.1391	1.000	.3252	.2686
2.150	6.520	7.058	-.5379	1.000	.2833	-.9929
2.320	10.00	8.934	1.066	1.000	.3262	2.063

```
CORRECTED SUM OF SQUARED OBSERVATIONS =  218.512
WEIGHTED CORRECTED SUM OF SQUARED OBSERVATIONS =  218.512
SUM OF SQUARED RESIDUALS =          11.9597
SUM OF WEIGHTED SQUARED RESIDUALS =  11.9597
S =  .611343     WITH    32 DEGREES OF FREEDOM
CORRELATION (OBSERVED,PREDICTED) =  .9728

AIC criteria =        101.81709
SC  criteria =        109.87168
```

PD8 – Sigmoidal growth models

Objectives

◆ To analyze data with the Gompertz, Logistic, Weibull, Richard, Hill and
Morgan-Mercer-Flodin models

◆ To obtain initial parameter estimates graphically

◆ To apply user specified models

Problem specification

This exercise highlights some issues of modeling sigmoidal curves. Data, which were obtained from Heyes and Brown [1956] were also analyzed in *Nonlinear Regression Modeling: A Unified Practical Approach* by Ratowsky [1983]. Specific emphasis will be put on obtaining *initial parameter estimates*, application of constant absolute *weights*, and assessing the *goodness-of-fit*. Asymmetric growth models are further explored at the end of this exercise.

The dataset is drawn in Figure 8.1 and is also presented in Table 8.1. The dataset is typical of the ones found in pharmacodynamic practice with respect to sample size and error structure.

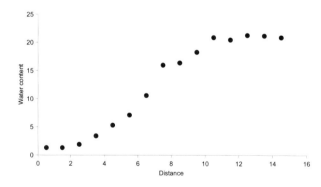

Figure 8.1 Observed water content (Y) versus distance (X). Data are taken from Heyes and Brown [1956] and Ratowsky [1983]. The dataset is typical of the ones found in pharmacodynamic practice with respect to sample size, error structure and complexity.

Table 8.1   Growth model data (Ratowsky [1983])

X	Y	X	Y
0.5	1.3	8.5	16.4
1.5	1.3	9.5	18.3
2.5	1.9	10.5	20.9
3.5	3.4	11.5	20.5
4.5	5.3	12.5	21.3
5.5	7.1	13.5	21.2
6.5	10.6	14.5	20.9
7.5	16	-	-

The six models that we will fit to the data in Table 8.1 are expressed below. The models contain either three parameters (Gompertz, Logistic) or four parameters (Weibull, Richards, Morgan-Mercer-Flodin and Hill type).
The Gompertz model

$$Y = \alpha \cdot e^{\left[-e(\beta - \gamma \cdot X)\right]} \tag{8:1}$$

The Logistic model

$$Y = \frac{\alpha}{1 + e^{(\beta - \gamma \cdot X)}} \tag{8:2}$$

The Weibull model

$$Y = \alpha - \beta \cdot e^{\left(-\gamma \cdot X^{\delta}\right)} \qquad \text{The ai}$$

mof □rt ol sxc p nxf-

$$Y = \frac{\alpha}{\left[1 + e^{(\beta - \gamma \cdot X)}\right]^{\frac{1}{\delta}}} \tag{8:4}$$

The Morgan-Mercer-Flodin model

$$Y = \frac{\beta \cdot \gamma + \alpha \cdot X^{\delta}}{\gamma + X^{\delta}} \tag{8:5}$$

The Hill model

$$Y = \alpha + \frac{\beta \cdot X^{\delta}}{\gamma^{\delta} + X^{\delta}} \tag{8:6}$$

Initial parameter estimates

Figure 8.2 demonstrates how to graphically obtain the initial estimates for some of the model parameters for the Hill and Gompertz models.

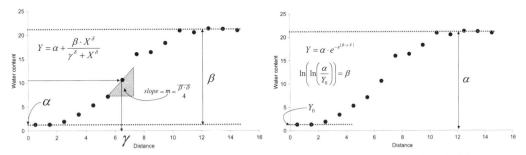

***Figure 8.2** Schematic illustration of initial parameter estimates for the Hill model (left) and of the Gompertz model (right). The δ parameter is more difficult to obtain graphically for the Gompertz model, but the parameter is related to the steepness of the curve.*

See sections 4.1 and 4.2 for a mechanistic interpretation for the Hill-parameters. The Hill parameters are usually denoted E_0 (α), E_{max} (β), EC_{50} (γ) and n (δ).

Interpretation of results and conclusions

So far, this exercise has dealt with obtaining *initial parameter estimates* for some commonly used growth models. It has also been shown how to implement the equations into *WinNonlin*.

Figure 8.3 contains the observed and Hill model-predicted water content *versus* distance. The plot demonstrates a high consistency between observed and model-predicted values. Tables 8.2 and 8.3 give the final parameter estimates and their precision (CV%) together with the *weighted residual sum of squares WRSS* of each fit.

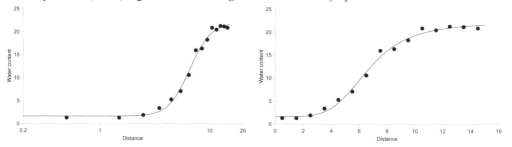

Figure 8.3 *The observed and Hill model-predicted data on a semi-logarithmic scale (left) and on a linear scale (right).*

Table 8.2   Parameter estimates of Gompertz, Logistic and Weibull Models

Parameter	Gompertz model		Logistic model		Weibull model	
	Estimate	CV%	Estimate	CV%	Estimate	CV%
α	22.5	3.7	21.5	1.9	21.1	1.8
β	2.11	11.2	3.95	6.7	19.8	3.1
γ	0.388	11.9	0.621	7.2	0.00181	57.2
δ					3.17	9.1
WRSS	12.59	-	6.21	-	5.45	

Table 8.3   Parameter estimates of Richards, Morgan-Mercer-Flodin and Hill Models

Parameter	Richards model		MMF model		Hill model	
	Estimate	CV%	Estimate	CV%	Estimate	CV%
α	21.2	2.1	22.1	2.9	1.65	26.4
β	5.71	33	1.64	26.6	20.4	4.2
γ	0.78	24	5300	98.1	6.63	2.6
δ	1.62	42	4.53	11.7	4.56	11.5
WRSS	5.52		6.37	-	6.37	-

Figure 8.4 compares the fit of the Hill and Gompertz models and Figure 8.5 the Richards and Logistic models. The Hill and Richards models showed generally a good fit to the data.

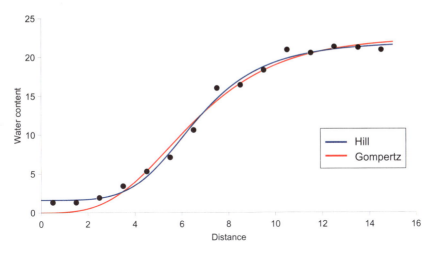

Figure 8.4 *Observed and model-predicted data using the Hill model (blue line) and Gompertz model (red line). Note the systematic deviation of the Gompertz model from the observed data at low values.*

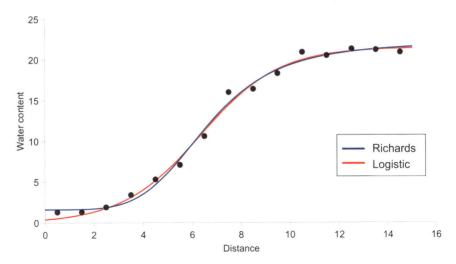

Figure 8.5 *Observed and model-predicted data using the Richards model (blue line) and Logistic model (red line). Note the systematic deviation of the Logistic model from the observed data at low values.*

Figure 8.6 shows the delta function of the Hill and Weibull models. The delta function takes the form

$$\Delta = \left[Hill\ model - Weibull\ model \right]^2 = \left[\left(\alpha + \frac{\beta \cdot X^\delta}{\gamma^\delta + X^\delta} \right) - \left(\alpha - \beta \cdot e^{-\gamma \cdot X^\delta} \right) \right]^2 \quad (8{:}7)$$

which is then plotted in Figure 8.6.

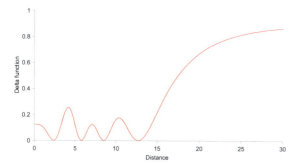

Figure 8.6 *Plot of the delta function of the Hill and Weibull models versus distance. This plot shows five (5) extreme (maximum) values where one would like to sample in future experiments to be able to discriminate between the models. However, obtaining more datasets is probably a more robust approach, since the deviation between the models is marginal.*

As seen from Table 8.2 and 8.3, the Gompertz model had the highest (12.59) *WRSS* and the Weibull model the lowest (5.45). The Gompertz model is rejected because of the systematic deviations between observed and predicted values. The Logistic model is also questionable for this partricular dataset because of systematic deviations from the observed values at lower values. However, both 3-parameter models (Gompertz and Logistic) demonstrated high parameter precision. In spite of the fact that the Hill and Weibull models are structurally very different, they had low *WRSS* and high parameter precision for at least three of their four parameters. One would ideally need more datasets to be able to fully discriminate between the models.

We highly recommend the work by Ratowsky [1983] on growth models. The reference also contains several other datasets that are analyzed by means of the models presented here, as well as other logistic models. Godfrey [1983] discusses the delta function in detail.

Appendix

We discuss the behavior of the Weibull equation briefly below. The Weibull model is known for its asymmetric characteristics. Figure 8.7 demonstrates some of the features of this model.

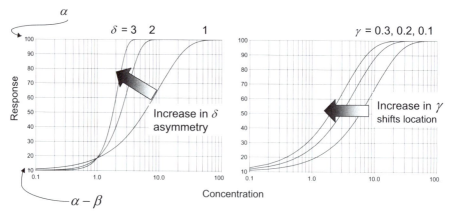

Figure 8.7 *The behavior of the Weibull model when the delta (left) and gamma (right) parameters are changed. Note the asymmetric characteristics of the model.*

Figure 8.8 demonstrates two curves (Weibull and Hill) fit to an asymmetric dataset.

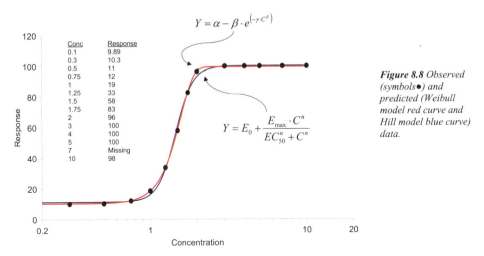

$$Y = \alpha - \beta \cdot e^{\left(-\gamma \cdot C^{\delta}\right)}$$

Conc	Response
0.1	9.89
0.3	10.3
0.5	11
0.75	12
1	19
1.25	33
1.5	58
1.75	83
2	96
3	100
4	100
5	100
7	Missing
10	98

$$Y = E_0 + \frac{E_{max} \cdot C^n}{EC_{50}^n + C^n}$$

Figure 8.8 Observed (symbols●) and predicted (Weibull model red curve and Hill model blue curve) data.

Figure 8.9 demonstrates the residuals obtained from the fitted Weibull and Hill models of Figure 8.8. As the Weibull model exhibits more randomness about the zero-line ans smaller residuals, it is the model of choice.

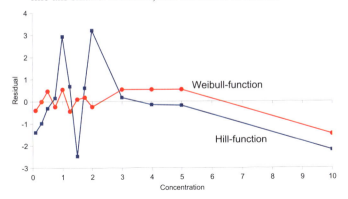

Weibull-function

Hill-function

Figure 8.9 Absolute residuals of the Weibull model (red line) and the Hill model (blue line) results shown in Figure 8.8. Note that the Weibull model exhibits more randomness about the zero-line than does the Hill model, and is associated with smaller residuals.

Solution I – Gompertz model

```
MODEL
 COMMANDS
NPAR 3
PNAM  'alpha', 'beta', 'gam'

END
     1: TEMPORARY
     2: C = X
     3: END
     4: FUNC 1
     5: DEL1 = -EXP(BETA-GAM*C)
     6: DEL2 = EXP(DEL1)
```

```
    7: F = ALPHA*DEL2
    8: END
    9: EOM
NOPAGE BREAKS
NVARIABLES 2
NPOINTS 100
XNUMBER 1
YNUMBER 2
METHOD 2   'Gauss-Newton (Levenberg and Hartley)
ITERATIONS 50
INITIAL 50,1,1
LOWER BOUNDS 0,0,0
UPPER BOUNDS 100,10,10
MISSING 'Missing'
NOBSERVATIONS 15
DATA 'WINNLIN.DAT'
BEGIN
```

PARAMETER	ESTIMATE	STANDARD ERROR	CV%	UNIVARIATE C.I.	
ALPHA	22.506955	.837396	3.72	20.682424	24.331486
BETA	2.105892	.236156	11.21	1.591351	2.620432
GAM	.388004	.046091	11.88	.287581	.488427

```
*** CORRELATION MATRIX OF THE ESTIMATES ***
PARAMETER  ALPHA       BETA        GAM
ALPHA      1.00000
BETA       -.64830  1.00000
GAM        -.79564   .95347   1.00000

  Condition_number=          97.87

    FUNCTION   1
```

X	OBSERVED Y	PREDICTED Y	RESIDUAL	WEIGHT	SE-PRED	STANDARDI RESIDUAL
.5000	1.300	.2594E-01	1.274	1.000	.3797E-01	1.245
1.500	1.300	.2285	1.071	1.000	.1839	1.063
2.500	1.900	.9999	.9001	1.000	.4246	.9656
3.500	3.400	2.722	.6781	1.000	.5823	.8047
4.500	5.300	5.369	-.6919E-01	1.000	.5665	-.8108E-01
5.500	7.100	8.513	-1.413	1.000	.4977	-1.578
6.500	10.60	11.64	-1.038	1.000	.4950	-1.157
7.500	16.00	14.39	1.613	1.000	.5063	1.811
8.500	16.40	16.61	-.2141	1.000	.4754	-.2360
9.500	18.30	18.32	-.1792E-01	1.000	.4179	-.1916E-01
10.50	20.90	19.57	1.328	1.000	.3800	1.396
11.50	20.50	20.47	.2836E-01	1.000	.3939	.2999E-01
12.50	21.30	21.11	.1947	1.000	.4502	.2117
13.50	21.20	21.55	-.3462	1.000	.5215	-.3927
14.50	20.90	21.85	-.9506	1.000	.5899	-1.135

```
CORRECTED SUM OF SQUARED OBSERVATIONS =  949.929
WEIGHTED CORRECTED SUM OF SQUARED OBSERVATIONS =  949.929
SUM OF SQUARED RESIDUALS =        12.5905
SUM OF WEIGHTED SQUARED RESIDUALS =  12.5905
S =  1.02431     WITH     12 DEGREES OF FREEDOM
CORRELATION (OBSERVED,PREDICTED) =  .9944
```

```
AIC criteria =          43.99418
SC  criteria =          46.11833
```

Solution II – Logistic model

```
MODEL
 COMMANDS
NPAR 3
PNAM  'alpha', 'beta', 'gam'
END
     1: TEMPORARY
     2: C = X
     3: END
     4: FUNC 1
     5: DEL1 = BETA-GAM*C
     6: DEL2 = 1 + EXP(DEL1)
     7: F = ALPHA/DEL2
     8: END
     9: EOM
NOPAGE BREAKS
NVARIABLES 2
NPOINTS 100
XNUMBER 1
YNUMBER 2
METHOD 2  'Gauss-Newton (Levenberg and Hartley)
ITERATIONS 50
INITIAL 50,1,1
LOWER BOUNDS 0,0,0
UPPER BOUNDS 100,10,10
MISSING 'Missing'
NOBSERVATIONS 15
DATA 'WINNLIN.DAT'
BEGIN
```

PARAMETER	ESTIMATE	STANDARD ERROR	CV%	UNIVARIATE C.I.	
ALPHA	21.511675	.415351	1.93	20.606702	22.416648
BETA	3.953035	.263032	6.65	3.379937	4.526133
GAM	.621493	.044797	7.21	.523889	.719097

```
*** CORRELATION MATRIX OF THE ESTIMATES ***
PARAMETER  ALPHA      BETA       GAM
ALPHA      1.00000
BETA       -.42994   1.00000
GAM        -.59814   .95341     1.00000

 Condition_number=        45.32

    FUNCTION   1
```

X	OBSERVED Y	PREDICTED Y	RESIDUAL	WEIGHT	SE-PRED	STANDARDI RESIDUAL
.5000	1.300	.5491	.7509	1.000	.1338	1.062
1.500	1.300	1.000	.2998	1.000	.1981	.4335
2.500	1.900	1.790	.1097	1.000	.2736	.1648
3.500	3.400	3.110	.2900	1.000	.3406	.4576

4.500	5.300	5.149	.1515	1.000	.3698	.2454
5.500	7.100	7.946	-.8463	1.000	.3592	-1.358
6.500	10.60	11.22	-.6216	1.000	.3651	-1.003
7.500	16.00	14.41	1.587	1.000	.3881	2.620
8.500	16.40	17.01	-.6111	1.000	.3634	-.9842
9.500	18.30	18.84	-.5350	1.000	.3039	-.8205
10.50	20.90	19.99	.9139	1.000	.2720	1.372
11.50	20.50	20.66	-.1644	1.000	.2882	-.2494
12.50	21.30	21.05	.2519	1.000	.3233	.3920
13.50	21.20	21.26	-.6017E-01	1.000	.3551	-.9617E-01
14.50	20.90	21.38	-.4758	1.000	.3779	-.7774

```
CORRECTED SUM OF SQUARED OBSERVATIONS =  949.929
WEIGHTED CORRECTED SUM OF SQUARED OBSERVATIONS =  949.929
SUM OF SQUARED RESIDUALS =          6.21037
SUM OF WEIGHTED SQUARED RESIDUALS =  6.21037
S = .719396      WITH      12 DEGREES OF FREEDOM
CORRELATION (OBSERVED,PREDICTED) = .9969

AIC criteria =        33.39331
SC  criteria =        35.51746
```

Solution III – Weibull model

```
MODEL
 COMMANDS
NPAR 4
PNAM  'alpha', 'beta', 'gam', 'delta'

END
    1: TEMPORARY
    2: C = X
    3: END
    4: FUNC 1
    5: DEL1 = -GAM*C**DELTA
    6: DEL2 = EXP(DEL1)
    7: F = ALPHA - BETA*DEL2
    8: END
    9: EOM
NOPAGE BREAKS
NVARIABLES 2
NPOINTS 100
XNUMBER 1
YNUMBER 2
METHOD 2  'Gauss-Newton (Levenberg and Hartley)
ITERATIONS 50
INITIAL 50,20,.002,1
LOWER BOUNDS 0,0,0,0
UPPER BOUNDS 100,50,1,10
MISSING 'Missing'
NOBSERVATIONS 15
DATA 'WINNLIN.DAT'
BEGIN
```

PARAMETER	ESTIMATE	STANDARD ERROR	CV%	UNIVARIATE C.I.	
ALPHA	21.108527	.378024	1.79	20.276499	21.940554
BETA	19.830399	.619996	3.13	18.465793	21.195005

GAM	.001808	.001033	57.15	-.000466	.004082
DELTA	3.169239	.288035	9.09	2.535276	3.803202

```
*** CORRELATION MATRIX OF THE ESTIMATES ***
PARAMETER  ALPHA      BETA        GAM        DELTA
ALPHA      1.00000
BETA        .71464   1.00000
GAM         .35904    .66391    1.00000
DELTA      -.42993   -.67739    -.99252    1.00000

  Condition_number=       8226.

    FUNCTION    1
```

X	OBSERVED Y	PREDICTED Y	RESIDUAL	WEIGHT	SE-PRED	STANDARDI RESIDUAL
.5000	1.300	1.282	.1789E-01	1.000	.4365	.3241E-01
1.500	1.300	1.407	-.1073	1.000	.4005	-.1854
2.500	1.900	1.922	-.2159E-01	1.000	.3301	-.3474E-01
3.500	3.400	3.090	.3099	1.000	.3159	.4929
4.500	5.300	5.074	.2256	1.000	.3765	.3794
5.500	7.100	7.834	-.7341	1.000	.4003	-1.268
6.500	10.60	11.08	-.4775	1.000	.3690	-.7968
7.500	16.00	14.32	1.676	1.000	.3721	2.805
8.500	16.40	17.08	-.6841	1.000	.3993	-1.181
9.500	18.30	19.06	-.7575	1.000	.3631	-1.257
10.50	20.90	20.23	.6709	1.000	.2939	1.049
11.50	20.50	20.80	-.2981	1.000	.2899	-.4649
12.50	21.30	21.02	.2798	1.000	.3328	.4512
13.50	21.20	21.09	.1113	1.000	.3631	.1846
14.50	20.90	21.11	-.2051	1.000	.3745	-.3443

```
CORRECTED SUM OF SQUARED OBSERVATIONS =   949.929
WEIGHTED CORRECTED SUM OF SQUARED OBSERVATIONS =  949.929
SUM OF SQUARED RESIDUALS =          5.44747
SUM OF WEIGHTED SQUARED RESIDUALS =  5.44747
S =  .703722     WITH     11 DEGREES OF FREEDOM
CORRELATION (OBSERVED,PREDICTED) =   .9971

AIC criteria =        33.42727
SC  criteria =        36.25947
```

Solution IV – Richards model

```
MODEL
 COMMANDS
NPAR 4
PNAM 'alpha', 'beta', 'gam', 'delta'
END
    1: TEMPORARY
    2: C = X
    3: END
    4: FUNC 1
    5: DEL1 = BETA - GAM*C
    6: DEL2 = EXP(DEL1)
    7: DEL3 = (1 + DEL2)**(1/DELTA)
    8: F = ALPHA/DEL3
```

```
   9: END
  10: EOM
NOPAGE BREAKS
NVARIABLES 2
NPOINTS 1000
XNUMBER 1
YNUMBER 2
METHOD 2   'Gauss-Newton (Levenberg and Hartley)
ITERATIONS 50
INITIAL 10,3,1,2
LOWER BOUNDS 0,0,0,.0001
UPPER BOUNDS 30,10,5,10
MISSING 'Missing'
NOBSERVATIONS 15
DATA 'WINNLIN.DAT'
BEGIN
```

PARAMETER	ESTIMATE	STANDARD ERROR	CV%	UNIVARIATE C.I.	
ALPHA	21.201315	.443708	2.09	20.224717	22.177914
BETA	5.705826	1.870607	32.78	1.588631	9.823020
GAM	.778712	.183057	23.51	.375805	1.181618
DELTA	1.622873	.675730	41.64	.135597	3.110150

```
*** CORRELATION MATRIX OF THE ESTIMATES ***
PARAMETER  ALPHA      BETA       GAM        DELTA
ALPHA      1.00000
BETA       -.601717   1.00000
GAM        -.657386   .990659    1.00000
DELTA      -.562503   .984335    .95638     1.00000

Condition_number=        173.3

FUNCTION    1
```

X	OBSERVED Y	PREDICTED Y	RESIDUAL	WEIGHT	SE-PRED	STANDARDI RESIDUAL
.5000	1.300	.7986	.5014	1.000	.2719	.7663
1.500	1.300	1.286	.1420E-01	1.000	.3246	.2255E-01
2.500	1.900	2.062	-.1618	1.000	.3558	-.2640
3.500	3.400	3.278	.1225	1.000	.3532	.1994
4.500	5.300	5.118	.1819	1.000	.3489	.2949
5.500	7.100	7.723	-.6232	1.000	.4057	-1.073
6.500	10.60	10.99	-.3907	1.000	.4395	-.7029
7.500	16.00	14.40	1.603	1.000	.4036	2.753
8.500	16.40	17.22	-.8221	1.000	.4237	-1.448
9.500	18.30	19.10	-.8041	1.000	.3987	-1.373
10.50	20.90	20.17	.7327	1.000	.3178	1.157
11.50	20.50	20.71	-.2098	1.000	.2922	-.3250
12.50	21.30	20.97	.3280	1.000	.3290	.5227
13.50	21.20	21.10	.1048	1.000	.3729	.1739
14.50	20.90	21.15	-.2524	1.000	.4038	-.4336

```
CORRECTED SUM OF SQUARED OBSERVATIONS =  949.929
WEIGHTED CORRECTED SUM OF SQUARED OBSERVATIONS =  949.929
SUM OF SQUARED RESIDUALS =          5.52296
SUM OF WEIGHTED SQUARED RESIDUALS =  5.52296
```

```
S = .708581      WITH    11 DEGREES OF FREEDOM
CORRELATION (OBSERVED,PREDICTED) =   .9971
```

```
AIC criteria =          33.63370
SC  criteria =          36.46590
```

Solution V – Morgan-Mercer-Flodin model

```
MODEL
 COMMANDS
NPAR 4
PNAM  'alpha', 'beta', 'gam', 'delta'
END
     1: TEMPORARY
     2: C = X
     3: END
     4: FUNC 1
     5: DEL1 = GAM + C**DELTA
     6: DEL2 = BETA*GAM + ALPHA*(C**DELTA)
     7: F = DEL2/DEL1
     8: END
     9: EOM
NOPAGE BREAKS
NVARIABLES 2
NPOINTS 100
XNUMBER 1
YNUMBER 2
METHOD 2  'Gauss-Newton (Levenberg and Hartley)
ITERATIONS 50
INITIAL 20,1,1000,5
LOWER BOUNDS 0,0,0,0
UPPER BOUNDS 100,10,10000,10
MISSING 'Missing'
NOBSERVATIONS 15
DATA 'WINNLIN.DAT'
BEGIN
```

PARAMETER	ESTIMATE	STANDARD ERROR	CV%	UNIVARIATE C.I.	
ALPHA	22.100085	.645292	2.92	20.679800	23.520370
BETA	1.640143	.436360	26.60	.679718	2.600568
GAM	5303.769524	5201.819862	98.08	-6145.402525	16752.941572
DELTA	4.532653	.528875	11.67	3.368602	5.696705

```
*** CORRELATION MATRIX OF THE ESTIMATES ***
PARAMETER ALPHA     BETA      GAM       DELTA
ALPHA     1.00000
BETA      -.26817  1.00000
GAM       -.70452  .53052  1.00000
DELTA     -.76200  .48181  .99280   1.00000
```

```
  Condition_number=        .1384E+06
```

```
    FUNCTION   1
```

X	OBSERVED Y	PREDICTED Y	RESIDUAL	WEIGHT	SE-PRED	STANDARDI RESIDUAL
.5000	1.300	1.640	-.3403	1.000	.4362	-.5457

1.500	1.300	1.664	-.3644	1.000	.4259	-.5776
2.500	1.900	1.883	.1727E-01	1.000	.3751	.2608E-01
3.500	3.400	2.709	.6906	1.000	.3489	1.021
4.500	5.300	4.647	.6532	1.000	.4505	1.065
5.500	7.100	7.770	-.6702	1.000	.4738	-1.125
6.500	10.60	11.40	-.8002	1.000	.4177	-1.258
7.500	16.00	14.65	1.354	1.000	.4182	2.129
8.500	16.40	17.08	-.6819	1.000	.3991	-1.052
9.500	18.30	18.74	-.4429	1.000	.3408	-.6509
10.50	20.90	19.83	1.068	1.000	.2964	1.524
11.50	20.50	20.54	-.3960E-01	1.000	.3000	-.5662E-01
12.50	21.30	21.00	.2956	1.000	.3388	.4337
13.50	21.20	21.31	-.1147	1.000	.3875	-.1751
14.50	20.90	21.53	-.6259	1.000	.4332	-1.000

```
CORRECTED SUM OF SQUARED OBSERVATIONS =  949.929
WEIGHTED CORRECTED SUM OF SQUARED OBSERVATIONS =  949.929
SUM OF SQUARED RESIDUALS =              6.37135
SUM OF WEIGHTED SQUARED RESIDUALS =  6.37135
S =  .761061      WITH      11 DEGREES OF FREEDOM
CORRELATION (OBSERVED,PREDICTED) =  .9966

AIC criteria =         35.77718
SC  criteria =         38.60938
```

Solution VI – Hill model

```
MODEL
 COMMANDS
NPAR 4
PNAM 'alpha', 'beta', 'gam', 'delta'
END
    1: TEMPORARY
    2: C = X
    3: END
    4: FUNC 1
    5: DEL1 = BETA*(C**GAM)
    6: DEL2 = DELTA**GAM + C**GAM
    7: F = ALPHA + DEL1/DEL2
    8: END
    9: EOM
NOPAGE BREAKS
NVARIABLES 2
NPOINTS 100
XNUMBER 1
YNUMBER 2
METHOD 2   'Gauss-Newton (Levenberg and Hartley)
ITERATIONS 50
INITIAL 1,20,2,10
LOWER BOUNDS 0,0,0,0
UPPER BOUNDS 10,50,10,50
MISSING 'Missing'
NOBSERVATIONS 15
DATA 'WINNLIN.DAT'
BEGIN
```

PARAMETER	ESTIMATE	STANDARD ERROR	CV%	UNIVARIATE C.I.	
ALPHA	1.648966	.434569	26.35	.692484	2.605449

```
BETA        20.424968         .864124        4.23      18.523036      22.326900

GAM          4.559471         .522484       11.46       3.409486       5.709455

DELTA        6.631007         .175292        2.64       6.245191       7.016823
```

*** CORRELATION MATRIX OF THE ESTIMATES ***

```
PARAMETER  ALPHA      BETA       GAM        DELTA
ALPHA      1.00000
BETA       -.69908  1.00000
GAM         .47783  -.80408   1.00000
DELTA       .31887   .27871   -.22183    1.00000
```

```
Condition_number=        10.65
```

FUNCTION   1

X	OBSERVED Y	PREDICTED Y	RESIDUAL	WEIGHT	SE-PRED	STANDARDI RESIDUAL
.5000	1.300	1.649	-.3491	1.000	.4345	-.5588
1.500	1.300	1.672	-.3722	1.000	.4247	-.5895
2.500	1.900	1.885	.1469E-01	1.000	.3754	.2220E-01
3.500	3.400	2.701	.6993	1.000	.3472	1.033
4.500	5.300	4.628	.6723	1.000	.4487	1.094
5.500	7.100	7.753	-.6535	1.000	.4758	-1.100
6.500	10.60	11.40	-.7972	1.000	.4188	-1.255
7.500	16.00	14.66	1.345	1.000	.4189	2.116
8.500	16.40	17.10	-.6951	1.000	.4003	-1.074
9.500	18.30	18.75	-.4536	1.000	.3414	-.6670
10.50	20.90	19.84	1.063	1.000	.2965	1.517
11.50	20.50	20.54	-.3936E-01	1.000	.3001	-.5628E-01
12.50	21.30	21.00	.3009	1.000	.3389	.4415
13.50	21.20	21.31	-.1053	1.000	.3875	-.1607
14.50	20.90	21.51	-.6131	1.000	.4329	-.9797

```
CORRECTED SUM OF SQUARED OBSERVATIONS =  949.929
WEIGHTED CORRECTED SUM OF SQUARED OBSERVATIONS =  949.929
SUM OF SQUARED RESIDUALS =            6.37002
SUM OF WEIGHTED SQUARED RESIDUALS =  6.37002
S =  .760982      WITH    11 DEGREES OF FREEDOM
CORRELATION (OBSERVED,PREDICTED) =  .9966
```

```
AIC criteria =       35.77405
SC  criteria =       38.60625
```

PD9 - Tolerance modeling I – Single intravenous infusion

Objectives

◆ To analyze the turnover of response after a constant intravenous infusion

◆ To write a multi-compartment model in terms of differential equations

◆ To obtain initial parameter estimates for the feedback and pool models

◆ To discuss various aspects of experimental design

Problem specification

The following exercise demonstrates how to build an indirect response model that also captures tolerance/rebound. Response-time data are given in Table 9.1.

Table 9.1 Response-time data.

Time (min)	Response (MHz)	Time (min)	Response (MHz)	Time (min)	Response (MHz)
0	18.85	16	5.51	32	18.44
0.5	19.91	18	6.68	34	20.5
1	19.51	20	5.96	36	20.09
1.5	18.67	20.5	6.24	38	23.7
2	19.03	21	5.68	40	21.76
2.5	16.04	21.5	6.34	45	22.96
3	15.83	22	6.63	50	24.04
3.5	11.1	22.5	6.13	55	19.39
4	10.8	23	8.6	65	19.74
4.5	9.91	23.5	11.48	70	20
5	8.72	24	9.92	90	18.5
6	7.61	24.5	12.77	100	20.49
7	8.82	25	12.48	110	18.8
8	8.48	26	13.69	200	18.6
9	7.75	27	16.31	230	20.82
10	7.34	28	13.26	260	18.84
12	7.76	29	16.73		
14	5.5	30	17.18		

We will estimate the determinants of the response following a 20 min constant intravenous infusion dose of 1440 units. A three-compartment pharmacokinetic model was fit to the plasma concentration-time data. The following pharmacokinetic parameter estimates (*WinNonlin* code names, see output) were obtained by fitting a three compartment physiological model to the plasma concentration-time data. The values below are the kinetic parameter values for the 3-compartment model used to drive the dynamics.

$VC = 8.6594,$ $CL = 2.6958,$ $CLD1 = 1.24486,$ $CLD2 = 0.25277,$

$VT1 = 13.5731,$ $VT2 = 101.8012$

These parameters correspond to the volume of the central compartment V_c, clearance Cl, inter-compartmental diffusion between central and shallow compartment Cl_{d1}, inter-compartmental diffusion between central and deep compartment Cl_{d2}, tissue volume of the shallow compartment V_{t1}, and tissue volume of the deep compartment V_{t2}. The pharmacokinetic parameters were estimated and served as (constant) input to the pharmacodynamic model. Model 1 was an ordinary indirect response model with inhibition of production of response. Model 2 (Equations 9:1 and 9:2) was a feedback model in which drug inhibits the production of response and the moderator stimulates the loss of response. Total inhibition of formation of response is possible with Model 2. Drug acts by stimulating the loss of response in Model 3 (Equations shown in program output) and moderator M inhibits the production of response. Model 4 (Equation 9:18 and 9:19) is a pool model with I_{max} set to unity (1). Model 5 (Equations shown in program output) is also a pool model in which I_{max} is a parameter to be estimated.

Fit a feedback model (Model 2 in Figure 9.2) and a pool (precursor) model (Figure 9.3) to the response-time data in Table 9.1 and characterize the response-time course.

Description of the feedback model

One needs to decide what each parameter means and what data are needed in order to distinguish between dynamic steady state, rebound effect, and return to the baseline response. We will demonstrate how to estimate I_{max} or E_{max}, IC_{50} or EC_{50}, k_{in}, k_{out}, k_{tol} and k_{loss}.

Start by plotting the response *versus* time and response *versus* concentration (Figure 9.1) relationship.

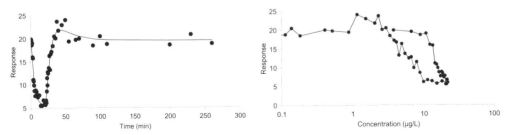

Figure 9.1 Observed response-time (left) and response-concentration (right, hysteresis) relationships.

In order to capture the tolerance and rebound phenomena that are evident from Figure 9.1, an expansion of the basic indirect response model was utilized, namely the feedback models and the pool models. Figure 9.2 displays schematically the turnover of response R and moderator M for the two feedback models (Models 2 and 3 in Figure 9.2). We believe that the assumptions we have made are reasonable in terms of a simple description of the behavior that might occur. The mechanism of action of drug X on R is not well defined, but you should fit Model 2 to the response-time data. Observations of the time course of its response shows no delay in response, although a rebound effect above the baseline level could be observed upon cessation of infusion.

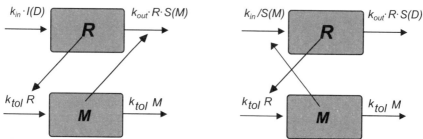

Figure 9.2 *Schematic illustration of the feedback models, where drug acts to inhibit the production of response (Model 2, left) or stimulate the loss of response (Model 3, right).*

The turnover of response in Model 2 (Figure 9.2 left) is written as

$$\frac{dR}{dt} = k_{in} \cdot H(t) - k_{out} \cdot R \cdot \left[1 + \frac{M}{M_{50}}\right] \tag{9:1}$$

$H(t)$ is the action (inhibition or stimulation) of drug on production of response. The term $1 + M/M_{50}$ serves as an amplifier to the loss of response. Since k_{out} and M_{50} partly become a ratio it will be hard to estimate them simultaneously with acceptable parameter precision. For simplicity, we assume M_{50} to be equal to unity and then the response function becomes

$$\frac{dR}{dt} = k_{in} \cdot H(t) - k_{out} \cdot R \cdot (1 + M) \tag{9:2}$$

The turnover of the moderator is

$$\frac{dM}{dt} = k_{tol} \cdot R - k_{tol} \cdot M \tag{9:3}$$

The turnover of the response at steady state is

$$\frac{dR}{dt} = k_{in} \cdot H(t) - k_{out} \cdot R_{ss} \cdot (1 + M_{ss}) = 0 \tag{9:4}$$

$H(t)$ is, in this case, inhibition of production of response. The turnover of moderator at steady state is

$$\frac{dM}{dt} = k_{tol} \cdot R_{ss} - k_{tol} \cdot M_{ss} = 0 \tag{9:5}$$

which rearranged becomes

$$k_{tol} \cdot R_{ss} - k_{tol} \cdot M_{ss} = 0 \tag{9:6}$$

and

$$k_{tol} \cdot R_{ss} = k_{tol} \cdot M_{ss} \qquad (9:7)$$

As k_{out} cancels out

$$R_{ss} = M_{ss} \qquad (9:8)$$

M_{ss} replaced by R_{ss} in Equation 9:4 gives

$$k_{in} \cdot H(t) - k_{out} \cdot R_{ss} \cdot (1 + R_{ss}) = 0 \qquad (9:9)$$

and rearranged gives

$$R_{ss}^2 + R_{ss} - \frac{k_{in}}{k_{out}} \cdot H(t) = 0 \qquad (9:10)$$

A second-order polynomial of the form

$$x^2 + x - a = 0 \qquad (9:11)$$

has the solution

$$x_{1,2} = -\frac{1}{2} \pm \sqrt{\frac{1}{4} + a} \qquad (9:12)$$

The solution for the R function with no drug present is then

$$R_0 = -\frac{1}{2} + \sqrt{\frac{1}{4} + \frac{k_{in}}{k_{out}}} \qquad (9:13)$$

Including the drug response, then the steady state solution becomes

$$R_{ss} = -\frac{1}{2} + \sqrt{\frac{1}{4} + \frac{k_{in}}{k_{out}} \cdot H(t)} \qquad (9:14)$$

If $H(t)$ is an E_{max} type of function, then R_{ss} is

$$R_{ss} = -\frac{1}{2} + \sqrt{\frac{1}{4} + \frac{k_{in}}{k_{out}} \cdot \left[1 + \frac{E_{max} \cdot C}{EC_{50} + C}\right]} \qquad (9:15)$$

The maximum value of R, R_{max} is then

$$R_{max} = -\frac{1}{2} + \sqrt{\frac{1}{4} + \frac{k_{in}}{k_{out}} \cdot [1 + E_{max}]}$$ (9:16)

If $H(t)$ is an I_{max} type of function, then the minimum value of R, R_{min} is

$$R_{min} = -\frac{1}{2} + \sqrt{\frac{1}{4} + \frac{k_{in}}{k_{out}} \cdot [1 - I_{max}]}$$ (9:17)

Description of the pool model

A pool model was also elaborated (Figure 9.3). This model incorporates k_{in}, which we still call the zero-order production, and k_{loss}, the first-order exit rate constant from the pool compartment.

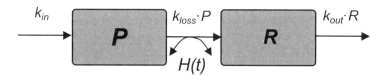

Figure 9.3 *Schematic illustration of the basic physiological pool model. H(t) is either stimulation or inhibition of the process. In this example, H(t) inhibits the loss from the pool compartment P.*

Turnover in the pool compartment P is

$$\frac{dP}{dt} = k_{in} - k_{loss} \cdot P$$ (9:18)

Turnover in the response compartment R is

$$\frac{dR}{dt} = k_{loss} \cdot P - k_{out} \cdot R$$ (9:19)

The baseline value of the pool compartment

$$P_0 = \frac{k_{in}}{k_{loss}}$$ (9:20)

and of the response compartment

$$R_0 = \frac{k_{in}}{k_{out}}$$ (9:21)

Inhibition of loss from the pool compartment

$$\frac{dP}{dt} = k_{in} - k_{loss} \cdot P \cdot H(t) \tag{9:22}$$

$$\frac{dR}{dt} = k_{loss} \cdot P \cdot H(t) - k_{out} \cdot R \tag{9:23}$$

The steady state condition of the pool compartment when drug is present becomes

$$P_{ss} = \frac{k_{in}}{k_{loss}} \cdot \frac{1}{H(t)} \tag{9:24}$$

The steady state condition of the response compartment with drug present is

$$R_{ss} = \frac{P_{ss} \cdot k_{loss}}{k_{out}} \cdot H(t) = \frac{P_{ss} \cdot k_{loss}}{k_{out}} \cdot \frac{k_{in}}{k_{loss} \cdot P_{ss}} = \frac{k_{in}}{k_{out}} \tag{9:25}$$

Initial parameter estimates

Derive initial parameter estimates graphically of k_{in}, k_{out}, I_{max}, IC_{50} and k_{tol}/k_{loss} for the feedback and pool models. The k_{out} parameter is obtained from the initial downswing of the time-course of the response

$$k_{out} = \frac{\ln(R_1 / R_2)}{t_1 - t_2} \tag{9:26}$$

According to the feedback model, k_{in} is obtained from the baseline value R_0 squared and then multiplied by k_{out} since

$$R_0 \propto \sqrt{\frac{k_{in}}{k_{out}}} \tag{9:27}$$

A good initial estimate of IC_{50} is obtained from the response *versus* concentration plot (Figure 9.1, right) when the concentration is declining and the response is increasing. The n and m parameters are obtained from the relationship below

$$slope_{IC_{50}} = m = \frac{R_1 - R_2)}{\ln(C_1 / C_2)} \approx \frac{I_{max} \cdot n}{4} \tag{9:28}$$

Interpretation of results and conclusions

We fit five different dynamic models to the data (Table 9.2). Model 1 was an ordinary indirect response model. Model 2 was a feedback model in which drug acts as inhibitor to the production of response and the moderator stimulates the loss of response. Total inhibition of formation of response is possible with Model 2. In Model 3, drug acts by stimulating the loss of R and M inhibits the production of response. Model 4 was a pool model with I_{max} set to unity (1), and in Model 5, I_{max} was a parameter to be estimated. Pertinent data on all models are shown in Table 9.2, including the weighted sum of squares *WRSS* and the *Akaike Information Criterion AIC*.

Table 9.2   Summary statistics from regressions of models 1-5.

Parameter	Model 1* mean ± CV%	Model 2 fb** mean ± CV%	Model 3 fb** mean ± CV%	Model 4 pool mean ± CV%	Model 5 pool mean ± CV%
k_{in}	9.57 ± 13	6.88 ± 11	47.8 ± 17	6.68 ± 13	6.81 ± 14
k_{out}	0.48 ± 13	0.017 ± 11	0.12 ± 17	0.35 ± 12	0.36 ± 13
EC_{50}/IC_{50}	12.5 ± 5.3	10.3 ± 12	10.6 ± 5.1	11.5 ± 6.1	8.44 ± 11
n	2.06 ± 13	1.57 ± 14	40.0 ± 441	1.81 ± 12	2.97 ± 35
k_{tol}/k_{loss}	–	0.023 ± 43	.022 ± 34	0.014 ± 31	0.009 ± 40
E_{max}/I_{max}	–	–	2.40 ± 9.6	1	0.76 ± 10
WRSS	146	80.7	62.0	96.8	90.5
AIC	277	238	226	248	246

* The program output from fitting Model 1 to data is not presented due to inadequate fit. **fb denotes feedback.

Model 3 had the lowest *WRSS* value, this being 77% of that for Model 2, and 64% of that for Model 4. However, Model 3 also estimated an exponent with a very low precision (CV ~440%). Model 5, which includes an I_{max} parameter, reduced *WRSS* to about 90.5. This is significantly better than Model 4 by an *F-test*. I_{max} for Model 5 was $0.76 \pm 10\%$, which was significantly different from 1. This means that full blockage by the drug could not be established. However, *WRSS* was not lower than for Models 2 and 3, and the parameter precision was lower than for Model 4 for n (35%, compared with 12%). Based on these results only, we would select Models 2 and 3.

Figure 9.4 shows the fits of Models 2 and 3 with ±2 times the predicted standard deviation. These two models are structurally completely different, which results in very different estimates of k_{in} and k_{out}. Since the parameters of Model 2 generally have good precision (except for k_{tol}) and also capture the rebound effect well, we would select that model as the 'best' of the ones fitted.

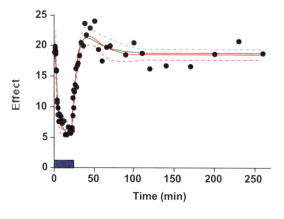

Figure 9.4 Schematic representation of observed and predicted response versus time. Solid curves are predicted response with Models 2 and 3. Dashed lines are ± 2SD

The pool model also mimics the data reasonably well. It captures the rebound effect and the parameters have good precision. The k_{in}, IC_{50}, n and k_{tol} are close to the estimates of Model 2, with similar parameter precision (except for k_{loss} which has a CV% of 31). However, it is difficult to select the feedback model as being superior to the pool model based on only one dataset. The mechanism behind the response needs to be elucidated, such as whether the response is determined by a pool of some endogenous substance, or counter-regulated by means of other biochemical substances. Several dose levels and multiple dosing would also be highly recommended. In other words, *Challenge your models!*

Application of the modified feedback model (Models 2 and 3) has enabled us to describe the important components of the observed response. By identifying and quantifying the influence of each of these, we can not only draw conclusions about the response of this drug, but also predict the time course of the treatment response over the period of a trial.

The key features of Models 2-5 capture:

> - **the response after multiple dosing – onset and loss of response**
> - **reversible/irreversible responses**
> - **tolerance and rebound phenomena**
> - **apparently time-variant systems that may contain time-invariant parameters**

Solution I - Feedback model (Model 2 – Inhibition of production)

```
MODEL
 COMM
   NDER X
   NFUN X
   NCON X
   NPAR X
   PNAMES 'Kin', 'Kout', 'IC50', 'g1', 'ktol'
END
   TEMP
   T=X
   D1 = CON(1)
   T1 = CON(2)
   VC = 8.6594
   CL = 2.6958
   CLD1= 1.24486
   CLD2= 0.25277
   VT1= 13.5731
   VT2= 101.8012

   KIN = P(1)
   KOUT = P(2)
   IC50 = P(3)
   G1 = P(4)
   KTOL = P(5)
   END
   START
   R0 = -0.5 + SQRT(0.25 + KIN/KOUT)
   Z(1)=R0
   Z(2)=0.0
   Z(3)=0.0
```

```
Z(4)=0.0
Z(5)=R0
END
DIFF
FI1 = 0.0
IF T LE T1 THEN
FI1 = D1/T1
ENDIF
IF Z(2) LE 0.0 THEN
ZZ2 = 0.0
ELSE
ZZ2 = Z(2)**G1
ENDIF
DRUG = (1. - ZZ2/(IC50**G1+ZZ2))
DZ(1) = KIN*DRUG - KOUT*Z(1)*(1. + Z(5))
DZ(2)=(FI1 - CL*Z(2)-CLD1*Z(2)-CLD2*Z(2)+CLD1*Z(3)+CLD2*Z(4))/VC
DZ(3)=(CLD1*Z(2) - CLD1*Z(3))/VT1
DZ(4)=(CLD2*Z(2) - CLD2*Z(4))/VT2
DZ(5) = KTOL*(Z(1) - Z(5))
END
FUNC 1
F=Z(1)
END
EOM
```

Solution - Model 4 – Precursor/pool model

```
MODEL
COMM
  NDER 5
  NFUN 1
  NCON 2
  NPARM 5
  PNAMES 'Kin', 'Kout', 'IC50', 'n', 'ktol'
END
TEMP
  T=X
  D1 = CON(1)
  T1 = CON(2)
  VC = 8.6594
  CL = 2.6958
  CLD1= 1.24486
  CLD2= 0.25277
  Vt1= 13.5731
  Vt2= 101.8012
  Kin  = P(1)
  Kout = P(2)
  IC50  = P(3)
  n     = P(4)
  Ktol = P(5)
END
START
  R0  = Kin/kout
  Z(1)= R0
  Z(2)= 0.
  Z(3)= 0.0
  Z(4)= 0.0
  Z(5)= Kin/ktol
END
DIFF
    Fi1 = 0.0
  IF T LE T1 THEN
```

```
   Fi1 = D1/T1
  ENDIF
  IF Z(2) LE 0.0 THEN
     ZZ2 = 0.0
  ELSE
     ZZ2 = Z(2)**n
  ENDIF
  Drug  = (1. - ZZ2/(IC50**n+ZZ2))
  DZ(1)  = ktol*Z(5)*Drug - kout*Z(1)
  DZ(2)=(Fi1 - CL*Z(2)-CLD1*Z(2)-CLD2*Z(2)+CLD1*Z(3)+CLD2*Z(4))/Vc
  DZ(3)=(CLD1*Z(2)  - CLD1*Z(3))/Vt1
  DZ(4)=(CLD2*Z(2)  - CLD2*Z(4))/Vt2
  DZ(5)  = Kin - ktol*Z(5)*Drug
END
func 1
 f=z(1)
end
EOM
```

PD10 - Tolerance modeling II – Repeated intravenous infusions

Objectives
◆ **To obtain initial estimates and fit a feedback model to data**

Problem specification

The following exercise will demonstrate how to build a turnover (indirect response) model that also captures tolerance. Two consecutive intravenous infusions, with a washout between, were administered to elucidate the time course of response when tolerance is observed. Figure 10.1 displays the observed data including the baseline response, tolerance development and rebound.

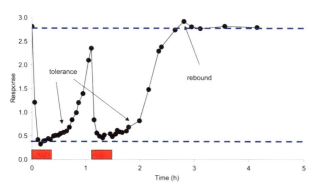

Figure 10.1 Schematic representation of the observed effect versus time and the principal parts including baseline, tolerance, and rebound. The two horizontal bars represent the infusion regimen.

One starts by plotting the data. What do the data in Figures 10.1 and 10.2 tell you?

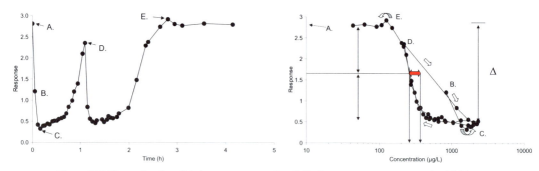

Figure 10.2 Observed and predicted response versus time (left). Response versus concentration (right). Arrows parallel to the predicted curve indicate the time order. Arrow on concentration axis indicates a range of initial estimates of EC_{50}.

Section 5.4.1.2 demonstrates how to obtain the initial estimates. It may be necessary to derive competing models. Then fit the model(s) to the experimental data and interpret and compare the output from several runs. Characterize the dynamics.

The negative feedback model was successfully fit to data. Note the rapidly induced tolerance, which is already acting during the infusion regimens. Note also that the trough value is not the same during the second infusion regimen as compared to the first. There is also a slight rebound after cessation of the second infusion regimen. Final parameter estimates are displayed in Table 10.1. The rate constant that governs the tolerance development (k_{tol}) is larger than the rate constant for the decline of response (k_{out}), which means that you will be able to observe tolerance development provided the experimental data have acceptable resolution.

Table 10.1 Comparisons of parameter estimates

Parameter	Basic turnover model mean ± CV%	Feedback model mean ± CV%	Precursor model mean ± CV%
k_{in}	21 ± 10	30 ± 5	94 ± 10
k_{out}	7.3 ± 10	2.9 ± 6	35 ± 9
EC_{50}/IC_{50}	390 ± 3	350 ± 1	270 ± 2
n	6.6 ± 10	7.4 ± 5	5.3 ± 7
k_{tol}	-	4.2 ± 10	0.05 ± 60
$E_{max} I_{max}$	4.8 ± 10	9.8 ± 4	0.84 ± 1
WRSS	0.33	0.083	0.28
AIC	−39.0	−97.5	−43.5
runs	15	23	

Figure 10.3 shows a simulation of the response-time course after a single constant rate intravenous infusion using the final parameter estimates from the regression. Note that the rebound is more pronounced.

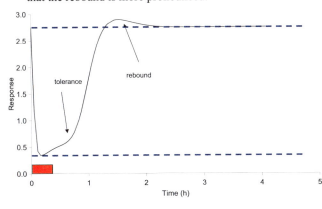

Figure 10.3 Simulation of the response time course after a single infusion using the final parameter estimates from the previous regression. Note that the rebound is more pronounced, since less of the drug from the terminal phase is controlling the response.

Notes for the reader

This exercise is aimed to demonstrate how to build an ordinary indirect response model and then extend it to incorporate compensatory mechanisms or a pool such as the feedback and pool models, respectively. Response-time data were obtained from a multiple infusion study of a new compound. Initial parameter estimates were taken from previous studies in the same animal species (see section 5.4). First, fit an ordinary indirect response model to the data and then the proposed tolerance model.

Solution – Basic turnover model

```
TITLE 1
Modelling of hysteresis data with adaptation
MODEL
 COMM
  NDER 4
  NFUN 1
  NCON 5
  NPARM 5
  PNAMES 'Kin', 'Kout', 'EC50', 'Emax', 'n'
END
     1: TEMP
     2: T=X
     3: D1 = CON(1)
     4: D2 = CON(2)
     5: TI1 = CON(3)
     6: TI2 = CON(4)
     7: T2 = CON(5)
     8: VC = 1.0
     9: CL = 6.5
    10: CLD1= 5.5
    11: CLD2= 0.90
    12: VT1= 1.75
    13: VT2= 3.45
    14: KIN = P(1)
    15: KOUT = P(2)
    16: EC50 = P(3)
    17: EMAX = P(4)
    18: N = P(5)
    19: R0 = KIN/KOUT
    20: END
    21: START
    22: Z(1)=R0
    23: Z(2)=0.0
    24: Z(3)=0.0
    25: Z(4)=0.0
    26: END
    27: DIFF
    28: FI1 = 0.0
    29: FI2 = 0.0
    30: IF T LE TI1 THEN
    31: FI1 = D1/TI1
    32: FI2 = 0.0
    33: ENDIF
    34: IF T GE T2 AND T LE T2+TI2 THEN
    35: FI1 = 0.0
    36: FI2 = D2/TI2
    37: ENDIF
    38: IF Z(2) LE 0.0 THEN
    39: ZZ2 = 0.0
    40: ELSE
    41: ZZ2 = Z(2)**N
    42: ENDIF
    43: DRUG = (1. + (EMAX*ZZ2/(EC50**N + ZZ2)))
```

```
    44: DZ(1) = KIN - DRUG*KOUT*Z(1)
    45: DZ(2)=(FI1+FI2-CL*Z(2)-CLD1*Z(2)-CLD2*Z(2)+CLD1*Z(3)+CLD2*Z(4))/VC
    46: DZ(3)=(CLD1*Z(2) - CLD1*Z(3))/VT1
    47: DZ(4)=(CLD2*Z(2) - CLD2*Z(4))/VT2
    48: END
    49: FUNC 1
    50: F=Z(1)
    51: END
    52: EOM
NVARIABLES 2
NPOINTS 500
XNUMBER 1
YNUMBER 2
CONSTANTS 7450,7350,.33,.28,1.1
METHOD 2  'Gauss-Newton (Levenberg and Hartley)
REWEIGHT -1
ITERATIONS 50
INITIAL 35,3.5,400,11,7.5
LOWER BOUNDS 1,.1,100,1,1
UPPER BOUNDS 45,10,550,25,15
NOBSERVATIONS 44
BEGIN
```

PARAMETER	ESTIMATE	STANDARD ERROR	CV%	UNIVARIATE C.I.	
KIN	20.968641	2.068861	9.87	16.783994	25.153288
KOUT	7.320378	.727977	9.94	5.847913	8.792843
EC50	386.315302	10.015277	2.59	366.057591	406.573013
EMAX	4.827919	.220508	4.57	4.381901	5.273937
N	6.558283	.644305	9.82	5.255060	7.861507

```
*** CORRELATION MATRIX OF THE ESTIMATES ***
```

PARAMETER	KIN	KOUT	EC50	EMAX	N
KIN	1.00000				
KOUT	.976426	1.00000			
EC50	-.289938	-.291380	1.00000		
EMAX	-.265068	-.398965	.503019	1.00000	
N	-.255197	-.179719	-.591191	-.367695	1.00000

```
 Condition_number=        128.8
```

X	OBSERVED Y	PREDICTED Y	RESIDUAL	WEIGHT	SE-PRED	STANDARDIZED RESIDUAL
.0000	2.811	2.864	-.5342E-01	.3491	.6161E-01	-.3748
.5000E-01	1.207	1.168	.3909E-01	.8559	.7786E-01	.6364
.1100	.4153	.5443	-.1290	1.837	.2100E-01	-2.004
.1600	.3213	.4978	-.1765	2.009	.1436E-01	-2.796
.2200	.3887	.4920	-.1033	2.033	.1446E-01	-1.647
.2500	.3957	.4916	-.9595E-01	2.034	.1450E-01	-1.531
.3100	.4392	.4915	-.5232E-01	2.035	.1451E-01	-.8350
.3500	.4216	.4915	-.6991E-01	2.035	.1451E-01	-1.116
.4100	.4936	.4918	.1821E-02	2.033	.1445E-01	.2905E-01
.4500	.5075	.4930	.1447E-01	2.028	.1425E-01	.2303
.5000	.5116	.4992	.1238E-01	2.003	.1374E-01	.1954
.5400	.5469	.5118	.3508E-01	1.954	.1391E-01	.5469
.6000	.5720	.5511	.2085E-01	1.814	.1771E-01	.3170
.6500	.6000	.6086	-.8568E-02	1.643	.2365E-01	-.1268

.7000	.6837	.6938	-.1009E-01	1.441	.2966E-01	-.1433
.7500	.8407	.8113	.2943E-01	1.233	.3395E-01	.3907
.8300	.9894	1.069	-.7992E-01	.9352	.3685E-01	-.9142
.8700	1.202	1.226	-.2442E-01	.8156	.3832E-01	-.2595
.9500	1.392	1.569	-.1770	.6375	.4493E-01	-1.674
1.050	2.097	1.984	.1133	.5041	.5284E-01	.9612
1.100	2.354	2.162	.1917	.4625	.5359E-01	1.549
1.150	.8394	.7730	.6640E-01	1.294	.4474E-01	.9893
1.200	.5576	.5249	.3270E-01	1.905	.1721E-01	.5093
1.250	.4911	.4955	-.4364E-02	2.018	.1438E-01	-.6932E-01
1.310	.4551	.4918	-.3671E-01	2.033	.1449E-01	-.5856
1.340	.5153	.4916	.2371E-01	2.034	.1451E-01	.3784
1.450	.5405	.4916	.4895E-01	2.034	.1450E-01	.7811
1.490	.4808	.4918	-.1100E-01	2.033	.1445E-01	-.1755
1.550	.5357	.4936	.4212E-01	2.026	.1417E-01	.6698
1.580	.6169	.4958	.1211	2.017	.1394E-01	1.920
1.620	.5841	.5009	.8320E-01	1.996	.1368E-01	1.311
1.670	.5691	.5120	.5715E-01	1.953	.1391E-01	.8908
1.750	.5992	.5455	.5370E-01	1.833	.1716E-01	.8193
1.790	.6854	.5731	.1123	1.745	.2032E-01	1.691
1.990	.8197	.9185	-.9879E-01	1.089	.3647E-01	-1.235
2.160	1.478	1.539	-.6083E-01	.6498	.4545E-01	-.5829
2.350	2.292	2.233	.5949E-01	.4479	.5485E-01	.4735
2.400	2.380	2.367	.1311E-01	.4225	.5312E-01	.1003
2.650	2.739	2.737	.2226E-02	.3654	.4887E-01	.1549E-01
2.810	2.919	2.816	.1035	.3552	.5387E-01	.7177
2.950	2.805	2.844	-.3858E-01	.3516	.5747E-01	-.2686
3.110	2.768	2.857	-.8858E-01	.3500	.5978E-01	-.6191
3.560	2.818	2.864	-.4635E-01	.3492	.6145E-01	-.3251
4.150	2.794	2.864	-.7070E-01	.3491	.6160E-01	-.4960

```
CORRECTED SUM OF SQUARED OBSERVATIONS =  35.1274
WEIGHTED CORRECTED SUM OF SQUARED OBSERVATIONS =  18.8391
SUM OF SQUARED RESIDUALS =           .267092
SUM OF WEIGHTED SQUARED RESIDUALS = .328258
S =  .917435E-01 WITH    39 DEGREES OF FREEDOM
CORRELATION (OBSERVED,PREDICTED) =  .9962

AIC criteria =          -39.01407
SC  criteria =          -39.55360
```

Solution - Feedback model

```
TITLE 1
Modelling of hysteresis data with adaptation
MODEL
 COMM
  NDER 5
  NFUN 1
  NCON 5
  NPARM 6
  PNAMES 'Kin', 'Kout', 'EC50', 'Emax', 'n', 'ktol'
END
    1: TEMP
    2: T=X
    3: D1 = CON(1)
    4: D2 = CON(2)
    5: TI1 = CON(3)
    6: TI2 = CON(4)
    7: T2 = CON(5)
```

```
   8: VC = 1.0
   9: CL = 6.5
  10: CLD1= 5.5
  11: CLD2= 0.90
  12: VT1= 1.75
  13: VT2= 3.45
  14: KIN = P(1)
  15: KOUT = P(2)
  16: EC50 = P(3)
  17: EMAX = P(4)
  18: N = P(5)
  19: KTOL = P(6)
  20: END
  21: START
  22: R0 = -0.5 + SQRT(0.25 + KIN/KOUT)
  23: Z(1)=R0
  24: Z(2)=0.0
  25: Z(3)=0.0
  26: Z(4)=0.0
  27: Z(5)=Z(1)
  28: END
  29: DIFF
  30: FI1 = 0.0
  31: FI2 = 0.0
  32: IF T LE TI1 THEN
  33: FI1 = D1/TI1
  34: FI2 = 0.0
  35: ENDIF
  36: IF T GE T2 AND T LE T2+TI2 THEN
  37: FI1 = 0.0
  38: FI2 = D2/TI2
  39: ENDIF
  40: IF Z(2) LE 0.0 THEN
  41: ZZ2 = 0.0
  42: ELSE
  43: ZZ2 = Z(2)**N
  44: ENDIF
  45: DRUG = (1. + (EMAX*ZZ2/(EC50**N + ZZ2)))
  46: DZ(1) = KIN/(1. + Z(5)) - DRUG*KOUT*Z(1)
  47: DZ(2)=(FI1+FI2-CL*Z(2)-CLD1*Z(2)-CLD2*Z(2)+CLD1*Z(3)+CLD2*Z(4))/VC
  48: DZ(3)=(CLD1*Z(2) - CLD1*Z(3))/VT1
  49: DZ(4)=(CLD2*Z(2) - CLD2*Z(4))/VT2
  50: DZ(5) = KTOL*(Z(1) - Z(5))
  51: END
  52: FUNC 1
  53: F=Z(1)
  54: END
  55: EOM
NVARIABLES 2
NPOINTS 500
XNUMBER 1
YNUMBER 2
CONSTANTS 7450,7350,.33,.28,1.1
METHOD 2   'Gauss-Newton (Levenberg and Hartley)
REWEIGHT -1
ITERATIONS 50
INITIAL 35,3.5,400,11,7.5,4.5
LOWER BOUNDS 1,.1,100,1,1,.1
UPPER BOUNDS 45,10,550,25,15,10
NOBSERVATIONS 44
BEGIN
```

PARAMETER	ESTIMATE	STANDARD ERROR	CV%	UNIVARIATE C.I.	
KIN	30.029140	1.470818	4.90	27.051640	33.006640
KOUT	2.888417	.161598	5.59	2.561281	3.215553
EC50	348.141199	4.145161	1.19	339.749802	356.532596
EMAX	9.749973	.396144	4.06	8.948025	10.551921
N	7.357972	.400665	5.45	6.546873	8.169071
KTOL	4.206192	.430563	10.24	3.334567	5.077817

Condition_number= 143.9

X	OBSERVED Y	PREDICTED Y	RESIDUAL	WEIGHT	SE-PRED	STANDARDIZ RESIDUAL
.0000	2.811	2.763	.4812E-01	.3619	.3281E-01	.6827
.5000E-01	1.207	1.187	.2081E-01	.8428	.3559E-01	.5710
.1100	.4153	.4276	-.1226E-01	2.339	.1425E-01	-.4530
.1600	.3213	.3465	-.2518E-01	2.886	.6258E-02	-.9394
.2200	.3887	.3646	.2411E-01	2.743	.6760E-02	.8794
.2500	.3957	.3839	.1182E-01	2.605	.7850E-02	.4237
.3100	.4392	.4249	.1431E-01	2.354	.9955E-02	.4966
.3500	.4216	.4509	-.2928E-01	2.218	.1095E-01	-.9949
.4100	.4936	.4858	.7832E-02	2.059	.1156E-01	.2570
.4500	.5075	.5061	.1412E-02	1.976	.1136E-01	.4516E-01
.5000	.5116	.5293	-.1771E-01	1.889	.1047E-01	-.5471
.5400	.5469	.5482	-.1338E-02	1.824	.9399E-02	-.4014E-01
.6000	.5720	.5832	-.1120E-01	1.715	.8261E-02	-.3225
.6500	.6000	.6251	-.2506E-01	1.600	.9362E-02	-.7007
.7000	.6837	.6864	-.2714E-02	1.457	.1248E-01	-.7399E-01
.7500	.8407	.7762	.6447E-01	1.288	.1615E-01	1.701
.8300	.9894	.9998	-.1041E-01	1.000	.2016E-01	-.2467
.8700	1.202	1.152	.4981E-01	.8681	.2107E-01	1.093
.9500	1.392	1.518	-.1265	.6587	.2336E-01	-2.402
1.050	2.097	2.005	.9209E-01	.4988	.2849E-01	1.541
1.100	2.354	2.222	.1320	.4500	.3042E-01	2.104
1.150	.8394	.8676	-.2819E-01	1.153	.2342E-01	-.7674
1.200	.5576	.5114	.4624E-01	1.956	.1403E-01	1.523
1.250	.4911	.4592	.3186E-01	2.177	.1101E-01	1.072
1.310	.4551	.4758	-.2066E-01	2.102	.9524E-02	-.6703
1.340	.5153	.4898	.2546E-01	2.041	.8941E-02	.8085
1.450	.5405	.5362	.4302E-02	1.865	.7959E-02	.1292
1.490	.4808	.5490	-.6820E-01	1.821	.7883E-02	-2.021
1.550	.5357	.5647	-.2900E-01	1.771	.7889E-02	-.8469
1.580	.6169	.5715	.4541E-01	1.750	.7944E-02	1.318
1.620	.5841	.5801	.4011E-02	1.724	.8092E-02	.1156
1.670	.5691	.5912	-.2212E-01	1.691	.8474E-02	-.6329
1.750	.5992	.6144	-.1517E-01	1.628	.9884E-02	-.4298
1.790	.6854	.6314	.5403E-01	1.584	.1113E-01	1.524
1.990	.8197	.8782	-.5850E-01	1.139	.2065E-01	-1.513
2.160	1.478	1.483	-.5103E-02	.6742	.2274E-01	-.9770E-01
2.350	2.292	2.293	-.2598E-03	.4362	.3193E-01	-.4111E-02
2.400	2.380	2.453	-.7293E-01	.4077	.3274E-01	-1.113
2.650	2.739	2.848	-.1093	.3511	.3220E-01	-1.517
2.810	2.919	2.881	.3825E-01	.3471	.3166E-01	.5255
2.950	2.805	2.859	-.5419E-01	.3497	.3145E-01	-.7467
3.110	2.768	2.823	-.5472E-01	.3542	.3174E-01	-.7612

```
  3.560     2.818     2.766     .5177E-01 .3616     .3330E-01 .7364
  4.150     2.794     2.761     .3231E-01 .3621     .3300E-01 .4592

CORRECTED SUM OF SQUARED OBSERVATIONS =  35.1274
WEIGHTED CORRECTED SUM OF SQUARED OBSERVATIONS =  19.1652
SUM OF SQUARED RESIDUALS =          .103188
SUM OF WEIGHTED SQUARED RESIDUALS =  .831172E-01
S =  .467685E-01 WITH    38 DEGREES OF FREEDOM
CORRELATION (OBSERVED,PREDICTED) =   .9985

 AIC criteria =         -97.45016
 SC  criteria =         -98.09759
```

Solution - Pol model

```
TITLE 1
Modelling of hysteresis data with adaptation
MODEL
 COMM
  NDER 5
  NFUN 1
  NCON 5
  NPARM 6
  PNAMES 'Kin', 'Kout', 'IC50', 'Imax', 'n', 'ktol'
END
    1: TEMP
    2: T=X
    3: D1 = CON(1)
    4: D2 = CON(2)
    5: TI1 = CON(3)
    6: TI2 = CON(4)
    7: T2 = CON(5)
    8: VC = 1.0
    9: CL = 6.5
   10: CLD1= 5.5
   11: CLD2= 0.90
   12: VT1= 1.75
   13: VT2= 3.45
   14: KIN = P(1)
   15: KOUT = P(2)
   16: IC50 = P(3)
   17: IMAX = P(4)
   18: N = P(5)
   19: KTOL = P(6)
   20: END
   21: START
   22: Z(1)=KIN/KOUT
   23: Z(2)=0.0
   24: Z(3)=0.0
   25: Z(4)=0.0
   26: Z(5)=KIN/KTOL
   27: END
   28: DIFF
   29: FI1 = 0.0
   30: FI2 = 0.0
   31: IF T LE TI1 THEN
   32: FI1 = D1/TI1
   33: FI2 = 0.0
   34: ENDIF
   35: IF T GE T2 AND T LE T2+TI2 THEN
```

```
36: FI1 = 0.0
37: FI2 = D2/TI2
38: ENDIF
39: IF Z(2) LE 0.0 THEN
40: ZZ2 = 0.0
41: ELSE
42: ZZ2 = Z(2)**N
43: ENDIF
44: DRUG = (1. - (IMAX*ZZ2/(IC50**N + ZZ2)))
45: DZ(1) = KTOL*Z(5)*DRUG - KOUT*Z(1)
46: DZ(2)=(FI1+FI2-CL*Z(2)-CLD1*Z(2)-CLD2*Z(2)+CLD1*Z(3)+CLD2*Z(4))/VC
47: DZ(3)=(CLD1*Z(2) - CLD1*Z(3))/VT1
48: DZ(4)=(CLD2*Z(2) - CLD2*Z(4))/VT2
49: DZ(5) = KIN - KTOL*Z(5)*DRUG
50: END
51: FUNC 1
52: F=Z(1)
53: END
54: EOM
NVARIABLES 2
NPOINTS 500
XNUMBER 1
YNUMBER 2
CONSTANTS 7450,7350,.33,.28,1.1
METHOD 2   'Gauss-Newton (Levenberg and Hartley)
REWEIGHT -1
ITERATIONS 50
INITIAL 80,30,300,.8,4,1
LOWER BOUNDS 1,.1,100,.001,1,.001
UPPER BOUNDS 150,100,550,1,15,10
NOBSERVATIONS 44
BEGIN
```

PARAMETER	ESTIMATE	STANDARD ERROR	CV%	UNIVARIATE C.I.	
KIN	94.826187	10.439532	11.01	73.692567	115.959807
KOUT	35.332853	3.196216	9.05	28.862486	41.803220
IC50	274.396480	5.306160	1.93	263.654775	285.138185
IMAX	.825393	.006946	.84	.811332	.839454
N	5.349918	.395551	7.39	4.549171	6.150665
KTOL	.048193	.030474	63.23	-.013498	.109884

```
*** CORRELATION MATRIX OF THE ESTIMATES ***
PARAMETER  KIN       KOUT      IC50      IMAX      N         KTOL
KIN        1.00000
KOUT       .937491   1.00000
IC50      -.350530  -.234725   1.00000
IMAX       .195353  -.029245  -.256894   1.00000
N          .019863   .085185   .419072  -.448340   1.00000
KTOL      -.583582  -.319503   .155976  -.382030  -.05047    1.00000

 Condition_number=        2563.
```

X	OBSERVED Y	PREDICTED Y	RESIDUAL	WEIGHT	SE-PRED	STANDARDI RESIDUAL
.0000	2.811	2.684	.1272	.3726	.1084	1.400

.5000E-01	1.207	1.111	.9652E-01	.9002	.6689E-01	1.564
.1100	.4153	.5478	-.1325	1.825	.2259E-01	-2.217
.1600	.3213	.4843	-.1630	2.065	.1497E-01	-2.801
.2200	.3887	.4736	-.8493E-01	2.111	.1552E-01	-1.481
.2500	.3957	.4731	-.7742E-01	2.114	.1558E-01	-1.351
.3100	.4392	.4737	-.3452E-01	2.111	.1545E-01	-.6014
.3500	.4216	.4744	-.5282E-01	2.108	.1529E-01	-.9190
.4100	.4936	.4763	.1732E-01	2.100	.1492E-01	.3003
.4500	.5075	.4796	.2792E-01	2.085	.1443E-01	.4812
.5000	.5116	.4901	.2147E-01	2.040	.1355E-01	.3645
.5400	.5469	.5071	.3977E-01	1.972	.1336E-01	.6626
.6000	.5720	.5517	.2033E-01	1.813	.1589E-01	.3271
.6500	.6000	.6093	-.9306E-02	1.641	.2037E-01	-.1448
.7000	.6837	.6891	-.5442E-02	1.451	.2551E-01	-.8123E-01
.7500	.8407	.7956	.4511E-01	1.257	.3009E-01	.6363
.8300	.9894	1.031	-.4151E-01	.9700	.3493E-01	-.5162
.8700	1.202	1.179	.2249E-01	.8480	.3692E-01	.2609
.9500	1.392	1.521	-.1292	.6575	.4417E-01	-1.333
1.050	2.097	1.959	.1378	.5104	.5642E-01	1.289
1.100	2.354	2.149	.2046	.4652	.5941E-01	1.830
1.150	.8394	.8238	.1564E-01	1.214	.4586E-01	.2461
1.200	.5576	.5446	.1302E-01	1.836	.1938E-01	.2146
1.250	.4911	.4975	-.6370E-02	2.010	.1415E-01	-.1075
1.310	.4551	.4898	-.3473E-01	2.042	.1454E-01	-.5921
1.340	.5153	.4896	.2570E-01	2.042	.1470E-01	.4385
1.450	.5405	.4914	.4908E-01	2.035	.1505E-01	.8371
1.490	.4808	.4929	-.1206E-01	2.029	.1506E-01	-.2053
1.550	.5357	.4977	.3805E-01	2.009	.1483E-01	.6440
1.580	.6169	.5020	.1149	1.992	.1465E-01	1.934
1.620	.5841	.5106	.7352E-01	1.959	.1452E-01	1.226
1.670	.5691	.5264	.4274E-01	1.900	.1494E-01	.7026
1.750	.5992	.5667	.3246E-01	1.764	.1784E-01	.5193
1.790	.6854	.5965	.8894E-01	1.677	.2042E-01	1.401
1.990	.8197	.9224	-.1027	1.084	.3701E-01	-1.384
2.160	1.478	1.532	-.5332E-01	.6529	.4715E-01	-.5561
2.350	2.292	2.280	.1260E-01	.4387	.5945E-01	.1086
2.400	2.380	2.422	-.4235E-01	.4128	.5793E-01	-.3492
2.650	2.739	2.782	-.4237E-01	.3595	.5220E-01	-.3157
2.810	2.919	2.846	.7315E-01	.3514	.5571E-01	.5435
2.950	2.805	2.867	-.6197E-01	.3488	.5804E-01	-.4618
3.110	2.768	2.876	-.1081	.3477	.5931E-01	-.8072
3.560	2.818	2.879	-.6132E-01	.3474	.5954E-01	-.4580
4.150	2.794	2.874	-.8055E-01	.3479	.5870E-01	-.6006

CORRECTED SUM OF SQUARED OBSERVATIONS = 35.1274
WEIGHTED CORRECTED SUM OF SQUARED OBSERVATIONS = 18.9053
SUM OF SQUARED RESIDUALS = .256248
SUM OF WEIGHTED SQUARED RESIDUALS = .283338
S = .863497E-01 WITH 38 DEGREES OF FREEDOM
CORRELATION (OBSERVED,PREDICTED) = .9963

AIC criteria = -43.48905
SC criteria = -44.13648

PD11 – Feedback modeling - Cortisol/*ACTH*

Objectives
◆ **To analyze a pharmacodynamic dataset exhibiting non-linear behavior**
◆ **To propose a model that captures tolerance and rebound**
◆ **To write a feedback model in terms of differential equations**
◆ **To obtain initial parameter estimates**

Problem specification

Adrenocorticotropic hormone (*ACTH*) stimulates the adrenal cortex to produce steroid hormones, of which cortisol is the principal one found in most mammalian species. Substantial efforts have been invested in utilizing the adrenocortical response to *ACTH* as a means of bioassaying *ACTH*. The dynamics of the adrenocortical secretory response to *ACTH* include several striking features. The first is the substantial transient overshoot in cortisol secretion rate following a stepwise increase in *ACTH* concentration. Urquhart and Li [1969] showed that a succession of overshooting responses could be elicited by a series of small stepwise increases in *ACTH* concentration. Data adapted from their study on cortisol-*ACTH* interaction (Urquhart and Li [1969]) indicate the very rapid decline in cortisol secretion rate when the *ACTH* concentration is reduced in a stepwise manner (Figure 11.1). Finally, they also show that there is a dip in the cortisol secretion below the basal level (rebound) when the *ACTH* concentration declines rapidly, after which the cortisol level returns to baseline.

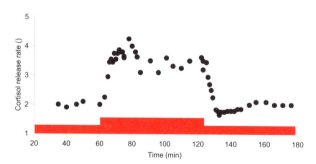

Figure 11.1 Time course of cortisol release rate following a stepwise increase in ACTH concentration from 1 to 2 µU/mL, and then a stepwise decrease back to 1 µU/mL. Data adapted from Urquhart and Li [1969]. The horizontal bar represents the step-function of ACTH that starts at 1 µU/mL, is increased to 2 µU/mL at 61 min (actually 60 min, but in this case modeled as 61 min), and then returns to 1 µU/mL at 122 min.

An initial model of the adrenocortical response to *ACTH* can be formulated as two parallel turnover models coupled together as shown in Figure 11.2. *ACTH (H(t))* is assumed to act on the production of cortisol release/secretion.

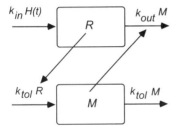

Figure 11.2 Schematic representation of the turnover of cortisol R and the moderator M. This model assumes that M directly acts on the loss of R.

The turnover of cortisol R (release rate or secretion rate) is modeled according to Equation 11:1.

$$\frac{dCort}{dt} = \frac{dR}{dt} = k_{in} \cdot H(t) - k_{out} \cdot M \tag{11:1}$$

where $H(t)$ is the stimulatory action of *ACTH* on production of cortisol.

ACTH was given as a step-function with a baseline value of 1 µU/mL between zero and 61 min; 2 µU/mL between 61 and 122 min and then 1 µU/mL between 122 and 180 min. The model we present embodies the assumption that the input and inhibitory feedback interact to determine the magnitude of response. An endogenous moderator M, which acts upon cortisol by means of a negative feedback mechanism, governs the reduction of cortisol secretion.

The turnover of the moderator (M) that balances the loss of cortisol is written as

$$\frac{dM}{dt} = k_{tol} \cdot R - k_{tol} \cdot M \tag{11:2}$$

The turnover of response at steady state is

$$\frac{dR}{dt} = k_{in} \cdot H(t) - k_{out} \cdot M_{ss} = 0 \tag{11:3}$$

and the turnover of the moderator at steady state is

$$\frac{dM}{dt} = k_{tol} \cdot R_{ss} - k_{tol} \cdot M_{ss} = 0 \tag{11:4}$$

which rearranged becomes

$$k_{tol} \cdot R_{ss} - k_{tol} \cdot M_{ss} = 0 \tag{11:5}$$

thus

$$k_{tol} \cdot R_{ss} = k_{tol} \cdot M_{ss} \tag{11:6}$$

and k_{tol} cancels out to give

$$R_{ss} = M_{ss} \tag{11:7}$$

R_{ss} then replaces M_{ss} in Equation 11:3, which after rearrangement gives

$$R_{ss} = \frac{k_{in}}{k_{out}} \cdot H(t) \tag{11:8}$$

If $H(t)$ is e.g., an E_{max} type of function, then R_{ss} is

$$R_{ss} = \frac{k_{in}}{k_{out}} \cdot \left[1 + \frac{E_{max} \cdot ACTH}{EC_{50} + ACTH} \right] \tag{11:9}$$

The maximum value of R (R_{max}) is then

$$R_{max} = \frac{k_{in}}{k_{out}} \cdot \left[1 + E_{max} \right] \tag{11:10}$$

The M function becomes

$$M_0 = \frac{k_{in}}{k_{out}} \tag{11:11}$$

The maximum difference Δ between the baseline value R_0 and R_{ss} when drug exhibits a stimulatory response is then

$$\Delta = \left| R_0 - R_{ss} \right| = \left| \frac{k_{in}}{k_{out}} \cdot \left[1 + E_{max} \right] - \frac{k_{in}}{k_{out}} \right| = \frac{k_{in}}{k_{out}} \cdot E_{max} \tag{11:12}$$

In this particular example Δ is

$$\Delta = \left| R_{ss} - R_0 \right| = \left| \frac{k_{in}}{k_{out}} \cdot \left[2^n - 1^n \right] \right| \approx 2 \cdot \left[2^n - 1^n \right] = 1.5 \tag{11:13}$$

since the $H(t)$ function takes the form

$$H(t) = ACTH_t^n \tag{11:14}$$

In the present example, $H(t)$ is modeled as either 1^n or 2^n at an $ACTH$ concentration of 1 and 2 μU/mL, respectively. The exponent n is added to give the function more flexibility.

Initial parameter estimates

Demonstrate how the initial parameter estimates of the model could be obtained graphically. The k_{out} parameter was derived from the initial slope of the upswing of the response-time curve after the largest dose. The k_{in} parameter is derived from k_{out} and the ratio of k_{in}/k_{out}.

$$k_{out} \sim 0.2 \text{ min}^{-1}$$
$$k_{in} = 0.4 \qquad\qquad R_0 \cdot 0.2$$
$$n = 1$$
$$k_{tol} \sim 0.1 \text{ min}^{-1} \qquad \text{This is set to a lower value than } k_{out}$$

Interpretations of results and conclusions

Figure 11.3 displays the observed and model predicted data. The model captures the asymmetric response-time course with the oscillations.

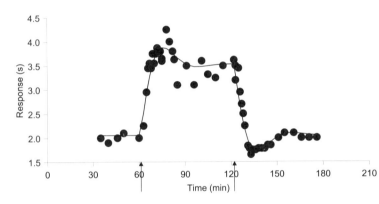

Figure 11.3 *Time course of observed and predicted cortisol release/secretion rate following a stepwise increase in ACTH concentration from 1 to 2 $\mu U/mL$, and then a stepwise decrease back to 1 $\mu U/mL$, as indicated by the two arrows. Data adapted from Urquhart and Li [1969].*

The half-lives of k_{out} and k_{tol} are 4.4 (0.693/0.1579) and 3.8 (0.693/0.1811) min, respectively.

This exercise has shown how to obtain initial parameter estimates, build a model that captures tolerance and rebound, and how to fit that model to experimental data. This exercise also demonstrates how to model dose-response-time data, when exposure data (*ACTH* concentrations or drug input rate) are known.

Solution

```
MODEL
  COMM
    NDER   2
    NFUN  1
    NCON  2
    NPAR   4
    PNAMES 'Kin', 'ktol','kout', 'n1'
  END
        1: TEMP
        2: T = X
        3: T1  = CON(1)
        4: T2  = CON(2)
        5: END
```

```
 6: START
 7: Z(1)= KIN/KOUT
 8: Z(2)= KIN/KOUT
 9: END
10: DIFF
11: ACTH = 1.
12: IF T GE T1 AND T LE T2 THEN
13: ACTH = 2.**N1
14: ELSE
15: ACTH = 1.0
16: ENDIF
17: DZ(1) = KIN*ACTH - KOUT*Z(2)
18: DZ(2) = KTOL*(Z(1) - Z(2))
19: END
20: FUNC 1
21: F = Z(1)
22: END
23: EOM
NVARIABLES 2
NPOINTS 100
XNUMBER 1
YNUMBER 2
CONSTANTS 61,122
METHOD 2  'Gauss-Newton (Levenberg and Hartley)
ITERATIONS 50
INITIAL .5,.1,.2,1
LOWER BOUNDS 0,0,0,0
UPPER BOUNDS 3,1,2,3
NOBSERVATIONS 54
BEGIN
```

PARAMETER	ESTIMATE	STANDARD ERROR	CV%	UNIVARIATE C.I.	
KIN	.324718	.018981	5.85	.286574	.362862
KTOL	.181135	.024619	13.59	.131663	.230608
KOUT	.157919	.007765	4.92	.142314	.173524
N1	.777572	.045732	5.88	.685670	.869474

```
*** CORRELATION MATRIX OF THE ESTIMATES ***
```

PARAMETER	KIN	KTOL	KOUT	N1
KIN	1.00000			
KTOL	-.090225	1.00000		
KOUT	.919748	.159238	1.00000	
N1	-.577177	.681519	-.252900	1.00000

```
Condition_number=        37.43
```

X	OBSERVED Y	PREDICTED Y	RESIDUAL	WEIGHT	SE-PRED	STANDARDI RESIDUAL
35.00	2.000	2.056	-.5623E-01	1.000	.4813E-01	-.2934
40.00	1.900	2.056	-.1562	1.000	.4813E-01	-.8152
46.00	2.000	2.056	-.5623E-01	1.000	.4813E-01	-.2934
50.00	2.100	2.056	.4377E-01	1.000	.4813E-01	.2284
60.00	2.000	2.056	-.5623E-01	1.000	.4813E-01	-.2934
63.00	2.250	2.514	-.2636	1.000	.4065E-01	-1.363
65.00	2.950	2.927	.2307E-01	1.000	.5016E-01	.1207
66.00	3.450	3.109	.3411	1.000	.5387E-01	1.794
67.00	3.550	3.272	.2785	1.000	.5600E-01	1.469
68.00	3.450	3.414	.3615E-01	1.000	.5658E-01	.1909

69.00	3.750	3.535	.2145	1.000	.5590E-01	1.132
70.00	3.550	3.637	-.8670E-01	1.000	.5446E-01	-.4564
71.00	3.750	3.718	.3178E-01	1.000	.5285E-01	.1669
72.00	3.870	3.781	.8881E-01	1.000	.5161E-01	.4656
73.00	3.800	3.827	-.2704E-01	1.000	.5117E-01	-.1417
74.00	3.800	3.857	-.5740E-01	1.000	.5170E-01	-.3010
75.00	3.600	3.874	-.2741	1.000	.5311E-01	-1.440
75.00	3.650	3.874	-.2241	1.000	.5311E-01	-1.177
78.00	4.250	3.860	.3899	1.000	.5923E-01	2.068
80.00	4.000	3.815	.1848	1.000	.6176E-01	.9846
82.00	3.800	3.756	.4367E-01	1.000	.6167E-01	.2326
83.00	3.630	3.725	-.9484E-01	1.000	.6073E-01	-.5043
85.00	3.100	3.663	-.5627	1.000	.5775E-01	-2.977
91.00	3.500	3.523	-.2335E-01	1.000	.5381E-01	-.1228
96.00	3.100	3.480	-.3795	1.000	.6089E-01	-2.019
101.0	3.600	3.482	.1181	1.000	.6404E-01	.6317
105.0	3.320	3.498	-.1776	1.000	.6111E-01	-.9453
110.0	3.250	3.517	-.2672	1.000	.5459E-01	-1.407
115.0	3.500	3.528	-.2849E-01	1.000	.4961E-01	-.1490
120.0	.	3.531	.	.0000	-.6400+152	.
122.0	3.620	3.531	.8889E-01	1.000	.4761E-01	.4635
123.0	3.500	3.304	.1960	1.000	.4205E-01	1.015
123.0	3.200	3.304	-.1040	1.000	.4205E-01	-.5387
125.0	3.450	2.864	.5864	1.000	.4387E-01	3.043
126.0	2.950	2.663	.2867	1.000	.4800E-01	1.496
127.0	2.700	2.480	.2197	1.000	.5187E-01	1.152
128.0	2.500	2.317	.1835	1.000	.5464E-01	.9661
129.0	2.250	2.173	.7687E-01	1.000	.5610E-01	.4057
131.0	1.830	1.948	-.1182	1.000	.5597E-01	-.6238
132.0	1.780	1.866	-.8575E-01	1.000	.5522E-01	-.4519
133.0	1.650	1.802	-.1519	1.000	.5463E-01	-.8000
134.0	1.750	1.755	-.5301E-02	1.000	.5456E-01	-.0279
136.0	1.750	1.707	.4297E-01	1.000	.5640E-01	.2269
138.0	1.780	1.707	.7337E-01	1.000	.5978E-01	.3895
140.0	1.780	1.740	.4041E-01	1.000	.6251E-01	.2156
142.0	1.780	1.793	-.1264E-01	1.000	.6307E-01	-.0675
144.0	1.850	1.855	-.4627E-02	1.000	.6119E-01	-.0246
146.0	1.850	1.917	-.6694E-01	1.000	.5768E-01	-.3542
151.0	2.000	2.040	-.4039E-01	1.000	.5160E-01	-.2117
155.0	2.100	2.091	.9035E-02	1.000	.5538E-01	.0476
161.0	2.100	2.102	-.2107E-02	1.000	.6138E-01	-.0112
166.0	2.000	2.084	-.8369E-01	1.000	.5868E-01	-.4435
171.0	2.000	2.064	-.6407E-01	1.000	.5252E-01	-.3363
176.0	2.000	2.053	-.5267E-01	1.000	.4782E-01	-.2747

```
CORRECTED SUM OF SQUARED OBSERVATIONS =   43.1952
WEIGHTED CORRECTED SUM OF SQUARED OBSERVATIONS =  35.4755
SUM OF SQUARED RESIDUALS =            1.91335
SUM OF WEIGHTED SQUARED RESIDUALS =  1.91335
S =  .197605     WITH    49 DEGREES OF FREEDOM
CORRELATION (OBSERVED,PREDICTED) =  .8815

AIC criteria =          42.38928
SC  criteria =          50.27045
```

PD12 - Oscillating response

Objectives

◆ **To analyze a response (temperature) dataset with oscillating baseline**

◆ **To obtain initial parameter estimates**

◆ **To implement a cosine-function for baseline**

◆ **To apply the indirect response model to temperature data**

◆ **To estimate k_{in} (mean), amplitude, peak shift, I_{max} and IC_{50}**

Problem specification

The aim of this exercise is to implement and fit a model that mimics oscillating data. Diurnal variation in body temperature of the rat was studied over three different 24 h periods, namely at baseline, after a placebo dose, and after a test dose (100 μg). All doses were given subcutaneously. The placebo data were modeled separately, with the dose as a parameter, which gave a value of about 10 μg. A cosine-function mimicking the oscillations in body temperature was simultaneously characterized by means of its average value k_{inm}, relative amplitude A, and peak shift t_0, together with the fractional turnover rate of temperature, k_{out}, and drug-specific parameters such as I_{max} and IC_{50}. The original experimental data were filtered to remove unexplainable spikes, and the resulting observational data were used for modeling (Figure 12.1).

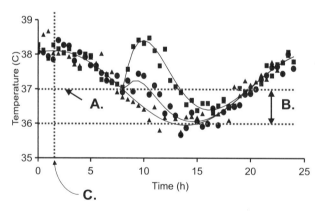

Figure 12.1 Observed (symbols) and predicted responses (solid lines) for baseline (▲), the placebo (10 μg) dose (●) and the test (100 μg) dose (■). The mean body temperature (A), amplitude (B) and peak shift (C) are superimposed on the observed and model-predicted data.

Initial parameter estimates

The doses were 100 and 10μg (The placebo response was modelled with a hypothetical dose as parameter). The rate constants of the first-order one-compartment model were 1.6 and 0.7 h^{-1}, and the volume approximately 1 L/kg. The k_{in} parameter is either estimated by taking the mean value (37 °C) of the placebo data and then using that value as a constant during the regression, or including it as a model parameter. We applied a circadian input process for the turnover rate according to

$$k_{in} = k_{inm} + A \cdot k_{inm} \cdot cosine[\omega \cdot (t - t_0)] \tag{12:1}$$

A is the relative amplitude and it was estimated as a fraction of k_{inm} the mean turnover rate. The latter was set as a constant, which was obtained by taking the average of the baseline data over 24 h. The ω parameter equals $2\Pi/24$ and t_0 was estimated to be 1.5 h. The differential equation model for the turnover of response employed Equation 12:2 as input according to

$$\frac{dR}{dt} = k_{in} - k_{out} \cdot I(t) \cdot R \tag{12:2}$$

Interpretation of output and conclusions

The experimental and predicted data were consistent (Figure 12.2, left), and the parameters had high precision (see program output). Figure 12.2 (right) shows the predicted response *versus* concentration course and Figure 12.3 the observed difference in response between the baseline and test dose data, and the baseline and placebo data.

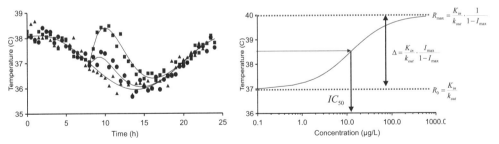

Figure 12.2 *Left hand figure shows observed (symbols) and predicted responses (solid lines) versus time for the baseline (▲), the placebo dose (●) and test dose (■). The right hand figure shows predicted response versus concentration of drug X. The mean baseline value (k_{in}/k_{out}) is 37 °C and R_{max}, potency (IC_{50}) and drug induced maximum response (Δ) are indicated.*

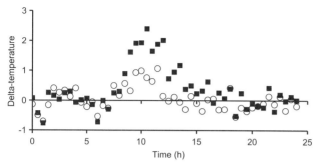

Figure 12.3 *Difference in response between the test dose (Dose 100 µg) and baseline data (■), and between the placebo and baseline data (○).*

See also Casati *et al* [1995] for information about regression analysis of hemodynamic responses obtained by telemetry.

Solution

```
MODEL
 COMM
   NDER   3
   NFUN   3
   NPARM  5
   NCON   8
   PNAMES    'kout', 'Imax', 'IC50', 't0', 'amplt'
END
     1: TEMP
     2: DOSE1=CON(1)
     3: DOSE2=CON(2)
     4: KA =CON(3)
     5: K =CON(4)
     6: TD =CON(5)
     7: CYCLE=CON(6)
     8: V =CON(7)
     9: KINM =CON(8)
    10: T=X
    11: KINT = KINM + AMPLT*KINM*COS((2*3.14/CYCLE)*(T - T0))
    12: IF T LE TD THEN
    13: C1 = 0.0
    14: C2 = 0.0
    15: ELSE
    16: C1 = (KA*DOSE1/(V*(KA - K)))*(DEXP(-K*(T-TD)) - DEXP(-KA*(T-TD)))
    17: C2 = (KA*DOSE2/(V*(KA - K)))*(DEXP(-K*(T-TD)) - DEXP(-KA*(T-TD)))
    18: ENDIF
    19: DRUG1 = 1 - IMAX*C1/(IC50 + C1)
    20: DRUG2 = 1 - IMAX*C2/(IC50 + C2)
    21: END
    22: START
    23: Z(1) = KINT/KOUT
    24: Z(2) = KINT/KOUT
    25: Z(3) = KINT/KOUT
    26: END
    27: DIFF
    28: DZ(1) = KINT - KOUT*DRUG1*Z(1)
    29: DZ(2) = KINT - KOUT*DRUG2*Z(2)
    30: DZ(3) = KINT - KOUT*Z(3)
    31: END
    32: FUNC 1
```

```
   33: F = Z(1)
   34: END
   35: FUNC 2
   36: F = Z(2)
   37: END
   38: FUNC 3
   39: F = Z(3)
   40: END
   41: EOM
NVARIABLES 3
FNUMBER 3
NPOINTS 100
XNUMBER 1
YNUMBER 2
CONSTANTS 10,100,2,.7,8,24,1,37
METHOD 2   'Gauss-Newton (Levenberg and Hartley)
INITIAL .9,.1,15,1.5,.04
LOWER BOUNDS 0,0,0,0,0
UPPER BOUNDS 2,.4,20,6,.1
DATA 'WINNLIN.DAT'
BEGIN
```

PARAMETER	ESTIMATE	STANDARD ERROR	CV%	UNIVARIATE C.I.	
KOUT	.999593	.000533	.05	.998539	1.000647
IMAX	.074462	.005901	7.92	.062796	.086127
IC50	11.417036	2.532108	22.18	6.411490	16.422582
T0	.954227	.090565	9.49	.775195	1.133259
AMPLT	.029733	.000778	2.62	.028195	.031271

```
*** CORRELATION MATRIX OF THE ESTIMATES ***
```

PARAMETER	KOUT	IMAX	IC50	T0	AMPLT
KOUT	1.00000				
IMAX	-.064049	1.00000			
IC50	-.344754	.837681	1.00000		
T0	-.139615	-.142948	.044959	1.00000	
AMPLT	.328364	-.110638	-.390242	-.14549	1.00000

```
  Condition_number=        5652.

     FUNCTION   1
```

X	OBSERVED Y	PREDICTED Y	RESIDUAL	WEIGHT	SE-PRED	STANDARDI RESIDUAL
.0000	38.07	38.08	-.1148E-01	1.000	.2960E-01	-.5725E-01
.5000	38.05	38.09	-.3770E-01	1.000	.2955E-01	-.1880
1.000	37.94	38.10	-.1536	1.000	.2938E-01	-.7656
1.500	37.85	38.10	-.2530	1.000	.2915E-01	-1.261
2.000	38.41	38.10	.3134	1.000	.2891E-01	1.562
2.500	38.27	38.08	.1928	1.000	.2870E-01	.9606
3.000	38.25	38.05	.2025	1.000	.2857E-01	1.009
3.500	37.81	38.00	-.1820	1.000	.2853E-01	-.9070
4.000	38.06	37.93	.1310	1.000	.2863E-01	.6528
4.500	37.92	37.85	.7220E-01	1.000	.2887E-01	.3598
5.000	37.68	37.76	-.7564E-01	1.000	.2928E-01	-.3771
5.500	37.67	37.65	.2464E-01	1.000	.2984E-01	.1229
6.000	37.46	37.53	-.6970E-01	1.000	.3054E-01	-.3478
6.500	37.24	37.41	-.1659	1.000	.3138E-01	-.8283
7.000	37.21	37.27	-.5986E-01	1.000	.3232E-01	-.2991
7.500	37.38	37.14	.2466	1.000	.3332E-01	1.233

8.000	36.91	37.00	-.8619E-01	1.000	.3436E-01	-.4314
8.500	37.24	37.10	.1411	1.000	.3339E-01	.7058
9.000	36.93	37.23	-.2955	1.000	.4460E-01	-1.494
9.500	37.44	37.23	.2042	1.000	.5378E-01	1.045
10.00	37.38	37.14	.2428	1.000	.5768E-01	1.249
10.50	36.84	36.99	-.1494	1.000	.5675E-01	-.7678
11.00	37.10	36.81	.2946	1.000	.5243E-01	1.504
11.50	36.85	36.62	.2266	1.000	.4648E-01	1.148
12.00	36.29	36.46	-.1707	1.000	.4058E-01	-.8593
12.50	36.50	36.32	.1808	1.000	.3594E-01	.9060
13.00	36.23	36.21	.2070E-01	1.000	.3312E-01	.1035
13.50	35.67	36.13	-.4570	1.000	.3195E-01	-2.283
14.00	35.87	36.08	-.2058	1.000	.3184E-01	-1.028
14.50	36.32	36.06	.2654	1.000	.3219E-01	1.326
15.00	36.06	36.06	-.5237E-02	1.000	.3257E-01	-.2618E-01
15.50	35.91	36.09	-.1802	1.000	.3280E-01	-.9010
16.00	35.99	36.14	-.1515	1.000	.3280E-01	-.7571
16.50	36.44	36.21	.2333	1.000	.3259E-01	1.166
17.00	35.95	36.29	-.3401	1.000	.3221E-01	-1.699
17.50	36.34	36.39	-.5566E-01	1.000	.3174E-01	-.2780
18.00	36.49	36.51	-.2031E-01	1.000	.3122E-01	-.1014
18.50	36.49	36.63	-.1388	1.000	.3072E-01	-.6928
19.00	36.54	36.76	-.2173	1.000	.3027E-01	-1.084
19.50	36.85	36.90	-.4460E-01	1.000	.2989E-01	-.2224
20.00	36.87	37.03	-.1681	1.000	.2960E-01	-.8380
20.50	37.00	37.17	-.1697	1.000	.2940E-01	-.8459
21.00	37.28	37.31	-.2662E-01	1.000	.2928E-01	-.1327
21.50	37.40	37.44	-.3846E-01	1.000	.2922E-01	-.1917
22.00	37.63	37.56	.6901E-01	1.000	.2919E-01	.3440
22.50	37.65	37.68	-.2527E-01	1.000	.2918E-01	-.1260
23.00	37.45	37.78	-.3318	1.000	.2917E-01	-1.654
23.50	38.01	37.87	.1407	1.000	.2913E-01	.7012
24.00	37.60	37.94	-.3450	1.000	.2907E-01	-1.720

```
CORRECTED SUM OF SQUARED OBSERVATIONS =  26.0150
WEIGHTED CORRECTED SUM OF SQUARED OBSERVATIONS =  26.0150
SUM OF SQUARED RESIDUALS =              1.79488
SUM OF WEIGHTED SQUARED RESIDUALS =   1.79488
S =  .201972     WITH     44 DEGREES OF FREEDOM
CORRELATION (OBSERVED,PREDICTED) =  .9660
```

FUNCTION   2

X	OBSERVED Y	PREDICTED Y	RESIDUAL	WEIGHT	SE-PRED	STANDARDI RESIDUAL
.0000	38.28	38.08	.1987	1.000	.2960E-01	.9909
.5000	38.12	38.09	.3680E-01	1.000	.2955E-01	.1835
1.000	37.88	38.10	-.2142	1.000	.2938E-01	-1.068
1.500	38.27	38.10	.1630	1.000	.2915E-01	.8127
2.000	38.16	38.10	.6574E-01	1.000	.2891E-01	.3276
2.500	38.05	38.08	-.3332E-01	1.000	.2870E-01	-.1660
3.000	38.16	38.05	.1165	1.000	.2857E-01	.5804
3.500	37.99	38.00	-.6730E-02	1.000	.2853E-01	-.3353E-01
4.000	37.59	37.93	-.3406	1.000	.2863E-01	-1.697
4.500	38.01	37.85	.1585	1.000	.2887E-01	.7899
5.000	37.96	37.76	.2056	1.000	.2928E-01	1.025
5.500	37.58	37.65	-.6606E-01	1.000	.2984E-01	-.3295
6.000	37.27	37.53	-.2570	1.000	.3054E-01	-1.282
6.500	37.41	37.41	.7607E-02	1.000	.3138E-01	.3798E-01
7.000	37.18	37.27	-.9756E-01	1.000	.3232E-01	-.4875
7.500	37.14	37.14	.1150E-02	1.000	.3332E-01	.5749E-02

8.000	37.05	37.00	.4841E-01	1.000	.3436E-01	.2423
8.500	37.57	37.63	-.5842E-01	1.000	.3731E-01	-.2932
9.000	38.23	38.11	.1172	1.000	.6677E-01	.6124
9.500	38.42	38.35	.7121E-01	1.000	.8597E-01	.3879
10.00	38.32	38.41	-.8502E-01	1.000	.9162E-01	-.4701
10.50	38.48	38.34	.1331	1.000	.8698E-01	.7270
11.00	38.11	38.19	-.7718E-01	1.000	.7695E-01	-.4115
11.50	37.67	37.98	-.3081	1.000	.6715E-01	-1.611
12.00	38.16	37.73	.4354	1.000	.6191E-01	2.256
12.50	37.25	37.47	-.2138	1.000	.6119E-01	-1.106
13.00	37.07	37.21	-.1383	1.000	.6157E-01	-.7161
13.50	36.90	36.98	-.7200E-01	1.000	.6018E-01	-.3719
14.00	36.55	36.78	-.2248	1.000	.5627E-01	-1.154
14.50	36.68	36.62	.6090E-01	1.000	.5058E-01	.3102
15.00	36.38	36.50	-.1199	1.000	.4436E-01	-.6060
15.50	36.59	36.43	.1604	1.000	.3876E-01	.8059
16.00	36.59	36.40	.1903	1.000	.3450E-01	.9527
16.50	36.31	36.40	-.9420E-01	1.000	.3175E-01	-.4705
17.00	36.52	36.44	.8248E-01	1.000	.3027E-01	.4115
17.50	36.71	36.50	.2050	1.000	.2962E-01	1.022
18.00	36.46	36.59	-.1228	1.000	.2939E-01	-.6124
18.50	36.52	36.69	-.1634	1.000	.2934E-01	-.8145
19.00	37.03	36.80	.2250	1.000	.2933E-01	1.122
19.50	36.93	36.93	.6240E-02	1.000	.2931E-01	.3111E-01
20.00	37.00	37.06	-.5496E-01	1.000	.2929E-01	-.2740
20.50	37.01	37.19	-.1808	1.000	.2926E-01	-.9013
21.00	37.16	37.32	-.1619	1.000	.2925E-01	-.8071
21.50	37.69	37.45	.2427	1.000	.2924E-01	1.210
22.00	37.39	37.57	-.1765	1.000	.2924E-01	-.8797
22.50	37.72	37.68	.3911E-01	1.000	.2924E-01	.1950
23.00	37.77	37.78	-.1085E-01	1.000	.2922E-01	-.5407E-01
23.50	38.08	37.87	.2116	1.000	.2918E-01	1.055
24.00	37.80	37.95	-.1507	1.000	.2910E-01	-.7510

CORRECTED SUM OF SQUARED OBSERVATIONS =   19.7577
  WEIGHTED CORRECTED SUM OF SQUARED OBSERVATIONS = 19.7577
  SUM OF SQUARED RESIDUALS =   1.30748
  SUM OF WEIGHTED SQUARED RESIDUALS =   1.30748
  S = .172382   WITH   44 DEGREES OF FREEDOM
  CORRELATION (OBSERVED,PREDICTED) = .9664

     FUNCTION   3

X	OBSERVED Y	PREDICTED Y	RESIDUAL	WEIGHT	SE-PRED	STANDARDI RESIDUAL
.0000	38.20	38.08	.1212	1.000	.2960E-01	.6044
.5000	38.55	38.09	.4588	1.000	.2955E-01	2.288
1.000	38.65	38.10	.5516	1.000	.2938E-01	2.750
1.500	38.01	38.10	-.9766E-01	1.000	.2915E-01	-.4868
2.000	38.00	38.10	-.9746E-01	1.000	.2891E-01	-.4857
2.500	38.01	38.08	-.6852E-01	1.000	.2870E-01	-.3414
3.000	37.91	38.05	-.1335	1.000	.2857E-01	-.6652
3.500	37.69	38.00	-.3097	1.000	.2853E-01	-1.543
4.000	37.65	37.93	-.2772	1.000	.2863E-01	-1.381
4.500	37.99	37.85	.1439	1.000	.2887E-01	.7172
5.000	37.89	37.76	.1390	1.000	.2928E-01	.6928
5.500	37.72	37.65	.7354E-01	1.000	.2984E-01	.3668
6.000	38.00	37.53	.4686	1.000	.3054E-01	2.338
6.500	37.42	37.41	.1091E-01	1.000	.3138E-01	.5446E-01
7.000	37.47	37.27	.1992	1.000	.3232E-01	.9956
7.500	36.89	37.14	-.2492	1.000	.3332E-01	-1.246

8.000	36.74	37.00	-.2530	1.000	.3436E-01	-1.266
8.500	36.68	36.86	-.1812	1.000	.3538E-01	-.9080
9.000	36.61	36.72	-.1097	1.000	.3637E-01	-.5500
9.500	36.50	36.59	-.8663E-01	1.000	.3727E-01	-.4347
10.00	36.39	36.47	-.7462E-01	1.000	.3807E-01	-.3748
10.50	36.08	36.35	-.2781	1.000	.3873E-01	-1.397
11.00	36.46	36.25	.2117	1.000	.3924E-01	1.064
11.50	35.78	36.16	-.3776	1.000	.3958E-01	-1.899
12.00	36.15	36.08	.6345E-01	1.000	.3974E-01	.3192
12.50	36.52	36.02	.4966	1.000	.3972E-01	2.498
13.00	36.12	35.98	.1358	1.000	.3951E-01	.6829
13.50	35.73	35.96	-.2296	1.000	.3914E-01	-1.154
14.00	36.17	35.95	.2232	1.000	.3861E-01	1.122
14.50	36.18	35.96	.2213	1.000	.3793E-01	1.111
15.00	36.15	35.99	.1606	1.000	.3715E-01	.8056
15.50	36.27	36.04	.2309	1.000	.3628E-01	1.158
16.00	35.96	36.10	-.1412	1.000	.3535E-01	-.7074
16.50	36.39	36.18	.2064	1.000	.3441E-01	1.033
17.00	36.26	36.27	-.1085E-01	1.000	.3347E-01	-.5429E-01
17.50	36.64	36.38	.2584	1.000	.3259E-01	1.291
18.00	36.07	36.50	-.4280	1.000	.3178E-01	-2.137
18.50	37.05	36.62	.4230	1.000	.3107E-01	2.111
19.00	36.79	36.76	.3269E-01	1.000	.3048E-01	.1631
19.50	37.22	36.89	.3257	1.000	.3001E-01	1.624
20.00	37.10	37.03	.7044E-01	1.000	.2967E-01	.3513
20.50	37.19	37.17	.1730E-01	1.000	.2943E-01	.8623E-01
21.00	37.39	37.31	.8580E-01	1.000	.2929E-01	.4277
21.50	37.27	37.44	-.1708	1.000	.2922E-01	-.8515
22.00	37.76	37.56	.2009	1.000	.2919E-01	1.001
22.50	37.53	37.68	-.1505	1.000	.2917E-01	-.7502
23.00	37.79	37.78	.1312E-01	1.000	.2916E-01	.6541E-01
23.50	37.96	37.87	.9500E-01	1.000	.2913E-01	.4735
24.00	37.79	37.94	-.1539	1.000	.2906E-01	-.7672

```
CORRECTED SUM OF SQUARED OBSERVATIONS =  30.6721
WEIGHTED CORRECTED SUM OF SQUARED OBSERVATIONS =  30.6721
SUM OF SQUARED RESIDUALS =            2.73326
SUM OF WEIGHTED SQUARED RESIDUALS =  2.73326
S =  .249238    WITH    44 DEGREES OF FREEDOM
CORRELATION (OBSERVED,PREDICTED) =  .9556
```

```
TOTALS FOR ALL CURVES COMBINED
   SUM OF SQUARED RESIDUALS =            5.83563
   SUM OF WEIGHTED SQUARED RESIDUALS =  5.83563
   S =  .202721    WITH   142 DEGREES OF FREEDOM
```

```
   AIC criteria =        269.30530
   SC  criteria =        284.25746
```

PD13 – Irreversible action – The dreker function

Objectives
- ◆ To analyze dynamic data obtained from irreversible drug action
- ◆ To write an irreversible drug action model in terms of differential equations
- ◆ To obtain initial parameter estimates of k_{in}, k_{out}, k_{irrev}.

Problem specification

The preliminary dynamics of drug X were characterized after two different oral doses (400 and 1600 mg). The response showed a negative trough shift with increasing dose, in that the time to the trough value decreased with the higher dose. This problem highlights some complexities in modeling irreversible response. Specific emphasis is put on obtaining *initial parameter estimates*, implementation of *differential equations* of the drekker (Equation 13:1) model, and assessing the *goodness-of-fit*.

Studying a new chemotherapeutic agent, the clinical investigator obtained the following response-time data after 400 mg and 1600 mg oral doses to a single subject (Figure 13.1). The high resolution of the data were required for research purposes.

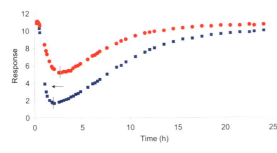

Figure 13.1 Observed response-time data obtained from a study where oral (400 mg dose • and 1600 mg ■) doses were administered to a single subject at two different occasions separated by a week. Note the shift to the left in the trough value (t_{max} decreases) with an increase in dose. The irreversible drug action is on the loss of response.

The response R is produced by a zero-order process k_{in} and eliminated by a first-order process $-k_{out} \cdot R$ when no drug is present. The drug concentration C acts via depletion of R, which can be modeled as a second-order process by means of $-k_{irrev} \cdot C \cdot R$. We will fit the following differential equation to the data. The turnover of response after an oral dose of drug X can be written mathematically as

$$\frac{dR}{dt} = k_{in} - k_{out} \cdot R - k_{irrev} \cdot C \cdot R \tag{13:1}$$

where k_{in}, k_{out}, k_{irrev}, and C are the turnover rate, fractional turnover rate, the second-order rate constant (time$^{-1} \cdot$concentration^{-1}) for the drug-response complex, and kinetics of drug (given in program output), respectively.

Initial parameter estimates

The kinetic parameter estimates are shown in rows 3-9 in the output of model code. Observe the lag-time of about 20 min. Then the slope of the downswing of the low dose curve in Figure 13.1 is approximated by

$$slope = -k_{out} = \frac{\ln(10.6) - \ln(5.9)}{0.5 - 1.75} \approx -0.5 h^{-1} \tag{13:2}$$

The k parameter then becomes $0.5 \cdot 11$ ($k_{out} \cdot$ baseline ≈ 6 units). We will set the initial estimate of k_{out} to 1. The second-order rate constant for the irreversible drug action k_{irrev} was derived from biochemical data on cell death ($0.03\ h^{-1}$).

Interpretation of results and conclusions

This exercise has dealt with obtaining *initial parameter estimates*, implementation of *differential equations*, and the *goodness-of-fit* of a nonlinear kinetic system. The observed and predicted data are shown in Figure 13.2.

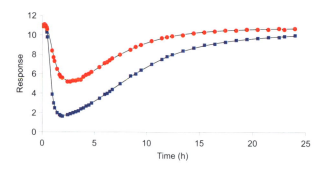

Figure 13.2. Observed (low dose ● and high dose ■) and model-predicted (solid lines) response-time courses of drug X after two oral doses.

You have
☐ analyzed an irreversible response system
☐ learned to derive the initial estimates
☐ characterized the pharmacodynamics after single dose administration to one subject

Next step in the characterization of dynamics would be to give multiple doses via the intended route of administration.

Solution
```
TITLE 1
  The Dreker model
Model
 COMM
  Nder 2
  Nfun 2
  NPAR 3
  PNAM 'kin', 'kout', 'Ke'
END
    1: TEMP
    2: DOSE1=CON(1)
```

```
 3: A=27.47/DOSE1
 4: B=0.894/DOSE1
 5: C=-1*(A+B)
 6: K01=1.934
 7: ALPHA=0.381
 8: BETA=0.0569
 9: TLAG=0.328
10: I=0
11: J=2
12: L=CON(2)
13: GREEN:
14: I=I+1
15: J=J+2
16: IF X <= CON(J) THEN GOTO RED
17: ELSE IF I < L THEN GOTO GREEN
18: ENDIF
19: ENDIF
20: I=I+1
21: RED:
22: L=I-1
23: SUM=0
24: I=0
25: J=2
26: BLUE:
27: I=I+1
28: J=J+2
29: T = X - CON(J) - TLAG
30: D=CON(J-1)
31: AMT=A*DEXP(-ALPHA*T) + B*DEXP(-BETA*T) + C*DEXP(-K01*T)
32: SUM=SUM + MAX(0,D*AMT)
33: IF I < L THEN GOTO BLUE
34: ENDIF
35: CP1 = SUM
36: CP2 = 4*SUM
37: END
38: START
39: Z(1) = KIN/KOUT
40: Z(2) = KIN/KOUT
41: END
42: DIFF
43: DZ(1) = KIN - KOUT*Z(1) - KE*CP1*Z(1)
44: DZ(2) = KIN - KOUT*Z(2) - KE*CP2*Z(2)
45: END
46: FUNC1
47: F = Z(1)
48: END
49: FUNC 2
50: F = Z(2)
51: END
52: EOM
NOPAGE BREAKS
NVARIABLES 3
FNUMBER 3
NPOINTS 1000
XNUMBER 1
YNUMBER 2
NCONSTANTS 4
CONSTANTS 400,1,400,0
METHOD 2  'Gauss-Newton (Levenberg and Hartley)
ITERATIONS 50
INITIAL 6,1,.03
```

```
LOWER BOUNDS 0,0,0
UPPER BOUNDS 20,5,.3
MISSING 'Missing'
DATA 'WINNLIN.DAT'
BEGIN


PARAMETER ESTIMATE      STANDARD    CV%        UNIVARIATE C.I.
                        ERROR
KIN        5.723620     .030153     .53     5.663758        5.783482


KOUT       .521694      .002844     .55      .516048         .527340


KE         .050307      .000262     .52      .049786         .050828



*** CORRELATION MATRIX OF THE ESTIMATES ***
PARAMETER  KIN          KOUT        KE
KIN        1.00000
KOUT       .98975       1.00000
KE         .85365       .80663      1.00000


  Condition_number=           262.7


    FUNCTION    1


X          OBSERVED    PREDICTED   RESIDUAL    WEIGHT    SE-PRED    STANDARD
           Y           Y                                            RESIDUAL
    .0000     11.00       10.97      .2878E-01   1.000    .8635E-02    .6677
    .1500     10.90       10.97     -.7122E-01   1.000    .8635E-02  -1.653
    .2500     11.10       10.97      .1288       1.000    .8635E-02   2.988
    .3500     11.00       10.97      .3393E-01   1.000    .8643E-02    .7874
    .4500     10.80       10.81     -.1053E-01   1.000    .8439E-02   -.2441
    .5000     10.60       10.67     -.6774E-01   1.000    .8313E-02  -1.570
   1.000      8.400        8.362     .3778E-01   1.000    .1113E-01    .8887
   1.150      7.700        7.696     .3519E-02   1.000    .1218E-01    .0833
   1.250      7.300        7.301    -.1083E-02   1.000    .1261E-01   -.0257
   1.500      6.500        6.495     .5477E-02   1.000    .1284E-01    .1303
   1.750      5.900        5.928    -.2846E-01   1.000    .1220E-01   -.6740
   1.850      5.700        5.760    -.5957E-01   1.000    .1180E-01  -1.407
   2.000      5.600        5.558     .4179E-01   1.000    .1113E-01    .9828
   2.500      5.200        5.228    -.2809E-01   1.000    .8914E-02   -.6527
   2.750      5.200        5.200    -.1036E-03   1.000    .8056E-02   -.0023
   3.000      5.300        5.232     .6830E-01   1.000    .7428E-02   1.577
   3.250      5.300        5.307    -.6808E-02   1.000    .7020E-02   -.1569
   3.500      5.400        5.414    -.1367E-01   1.000    .6795E-02   -.3149
   3.750      5.400        5.544    -.1436       1.000    .6707E-02  -3.307
   4.000      5.700        5.690     .9716E-02   1.000    .6711E-02    .2237
   4.250      5.900        5.849     .5124E-01   1.000    .6770E-02   1.180
   4.500      6.000        6.015    -.1538E-01   1.000    .6856E-02   -.3543
   4.750      6.200        6.187     .1270E-01   1.000    .6950E-02    .2927
   5.500      6.700        6.715    -.1467E-01   1.000    .7183E-02   -.3384
   6.000      7.100        7.062     .3782E-01   1.000    .7256E-02    .8724
   6.250      7.200        7.232    -.3187E-01   1.000    .7266E-02   -.7352
   6.500      7.400        7.398     .2035E-02   1.000    .7259E-02    .0469
   6.750      7.600        7.560     .4003E-01   1.000    .7236E-02    .9233
   7.500      8.000        8.018    -.1783E-01   1.000    .7096E-02   -.4112
   7.500      8.000        8.018    -.1783E-01   1.000    .7096E-02   -.4112
   8.500      8.500        8.555    -.5496E-01   1.000    .6810E-02  -1.266
   9.000      8.800        8.791     .9240E-02   1.000    .6660E-02    .2127
   9.500      9.000        9.005    -.5057E-02   1.000    .6521E-02   -.1163
  10.50       9.400        9.373     .2743E-01   1.000    .6314E-02    .6306
```

```
11.50      9.700      9.667      .3333E-01  1.000      .6231E-02   .7661
12.00      9.800      9.790      .1036E-01  1.000      .6235E-02   .2381
12.50      9.900      9.898      .1519E-02  1.000      .6265E-02   .0349
13.50      10.00      10.08     -.7932E-01  1.000      .6386E-02  -1.824
14.50      10.20      10.22     -.1959E-01  1.000      .6556E-02  -.4508
15.50      10.30      10.33     -.2825E-01  1.000      .6743E-02  -.6506
16.50      10.40      10.41     -.1271E-01  1.000      .6926E-02  -.2928
17.50      10.50      10.48      .2115E-01  1.000      .7094E-02   .4876
18.50      10.50      10.53     -.3126E-01  1.000      .7242E-02  -.7211
19.50      10.60      10.57      .2658E-01  1.000      .7369E-02   .6135
20.50      10.60      10.61     -.7937E-02  1.000      .7478E-02  -.1833
21.50      10.60      10.64     -.3675E-01  1.000      .7572E-02  -.8488
22.50      10.70      10.66      .3873E-01  1.000      .7652E-02   .8948
23.00      10.60      10.67     -.7226E-01  1.000      .7688E-02  -1.670
24.00      10.70      10.69      .7809E-02  1.000      .7753E-02   .1805
```

```
CORRECTED SUM OF SQUARED OBSERVATIONS =   213.760
WEIGHTED CORRECTED SUM OF SQUARED OBSERVATIONS =  213.760
SUM OF SQUARED RESIDUALS =              .933200E-01
SUM OF WEIGHTED SQUARED RESIDUALS =    .933200E-01
S =  .450411E-01 WITH     46 DEGREES OF FREEDOM
CORRELATION (OBSERVED,PREDICTED) =   .9998
```

```
   FUNCTION    2
```

X	OBSERVED Y	PREDICTED Y	RESIDUAL	WEIGHT	SE-PRED	STANDARD RESIDUAL
.0000	11.00	10.97	.2878E-01	1.000	.8635E-02	.6677
.1500	10.90	10.97	-.7122E-01	1.000	.8635E-02	-1.653
.2500	11.10	10.97	.1288	1.000	.8635E-02	2.988
.3500	10.85	10.95	-.1006	1.000	.8668E-02	-2.335
.4500	10.30	10.34	-.4261E-01	1.000	.8249E-02	-.9870
.5000	9.800	9.808	-.7628E-02	1.000	.8725E-02	-.1771
1.000	3.900	3.884	.1619E-01	1.000	.1624E-01	.3965
1.150	3.000	2.963	.3684E-01	1.000	.1364E-01	.8817
1.250	2.500	2.548	-.4784E-01	1.000	.1167E-01	-1.129
1.500	2.000	1.964	.3555E-01	1.000	.7548E-02	.8210
1.750	1.800	1.746	.5390E-01	1.000	.5225E-02	1.235
1.850	1.700	1.712	-.1208E-01	1.000	.4727E-02	-.2765
2.000	1.650	1.695	-.4474E-01	1.000	.4293E-02	-1.023
2.500	1.800	1.791	.8995E-02	1.000	.4100E-02	.2056
2.750	1.900	1.884	.1627E-01	1.000	.4240E-02	.3718
3.000	2.000	1.993	.7418E-02	1.000	.4423E-02	.1696
3.250	2.100	2.113	-.1344E-01	1.000	.4624E-02	-.3075
3.500	2.200	2.244	-.4397E-01	1.000	.4837E-02	-1.007
3.750	2.400	2.383	.1728E-01	1.000	.5054E-02	.3958
4.000	2.500	2.529	-.2865E-01	1.000	.5275E-02	-.6565
4.250	2.700	2.681	.1903E-01	1.000	.5494E-02	.4363
4.500	2.800	2.839	-.3904E-01	1.000	.5712E-02	-.8959
4.750	3.000	3.002	-.2262E-02	1.000	.5925E-02	-.0519
5.500	3.500	3.517	-.1738E-01	1.000	.6524E-02	-.3999
6.000	3.900	3.877	.2332E-01	1.000	.6875E-02	.5373
6.250	4.000	4.059	-.5948E-01	1.000	.7032E-02	-1.371
6.500	4.200	4.244	-.4367E-01	1.000	.7176E-02	-1.007
6.750	4.400	4.429	-.2870E-01	1.000	.7305E-02	-.6622
7.500	5.000	4.984	.1643E-01	1.000	.7603E-02	.3795
7.500	5.000	4.984	.1643E-01	1.000	.7603E-02	.3795
8.500	5.800	5.704	.9557E-01	1.000	.7779E-02	2.209
9.000	6.000	6.049	-.4888E-01	1.000	.7775E-02	-1.130
9.500	6.400	6.379	.2107E-01	1.000	.7716E-02	.4870
10.50	7.000	6.988	.1209E-01	1.000	.7460E-02	.2791

11.50	7.500	7.521	-.2120E-01	1.000	.7080E-02	-.4887
12.00	7.800	7.759	.4148E-01	1.000	.6868E-02	.9555
12.50	8.000	7.977	.2335E-01	1.000	.6654E-02	.5374
13.50	8.400	8.358	.4178E-01	1.000	.6252E-02	.9605
14.50	8.700	8.674	.2641E-01	1.000	.5923E-02	.6065
15.50	9.000	8.932	.6777E-01	1.000	.5693E-02	1.555
16.50	9.100	9.144	-.4389E-01	1.000	.5562E-02	-1.007
17.50	9.300	9.318	-.1763E-01	1.000	.5517E-02	-.4044
18.50	9.500	9.461	.3866E-01	1.000	.5535E-02	.8868
19.50	9.600	9.582	.1844E-01	1.000	.5596E-02	.4230
20.50	9.700	9.684	.1644E-01	1.000	.5681E-02	.3772
21.50	9.800	9.771	.2854E-01	1.000	.5780E-02	.6550
22.50	9.800	9.848	-.4841E-01	1.000	.5885E-02	-1.112
23.00	9.900	9.884	.1644E-01	1.000	.5938E-02	.3776
24.00	10.00	9.948	.5162E-01	1.000	.6044E-02	1.186

```
CORRECTED SUM OF SQUARED OBSERVATIONS =  538.286
WEIGHTED CORRECTED SUM OF SQUARED OBSERVATIONS =   538.286
SUM OF SQUARED RESIDUALS =              .901862E-01
SUM OF WEIGHTED SQUARED RESIDUALS =    .901862E-01
S =  .442783E-01 WITH    46 DEGREES OF FREEDOM
CORRELATION (OBSERVED,PREDICTED) =   .9999

TOTALS FOR ALL CURVES COMBINED
SUM OF SQUARED RESIDUALS =              .183506
SUM OF WEIGHTED SQUARED RESIDUALS =    .183506
S =  .439505E-01 WITH    95 DEGREES OF FREEDOM

AIC criteria =         -160.15964
SC  criteria =         -152.40473
```

PD14 - Composite model I - I_{max}

Objectives
◆ To analyze a composite concentration-response relationship
◆ To obtain initial parameter estimates for the composite I_{max} model

Problem specification

In the following exercise, we elaborate a composite I_{max} model originally proposed by Lundström *et al* [1992]. The relationship between response (in this case, spontaneous locomotor activity) and plasma concentration of a new dopaminergic compound was studied in rats. The results showed that the compound produced an interesting response-concentration relationship (Figure 14.1). We will use the original parameter estimates from the article and then generate a new dataset with a 10% *CV* added to data. Taking our simulated data, we will re-fit Equation 14:1 to the data and analyze the output.

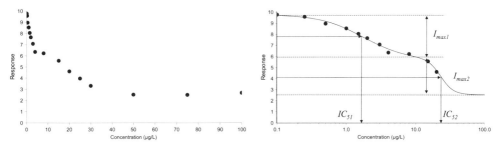

Figure 14.1 *The relationship between observed response and the steady state concentration of a new experimental agent, on a Cartesian scale (left) and semi-logarithmic scale (right). The right hand plot also includes the model-predicted response-concentration profile.*

The IC_{51} and IC_{52} represent IC_{50} for the first and second phase of the response, respectively. In the reference, these were estimated to be 1.8 and 23 µg/L, respectively, and I_{max1} and I_{max2} were 4 and 3.2 units, respectively. The corresponding exponents were 1.4 and 4.7. Presumably, the first phase (I_1) corresponded to the predominantly autoreceptor-mediated responses, whereas the later phase (I_2) reflected a postsynaptic dopamine receptor blockade (Lundström *et al* [1992]).

$$E = E_0 - \frac{I_{max1} \cdot C^{n_1}}{IC_{51}^{n_1} + C^{n_1}} - \frac{I_{max2} \cdot C^{n_2}}{IC_{52}^{n_2} + C^{n_2}} \qquad (14:1)$$

Initial parameter estimates

The initial parameter estimates of I_{max1} and I_{max2} are 4-5 and 3-4, respectively, and of IC_{51} and IC_{52} about 1.5 and 25, respectively (Figures 14.1 and 14.3). The n_1 and n_2 parameters are 2 and 5, respectively, and the baseline value E_0 is 10-11.

Interpretation of results and conclusions

The data in Figure 14.1 were simulated by means of the parameter estimates given by the authors (Lundström *et al* [1992]). A constant coefficient of variation error model was used for the generation of data. Figure 14.2 shows the observed and predicted data of the composite I_{max} model.

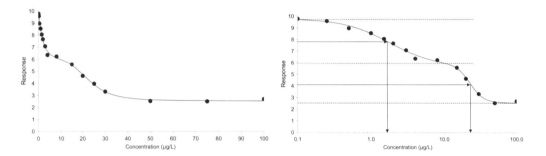

Figure 14.2 *The observed and model-predicted relationship between response and steady state concentration of a new experimental agent.*

We may conclude that Equation 14:1 is a flexible system that can summarize data where we know the underlying receptor model (Lundström *et al* [1992]). Figure 14.3 shows schematically the composite I_{max} model.

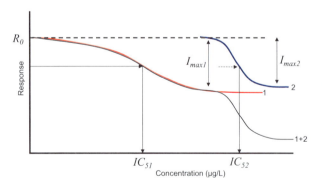

Figure 14.3 *Schematic representation of the composite I_{max} model. The red line (No. 1) is characterized by means of I_{max1} and IC_{51} and the blue line (No. 2) by means of I_{max2} and IC_{52}. The combined function (No. 1+2) is the black line.*

The situation of combined drug action arises either when two or more active compounds exert an response at a certain biological system, or when a single drug acts simultaneously at two different receptors. This approach, of modeling the responses from either competing agonists or multiple receptor actions of one compound, has been described by several people (Ariens [1964], Gero [1971], Paalzow and Edlund [1979], Ebling *et al* [1991], Lundström *et al* [1992]).

Solution

```
MODEL
 COMM
 NPARM 7
 PNAMES 'EMAX1', 'EC51', 'n1','EMAX2', 'EC52', 'n2', 'E0'
END
    1: TEMP
    2: C=X
    3: EMAX1 = P(1)
    4: EC51 = P(2)
    5: N1 = P(3)
    6: EMAX2 = P(4)
    7: EC52 = P(5)
    8: N2 = P(6)
    9: E0 = P(7)
   10: END
   11: FUNC1
   12: E1=EMAX1*(ABS(C)**N1)/(EC51**N1 + (ABS(C)**N1))
   13: E2=EMAX2*(ABS(C)**N2)/(EC52**N2 + (ABS(C)**N2))
   14: F = E0 - E1 - E2
   15: END
   16: EOM
NVARIABLES 2
NPOINTS 1000
XNUMBER 1
YNUMBER 2
METHOD 2  'Gauss-Newton (Levenberg and Hartley)
ITERATIONS 50
INITIAL 5,1.5,2,4,25,5,11
LOWER BOUNDS .1,.01,.1,.1,1,.1,1
UPPER BOUNDS 10,5,5,6,50,10,20
MISSING '-'
NOBSERVATIONS 16
DATA 'WINNLIN.DAT'
```

PARAMETER	ESTIMATE	STANDARD ERROR	CV%	UNIVARIATE C.I.	
EMAX1	4.084678	.559734	13.70	2.818460	5.350896
EC51	1.755838	.293871	16.74	1.091049	2.420626
N1	1.399406	.322179	23.02	.670580	2.128233
EMAX2	3.178822	.438450	13.79	2.186971	4.170674
EC52	22.886482	1.202129	5.25	20.167054	25.605910
N2	4.704920	1.085905	23.08	2.248410	7.161429
E0	9.778641	.192535	1.97	9.343093	10.214189

```
 Condition_number=      47.73

    FUNCTION    1
```

X	OBSERVED Y	PREDICTED Y	RESIDUAL	WEIGHT	SE-PRED	STANDARDI RESIDUAL
.1000	9.772	9.706	.6611E-01	1.000	.1415	.7164
.2500	9.576	9.528	.4787E-01	1.000	.9845E-01	.3486
.5000	8.962	9.178	-.2163	1.000	.1071	-1.655
1.000	8.535	8.502	.3341E-01	1.000	.1063	.2544
1.500	8.046	7.960	.8557E-01	1.000	.9079E-01	.6005
2.000	7.656	7.551	.1057	1.000	.9251E-01	.7475
3.000	7.076	7.005	.7187E-01	1.000	.9884E-01	.5244
4.000	6.352	6.674	-.3220	1.000	.9745E-01	-2.333
8.000	6.227	6.108	.1188	1.000	.1410	1.277
15.00	5.561	5.504	.5668E-01	1.000	.1229	.4888
20.00	4.609	4.724	-.1143	1.000	.1312	-1.074
25.00	3.955	3.876	.7889E-01	1.000	.1143	.6339
30.00	3.303	3.286	.1750E-01	1.000	.1282	.1590
50.00	2.517	2.631	-.1137	1.000	.9026E-01	-.7960
75.00	2.490	2.548	-.5867E-01	1.000	.1017	-.4348
100.0	2.680	2.532	.1475	1.000	.1052	1.116

```
CORRECTED SUM OF SQUARED OBSERVATIONS =   99.4250
WEIGHTED CORRECTED SUM OF SQUARED OBSERVATIONS =   99.4250
SUM OF SQUARED RESIDUALS =              .256934
SUM OF WEIGHTED SQUARED RESIDUALS =    .256934
S =  .168962      WITH      9 DEGREES OF FREEDOM
CORRELATION (OBSERVED,PREDICTED) =   .9987

AIC criteria =        -7.74299
SC  criteria =        -12.03893
```

PD15 - Composite model II - E_{max} / I_{max}

Objectives
◆ **To analyze a composite concentration-response relationship**
◆ **To obtain initial parameter estimates for the composite E_{max} - I_{max} model**

Problem specification

In the following exercise, we elaborate a composite E_{max}/I_{max} model originally proposed by Gero [1970], Paalzow and Edlund [1979], van Boxtel _et al_ [1992], and Lundström _et al_ [1992]. When you design an experiment of a compound that exhibits a bell-shaped response-concentration relationship, it is important to not only sample the data correctly, but also to reduce the experimental noise and obtain good initial estimates. Figure 15.1 shows the steady state plasma concentration _versus_ pharmacological response. Due to the large number of observations, data are only listed in the output.

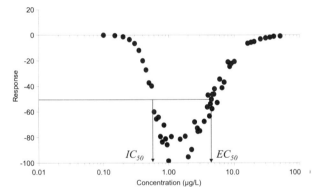

Figure 15.1 _Observed and predicted response versus steady state concentration._

Analyze the concentration-response data using the following model

$$E = \frac{E_{max} \cdot C^{n_1}}{EC_{50}^{n_1} + C^{n_1}} - \frac{I_{max} \cdot C^{n_2}}{IC_{50}^{n_2} + C^{n_2}} \tag{15:1}$$

In Step No. 1, we will fit the structural model assuming a constant absolute error (_homoscedastic error_) model. We will observe any deviations of the parameter estimates, parameter precision and/or parameter correlation.

In Step No. 2, we will fit the same structural model using a constant _CV_ error model. We will then observe any differences between the two fits. This type of model (data) is a good test of the robustness of the algorithm of the program. We recommend that the interested reader run the program with different sets of initial estimates.

What needs to be done?
- Plot data - What do you see?
- Derive the necessary pharmacodynamic relationships
- Obtain initial estimates
- Fit the model to the data
- Interpret the output and refine the model

Initial parameter estimates

Initial estimates are obtained graphically from Figures 15.2 and 15.3.

$$
\begin{aligned}
E_{max} &= 100 \ \text{(units)} \\
EC_{50} &= 10.0 \ (\mu g/L) \\
n_1 &= 4.0 \\
I_{max} &= 100 \ \text{(units)} \\
IC_{50} &= 0.4 \ (\mu g/L) \\
n_2 &= 2.0
\end{aligned}
$$

Interpretation of results and conclusions

A high correlation is generally seen between two or more parameters, even though there is little scatter in the data and the data are well-spaced. If the EC_{50}/IC_{50} values of the model are not sufficiently separated, a clear maximal activation plateau is not reflected in the data. The absence of an activation plateau makes it difficult to estimate the two separate E_{max}/I_{max} parameters. In this example, the model could have been reduced to only one E_{max}/I_{max} parameter.

The model implicitly relates the observed response E, to a plasma, tissue, or effect compartment concentration. There are inherent limitations of this model unless effect data exhibit a low variability, and depending on the separation of the EC_{50} and IC_{50} values. At high plasma concentrations ($C \gg EC_{50}$ and IC_{50})

$$
E \rightarrow E_{max} - I_{max} \tag{15:2}
$$

At low plasma concentrations ($C \ll EC_{50}$ and IC_{50})

$$
E \rightarrow E_{max} \cdot \frac{C^{n_1}}{EC_{50}^{n_1}} - I_{max} \cdot \frac{C^{n_2}}{IC_{50}^{n_2}} \tag{15:3}
$$

At intermediate plasma concentrations ($IC_{50} \ll C \ll EC_{50}$)

$$
E \rightarrow E_{max} - I_{max} \cdot \frac{C^{n_2}}{IC_{50}^{n_2}} \tag{15:4}
$$

Since E_{max} and I_{max} appear as a sum on all occasions, they will often be highly correlated with each other. In the presence of these difficulties, physiological interpretation of the model parameters would be speculative. For additional reading, see Gero [1970], Paalzow and Edlund [1979], van Boxtel *et al* [1992] and Lundström *et al* [1992].

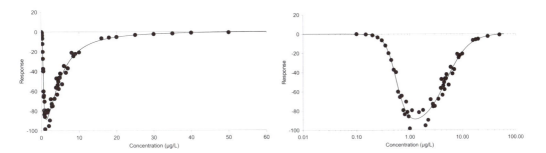

Figure 15.2 *Observed and predicted response versus concentration data on a linear scale (left) and semi-logarithmic scale (right).*

Figure 15.3 shows a schematic illustration of the composite E_{max}/I_{max} model. It shows clearly how the composite curve is made up of two individual curves.

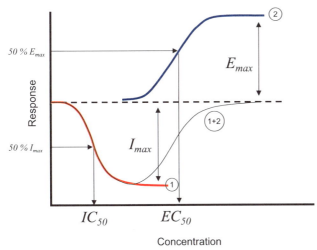

Figure 15.3 *Schematic illustration of the composite I_{max} model. The red line (No. 1) is characterized by means of I_{max} and IC_{50} and the blue line (No. 2) by means of E_{max} and EC_{50}. The combined function (No. 1+2) is the black line.*

The relative residual plots of the weighted fit can be seen in Figure 15.4. In Figure 15.4 (left) there is a cone-shaped pattern, which is largely eliminated by means of weighting. In Figure 15.4 (right) there is a more horizontal uniform band in the scatter, which indicates a more correct structural model and weighting scheme.

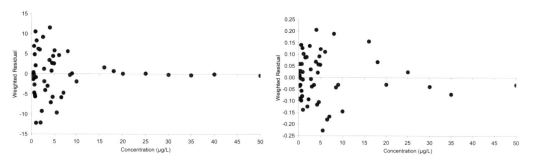

Figure 15.4 *Relative residual versus the predicted effect using a constant absolute (left) and constant relative (right) variance model.*

We may conclude that the sum of the E_{max} functions is a flexible system. We have already demonstrated its usefulness where the underlying receptor model is known (PD14). Here, we show it can be used in more empirical situations, such as when there are bell-shaped effect-concentration data, and the same is true for u-shaped effect-concentration data. Finally, we would like to point out that this dataset is good for discrimination between programs with respect to algorithm robustness.

Solution I - Constant absolute variance model

```
TITLE 1
Multiple effect sites
MODEL
 COMM
NPARM 6
PNAMES 'EMAX1', 'EC51', 'n1','EMAX2', 'EC52', 'n2'
END
     1: TEMP
     2: CP=X
     3: EMAX1=P(1)
     4: EC51=P(2)
     5: N1=P(3)
     6: EMAX2=P(4)
     7: EC52=P(5)
     8: N2=P(6)
     9: END
    10: FUNC1
    11: E1=EMAX1*(ABS(CP)**N1)/((EC51**N1) + (ABS(CP)**N1))
    12: E2=EMAX2*(ABS(CP)**N2)/((EC52**N2) + (ABS(CP)**N2))
    13: F=E1-E2
    14: END
    15: EOM
NVARIABLES 2
NPOINTS 100
XNUMBER 1
YNUMBER 2
METHOD 2   'Gauss-Newton (Levenberg and Hartley)
ITERATIONS 50
INITIAL 100,10,4,100,.4,2
LOWER BOUNDS .1,.1,.1,.1,.01,.01
UPPER BOUNDS 200,50,9,200,5,9
NOBSERVATIONS 54
```

```
BEGIN
```

PARAMETER	ESTIMATE	STANDARD ERROR	CV%	UNIVARIATE C.I.	
EMAX1	97.672163	8.423311	8.62	80.736001	114.608324
EC51	4.943094	.368587	7.46	4.202003	5.684185
N1	1.993429	.301372	15.12	1.387482	2.599376
EMAX2	97.270600	6.565479	6.75	84.069851	110.471348
EC52	.553966	.020059	3.62	.513634	.594297
N2	4.265895	.530867	12.44	3.198517	5.333272

```
  *** CORRELATION MATRIX OF THE ESTIMATES ***
PARAMETER  EMAX1       EC51        N1         EMAX2       EC52       N2
EMAX1      1.00000
EC51       -.630365    1.00000
N1         -.903283    .512573    1.00000
EMAX2      .949659    -.815313    -.839027    1.00000
EC52       .744919    -.697213    -.597619    .812572    1.00000
N2         -.692737    .640841    .549742    -.755439    -.56058    1.00000

  Condition_number=        962.8

   FUNCTION    1
```

X	OBSERVED Y	PREDICTED Y	RESIDUAL	WEIGHT	SE-PRED	STANDARDI RESIDUAL
.1000	-.6000E-01	-.2448E-01	-.3552E-01	1.000	.5176E-01	-.6520E-02
.1500	-.4400	-.2761	-.1639	1.000	.2023	-.3010E-01
.2000	-1.360	-1.081	-.2787	1.000	.5514	-.5142E-01
.2500	-3.540	-2.905	-.6348	1.000	1.102	-.1190
.3000	-6.570	-6.258	-.3116	1.000	1.749	-.6039E-01
.3500	-12.00	-11.53	-.4731	1.000	2.296	-.9574E-01
.4000	-20.08	-18.76	-1.316	1.000	2.559	-.2736
.4500	-27.20	-27.57	.3675	1.000	2.509	.7599E-01
.5000	-37.30	-37.17	-.1350	1.000	2.326	-.2740E-01
.5500	-39.76	-46.68	6.919	1.000	2.226	1.391
.6000	-60.09	-55.40	-4.690	1.000	2.228	-.9433
.6500	-65.62	-62.92	-2.697	1.000	2.217	-.5418
.7000	-64.30	-69.13	4.830	1.000	2.132	.9633
.7500	-79.33	-74.09	-5.240	1.000	1.997	-1.034
.8000	-83.59	-77.97	-5.624	1.000	1.871	-1.099
.8500	-70.42	-80.95	10.53	1.000	1.802	2.047
.9000	-81.12	-83.21	2.086	1.000	1.804	.4057
.9500	-85.74	-84.90	-.8418	1.000	1.863	-.1644
1.000	-98.40	-86.15	-12.25	1.000	1.953	-2.409
1.100	-79.37	-87.67	8.300	1.000	2.141	1.657
1.500	-81.38	-87.61	6.227	1.000	2.271	1.257
1.800	-79.10	-85.13	6.035	1.000	2.013	1.192
2.000	-95.22	-83.05	-12.17	1.000	1.844	-2.373
2.250	-89.43	-80.19	-9.240	1.000	1.697	-1.785
2.500	-67.96	-77.15	9.188	1.000	1.627	1.767
2.750	-72.78	-74.01	1.233	1.000	1.605	.2368
2.800	-75.25	-73.38	-1.870	1.000	1.604	-.3591
3.000	-74.91	-70.84	-4.067	1.000	1.600	-.7808
3.500	-67.57	-64.57	-3.002	1.000	1.570	-.5754
3.900	-56.32	-59.74	3.418	1.000	1.510	.6529

4.000	-47.04	-58.57	11.53	1.000	1.493	2.200
4.200	-63.37	-56.28	-7.092	1.000	1.459	-1.351
4.350	-53.84	-54.61	.7678	1.000	1.435	.1461
4.500	-50.30	-52.98	2.680	1.000	1.414	.5093
4.600	-57.66	-51.92	-5.742	1.000	1.403	-1.091
4.700	-46.44	-50.88	4.436	1.000	1.395	.8423
4.900	-46.27	-48.85	2.582	1.000	1.385	.4900
5.000	-42.06	-47.87	5.809	1.000	1.384	1.102
5.500	-52.88	-43.25	-9.628	1.000	1.421	-1.830
6.000	-34.48	-39.11	4.635	1.000	1.507	.8851
6.500	-41.25	-35.43	-5.825	1.000	1.614	-1.119
7.000	-36.84	-32.15	-4.695	1.000	1.720	-.9081
8.000	-21.05	-26.65	5.596	1.000	1.880	1.094
8.500	-24.61	-24.35	-.2628	1.000	1.929	-.5157E-01
9.000	-22.22	-22.30	.8166E-01	1.000	1.958	.1606E-01
10.00	-20.74	-18.85	-1.891	1.000	1.969	-.3723
16.00	-6.560	-8.169	1.609	1.000	1.711	.3110
18.00	-5.800	-6.502	.7020	1.000	1.694	.1356
20.00	-5.230	-5.270	.4016E-01	1.000	1.720	.7767E-02
25.00	-3.220	-3.311	.9108E-01	1.000	1.876	.1781E-01
30.00	-2.410	-2.210	-.2000	1.000	2.055	-.3964E-01
35.00	-1.850	-1.533	-.3172	1.000	2.210	-.6370E-01
40.00	-1.160	-1.088	-.7240E-01	1.000	2.335	-.1471E-01
50.00	-.9200	-.5582	-.3618	1.000	2.514	-.7486E-01

```
CORRECTED SUM OF SQUARED OBSERVATIONS =  50057.4
WEIGHTED CORRECTED SUM OF SQUARED OBSERVATIONS =  50057.4
SUM OF SQUARED RESIDUALS =              1424.90
SUM OF WEIGHTED SQUARED RESIDUALS =    1424.90
S =  5.44842      WITH      48 DEGREES OF FREEDOM
CORRELATION (OBSERVED,PREDICTED) =  .9857

AIC criteria =        404.14010
SC  criteria =        404.10706
```

Solution II - Constant CV-model

```
TITLE 1
Multiple effect sites
MODEL
 COMM
NPARM 6
PNAMES 'EMAX1', 'EC51', 'n1','EMAX2', 'EC52', 'n2'
END
    1: TEMP
    2: CP=X
    3: EMAX1=P(1)
    4: EC51=P(2)
    5: N1=P(3)
    6: EMAX2=P(4)
    7: EC52=P(5)
    8: N2=P(6)
    9: END
   10: FUNC1
   11: E1=EMAX1*(CP**N1)/((EC51**N1) + CP**N1)
   12: E2=EMAX2*(CP**N2)/((EC52**N2) + CP**N2)
   13: IF E2 > E1 THEN
   14: F = E2 - E1
   15: WT = 1 / (F*F)
   16: ELSE
```

```
   17: F = 0
   18: ENDIF
   19: F=E1-E2
   20: END
   21: EOM
NVARIABLES 2
NPOINTS 100
XNUMBER 1
YNUMBER 2
METHOD 2   'Gauss-Newton (Levenberg and Hartley)
ITERATIONS 50
INITIAL 100,10,4,100,.4,2
LOWER BOUNDS .1,.1,.1,.1,.01,.01
UPPER BOUNDS 200,50,9,200,5,9
NOBSERVATIONS 54
BEGIN
```

PARAMETER	ESTIMATE	STANDARD ERROR	CV%	UNIVARIATE C.I.	
EMAX1	98.649243	1.375252	1.39	95.884121	101.414365
EC51	4.853153	.148105	3.05	4.555367	5.150938
N1	2.075233	.059791	2.88	1.955016	2.195450
EMAX2	98.767708	1.741123	1.76	95.266954	102.268462
EC52	.561584	.009225	1.64	.543037	.580132
N2	4.037894	.067118	1.66	3.902946	4.172843

```
*** CORRELATION MATRIX OF THE ESTIMATES ***
PARAMETER  EMAX1      EC51          N1         EMAX2      EC52         N2
EMAX1    1.00000
EC51     -.722693    1.00000
N1        .382544     .046155    1.00000
EMAX2     .996093    -.715679     .446254    1.00000
EC52      .472247    -.380504     .108361     .466416    1.00000
N2       -.105442     .335299     .498949    -.067341    -.67921    1.00000

 Condition_number=        495.1

   FUNCTION   1
```

X	OBSERVED Y	PREDICTED Y	RESIDUAL	WEIGHT	SE-PRED	STANDARDI RESIDUAL
.1000	-.6000E-01	-.6166E-01	.1663E-02	268.6	.5399E-02	.5068
.1500	-.4400	-.4034	-.3662E-01	6.223	.1970E-01	-1.002
.2000	-1.360	-1.373	.1296E-01	.5353	.5561E-01	.9961E-01
.2500	-3.540	-3.415	-.1252	.8633E-01	.1274	-.3811
.3000	-6.570	-6.971	.4011	.2068E-01	.2555	.5959
.3500	-12.00	-12.33	.3285	.6602E-02	.4547	.2759
.4000	-20.08	-19.46	-.6208	.2647E-02	.7164	-.3301
.4500	-27.20	-27.96	.7574	.1281E-02	1.002	.2791
.5000	-37.30	-37.14	-.1641	.7256E-03	1.260	-.4518E-01
.5500	-39.76	-46.24	6.483	.4677E-03	1.450	1.421
.6000	-60.09	-54.67	-5.420	.3345E-03	1.561	-.9960
.6500	-65.62	-62.05	-3.566	.2596E-03	1.604	-.5730
.7000	-64.30	-68.26	3.965	.2145E-03	1.606	.5759
.7500	-79.33	-73.34	-5.993	.1858E-03	1.590	-.8070
.8000	-83.59	-77.39	-6.199	.1669E-03	1.573	-.7887

.8500	-70.42	-80.58	10.16	.1539E-03	1.561	1.240
.9000	-81.12	-83.07	1.946	.1449E-03	1.557	.2300
.9500	-85.74	-84.97	-.7653	.1384E-03	1.560	-.8836E-01
1.000	-98.40	-86.42	-11.98	.1338E-03	1.569	-1.359
1.100	-79.37	-88.30	8.930	.1282E-03	1.591	.9917
1.500	-81.38	-89.00	7.620	.1262E-03	1.643	.8403
1.800	-79.10	-86.71	7.612	.1330E-03	1.632	.8622
2.000	-95.22	-84.66	-10.56	.1395E-03	1.611	-1.225
2.250	-89.43	-81.77	-7.662	.1496E-03	1.577	-.9211
2.500	-67.96	-78.65	10.69	.1617E-03	1.535	1.336
2.750	-72.78	-75.40	2.617	.1759E-03	1.490	.3415
2.800	-75.25	-74.74	-.5119	.1790E-03	1.480	-.6739E-01
3.000	-74.91	-72.09	-2.821	.1924E-03	1.442	-.3853
3.500	-67.57	-65.50	-2.072	.2331E-03	1.343	-.3117
3.900	-56.32	-60.41	4.086	.2741E-03	1.265	.6671
4.000	-47.04	-59.17	12.13	.2856E-03	1.247	2.023
4.200	-63.37	-56.76	-6.613	.3104E-03	1.211	-1.150
4.350	-53.84	-55.00	1.156	.3306E-03	1.184	.2075
4.500	-50.30	-53.28	2.980	.3523E-03	1.159	.5525
4.600	-57.66	-52.16	-5.498	.3675E-03	1.143	-1.042
4.700	-46.44	-51.07	4.625	.3835E-03	1.127	.8954
4.900	-46.27	-48.94	2.666	.4176E-03	1.097	.5389
5.000	-42.06	-47.90	5.843	.4358E-03	1.082	1.207
5.500	-52.88	-43.07	-9.815	.5392E-03	1.014	-2.260
6.000	-34.48	-38.75	4.271	.6659E-03	.9548	1.096
6.500	-41.25	-34.93	-6.324	.8198E-03	.9015	-1.806
7.000	-36.84	-31.55	-5.294	.1005E-02	.8527	-1.679
8.000	-21.05	-25.93	4.881	.1487E-02	.7643	1.897
8.500	-24.61	-23.61	-1.003	.1794E-02	.7234	-.4297
9.000	-22.22	-21.55	-.6695	.2153E-02	.6842	-.3152
10.00	-20.74	-18.11	-2.631	.3049E-02	.6106	-1.484
16.00	-6.560	-7.772	1.212	.1656E-01	.3015	1.624
18.00	-5.800	-6.215	.4147	.2589E-01	.2424	.6958
20.00	-5.230	-5.078	-.1524	.3879E-01	.1996	-.3133
25.00	-3.220	-3.299	.7876E-01	.9190E-01	.1406	.2530
30.00	-2.410	-2.319	-.9073E-01	.1859	.1142	-.4296
35.00	-1.850	-1.727	-.1234	.3355	.9450E-01	-.8135
40.00	-1.160	-1.342	.1822	.5551	.7624E-01	1.568
50.00	-.9200	-.8922	-.2783E-01	1.256	.8773E-01	-.9622

```
CORRECTED SUM OF SQUARED OBSERVATIONS =  50057.4
WEIGHTED CORRECTED SUM OF SQUARED OBSERVATIONS =  52.1526
SUM OF SQUARED RESIDUALS =            1453.22
SUM OF WEIGHTED SQUARED RESIDUALS = .514663
S = .103548      WITH     48 DEGREES OF FREEDOM
CORRELATION (OBSERVED,PREDICTED) =  .9854

AIC criteria =        -23.86916
SC  criteria =        -23.90221
```

PD16 - Enantiomer interaction

Objectives
◆ To model the enantiomer interaction between (+)- and (-)-ketamine
◆ To implement an agonist/partial agonist model

Problem specification

In this exercise, we illustrate the complexity of pharmacodynamic interactions between enantiomers with a study with ketamine where quantitative electroencephalogram (*EEG*) parameters were used as the pharmacodynamic measurements. The pharmacodynamics of a racemic mixture of RS(±)-ketamine and of each enantiomer (S(+)-ketamine and R(-)-ketamine) were studied in five volunteers (Schüttler *et al* [1987]). The median frequency of the *EEG* power spectrum, which provides a continuous non-invasive measure of the degree of central nervous system depression (pharmacodynamics), was related to the serum concentration of the drug (pharmacokinetics).

The data generated here use information from the Schüttler analysis. In part I, we fit the racemic effect data (Equation 16:2, Figure 16.1 left), and in part II the S(+)- and R(-)-ketamine data simultaneously with two different functions (Equation 16:2, Figure 16.1 right). In part III, we fit an extended dataset (Figure 16.2) by means of Equation 16:1.

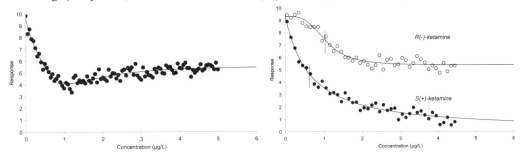

Figure 16.1 *Observed (symbols) and predicted (curves) effects of the racemic mixture of ketamine (left), and of the individual enantiomers (right). Vertical lines in the right hand graph indicate the respective IC_{50} values.*

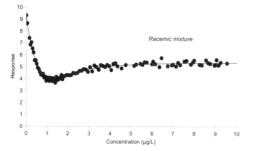

Figure 16.2 *Plot of the observed response versus the concentration of the racemic mixture of ketamine, using an extended dataset with an added constant error.*

Part I

The concentration-effect relationship was described by an inhibitory sigmoid I_{max} model, yielding estimates of both maximal inhibitory effect I_{max} and potency IC_{50} of the racemic (Equation 16:1) and enantiomeric forms (Equation 16:2) of ketamine. Equation 16:1 is the agonist/partial agonist model for the racemate. It includes a baseline effect E_0, the maximum drug-induced effect I_{max1}, the potency IC_{51} and the sigmoidicity factor n_1 corresponding to the S(+)-form and the corresponding parameters (I_{max2}, IC_{52}, and n_2) for the R(-)-form of ketamine. Racemic data (Figure 16.1) were fit using this equation

$$E = E_0 - \frac{I_{max1} \cdot \left[\dfrac{C}{IC_{51}}\right]^{n1} + I_{max2} \cdot \left[\dfrac{C}{IC_{52}}\right]^{n2}}{1 + \left[\dfrac{C}{IC_{51}}\right]^{n1} + \left[\dfrac{C}{IC_{52}}\right]^{n2}} \tag{16:1}$$

Part II

Figure 16.1 (right) includes enantiomeric data of R(-)- and S(+)-ketamine. The plasma concentration-time range is similar to the racemic data. Enantiomeric data were then fit by means of Equation 16:2, including one function for the S(+)-ketamine data and a separate function for the R(-)-ketamine data.

$$E(R(-)\text{- or }S(+)\text{- }ketamine) = E_0 - \frac{I_{max1} \cdot C^n}{IC_{50}^n + C^n} \tag{16:2}$$

Part III

New data, covering a larger concentration range (0 - 10 μg/L), were then generated, and an error with a constant (3%) coefficient of variation was added to data (Figure 16.2). The combined agonist/partial agonist model (i.e., Equation 16:1) was then re-fitted to the new dataset. Note the dip on the curve, with a minimum effect of about 4 units at a plasma concentration of 1.2 - 1.3 μg/L. The initial dip stems from the impact of the more potent S(+)-ketamine which also exhibits a larger intrinsic activity than R(-)-ketamine. R(-)-ketamine, with a higher IC_{50} value and a lower E_{max} *dilutes* the overall effect at higher plasma concentrations. We will apply the *IRLS* method by means of setting $WT = 1/(C\text{-}hat)^2$.

Because of the large number of observations (100 per dataset), data are only given in

Figures 16.1 and 16.2, and in the program output.

Initial parameter estimates

Racemic model	Enantiomeric models
I_{max1} = 10	I_{max1} = 10
IC_{51} = 1.0	IC_{51} = 1.0
n_1 = 2.0	n_1 = 2.0
I_{max2} = 5	I_{max2} = 5
IC_{52} = 1.0	IC_{52} = 1.0
n_2 = 2.0	n_2 = 4.0
E_0 = 1	E_0 = 2

Interpretation of results and conclusions

We did not succeed in fitting the combined racemic model to the simulated original data (Part I) since the concentration span was too limited. However, it is interesting to note that the effect-concentration course reached a minimum and then increased slightly because of the *diluting* effect of the R(-)-isomer (compare Figures 16.1 left and 16.2). In the Schüttler data there was no obvious dip followed by a slight increase in effect, such as we saw when we simulated the agonist/partial agonist model over a larger concentration range (Part III).

In part II, we could successfully fit the separated models to the R(-)- and S(+)-ketamine data with a high parameter precision and low correlation between parameters. A new dataset with 3% *CV* error was then generated (Part III and Figure 16.2). The concentration range was increased to 10 µg/L. With this dataset, we could then estimate the parameters correctly.

The most common approach when judging the validity of a model is to examine the residuals. Figure 16.3 shows an example of a residual plot of the relative residual against the independent variable (*C*) from the last fit (Part III).

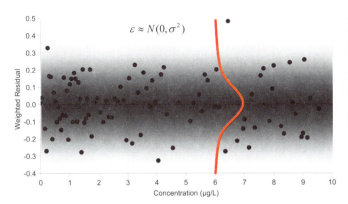

Figure 16.3 Scatter plot of weighted residual against the independent variable (concentration). The shaded background represents the density of the residuals provided an infinite number of observations were fitted by the model. The red line symbolizes the distribution with mean weighted residual of zero and variance σ^2.

The purpose of these plots is to examine the validity of the assumption of constant relative variance. It can easily be seen that there is a random scatter about the predicted curve (horizontal line through zero). In other words, there is no obvious trend in the scatter. Therefore, it is important that simple residual plots be examined routinely by plotting absolute and weighted residuals, or standardized residuals as a function of the independent variable or of the predicted value of the dependent variable. In general, an overall horizontal band of residuals around the value zero, displaying no systematic tendencies to be positive or negative, suggests random error in the data and an appropriate model and weighting scheme.

Solution I – Part I (Classical agonist/partial agonist interaction model)

```
TITLE 1
Ketamine model: data taken from Schuttler et al 1987.
MODEL
 COMM
NFUN   1
NPARM 7
PNAMES 'Emax1','Emax2','EC51','EC52','n1','n2','E0'
END
    1: TEMP
    2: EMAX1=P(1)
    3: EMAX2=P(2)
    4: EC51=P(3)
    5: EC52=P(4)
    6: N1=P(5)
    7: N2=P(6)
    8: E0 =P(7)
    9: CP=X
   10: END
   11: FUNC1
   12: A = (CP/EC51)**N1
   13: B = (CP/EC52)**N2
   14: F = E0 - (EMAX1*A + EMAX2*B)/(1. + A + B)
   15: END
   16: EOM
NVARIABLES 2
NPOINTS 100
XNUMBER 1
YNUMBER 2
METHOD 2   'Gauss-Newton (Levenberg and Hartley)
ITERATIONS 50
INITIAL 10,5,1,1,2,2,10
LOWER BOUNDS 0,0,0,0,0,0,0
UPPER BOUNDS 25,15,5,5,5,5,20
MISSING '-'
NOBSERVATIONS 101
DATA 'WINNLIN.DAT'
BEGIN
```

PARAMETER	ESTIMATE	STANDARD ERROR	CV%	UNIVARIATE C.I.	
EMAX1	22.639465	72.553086	320.47	-121.417164	166.696095
EMAX2	4.066760	.386678	9.51	3.298998	4.834522
EC51	3.010063	16.846546	559.67	-30.439329	36.459454
EC52	1.013819	.130154	12.84	.755394	1.272243

N1	.833351	.445938	53.51	-.052074	1.718776
N2	3.109542	1.189787	38.26	.747180	5.471904
E0	9.633181	.335162	3.48	8.967705	10.298658

```
*** CORRELATION MATRIX OF THE ESTIMATES ***
PARAMETER EMAX1     EMAX2      EC51        EC52     N1         N2         E0
EMAX1     1.00000
EMAX2     -.04892  1.00000
EC51       .99917  -.02620   1.00000
EC52       .14438   .52539    .18158    1.00000
N1        -.95935  -.17607   -.96889    -.36701  1.00000
N2        -.91206   .26771   -.90055     .16637   .81726   1.00000
E0         .36472   .81258    .37880     .39467  -.54004  -.26969   1.0000
```

```
Condition_number=      5539.

CORRECTED SUM OF SQUARED OBSERVATIONS =  99.1960
WEIGHTED CORRECTED SUM OF SQUARED OBSERVATIONS =  99.1960
SUM OF SQUARED RESIDUALS =          10.8356
SUM OF WEIGHTED SQUARED RESIDUALS =  10.8356
S =  .339518      WITH     94 DEGREES OF FREEDOM
CORRELATION (OBSERVED,PREDICTED) =  .9438

AIC criteria =      254.66677
SC  criteria =      256.81969
```

Solution II – Part II (Separate E_{max} models)

```
TITLE 1
Ketaminmodell with data from Schuttler et al 1987.
MODEL
 COMM
NFUN   2
NPARM  7
PNAMES 'Emax1','Emax2','EC51','EC52','n1','E0','n2'
END
    1: TEMP
    2: EMAX1=P(1)
    3: EMAX2=P(2)
    4: EC51=P(3)
    5: EC52=P(4)
    6: N1=P(5)
    7: E0 =P(6)
    8: N2=P(7)
    9: CP=X
   10: END
   11: FUNC 1
   12: F = E0 - EMAX1*(CP**N1)/(EC51**N1 +CP**N1)
   13: END
   14: FUNC 2
   15: F = E0 - EMAX2*(CP**N2)/(EC52**N2 +CP**N2)
   16: END
   17: EOM
NVARIABLES 3
FNUMBER 3
NPOINTS 100
XNUMBER 1
```

```
YNUMBER 2
METHOD 2  'Gauss-Newton (Levenberg and Hartley)
ITERATIONS 50
INITIAL 10,5,1,1,2,2,4
LOWER BOUNDS 0,0,0,0,0,0,0
UPPER BOUNDS 15,15,10,10,10,15,10
MISSING '-'
DATA 'WINNLIN.DAT'
BEGIN
```

PARAMETER	ESTIMATE	STANDARD ERROR	CV%	UNIVARIATE C.I.	
EMAX1	9.270923	.372923	4.02	8.530471	10.011375
EMAX2	3.959803	.185028	4.67	3.592424	4.327182
EC51	.615282	.049500	8.05	.516998	.713566
EC52	1.050176	.048628	4.63	.953624	1.146729
N1	1.046080	.092291	8.82	.862833	1.229328
E0	9.313265	.151557	1.63	9.012344	9.614186
N2	3.498778	.487245	13.93	2.531337	4.466219

```
*** CORRELATION MATRIX OF THE ESTIMATES ***
PARAMETER EMAX1    EMAX2      EC51      EC52       N1        E0        N2
EMAX1  1.00000
EMAX2   .54516   1.0000
EC51    .51006   -.24037    1.0000
EC52   -.36156   -.41260     .15942   1.0000
N1     -.89558   -.33471    -.47986    .22199   1.0000
E0      .60002    .90856    -.26456   -.60257   -.36840   1.00000
N2     -.29290   -.65968     .12914    .28492    .17983   -.48814   1.0000

 Condition_number=        37.63

  FUNCTION   1

 CORRECTED SUM OF SQUARED OBSERVATIONS =  166.902
 WEIGHTED CORRECTED SUM OF SQUARED OBSERVATIONS =  166.902
 SUM OF SQUARED RESIDUALS =          4.45217
 SUM OF WEIGHTED SQUARED RESIDUALS =  4.45217
 S =  .321774    WITH     43 DEGREES OF FREEDOM
 CORRELATION (OBSERVED,PREDICTED) =  .9866

   FUNCTION   2

 CORRECTED SUM OF SQUARED OBSERVATIONS =  95.5729
 WEIGHTED CORRECTED SUM OF SQUARED OBSERVATIONS =  95.5729
 SUM OF SQUARED RESIDUALS =          5.37994
 SUM OF WEIGHTED SQUARED RESIDUALS =  5.37994
 S =  .349673    WITH     44 DEGREES OF FREEDOM
 CORRELATION (OBSERVED,PREDICTED) =  .9715


 TOTALS FOR ALL CURVES COMBINED
 SUM OF SQUARED RESIDUALS =          9.83211
 SUM OF WEIGHTED SQUARED RESIDUALS =  9.83211
```

```
    S =   .323414      WITH      94 DEGREES OF FREEDOM
```

Solution III – Part III (Agonist/partial agonist model - extended data)

```
TITLE 1
Ketamine model: data taken from Schuttler et al 1987.
MODEL
 COMM
NPARM 7
PNAMES 'Emax1','Emax2','EC51','EC52','n1','n2','E0'
END
    1: TEMP
    2: EMAX1=P(1)
    3: EMAX2=P(2)
    4: EC51=P(3)
    5: EC52=P(4)
    6: N1=P(5)
    7: N2=P(6)
    8: E0 =P(7)
    9: CP=X
   10: END
   11: FUNC 1
   12: A = (CP/EC51)**N1
   13: B = (CP/EC52)**N2
   14: F = E0 - (EMAX1*A + EMAX2*B)/(1. + A + B)
   15: END
   16: EOM
NVARIABLES 2
NPOINTS 100
XNUMBER 1
YNUMBER 2
METHOD 2  'Gauss-Newton (Levenberg and Hartley)
ITERATIONS 50
INITIAL 10,5,1,1,2,2,10
LOWER BOUNDS 0,0,0,0,0,0,0
UPPER BOUNDS 25,15,5,5,5,5,20
MISSING '-'
NOBSERVATIONS 96
DATA 'WINNLIN.DAT'
BEGIN
```

PARAMETER	ESTIMATE	STANDARD ERROR	CV%	UNIVARIATE C.I.	
EMAX1	11.569170	5.888893	50.90	-.131991	23.270330
EMAX2	3.907497	.163172	4.18	3.583277	4.231717
EC51	.957010	.925273	96.68	-.881496	2.795516
EC52	1.071459	.066813	6.24	.938702	1.204215
N1	.930283	.203612	21.89	.525709	1.334858
N2	3.186737	.481768	15.12	2.229470	4.144004
E0	9.303580	.157724	1.70	8.990184	9.616976

```
*** CORRELATION MATRIX OF THE ESTIMATES ***
PARAMETER  EMAX1      EMAX2    EC51        EC52       N1        N2        E0
```

```
EMAX1    1.0000
EMAX2    .18603   1.0000
EC51     .99823   .15633   1.00000
EC52    -.04889   .37331  -.01851   1.00000
N1      -.96058  -.35759  -.96159  -.17884   1.00000
N2      -.92083  -.02484  -.91035   .34050   .83313   1.00000
E0       .38692   .93400   .35106   .22693  -.52280  -.30359   1.0000


    Condition_number=          707.6

    CORRECTED SUM OF SQUARED OBSERVATIONS =   78.7241
    WEIGHTED CORRECTED SUM OF SQUARED OBSERVATIONS =   78.7241
    SUM OF SQUARED RESIDUALS =           2.30938
    SUM OF WEIGHTED SQUARED RESIDUALS =  2.30938
    S =  .161084     WITH      89 DEGREES OF FREEDOM
    CORRELATION (OBSERVED,PREDICTED) =  .9852

AIC criteria =        94.35014
SC  criteria =        96.32536
```

PD17 – Effect-compartment model I – Intravenous bolus dosing

Objectives

◆ **To fit an effect-compartment model to intravenous bolus dose data**

◆ **To simulate the response-time curves following different doses**

◆ **To simulate the response-time curves of different k_{e0} values**

Problem specification

The effect-compartment model (also called the link model) attempts to account for distributional processes that delay the onset and loss of response (Figure 17.1)

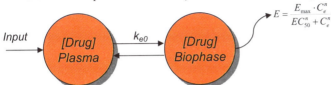

$$E = \frac{E_{\max} \cdot C_e^n}{EC_{50}^n + C_e^n}$$

Figure 17.1 Schematic illustration of the effect-compartment model.

Figure 17.2 shows schematically the relationship between the time course of plasma kinetics linked to the time course of a response. The goal of fitting the effect-compartment model to response-time data is to obtain estimates of the time delay (k_{e0}) between plasma concentration C and the effect-compartment concentration C_e and the pharmacodynamics (E_{max}, EC_{50} and n).

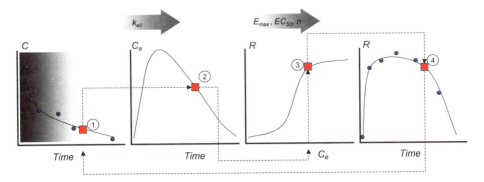

Figure 17.2 *Schematic representation of the effect-compartment model. The drug is given as an intravenous bolus dose. The concentration at time point ① in plasma corresponds to the effect-compartment concentration at ②. The k_{e0} parameter will enable us to generate the relationship between C_e and time course. The k_{e0} parameter gives us the time delay between C and C_e. The C_e at ② will generate a response corresponding to ③ provided response R versus time data are available ④ for estimation of E_{max}, EC_{50} and n.*

The plasma kinetics of a new analgesic can be described by Equation 17:1 after an intravenous bolus dose of 45 µg/kg, with $C^0 = 45.0$, $V = 1$ L/kg and $K = 0.50$ h^{-1}

$$C = \frac{Dose_{iv}}{V} \cdot e^{-K \cdot t} = 45.0 \cdot e^{-0.50 \cdot t} \tag{17:1}$$

The surrogate response was recorded for 80 h as shown in Figure 17.3. It was proposed that a kinetic/dynamic analysis should be carried out in order to refine the design of a planned extended exposure study in patients. The potency EC_{50}, intrinsic activity E_{max} and half-life (or rate-constant k_{e0}) of the response were sought.

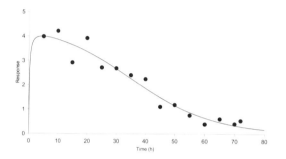

Figure 17.3 *Observed (●) and predicted (solid line) effect-time data after an intravenous bolus dose of 45 µg/kg.*

The effect-compartment (link) model allows estimation of the *in vivo* dynamic relationship from non steady state effect *versus* time and concentration *versus* time data (Segre [1968], Dahlström *et al* [1978] and Sheiner *et al* [1979]). The rate of change of the amount of drug A_e in the hypothetical effect compartment can be expressed as

$$\frac{dA_e}{dt} = k_{1e} \cdot A_1 - k_{e0} \cdot A_e \tag{17:2}$$

A_1 is the amount of drug in the central compartment of a pharmacokinetic model linked to the effect compartment, with first-order rate constants k_{1e} and k_{e0}. The concentration of drug in the effect compartment C_e, is obtained by dividing A_e by the effect compartment volume V_e

$$C_e = \frac{k_{1e} \cdot D}{V_e \cdot (k_{e0} - K)} \cdot \left[e^{-K \cdot t} - e^{-k_{e0} \cdot t} \right] \tag{17:3}$$

At steady state, C will be equal to C_e/K_p by definition and V_e will be equal to V_c, which gives

$$C_e = \frac{k_{1e} \cdot D}{V_c \cdot (k_{e0} - K)} \cdot \left[e^{-K \cdot t} - e^{-k_{e0} \cdot t} \right] \tag{17:4}$$

The effect-compartment model relates the kinetics of the drug in plasma to the kinetics of the drug in the effect compartment. This model may be used together with the E_{max} model for estimation of the maximal drug-induced effect E_{max}, the plasma concentration at half-maximal effect EC_{50}, and the rate constant of the disappearance of the effect k_{e0}.

$$E = \frac{E_{max} \cdot C_e^n}{EC_{50}^n + C_e^n}$$

(17:5)

We will fit Equations 17:4 and 17:5 to the effect data, and estimate k_{e0}, EC_{50}, and E_{max}. Assume the sigmoidicity factor n to be equal to unity.

At steady state, C_e is directly proportional to the plasma concentration C since C_e is equal to $K_p \cdot C$. We assume that the partition coefficient between C and C_e is equal to unity. Consequently, the obtained potency EC_{50} by regressing Equations 17:4 and 17:5 to the data, represents the steady state plasma concentration producing 50% of E_{max}. The effect equilibration rate constant k_{e0} may be viewed as a first-order distribution rate constant. It can also be thought of in terms of the rate of presentation of a drug to a specific tissue, determined by, e.g., tissue perfusion rate, the apparent volume of the tissue and eventual diffusion into the tissue.

The rate of change of the amount of drug C_e in the hypothetical effect-compartment may also be expressed as depicted in Figure 17:1

$$\frac{dC_e}{dt} = k_{e0} \cdot C_p - k_{e0} \cdot C_e$$

(17:6)

In this case, we assume that k_{1e} and k_{e0} are equal. Note however, that there is no net loss of compound from the central to the effect-compartment. The arrow pointing from the effect-compartment to the plasma compartment could equally well point from effect-compartment and out.

Initial parameter estimates

E_{max} = 3.0 (units)   or 4-5

EC_{50} = 1.5 (μg/L)

k_{e0} = 0.1 (h^{-1}) $-slope = (ln(E_1/E_2))/(t_1-t_2) = k_{e0}$

The k_{e0} parameter is estimated from the terminal linear portion of a semi-logarithmic plot of response *versus* time.

Interpretation of results and conclusions

We started by fitting the pharmacokinetic model to the data and then used the effect-compartment library to input *Dose* (45 µg/kg), volume V (1 L/kg) and elimination rate constant K (0.5 h^{-1}) as constant values into the program. For most drugs the kinetics 'drive' the dynamics but not *vice versa*. In the second step, we specified the type of E_{max} model to be fitted together with the effect-compartment model to the effect-time data. By doing so we obtained estimates of E_{max}, EC_{50}, and k_{e0}. The correlation between E_{max} and EC_{50} was high and the precision of EC_{50} was low. A possible explanation for the poorly estimated EC_{50} could be that the data only contained information from a single dose. Two or more dose levels would improve the parameter precision. Figure 17.4 shows three simulations of the response-time course after three different doses.

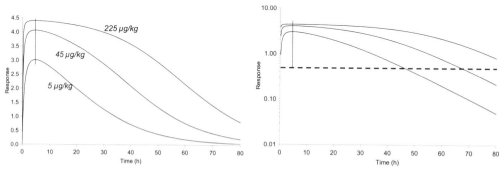

Figure 17.4 *Simulated response-time curves of the different intravenous bolus doses (5, 45 and 225 µg/kg) presented in a linear plot (left) and semi-logarithmic plot (right). Note that the time courses peak at the same t_{max} (left) and fall off linearly at plasma concentrations far below the EC_{50}/IC_{50} value. The level below which the response-time curve decline linearly on a logarithmic scale is indicated by the dashed horizontal line.*

The half-life of response (i.e., 0.693/0.5 or 1.4 h) will determine the onset and duration of the response, provided plasma concentration is substantially below the EC_{50} value. Figure 17.5 shows a simulation of the response-time curve with three different values of k_{e0} (0.1, 0.5 and 2.5 h^{-1}) corresponding to response half-lives of 0.3, 1.4 and 6.9 h, respectively. The half-life of the drug in plasma is 1.4 h ($K \sim 0.5$ h^{-1}), which means that the observed half-life of the response cannot be shorter than that.

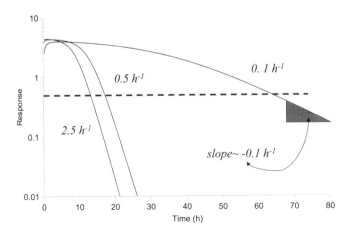

Figure 17.5 *Simulated response-time curves of different k_{e0} values (0.1, 0.5 and 2.5 h^{-1}). Note that the time course of the lowest k_{e0} corresponds to the effect data in this exercise. A k_{e0} of 0.5 h^{-1} will result in a response-time curve that falls in parallel with the slope of the intravenous bolus concentration when $C < EC_{50}$. The approximate response level at which the response declines linearly on a semi-logarithmic scale is indicated by the dashed horizontal line.*

This exercise has shown you how to implement and fit an effect-compartment model. Simulations were also done to demonstrate the response-time curve on linear and semi-logarithmic response-time scales, and to show how a change in k_{e0} affects the response time course. You may be interested to study more complex models, and we would then recommend Colburn [1981].

Solution

```
TITLE 1
Link model: One-cmpt iv bolus
MODEL
 COMM
   NCON 2
   NPAR 3
   PNAMES 'Emax','EC50','Keo'
END
     1: TEMP
     2: K=CON(1)
     3: C0 =CON(2)
     4: T=X
     5: END
     6: FUNC1
     7: CE = (C0*KEO/(KEO - K))*(EXP(-K*T) - EXP(-KEO*T))
     8: F= EMAX*(CE/(CE + EC50))
     9: END
    10: EOM
NOPAGE BREAKS
NVARIABLES 2
NPOINTS 1000
XNUMBER 1
YNUMBER 2
CONSTANTS .5,45
METHOD 2  'Gauss-Newton (Levenberg and Hartley)
ITERATIONS 50
INITIAL 3,1.5,.1
LOWER BOUNDS 0,0,0
UPPER BOUNDS 10,5,1
```

```
MISSING 'Missing'
NOBSERVATIONS 15
DATA 'WINNLIN.DAT'
BEGIN
```

PARAMETER	ESTIMATE	STANDARD ERROR	CV%	UNIVARIATE C.I.	
EMAX	4.512346	.630860	13.98	3.137819	5.886873
EC50	.609054	.395896	65.00	-.253531	1.471638
KEO	.071507	.015745	22.02	.037201	.105813

```
*** CORRELATION MATRIX OF THE ESTIMATES ***
PARAMETER  EMAX       EC50        KEO
EMAX       1.00000
EC50        .95375    1.00000
KEO        -.85813    -.94210    1.00000

 Condition_number=        152.8

   FUNCTION   1
```

X	OBSERVED Y	PREDICTED Y	RESIDUAL	WEIGHT	SE-PRED	STANDARDI RESIDUAL
5.000	3.980	3.988	-.8351E-02	1.000	.2390	-.3160E-01
10.00	4.210	3.863	.3471	1.000	.1990	1.174
15.00	2.920	3.646	-.7265	1.000	.1563	-2.269
20.00	3.920	3.370	.5501	1.000	.1478	1.697
25.00	2.720	3.039	-.3193	1.000	.1618	-1.006
30.00	2.690	2.665	.2460E-01	1.000	.1705	.7863E-01
35.00	2.410	2.267	.1433	1.000	.1639	.4531
40.00	2.250	1.867	.3828	1.000	.1493	1.183
45.00	1.110	1.491	-.3815	1.000	.1408	-1.165
50.00	1.190	1.158	.3178E-01	1.000	.1430	.9739E-01
55.00	.7600	.8778	-.1178	1.000	.1480	-.3634
60.00	.4000	.6520	-.2520	1.000	.1477	-.7773
65.00	.6100	.4768	.1332	1.000	.1396	.4065
70.00	.4100	.3444	.6562E-01	1.000	.1256	.1968
72.00	.5400	.3016	.2384	1.000	.1190	.7100

```
CORRECTED SUM OF SQUARED OBSERVATIONS =  26.6990
WEIGHTED CORRECTED SUM OF SQUARED OBSERVATIONS =  26.6990
SUM OF SQUARED RESIDUALS =          1.52343
SUM OF WEIGHTED SQUARED RESIDUALS =  1.52343
S =  .356304     WITH     12 DEGREES OF FREEDOM
CORRELATION (OBSERVED,PREDICTED) =  .9711

AIC criteria =        12.31451
SC  criteria =        14.43866
```

PD18 – Effect-compartment model II – Oral dosing

Objectives

◆ To derive an effect-compartment model for oral input

◆ To discriminate between the ordinary E_{max} model and the sigmoid E_{max} model

Problem specification

The pharmacokinetics of an orally administered Ca^{2+}-ion channel blocker was determined and could be described by a two-compartment model with first-order input. A plot of the effect-time data is given in Figure 18.1. The oral dose was 200 units and the pharmacokinetic parameters were determined separately by means of a micro-constant library model and found to be: $V_c = 2.44$ L, $K_a = 0.92$ h^{-1}, $k_{10} = 0.44$ h^{-1}, $k_{12} = 0.36$ h^{-1}, $k_{21} = 0.24$ h^{-1}.

The baseline effect E_0 was manually estimated to be 157 mmHg. We will fit a pharmacodynamic link-model to the observed effect-time data and estimate E_{max}, EC_{50}, k_{e0} and eventually n.

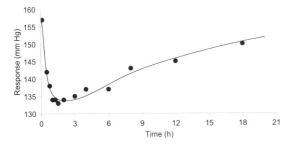

Figure 18.1 *Observed and predicted effect versus time data.*

The kinetics of the drug in plasma could be described by

$$C = \frac{K_a \cdot F \cdot D_{po}}{V_c} \cdot \left[\frac{(k_{21} - \alpha) \cdot e^{-\alpha \cdot t}}{(K_a - \alpha) \cdot (\beta - \alpha)} + \frac{(k_{21} - \beta) \cdot e^{-\beta \cdot t}}{(K_a - \beta) \cdot (\alpha - \beta)} + \frac{(k_{21} - K_a) \cdot e^{-K_a \cdot t}}{(\alpha - K_a) \cdot (\beta - K_a)} \right] \quad (18:1)$$

The concentration of the drug in the effect-compartment is

$$C = \frac{k_{1e} \cdot K_a \cdot F \cdot D_{po}}{V_c} \cdot \left[\frac{(k_{21} - \alpha) \cdot e^{-\alpha \cdot t}}{(K_a - \alpha) \cdot (\beta - \alpha) \cdot (k_{e0} - \alpha)} + \frac{(k_{21} - \beta) \cdot e^{-\beta \cdot t}}{(K_a - \beta) \cdot (\alpha - \beta) \cdot (k_{e0} - \beta)} + \right.$$

$$\left. + \frac{(k_{21} - K_a) \cdot e^{-K_a \cdot t}}{(\alpha - K_a) \cdot (\beta - K_a) \cdot (k_{e0} - K_a)} + \frac{(k_{21} - k_{e0}) \cdot e^{-k_{e0} \cdot t}}{(\alpha - k_{e0}) \cdot (\beta - k_{e0}) \cdot (k_{e0} - k_{e0})} \right] \quad (18:2)$$

The predicted effect is described by the Hill equation

$$E = E_0 - \frac{E_{max} C_e^n}{C_e^n + EC_{50}^n} \tag{18:3}$$

n is either 1 (ordinary E_{max} model) or greater than 1 (sigmoid E_{max} model).

What needs to be done and how can this be accomplished?
- Fit the kinetic data by means of a kinetic library model
- Fix kinetic parameters and let the kinetic model 'drive' the dynamic model
- Select an effect-compartment model with the appropriate effect (Hill) equation
- Fit the dynamic model to the effect-time data

Initial parameter estimates

E_{max} = 50 (mm Hg)
EC_{50} = 2 (ng/mL)
k_{e0} = 2 (1/h)

Interpretation of results and conclusions

The model fitted the data well and E_{max}, EC_{50} and k_{e0} were estimated to be 29 mm Hg, 5.0 µg/L and 1.3 h^{-1}, respectively.

Table 18.1   *F*-statistics

v_1 $\Delta df \rightarrow$	1	2	3	4
v_2 1	161.45	199.50	215.71	224.58
2	18.513	19.000	19.164	19.247
3	10.128	9.5521	9.2766	9.1172
4	7.7086	6.9443	6.5914	6.3882
5	6.6079	5.7861	5.4095	5.1922
6	5.9874	5.1433	4.7571	4.5337
7	5.5914	4.7374	4.3868	4.1203
8	5.3177	4.4590	4.0662	3.8379
9	→ *5.1174*	4.2565	3.8626	3.6331
10	4.9646	4.1028	3.7083	3.4780
11	4.8443	3.9823	3.5874	3.3567

The half-life of k_{e0} was about 0.5 h^{-1}. The *AIC* is 43.2 for the reduced (ordinary E_{max}) model and 44.2 for the full (sigmoid E_{max}) model. This negligible difference says little about which is the superior model. The condition number is 36.07 for the full model and 20.52 for the reduced model.

We will find that the sigmoid E_{max} model is not superior to the ordinary E_{max} model if we apply an *F*-test. The *F*-test is a measure of whether n is different from unity or not. In the full model, n is not significantly different from 1 (unity), since confidence interval of n also includes 1 *(Δdf = 1 = column 1 and Δf_2 = 9 = row 9)*. See *UNIVARIATE* confidence interval in output.

The calculated F^* value is obtained according to

$$F^* = \frac{\dfrac{\left|WRSS_1 - WRSS_2\right|}{\left|df_1 - df_2\right|}}{\dfrac{WRSS_2}{df_2}} = 0.7124 \tag{18:4}$$

The F^* value is less than F_{table} (= 5.1174, P = 0.05). $|df_1 - df_2| = \Delta df = v_2 = 1$ and df_2 = $v_2 = 9$. In other words, we were not able to discriminate between the two models. Thus, the ordinary E_{max} model is the model of choice because it is the simpler model. **Note**: It is only appropriate to apply the F-test when equal weights are used for both models.

This exercise has shown how to fit an effect-compartmentmodel (reduced or ordinary E_{max} model and full or sigmoid E_{max} model) to effect-time data and how to interpret the output. We have also shown that the sigmoid E_{max} model is not superior to the ordinary E_{max} model for this particular dataset.

Solution I - Ordinary E_{max} model

```
TITLE 1
2-CMPT ORAL Link-model
MODEL
  COMM
    NPARM 3
    NCON 1
    PNAMES 'Emax','EC50','Ke0'
END
     1: TEMP
     2: T=X
     3: V=2.44
     4: K01=0.92
     5: K10=0.44
     6: K12=0.36
     7: K21=0.24
     8: E0=157.
     9: EMAX=P(1)
    10: EC50=P(2)
    11: KE0=P(3)
    12: D=CON(1)
    13: R1=DSQRT((K12+K21+K10)**2 - (4*K21*K10))
    14: ALPHA=((K12+K21+K10) + R1)/2
    15: BETA=((K12+K21+K10) - R1)/2
    16: A=(D/V)*K01*(K21-ALPHA)/(ALPHA-BETA)/(ALPHA-K01)/(KE0-ALPHA)
    17: B=-1*(D/V)*K01*(K21-BETA)/(ALPHA-BETA)/(BETA-K01)/(KE0-BETA)
    18: C=(D/V)*K01*(K21-K01)/(BETA-K01)/(ALPHA-K01)/(KE0-K01)
    19: D=(D/V)*K01*(K21-KE0)/(BETA-KE0)/(ALPHA-KE0)/(K01-KE0)
    20: END
    21: FUNC1
    22: CE=A*DEXP(-ALPHA*T)+B*DEXP(-BETA*T)+C*DEXP(-K01*T)+D*DEXP(-KE0*T)
    23: F=E0 - EMAX*CE/(CE + EC50)
    24: END
    25: EOM
NVARIABLES 2
NPOINTS 100
```

```
XNUMBER 1
YNUMBER 2
CONSTANTS 200
METHOD 2  'Gauss-Newton (Levenberg and Hartley)
ITERATIONS 50
INITIAL 50,2,2
LOWER BOUNDS 0,0,0
UPPER BOUNDS 100,8,10
MISSING '-'
NOBSERVATIONS 13
DATA 'WINNLIN.DAT'
BEGIN
```

PARAMETER	ESTIMATE	STANDARD ERROR	CV%	UNIVARIATE C.I.	
EMAX	28.950375	1.554306	5.37	25.487143	32.413607
EC50	4.956502	1.176276	23.73	2.335579	7.577426
KE0	1.304965	.214638	16.45	*.826719*	*1.783212*

```
*** CORRELATION MATRIX OF THE ESTIMATES ***
PARAMETER  EMAX        EC50        KE0
EMAX       1.00000
EC50        .795227    1.00000
KE0        -.208996    -.695793    1.00000

Condition_number=      20.52
```

X	OBSERVED Y	PREDICTED Y	RESIDUAL	WEIGHT	SE-PRED	STANDARDI RESIDUAL
.1000	157.0	155.1	1.863	1.000	.3296	1.454
.5000	142.0	141.5	.5016	1.000	.9528	.5468
.7500	138.0	137.6	.3789	1.000	.6616	.3309
1.000	134.0	135.6	-1.637	1.000	.5357	-1.353
1.250	134.0	134.6	-.5558	1.000	.5320	-.4590
1.500	133.0	134.0	-.9525	1.000	.5635	-.7960
2.000	134.0	133.5	.5124	1.000	.6132	.4372
3.000	135.0	133.9	1.079	1.000	.6156	.9220
4.000	137.0	135.2	1.766	1.000	.5972	1.497
6.000	137.0	138.8	-1.792	1.000	.6670	-1.569
8.000	143.0	141.8	1.151	1.000	.7008	1.026
12.00	145.0	145.8	-.8045	1.000	.7068	-.7196
18.00	150.0	150.1	-.5320E-01	1.000	.5998	-.4513E-01

```
CORRECTED SUM OF SQUARED OBSERVATIONS =   630.923
WEIGHTED CORRECTED SUM OF SQUARED OBSERVATIONS =  630.923
SUM OF SQUARED RESIDUALS =              17.4953
SUM OF WEIGHTED SQUARED RESIDUALS =    17.4953
S =  1.32270      WITH     10 DEGREES OF FREEDOM
CORRELATION (OBSERVED,PREDICTED) =   .9871

AIC criteria =        43.20509
SC  criteria =        41.05252
```

Solution II - Sigmoid E_{max} model

```
TITLE 1
2-CMPT ORAL Link-model
```

```
MODEL
 COMM
NPARM 4
NCON 1
PNAMES 'Emax','EC50','Ke0','n'
END
    1: TEMP
    2: T=X
    3: V=2.44
    4: K01=0.92
    5: K10=0.44
    6: K12=0.36
    7: K21=0.24
    8: E0=157.
    9: EMAX=P(1)
   10: EC50=P(2)
   11: KE0=P(3)
   12: N=P(4)
   13: D=CON(1)
   14: R1=DSQRT((K12+K21+K10)**2 - (4*K21*K10))
   15: ALPHA=((K12+K21+K10) + R1)/2
   16: BETA=((K12+K21+K10) - R1)/2
   17: A=(D/V)*K01*(K21-ALPHA)/(ALPHA-BETA)/(ALPHA-K01)/(KE0-ALPHA)
   18: B=-1*(D/V)*K01*(K21-BETA)/(ALPHA-BETA)/(BETA-K01)/(KE0-BETA)
   19: C=(D/V)*K01*(K21-K01)/(BETA-K01)/(ALPHA-K01)/(KE0-K01)
   20: D=(D/V)*K01*(K21-KE0)/(BETA-KE0)/(ALPHA-KE0)/(K01-KE0)
   21: END
   22: FUNC1
   23: CE=A*DEXP(-ALPHA*T)+B*DEXP(-BETA*T)+C*DEXP(-K01*T)+D*DEXP(-KE0*T)
   24: IF CE LE 0 THEN
   25: ZZE = 0.0
   26: ELSE
   27: ZZE = CE**N
   28: ENDIF
   29: F=E0 - EMAX*ZZE/(ZZE + EC50**N)
   30: END
   31: EOM
NVARIABLES 2
NPOINTS 100
XNUMBER 1
YNUMBER 2
CONSTANTS 200
METHOD 2  'Gauss-Newton (Levenberg and Hartley)'
ITERATIONS 50
INITIAL 50,2,2,2
LOWER BOUNDS 0,0,0,0
UPPER BOUNDS 100,8,10,8
MISSING '-'
NOBSERVATIONS 13
DATA 'WINNLIN.DAT'
BEGIN
```

PARAMETER	ESTIMATE	STANDARD ERROR	CV%	UNIVARIATE C.I.	
EMAX	26.534584	2.578915	9.72	20.700625	32.368544
EC50	4.233479	1.078996	25.49	1.792599	6.674359
KE0	1.270494	.213662	16.82	.787154	1.753834
N	1.190105	.250661	21.06	.623066	1.757143

```
*** CORRELATION MATRIX OF THE ESTIMATES ***
PARAMETER  EMAX         EC50         KE0          N
EMAX      1.00000
EC50       .746538     1.00000
KE0        .028513     -.565106    1.00000
N         -.896926     -.529901    -.141245     1.00000

Condition_number=        36.07
```

X	OBSERVED Y	PREDICTED Y	RESIDUAL	WEIGHT	SE-PRED	STANDARDI RESIDUAL
.1000	157.0	155.7	1.262	1.000	.7234	1.117
.5000	142.0	141.3	.6512	1.000	1.049	.7780
.7500	138.0	137.4	.5890	1.000	.7168	.5191
1.000	134.0	135.6	-1.579	1.000	.5326	-1.282
1.250	134.0	134.6	-.6449	1.000	.5352	-.5240
1.500	133.0	134.1	-1.143	1.000	.5984	-.9516
2.000	134.0	133.8	.2394	1.000	.6815	.2070
3.000	135.0	134.1	.9240	1.000	.6357	.7817
4.000	137.0	135.1	1.866	1.000	.5880	1.546
6.000	137.0	138.4	-1.378	1.000	.8105	-1.288
8.000	143.0	141.5	1.482	1.000	.8130	1.387
12.00	145.0	145.9	-.8753	1.000	.7706	-.7966
18.00	150.0	150.6	-.6162	1.000	1.007	-.6942

```
CORRECTED SUM OF SQUARED OBSERVATIONS =  630.923
WEIGHTED CORRECTED SUM OF SQUARED OBSERVATIONS =  630.923
SUM OF SQUARED RESIDUALS =           16.2120
SUM OF WEIGHTED SQUARED RESIDUALS =  16.2120
S =  1.34214      WITH      9 DEGREES OF FREEDOM
CORRELATION (OBSERVED,PREDICTED) =   .9873

AIC criteria =        44.21478
SC  criteria =        41.34467
```

PD19 – Effect-compartment model III – Single infusion

> ## Objectives
> ◆ To analyze plasma concentration data after an intravenous infusion
> ◆ To analyze response data after an intravenous infusion
> ◆ To obtain initial parameter estimates
> ◆ To use a library effect-compartment model
> ◆ To simulate the concentration-time course of the effect-compartment

Problem specification

This problem highlights some complexities in modeling nonlinear time-delayed dynamics. Specific emphasis is put on obtaining *initial parameter estimates*, applying the *effect-compartment library*, and discussing *experimental design*.

A total dose of 69mg of a new CNS compound was infused over 15 minutes. Concentration and response were measured over a 12 h period (720 min). Plot and analyze the plasma concentration *versus* time, effect *versus* time, and effect *versus* plasma concentration relationships. Obtain initial parameter estimates for the pharmacokinetic and pharmacodynamic models. Propose an adequate model and then use the *WinNonlin* model library.

Table 19.1 Concentration and response-time data

Time (min)	Concentration (μg/L)	Response	Time (min)	Concentration (μg/L)	Response
0	-	-	150	8018	98.4
.5	4885	1.12	180	7109	92
1	10420	16	240	4506	88
2	20740	72	300	3551	84
3	25710	80	360	2295	80
7	52210	96	420	1532	68
15	81420	95.2	480	1270	54.4
20	48300	96	540	816.5	39.2
30	21780	100	600	713.0	24.8
45	15460	96	660	455.4	14.4
75	11240	95.2	720	305.9	8.0

Plot the data according to the three plots in Figure 19.1. Obtain the initial parameter estimates graphically. Characterize the pharmacokinetics and pharmacodynamics of this compound. E_{max} and EC_{50} are obtained graphically from Figure 19.1 (lower, left) and k_{e0} is calculated from the terminal portion of the plot of log-response *versus* time curve.

$$slope = -k_{e0} = \frac{\ln E_1 - \ln E_2}{t_1 - t_2} = \frac{\ln 14.4 - \ln 8.0}{660 - 720} = -0.01\,\text{min}^{-1} \qquad (19:1)$$

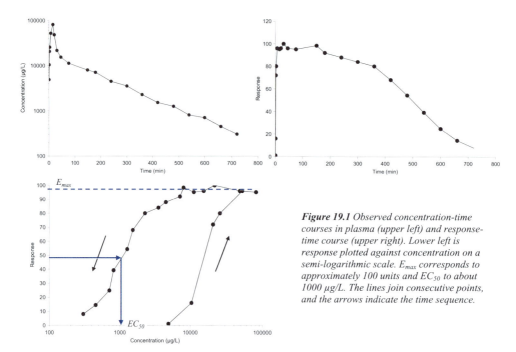

Figure 19.1 *Observed concentration-time courses in plasma (upper left) and response-time course (upper right). Lower left is response plotted against concentration on a semi-logarithmic scale. E_{max} corresponds to approximately 100 units and EC_{50} to about 1000 µg/L. The lines join consecutive points, and the arrows indicate the time sequence.*

How long does it take to achieve pharmacodynamic steady state with an instantaneously constant plasma concentration upon administration of the drug? Figure 19.2 shows three simulations with an effect-compartment model using a k_{e0} of 0.05, 0.15 and 0.5 time^{-1}. The plasma concentration Cp is held constant (set to unity) and Equation 19:2 is used to simulate the system.

$$\frac{dC_e}{dt} = k_{e0} \cdot \left[C_p - C_e \right]$$
(19:2)

C_p and C_e will reach the same steady state level, since the partition coefficient between C_p and C_e is set to unity. The time to steady state in the effect-compartment is governed by means of k_{e0}, provided C_p is much less than EC_{50}.

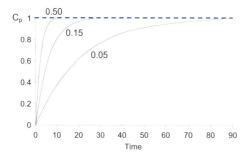

Figure 19.2 *Simulated plots of concentration versus time in the effect-compartment (solid lines) using three different k_{e0}s of 0.05, 0.15 and 0.5 time^{-1}. The plasma concentration remained constant at unity (1; dashed horizontal line). Note that the time to steady state in the effect-compartment is governed by the half-life of k_{e0}, but the actual level of C_e at steady state will be equal to C_p, unless the partition coefficient between C_e and C_p is different from 1.*

Initial parameter estimates

E_{max} = 100 (units)
EC_{50} = 1000 (µg/L)
k_{e0} = 0.1 (min^{-1})   Note: different from *-slope =(ln(E$_1$/E$_2$))/(t$_1$-t$_2$) = .01*

The k_{e0} parameter is estimated from the terminal linear portion of a semi-logarithmic plot of response *versus* time. Note that we used a value 10 times larger than the manually calculated value.

Interpretation of results and conclusions

In this exercise, we used a dose that resulted in plasma concentrations that far exceeded the EC_{50} value. In such a case, the half-life of k_{e0} ($t_{1/2ke0}$) is no longer valid, since the system does not behave like a linear system but, rather, nonlinearly. The k_{e0} parameter was estimated to 0.1 h^{-1}, which is one order of magnitude away from the calculated initial estimate (0.01 min^{-1}). E_{max} was estimated to 96.3 units, and EC_{50} to 1163 µg/L. The sigmoidicity factor (exponent *n*) in the Hill equation is 1.9. Figure 19.3 shows the high consistency between observed and predicted response-time data, which means that the model is compatible with the data.

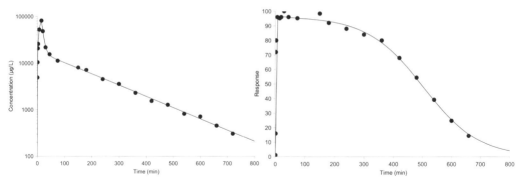

Figure 19.3 *Observed and predicted kinetic data (left). Observed and predicted dynamic data (right).*

We have also simulated the time course of drug in the effect-compartment C_e and demonstrated how k_{e0} determines the time to steady state provided plasma concentration C_p is held constant. Finally we have shown that C_e equals C_p at steady state provided the partitioning between C_e and C_p is 1.

Solution I - kinetics

```
MODEL 10
NOPAGE BREAKS
NVARIABLES 3
NPOINTS 1000
XNUMBER 1
YNUMBER 2
```

```
NCONSTANTS 4
CONSTANTS 1,69000,0,15
METHOD 2   'Gauss-Newton (Levenberg and Hartley)
REWEIGHT -1
ITERATIONS 50
MISSING 'Missing'
NOBSERVATIONS 21
DATA 'WINNLIN.DAT'
BEGIN
```

The following default parameter boundaries were generated.

Parameter	Lower Bound	Upper Bound
Volume	.0000	3.970
K21	.0000	.1881
Alpha	.0000	1.400
Beta	.0000	.5614E-01

PARAMETER	ESTIMATE	STANDARD ERROR	CV%	UNIVARIATE C.I.	
Volume	.416572	.008524	2.05	.398587	.434556
K21	.019077	.000895	4.69	.017190	.020965
Alpha	.136651	.004734	3.46	.126664	.146638
Beta	.005551	.000173	3.11	.005187	.005916

```
*** CORRELATION MATRIX OF THE ESTIMATES ***
PARAMETER  Volume     K21        Alpha      Beta
Volume     1.00000
K21        -.27865    1.00000
Alpha      -.75205    .71086     1.00000
Beta       -.17457    .78914     .39844     1.00000

 Condition_number=        108.7

    FUNCTION   1
```

X	OBSERVED Y	PREDICTED Y	RESIDUAL	WEIGHT	SE-PRED	STANDARDIZ RESIDUAL
.5000	4885.	5355.	-469.9	.1867E-03	105.2	-1.155
1.000	.1042E+05	.1039E+05	28.79	.9622E-04	196.1	.0521
2.000	.2074E+05	.1960E+05	1147.	.5103E-04	342.1	1.576
3.000	.2571E+05	.2776E+05	-2049.	.3602E-04	450.3	-2.426
7.000	.5221E+05	.5245E+05	-238.6	.1907E-04	679.4	-.2118
15.00	.8142E+05	.7954E+05	1884.	.1257E-04	936.8	1.425
20.00	.4830E+05	.4783E+05	468.4	.2091E-04	885.7	.5257
30.00	.2178E+05	.2322E+05	-1436.	.4308E-04	533.0	-2.069
45.00	.1546E+05	.1493E+05	532.1	.6702E-04	402.3	.9256
75.00	.1124E+05	.1177E+05	-533.9	.8503E-04	362.1	-1.053
150.0	8018.	7750.	268.0	.1292E-03	199.0	.5768
180.0	7109.	6561.	548.4	.1526E-03	164.8	1.261
240.0	4506.	4702.	-196.6	.2130E-03	127.5	-.5279
300.0	3551.	3370.	181.0	.2973E-03	109.4	.5753
360.0	2295.	2415.	-120.3	.4150E-03	96.23	-.4537
420.0	1532.	1731.	-199.4	.5792E-03	83.64	-.8918
480.0	1270.	1241.	28.86	.8084E-03	71.22	.1526
540.0	816.5	889.3	-72.75	.1128E-02	59.45	-.4537
600.0	713.0	637.3	75.66	.1575E-02	48.77	.5551

```
   660.0      455.4      456.8      -1.386     .2198E-02  39.45     -.0119
   720.0      305.9      327.4      -21.48     .3068E-02  31.52     -.2174
```

```
CORRECTED SUM OF SQUARED OBSERVATIONS = .894996E+10
WEIGHTED CORRECTED SUM OF SQUARED OBSERVATIONS =  284356.
SUM OF SQUARED RESIDUALS =           .127016E+08
SUM OF WEIGHTED SQUARED RESIDUALS =  561.297
S =  5.74609     WITH     17 DEGREES OF FREEDOM
CORRELATION (OBSERVED,PREDICTED) =  .9994
```

```
AIC criteria =          140.93527
SC  criteria =          145.11336
```

Solution II - dynamics
```
MODEL LINK 10,105
```
PKVAL .416572,.019077,.136651,.005551
```
NOPAGE BREAKS
NVARIABLES 6
NPOINTS 100
XNUMBER 1
YNUMBER 3
NCONSTANTS 4
CONSTANTS 1,69000,0,15
METHOD 2  'Gauss-Newton (Levenberg and Hartley)
ITERATIONS 50
INITIAL 100,1000,1.5,.1
LOWER BOUNDS 50,100,1,.01
UPPER BOUNDS 150,10000,3,1
MISSING 'Missing'
NOBSERVATIONS 21
DATA 'WINNLIN.DAT'
BEGIN
```

PARAMETER	ESTIMATE	STANDARD ERROR	CV%	UNIVARIATE C.I.	
Emax	96.288200	1.011477	1.05	94.154184	98.422217
EC50	1163.504326	37.093339	3.19	1085.244682	1241.763970
Gamma	1.875116	.091426	4.88	1.682225	2.068007
KEO	.098703	.006217	6.30	.085587	.111819

```
*** CORRELATION MATRIX OF THE ESTIMATES ***
PARAMETER  Emax       EC50       Gamma      KEO
Emax       1.00000
EC50        .51518    1.00000
Gamma      -.49886     -.21581    1.00000
KEO         .00166      .29239    -.06895    1.00000
```

```
 Condition_number=        6380.
```

```
   FUNCTION    1
```

X	OBSERVED Y	PREDICTED Y	RESIDUAL	WEIGHT	SE-PRED	STANDARDIZ RESIDUAL
.5000	1.120	1.584	-.4640	1.000	.3562	-.1755
1.000	16.00	16.72	-.7208	1.000	1.853	-.3757
2.000	72.00	68.53	3.472	1.000	2.145	2.190
3.000	80.00	87.44	-7.435	1.000	1.085	-3.051
7.000	96.00	95.61	.3887	1.000	.9148	.1551

15.00	95.20	96.19	-.9879	1.000	.9911	-.3989
20.00	96.00	96.21	-.2087	1.000	.9948	-.0843
30.00	100.0	96.15	3.847	1.000	.9851	1.552
45.00	96.00	95.90	.1013	1.000	.9484	.0406
75.00	95.20	95.25	-.4523E-01	1.000	.8799	-.0179
150.0	98.40	93.88	4.517	1.000	.8006	1.775
180.0	92.00	93.03	-1.031	1.000	.7811	-.4042
240.0	88.00	90.38	-2.379	1.000	.8064	-.9356
300.0	84.00	85.81	-1.811	1.000	.9618	-.7280
360.0	80.00	78.41	1.589	1.000	1.186	.6649
420.0	68.00	67.54	.4642	1.000	1.326	.2006
480.0	54.40	53.64	.7577	1.000	1.329	.3275
540.0	39.20	38.75	.4454	1.000	1.327	.1924
600.0	24.80	25.53	-.7257	1.000	1.331	-.3139
660.0	14.40	15.59	-1.189	1.000	1.206	-.4997
720.0	8.000	9.027	-1.027	1.000	.9616	-.4127

```
CORRECTED SUM OF SQUARED OBSERVATIONS =   23805.5
WEIGHTED CORRECTED SUM OF SQUARED OBSERVATIONS =   23805.5
SUM OF SQUARED RESIDUALS =              120.978
SUM OF WEIGHTED SQUARED RESIDUALS =   120.978
S =  2.66765      WITH     17 DEGREES OF FREEDOM
CORRELATION (OBSERVED,PREDICTED) =   .9975

AIC criteria =        108.70784
SC  criteria =        112.88593
```

PD20 - Logistic regression I – Single stimulus

Objectives
◆ **To build a dichotomous response model**
◆ **To include continuous and dichotomous variables**

Problem specification

Regression methods have become an integral component of any data analysis concerned with describing the relationship between a response variable and one or more explanatory variables. It is often the case that the outcome variable is discrete, taking two or more possible distinct values. The goal of this analysis is to find the best fitting and most parsimonious yet biologically reasonable model to describe the relationship between the outcome and a set of independent variables. These independent variables are often called covariates. The most common example of modeling is the linear regression model where the outcome variable is assumed to be continuous (Hosmer and Lemeshov [1989]). The thing that distinguishes a logistic regression model from the linear regression model is that the outcome variable in logistic regression is binary or dichotomous.

The above difference between logistic and linear regression models is reflected both in the choice of parametric model and in the assumptions. The specific form of the logistic regression model we will use is shown in Equation 20:1. $\Pi(C)$ denotes the probability of a response as a function of concentration (and possibly other covariates)

$$\Pi(C) = \frac{e^L}{1 + e^L} \qquad (20:1)$$

The transformation of $\Pi(C)$ that will be central to our study of logistic regression is the logit transformation. This transformation is defined, in terms of $\Pi(C)$, as follows

$$L = \text{logit}(\Pi(C)) = \frac{\Pi(C)}{1 - \Pi(C)} \qquad (20:2)$$

where the general (multiple linear regression) functional form is

$$L = \sum x_i \cdot \alpha_i \qquad (20:3)$$

and for this particular example

$$L = \alpha_1 \cdot S_1 - \alpha_2 \cdot C \qquad (20:4)$$

The logit L is linear in its parameters and ranges from $-\infty$ to $+\infty$ depending on α_2 and C. The conditional distribution of the outcome variable has a mean of zero and a variance equal to $\Pi(C)\cdot[1 - \Pi(C)]$. The conditional distribution of the outcome variable follows a binomial distribution, with probability given by the conditional mean $\Pi(C)$. This logistic regression model was used to analyze the impact of one stimulus S_1 on a dichotomous response (Figure 20.1). In the following plot, subjects who responded were coded as one, non-responders as zero. Note that the stimulus was also coded as a zero (no stimulus applied) or one (stimulus applied). It can be seen that the probability of a response decreased as concentration increased.

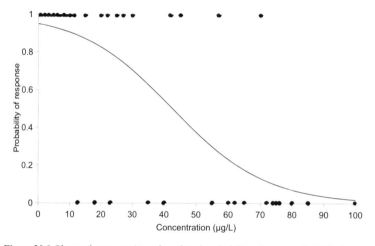

Figure 20.1 *Observed response (●) and predicted probability of response (solid line) versus plasma concentration.*

C is the plasma concentration of the drug. The probability of <u>no</u> response to a stimulus is

$$Q(C) = 1 - \Pi(C) = \frac{1}{1 + e^L}$$

(20:5)

The *odds ratio* (relative risk) of a response-to-no-response is given by

$$odds\ ratio = \frac{\Pi(C)}{Q(C)} = \frac{\dfrac{e^L}{1 + e^L}}{\dfrac{1}{1 + e^L}} = e^L$$

(20:6)

This expression can also be written as the ln*(odds ratio)* or log-*odds ratio*

$$\ln(odds\ ratio) = L = \alpha_1 \cdot S_1 - \alpha_2 \cdot C$$

(20:7)

We will calculate the *ln(odds ratio)* (LogODDS in program code) and relative risk (Rra50 in program code) of a response at a plasma concentration of 50 µg/L with stimulation as secondary parameters.

Initial parameter estimates

$$\alpha = 1$$
$$\beta = 0.1$$

The initial parameter estimates are set to zero for both parameters. *ALPHA* and *BETA* in program code corresponds to α_1 and α_2, respectively, in Equation 20:4. See also section 4.10 for alternative logistic regression models.

Interpretation of results and conclusions

This exercise has shown you how to fit a logistic model to dichotomous data. The condition number was low, and parameter precision was acceptable. The relative risk of a response at 50 µg/L is about 60% with stimulation.

We maximized the log-likelihood function in this example (*MEAN 1, METHOD 3*) which is not necessarily the same as minimizing the *WRSS*. The log-likelihood function is maximized when the weighted residual sum of squares (using the specified weighting scheme) no longer changes from iteration to iteration. Note that for these types of maximum likelihood models, the weighted residual sum of squares is also a χ^2 (chi-square) statistic for lack of fit *LOF* of the model, with degrees of freedom *df* equal to the number of observations minus the number of estimated parameters. Thus, for this example, $\chi^2_{LOF} = 28.43$ with 42 *df*, which is not statistically significant. Thus, there is insufficient evidence to reject the model.

Observe how *WRSS* decreases from 44 to 28.49 and then increases to 33.56 and finally reaches 36.33, where it remained constant.

Solution

```
TITLE 1
Logistic regression of thiopental data
MODEL
 COMM
NPARM 2
NSEC  2
PNAMES 'Alpha1','Beta'
SNAMES 'LogODDS', 'RRa50'
NVAR 3
END
    1: TEMP
    2: ALPHA=P(1)
    3: BETA=P(2)
    4: CP=DTA(1)
    5: S1=1.0
    6: END
    7: FUNC1
    8: AL = ALPHA*S1 - BETA*CP
    9: EL = DEXP(AL)
   10: PHAT = EL/(1 + EL)
```

```
11: F = PHAT
12: IF F NE 0 AND PHAT NE 1 THEN
13: WT = 1/(F*(1. - PHAT))
14: ELSE
15: WT = 0
16: ENDIF
17: END
18: SECO
19: S(1)=ALPHA - BETA*50
20: S(2)=EXP(S(1))
21: END
22: EOM
NVARIABLES 3
FNUMBER 3
NPOINTS 100
XNUMBER 1
YNUMBER 2
METHOD 3   'Gauss-Newton
CONVERGENCE 0
ITERATIONS 10
MEANSQUARE 1
INITIAL 0,0
BEGIN
 Parameter     Lower Bound    Upper Bound
 ALPHA            -10.00        10.00
 BETA             -10.00        10.00
 ITERATION WEIGHTED_SS  ALPHA     BETA
     0      44.0000    .0000    .0000    RANK =  2 CONDITION NO. = 62.90
     1      28.4960    1.967    .4647E-01 RANK =  2 CONDITION NO. = 69.70
     2      33.5619    2.693    .6345E-01 RANK =  2 CONDITION NO. = 77.47
     3      36.1099    2.912    .6839E-01 RANK =  2 CONDITION NO. = 80.71
     4      36.3328    2.930    .6877E-01 RANK =  2 CONDITION NO. = 81.01
     5      36.3344    2.930    .6878E-01 RANK =  2 CONDITION NO. = 81.01
     6      36.3344    2.930    .6878E-01 RANK =  2 CONDITION NO. = 81.01
     7      36.3344    2.930    .6878E-01 RANK =  2 CONDITION NO. = 81.01
     8      36.3344    2.930    .6878E-01 RANK =  2 CONDITION NO. = 81.01
     9      36.3344    2.930    .6878E-01 RANK =  2 CONDITION NO. = 81.01
    10      36.3344    2.930    .6878E-01
```

THE SPECIFIED VALUE OF THE MEANSQUARE ERROR 1.00000
WAS USED IN COMPUTING ALL OF THE FOLLOWING VARIANCES.

PARAMETER	ESTIMATE	STANDARD ERROR	CV%	UNIVARIATE C.I.	
ALPHA	2.929872	.788619	26.92	1.338381	4.521363
BETA	.068778	.018534	26.95	.031374	.106181

Condition_number=   75.33

X	OBSERVED Y	PREDICTED Y	RESIDUAL	WEIGHT	SE-PRED	STANDARDI RESIDUAL
.0000	1.000	.9493	.5070E-01	20.78	.3795E-01	.2346
.1000E-01	1.000	.9493	.5073E-01	20.77	.3797E-01	.2347
.2000E-01	1.000	.9492	.5076E-01	20.75	.3799E-01	.2348
.8000E-01	1.000	.9490	.5096E-01	20.68	.3808E-01	.2353
.1000	1.000	.9490	.5103E-01	20.65	.3811E-01	.2354
1.000	1.000	.9459	.5411E-01	19.54	.3959E-01	.2429
2.000	1.000	.9423	.5774E-01	18.38	.4126E-01	.2515
3.000	1.000	.9384	.6160E-01	17.30	.4298E-01	.2604
4.000	1.000	.9343	.6570E-01	16.29	.4473E-01	.2696

5.000	1.000	.9300	.7005E-01	15.35	.4652E-01	.2791
6.000	1.000	.9253	.7466E-01	14.47	.4835E-01	.2889
6.500	1.000	.9229	.7707E-01	14.05	.4927E-01	.2940
7.000	1.000	.9204	.7955E-01	13.65	.5021E-01	.2991
8.000	1.000	.9153	.8474E-01	12.89	.5209E-01	.3097
8.500	1.000	.9126	.8744E-01	12.53	.5305E-01	.3151
9.000	1.000	.9098	.9023E-01	12.18	.5401E-01	.3206
10.00	1.000	.9040	.9603E-01	11.51	.5595E-01	.3319
11.00	1.000	.8978	.1022	10.89	.5790E-01	.3436
12.50	.0000	.8880	-.8880	10.05	.6087E-01	-2.868
15.00	1.000	.8697	.1303	8.817	.6586E-01	.3946
18.00	.0000	.8445	-.8445	7.607	.7186E-01	-2.376
20.00	1.000	.8255	.1745	6.937	.7583E-01	.4690
22.00	1.000	.8048	.1952	6.360	.7975E-01	.5025
23.00	.0000	.7938	-.7938	6.104	.8170E-01	-2.002
25.00	1.000	.7704	.2296	5.648	.8554E-01	.5573
27.00	1.000	.7451	.2549	5.261	.8934E-01	.5972
30.00	1.000	.7040	.2960	4.795	.9496E-01	.6626
35.00	.0000	.6278	-.6278	4.277	.1042	-1.329
40.00	.0000	.5446	-.5446	4.031	.1128	-1.123
42.00	1.000	.5103	.4897	4.001	.1158	1.007
45.00	1.000	.4588	.5412	4.028	.1196	1.119
55.00	.0000	.2988	-.2988	4.780	.1211	-.6775
57.00	1.000	.2708	.7292	5.073	.1189	1.705
60.00	.0000	.2320	-.2320	5.624	.1142	-.5717
62.00	.0000	.2084	-.2084	6.076	.1102	-.5339
65.00	.0000	.1764	-.1764	6.902	.1032	-.4815
70.00	1.000	.1319	.8681	8.767	.8975E-01	2.666
72.00	.0000	.1169	-.1169	9.724	.8410E-01	-.3777
74.00	.0000	.1034	-.1034	10.83	.7843E-01	-.3523
75.00	.0000	.9722E-01	-.9722E-01	11.44	.7561E-01	-.3402
76.00	.0000	.9135E-01	-.9135E-01	12.10	.7282E-01	-.3285
80.00	.0000	.7094E-01	-.7094E-01	15.24	.6206E-01	-.2855
85.00	.0000	.5136E-01	-.5136E-01	20.63	.4988E-01	-.2395
100.0	.0000	.1893E-01	-.1893E-01	54.20	.2374E-01	-.1415

```
CORRECTED SUM OF SQUARED OBSERVATIONS =  10.1818
WEIGHTED CORRECTED SUM OF SQUARED OBSERVATIONS =  123.275
SUM OF SQUARED RESIDUALS =            5.29181
SUM OF WEIGHTED SQUARED RESIDUALS =  36.3514
S =  .930327     WITH     42 DEGREES OF FREEDOM
CORRELATION (OBSERVED,PREDICTED) =   .6932

AIC criteria =        162.10220
SC  criteria =        161.88638
```

SUMMARY OF ESTIMATED SECONDARY PARAMETERS

PARAMETER	ESTIMATE	STANDARD ERROR	CV%
LOGODDS	-.509021	.524165	102.98
RRA50	.601084	.314599	52.34

PD21 - Logistic regression II – Multiple stimulii

Objectives
◆ To build a dichotomous response model for multiple stimuli
◆ To include continuous and dichotomous variables

Problem specification

A logistic regression model was used to analyze the impact of four different stimuli (tetanus, trapezius squeeze, laryngoscopy and intubation) on three different dichotomous responses (movement, heart rate, and blood pressure). Data are taken from Hung _et al_ [1990] and were then adapted.

C is the plasma concentration of the compound Sedate®. The logit L is linear in its parameters and ranges from $-\infty$ to $+\infty$ depending on θ_5 and C. The parameters (θ_i, i = 1,4 corresponding to the four different stimuli: tetanus, trapezius squeeze, laryngoscopy and intubation) expressing the logit of the probability that a certain stimulus (S_i, i = 1, 4 which is either 1 or 0) yields a specific response such as movement, or change in blood pressure or heart rate). S_1 corresponds to tetanus, S_2 to trapezius squeeze, S_3 to laryngoscopy and S_4 to intubation. The model included five parameters, one for each stimulus corresponding to a specific response (movement) plus an additional parameter relating C to the probability of movement. A logistic regression model is of the form

$$\Pi(C) = \frac{e^L}{1+e^L} \tag{21:1}$$

where

$$L = \theta_1 \cdot S_1 + \theta_2 \cdot S_2 + \theta_3 \cdot S_3 + \theta_4 \cdot S_4 - \theta_5 \cdot C \tag{21:2}$$

We will estimate the parameters θ_i that relate a certain stimulus to the logit.

Initial parameter estimates

_P1_	= 1.8
_P2_	= 2.5
_P3_	= 3.0
_P4_	= 5.0
_P5_	= 0.05

_P1_ to _P5_ in the program code correspond to θ_1 to θ_5 in Equation 21:2. This type of logistic model is stochastic. We need maximum likelihood estimates (_MLE_) which can be obtained by iteratively reweighted least squares _IRLS_. Weight is equal to one divided by the product of the probability of a response $\Pi(C)$, and the probability of no response

(1-Π(C) or *Q(C)).* Convergence is turned off and the maximum number of iterations is set to 10. In addition to this, Method 3 (Gauss-Newton) is required to obtain *MLEs* of the parameters. We maximize the likelihood function in this example (*MEAN 1, METHOD 3*), which is not necessarily the same as minimizing the weighted residual sum of squares, WRSS.

MEAN 1 allows a specified value (argument) other than the residual mean square to be used in the estimates of variances and standard deviations. The argument *1* replaces *WRSS* in the estimates of variances and standard deviations. A value of 1 is used in maximum likelihood estimation. Remember that *WRSS* in least squares regression is used for the prediction of the mean variance of the predicted function, $S^2 = WRSS/df$, where *df* stands for the degrees of freedom, which is the number of observations minus the number of parameters.

Interpretation of results and conclusions

This exercise shows how to fit a logistic model to dichotomous data. Observed and predicted data are shown in Figure 21.1. The model is an expansion of the logistic model in problem PD20. The parameter precision was good and the parameters were poorly correlated. The condition number was also low. The $\chi^2_{LOF} = 100.41$ with 120 degrees of freedom, which is not statistically significant. Thus, there is insufficient evidence to reject the model.

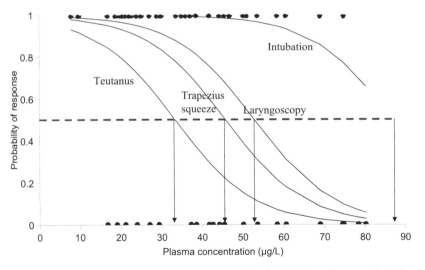

Figure 21.1 *Observed (symbols) response and predicted probability of response (solid lines) to each stimulus, versus plasma concentration.*

This exercise has shown how to simultaneously fit a logistic model to multiple dichotomous response data.

Solution

```
TITLE 1
Logistic regression of thiopental data
MODEL
 COMM
NPARM 5
PNAMES 'p1','p2','p3','p4','pcp'
NVAR 6
END
    1: TEMP
    2: P1=P(1)
    3: P2=P(2)
    4: P3=P(3)
    5: P4=P(4)
    6: PCP=P(5)
    7: CP=DTA(1)
    8: S1=DTA(2)
    9: S2=DTA(3)
   10: S3=DTA(4)
   11: S4=DTA(5)
   12: END
   13: FUNC1
   14: AL = P1*S1 +P2*S2 + P3*S3 +P4*S4 - PCP*CP
   15: EL = DEXP(AL)
   16: PHAT = EL/(1 + EL)
   17: F = PHAT
   18: WT = 1/(F*(1. - PHAT))
   19: END
   20: EOM
NVARIABLES 6
NPOINTS 100
XNUMBER 1
YNUMBER 6
METHOD 3   'Gauss-Newton
CONVERGENCE 0
ITERATIONS 10
MEANSQUARE 1
INITIAL 1.8,2.5,3,5,.05
MISSING 'Missing'
NOBSERVATIONS 125
DATA 'WINNLIN.DAT'
BEGIN
```

The following default parameter boundaries were generated.

Parameter	Lower Bound	Upper Bound
P1	.0000	18.00
P2	.0000	25.00
P3	.0000	30.00
P4	.0000	50.00
PCP	.0000	.5000

ITERATION	WEIGHTED_SS	P1	P2	P3	P4	PCP
0	82.1793	1.800	2.500	3.000	5.000	.5000E-01
		RANK = 5 CONDITION NO. = 5.964				
1	92.0244	3.125	4.261	4.882	7.914	.9341E-01
		RANK = 5 CONDITION NO. = 7.434				
2	99.7686	3.402	4.666	5.387	8.793	.1018
		RANK = 5 CONDITION NO. = 7.756				
3	100.411	3.426	4.698	5.424	8.876	.1026

```
                    RANK =   5 CONDITION NO. = 7.784
4      100.414  3.426    4.698       5.424        8.876        .1026
                    RANK =   5 CONDITION NO. = 7.784
5      100.414  3.426    4.698       5.424        8.876        .1026
                    RANK =   5 CONDITION NO. = 7.784
6      100.414  3.426    4.698       5.424        8.876        .1026
                    RANK =   5 CONDITION NO. = 7.784
7      100.414  3.426    4.698       5.424        8.876        .1026
                    RANK =   5 CONDITION NO. = 7.784
8      100.414  3.426    4.698       5.424        8.876        .1026
                    RANK =   5 CONDITION NO. = 7.784
9      100.414  3.426    4.698       5.424        8.876        .1026
                    RANK =   5 CONDITION NO. = 7.784
10      100.414  3.426   4.698       5.424        8.876        .1026
```

THE SPECIFIED VALUE OF THE MEANSQUARE ERROR 1.00000
WAS USED IN COMPUTING ALL OF THE FOLLOWING VARIANCES.

PARAMETER	ESTIMATE	STANDARD ERROR	CV%	UNIVARIATE C.I.	
P1	3.425800	.775058	22.62	1.891227	4.960374
P2	4.697747	.922294	19.63	2.871655	6.523839
P3	5.424346	1.021677	18.84	3.401482	7.447210
P4	8.876236	1.750594	19.72	5.410154	12.342318
PCP	.102579	.019942	19.44	.063094	.142064

Condition_number=   356.3

X	OBSERVED Y	PREDICTED Y	RESIDUAL	WEIGHT	SE-PRED	STANDARDI RESIDUAL
7.950	1.000	.9901	.9865E-02	102.3	.8673E-02	.1002
7.950	1.000	.9315	.6848E-01	15.67	.4125E-01	.2747
7.950	1.000	.9798	.2019E-01	50.51	.1558E-01	.1444
9.820	1.000	.9756	.2435E-01	42.05	.1800E-01	.1590
9.820	1.000	.9182	.8177E-01	13.31	.4645E-01	.3027
9.820	1.000	.9881	.1193E-01	84.78	.1011E-01	.1103
16.70	1.000	.9761	.2386E-01	42.86	.1751E-01	.1573
16.70	.0000	.8472	-.8472	7.715	.6792E-01	-2.396
16.70	1.000	.9519	.4812E-01	21.80	.2987E-01	.2269
19.00	.0000	.8141	-.8141	6.599	.7533E-01	-2.132
19.00	1.000	.9700	.3002E-01	34.28	.2092E-01	.1771
19.00	1.000	.9398	.6016E-01	17.66	.3503E-01	.2556
19.70	1.000	.9678	.3218E-01	32.04	.2208E-01	.1836
19.70	1.000	.8030	.1970	6.313	.7754E-01	.5047
19.70	1.000	.9357	.6435E-01	16.58	.3672E-01	.2650
20.00	1.000	.9668	.3316E-01	31.13	.2259E-01	.1865
20.00	1.000	.7981	.2019	6.197	.7847E-01	.5126
20.00	1.000	.9338	.6623E-01	16.14	.3747E-01	.2691
20.40	1.000	.7914	.2086	6.049	.7971E-01	.5233
20.40	1.000	.9312	.6881E-01	15.58	.3848E-01	.2748
21.10	.0000	.7793	-.7793	5.807	.8183E-01	-1.915
21.10	1.000	.9630	.3697E-01	28.03	.2455E-01	.1974
21.10	1.000	.9264	.7355E-01	14.65	.4030E-01	.2849
22.00	1.000	.9199	.8010E-01	13.55	.4272E-01	.2985
22.00	1.000	.7630	.2370	5.523	.8447E-01	.5683
22.00	1.000	.9596	.4040E-01	25.74	.2627E-01	.2068
22.70	1.000	.7497	.2503	5.324	.8645E-01	.5892

22.70	1.000	.9144	.8555E-01	12.76	.4468E-01	.3095
22.70	1.000	.9567	.4328E-01	24.10	.2768E-01	.2145
23.90	.0000	.9043	-.9043	11.53	.4816E-01	-3.113
23.90	.0000	.7259	-.7259	5.021	.8967E-01	-1.661
23.90	1.000	.9513	.4867E-01	21.55	.3024E-01	.2282
25.00	1.000	.8941	.1059	10.54	.5150E-01	.3487
25.00	1.000	.7029	.2971	4.784	.9239E-01	.6634
25.00	1.000	.9458	.5417E-01	19.47	.3277E-01	.2416
25.10	1.000	.7008	.2992	4.764	.9262E-01	.6669
25.10	1.000	.8931	.1069	10.45	.5181E-01	.3505
25.10	1.000	.9453	.5470E-01	19.29	.3301E-01	.2428
27.00	.0000	.6584	-.6584	4.442	.9667E-01	-1.417
27.00	.0000	.8730	-.8730	9.003	.5789E-01	-2.660
27.00	1.000	.9343	.6570E-01	16.25	.3782E-01	.2680
29.30	.0000	.6035	-.6035	4.177	.1004	-1.260
29.30	1.000	.6035	.3965	4.177	.1004	.8278
29.30	1.000	.8445	.1555	7.600	.6569E-01	.4358
29.30	1.000	.8445	.1555	7.600	.6569E-01	.4358
29.30	.0000	.9182	-.9182	13.29	.4440E-01	-3.392
29.30	1.000	.9182	.8175E-01	13.29	.4440E-01	.3020
30.00	1.000	.5862	.4138	4.120	.1013	.8582
30.00	1.000	.8348	.1652	7.238	.6814E-01	.4520
30.00	1.000	.9127	.8731E-01	12.52	.4657E-01	.3132
33.60	1.000	.7775	.2225	5.769	.8104E-01	.5449
33.60	1.000	.4948	.5052	4.001	.1034	1.033
33.60	1.000	.8784	.1216	9.340	.5899E-01	.3777
35.00	1.000	.4590	.5410	4.028	.1032	1.110
35.00	1.000	.8622	.1378	8.397	.6438E-01	.4063
35.00	1.000	.7517	.2483	5.347	.8607E-01	.5860
35.00	1.000	.9950	.5037E-02	198.8	.6489E-02	.0713
35.90	1.000	.8509	.1491	7.862	.6800E-01	.4259
35.90	1.000	.7340	.2660	5.113	.8925E-01	.6141
35.90	1.000	.4362	.5638	4.068	.1027	1.162
37.20	1.000	.9937	.6304E-02	159.0	.7978E-02	.0791
37.20	.0000	.4037	-.4037	4.157	.1016	-.8413
37.20	1.000	.7072	.2928	4.822	.9374E-01	.6570
37.20	1.000	.8332	.1668	7.177	.7343E-01	.4558
38.50	1.000	.9928	.7196E-02	139.4	.9013E-02	.0854
38.50	.0000	.3720	-.3720	4.285	.1000	-.7872
38.50	1.000	.6788	.3212	4.580	.9804E-01	.7030
38.50	1.000	.8138	.1862	6.584	.7905E-01	.4878
41.70	1.000	.7589	.2411	5.454	.9347E-01	.5769
41.70	.0000	.6035	-.6035	4.176	.1074	-1.264
41.70	.0000	.2991	-.2991	4.779	.9410E-01	-.6681
41.70	1.000	.9900	.9965E-02	100.9	.1217E-01	.1009
42.60	.0000	.2801	-.2801	4.969	.9198E-01	-.6379
42.60	.0000	.7416	-.7416	5.208	.9758E-01	-1.736
42.60	.0000	.5812	-.5812	4.106	.1095	-1.208
44.20	1.000	.9872	.1284E-01	78.55	.1538E-01	.1149
44.20	.0000	.2482	-.2482	5.371	.8779E-01	-.5875
44.20	1.000	.7090	.2910	4.837	.1048	.6578
44.20	1.000	.5409	.4591	4.025	.1128	.9457
45.40	.0000	.5102	-.5102	4.001	.1146	-1.048
45.40	1.000	.9855	.1450E-01	69.68	.1721E-01	.1223
45.40	.0000	.2260	-.2260	5.732	.8436E-01	-.5524
45.40	.0000	.6829	-.6829	4.610	.1100	-1.509
46.70	1.000	.9835	.1653E-01	61.22	.1944E-01	.1309
46.70	1.000	.2035	.7965	6.187	.8042E-01	2.022
46.70	1.000	.4768	.5232	4.010	.1159	1.077
46.70	1.000	.6534	.3466	4.409	.1154	.7502
46.80	1.000	.2018	.7982	6.225	.8011E-01	2.032

46.80	1.000	.4743	.5257	4.012	.1160	1.083
46.80	1.000	.6511	.3489	4.395	.1158	.7541
46.80	1.000	.9833	.1670E-01	60.62	.1962E-01	.1316
50.10	.0000	.5708	-.5708	4.079	.1273	-1.193
50.10	.0000	.1527	-.1527	7.756	.6942E-01	-.4335
50.10	.0000	.3914	-.3914	4.203	.1161	-.8261
50.10	1.000	.9767	.2327E-01	43.78	.2668E-01	.1564
50.80	1.000	.3744	.6256	4.275	.1154	1.332
50.80	1.000	.9750	.2496E-01	40.89	.2846E-01	.1623
50.80	1.000	.1437	.8563	8.159	.6710E-01	2.492
50.80	1.000	.5531	.4469	4.043	.1293	.9306
53.60	.0000	.1118	-.1118	10.11	.5792E-01	-.3617
53.60	.0000	.3099	-.3099	4.686	.1109	-.6911
53.60	1.000	.9670	.3299E-01	31.19	.3684E-01	.1883
53.60	1.000	.4815	.5185	4.006	.1346	1.078
58.40	.0000	.7144E-01	-.7144E-01	15.15	.4339E-01	-.2821
58.40	.0000	.2154	-.2154	5.938	.9674E-01	-.5400
58.40	1.000	.9471	.5287E-01	19.86	.5687E-01	.2436
60.60	.0000	.5784E-01	-.5784E-01	18.45	.3752E-01	-.2517
60.60	1.000	.9346	.6538E-01	16.28	.6901E-01	.2746
60.60	.0000	.3118	-.3118	4.672	.1287	-.7015
60.60	.0000	.1797	-.1797	6.812	.8853E-01	-.4820
68.90	.0000	.8550E-01	-.8550E-01	12.87	.5636E-01	-.3131
68.90	.0000	.1620	-.1620	7.401	.9527E-01	-.4563
68.90	.0000	.2553E-01	-.2553E-01	40.46	.2055E-01	-.1638
68.90	1.000	.8592	.1408	8.224	.1356	.4383
74.40	.0000	.1469E-01	-.1469E-01	69.62	.1334E-01	-.1233
74.70	1.000	.7709	.2291	5.639	.1994	.6176
74.70	.0000	.9637E-01	-.9637E-01	11.56	.6840E-01	-.3368
74.70	.0000	.4904E-01	-.4904E-01	21.59	.3797E-01	-.2315
78.30	.0000	.3442E-01	-.3442E-01	30.32	.2907E-01	-.1920
78.30	.0000	.9892E-02	-.9892E-02	102.9	.9714E-02	-.1008
78.30	.0000	.6865E-01	-.6865E-01	15.75	.5379E-01	-.2789
80.10	.0000	.6592	-.6592	4.440	.2577	-1.654
80.10	.0000	.5775E-01	-.5775E-01	18.51	.4734E-01	-.2538
80.10	.0000	.2878E-01	-.2878E-01	36.05	.2531E-01	-.1748
80.10	.0000	.8238E-02	-.8238E-02	123.4	.8369E-02	-.0919

```
CORRECTED SUM OF SQUARED OBSERVATIONS =  28.2080
WEIGHTED CORRECTED SUM OF SQUARED OBSERVATIONS =  517.735
SUM OF SQUARED RESIDUALS =        15.9633
SUM OF WEIGHTED SQUARED RESIDUALS =  100.356
S =  .914496    WITH   120 DEGREES OF FREEDOM
CORRELATION (OBSERVED,PREDICTED) =  .6590

AIC criteria =       586.09082
SC  criteria =       588.16160
```

PD22 - Dose-response-time analysis I – Surrogate marker

Objectives

◆ **To model dose-response-time data**

◆ **To estimate I_{max}, ID_{50}, K_a and K of drug**

◆ **To predict the dose-response relationship**

Problem specification

The analysis presented here exemplifies the use of dose-response-time data when there are no supportive drug concentrations. This technique is suitable for both directly and indirectly acting drugs. The example also illustrates the feasibility of using only effect-time data, e.g., in an early preclinical setting, when there is no available bioanalytical method and/or when there are no systemic exposure data.

The miotic response in the cat eye after application of latanoprost is assumed to mirror an interaction with latanoprost receptors in the smooth muscle of the iris. Thus, in a screening program of latanoprost for the treatment of glaucoma, the miotic response in the cat eye was one of the models used to evaluate potential drug candidates.

Domestic cats were specially trained to receive eye drops and to allow measurement of the pupillary response. Six animals were used in each of three dose groups. The horizontal diameter of the pupil was measured. The precision of the measurements of the pupil diameter was 1 mm ($\approx 10\%$). Figure 22.1 shows the observed effect-time data. Data are also shown in the program output.

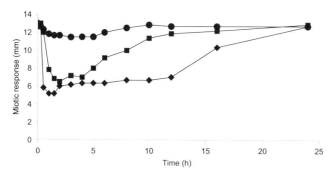

Figure 22.1 Mean observed effect versus time for latanoprost at the 0.1 (circles), 1.0 (squares) and 10 (diamonds) µg dose level.

The model is schematically illustrated in Figure 22.2.

Biophase kinetics

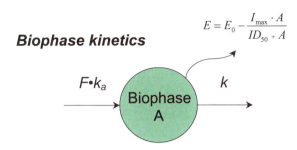

$$E = E_0 - \frac{I_{max} \cdot A}{ID_{50} + A}$$

Figure 22.2 *Schematic illustration of the biophase kinetics of the response model. The model is based on a first-order input-output process, and the biophase amount is directly linked to the response function.*

The kinetics of latanoprost in the biophase was assumed to be described by a first-order input-output model including a lag-time

$$A_{ev} = \frac{K_a FD_{ev}}{K_a - K} \cdot \left[e^{-K \cdot (t - t_{lag})} - e^{-K_a \cdot (t - t_{lag})} \right] \tag{22:1}$$

D_{ev}, K_a, K, and t_{lag} are the actual extravascular dose applied on the cornea, the first order input rate constant, the first-order elimination rate constant (assuming $K_a \gg K$), and the lag-time during the input of drug into the effect-compartment, respectively. We also assume that the biophase availability and volume of the biophase compartment are set equal to unity. The biophase function (Equation 22:1) is then directly *driving* the response

$$E = E_0 - \frac{I_{max} \cdot A^n}{ID_{50}^n + A^n} \tag{22:2}$$

including the baseline value of the contralateral control eye E_0, the maximum effect I_{max}, the concentration (dose) at half-maximal effect ID_{50}, and the sigmoidicity factor n. K_a, K, t_{lag}, I_{max}, ID_{50} and n were then estimated by simultaneously fitting Equations 22:1 and 22:2 to the mean values of the observed effect-time data obtained from each dose.

Initial parameter estimates

K_a	= 0.05 (min^{-1})	
K	= 0.01 (min^{-1})	K_e in model code
t_{lag}	= 10.0 (min)	
ID_{50}	= 1 (µg)	
I_{max}	= 10 (mm)	
n	= 1.0	

Interpretation of results and conclusions

The data treatment presented here exemplifies the use of a surrogate marker (response) when there are no drug concentration data for kinetic/dynamic analyses. This technique is suitable for indirectly acting drugs, for drugs with multi-compartment characteristics, or when absorption is not instantaneous. In this exercise, our aim is to demonstrate the feasibility of using dose-response-time data on its own (e.g., in an early preclinical setting when there is no bioanalytical method available and/or when there is no systemic exposure).

We assumed a first-order input-output kinetic model that drives the dynamics. The kinetic and dynamic models were then fit simultaneously to the data in order to estimate K_a, K, t_{lag}, I_{max}, ID_{50} and n (Table 22.1).

Table 22.1 Final parameter estimates (mean ± CV%)

	$t_{1/2Ka}$ (min)	$t_{1/2}$ (min)	t_{lag} (min)	I_{max} (mm)	ID_{50} (µg)	n
Mean ± CV%	35 ± 30	140 ± 7	26 ± 5	7.2 ± 2	0.17 ± 10	1.3 ± 7

When the kinetic/dynamic model was fit to the pooled mean data from each dose-effect level, the fit was generally good (Figure 22.3). There were some runs in the residuals, particularly for the lowest dose. This may be because of the averaged categorical data. As the resolution of the method was 1 mm, 12.8, 12.6, and 12.5 mm became 13 mm in the mean curve, 12.4, 12.0 and 11.6 became 12 mm, and so on.

The maximum concentration (amount) within the biophase compartment C_{max} and the corresponding time t_{max} is approximately $0.6 \cdot Dose_i$ and 120 min, respectively. The estimated half-lives of the absorption and elimination phases are 35 and 140 min.

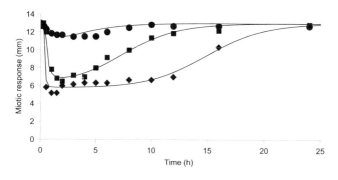

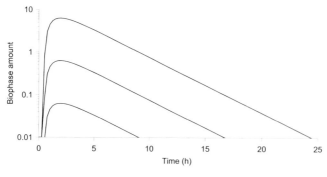

Figure 22.3 The upper plot shows observed (symbols) and predicted miotic effect (mm) versus time following three different doses of latanoprost (0.1 µg ●, 1.0 µg ■, 10 µg ◆). The observed effects are mean values with a coefficient of variation of approximately 30 % for each observation. The lower plot shows the kinetics of latanoprost in the biophase compartment (0.1 µg lower curve, 1.0 µg middle curve and 10 µg upper curve).

The predicted dose-response curve shown in Figure 22.4 was generated by means of Equation 22:3

$$E = 13 - \frac{7.2 \cdot A^{1.3}}{0.17^{1.3} + A^{1.3}}$$ (22:3)

using the final parameter estimates from the model fitting.

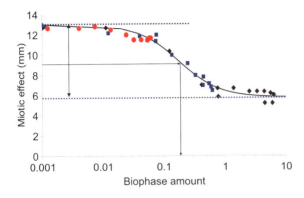

Figure 22.4 *Dose-response plot including all observed values (0.1 μg ●, 1.0 μg ■ and 10.0 μg ♦), superimposed on the corresponding predicted relationship (solid line). I_{max} is indicated by the double arrow to the left and ID_{50} by the single arrow to the right.*

For this particular example, the fit of the simple biophase response model (Equations 22:1 and 22:2) was superior to the indirect response (turnover) model (see exercise PD5) with respect to parameter precision and goodness-of-fit. Equations 22:1 and 22:2 combined captured the early effect-time points, whereas the indirect response model did not.

This analysis has shown that a well-designed experiment that includes only response-time data lends itself to estimation of the underlying kinetic processes without any additional concentration-time measurements. It also shows that absorption does not have to be instantaneous to enable estimation of biophase kinetics and dynamics. The results from such an exercise could be used for more refined recommendations regarding dose, dosing interval, and concentration-response sampling times in future (pre)clinical studies.

Solution

```
TITLE 1
Effect-time data following three extra-vascular doses
MODEL
 COMM
NFUN 3
NPARM 6
NCON 4
REMA  ************************************************************
REMA  Model assumes same EC50 and Emax for all doses and a
REMA  first order input rate model is assumed for the kinetics.
REMA  ************************************************************
REMA
PNAMES 'Ka', 'Ke', 'EC50', 'Emax', 'Tlag', 'N'
END
     1: TEMP
     2: DOSE1=CON(1)
     3: DOSE2=CON(2)
     4: DOSE3=CON(3)
```

```
 5: E0=CON(4)
 6: KA=P(1)
 7: KE=P(2)
 8: EC50=P(3)
 9: EMAX=P(4)
10: TLAG=P(5)
11: N =P(6)
12: NICK=N
13: T=X
14: TL = T - TLAG
15: END
16: FUNC1
17: A = DOSE1*KA/(KA - KE)
18: IF T LE TLAG THEN
19: C = 0.0
20: ELSE
21: C = A*DEXP(-KE*TL) - A*DEXP(-KA*TL)
22: ENDIF
23: F= E0 - (EMAX*(C**NICK))/((EC50**NICK)+C**NICK)
24: END
25: FUNC2
26: A = DOSE2*KA/(KA - KE)
27: IF T LE TLAG THEN
28: C = 0.0
29: ELSE
30: C = A*DEXP(-KE*TL) - A*DEXP(-KA*TL)
31: ENDIF
32: F= E0 - (EMAX*(C**NICK))/((EC50**NICK)+C**NICK)
33: END
34: FUNC3
35: A = DOSE3*KA/(KA - KE)
36: IF T LE TLAG THEN
37: C = 0.0
38: ELSE
39: C = A*DEXP(-KE*TL) - A*DEXP(-KA*TL)
40: ENDIF
41: F= E0 - (EMAX*(C**NICK))/((EC50**NICK)+C**NICK)
42: END
43: EOM
NOPAGE BREAKS
NVARIABLES 4
FNUMBER 3
NPOINTS 100
XNUMBER 1
YNUMBER 2
CONSTANTS .1,1,10,13
METHOD 2  'Gauss-Newton (Levenberg and Hartley)
REWEIGHT -1
ITERATIONS 50
INITIAL .05,.01,1,10,10,1
LOWER BOUNDS 0,0,0,0,0,0
UPPER BOUNDS 1,.1,10,20,60,2
MISSING 'Missing'
DATA 'WINNLIN.DAT'
BEGIN
```

PARAMETER	ESTIMATE	STANDARD ERROR	CV%	UNIVARIATE C.I.	
KA	.019847	.005304	26.73	.009143	.030551
KE	.004996	.000336	6.73	.004318	.005674

```
EC50        .174858      .020944     11.98        .132592          .217124

EMAX       7.244990      .149704      2.07       6.942876         7.547103

TLAG      25.951002     1.359098      5.24      23.208242        28.693761

N          1.296679      .096788      7.46       1.101353         1.492004

*** CORRELATION MATRIX OF THE ESTIMATES ***
PARAMETER  KA           KE          EC50        EMAX       TLAG        N
KA       1.00000
KE       -.700087     1.00000
EC50      .422936     -.751940     1.00000
EMAX     -.035070     -.002272      .403359    1.00000
TLAG      .687343     -.294392      .051123     .002710   1.00000
N         .055817     -.121767     -.054619    -.581318   -.03760    1.00000

  Condition_number=       8996.

   FUNCTION    1

   X            OBSERVED     PREDICTED    RESIDUAL      WEIGHT    SE-PRED   STANDARDI
                Y            Y                                              RESIDUAL
   .1000E-02   13.00         13.00        .0000        .7692E-01
   5.000       12.67         13.00       -.3300        .7692E-01
   15.00       12.67         13.00       -.3300        .7692E-01
   30.00       12.33         12.88       -.5470        .7765E-01  .4678E-01  -1.081
   60.00       11.83         11.94       -.1132        .8372E-01  .2094      -.2560
   90.00       11.67         11.56        .1071        .8647E-01  .2135       .2482
   120.0       11.67         11.48        .1897        .8709E-01  .1920       .4314
   180.0       11.47         11.66       -.1924        .8573E-01  .1738      -.4264
   240.0       11.50         11.96       -.4638        .8357E-01  .1609      -1.003
   300.0       11.50         12.24       -.7416        .8168E-01  .1370      -1.558
   360.0       12.00         12.46       -.4613        .8024E-01  .1096      -.9461
   480.0       12.50         12.74       -.2402        .7849E-01  .6474E-01  -.4792
   600.0       12.84         12.88       -.3807E-01    .7765E-01  .3670E-01  -.7512E-01
   720.0       12.67         12.94       -.2734        .7726E-01  .2031E-01  -.5372
   960.0       12.67         12.99       -.3180        .7699E-01  .5843E-02  -.6232
   1440.       12.67         13.00       -.3295        .7693E-01  .4076E-03  -.6454

CORRECTED SUM OF SQUARED OBSERVATIONS =   4.49938
WEIGHTED CORRECTED SUM OF SQUARED OBSERVATIONS =   .364494
SUM OF SQUARED RESIDUALS =            1.93571
SUM OF WEIGHTED SQUARED RESIDUALS =   .154884
S =   .124452     WITH    10 DEGREES OF FREEDOM
CORRELATION (OBSERVED,PREDICTED) =   .9080

   FUNCTION    2

   X            OBSERVED     PREDICTED    RESIDUAL      WEIGHT    SE-PRED   STANDARDI
                Y            Y                                              RESIDUAL
   .1000E-02   13.00         13.00        .0000        .7692E-01
   5.000       13.00         13.00        .0000        .7692E-01
   15.00       13.00         13.00        .0000        .7692E-01
   30.00       12.00         11.15        .8461        .8964E-01  .3852       3.085
   60.00       7.830         7.409        .4214        .1350      .2560       1.463
   90.00       6.840         6.983       -.1431        .1432      .1690      -.4287
   120.0       6.500         6.913       -.4132        .1447      .1357      -1.192
   180.0       7.170         7.076        .9376E-01    .1414      .1325       .2660
   240.0       7.000         7.438       -.4380        .1345      .1608      -1.248
```

300.0	8.000	7.940	.5955E-01	.1260	.1872	.1690
360.0	9.170	8.552	.6181	.1169	.1985	1.701
480.0	10.00	9.927	.7287E-01	.1007	.1861	.1797
600.0	11.37	11.17	.2040	.8955E-01	.1724	.4630
720.0	11.87	12.02	-.1533	.8316E-01	.1472	-.3273
960.0	12.17	12.77	-.5991	.7831E-01	.6736E-01	-1.195
1440.	12.87	12.99	-.1194	.7699E-01	.6176E-02	-.2340

```
CORRECTED SUM OF SQUARED OBSERVATIONS =  97.7616
WEIGHTED CORRECTED SUM OF SQUARED OBSERVATIONS =   10.2211
SUM OF SQUARED RESIDUALS =               2.11454
SUM OF WEIGHTED SQUARED RESIDUALS =  .223374
S =  .149457      WITH      10 DEGREES OF FREEDOM
CORRELATION (OBSERVED,PREDICTED) =   .9892
```

FUNCTION   3

X	OBSERVED Y	PREDICTED Y	RESIDUAL	WEIGHT	SE-PRED	STANDARDI RESIDUAL
.1000E-02	13.00	13.00	.0000	.7692E-01		
5.000	13.00	13.00	.0000	.7692E-01		
15.00	12.67	13.00	-.3300	.7692E-01		
30.00	5.830	6.687	-.8574	.1496	.2409	-3.110
60.00	5.170	5.862	-.6916	.1706	.1251	-2.167
90.00	5.170	5.829	-.6589	.1716	.1306	-2.086
120.0	6.000	5.824	.1760	.1717	.1315	.5582
180.0	6.170	5.836	.3343	.1714	.1294	1.056
240.0	6.330	5.864	.4659	.1705	.1246	1.459
300.0	6.340	5.910	.4303	.1692	.1181	1.331
360.0	6.340	5.978	.3620	.1673	.1108	1.104
480.0	6.670	6.220	.4501	.1608	.1009	1.330
600.0	6.670	6.695	-.2454E-01	.1494	.1232	-.7115E-01
720.0	7.000	7.528	-.5285	.1329	.1953	-1.574
960.0	10.33	10.14	.1887	.9860E-01	.3575	.6866
1440.	12.67	12.80	-.1253	.7815E-01	.8897E-01	-.2512

```
CORRECTED SUM OF SQUARED OBSERVATIONS =  139.845
WEIGHTED CORRECTED SUM OF SQUARED OBSERVATIONS =   13.3629
SUM OF SQUARED RESIDUALS =               2.96637
SUM OF WEIGHTED SQUARED RESIDUALS =  .463698
S =  .215336      WITH      10 DEGREES OF FREEDOM
CORRELATION (OBSERVED,PREDICTED) =   .9896


TOTALS FOR ALL CURVES COMBINED
SUM OF SQUARED RESIDUALS =               7.01662
SUM OF WEIGHTED SQUARED RESIDUALS =  .841955
S =  .141586      WITH      42 DEGREES OF FREEDOM

AIC criteria =        3.74264
SC  criteria =       14.96985
```

PD23 – Dose-response-time analysis II - Irreversible response

Objectives

◆ **To practise dose-response-time data analysis**

◆ **To characterize the turnover parameters of a bacterial growth model**

◆ **To obtain initial estimates graphically**

◆ **To implement and fit a bacterial growth and death model**

Problem specification

This problem highlights some complexities in modeling *dose-response-time data*. Specific emphasis is put on analyzing a *bacterial growth* and *death* model, obtaining *initial parameter estimates*, and implementation of *differential equations*.

A new potent antibacterial compound was being developed. To establish its potency in a resistant bacterial strain, a 10000 unit dose of *U-FU* bacteria was injected into the blood stream of four groups of Wistar rats (250 g). A dose of 1, 2, 4 or 8 μg of the antibiotic was given to each of the groups. Blood (10 μL) was drawn at selected time points for bacterial count.

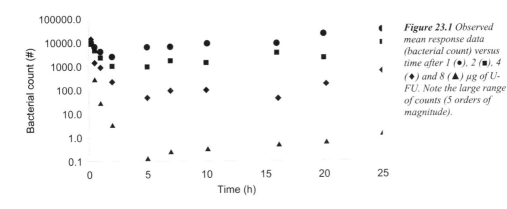

Figure 23.1 Observed mean response data (bacterial count) versus time after 1 (●), 2 (■), 4 (♦) and 8 (▲) μg of U-FU. Note the large range of counts (5 orders of magnitude).

We will fit the following differential equation to the data.

$$\frac{dN}{dt} = k_1 \cdot N - k_{out} \cdot D(t) \cdot N \qquad (23:1)$$

Where N is the number of counts, $D(t)$ is the dosing function, k_1 is the first-order growth constant, and k_{out} the second order bacterial killing. The drug function $D(t)$ is written as

$$D(t) = Dose_i \cdot e^{-K \cdot t} \qquad (23:2)$$

Initial parameter estimates

Derive initial parameter estimates of k_1, k_{out} and K. Table 23.1 contains *Dose*, mean bacterial count and time data. Note that the bacterial count is less than 1 in the 8 µg dose group for the 5 to 25 h time interval. That is due to the fact that data were averaged. Ideally, when the bacterial count is less than one all bacteria are gone, and the state variable N becomes zero (i.e., complete kill). In order to balance the data, a weight according to $1/Y^2$ is recommended.

Table 23.1 Bacterial count and function number

		Bacterial count		
Time (h)	Dose 1 µg # 1	Dose 2 µg # 2	Dose 4 µg # 3	Dose 8 µg # 4
0.2	10000	8730	13766	10273
0.5	6548	4662	1409	262
1	4131	2317	889	26
2	2517	1024	220	3.1
5	6235	940	46	0.12
7	6394	1674	90	0.23
10.1	8747	1378	95	0.3
16	8701	3515	42	0.45
20	21818	2184	171	0.57
25	32534	9120	591	1.28

Interpretation of results and conclusions

Figure 23.2 demonstrates that the observed and predicted data were consistent and the program output shows that the parameter estimates had high precision in both the two, and three parameter models.

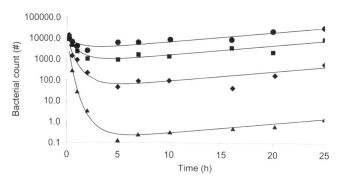

Figure 23.2 Plot of observed (symbols) and model predicted (solid lines) response data versus time for the 1, 2, 4 and 8 µg dose.

This analysis has shown that a well-designed experiment that includes only effect-time data lends itself to estimation of the underlying kinetic processes without any additional concentration-time measurements. The results from such an exercise could be used for more refined recommendations regarding dose, dosing interval, and effect-concentration sampling times in future preclinical studies. The reduced model is not discussed since the full model gave high parameter precision.

Solution I – Reduced model

```
MODEL
 COMM
  NDER   4
  NFUN   4
  NCON   5
  NPAR   2
  PNAMES 'K1', 'Ke'
END
    1: TEMP
    2: DOSE1 = CON(1)
    3: DOSE2 = CON(2)
    4: DOSE3 = CON(3)
    5: DOSE4 = CON(4)
    6: B0 = CON(5)
    7: T=X
    8: K1 = P(1)
    9: KE = P(2)
   10: END
   11: START
   12: Z(1) = B0
   13: Z(2) = B0
   14: Z(3) = B0
   15: Z(4) = B0
   16: END
   17: DIFF
   18: DRUG1 = DOSE1*DEXP(-KE*T)
   19: DRUG2 = DOSE2*DEXP(-KE*T)
   20: DRUG3 = DOSE3*DEXP(-KE*T)
   21: DRUG4 = DOSE4*DEXP(-KE*T)
   22: DZ(1) = K1*Z(1)  - DRUG1*Z(1)
   23: DZ(2) = K1*Z(2)  - DRUG2*Z(2)
   24: DZ(3) = K1*Z(3)  - DRUG3*Z(3)
   25: DZ(4) = K1*Z(4)  - DRUG4*Z(4)
   26: END
   27: FUNC 1
   28: F = Z(1)
   29: END
   30: FUNC 2
   31: F = Z(2)
   32: END
   33: FUNC 3
   34: F = Z(3)
   35: END
   36: FUNC 4
   37: F = Z(4)
   38: END
   39: EOM
NVARIABLES 3
FNUMBER 3
NPOINTS 100
XNUMBER 1
YNUMBER 2
CONSTANTS 1,2,4,8,10000
METHOD 2  'Gauss-Newton (Levenberg and Hartley)
REWEIGHT -2
ITERATIONS 50
INITIAL .2,1
LOWER BOUNDS 0,0
UPPER BOUNDS 5,4
BEGIN
```

PARAMETER	ESTIMATE	STANDARD ERROR	CV%	UNIVARIATE C.I.	
K1	.104157	.013529	12.99	.076768	.131545
KE	.697819	.022602	3.24	.652064	.743574

*** CORRELATION MATRIX OF THE ESTIMATES ***

PARAMETER	K1	KE
K1	1.00000	
KE	-.731248	1.00000

FUNCTION   1

X	OBSERVED Y	PREDICTED Y	RESIDUAL	WEIGHT	SE-PRED	STANDARDI RESIDUAL
.2000	.1000E+05	8472.	1528.	.1393E-07	20.51	.2549
.5000	6548.	6907.	-359.3	.2096E-07	36.93	-.7349E-01
.6000	.	6520.	.	.0000	-.6400+152	.
.8000	.	5888.	.	.0000	-.6400+152	.
1.000	4131.	5403.	-1272.	.3426E-07	51.95	-.3326
1.250	.	4947.	.	.0000	-.6400+152	.
1.500	.	4613.	.	.0000	-.6400+152	.
1.750	.	4368.	.	.0000	-.6400+152	.
2.000	2517.	4190.	-1673.	.5695E-07	77.45	-.5644
2.500	.	3976.	.	.0000	-.6400+152	.
3.000	.	3891.	.	.0000	-.6400+152	.
3.500	.	3891.	.	.0000	-.6400+152	.
4.000	.	3951.	.	.0000	-.6400+152	.
4.500	.	4056.	.	.0000	-.6400+152	.
5.000	6235.	4196.	2039.	.5680E-07	197.7	.6882
6.000	.	4555.	.	.0000	-.6400+152	.
7.000	6394.	5000.	1394.	.4000E-07	346.4	.3957
8.000	.	5519.	.	.0000	-.6400+152	.
9.000	.	6108.	.	.0000	-.6400+152	.
10.10	8747.	6840.	1907.	.2137E-07	736.7	.3986
11.00	.	7508.	.	.0000	-.6400+152	.
12.00	.	8329.	.	.0000	-.6400+152	.
13.00	.	9242.	.	.0000	-.6400+152	.
16.00	8701.	.1263E+05	-3929.	.6269E-08	2342.	-.4555
17.00	.	.1402E+05	.	.0000	-.6400+152	.
19.00	.	.1726E+05	.	.0000	-.6400+152	.
20.00	.2182E+05	.1916E+05	2661.	.2725E-08	4579.	.2085
21.00	.	.2126E+05	.	.0000	-.6400+152	.
22.00	.	.2359E+05	.	.0000	-.6400+152	.
23.00	.	.2618E+05	.	.0000	-.6400+152	.
24.00	.	.2906E+05	.	.0000	-.6400+152	.
25.00	.3253E+05	.3225E+05	285.7	.9616E-09	9880.	.1389E-01

CORRECTED SUM OF SQUARED OBSERVATIONS = .157077E+10
WEIGHTED CORRECTED SUM OF SQUARED OBSERVATIONS = 2.61305
SUM OF SQUARED RESIDUALS =   .392177E+08
SUM OF WEIGHTED SQUARED RESIDUALS = .757902
S = .307795   WITH   8 DEGREES OF FREEDOM
CORRELATION (OBSERVED,PREDICTED) = .6845

FUNCTION   2

X	OBSERVED Y	PREDICTED Y	RESIDUAL	WEIGHT	SE-PRED	STANDARDI RESIDUAL
.2000	8730.	7029.	1701.	.2024E-07	15.30	.3419
.5000	4662.	4529.	133.1	.4875E-07	21.00	.4152E-01

.7500	.	3363.	.	.0000	-.6400+152	.
.8500	.	3031.	.	.0000	-.6400+152	.
1.000	2317.	2630.	-313.1	.1446E-06	27.00	-.1682
1.250	.	2148.	.	.0000	-.6400+152	.
1.650	.	1673.	.	.0000	-.6400+152	.
2.000	1024.	1426.	-401.7	.4920E-06	36.73	-.3984
2.500	.	1219.	.	.0000	-.6400+152	.
3.000	.	1108.	.	.0000	-.6400+152	.
4.000	.	1029.	.	.0000	-.6400+152	.
5.000	940.0	1046.	-105.8	.9143E-06	57.95	-.1434
6.000	.	1111.	.	.0000	-.6400+152	.
6.500	.	1155.	.	.0000	-.6400+152	.
7.000	1674.	1206.	468.0	.6876E-06	81.39	.5509
8.000	.	1324.	.	.0000	-.6400+152	.
8.500	.	1390.	.	.0000	-.6400+152	.
9.000	.	1461.	.	.0000	-.6400+152	.
10.00	1378.	1617.	-239.3	.3823E-06	149.7	-.2109
11.00	.	1792.	.	.0000	-.6400+152	.
12.00	.	1988.	.	.0000	-.6400+152	.
13.00	.	2205.	.	.0000	-.6400+152	.
15.00	.	2715.	.	.0000	-.6400+152	.
16.00	3515.	3013.	501.6	.1101E-06	487.2	.2416
17.00	.	3344.	.	.0000	-.6400+152	.
18.00	.	3711.	.	.0000	-.6400+152	.
19.00	.	4119.	.	.0000	-.6400+152	.
20.00	2184.	4571.	-2387.	.4787E-07	971.8	-.7735
21.00	.	5072.	.	.0000	-.6400+152	.
22.00	.	5629.	.	.0000	-.6400+152	.
23.00	.	6247.	.	.0000	-.6400+152	.
24.00	.	6933.	.	.0000	-.6400+152	.
25.00	9120.	7694.	1426.	.1689E-07	2140.	.2848

```
CORRECTED SUM OF SQUARED OBSERVATIONS =  .171964E+09
WEIGHTED CORRECTED SUM OF SQUARED OBSERVATIONS =  3.51122
SUM OF SQUARED RESIDUALS =           .114386E+08
SUM OF WEIGHTED SQUARED RESIDUALS =  .670447
S =  .289492     WITH     8 DEGREES OF FREEDOM
CORRELATION (OBSERVED,PREDICTED) =  .6581
```

FUNCTION   3

X	OBSERVED Y	PREDICTED Y	RESIDUAL	WEIGHT	SE-PRED	STANDARDI RESIDUAL
.2000	.1377E+05	4839.	8927.	.4270E-07	9.074	2.606
.5000	1409.	1947.	-538.0	.2638E-06	11.94	-.3904
.6500	.	1323.	.	.0000	-.6400+152	.
.8500	.	840.7	.	.0000	-.6400+152	.
1.000	889.0	623.3	265.7	.2574E-05	13.11	.6025
1.300	.	375.2	.	.0000	-.6400+152	.
1.700	.	222.6	.	.0000	-.6400+152	.
2.000	220.0	165.0	54.96	.3671E-04	9.695	.4721
3.000	.	89.77	.	.0000	-.6400+152	.
3.500	.	76.80	.	.0000	-.6400+152	.
4.000	.	69.86	.	.0000	-.6400+152	.
5.000	46.00	64.97	-18.97	.2369E-03	7.811	-.4187
5.500	.	65.01	.	.0000	-.6400+152	.
6.000	.	66.04	.	.0000	-.6400+152	.
7.000	90.00	70.15	19.85	.2032E-03	8.856	.4063
8.000	.	76.17	.	.0000	-.6400+152	.
9.000	.	83.62	.	.0000	-.6400+152	.
10.00	95.00	92.31	2.694	.1174E-03	11.62	.4191E-01

```
11.00       .           102.2       .         .0000    -.6400+152      .
12.00       .           113.2       .         .0000    -.6400+152      .
13.00       .           125.6       .         .0000    -.6400+152      .
14.00       .           139.3       .         .0000    -.6400+152      .
16.10      42.00        173.3     -131.3      .3329E-04  26.18      -1.096
17.00       .           190.4       .         .0000    -.6400+152      .
18.00       .           211.2       .         .0000    -.6400+152      .
19.00       .           234.4       .         .0000    -.6400+152      .
20.20     171.0         265.6      -94.65     .1417E-04  49.71       -.5220
21.00       .           288.7       .         .0000    -.6400+152      .
22.00       .           320.4       .         .0000    -.6400+152      .
23.00       .           355.6       .         .0000    -.6400+152      .
24.00       .           394.6       .         .0000    -.6400+152      .
25.00     591.0         438.0      153.0      .5214E-05 104.7         .5246
```

CORRECTED SUM OF SQUARED OBSERVATIONS = .183353E+09
WEIGHTED CORRECTED SUM OF SQUARED OBSERVATIONS = 12.6493
SUM OF SQUARED RESIDUALS =   .801026E+08
SUM OF WEIGHTED SQUARED RESIDUALS = 4.76128
S = .771466   WITH   8 DEGREES OF FREEDOM
CORRELATION (OBSERVED,PREDICTED) = .9033

FUNCTION   4

```
X          OBSERVED    PREDICTED   RESIDUAL    WEIGHT    SE-PRED    STANDARDI
           Y           Y                                            RESIDUAL
.2000     .1027E+05    2293.       7980.      .1901E-06  5.202       4.916
.5000     262.0        359.9      -97.85      .7722E-05  4.974       -.3843
.6500       .          163.6       .          .0000     -.6400+152      .
.8500       .          64.69       .          .0000     -.6400+152      .
1.000      26.00       35.01      -9.010      .8159E-03  1.702       -.3645
1.300       .          12.30       .          .0000     -.6400+152      .
1.700       .          4.153       .          .0000     -.6400+152      .
2.000      3.100       2.212       .8883      .2044      .2937        .5777
2.500       .          1.010       .          .0000     -.6400+152      .
3.000       .          .5896       .          .0000     -.6400+152      .
3.500       .          .4096       .          .0000     -.6400+152      .
4.000       .          .3217       .          .0000     -.6400+152      .
5.000      .1200       .2508      -.1308      15.90      .6926E-01   -.8003
5.500       .          .2383       .          .0000     -.6400+152      .
6.000       .          .2334       .          .0000     -.6400+152      .
6.500       .          .2336       .          .0000     -.6400+152      .
6.800       .          .2355       .          .0000     -.6400+152      .
7.000      .2300       .2374      -.7364E-02  17.75      .6986E-01   -.4820E-01
7.500       .          .2437       .          .0000     -.6400+152      .
8.000       .          .2522       .          .0000     -.6400+152      .
9.000       .          .2739       .          .0000     -.6400+152      .
10.10      .3000       .3036      -.3616E-02  10.85      .8687E-01   -.1840E-01
11.00       .          .3319       .          .0000     -.6400+152      .
12.00       .          .3674       .          .0000     -.6400+152      .
13.00       .          .4071       .          .0000     -.6400+152      .
14.00       .          .4515       .          .0000     -.6400+152      .
15.00       .          .5009       .          .0000     -.6400+152      .
16.15      .4500       .5646      -.1146      3.137      .1469       -.3083
17.00       .          .6168       .          .0000     -.6400+152      .
17.50       .          .6498       .          .0000     -.6400+152      .
18.00       .          .6845       .          .0000     -.6400+152      .
19.00       .          .7596       .          .0000     -.6400+152      .
20.20      .5700       .8607      -.2907      1.350      .2190       -.5115
21.00       .          .9355       .          .0000     -.6400+152      .
22.00       .          1.038       .          .0000     -.6400+152      .
```

23.00	.	1.152	.	.0000	-.6400+152	.
24.00	.	1.279	.	.0000	-.6400+152	.
25.00	1.280	1.419	-.1390	.4966	.3730	-.1491

```
CORRECTED SUM OF SQUARED OBSERVATIONS =  .102665E+09
WEIGHTED CORRECTED SUM OF SQUARED OBSERVATIONS =  23.8855
SUM OF SQUARED RESIDUALS =           .636832E+08
SUM OF WEIGHTED SQUARED RESIDUALS =  12.8450
S =  1.26713      WITH       8 DEGREES OF FREEDOM
CORRELATION (OBSERVED,PREDICTED) =  .9509

TOTALS FOR ALL CURVES COMBINED
SUM OF SQUARED RESIDUALS =           .194442E+09
SUM OF WEIGHTED SQUARED RESIDUALS =  19.0347
S =  .707751      WITH      38 DEGREES OF FREEDOM

AIC criteria =        319.24998
SC  criteria =        324.59564
```

Solution II – Full model

```
MODEL
 COMM
  NDER   4
  NFUN   4
  NCON   5
  NPAR   3
  PNAMES 'K1', 'K', 'kout'
END
    1: TEMP
    2: DOSE1 = CON(1)
    3: DOSE2 = CON(2)
    4: DOSE3 = CON(3)
    5: DOSE4 = CON(4)
    6: B0 = CON(5)
    7: T=X
    8: K1 = P(1)
    9: K = P(2)
   10: KOUT= P(3)
   11: END
   12: START
   13: Z(1) = B0
   14: Z(2) = B0
   15: Z(3) = B0
   16: Z(4) = B0
   17: END
   18: DIFF
   19: DRUG1 = DOSE1*DEXP(-K*T)
   20: DRUG2 = DOSE2*DEXP(-K*T)
   21: DRUG3 = DOSE3*DEXP(-K*T)
   22: DRUG4 = DOSE4*DEXP(-K*T)
   23: DZ(1) = K1*Z(1)  - DRUG1*Z(1)*KOUT
   24: DZ(2) = K1*Z(2)  - DRUG2*Z(2)*KOUT
   25: DZ(3) = K1*Z(3)  - DRUG3*Z(3)*KOUT
   26: DZ(4) = K1*Z(4)  - DRUG4*Z(4)*KOUT
   27: END
   28: FUNC 1
   29: F = Z(1)
   30: END
   31: FUNC 2
```

```
      32: F = Z(2)
      33: END
      34: FUNC 3
      35: F = Z(3)
      36: END
      37: FUNC 4
      38: F = Z(4)
      39: END
      40: EOM
NVARIABLES 3
FNUMBER 3
NPOINTS 100
XNUMBER 1
YNUMBER 2
CONSTANTS 1,2,4,8,10000
METHOD 2   'Gauss-Newton (Levenberg and Hartley)
REWEIGHT -2
ITERATIONS 50
INITIAL .2,1,1
LOWER BOUNDS 0,0,0
UPPER BOUNDS 5,4,2
BEGIN
```

PARAMETER	ESTIMATE	STANDARD ERROR	CV%	UNIVARIATE C.I.	
K1	.106786	.012252	11.47	.081961	.131610
K	.618214	.060823	9.84	.494976	.741453
KOUT	.896651	.077101	8.60	.740431	1.052871

```
*** CORRELATION MATRIX OF THE ESTIMATES ***
PARAMETER        K1             K            KOUT
K1            1.00000
K            -.361693      1.00000
KOUT         -.152509       .955144      1.00000
```

```
   FUNCTION   1
```

X	OBSERVED Y	PREDICTED Y	RESIDUAL	WEIGHT	SE-PRED	STANDARDI RESIDUAL
.2000	.1000E+05	8630.	1370.	.1343E-07	121.6	.2534
.5000	6548.	7173.	-625.0	.1944E-07	209.5	-.1392
.6000	.	6802.	.	.0000	-.6400+152	.
.8000	.	6185.	.	.0000	-.6400+152	.
1.000	4131.	5701.	-1570.	.3077E-07	244.3	-.4405
1.250	.	5235.	.	.0000	-.6400+152	.
1.500	.	4885.	.	.0000	-.6400+152	.
1.750	.	4621.	.	.0000	-.6400+152	.
2.000	2517.	4424.	-1907.	.5110E-07	209.8	-.6899
2.500	.	4172.	.	.0000	-.6400+152	.
3.000	.	4053.	.	.0000	-.6400+152	.
3.500	.	4025.	.	.0000	-.6400+152	.
4.000	.	4062.	.	.0000	-.6400+152	.
4.500	.	4148.	.	.0000	-.6400+152	.
5.000	6235.	4272.	1963.	.5480E-07	194.7	.7354
6.000	.	4611.	.	.0000	-.6400+152	.
7.000	6394.	5047.	1347.	.3925E-07	315.4	.4280
8.000	.	5567.	.	.0000	-.6400+152	.
9.000	.	6165.	.	.0000	-.6400+152	.
10.10	8747.	6914.	1833.	.2092E-07	665.4	.4282

```
11.00          .        7602.      .       .0000    -.6400+152      .
12.00          .        8453.      .       .0000    -.6400+152      .
13.00          .        9402.      .       .0000    -.6400+152      .
16.00       8701.     .1295E+05  -4246.    .5966E-08  2152.       -.5429
17.00          .      .1441E+05    .       .0000    -.6400+152      .
19.00          .      .1783E+05    .       .0000    -.6400+152      .
20.00      .2182E+05  .1984E+05  1974.     .2539E-08  4261.        .1690
21.00          .      .2208E+05    .       .0000    -.6400+152      .
22.00          .      .2457E+05    .       .0000    -.6400+152      .
23.00          .      .2734E+05    .       .0000    -.6400+152      .
24.00          .      .3042E+05    .       .0000    -.6400+152      .
25.00      .3253E+05  .3385E+05  -1313.    .8729E-09  9331.       -.6893E-01
```

```
CORRECTED SUM OF SQUARED OBSERVATIONS = .157077E+10
WEIGHTED CORRECTED SUM OF SQUARED OBSERVATIONS = 2.40835
SUM OF SQUARED RESIDUALS =          .410396E+08
SUM OF WEIGHTED SQUARED RESIDUALS = .766005
S =  .330801     WITH     7 DEGREES OF FREEDOM
CORRELATION (OBSERVED,PREDICTED) = .6839
```

```
   FUNCTION   2
```

X	OBSERVED Y	PREDICTED Y	RESIDUAL	WEIGHT	SE-PRED	STANDARDI RESIDUAL
.2000	8730.	7290.	1440.	.1881E-07	200.7	.3154
.5000	4662.	4878.	-215.7	.4203E-07	276.9	-.7086E-01
.7500	.	3693.	.	.0000	-.6400+152	.
.8500	.	3346.	.	.0000	-.6400+152	.
1.000	2317.	2921.	-603.7	.1172E-06	240.5	-.3328
1.250	.	2398.	.	.0000	-.6400+152	.
1.650	.	1866.	.	.0000	-.6400+152	.
2.000	1024.	1581.	-556.6	.4003E-06	138.1	-.5676
2.500	.	1333.	.	.0000	-.6400+152	.
3.000	.	1193.	.	.0000	-.6400+152	.
4.000	.	1076.	.	.0000	-.6400+152	.
5.000	940.0	1070.	-129.9	.8736E-06	57.44	-.1945
6.000	.	1120.	.	.0000	-.6400+152	.
6.500	.	1160.	.	.0000	-.6400+152	.
7.000	1674.	1206.	467.6	.6871E-06	71.96	.6215
8.000	.	1319.	.	.0000	-.6400+152	.
8.500	.	1384.	.	.0000	-.6400+152	.
9.000	.	1454.	.	.0000	-.6400+152	.
10.00	1378.	1609.	-231.1	.3862E-06	131.9	-.2312
11.00	.	1785.	.	.0000	-.6400+152	.
12.00	.	1984.	.	.0000	-.6400+152	.
13.00	.	2206.	.	.0000	-.6400+152	.
15.00	.	2729.	.	.0000	-.6400+152	.
16.00	3515.	3036.	479.0	.1085E-06	435.7	.2587
17.00	.	3378.	.	.0000	-.6400+152	.
18.00	.	3758.	.	.0000	-.6400+152	.
19.00	.	4182.	.	.0000	-.6400+152	.
20.00	2184.	4653.	-2469.	.4619E-07	881.1	-.8885
21.00	.	5177.	.	.0000	-.6400+152	.
22.00	.	5761.	.	.0000	-.6400+152	.
23.00	.	6410.	.	.0000	-.6400+152	.
24.00	.	7133.	.	.0000	-.6400+152	.
25.00	9120.	7936.	1184.	.1588E-07	1972.	.2593

```
CORRECTED SUM OF SQUARED OBSERVATIONS = .171964E+09
WEIGHTED CORRECTED SUM OF SQUARED OBSERVATIONS = 3.25100
SUM OF SQUARED RESIDUALS =          .108091E+08
```

```
SUM OF WEIGHTED SQUARED RESIDUALS =  .722047
S =  .321169     WITH     7 DEGREES OF FREEDOM
CORRELATION (OBSERVED,PREDICTED) =  .6579
```

   FUNCTION   3

X	OBSERVED Y	PREDICTED Y	RESIDUAL	WEIGHT	SE-PRED	STANDARDI RESIDUAL
.2000	.1377E+05	5203.	8563.	.3694E-07	283.9	2.637
.5000	1409.	2255.	-846.5	.1966E-06	253.3	-.6088
.6500	.	1572.	.	.0000	-.6400+152	.
.8500	.	1022.	.	.0000	-.6400+152	.
1.000	889.0	766.7	122.3	.1701E-05	124.6	.2637
1.300	.	466.3	.	.0000	-.6400+152	.
1.700	.	275.5	.	.0000	-.6400+152	.
2.000	220.0	201.8	18.21	.2456E-04	34.64	.1497
3.000	.	103.2	.	.0000	-.6400+152	.
3.500	.	85.55	.	.0000	-.6400+152	.
4.000	.	75.58	.	.0000	-.6400+152	.
5.000	46.00	67.12	-21.12	.2220E-03	7.327	-.5099
5.500	.	66.00	.	.0000	-.6400+152	.
6.000	.	66.13	.	.0000	-.6400+152	.
7.000	90.00	68.91	21.09	.2106E-03	7.638	.4961
8.000	.	74.02	.	.0000	-.6400+152	.
9.000	.	80.81	.	.0000	-.6400+152	.
10.00	95.00	89.00	6.000	.1262E-03	10.23	.1095
11.00	.	98.48	.	.0000	-.6400+152	.
12.00	.	109.3	.	.0000	-.6400+152	.
13.00	.	121.4	.	.0000	-.6400+152	.
14.00	.	134.9	.	.0000	-.6400+152	.
16.10	42.00	168.7	-126.7	.3512E-04	22.83	-1.228
17.00	.	185.7	.	.0000	-.6400+152	.
18.00	.	206.6	.	.0000	-.6400+152	.
19.00	.	229.9	.	.0000	-.6400+152	.
20.20	171.0	261.4	-90.36	.1464E-04	43.37	-.5722
21.00	.	284.7	.	.0000	-.6400+152	.
22.00	.	316.7	.	.0000	-.6400+152	.
23.00	.	352.4	.	.0000	-.6400+152	.
24.00	.	392.2	.	.0000	-.6400+152	.
25.00	591.0	436.3	154.7	.5252E-05	92.34	.6010

```
CORRECTED SUM OF SQUARED OBSERVATIONS =  .183353E+09
WEIGHTED CORRECTED SUM OF SQUARED OBSERVATIONS =  10.6764
SUM OF SQUARED RESIDUALS =           .741095E+08
SUM OF WEIGHTED SQUARED RESIDUALS =  3.88981
S =  .745444     WITH     7 DEGREES OF FREEDOM
CORRELATION (OBSERVED,PREDICTED) =  .8944
```

   FUNCTION   4

X	OBSERVED Y	PREDICTED Y	RESIDUAL	WEIGHT	SE-PRED	STANDARDI RESIDUAL
.2000	.1027E+05	2650.	7623.	.1424E-06	287.9	4.663
.5000	262.0	482.3	-220.3	.4300E-05	107.8	-.7803
.6500	.	230.5	.	.0000	-.6400+152	.
.8500	.	95.45	.	.0000	-.6400+152	.
1.000	26.00	52.83	-26.83	.3583E-03	17.07	-.9459
1.300	.	18.93	.	.0000	-.6400+152	.
1.700	.	6.329	.	.0000	-.6400+152	.
2.000	3.100	3.289	-.1890	.9244E-01	1.124	-.1094
2.500	.	1.416	.	.0000	-.6400+152	.

```
3.000        .          .7738         .         .0000     -.6400+152        .
3.500        .          .5036         .         .0000     -.6400+152        .
4.000        .          .3727         .         .0000     -.6400+152        .
5.000      .1200        .2641      -.1441       14.34      .6446E-01    -.9455
5.500        .          .2421         .         .0000     -.6400+152        .
6.000        .          .2304         .         .0000     -.6400+152        .
6.500        .          .2254         .         .0000     -.6400+152        .
6.800        .          .2246         .         .0000     -.6400+152        .
7.000      .2300        .2249      .5097E-02    19.77      .5866E-01    .3978E-01
7.500        .          .2278         .         .0000     -.6400+152        .
8.000        .          .2332         .         .0000     -.6400+152        .
9.000        .          .2498         .         .0000     -.6400+152        .
10.10      .3000        .2748      .2519E-01    13.24      .7359E-01    .1618
11.00        .          .2996         .         .0000     -.6400+152        .
12.00        .          .3314         .         .0000     -.6400+152        .
13.00        .          .3676         .         .0000     -.6400+152        .
14.00        .          .4083         .         .0000     -.6400+152        .
15.00        .          .4539         .         .0000     -.6400+152        .
16.15      .4500        .5129      -.6292E-01   3.801      .1265        -.2130
17.00        .          .5615         .         .0000     -.6400+152        .
17.50        .          .5923         .         .0000     -.6400+152        .
18.00        .          .6247         .         .0000     -.6400+152        .
19.00        .          .6951         .         .0000     -.6400+152        .
20.20      .5700        .7901      -.2201       1.602      .1884        -.4808
21.00        .          .8605         .         .0000     -.6400+152        .
22.00        .          .9575         .         .0000     -.6400+152        .
23.00        .          1.065         .         .0000     -.6400+152        .
24.00        .          1.185         .         .0000     -.6400+152        .
25.00      1.280        1.319      -.3903E-01   .5748      .3191        -.5119E-01
```

```
CORRECTED SUM OF SQUARED OBSERVATIONS =  .102665E+09
WEIGHTED CORRECTED SUM OF SQUARED OBSERVATIONS =  17.5174
SUM OF SQUARED RESIDUALS =          .581639E+08
SUM OF WEIGHTED SQUARED RESIDUALS =  9.14728
S =  1.14313     WITH     7 DEGREES OF FREEDOM
CORRELATION (OBSERVED,PREDICTED) =  .9474

TOTALS FOR ALL CURVES COMBINED
SUM OF SQUARED RESIDUALS =          .184122E+09
SUM OF WEIGHTED SQUARED RESIDUALS =  14.5251
S =  .626555     WITH    37 DEGREES OF FREEDOM

AIC criteria =       292.31925
SC  criteria =       300.33773
```

PD24 - Dose-response-time analysis III – Different routes

Problem specification

In this example, we use pharmacological data obtained from different routes of administration to model and estimate pertinent screening parameters such as ED_{50}/ID_{50}, bioavailability, and the turnover characteristics of a response, such as half-life. As in PD22 and PD23, the underlying assumption is that pharmacodynamic data contain information about the kinetics of the drug in the *biophase*. In other words, apparent half-life, bioavailability and potency can be obtained simultaneously from dose-response-time data.

In the following example, a potential analgesic drug was given via the intravenous and subcutaneous routes. The time (in seconds) to respond to a noxious stimulus was determined at different times after dosing. Response-time data were obtained from each dose level after a single dose (Figure 24.1).

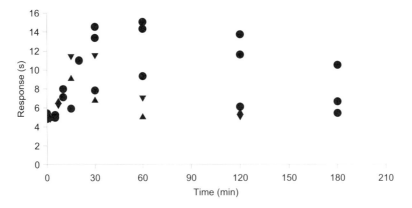

Figure 24.1 *The response(antinociception)-time courses of drug X after different doses given by the intravenous (▼ ▲) and subcutaneous (●) routes. The three subcutaneous doses at each time point are represented by solid circles.*

The underlying kinetic model is shown in Figure 24.2A. A sum of exponentials model (Equation 24:1) and a first-order input-output model (Equation 24:2) are assumed for the systemic and extravascular dosing, respectively. The kinetic model then drives the inhibition function (Equation 24:3) of the dynamic model, with drug inhibiting the loss of response.

The response (antinociception) is described by a turnover model, which is shown in Figure 24.2B. A zero-order input and first-order input-output model governs the turnover of the response.

The data and model were taken from Gabrielsson and Weiner [1994] and this analysis is also reviewed in Gabrielsson *et al* [2000].

Biophase kinetics

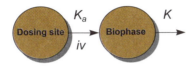

Dynamics

Figure 24.2 Schematic illustration of A. the biophase and B. the response turnover model. The drug in the biophase affects the turnover of response via inhibition of the loss of response (inhibition of the fractional turnover rate)

The amount of drug in the biophase compartment after an intravenous dose is modeled with mono-exponential decline and assuming the volume is unity (Equation 24:1)

$$A_s = \sum_{i=1}^{n} A_i e^{-a_i \cdot t} \tag{24:1}$$

The amount of drug in the biophase compartment after an extravascular dose (in this case subcutaneous) is modeled with first-order kinetics assuming the volume is unity (Equation 24:2)

$$A_{ev} = \frac{k_a F D_{ev}}{k_a - k} \left[e^{-k_a (t - t_{lag})} - e^{-k (t - t_{lag})} \right] \tag{24:2}$$

The inhibitory function driven by Equations 24:1 and 24:2 is

$$H(t) = 1 - \frac{I_{max} A}{ID_{50} + A} \tag{24:3}$$

The turnover function of the response where drug acts on loss (k_{out}) of response is given by

$$\frac{dR}{dt} = k_{in} - k_{out} \cdot R \cdot H(t) \tag{24:4}$$

The maximum drug-induced change Δ is obtained from

$$\Delta = \frac{k_{in}}{k_{out}} \cdot \frac{I_{max}}{1 - I_{max}} \tag{24:5}$$

Equation 24:5 rearranged gives the I_{max} parameter.

$$I_{max} = \frac{1}{\dfrac{k_{in}}{k_{out}} + 1} = \frac{1}{\dfrac{R_0}{\Delta} + 1} = \frac{1}{\dfrac{5}{10} + 1} = \frac{2}{3} \tag{24:6}$$

We will use I_{max} as a constant of 0.67 in this run, but one may also apply a model where it is a parameter.

Initial parameter estimates

One can either obtain the initial estimates graphically from Figure 24.1 or, as an alternative estimate bioavailability by means of the area under the response-time course following intravenous (AUC_{Eiv}) and subcutaneous (AUC_{Esc}) administration. An approximate estimate of F can be obtained by AUC_{Esc}/AUC_{Eiv} and then corrected for differences in doses, although this estimate of biophase availability will be less than 20%. The half-life of drug in the biophase after an intravenous bolus was guessed to be less than 30 min (actually 15 min). The k_{out} parameter was derived from the initial slope of the upswing of the effect-time curve after the largest dose. The k_{in} parameter was derived from k_{out} and the ratio of k_{in}/k_{out}. ID_{50} was obtained by simulating the model with a number of reasonable values of ID_{50}. These should be used as initial estimates for the regression program.

K_a	$= 0.02 \ (\text{min}^{-1})$	
$ID_{50}= 2$		ID_{50} in model code
n	$= 2$	Nick in model code
k_{in}	$= 2$	
k_{out}	$= 0.3$	
F	$= 0.6$	BIO in model code
K	$= 0.05$	

Interpretation of results and conclusions

Data were successfully fit by means of an indirect pharmacodynamic model using the hypothetical biophase function as input (Figure 24.3).

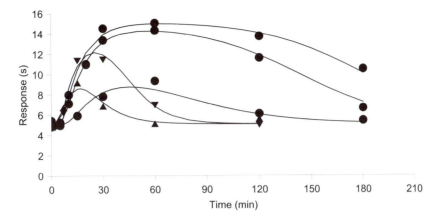

Figure 24.3 *The observed (symbols) and predicted (solid lines) response time courses after different doses given by the intravenous (▼▲) and subcutaneous (●) routes. Note the baseline value of the response at approximately 5 s. The three subcutaneous doses are represented by solid circles at each time point. An increase in response time is indicative of antinociception.*

The half-life of absorption was about 36 min and the half-life of elimination 15 min. This indicates a typical flip-flop situation (see also Figure 24.6). ID_{50} was estimated to be 1.1 µg, biophase availability 66% and the predicted baseline value was 5.1 s. The estimated half-life of effect was about 3 min. Note the difference between the model-predicted biophase availability of 66% (which is close to the systemic availability of 70% predicted from the plasma concentrations) and the AUC_E-predicted availability (18%). The poor predictability of the AUC_E approach is due to the fact that the reference dose (intravenous) produces a response that covers concentrations up to the ID_{50} value, with the effect being basically linearly proportional to the concentrations below ID_{50}. The test dose(s) (subcutaneous) reached the effect-maximum. In other words, the response behaved in a nonlinear manner. In order to use the AUE-approach, the reference dose (e.g., intravenous) and test dose must cover basically the same concentration range (see Figure 24.4).

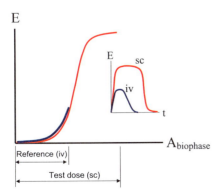

Figure 24.4 *Schematic illustration of the limitation of predicting biophase availability from AUE measurements when the test dose (iv) and reference dose (sc) cover different response (antinociception)-concentration ranges.*

The parameter precision (see program output) was good, in that no parameter had a CV larger than 17% and, on average, it was < 10%. The predicted dose-response curve is shown in Figure 24.5

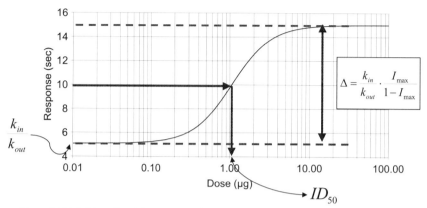

Figure 24.5 *Model-predicted dose-response (antinociception) curve, showing the baseline value (k_{in}/k_{out}), potency (ID_{50}) and delta value (Δ).*

The predicted biophase-time courses and the hysteresis plots of sequential response *versus* biophase amount are shown in Figure 24.6. Note the flip-flop situation in Figure 24.6 (left), where the decline of response is more rapid after intravenous than subcutaneous dosing.

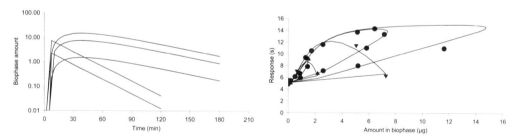

Figure 24.6 *The predicted biophase amount-time courses (left) and response-biophase amount plotted in time sequence (right) after different doses given by the intravenous (▼) and subcutaneous (●) routes.*

Solution

```
MODEL
 COMM
   nder    5
   NFUN    5
   NPARM   7
   NCON    8
   REMA
   REMA     * * * * * * * * * * * * * * * * * * * * * * * * * * * * * * * * * * * * * * * * * * * * * * * * * * * *
   REMA     Model assumes same ID50 and Imax for all doses and a
```

```
   REMA    first order input rate model is assumed for the kinetics.
   REMA    ************************************************************
   REMA
   PNAMES  'Ka', 'ID50', 'Nick', 'Kin', 'kout', 'Bio', 'K'
END
    1: TEMP
    2: DOSE1=CON(1)
    3: DOSE2=CON(2)
    4: DOSE3=CON(3)
    5: DOSE4=CON(4)
    6: DOSE5=CON(5)
    7: E0=CON(6)
    8: IMAX=CON(7)
    9: TLAG=CON(8)
   10: KA=P(1)
   11: ID50=P(2)
   12: NICK=P(3)
   13: KIN =P(4)
   14: KOUT=P(5)
   15: BIO =P(6)
   16: K =P(7)
   17: T=X
   18: TL = T - TLAG
   19: END
   20: START
   21: Z(1) = KIN/KOUT
   22: Z(2) = KIN/KOUT
   23: Z(3) = KIN/KOUT
   24: Z(4) = KIN/KOUT
   25: Z(5) = KIN/KOUT
   26: END
   27: DIFF
   28: IF T LE TLAG THEN
   29: C1 = 0.0
   30: DZ(1) = 0.
   31: C2 = 0.0
   32: DZ(2) = 0.0
   33: C3 = 0.0
   34: DZ(3) = 0.0
   35: C4 = 0.0
   36: DZ(4) = 0.0
   37: C5 = 0.0
   38: DZ(5) = 0.0
   39: ELSE
   40: C1 = (BIO*DOSE1*KA/(KA-K))*(DEXP(-K*TL) - DEXP(-KA*TL))
   41: DRUG1= 1. - IMAX*(C1**NICK)/((ID50**NICK) + C1**NICK)
   42: DZ(1) = KIN - KOUT*DRUG1*Z(1)
   43: C2 = (BIO*DOSE2*KA/(KA-K))*(DEXP(-K*TL) - DEXP(-KA*TL))
   44: DRUG2= 1. - IMAX*(C2**NICK)/((ID50**NICK) + C2**NICK)
   45: DZ(2) = KIN - KOUT*DRUG2*Z(2)
   46: C3 = (BIO*DOSE3*KA/(KA-K))*(DEXP(-K*TL) - DEXP(-KA*TL))
   47: DRUG3 = 1. - IMAX*(C3**NICK)/((ID50**NICK) + C3**NICK)
   48: DZ(3) = KIN - KOUT*DRUG3*Z(3)
   49: C4 = DOSE4*DEXP(-K*T)
   50: DRUG4= 1. - IMAX*(C4**NICK)/((ID50**NICK) + C4**NICK)
   51: DZ(4) = KIN - KOUT*DRUG4*Z(4)
   52: C5 = DOSE5*DEXP(-K*T)
   53: DRUG5 = 1. - IMAX*(C5**NICK)/((ID50**NICK) + C5**NICK)
   54: DZ(5) = KIN - KOUT*DRUG5*Z(5)
   55: ENDIF
   56: END
```

```
57: FUNC 1
58: F = Z(1)
59: END
60: FUNC 2
61: F = Z(2)
62: END
63: FUNC 3
64: F = Z(3)
65: END
66: FUNC 4
67: F = Z(4)
68: END
69: FUNC 5
70: F = Z(5)
71: END
72: EOM
NOPAGE BREAKS
NVARIABLES 3
FNUMBER 3
NPOINTS 100
XNUMBER 1
YNUMBER 2
CONSTANTS 10,50,100,3,10,5.2, &
          .67,5.2
METHOD 2   'Gauss-Newton (Levenberg and Hartley)
ITERATIONS 50
INITIAL .02,2,2,2,.3,.6,.05
LOWER BOUNDS 0,0,0,0,0,0,0
UPPER BOUNDS .2,20,10,10,1,1,.5
MISSING 'Missing'
DATA 'WINNLIN.DAT'
BEGIN
```

PARAMETER	ESTIMATE	STANDARD ERROR	CV%	UNIVARIATE C.I.	
KA	.018781	.001301	6.93	.016123	.021438
ID50	1.064434	.177923	16.72	.701068	1.427799
NICK	1.756721	.127275	7.25	1.496792	2.016649
KIN	1.443095	.122114	8.46	1.193706	1.692484
KOUT	.283777	.026281	9.26	.230104	.337450
BIO	.663651	.061906	9.33	.537224	.790079
K	.046120	.005287	11.46	.035322	.056917

```
*** CORRELATION MATRIX OF THE ESTIMATES ***
PARAMETER  KA        ID50      NICK      KIN       KOUT      BIO        K
KA         1.00000
ID50        .56752   1.00000
NICK        .13206    .10195   1.00000
KIN         .08921    .37815   -.161881  1.00000
KOUT        .11760    .34087   -.083878   .98492   1.00000
BIO         .36279    .64334   -.101628   .20329    .20259   1.00000
K          -.68296   -.86016   -.072334  -.49529   -.49644  -.38560   1.00000

   Condition_number=         252.1

     FUNCTION    1
```

X	OBSERVED Y	PREDICTED Y	RESIDUAL	WEIGHT	SE-PRED	STANDARDI RESIDUAL
.0000	5.077	5.085	-.8318E-02	1.000	.8785E-01	-.1966E-01
5.000	5.037	5.085	-.4832E-01	1.000	.8785E-01	-.1142
15.00	5.920	6.221	-.3018	1.000	.1672	-.7575
30.00	7.829	8.230	-.4009	1.000	.2668	-1.180
60.00	9.345	8.474	.8712	1.000	.2588	2.518
120.0	6.103	6.040	.6285E-01	1.000	.1602	.1566
180.0	5.434	5.236	.1980	1.000	.1044	.4722

```
CORRECTED SUM OF SQUARED OBSERVATIONS =  15.5747
WEIGHTED CORRECTED SUM OF SQUARED OBSERVATIONS =  15.5747
SUM OF SQUARED RESIDUALS =          1.05643
SUM OF WEIGHTED SQUARED RESIDUALS =  1.05643
S =  .000000    WITH    0 DEGREES OF FREEDOM
CORRELATION (OBSERVED,PREDICTED) =  .9686
```

FUNCTION   2

X	OBSERVED Y	PREDICTED Y	RESIDUAL	WEIGHT	SE-PRED	STANDARDI RESIDUAL
.0000	4.841	5.085	-.2441	1.000	.8785E-01	-.5770
5.000	5.253	5.085	.1673	1.000	.8785E-01	.3954
10.00	7.124	7.015	.1091	1.000	.1188	.2626
20.00	11.02	11.23	-.2098	1.000	.1883	-.5395
30.00	13.37	13.17	.2026	1.000	.1572	.5034
60.00	14.32	14.27	.5104E-01	1.000	.2015	.1335
120.0	11.61	11.86	-.2496	1.000	.2608	-.7244
180.0	6.655	7.234	-.5782	1.000	.2697	-1.713

```
CORRECTED SUM OF SQUARED OBSERVATIONS =  98.0395
WEIGHTED CORRECTED SUM OF SQUARED OBSERVATIONS =  98.0395
SUM OF SQUARED RESIDUALS =          .583748
SUM OF WEIGHTED SQUARED RESIDUALS =  .583748
S =  .764034    WITH    1 DEGREES OF FREEDOM
CORRELATION (OBSERVED,PREDICTED) =  .9975
```

FUNCTION   3

X	OBSERVED Y	PREDICTED Y	RESIDUAL	WEIGHT	SE-PRED	STANDARDI RESIDUAL
.0000	5.423	5.085	.3375	1.000	.8785E-01	.7977
5.000	4.969	5.085	-.1167	1.000	.8785E-01	-.2759
10.00	7.992	7.788	.2031	1.000	.1326	.4938
20.00	10.98	12.12	-1.135	1.000	.2133	-3.019
30.00	14.54	13.92	.6161	1.000	.1656	1.544
60.00	15.05	15.00	.4935E-01	1.000	.2008	.1290
120.0	13.73	14.06	-.3299	1.000	.2142	-.8790
180.0	10.52	10.08	.4436	1.000	.3358	1.631

```
CORRECTED SUM OF SQUARED OBSERVATIONS =  110.230
WEIGHTED CORRECTED SUM OF SQUARED OBSERVATIONS =  110.230
SUM OF SQUARED RESIDUALS =          2.14352
SUM OF WEIGHTED SQUARED RESIDUALS =  2.14352
S =  1.46408    WITH    1 DEGREES OF FREEDOM
CORRELATION (OBSERVED,PREDICTED) =  .9905
```

FUNCTION   4

X	OBSERVED	PREDICTED	RESIDUAL	WEIGHT	SE-PRED	STANDARDI

```
              Y            Y                                              RESIDUAL
    .0000      5.384        5.085        .2987       1.000      .8785E-01    .7060
   2.000       4.840        5.085       -.2458       1.000      .8785E-01   -.5810
   7.000       6.629        6.311        .3183       1.000      .8957E-01    .7531
   15.00       9.077        8.597        .4799       1.000      .2815       1.464
   30.00       6.762        7.264       -.5021       1.000      .2630      -1.465
   60.00       5.006        5.312       -.3059       1.000      .1109      -.7324
   120.0       5.477        5.087        .3899       1.000      .8781E-01    .9216
```

CORRECTED SUM OF SQUARED OBSERVATIONS = 13.2341
WEIGHTED CORRECTED SUM OF SQUARED OBSERVATIONS = 13.2341
SUM OF SQUARED RESIDUALS = .978997
SUM OF WEIGHTED SQUARED RESIDUALS = .978997
S = .000000   WITH   0 DEGREES OF FREEDOM
CORRELATION (OBSERVED,PREDICTED) = .9641

   FUNCTION   5

```
   X          OBSERVED     PREDICTED    RESIDUAL    WEIGHT     SE-PRED    STANDARDI
              Y            Y                                              RESIDUAL
    .0000      5.100        5.085        .1488E-01   1.000      .8785E-01    .3518E-01
   2.000       4.865        5.085       -.2204       1.000      .8785E-01   -.5210
   7.000       6.307        6.635       -.3286       1.000      .1129      -.7878
   15.00       11.46        10.78        .6849       1.000      .2189       1.838
   30.00       11.56        11.95       -.3866       1.000      .2855      -1.192
   60.00       7.071        6.820        .2512       1.000      .3946       1.427
   120.0       5.095        5.100       -.5425E-02   1.000      .8755E-01   -.1282E-01
```

CORRECTED SUM OF SQUARED OBSERVATIONS = 52.1374
WEIGHTED CORRECTED SUM OF SQUARED OBSERVATIONS = 52.1374
SUM OF SQUARED RESIDUALS = .838491
SUM OF WEIGHTED SQUARED RESIDUALS = .838491
S = .000000   WITH   0 DEGREES OF FREEDOM
CORRELATION (OBSERVED,PREDICTED) = .9922

TOTALS FOR ALL CURVES COMBINED
SUM OF SQUARED RESIDUALS = 5.60119
SUM OF WEIGHTED SQUARED RESIDUALS = 5.60119
S = .432095   WITH   30 DEGREES OF FREEDOM

AIC criteria = 77.75020
SC criteria = 89.02663

PD25 – Dose-response-time analysis IV – Δ response

Objectives

◆ **To analyze four response-time datasets after subcutaneous dosing**

◆ **To write an indirect response model in terms of differential equations**

◆ **To model change from baseline (Δ-temperature) and fix k_{in} to a constant value**

◆ **To obtain initial parameter estimates including maximum obtainable response**

◆ **To characterize the dose-response relationship and turnover of response**

◆ **To simulate the response-time course at a constant exposure of ID_{50}**

◆ **To apply allometric scaling to the turnover parameters**

Problem specification

This problem highlights some complexities in modeling *dose-response-time data* when the measured response is a difference and the baseline value is taken as zero. Specific emphasis is put on obtaining *initial parameter estimates*, implementing *differential equations*, and practicing *indirect response modeling* when the baseline effect is set to zero (delta response).

The primary goal of this exercise is to obtain a dose-response relationship from which one can select an appropriate dose level for a subsequent repeated dose study. Five groups of rats received a dose of 0, 30, 125, 500 or 2000 µg of *8-OH-DPAT* subcutaneously. Body temperature was measured by telemetry for 60 min. No plasma concentrations were available for *8-OH-DPAT* since the handling of the animals would disturb the temperature measurements. The decrease in body temperature from the baseline curve was obtained (see Table 25.1) and is plotted in Figure 25.1 as mean values against time. Data were obtained from Deveney *et al* [1999].

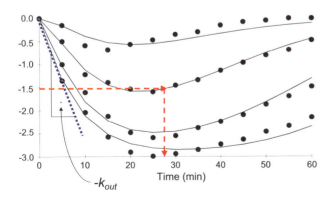

Figure 25.1 Observed (symbols) mean telemetric response-time data of 8-OH-DPAT following four different subcutaneous doses (30, 125, 500 and 2000 µg). Response is expressed as change from baseline response (0 µg) in each group. Vertical arrow indicates t_{ID50} for the 125 µg dose.

Table 25.1   Time, response and dose data

Time (min)	Response	Dose (µg)	Time (min)	Response	Dose (µg)
0	0	30	0	0	500
5	-0.25		5	-1.00	
10	-0.59		10	-1.61	
15	-0.68		15	-2.12	
20	-0.56		20	-2.48	
25	-0.48		25	-2.59	
30	-0.35		30	-2.55	
35	-0.30		35	-2.38	
40	-0.15		40	-2.24	
45	-0.078		45	-2.30	
50	-0.048		50	-1.87	
55	0		55	-1.69	
60	0		60	-1.48	
0	0	125	0	0	2000
5	-0.50		5	-1.36	
10	-1.22		10	-2.04	
15	-1.39		15	-2.58	
20	-1.54		20	-2.90	
25	-1.59		25	-2.99	
30	-1.45		30	-2.94	
35	-1.34		35	-2.85	
40	-1.13		40	-2.73	
45	-0.96		45	-2.63	
50	-0.80		50	-2.46	
55	-0.69		55	-2.34	
60	-0.47		60	-2.15	

We assume that we can *approximate* the biophase compartment kinetics by a first-order input-output model, where the input rate constant is equal to the output rate constant. This reduces the number of model parameters. This simple model is assumed because it is not possible to estimate more than one biophase rate constant with the present data. The amount of drug in the biophase becomes

$$A = Dose_i \cdot K' \cdot t \cdot e^{-k \cdot t} \tag{25:1}$$

The inhibitory function is written as

$$H(t) = 1 - \frac{I_{max} A^n}{ID_{50}^n + A^n} \tag{25:2}$$

The turnover of the response (R) can be written as:

$$\frac{dR}{dt} = k_{in} \cdot H(t) - k_{out} \cdot R \tag{25:3}$$

As can be seen, the antagonistic effect of the drug is assumed to act on the turnover rate (k_{in}) rather than to stimulate the loss of response. Since Δ-temperature (change from baseline) is modeled, we write the *WinNonlin* code of the equation to be fitted to the data as

       FUNC 1
$$F = Z(1) - R0$$
       End

R_0 is the baseline response defined as k_{in}/k_{out}, and $Z(1)$ is R in Equation 25:3. The combined response (Equation 25:3) and inhibitory (Equation 25:2) function at dynamic steady state becomes

$$R_{ss} = R_0 \cdot H(t) = \frac{k_{in}}{k_{out}} \cdot \left[1 - \frac{I_{max} A^n}{ID_{50}^n + A^n} \right] \tag{25:4}$$

This relationship is the dose-response function for *8-OH-DPAT* with respect to body temperature in the rat. The function starts at a value equal to k_{in}/k_{out} and then decreases with increasing values of A (or *Dose*). The maximum obtainable inhibition from baseline (Δ_{max}) becomes

$$\Delta_{max} = R_0 - R_0 \cdot H(t) = \frac{k_{in}}{k_{out}} - \frac{k_{in}}{k_{out}} \cdot \left[1 - \frac{I_{max} A_\infty^n}{ID_{50}^n + A_\infty^n} \right] = \frac{k_{in}}{k_{out}} \cdot I_{max} \tag{25:5}$$

Initial parameter estimates

Figures 25.1 and 25.3 demonstrate how the initial parameter estimates of the model can be obtained graphically.

     k_{in} = was set to a constant (e.g., 1) since we model a change from baseline.
     k_{out} = 0.15 min^{-1}
     K' = 0.15 min^{-1}

     I_{max} = 0.45 $I_{max} = \dfrac{k_{out}}{k_{in}} \cdot \Delta$ for the highest dose

     ID_{50} = 0.125 · 0.15 · 27 e$^{-.15 \cdot 27}$ = 0.01 mg   (Equation 25:1)
     n = 2

     The maximum value Δ for the highest dose is approximately 3 °C. According to the relationship above, this gives $I_{max} = (k_{out}/k_{in}) \cdot \Delta = (0.15/1) \cdot 3 = 0.45$. It may not be necessary to apply an exponent (n) different from unity (1), but we will also try to estimate that parameter. The initial parameter estimates for I_{max}, ID_{50}, n, k_{out} and K' (0.45, 0.02 mg, 1.5, 2, 0.15 min^{-1} and 0.15 min^{-1}), were obtained graphically and by using previous study information. Since we are modeling a change in response from the baseline, k_{in} was set to unity and the only turnover parameter to be estimated is the fractional turnover rate constant, k_{out}. The latter parameter is also easily obtained from the initial slope (downswing) of the response-time data. In the program output, *FUNCTIONS* 1-4 correspond to the doses 30, 125, 500 and 2000 µg, respectively.

Interpretation of results and conclusions

This exercise has dealt with implementing of *differential equations* and practicing *dose-response-time data analysis*. It is of considerable value to simultaneously model several dose-response-time data sets in order to get robust and precise parameter estimates. The telemetry system enables one to record body temperature continuously in freely moving animals and the lack of stressful conditions makes it possible to obtain a more accurate measure of temperature turnover. Figure 25.1 contains the superimposed values of observed and predicted concentrations of *8-OHDPAT*. Note the shift in the trough to the right with increasing dose. Table 25.2 shows the final parameter estimates.

Table 25.2   Parameter estimates

Parameter	Mean	CV%
I_{max}	0.41	6.6
ID_{50}	0.033	7.8
k_{out}	0.13	7.5
K'	0.085	4.1
n	1.14	5.4

The half-life of the response was about 8 min and therefore the time to dynamic steady state was about 30 min. What kind of design would you propose in order to increase the parameter precision? Simulate the model with the final parameter estimates during a 3 h time course. Plot the partial derivatives of the response (Figure 25.4) with respect to ID_{50}, k_{out} and I_{max}. What do you see?

Figure 25.2 shows the dose-delta response curve. From this curve, one can easily obtain the ID_{50} value or the delta value as shown. The latter is the maximum obtained effect of *8-OHDPAT*, which, in this particular example, will be 3°C (I_{max}/k_{out}), since k_{in} was set to 1 (unity). This is also confirmed by the simulation in Figure 25.3 where a constant exposure (A) equal to ID_{50} ($A=33$ µg) and to 3000 µg were simulated and are presented as the response-time courses. Exposure equal to ID_{50} and to 3000 µg gave a 1.5 and 3.0°C reduction in body temperature, respectively (Figure 25.3).

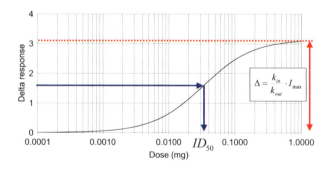

$$\Delta = \frac{k_{in}}{k_{out}} \cdot I_{max}$$

Figure 25.2 *Simulated delta response versus dose (mg) profile. The delta response corresponds to the change from the baseline value. Delta (Δ) is equal to 0.41/0.13 since k_{in} is set to 1.*

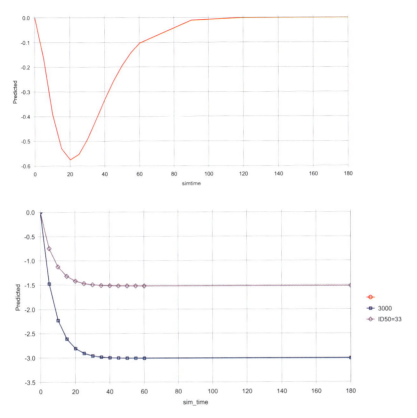

Figure 25.3 *Simulated delta response versus time profile of a single dose corresponding to ID$_{50}$ (upper). Simulated delta response versus time profile (lower) of a constant exposure corresponding to a dose equal to ID$_{50}$ (purple) and a dose equal to 3000 µg (blue). Note that a single dose corresponding to ID$_{50}$ does not necessarily reach half-maximal effect (upper panel). However, a constant body load corresponding to ID$_{50}$ will reduced the response to half-maximal effect (lower panel).*

The application of partial derivatives with respect to the model parameters may be used to obtain information about information-rich sampling domains for the model parameters (Figure 25.4). When the derivative reaches a maximum or minimum, the region is *information rich* with respect to that particular parameter. For example, if one were to sample at about 20 min, one would obtain correspondingly more information about the *ID$_{50}$* parameter. On the other hand, the kinetic rate constant for the biophase *K* has two extreme values, namely one at about 10 min and the other at 40 min.

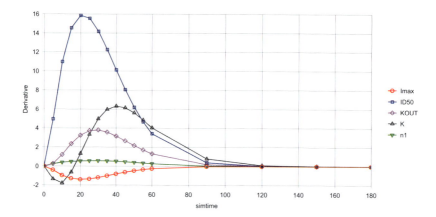

Figure 25.4 *Simulated partial derivatives of the response with respect to each model parameter.*

Allometric scaling of turnover parameters

If the half-life of the response $(ln(2)/k_{out})$ is proportional to the turnover time of the substance causing the effect, then the principles of allometric scaling are applicable to turnover parameters or pharmacodynamic response. All rate processes like flows, clearances, secretion rates, and synthesis rates require energy, and scale generally proportionally to body weight raised to the power b, where b has a value of 0.6-0.8. Turnover rate k_{in} is scaled according to Equation 25:6

$$Turnover\ rate = k_{in} = a \cdot BW^{b} \tag{25:6}$$

The fractional turnover rate k_{out} scales like any first-order rate constant or frequency proportionally to body weight raised to the power of ~ -0.25

$$k_{out} = d \cdot BW^{-0.25} \tag{25:7}$$

The turnover time is derived as the ratio of A_{ss} to k_{in} and scales proportionally with body weight according to

$$Turnover\ time = \frac{A_{ss}}{k_{in}} = \frac{c \cdot BW^{1}}{a \cdot BW^{0.75}} = e \cdot BW^{0.25} \tag{25:8}$$

$$t_{1/2kout} = \frac{\ln(2)}{k_{out}} = \ln(2) \cdot e \cdot BW^{0.25} \tag{25:9}$$

For scaling the half-life of a response from rat to man, one can use Equation 25:10

$$t_{1/2\,response}(man) = t_{1/2\,response}(rat) \cdot \left[\frac{BW_{man}}{BW_{rat}}\right]^{0.25} \qquad (25:10)$$

$$t_{1/2\,response}(man) = \frac{\ln(2)}{k_{out}} \cdot \left[\frac{70}{0.25}\right]^{0.25} = 7.69 \cdot 4.09 \approx 32 \, \text{min} \qquad (25:11)$$

The simulation shown in Figure 25.5, gives a schematic illustration of the fictitious response time courses in rat and man with baseline values normalized and the only difference being k_{out}.

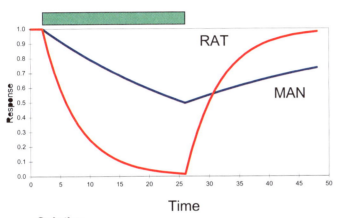

Figure 25.5 *Schematic illustration of the response time courses in the rat (red line) and man (blue line) after having scaled animal data to man by means of Equations 25:6 and 25:7. The green horizontal bar represents drug exposure.*

Solution

```
MODEL
  COMM
    NFUN   4
    NDER   4
    NCON   4
    NPARM  5
    PNAMES 'Imax', 'ID50', 'KOUT', 'K', 'n1'
  END
       1:  TEMP
       2:  T=X
       3:  DOSE1=CON(1)
       4:  DOSE2=CON(2)
       5:  DOSE3=CON(3)
       6:  DOSE4=CON(4)
       7:  END
       8:  START
       9:  KIN = 1
      10:  RO=KIN/KOUT
      11:  Z(1) = RO
      12:  Z(2) = RO
      13:  Z(3) = RO
      14:  Z(4) = RO
      15:  END
      16:  DIFF
      17:  CP1 = DOSE1*K*T*DEXP(-K*T)
      18:  CP2 = DOSE2*K*T*DEXP(-K*T)
```

```
19: CP3 = DOSE3*K*T*DEXP(-K*T)
20: CP4 = DOSE4*K*T*DEXP(-K*T)
21: INHIB1 = 1 - IMAX*CP1**N1/(ID50**N1 + CP1**N1)
22: INHIB2 = 1 - IMAX*CP2**N1/(ID50**N1 + CP2**N1)
23: INHIB3 = 1 - IMAX*CP3**N1/(ID50**N1 + CP3**N1)
24: INHIB4 = 1 - IMAX*CP4**N1/(ID50**N1 + CP4**N1)
25: DZ(1)= KIN*INHIB1 - KOUT*Z(1)
26: DZ(2)= KIN*INHIB2 - KOUT*Z(2)
27: DZ(3)= KIN*INHIB3 - KOUT*Z(3)
28: DZ(4)= KIN*INHIB4 - KOUT*Z(4)
29: END
30: FUNC 1
31: F = Z(1) - R0
32: END
33: FUNC 2
34: F = Z(2) - R0
35: END
36: FUNC 3
37: F = Z(3) - R0
38: END
39: FUNC 4
40: F = Z(4) - R0
41: END
42: EOM
NVARIABLES 3
FNUMBER 3
NPOINTS 100
XNUMBER 1
YNUMBER 2
CONSTANTS .03,.125,.5,2
METHOD 2  'Gauss-Newton (Levenberg and Hartley)
ITERATIONS 50
INITIAL .45,.03,.15,.05,1.5
LOWER BOUNDS 0,0,0,0,0
UPPER BOUNDS 2,.3,1,.3,3
DATA 'WINNLIN.DAT'
BEGIN
```

PARAMETER	ESTIMATE	STANDARD ERROR	CV%	UNIVARIATE C.I.	
IMAX	.407102	.026833	6.59	.353122	.461083
ID50	.032767	.002566	7.83	.027605	.037929
KOUT	.134430	.010058	7.48	.114196	.154664
K	.084693	.003450	4.07	.077751	.091634
N1	1.142635	.062152	5.44	1.017600	1.267669

```
*** CORRELATION MATRIX OF THE ESTIMATES ***
PARAMETER  IMAX       ID50       KOUT       K          N1
IMAX       1.00000
ID50        .51453    1.00000
KOUT        .94773     .30934    1.00000
K          -.73885    -.57864    -.76988    1.0000
N1         -.20094    -.50277     .04154    -.01154    1.00000

     FUNCTION   1

     X       OBSERVED   PREDICTED   RESIDUAL    WEIGHT   SE-PRED   STANDARDI
```

X	OBSERVED Y	PREDICTED Y	RESIDUAL	WEIGHT	SE-PRED	STANDARDI RESIDUAL
.0000	.0000	.0000	.0000	1.000		
5.000	-.2500	-.1621	-.8786E-01	1.000	.1576E-01	-.7858
10.00	-.5875	-.3882	-.1993	1.000	.2850E-01	-1.825
15.00	-.6813	-.5271	-.1541	1.000	.3384E-01	-1.431
20.00	-.5638	-.5726	.8872E-02	1.000	.3489E-01	.8262E-01
25.00	-.4688	-.5508	.8202E-01	1.000	.3393E-01	.7616
30.00	-.3488	-.4895	.1408	1.000	.3221E-01	1.301
35.00	-.2963	-.4110	.1148	1.000	.3018E-01	1.055
40.00	-.1500	-.3304	.1804	1.000	.2784E-01	1.648
45.00	-.7750E-01	-.2564	.1789	1.000	.2509E-01	1.625
50.00	-.4750E-01	-.1934	.1459	1.000	.2200E-01	1.318
55.00	.0000	-.1425	.1425	1.000	.1872E-01	1.280
60.00	.0000	-.1030	.1030	1.000	.1549E-01	.9209

CORRECTED SUM OF SQUARED OBSERVATIONS = .722561
WEIGHTED CORRECTED SUM OF SQUARED OBSERVATIONS = .722561
SUM OF SQUARED RESIDUALS = .227790
SUM OF WEIGHTED SQUARED RESIDUALS = .227790
S = .168742   WITH   8 DEGREES OF FREEDOM
CORRELATION (OBSERVED,PREDICTED) = .8595

FUNCTION   2

X	OBSERVED Y	PREDICTED Y	RESIDUAL	WEIGHT	SE-PRED	STANDARDI RESIDUAL
.0000	.0000	.0000	.0000	1.000		
5.000	-.5000	-.5434	.4341E-01	1.000	.2589E-01	.3950
10.00	-1.219	-1.121	-.9745E-01	1.000	.3909E-01	-.9199
15.00	-1.391	-1.451	.6009E-01	1.000	.4162E-01	.5725
20.00	-1.536	-1.583	.4643E-01	1.000	.4083E-01	.4410
25.00	-1.594	-1.577	-.1719E-01	1.000	.3843E-01	-.1619
30.00	-1.453	-1.479	.2630E-01	1.000	.3451E-01	.2446
35.00	-1.338	-1.323	-.1438E-01	1.000	.3062E-01	-.1323
40.00	-1.133	-1.136	.3762E-02	1.000	.2931E-01	.3450E-01
45.00	-.9613	-.9401	-.2113E-01	1.000	.3119E-01	-.1947
50.00	-.8025	-.7518	-.5069E-01	1.000	.3399E-01	-.4708
55.00	-.6875	-.5830	-.1045	1.000	.3543E-01	-.9745
60.00	-.4738	-.4400	-.3379E-01	1.000	.3469E-01	-.3145

CORRECTED SUM OF SQUARED OBSERVATIONS = 2.84176
WEIGHTED CORRECTED SUM OF SQUARED OBSERVATIONS = 2.84176
SUM OF SQUARED RESIDUALS = .334292E-01
SUM OF WEIGHTED SQUARED RESIDUALS = .334292E-01
S = .646425E-01 WITH   8 DEGREES OF FREEDOM
CORRELATION (OBSERVED,PREDICTED) = .9950

FUNCTION   3

X	OBSERVED Y	PREDICTED Y	RESIDUAL	WEIGHT	SE-PRED	STANDARDI RESIDUAL
.0000	.0000	.0000	.0000	1.000		
5.000	-1.001	-1.039	.3820E-01	1.000	.3802E-01	.3593
10.00	-1.606	-1.813	.2071	1.000	.4364E-01	1.989
15.00	-2.123	-2.225	.1025	1.000	.3442E-01	.9534
20.00	-2.479	-2.418	-.6081E-01	1.000	.2864E-01	-.5568
25.00	-2.591	-2.479	-.1120	1.000	.2961E-01	-1.028
30.00	-2.549	-2.455	-.9397E-01	1.000	.3212E-01	-.8681
35.00	-2.378	-2.367	-.1061E-01	1.000	.3331E-01	-.9833E-01
40.00	-2.243	-2.226	-.1632E-01	1.000	.3352E-01	-.1514
45.00	-2.096	-2.039	-.5689E-01	1.000	.3484E-01	-.5297

```
50.00        -1.871       -1.814       -.5679E-01  1.000      .3937E-01  -.5366
55.00        -1.693       -1.564       -.1289      1.000      .4684E-01  -1.254
60.00        -1.483       -1.303       -.1796      1.000      .5463E-01  -1.818
```

CORRECTED SUM OF SQUARED OBSERVATIONS = 6.36255
WEIGHTED CORRECTED SUM OF SQUARED OBSERVATIONS = 6.36255
SUM OF SQUARED RESIDUALS =   .135660
SUM OF WEIGHTED SQUARED RESIDUALS = .135660
S = .130221   WITH   8 DEGREES OF FREEDOM
CORRELATION (OBSERVED,PREDICTED) = .9901

   FUNCTION   4

X	OBSERVED Y	PREDICTED Y	RESIDUAL	WEIGHT	SE-PRED	STANDARDI RESIDUAL
.0000	.0000	.0000	.0000	1.000		
5.000	-1.355	-1.330	-.2473E-01	1.000	.5212E-01	-.2469
10.00	-2.039	-2.115	.7670E-01	1.000	.5588E-01	.7818
15.00	-2.576	-2.521	-.5564E-01	1.000	.4632E-01	-.5403
20.00	-2.903	-2.723	-.1794	1.000	.4170E-01	-1.710
25.00	-2.993	-2.817	-.1759	1.000	.4254E-01	-1.681
30.00	-2.939	-2.849	-.8993E-01	1.000	.4401E-01	-.8648
35.00	-2.851	-2.842	-.8930E-02	1.000	.4363E-01	-.8575E-01
40.00	-2.734	-2.806	.7251E-01	1.000	.4099E-01	.6892
45.00	-2.630	-2.742	.1121	1.000	.3680E-01	1.050
50.00	-2.461	-2.647	.1856	1.000	.3322E-01	1.720
55.00	-2.336	-2.516	.1793	1.000	.3457E-01	1.669
60.00	-2.151	-2.344	.1924	1.000	.4417E-01	1.851

CORRECTED SUM OF SQUARED OBSERVATIONS = 8.22816
WEIGHTED CORRECTED SUM OF SQUARED OBSERVATIONS = 8.22816
SUM OF SQUARED RESIDUALS =   .202333
SUM OF WEIGHTED SQUARED RESIDUALS = .202333
S = .159034   WITH   8 DEGREES OF FREEDOM
CORRELATION (OBSERVED,PREDICTED) = .9880

TOTALS FOR ALL CURVES COMBINED
SUM OF SQUARED RESIDUALS =   .599213
SUM OF WEIGHTED SQUARED RESIDUALS = .599213
S = .112912   WITH   47 DEGREES OF FREEDOM

AIC criteria =   -16.63115
SC criteria =   -6.87493

PD26 – Synergy modeled via hyperbolic functions

Objectives

◆ **To analyze two nested sets of receptor binding/stimulus/response data**

◆ **To obtain initial parameter estimates of K_d, n_1, beta and n_2**

◆ **To apply a receptor binding/stimulus/response model**

Problem specification

This exercise highlights how to *model binding/stimulus/response data*. For further reading about the different synergistic models, see also Sections 4.12 and 4.13.

Two sets of binding/stimulus/response data were obtained experimentally. Data are shown in Figure 26.1, Table 26.1 and the program output. Synergistic or cascade effects may be predicted from a nested system of hyperbolic functions (see Figure 4.4 in section 4.2 for a schematic illustration of nested hyperbolic functions). Equation 26:1 'drives' Equation 26:2 via the stimulus variable.

$$Stimulus\ I = \frac{B_{max} \cdot C^n}{K_d^n + C^n} \tag{26:1}$$

which then enters

$$Response = R = \frac{Stim_{max} \cdot Stimulus\ I^n}{beta^n + Stimulus\ I^n} \tag{26:2}$$

This system (Equations 26:1 and 26:2) is fit to the experimental data shown in Figure 26.1 and the resulting parameter estimates are shown in the figure text. The model parameters were n_1, K_d, n_2 and *beta*, but B_{max} and $Stim_{max}$ were constants, since *Stimulus I* and *Response* were normalized to 100 per cent, respectively.

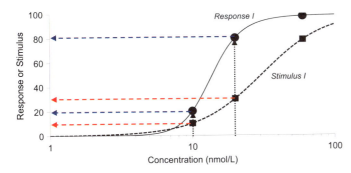

Figure 26.1 *Response/Stimulus I versus drug concentration. By doubling the drug concentration from 10 to 20 nmol/L Stimulus I increases from 10 to 30 units (3-fold increase) and Response from 20 to 80 units (4-fold increase). K_d and n_1 were 30.9 ± 0.17 % (mean ± CV %) nmol/L and 1.95 ± 0.27 , respectively. Beta and n_2 were 17.3 ± 0.43 nmol/L and 2.52 ± 0.44, respectively.*

Another way of viewing this synergistic system is to study Figure 26.2 where the transduction function for ligand concentration-stimulus is separated from stimulus-response.

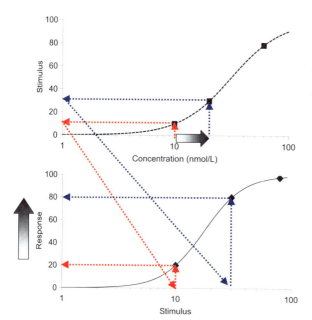

Figure 26.2 *Ligand-concentration versus stimulus (upper figure) and stimulus versus response (bottom figure). The doubling of the ligand concentration from 10 to 20 nmol/L(2-fold increase shown by the gray horizontal arrow) increases the response from 20 to 40 units (4-fold increase shown by the gray vertical arrow). Solid symbols represent measured data of ligand concentration-stimulus (∎) and stimulus-response (●). Solid lines are model predicted. The 10 nmol/l ligand concentration-response relationship is shown by the dotted red line. The 20 nmol/L ligand concentration-response relationship is shown by the dotted blue line.*

The magnitude of amplification throughout the cascade is, of course, very much dependent on the individual *Stim_{max}* values (Equation 26:2). The relative magnitude of two different *Stim_{max}* values may differ substantially.

Table 26.1 Concentration-stimulus-response data of compound X

Conc (nM)	Stimulus (%)	Response (%)
10	10	20
20	30	80
60	78	97
100	-	-

Initial parameter estimates

Obtain the initial parameter estimates graphically from Figures 26.1 and 26.2. For the binding/stimulus model (Equation 26:1), concentration and stimulus are the independent and dependent variables, respectively. For the stimulus/response model (Equation 26:2), the concentration and response are the independent and dependent variables, respectively, as the model code is setup.

Interpretation of results and conclusions
We succeeded in fitting a nested binding/stimulus/response model to the experimental data. The final parameter estimates for this model are shown in the program output together with their precision.

You have
- analyzed a model for synergy by means of nested hyperbolic functions
- learned to graphically obtain the initial estimates
- characterised the parameters for synergy

The next step in the characterisation of dynamics would be to analyse *in vivo* data obtained from the intended route of administration after repeated dosing.

Solution
```
MODEL
 COMM
  nfun   2
  NPAR   4
  PNAM 'kd', 'n1', 'beta', 'n2'
END
    1: TEMP
    2: C = X
    3: STIM1 = 100*C**N1/(KD**N1 + C**N1)
    4: END
    5: FUNC 1
    6: F = 100*C**N1/(KD**N1 + C**N1)
    7: END
    8: FUNC 2
    9: F = 100*STIM1**N2/(BETA**N2 + STIM1**N2)
   10: END
   11: EOM
NOPAGE BREAKS
NVARIABLES 3
FNUMBER 3
NPOINTS 100
XNUMBER 1
YNUMBER 2
METHOD 2   'Gauss-Newton (Levenberg and Hartley)
ITERATIONS 50
INITIAL 40,2,10,3
LOWER BOUNDS 0,0,0,0
UPPER BOUNDS 100,5,25,7
MISSING 'Missing'
DATA 'WINNLIN.DAT'
BEGIN
```

PARAMETER	ESTIMATE	STANDARD ERROR	CV%	UNIVARIATE C.I.	
KD	30.877097	.051921	.17	30.652404	31.101790
N1	1.950039	.005277	.27	1.927201	1.972877
BETA	17.316460	.073907	.43	16.996622	17.636297
N2	2.517757	.010976	.44	2.470255	2.565258

```
*** CORRELATION MATRIX OF THE ESTIMATES ***
```

```
PARAMETER  KD          N1          BETA        N2
KD       1.00000
N1       -.182761    1.00000
BETA     -.463347    -.70458     1.00000
N2       -.004595    -.74664     .61842      1.00000

  Condition_number=        40.63

   FUNCTION   1

   X          OBSERVED    PREDICTED   RESIDUAL    WEIGHT    SE-PRED     STANDAR
              Y           Y                                            RESIDUAL
   10.00      10.00       9.988       .1176E-01   1.000     .5611E-01   .1622
   20.00      30.00       30.01       -.9035E-02  1.000     .7642E-01   -.1786
   60.00      78.50       78.51       -.7145E-02  1.000     .8806E-01   -.2810
   100.0      .           90.82       .           .0000     -.6400+152  .

CORRECTED SUM OF SQUARED OBSERVATIONS =  3651.69
WEIGHTED CORRECTED SUM OF SQUARED OBSERVATIONS =  2481.50
SUM OF SQUARED RESIDUALS =              .270924E-03
SUM OF WEIGHTED SQUARED RESIDUALS =  .270924E-03
S =  .000000     WITH     -1 DEGREES OF FREEDOM
CORRELATION (OBSERVED,PREDICTED) =  .8243

   FUNCTION   2

   X          OBSERVED    PREDICTED   RESIDUAL    WEIGHT    SE-PRED     STANDARD
              Y           Y                                            RESIDUAL
   10.00      20.00       20.01       -.1443E-01  1.000     .9109E-01   -1.412
   20.00      80.00       79.97       .3087E-01   1.000     .8900E-01   1.409
   60.00      97.70       97.82       -.1240      1.000     .2537E-01   -1.408
   100.0      .           98.48       .           .0000     -.6400+152  .

CORRECTED SUM OF SQUARED OBSERVATIONS =  6573.97
WEIGHTED CORRECTED SUM OF SQUARED OBSERVATIONS =  3316.86
SUM OF SQUARED RESIDUALS =              .165314E-01
SUM OF WEIGHTED SQUARED RESIDUALS =  .165314E-01
S =  .000000     WITH     -1 DEGREES OF FREEDOM
CORRELATION (OBSERVED,PREDICTED) =  .7103

TOTALS FOR ALL CURVES COMBINED
SUM OF SQUARED RESIDUALS =              .168023E-01
SUM OF WEIGHTED SQUARED RESIDUALS =  .168023E-01
S =  .916578E-01 WITH      2 DEGREES OF FREEDOM
```

PD27 – Incomplete response data

Objectives

◆ To analyze data with an *m·log(C)* model

◆ To analyze data with an ordinary E_{max} model

◆ To discriminate between competing models

Problem specification

The goals of this exercise are to demonstrate some ways of discriminating between rival response models for incomplete concentration-effect data with respect to the upper asymptote. We will also propose an alternative design in order to select the most appropriate model. See sections 4.4.2 and 4.4.3 for more details.

The effect is the plasma concentration of a surrogate marker and covers an interval between 20 (approximate baseline level E_0) and 180. The data are taken from a study involving measurements at steady state, and are shown in Figure 27.1 and in the program output. We will model the response by using the ordinary E_{max} and $m·logC$ models. We observe E_0 when no drug is present. By increasing the plasma concentration of drug A, the response is increasing. The maximum drug induced effect is not at a plateau.

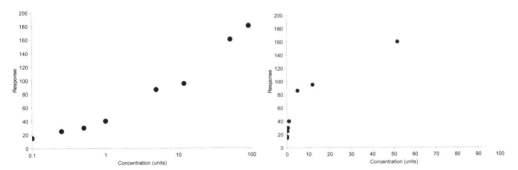

Figure 27.1 *Semi-logarithmic plot (left) and linear plot (right) of response versus steady state plasma concentration of drug X.*

We plot the effect-concentration data and according to Figure 27.1. The models are shown in Equation 27:1 (2 parameters) and Equation 27:2 (3 parameters).

$$Model_1 = m \cdot \ln(C + C_0) \tag{27:1}$$

$$Model_2 = E_0 + \frac{E_{max} \cdot C}{EC_{50} + C} \tag{27:2}$$

We then discriminate between them using appropriate statistical methods such as the residuals, *WRSS* and parameter precision.

Initial parameter estimates

The initial parameter estimates for E_0 (baseline), E_{max}, EC_{50} and m were obtained graphically according to Figure 27.2. E_{max} is estimated to about 200 units, EC_{50} is about 10 concentration units, E_0 is 20. To obtain initial estimates for m and C_0 in Equation 27:1 we use two concentration-effect measurements

$$20 \approx m \cdot \ln(0 + C_0) \tag{27:3}$$

$$140 \approx m \cdot \ln(30 + C_0) \tag{27:4}$$

The m and C_0 parameters are then solved algebraically by rearranging Equation 27:3 and solve for m, and then insert that new expression into Equation 27:4. C_0 and m are then approximately 1.6 and 40, respectively.

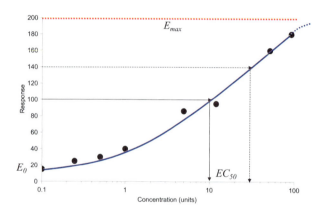

Figure 27.2 *Schematic illustration of how to obtain the initial parameter estimates of E_0 (intercept of the Y axis), E_{max} (horizontal dashed red line) and EC_{50} (vertical solid arrow). The blue curve is drawn by eyeballing the data. The dashed hairthin arrows corresponds to the data in Equation 27:4. The same is done for Equation 27:3 where C is approximated to be much less than C_0.*

Interpretation of results and conclusions

Comparisons of the observed and model-predicted data of both models are displayed in Figure 27.3, and parameter estimates and their precision are shown in Table 27.1.

Although there are no absolute criteria regarding the selection of a model, two goals are generally sought when selection is for kinetic/dynamic data. Firstly, it is necessary that the scatter of the residuals is randomly distributed around the fitted model, and secondly, that the sums of the weighted squared residuals (*WRSS*) have been sufficiently reduced to justify fitting with additional parameters.

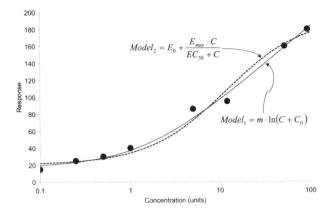

Figure 27.3 *Observed and predicted effects obtained using two models. The predicted effects of the E_{max} and log-C models are indicated by arrows. Note the marginal difference between the models within the studied concentration-response range.*

Table 27.1 Final parameter estimates ± CV% compared with initial estimates

	$m \cdot \log C$ model	E_{max} model	Initial estimates
m	39.9 ± 2.2	-	40
C_0	1.54 ± 8.1	-	1.6
E_0	-	20.5 ± 23	20
E_{max}	-	174 ± 6.6	200
EC_{50}	-	11.8 ± 26	10
WRSS	244	503	-
+/-	4	3	-

The two *WRSS* of the ordinary E_{max} and $m \cdot \log C$ models were 503 and 244, respectively. In spite of the fact that a third parameter was added to the ordinary E_{max} model, it did not reduce the *WRSS*. Figure 27.4 shows the weighted (absolute, w = 1) residual plotted against the concentration. Note that the amplitude of the scatter is larger for the ordinary E_{max} model than for the $m \cdot \log C$ model. The changes in the sign are more frequent for the $m \cdot \log C$ model.

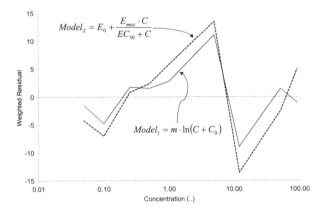

Figure 27.4 *Weighted (i.e., absolute residual residual since weight is 1) versus steady state concentration.*

The analysis has shown that residual analysis, *WRSS* and precision of the parameter estimates favor the *m·logC* model with the actual dataset. We propose a design with sampling points at higher concentrations at the asymptote of the response in order to fully discriminate between the models.

Solution I - *m·logC* model

```
MODEL
   COMM
   NPAR 2
   PNAM 'm', 'C0'
END
     1: FUNC 1
     2: C = X
     3: F = M*LOGE(C + C0)
     4: END
     5: EOM
NOPAGE BREAKS
NVARIABLES 3
NPOINTS 100
XNUMBER 2
YNUMBER 3
METHOD 2   'Gauss-Newton (Levenberg and Hartley)
ITERATIONS 50
INITIAL 40,1.6
LOWER BOUNDS 10,.5
UPPER BOUNDS 100,5
MISSING 'Missing'
NOBSERVATIONS 13
DATA 'WINNLIN.DAT'
BEGIN
```

PARAMETER	ESTIMATE	STANDARD ERROR	CV%	UNIVARIATE C.I.	
M	39.890026	.875087	2.19	37.820758	41.959294
C0	1.541506	.125237	8.12	1.245364	1.837648

```
*** CORRELATION MATRIX OF THE ESTIMATES ***
PARAMETER    M           C0
M       1.00000
C0     -.26129     1.00000

Condition_number=       7.249

   FUNCTION   1
```

X	OBSERVED Y	PREDICTED Y	RESIDUAL	WEIGHT	SE-PRED	STANDARDIZED RESIDUAL
.5000E-01	17.00	18.54	-1.536	1.000	3.057	-.3039
.1000	15.00	19.77	-4.770	1.000	2.958	-.9331
.2500	25.00	23.26	1.742	1.000	2.699	.3316
.5000	30.00	28.47	1.531	1.000	2.361	.2828
1.000	40.00	37.21	2.792	1.000	1.921	.4999
5.000	86.00	74.92	11.08	1.000	1.621	1.951
12.00	95.00	103.9	-8.944	1.000	2.213	-1.633
23.00	.	127.7	.	.0000	-.6400+152	.
51.70	160.0	158.6	1.444	1.000	3.455	.3013
65.00	.	167.5	.	.0000	-.6400+152	.
75.00	.	173.0	.	.0000	-.6400+152	.

| 92.00 | 180.0 | 181.0 | -1.037 | 1.000 | 3.958 | -.2366 |
| 100.0 | . | 184.3 | . | .0000 | -.6400+152 | . |

```
CORRECTED SUM OF SQUARED OBSERVATIONS =  45759.7
WEIGHTED CORRECTED SUM OF SQUARED OBSERVATIONS =  31404.0
SUM OF SQUARED RESIDUALS =          244.203
SUM OF WEIGHTED SQUARED RESIDUALS =  244.203
S =  5.90644      WITH      7 DEGREES OF FREEDOM
CORRELATION (OBSERVED,PREDICTED) =  .8252
```

```
AIC criteria =          53.48198
SC  criteria =          53.87643
```

Solution II - Ordinary E_{max} model

```
MODEL
 COMM
  NPAR 3
  PNAM 'EMAX', 'c50', 'E0'
END
     1: FUNC 1
     2: C = X
     3: F = E0 + EMAX*C/(C + C50)
     4: END
     5: EOM
NOPAGE BREAKS
NVARIABLES 3
NPOINTS 100
XNUMBER 2
YNUMBER 3
METHOD 2   'Gauss-Newton (Levenberg and Hartley)
ITERATIONS 50
INITIAL 200,10,20
LOWER BOUNDS 10,1,1
UPPER BOUNDS 500,100,50
MISSING 'Missing'
NOBSERVATIONS 13
DATA 'WINNLIN.DAT'
BEGIN
```

PARAMETER	ESTIMATE	STANDARD ERROR	CV%	UNIVARIATE C.I.	
EMAX	174.307640	11.417862	6.55	146.369117	202.246163
C50	11.766812	3.046067	25.89	4.313348	19.220276
E0	20.536384	4.639542	22.59	9.183827	31.888942

```
*** CORRELATION MATRIX OF THE ESTIMATES ***
PARAMETER  EMAX         C50         E0
EMAX     1.00000
C50       .60635   1.00000
E0       -.16652    .48300    1.00000
```

```
 Condition_number=          7.893
```

```
    FUNCTION   1
```

X	OBSERVED Y	PREDICTED Y	RESIDUAL	WEIGHT	SE-PRED	STANDARDIZED RESIDUAL

```
.5000E-01   17.00      21.27     -4.274    1.000     4.542     -.5374
.1000       15.00      22.01     -7.005    1.000     4.449     -.8751
.2500       25.00      24.16      .8373    1.000     4.204      .1029
.5000       30.00      27.64     2.359     1.000     3.903      .2847
1.000       40.00      34.19     5.810     1.000     3.663      .6922
5.000       86.00      72.52     13.48     1.000     5.971     1.942
12.00       95.00     108.5     -13.55     1.000     6.604    -2.135
23.00         .       135.8        .       .0000    -.6400+152    .
51.70      160.0      162.5     -2.527     1.000     5.718     -.3533
65.00         .       168.1        .       .0000    -.6400+152    .
75.00         .       171.2        .       .0000    -.6400+152    .
92.00      180.0      175.1     4.922      1.000     7.275      .8846
100.0         .       176.5        .       .0000    -.6400+152    .
```

```
CORRECTED SUM OF SQUARED OBSERVATIONS =  45759.7
WEIGHTED CORRECTED SUM OF SQUARED OBSERVATIONS =  31404.0
SUM OF SQUARED RESIDUALS =              503.262
SUM OF WEIGHTED SQUARED RESIDUALS =    503.262
S =  9.15844     WITH     6 DEGREES OF FREEDOM
CORRELATION (OBSERVED,PREDICTED) =  .8218

AIC criteria =        61.99000
SC  criteria =        62.58168
```

References

Pharmacokinetics-pharmacodynamics

Abramson FP. Kinetic models of induction: I. Persistence of inducing agent. *J. Pharm. Sci.* **75**:pp 223 (1986).

Abramson FP. Kinetic models of induction: II. Decreased turnover of a product or its precursor. *J. Pharm. Sci.* **75**:pp 229 (1986).

Ackerman E, Rosevear JW, McGuckin WF. A mathematical model of the glucose-tolerance test. *Phys. Med. Biol.* **9**: pp 203 (1964).

Andersson JM, Kim SW, Kopecek J, Knutson K. Proceedings of the sixth symposium on recent advances in drug delivery systems. *J. Controlled Release.* **28**(1-3):pp 1 (1994).

Andersson A. Analytical and pharmacokinetic aspects of the antineoplastic drug cisplatin and its monohydrated complex, Thesis, *Faculty of Pharmacy, Uppsala University* (1995).

Ariens EJ. Affinity and intrinsic activity in the theory of competitive inhibition. *Arch. Int. Pharmacodyn. Ther.* **99**:pp 32 (1954).

Ariens EJ, Simonis AM. A molecular basis for drug action: the interaction of one or more drugs with different receptors. *J. Pharm. Pharmacol.* **16**:pp 289 (1964).

Arimori K, Nakano M. Drug exsorption from blood into the gastrointestinal tract. *Pharm. Res.* **15**:pp 371 (1998).

von Bahr C, Steiner E, Koike U, Gabrielsson J. Time course of enzyme induction in humans: Effect of pentobarbital on nortriptyline metabolism. *Clin. Pharmacol. Ther.*, **64**:pp 18 (1998).

Barlow RB. Use of an antagonist for estimating the degree of agonist stimulation during physiological release. *TiPS*, **16**:pp 114 (1995).

Bass L, Keiding S. Physiologically based models and strategic experiments in hepatic pharmacology. *Biochem. Pharmacol.* **37**:pp 1425 (1988).

Benet LZ. General treatment of linear mammillary models with elimination from any compartment as used in pharmacokinetics. *J. Pharm. Sci.* **61**:pp 536 (1972).

Benet LZ, Galeazzi RL. Noncompartmental determination of the steady-state volume of distribution. *J. Pharm. Sci.* **48**:pp 1071 (1979).

Benet LZ, Massoud N, Gambertoglio JG. Pharmacokinetic basis for drug treatment,

Raven Press, New York (1985).

Berlin CM, Schimke RT. Influence of turnover rates on the responses of enzymes to cortisone. *Mol. Pharmacol.* **1**:pp 149 (1965).

Berman M. Kinetic analysis of turnover data. *Prog. Biochem. Pharmacol.* **15**:pp 67 (1979).

de Boer AG. Drug absorption enhancements: Concepts, possibilities, limitations and trends. *Harwood academic publishers* (1994).

Black JW, Leff P. Operational models of pharmacological agonism. *Proc. R. Soc. Lond. B* **220**:pp 141 (1983).

Bonate P, Howard X. Critique of prospective allometric scaling: Does the emperor have clothes?. *J. Clin. Pharmacol.* **40**:pp 335 (2000).

Boxenbaum H. Interspecies scaling, allometry, physiological time and the ground plan for pharmacokinetics. *J. Pharmacokin. Biopharm.* **10**:pp 201 (1982).

Boxenbaum H, D'Souza RW. Interspecies pharmacokinetic scaling: Biological design and neoteny. *Adv. Drug Research* **19**:pp 139 (1990).

Boxenbaum H. Pharmacokinetics: Philosophy of Modeling. *Drug Metabolism Rev.* **24**(1):pp 89 (1992).

Boxenbaum H, DiLea C. First-time-in-human dose selection: Allometric thoughts and perspectives. *J. Clin. Pharmacol.* **35**:pp 957 (1995).

Boxenbaum H. Interspecies pharmacokinetic scaling: Desultory Reflections, In: Pharmacokinetic/Pharmacodynamic analysis: Accelerating drug discovery and development. *Biomedical Library Series, IBC, Inc.* (1996).

Breimer DD, Danhof M. Relevance of the application of pharmacokinetic-pharmaco-dynamic modelling concepts in drug development. *Clin. Pharmacokinet.* **32**:pp 259 (1997).

Brody S. Bioenergetics and Growth. *Reinhold*, New York (1945).

Bäärnhielm C, Dahlbäck H, Skånberg I. In vivo pharmacokinetics of felodipine predicted from in vitro studies in rat, dog, and man. *Acta Pharm. Toxicol.* **59**: pp113 (1986).

Calabrese EJ. Principles of animal extrapolation. *Lewis Publishers*, Chelsea MI (1991).

Calder III WA. Scaling of physiological processes in homeothermic animals. *Ann. Rev.*

Physiol. **43**:pp 301 (1981).

Casati C, Monopoli A, Forlani A, Bonizzoni E, Ongini E. Telemetry monitoring of hemodynamic effects induced over time by adenosine agonists in spontaneously hypertensive rats.*J. Pharmacol. Exp. Ther.*, **275**:pp 914 (1995).

Ceresa F, Ghemi F, Martini PF, Martino P, Segre G and Vitelli A. Control of blood glucose in normal, diabetic and obese subjects: Studies by compartmental analysis and digital computer technics. *Diabetes.* **17**:pp 570 (1968).

Chakraborty A, Krzyzanski W, Jusko WJ. Mathematical modeling of circadian cortisol concentrations using indirect response models: Comparison of several methods. *J. Pharmacokin. Biopharm.* **27**:pp 23 (1999).

Cheng Y-C, Prusoff WH. *Biochem. Pharmacol.* **22**:pp 3099 (1973).

Cheng H, Jusko W. Noncompartmental Determination of the Mean Residence Time and Steady-State Volume of Distribution during Multiple Dosing, *J. Pharm. Sci.* **80**:pp 202 (1991).

Cheung WK, Levy G. Comparative pharmacokinetics of coumarin anticoagulants XLIX: Nonlinear tissue distribution of S-warfarin in rats, *J. Pharm. Sci.* **78**:pp 541 (1989).

Cobelli C, Federspiel G, Pacini G, Salvan A, Scandellari C. An integrated mathematical model of the dynamics of blood glucose and its hormonal control. *Mathematical Biosciencees.* **58**:pp 605 (1981).

Colburn WA. Simultaneous pharmacokinetic and pharmacodynamic modeling. *J. Pharmacokin. Biopharm.* **9**:pp 367 (1981).

Colburn WA. A time-dependent volume of distribution term used to describe linear concentration-time profiles. *J. Pharmacokin. Biopharm.* **11**:pp 389 (1983).

Colburn WA, Eldon MA. Simultaneous pharmacokinetic/Pharmacodynamic modeling. In: Pharmacodynamics and drug development: Perspectives in clinical pharmacology. Edited by Cutler NR, Sramek JJ, Narang PK. *Wiley*, New York (1994).

Colburn WA. Clinical markers and endpoints in bioequivalence assessment. *Drug Information Journal.* **29**:pp 917 (1995).

Collier HOJ. Tolerance, physical dependence and receptors: A Theory of the genesis of tolerance and physical dependence through drug-induced changes in the number of receptors. *Adv. Drug Res.* **3**:pp 171 (1966); In Scientific basis of drug dependence (ed. Steinberg, pp 49, *Churchill*, London (1969).

Collier PS. Some considerations on the estimation of steady state apparent volume of distribution and the relationships between volume terms *J. Pharmacokin. Biopharm.* **11**:pp 93 (1983)

Creasy GW, Jaffe ME. Endocrine/reproductive pulsatile delivery systems. *Advanced Drug Delivery Reviews.* **6**:pp 51 (1991).

Cutler NR, Sramek JJ and Narang PK. Pharmacodynamics and drug development: Perspectives in clinical pharmacology, 1st ed. *John Wiley and Sons,* Chichester (1994).

Dahlström B, Paalzow LK, Segre G *et al.* Relation between morphine pharmaco-kinetics and analgesia. *J. Pharmacokin. Biopharm.* **6**:pp 41 (1978).

Davies B, Morris B. Physiological parameters in laboratory animals and humans. *Pharm.Res.* **10**:pp1093 (1993).

Davson H, Welch K, Segal MB. Secretion of cerebrospinal fluid. In: The Physiology and pathophysiology of the cerebrospinal fluid. pp 201, *Churchill Livingstone*, London (1987).

Dayneka NL, Garg V and Jusko W, Comparison of four basic models of indirect pharmacodynamic responses, *J. Pharmacokin. Biopharm.* **21**:pp 457 (1993).

Dedrick RL. Animal Scale-up. In: *Pharmacology and Pharmacokinetics.* Edited by Teorell T, Dedrick RL, Cundliffe PG *Plenum Press,* New York (1974).

Deveney AM, Kjellström Å, Forsberg T, Jackson DM. A Pharmacological validation of radiotelemetry in conscious, freely moving rats. *J. Pharm. Toxicol. Meth.* **40**:pp 71 (1998).

Drayer DE. Pharmacodynamic and pharmacokinetic differences between drug enantiomers in humans: An overview *Clin. Pharm. Ther.* **40**:pp 125 (1986).

Dutta S, Matsumoto Y, Ebling WF. Is it possible to estimate the parameters of the sigmoid E_{max} model with truncated data typical of clinical studies?. *J. Pharm. Sci.* **85**:pp 232 (1996).

Ebling WF, Danhof M, Stanski DR. Pharmacodynamic characterization of the electroencephalographic effects of thiopental in rats. *J. Pharmacokin. Biopharm.* **19**:pp 123 (1991).

Ebling WF, Jusko W, Determination of essential clearance, volume, and residence time parameters of recirculating metabolic systems: The reversible metabolism of methylprednisolone and methylprednisone in rabbits. *J. Pharmacokin. Biopharm.* **14**:pp 557 (1986).

Ebling WF. Lecture material from 'Course on Pharmacokinetic-Pharmacodynamic Modeling: Concepts and applications'. Buffalo, May 21-24 (1995).

Ekblad EBM, Licko V. A model eliciting transient response. *Am.J.Physiol.* pp 246 (1984).

Evans WE, Schentag JJ, Jusko WJ. Applied Pharmacokinetics: Principles of therapeutic drug monitoring, third edition. *Applied Therapeutics, Inc.* Vancouver WA (1980, 1986, 1992).

Finkelstein L and Carson ER. Mathematical modelling of dynamic biological systems. *Research Studies Press Ltd.* (1985).

Gabrielsson J, *et al*. Constant rate of infusion – Improvement of tests for teratogenicity and embryotoxicity. *Life Sciences.* **37**:pp 2275 (1985).

Gabrielsson J, Larsson KS. Proposals for improving risk assessment in reproductive toxicology. *Pharmacol. Toxicol.* 66: pp 10 (1990).

Gabrielsson J. Lecture material from 'Advanced Pharmacokinetic-Pharmacodynamic Modeling: Concepts and applications'. Stockholm, December (1996).

Gabrielsson J, Weiner D. Pharmacokinetic and pharmacodynamic data analysis: Concepts and applications, 1st and 2nd ed., *Swedish Pharmaceutical Press*, Stockholm (1994, 1997).

Gabrielsson J, Wallenbeck I, Larsson G, Birgersson L, Heimer G. New kinetic data on estradiol in light of the vaginal ring concept. *Maturitas*, **22**:S35 (1995).

Gabrielsson J, Jusko W, Alari L. Modeling of dose-response-time data: Four examples of estimating the turnover parameters and generating kinetic functions from response profiles. *Biopharm. Drug Disposition.* **(in press)** (2000).

Gerlowski LE, Jain KJ. Physiologically based pharmacokinetic modeling: Principles and applications. *J.Pharm.Sci.* **72**:pp 1103 (1983).

Gero A. Intimate study of drug action III: Mechanisms of molecular drug action. In: Drill's Pharmacology in Medicine. *McGraw-Hill* (1971).

Gibaldi M, Perrier D. Pharmacokinetics, 2nd ed. Revised and Expanded. *Marcel Dekker Inc.*, New York (1982).

Gillespie WR. Noncompartmental versus compartmental modelling in clinical pharmacokinetics. *Clin. Pharmacokinet.* **20**:pp 253 (1991).

Godfrey K. Compartmental models and their application. *Academic Press*, London

(1983).

Gorrod JW, Wahren J. Nicotine and related alcaloids. *Chapman and Hall*, London (1993).

Gries JM, Benowitz N, Verotta D. Chronopharmacokinetics of nicotine. *Clin. Pharmacol. Ther.* **60**(4):pp 385 (1996).

Goyette A, Pharmacokinetics: Statistical Moment Calculations, *Arzneimittel-forschung-Drug Research* **33**(1):pp 173 (1983).

Van der Graaf PH, Schoemaker RC. Analysis of asymmetry of agonist concentration-effect curves. *J. Pharmaco. Toxicol.* **41**:pp 107 (1999).

Günther B. In Biogenesis-Evolution-Homeostasis, pp 127, Locker A ed. *Springer Verlag*, Heidelberg (1973).

Hill AV. The possible effects of the aggregation of the molecules of haemoglobin on its dissociation curves. *J.Physiol (Lond.)* **40**:iv (1910).

Holford NHG. Input from the deep south compartment: A personal viewpoint. *Clin. Pharmacokinet.* **29**: pp139 (1995).

Holford NHG, Gabrielsson J, Sheiner LB, Benowitz N, Jones R. A Physiological pharmacodynamic model for tolerance to cocaine effects on systolic blood pressure, heart-rate and euphoria in human volunteers. *Poster at Measurement and kinetics of in vivo drug effects: Advances in simultaneous pharmacokinetic/pharmacodynamic modelling. 28-30 June, 1990 Nordwijk, The Netherlands.*

Houston JB. Kinetics of disposition of xenobiotics and their metabolites. *Drug Metab. Drug Interact.* **6**:47 (1994)

Hung O, Varvel JR, Shafer SL, Stanski DR. Quantitation of thiopental anesthetic depth with clinical stimuli. *Can. J. Anesth.* **18S**:pp 37 (1990).

Iwatsubo T, Hirota N, Ooie T, Suzuki H, Sugiyama Y. Prediction of in vivo disposition from in vitro data based on physiological pharmacokinetics. *Biopharm. Drug Disposition.* **17**:pp 273 (1996).

Jenkinson DH, Barnard EA, Hoyer D, *et al* International union of pharmacology committee on receptor nomenclature and drug classification. IX. Recommendations on terms and symbols in quantitative pharmacology. *Pharmacol. Rev.* **47**:pp 225 (1995).

Jusko WJ. Pharmacodynamics of chemotherapeutic effects: Dose-time-response relationships for phase-non-specific agents. *J.Pharm.Sci.* **60**:pp 892 (1971).

Jusko WJ. Guidelines for collection and analysis of pharmacokinetic data, In: Applied Pharmacokinetics: Principles of Therapeutic Drug Monitoring, 3rd ed. Evans WE, Schentag JJ and Jusko WJ eds. *Applied Therapeutics*, Spokane WA (1980, 1992).

Jusko WJ. Corticosteroid pharmacodynamics: models for a broad array of receptor-mediated pharmacologic effects. *J. Clin. Pharmacol.* **30**:pp 303 (1990).

Jusko WJ. Lecture material from 'Course on Pharmacokinetic-Pharmacodynamic Modeling: Concepts and applications'. Buffalo, May 21-24 (1995).

Jusko WJ, Ko HC . Physiological indirect response models characterize diverse types of pharmacodynamic effects. *Clin. Pharmacol. Ther.* **56**:pp 406 (1994).

Kakkar T, Boxenbaum H, Mayersohn M. Estimation of K_i in a competitive enzyme-inhibition model: Comparisons among three methods of data analysis. *Drug Metab. Dispos.* **27**:pp 756 (1999).

Kakkar T, Pak Y, Mayersohn M. Evaluation of a minimal experimental design for determination of enzyme kinetic parameters and inhibition mechanism. *J. Pharmacol. Exp. Ther.* **293**:pp 861 (1999).

Keener J, Sneyd J. Mathematical Physiology. *Springer-Verlag,* New York (1998).

Kenakin T. Pharmacologic analysis of drug-receptor interaction, second ed. *Raven Press*, New York (1993).

Kenakin T. Molecular pharmacology: A short course. *Blackwell Science*, Cambridge (1996).

Kleiber M. Metabolic turnover rate: A physiological meaning of the metabolic rate per unit body weight. *J. Theor. Biol.* **53**:pp 199 (1975).

Lassen NA, Perl W. Tracer kinetic methods in medical physiology. *Raven Press,* New York (1979).

Labrecque G, Belanger PM. Time-dependency in the pharmacokinetics and disposition of drugs. In: Topics in Pharmaceutical sciences. Edited by Breimer DD, Speiser P. *Elsevier*, Amsterdam (1985).

Lave T, Levet-Trafit B, Schmitt-Hoffmann, Morgenroth B, Richter W, and Chou R. Interspecies scaling of interferon disposition and comparison of allometric scaling with concentration-time transformations. *J. Pharm. Sci.* **84**:pp 1285 (1995).

Lebel L. Turnover of circulating hyaluronan: Studies in man and experimental animals. *Acta Universitatis Upsaliensis.* Comprehensive summaries of Uppsala Dissertations from the Faculty of Medicine. **217** (1989).

Lemmer B. Chronopharmacological aspects of PK/PD modelling. *Int. J. Clin. Pharm. Ther.* **35**:pp 458 (1997).

Lesko LJ and Williams RL. Regulatory Perspectives: The role of pharmacokinetics and pharmacodynamics. In: Pharmacodynamics and drug development: Perspectives in clinical pharmacology, 1st ed. Edited by Cutler NR, Sramek JJ and Narang PK. *John Wiley and Sons*, Chichester (1994).

Levy G. Relationship between elimination rate of drugs and rate of decline of their pharmacologic effects. *J Pharm Sci.* **53**:pp 342 (1964).

Levy G. Kinetics of pharmacological effects. *Clin. Pharm. Ther.* **7**:pp 362 (1966).

Levy G. Effect of protein binding on the pharmacologic activity of drugs as exemplified by warfarin. In Drug-protein binding. Edited by Reidenberg MM, Errill S . *Praeger* New York (1986).

Levy G. The case for preclinical pharmacodynamics, In Integration of pharmacokinetics, pharmacodynamics, and toxicokinetics in rational drug development. Edited by Yacobi A, Shah VP, Skelley JP, Benet LZ. pp7 *Plenum Press* New York (1993).

Levy G. Mechanism-based pharmacodynamic modeling. *Clin. Pharmacol. Ther.* **56**:pp 356 (1994).

Levy G. Lecture material from 'Course on Pharmacokinetic-Pharmacodynamic Modeling: Concepts and applications', Buffalo, May 21-24 (1995).

Levy RH. Dumain MS. Time-dependent kinetics VI: direct relationship between equations from drug levels during induction and those involving constant clearance. *J. Pharm. Sci.* **68**: pp 398 (1979).

Levy RH. Time-dependent pharmacokinetics In: *Pharmacokinetics: Theory and Methodology* Edited by Rowland M, Tucker G. *Pergamon Press* (1982).

Lew KH, Ludwig EA, Milad MA, Donovan K, Middleton E, Ferry JJ, Jusko WJ. Gender-based effects on methylprednisolone pharmacokinetics and pharmacodynamics. *Clin. Pharmacol. Ther.,* **54**:pp 402 (1993).

Lundström J, Lindgren J-E, Ahlenius S, Hillegaart V. Relationship between brain level of 3-(3-hydroxyphenyl)-N-n-propylpiperidine HCL enantiomers and effects on locomotor activity in rats. *J. Pharmacol. Exp. Ther.* **262**:pp 41 (1992).

Matsumoto M *et al.* Regional blood flows measured in Mongolian gerbil by a modified microsphere method. *Am. J. Physiol.* **242**:H990 (1982).

Matthews JC, Fundamentals of receptor, enzyme and transport kinetics, *CRC Press,* Bocan Raton (1993).

McMahon TA and Bonner JT. On size and Life. *Scientific American Library*, New York (1983).

Moore-Ede MC, Sulzman FM, Fuller CA. The clocks that time us: Physiology of the circadian timing system. *Harvard University Press*, Cambridge Massachusetts (1982).

Morgan DJ, Smallwood RA. Clinical significance of pharmacokinetic models of hepatic elimination. *Clin. Pharmacokinet.* **18**:pp 61 (1990).

Nagashima R, O'Reilly RA, Levy G. Kinetics of pharmacologic effects in man: The anticoagulant action of warfarin. *Clin. Pharmacol. Ther.* **10**:pp 22 (1969).

Nakashima E, Benet LZ. General treatment of mean residence time, clearance, and volume parameters in linear mamillary models with elimination from any compartment. *J. Pharmacokin. Biopharm.* **16**:pp 475 (1988).

Nakashima E, Benet LZ. An integrated approach to pharmacokinetic analysis for linear mammillary systems in which input and exit may occur in/from any compartment. *J. Pharmacokin. Biopharm.* **17**:pp 673 (1989).

Nau H. *et al.* A New model for embryotoxicity testing: Teratogenicity and pharmacokinetics of valproic acid following constant-rate administration in the mouse using human therapeutic drug and metabolite concentrations. *Life Sciences.* **29**:pp 2803 (1981).

Nimmo IA, Atkins GL. Methods for fitting equations with two or more non-linear parameters. Biochem. J. **157**:pp 489 (1976).

Norberg Å, Gabrielsson J, Jones AW, Hahn RG. Within-and between-subject variations in pharmacokinetic parameters of ethanol by analysis of breath, venous blood and urine. *Br. J. Clin. Pharmacol.* **49**:pp 399 (2000).

Obach RS. Predictions of human pharmacokinetics using *In vitro-in vivo* correlations, In Pharmacokinetic/Pharmacodynamic analysis: Accelerating drug discovery and development. *Biomedical Library Series, IBC, Inc.* (1996).

Paalzow L, Edlund PO. Multiple receptor response: a new concept to describe the relationship between pharmacological effects and pharmacokinetics of a drug in studies on clonidine in the rat and the cat. *J. Pharmacokin. Biopharm.* **7**:pp 495 (1979).

Pang KS. A review of metabolic kinetics. *J. Pharmacokin. Biopharm.* **13**:pp 633 (1985).

Park BK, Kitteringham NR, Pirmohammed M, Tucker GT. Relevance of induction of human drug-metabolizing enzymes: Pharmacological and toxicological implications. *Br. J. Clin. Pharmacol.* **41**:pp 477 (1996)

Paton WDM. A Theory of drug action based on the rate of drug-receptor combination. *Proc. R. Soc. London* [Biol.] **152**:pp 21 (1961)

Perrier D and Mayersohn M. Noncompartmental determination of steady-state volume of distribution for any mode of administration. *J. Pharm. Sci.* **71**:pp 372 (1982).

Peters RH. The ecological implications of body size. *Cambridge University Press,* Cambridge (1983).

Peeters LLH, Grutters G, Martin CB. Distribution of cardiac output in the unstressed pregnant guinea pig. *Am.J.Obstet.Gynecol.* **138**:pp 1177 (1980).

Pidgeon C, Pitlick WH. Unique approach for calculation of first-order absorption rate constants from blood or urine. *J. Pharmacokin. Biopharm.* **8**:pp 203 (1980)

Pitlick WH, Levy RH. Time-dependent pharmacokinetics: I. Exponential autoinduction of carbamazepin in monkeys. *J. Pharm. Sci.* **66**:pp 647 (1977).

Pitsui M, Parker E. Aarons L and Rowland M. Population pharmacokinetics and pharmacodynamics of warfarin in healthy young adults. *Eur. J. Pharm. Sci.* **1**:pp 151 (1993).

Posen S. Turnover of circulating Enzymes. *Clinical Chemistry.* **16**:pp 71 (1970).

Purves RD. Anomalous parameter estimates in the one-compartment model with first-order absorption. *J. Pharm. Pharmacol.* **45**:pp 934 (1993).

Rescigno A, Segre G. Drug and tracer kinetics. *Blaisdell Publishing Company,* London (1961, 1966).

Riggs DS. Mathematical approach to physiological problems. *Williams and Wilkins,* Baltimore, pp 208 (1963).

Roberts MS, Rowland M. Hepatic elimination-dispersion model. *J.Pharm.Sci.* **74**:pp 585 (1985).

Roberts MS, Rowland M. A dispersion model of hepatic elimination. 1. Formulation of the model and bolus considerations. *J. Pharmacokin. Biopharm.* **14**:pp 226 (1986).

Roberts MS, Rowland M. A dispersion model of hepatic elimination. 2. Steady-state considerations. Influence of blood flow, protein binding and hepatocellular enzymatic activity. *J. Pharmacokin. Biopharm.* **14**:pp 261 (1986).

Roberts MS, Rowland M. A dispersion model of hepatic elimination. 3. Application to metabolite formation and elimination kinetics. *J. Pharmacokin. Biopharm.* **14**:pp 289 (1986).

Roberts MS, Donaldson JD, Rowland M. Models of hepatic elimination: Comparison of stochastic models to describe residence time distributions and to predict the influence of drug distribution, enzyme heterogeneity, and systemic recycling on hepatic elimination. *J. Pharmacokin. Biopharm.* **16**:pp 41 (1988).

Robinson ICAF. Chronopharmacology of growth hormone and related peptides. *Advanced Drug Delivery Reviews.* **6**:pp 57 (1991).

Ross EM. Pharmacodynamics: Mechanisms of drug action and the relationship between drug concentration and effect. In Goodman Gilman's The Pharmacological basis of therapeutics. 10th edition. *Pergamon Press*, New york (1996).

Rovati GE. Ligand-binding studies: old beliefs and new strategies. *TiPS.* **19**:pp 365 (1998).

Rowland M and Tozer T. Clinical Pharmacokinetics: Concepts and Applications, 3rd ed. *Lea & Febiger* New York (1995).

Sacher GA. Relation of lifespan to brain weight and body weight in mammals, In: The Lifespan of animals, Edited by Wolstenholme GEW, O'Connor M. pp 115. *Little Brown & Co.*, Boston (1959).

Schoenwald RD, Smolen VF. Drug-Absorption analysis from pharmacological data II: Transcorneal biophasic availability of tropicamide. *J. Pharm. Sci.* **60**:pp 1039 (1971).

Schmidt-Nielsen K. Why is animal size so important? *Cambridge University Press*, Cambridge (1996).

Schwilden H, Stoeckel H, Schüttler J, Lauven PM. Pharmacological models and their use in clinical anaesthesia. *Eur. J. Anaesth.* **3**:pp 175 (1986).

Schüttler J, Stanski DR, White PF, White PF, Trevor AJ, Horai Y, Verotta D, Sheiner LB. Pharmacodynamic modeling of the EEG effects of ketamine and its enantiomers in man. *J. Pharmacokin. Biopharm.* **15**:pp 241 (1987).

Segre G. Kinetics of interaction between drugs and biological systems. *Il Farmaco* **23**:pp 907 (1968).

Segre G, Turco GL and Vercellone G. Modeling blood glucose and insulin kinetics in normal and diabetic subjects. *Diabetes.* **22**:pp 94 (1970).

Shahin M, Iyengar SS, Rao RM. Computers in simulation and modeling of complex biological systems. *CRC Press*, Boca Raton (1985).

Sharma A, Ebling WF, Jusko WJ. Indirect pharmacodynamic response models for tolerance rebound phenomena. *Pharm. Res.* **12**: S364 (1995).

Sheiner LB, Stanski DR, Vozeh S, Miller RD and Ham J. Simultaneous modelling of pharmacokinetics and pharmacodynamics: Application to d-tubocurarine. *Clin. Pharmacol. Ther.* **25**:pp 358 (1979).

Smolen VF. Quantitative determination of drug bioavailability and biokinetic behavior from pharmacological data for ophtalmic and oral administration of a mydriatic drug. *J. Pharm. Sci.* **60**:pp 354 (1971).

Smolen VF and Weigand WA. Drug bioavailability and pharmacokinetic analysis from pharmacological data. *J. Pharmacokin. Biopharm.* **1**:pp 329 (1973).

Smolen VF. Theoretical and computational basis for drug bioavailability determinations using pharmacological data I: General considerations and procedures, *J. Pharmacokin. Biopharm.* **4**:pp 337 (1976).

Stephenson RP. A modification of receptor theory. *Br. J. Pharmacol.* **11**:pp 379 (1956).

Struyker_Boudier HAJ, Smits JFM, Schoemaker RG, Janssen BJA. Dynamics of cardiovascular drug action. *Lecture at Measurement and kinetics of in vivo drug effects: Advances in simultaneous pharmacokinetic/pharmacodynamic modelling. 28-30 June, 1990 Nordwijk, The Netherlands.*

Sturis J. Polonsky KS, Mosekilde E, van Cauter E. Computer model for mechanisms underlying ultradian oscillations of insulin and glucose. *Am. J. Physiol.* **260**:pp E801 (1991).

Sun Y-N, Jusko WJ. Role of baseline parameters in determining indirect pharmacodynamic response. *J. Pharm. Sci.* **88**:pp 987 (1999).

Takada K, Asada S. A model independent approach to describe the blood disappearance profile of intravenously administered drugs. *Chem. Pharm. Bull.* **29**:pp 1462 (1981).

Taylor JL, Mayer RT, Himel CM. Conformers of acetylcholinesterase: A mechanism of allosteric control. *Molec. Pharmacol.* **45**:pp 74 (1994).

Tozer TN, Winter ME. Phenytoin, In: Applied Pharmacokinetics: Principles of Therapeutic Drug Monitoring, 3rd ed. Edited by Evans WE, Schentag JJ, Jusko WJ. *Applied Therapeutics*, Spokane WA (1980, 1986, 1992).

Urquhart J, Li CC. Dynamic testing and modeling of adrenocortical secretory function. *Ann. New York Acad. Sci.* **156**:pp 756 (1969).

Urquhart J. Physiological actions of adrenocorticotropic hormone. In The Pituitary gland and its neuroendocrine control, Part 2. ed. Knobhil E and Sawyer WH. *Am. Physiol. Soc.* Washington DC. pp 133 (1974).

Van Boxtel CJ, Holford NHG, Danhof M (eds). The *in vivo* study of drug action. *Elsevier* (1992).

Van der Graaf PH, Danhof M. On the reliability of affinity and efficacy estimates obtained by direct operational fitting of agonist concentration-effect curves following irreversible receptor inactivation. *J. Pharmacol. Toxicol. Meth.* **38**:pp 81 (1997).

Van Rossum JM, van Lingen G. Dose-dependent pharmacokinetics, *Pharmacol. Ther.* **21**:pp 77 (1983).

Vaughan DP. A model independent method for estimating the *in vivo* release rate constant of a drug from its oral formulation. *J. Pharm. Pharmacol.* **28**:pp 505 (1976).

Verotta D, Sheiner LB. Semiparametric analysis of non-steady-state pharmacodynamic data. *J. Pharmacokin. Biopharm.* **19**:pp 691 (1991).

Wagner JG. Linear pharmacokinetic equations allowing direct calculation of many needed pharmacokinetic parameters from coefficients and exponents of poly-exponential equations which have been fitted to data. *J. Pharmacokin. Biopharm.* **4**:pp 443 (1976).

Wagner JG. Kinetics of pharmacological response I. Proposed relationships between response and drug concentration in the intect animal and man. *J. Theor. Biol.* **20**:pp 171 (1968).

Wagner JG. Aghajanian GK, Bing OHL. Correlation of performance test scores with tissue concentration of lysergic acid diethylamide in human subjects. *Clin. Pharmacol. Ther.* **9**:pp 635 (1968).

Wagner JG. Pharmacokinetics for the pharmaceutical scientist. *Technomic Publishing* (1993).

Wakelkamp M, Alvan G, Gabrielsson J, Paintaud G. Pharmacodynamic modeling of furosemide tolerance after multiple intravenous administration. *Clin. Pharmacol. Ther.* **60**:pp 75 (1996).

Waldmann TA, Strober W, Blaese RM. Variations in the metabolism of immunoglobulins measured by turnover rates, In: Immunoglobulins. Biological aspects and clinical uses. *Natl. Acad. Sci.* pp 33 (1970).

Wang Y *et al.* A Double-peak in the pharmacokinetics of alprazolam after oral administration. *Drug Metab. Dispos.* **27**:pp 855 (1999).

Wang Y-M C, Reuning RH. An experimental design strategy for quantitating complex pharmacokinetic models: Enterohepatic circulation with time-varying gallbladder emptying as an example. *Pharm. Res.* **9**:pp 169 (1992).

Watari N, Benet LZ. Determination of mean input time, mean residence time, and steady-state volume of distribution with multiple drug inputs. *J. Pharmacokin. Biopharm.* **17**(1):pp 593 (1989).

Welling P, Tse F. Pharmacokinetics: Regulatory, Industrial, Academic Perspectives, *Marcel Dekker,* New York (1988).

Wenthold RJ, Mahler HR, Moore WJ. The half-life of acetylcholinesterase in mature rat brain. *J. Neurochemistry.* **22**:pp 941 (1974).

Whitlock JP, Denison MS. Induction of cytochrome P450 enzymes that metabolize xenobiotics, In Ortiz de Montellano PR. Cytochrome P450: Structure, Mechanisms, and Biochemistry, 2nd ed. *Plenum Press,* New York (1995).

Wilkinson GR, Shand DG. A physiological approach to hepatic drug clearance. *Clin. Pharm. Ther.* **18**:pp 377 (1975).

Wilkinson PK, Sedman AJ, Sakmar E, Earhart RH, Weidler DJ, Wagner JG. Blood ethanol concentrations during and following constant-rate intravenous infusion of alcohol. *Clin. Pharmacol. Ther.* **19**:pp 213 (1976).

Wills RJ. Basic pharmacodynamic concepts and models, In: Pharmacodynamics and drug development: Perspectives in clinical pharmacology, 1st ed. Edited by Cutler NR, Sramek JJ and Narang PK. *John Wiley and Sons,* Chichester (1994).

Winter ME. Basic clinical pharmacokinetics, 3rd ed. *Applied Therapeutics*, Vancouver WA (1994).

Data analysis

Akaike A. Posterior Probabilities for choosing a regression model, *Annal. Instit. Math. Stats.* **30**:pp A9 (1978).

Allen D, Cady F. Analyzing experimental data by regression, *Lifetime Learning Publications*, Belmont, CA (1982).

Allen DM. Parameter estimation for nonlinear models with emphasis on compartmental models, *Biometrics,* **39**:pp 629 (1981).

Anscombe FJ, Tukey JN. The examination and analysis of residuals, *Technometrics,* **5**:pp 141 (1960).

Atkins GL. Weighting functions and data truncation in the fitting of multi-exponential functions, *Biochem. J.* **138**:pp 125 (1974).

Atkinson AC. A note on the generalized information criterion for choice of a model, *Biometrika.* **67**:pp 2 (1980).

Atkinson AC. Plots, transformations and regression, *Clarendon Press,* Oxford (1985).

Barlow RB. Foundations of Pharmacology. Ash Lea Cottage (1991).

Bard Y. Nonlinear Parameter Estimation, *Academic Press*, NY (1974).

Bates D, Watts D. Nonlinear Regression Analysis and Its Applications, *John Wiley & Sons*, New York, 1988

Beal SL, Sheiner LB. Heteroscedastic nonlinear regression, *Technometrics,* **30**:pp 327 (1988).

Beck JJ, Arnold KJ. Parameter Estimation, *John Wiley & Sons*, New York (1977).

Belsey DA, Kuh E and Welsch RE. Regression diagnostics, *John Wiley & Sons*, New York (1982).

Carroll RJ, Rupert D. Transformation and weighting in regression, *John Wiley & Sons*, New York (1988).

Carson ER, Cobelli C, Finkelstein L. The mathematical modeling of metabolic and endocrine systems. *Wiley*, New York (1983).

Chambers JM. Fitting nonlinear models: Numerical techniques, *Biometrika,* **60**:pp 1 (1973).

Chamberlain R. Computer systems validation for the pharmaceutical and medical device industries, Second edition, *ALAREN Press,* Libertyville (1994).

Christopoulos A. Assessing the distribution of parameters in models of ligand-receptor interaction: to log or not to log. *TiPS.* **19**:pp 351 (1998).

Cobelli C, DiStefano JJ III. Parameters and structural identifiability concepts and ambiguities: A critical review and analysis, *Am. J. Physiol.* **239**:pp R7 (1980).

Cook RD, Weisberg S. Residuals and influence in regression, *Chapman & Hall,* New York (1982).

Cook RD, Weisberg S. An Introduction to regression graphics, *John Wiley & Sons*, New York (1994).

Daniel C, Woods FS. Fitting equations to data. *Wiley Interscience,* New York 1971.

DiStefano JJ III, Landaw EM. Multiexponential, multicompartmental and noncompartmental modelling I. Methodological limitations and physiological interpretations. *Am. J. Physiol.* **246**: R 651 (1984)

Double ME, McKendry M. Computer validation compliance: A quality assurance perspective, *Interpharm Press,* Buffalo Grove (1994).

Draper NR, Smith H. Applied regression analysis, 3rd Ed., *John Wiley & Sons,* New York (1998).

Ebling WF. Lecture material from 'Course on Pharmacokinetic-Pharmacodynamic Modeling: Concepts and applications', Buffalo, May 21-24 (1995).

Endrenyi L. Kinetic data analysis: design and analysis of enzyme and pharmacokinetic experiments. *Plenum Press*, New York (1981).

Fletcher R. Practical methods of optimization, 2nd Ed., *John Wiley & Sons*, New York (1981)

Fox T, Hinkley D, Lantz K. Jacknifing in nonlinear regression, *Technometrics,* **22**:pp 29 (1980).

Gill PE, Murray W, Wright MH. Practical optimization, *Academic Press* (1981).

Giltinan DH, Ruppert D. Fitting heteroscedastic regression models to individual pharmacokinetic data using standard statistical software, *J. Pharmacokin. Biopharm.* **17**:pp 601 (1989).

Haefner JW. Modeling biological systems: Principles and applications. *Chapman and Hall* , London (1996).

Hartley H. The Modified Gauss-Newton Method for the Fitting of Nonlinear Regression Functions by Least Squares, *Technometrics,* **3**:pp 269 (1961)

Hocking RR. The analysis and selection of variables in linear regression, *Biometrics,* **32**:pp 1 (1976).

Hosmer DW, Lemeshow S. Applied Logistic Regression, *John Wiley & Sons*, New York (1989).

Jennrich RI, Moore RH. Maximum likelihood estimation by means of nonlinear least squares, *Amer. Stat. Assoc. Proceedings Statistical Computing Section*: pp 57 (1975).

Jennrich RI, Ralston ML. Fitting nonlinear models to data, *Ann. Rev. Biophys. Bioeng.* **8**:pp 195 (1979).

Johnson LE. Computers, models and optimization in physiological kinetics, *CRC-Critical reviews in Bioengineering*, **2**:pp 1 (1974).

Landaw EM, DiStefano JJ III. Multiexponential, multicompartmental and non-compartmental modelling II. Data analysis and statistical considerations, *Am. J. Physiol.* **246**:R 665 (1984)

Landaw EM. Optimal multicompartmental sampling designs for parameter estimation: Practical aspects of the identification problem, *Math. Comput. Simul.* **24**:pp 525 (1982).

Langenbücher. Numerical convolution/deconvolution as a tool for correlating in vitro with in vivo drug availability. *Pharm. Ind.* **44**:pp 1166 (1982).

Langenbücher. Improved understanding of convolution algorithms correlating body response with drug input. *Pharm. Ind.* **44**:pp 1275 (1982).

Langenbücher, Möller. Correlation of in vitro drug release with in vivo response kinetics. *Pharm. Ind.* **45**:pp 623 (1983).

Lawson CL, Hanson RJ. Solving least squares problems, *Prentice Hall*, New Jersey (1974).

Leatherbarrow RJ. Using linear and nonlinear regression to fit biochemical data. *Trends in Biochemical Sciences*, **15**:pp 455 (1990).

Madden FN, Godfrey KR, Chappell MJ, Hovroka R, Bates R. A comparison of six deconvolution techniques, *J. Pharmacokin. Biopharm.* **24**:pp 282 (1996).

Mannervik B. Design and analysis of kinetic experiments for discrimination between rival models. In: Kinetic Data Analysis: Design and Analysis of Enzyme and Pharmacokinetic Experiments, Edited by Endrenyi L., *Plenum Press*, New York (1981).

Metzler C, Tong DDM. Computational problems of compartmental models with Michaelis-Menten-type elimination. *J. Pharm. Sci.* **70**:pp 733 (1981).

Metzler C. Extended Least Squares (ELS) for Pharmacokinetic Models, *J. Pharm. Sci.* **76**(7) (1987).

Metzler CM. Estimation of pharmacokinetic parameters: Statistical considerations, In 'Pharmacokinetics: Theory and Methodology' Edited by Rowland M and Tucker GT. Chapter **15**:pp 407, *Pergamon Press*, Oxford (1986)

Nelder J, Mead R. A Simplex Method for Function Minimization, *Computing Journal*,

7:pp 308 (1965)

Peck CC, D'Argenio DZ, Rodman JH. Analysis of clinical pharmacokinetic data for individualizing drug dosage regimens, In: Applied Pharmacokinetics: Principles of Therapeutic Drug Monitoring, 3rd ed. Edited by Evans WE, Schentag JJ, Jusko WJ. *Applied Therapeutics*, Spokane WA (1980, 1992).

Peck CC, Sheiner LB, Nichols AI. Extended least squares nonlinear regression: A possible solution to the 'choice of weights' problem in analysis of individual pharmacokinetic data. *J. Pharmacokin. Biopharm.* **12**:pp 545 (1984).

Peck CC, Sheiner LB, Nichols AI. The problem of choosing weights in nonlinear regression analysis of pharmacokinetic data. *Drug. Metab. Rev.* **15**:pp 133 (1984).

Purves RD. Anomalous parameter estimates in the one-compartment model with first-order absorption. *J. Pharm. Pharmacol.* **45**:pp 934 (1993).

QWERT®: A convolution/deconvolution software package for the personal computer, *SI-Computing*, Uppsala, Sweden (1994)

Ratkowski DA. Nonlinear regression modelling, *Marcel Dekker Inc.*, New York (1983).

Schwarz G. Estimating the dimension of a model. *Annals. Stats.* **6**:pp 461 (1978).

Serber GAF, Wild CJ. Chapter 3.4. Commonly encountered problems-Identifiability and Ill-Conditioning. In: Nonlinear Regression. *John Wiley and sons.* New York (1989).

Sheiner LB, Beal S. Pharmacokinetic Parameter estimates from several least squares Procedures: Superiority of extended least squares, *J. Pharmacokin. Biopharm.* **13**:pp 185 (1985).

Snee RD. Validation of regression models: Methods and examples, *Technometrics,* **18**:pp 415 (1977).

Stokes T, Branning RC, Chapman KG, Hambloch H, Trill AJ. Good computer validation practice; Common sense implementation, *Interpharm Press,* Buffalo Grove (1994).

Tukey JW. Exploratory data analysis, *Addison-Wesley*, Massachusetts (1977).

Van Houwelingen J. Use and abuse of variance models in regression, *Biometrics* **44**:pp1073 (1988).

Index

Symbols and their definitions

Symbol	Definition or Explanation (units)
a, b, c, d	Allometric constants (arbitrary units)
A	Amplitude
A	Ligand concentration (e.g., µmol/L)
A, B, C	Coefficients in the sum of exponentials (e.g., µmol/L)
α, β, γ	Slope factors or exponents (e.g., h^{-1})
A_b	Amount in body (e.g., µmol)
A_e	Cumulative amount excreted unchanged into urine (e.g., µmol)
A_e	Amount in effect-compartment (e.g., µmol)
A_{ev}	Amount in biophase compartment after extravascular dosing (e.g., µmol)
α_i	The logit of the probability that a certain stimulus yields a response
Amount	e.g., µg, mg, g or µmol, mmol, mol
A_p	Amount in plasma (e.g., µmol)
A_s	Shifted ligand concentration (e.g., µmol/L)
A_{ss}	Amount at steady state (e.g., µmol)
A_t	Amount in tissue (e.g., µmol)
A_u	Amount in urine (e.g., µmol)
AIC	Akaike Information Criterion
AUC	Area Under the (zero moment curve) Curve (e.g., µmol · h/L)
AUC_{extr}	Extrapolated Area Under the Curve (e.g., µmol · h/L)
AUC_E	Area Under the effect Curve (arbitrary units)
AUC_R	Area Under the Response Curve (arbitrary units) or Area due to rebound
AUC_0^∞	Area Under the Curve from zero to infinity (e.g., µmol · h/L)
$AUMC$	Area Under the first Moment Curve (e.g., µmol · h^2/L)
$AUMC_0^\infty$	Area Under the first Moment Curve from zero to infinity (e.g., µmol · h^2/L)
$AUMC_{extr}$	Extrapolated Area Under the first Moment Curve (e.g., µmol · h^2/L)
b	Allometric exponent
B_{max}	Maximum amount bound
bid	bis in diem, twice a day
BW	Body weight (usually g or kg)
BZP	Benzodiazepine
C	Concentration (e.g., µmol/L)
$C(t)$	Concentration at time t (e.g., µmol/L)
$C_1 (= C)$	Concentration in central compartment (e.g., µmol/L)
C_{ave}	Average concentration (e.g., µmol/L)
C_e	Effect-compartment concentration (e.g., µmol/L)
C_{in}	Incoming concentration (e.g., µmol/L)
C_{iv}	Concentration due to intravenous dose (e.g., µmol/L)
C_{last}	Terminal concentration (e.g., µmol/L)
C_M	Metabolite concentration (e.g., µmol/L)

C_{max}	Maximum concentration (e.g., µmol/L)
C_{out}	Outgoing blood concentration (e.g., µmol/L)
C_P	Plasma concentration
C_{po}	Concentration due to oral dosing (e.g., µmol/L)
C_{SS}	Concentration at steady state (e.g., µmol/L)
C_t	Tissue concentration (e.g., µmol/L)
C_u	Unbound concentration (e.g., µmol/L)
C_{ub}	Unbound blood concentration (e.g., µmol/L)
C_{uss}	Unbound concentration at steady state (e.g., µmol/L)
C^I	Back-extrapolated intercept on the concentration axis (e.g., µmol/L)
C^0	Concentration at time zero (e.g., µmol/L)
CATD	Computer Assisted Trial Design
Cl	Clearance (e.g., $L \cdot h^{-1}$)
Cl_d	Inter-compartmental distribution (e.g., $L \cdot h^{-1}$)
Cl_E	Denotes clearance in the induction-feedback model (e.g., $L \cdot h^{-1}$)
Cl_H	Hepatic organ clearance (e.g., $L \cdot h^{-1}$)
Cl_{int}	Intrinsic clearance (e.g., $L \cdot h^{-1}$)
Cl_M	Metabolic clearance also formation clearance of metabolite (e.g., $L \cdot h^{-1}$)
Cl_{ME}	Metabolic clearance of the metabolite (e.g., $L \cdot h^{-1}$)
Cl_o	Oral clearance (e.g., $L \cdot h^{-1}$)
Cl_R	Renal clearance (e.g., $L \cdot h^{-1}$)
Cl_u	Unbound clearance (e.g., $L \cdot h^{-1}$)
Cl_{ub}	Unbound blood clearance (e.g., $L \cdot h^{-1}$)
CSV	Computer System Validation
CV	Coefficient of variation also called relative standard deviation
Δ	Difference
D	Dose or daily
D_{ev}	Extravascular (e.g., oral, subcutaneous) dose (e.g., µmol)
D_{inf}	Infusion dose (e.g., µmol)
D_{iv}	Intravenous dose (e.g., µmol)
D_{po}	Oral dose (e.g., µmol)
D_{sc}	Subcutaneous dose (e.g., µmol)
df	Degrees of freedom ($N_{obs} - N_{par}$)
ε	Residual error or measurement error or random unobservable error
ε	Intrinsic activity
E	Effect variable or Extraction ration
E_0	Baseline effect
E_H	Extraction ratio across the liver
E_{max}	Maximum drug-induced effect (arbitrary units)
EA_{50}	Concentration of competing agonist at 50% of maximal effect (e.g., µmol/L)
EC_{50}	Concentration at 50% of maximal effect (e.g., µmol/L)
EC_{51} and EC_{52}	EC_{50} at two different receptors or two different compounds (e.g., µmol/L)

EDA	Exploratory Data Analysis
EEG	Electroencephalogram
ELS	Extended least squares
E_v	Extravascular
F	Bioavailability
f_e	Fraction of dose excreted via urine
F_H	Bioavailability across the liver
F_i	Fraction of binding to the ith site
F_{inf}	Infusion rate (e.g., µmol/min or µmol/h)
f_m	Fraction metabolized
F_{table}	F-value obtained from the F-table by means of df and Δdf
f_u	Free fraction
F^*	Calculated F-value from $WRSS$, df, and Δdf
FDA	Food and Drug Administration
GLS	Generalized Least Squares
H	Hepatic
$H(A)$	Inhibition or stimulation function based on the biophase amount
$H(t)$	Inhibition or stimulation function
i	Measurement number e.g., C_1 = first concentration measurement
$I(C)$	Inhibition function, also written as $I(D)$ or $I(t)$
I_{max}	Maximum drug induced inhibition
IC_{50}	Concentration producing 50% of maximal inhibition (e.g., µmol/L)
IC_{51} and IC_{52}	IC_{50} at two different receptors or for two different compounds (e.g., µmol/L)
ICH	International Conference on Harmonization
ID_{50}	Dose producing 50% of maximal inhibition (e.g., µmol)
In	Input or incoming (arbitrary units)
$IRLS$	Iterative reweighted least squares
IRP	Indirect response
iv	Intravenous
K	Elimination rate constant (e.g., h^{-1}) also called K_e
K	Rate constant (e.g., $h^{-1} C^{-1}$) for bacterial killing in $-K \cdot D \cdot R$
k'	First-order absorption/elimination rate constant when $K_a = K$ (e.g., h^{-1})
K_1	Affinity constant (e.g., $L \cdot$ µmol)
k_{10}	Fractional rate constant central compartment $\Rightarrow$ out (e.g., h^{-1})
k_{12}	Fractional rate constant central $\Rightarrow$ peripheral compartment (e.g., h^{-1})
k_{21}	Fractional rate constant peripheral $\Rightarrow$ central compartment (e.g., h^{-1})
k_{1e}	Rate constant from plasma $\Rightarrow$ effect-compartment (e.g., h^{-1})
K_a	Absorption rate constant (t^{-1}) or apparent affinity constant (e.g., $L \cdot$µmol)
k_d	Fractional turnover rate of Prothrombin Complex Activity (e.g., h^{-1})
K_d	Fractional rate constant for chemical degradation of drug (e.g., h^{-1})
K_d	Apparent dissociation constant (e.g., h^{-1})
k_{em}	Elimination rate constant for the metabolite (e.g., h^{-1})
k_{e0}	Rate constant from effect-compartment to out (e.g., h^{-1})
k_{in}	Zero-order turnover rate for production of response
k_{inm}	Mean turnover rate

k_{irrev}	Second-order rate constant for irreversible drug action
k_m	Fractional metabolic rate constant (e.g., h^{-1})
K_m	Michaelis-Menten constant (e.g., $\mu mol/L$)
k_{ME}	First-order rate constant for elimination of metabolite (e.g., h^{-1})
k_{out}	Fractional turnover rate (e.g., h^{-1})
K_p	Partition coefficient
K_t	Distribution rate constant (e.g., h^{-1})
k_{tol}	First-order rate constant for tolerance development (e.g., h^{-1})
k_u	Fractional urinary rate constant (e.g., h^{-1})
L	Length also Radiolabeled ligand
$[L]$	Ligand concentration (e.g., $\mu mol/L$)
$L()$	Logit
λ	Weighting exponent
λ_z	Terminal slope (e.g., h^{-1})
LOF	Lack of fit
LOQ	Limit of detection (e.g., $\mu mol/L$)
m	Slope at EC_{50} on effect vs. log-concentration curve
M	Metabolite or endogenous modulator
M_{50}	50% of maximal effect of the modulator (e.g., $\mu mol/L$)
MAT	Mean Absorption Time (e.g., h)
MIT	Mean Input Time (e.g., h)
ML	Maximum likelihood
MLP	Maximum lifespan potential (e.g., h)
MRT	Mean Residence Time (e.g., h)
MST	Mean Sojourn Time (e.g., h)
MTT	Mean Transit Time (e.g., h)
μ	Availability of growth medium
n	Sigmoidicity factor also called the Hill exponent
N and N_{obs}	Number of observations
N_{par} and NP	Number of parameters
N_{tot}	Total number of binding sites or receptors
NAD	Naive Averaged Data
NCA	Non-Compartmental Analysis
NL	Nonlinear
NP	Number of parameters
NPD	Naive Pooled Data
NSB	Non-specific binding
ω	Omega (used as wave-length in the cosine function)
OLS	Ordinary least squares
Π	Pi (used to convert the frequency in the cosine function) also called π
$\Pi(C)$	Probability of response as a function of C
P	Plasma or Precursor or Pool
P_{ss}	Prothrombin activity at steady state or condition of Pool compartment at steady state
$PBPK$	Physiologically Based Pharmacokinetics
PCA	Prothrombin Complex Activity

PD	Pharmacodynamics
PD_{ss}	Pharmacodynamics at steady state
PE	Excretory Pathways
PK	Pharmacokinetics
po	*per os*, oral
Q	Flow or organ perfusion (e.g., L/h)
Q(C)	Probability of no response as a function of *C*
Q_H	Hepatic blood flow (e.g., L/h)
qd	*quaque diem*, daily, per day
r	Response, equivalent to pharmacological effect (E)
R	Receptor concentration (arbitrary units)
R	Response, equivalent to pharmacological effect (E)
R_{in}	Turnover rate (e.g., $\mu mol \cdot h^{-1}$) such as secretion or synthesis
r(t)	Response function for a unit intravenous (bolus) dose or Response function in convolution/deconvolution
R_a	Active unbound enzyme
R_{ED50}	Response at ED_{50}
R_i	Inactive unbound enzyme
R_i	Observed or predicted maximum response
R_{in}	Turnover rate (arbitrary units/h) or constant rate of infusion
R_{in}	Build-up of Response such as secretion or synthesis (arbitrary units/h)
R_{inf}	Rate of infusion
R_{max}	Observed or predicted maximum response
R_{min}	Observed or predicted minimum response
R_0	Dosing rate (e.g., $\mu mol/h$) or baseline of response (arbitrary units)
R_t	Receptor concentration (e.g., $\mu mol/L$)
R_{ss}	Steady state value of response (arbitrary units)
RSD	Relative standard deviation
RSS	Residual sum of squares also called *WSS* or *WRSS*
$\hat{\sigma}$	Mean predicted standard deviation
$\hat{\sigma}^2$	Mean predicted variance
S	Slope factor (arbitrary units)
S(C)	Stimulation function also denoted *S(D)* or *S(t)*
S_i	Stimulus *I*
sc	Subcutaneous
SC	Schwarz Criterion
SD	Standard Deviation
SE	Standard Error
SOP	Standard Operating Procedure(s)
SS	Steady state or Sum of squares also called *WRSS*, *RSS*
t	Time (seconds, minutes, hours)
t'	Time after end of infusion (e.g., h)
t_0	Shifted peak time in the *cosine* function
$t_{1/2}$	Half-life (e.g., h)
$t_{1/2d}$	Distribution half-life (e.g., h)
$t_{1/2eq}$	Equilibration half-life (e.g., h)

$t_{1/2Ka}$	Absorption half-life (e.g., h)
$t_{1/2z}$	Terminal half-life (e.g., h)
T_{abs}	Absorption time (e.g., h)
t_d	Duration of response (e.g., h)
T_{fst}	Rapid infusion time (e.g., h)
T_{inf}	Infusion time (e.g., h)
t_{lag}	Lag-time or time delay (e.g., h)
t_{max}	Time to maximum concentration or effect (e.g., h)
T_{ss}	Time to steady state with respect to concentration or response (e.g., h)
t_t	Turnover time (e.g., h)
tid	*ter in diem*, three times a day
TS	Number of Tissue Spaces or binding proteins
V	Volume (e.g., L) or volume of distribution
V_c	Volume of the central compartment (e.g., L)
V_{cM}	Volume of the central compartment for the metabolite (e.g., L)
V_e	Volume of the effect-compartment (e.g., L)
V_M	Volume of distribution of metabolite (e.g., L)
V_{max}	Maximum metabolic rate (e.g., $\mu mol \cdot h^{-1}$)
V_{max} *and* K_d	Takada distribution model parameters; Maximum tissue volume (e.g., L) and time to establish half-maximal distribution equilibrium (e.g., h)
V_{max} *and* K_d	Colburn distribution model parameters; Maximum tissue volume (e.g., L) and first-order distribution rate constant (e.g., h^{-1})
V_{ss}	Volume of distribution at steady state (e.g., L)
V_t	Volume of the tissue compartment (e.g., L)
V_u	Unbound volume of distribution (e.g., L)
V_z	Volume of distribution based on the terminal phase (e.g., L) also called $V_{d\beta}$ for a bi-exponential system
VIF	Variance Inflation Factor
W	Weight ad organ weight (e.g., kg)
w(t)	Weighting function in convolution/deconvolution
WLS	Weighted least squares
WRSS	Weighted residual sum of squares also called *WSS*
WT	Weighting function in the *WinNonlin* code also called *W*
X	Independent variable or any parameter
X_u	Cumulative amount in urine (e.g., μmol)
X_{u0}^{∞}	Cumulative amount in urine from zero to infinity (e.g., μmol)
Y	Any measure quantity
$\hat{Y}$	Predicted value also called Y-hat
Y_{calc}	Predicted or calculated value also called Y-hat ($= \hat{Y}$)
Y_{obs}	Observed value
*	Convolution operator
$\bar{}$	A vector (e.g., $\underline{X}$ = vector of independent variable)
$\wedge$	Predicted (e.g., $\hat{C}$ = predicted concentration)
//	Deconvolution operator